Nanoarchitectonics in Biomedicine

Nanoarchitectonics in Biomedicine

Edited by

Alexandru Mihai Grumezescu
Faculty of Applied Chemistry and Materials Science,
University Politehnica of Bucharest,
Bucharest, Romania

ELSEVIER

Elsevier
Radarweg 29, PO Box 211, 1000 AE Amsterdam, Netherlands
The Boulevard, Langford Lane, Kidlington, Oxford OX5 1GB, United Kingdom
50 Hampshire Street, 5th Floor, Cambridge, MA 02139, United States

British Library Cataloguing-in-Publication Data
A catalogue record for this book is available from the British Library

Library of Congress Cataloging-in-Publication Data
A catalog record for this book is available from the Library of Congress

ISBN: 978-0-12-816200-2

Publisher: Matthew Deans
Acquisition Editor: Simon Holt
Editorial Project Manager: Andrae Akeh
Production Project Manager: Anitha Sivaraj
Cover Designer: Matthew Limbert

Typeset by MPS Limited, Chennai, India

Contents

List of Contributors

Lucero G. Aceves-Serrano
School of Engineering and Sciences, Tecnologico de Monterrey, Monterrey, Mexico

K. Thanigai Arul
Department of Physics, AMET (Deemed to be University), Kanathur, Chennai, India

Gurjit Kaur Bhatti
UGC Centre of Excellence in Nano Applications, University Institute of Pharmaceutical Sciences, Panjab University, Chandigarh, India; University Institute of Pharmacy Sciences, Chandigarh University, Mohali, Punjab, India

Jasvinder Singh Bhatti
Department of Biotechnology, Sri Guru Gobind Singh College, Chandigarh, India

Ziyad Binkhathlan
Nanomedicine Research Unit, Department of Pharmaceutics, King Saud University, Riyadh, Saudi Arabia; King Abdullah Institute for Nanotechnology, King Saud University, Riyadh, Saudi Arabia; Faculty of Pharmacy and Pharmaceutical Sciences, University of Alberta, Edmonton, AB, Canada

Asuman Bozkır
Department of Pharmaceutical Technology, Ankara University Faculty of Pharmacy, Ankara, Turkey

Naseem Ahmad Charoo
Zeino Pharma LLC, Khalifa Industrial Zone, Abu Dhabi, United Arab Emirates

Cristina Chircov
Faculty of Medical Engineering, University Politehnica of Bucharest, Bucharest, Romania

Ishwar Das
Department of Chemistry, DDU Gorakhpur University, Gorakhpur, India

Venugopal Dhananjayan
Industrial Hygiene & Toxicology Division, ICMR-Regional Occupational Health Centre (Southern), Indian Council of Medical Research, Bangalore, India

M. Ezgi Durgun
Faculty of Pharmacy, Department of Pharmaceutical Technology, Istanbul University, Istanbul, Turkey

Poopak Farnia
Mycobacteriology Research Centre (MRC), National Research Institute of Tuberculosis and Lung Disease (NRITLD), Shahid Beheshti University of Medical Sciences, Tehran, Iran; Department of Biotechnology, Advanced Technologies in Medicine, Shahid Beheshti University of Medical Sciences, Tehran, Iran

Anton Ficai
Department of Science and Engineering of Oxide Materials and Nanomaterials, Faculty of Applied Chemistry and Materials Science, Politehnica University of Bucharest, Bucharest, Romania

Jalaledin Ghanavi
Mycobacteriology Research Centre (MRC), National Research Institute of Tuberculosis and Lung Disease (NRITLD), Shahid Beheshti University of Medical Sciences, Tehran, Iran

Oana Gherasim
Department of Science and Engineering of Oxide Materials and Nanomaterials, Faculty of Applied Chemistry and Materials Science, Politehnica University of Bucharest, Bucharest, Romania; Lasers Department, National Institute for Lasers, Plasma and Radiation Physics, Magurele, Romania

Alexandru Mihai Grumezescu
Faculty of Applied Chemistry and Materials Science, University Politehnica of Bucharest, Bucharest, Romania; ICUB—Research Institute of University of Bucharest, University of Bucharest, Bucharest, Romania

Valentina Grumezescu
Department of Science and Engineering of Oxide Materials and Nanomaterials, Faculty of Applied Chemistry and Materials Science, Politehnica University of Bucharest, Bucharest, Romania; Lasers Department, National Institute for Lasers, Plasma and Radiation Physics, Magurele, Romania

Sevgi Güngör
Faculty of Pharmacy, Department of Pharmaceutical Technology, Istanbul University, Istanbul, Turkey

Hyundoo Hwang
BBB Inc., Seoul, Republic of Korea

Josef Jampílek
Regional Centre of Advanced Technologies and Materials, Palacky University, Olomouc, Czech Republic

Emine Kahraman
Faculty of Pharmacy, Department of Pharmaceutical Technology, Istanbul University, Istanbul, Turkey

Suresh Kumar Kailasa
Department of Applied Chemistry, S. V. National Institute of Technology, Surat, India; Department of Chemistry, Institute of Interdisciplinary Convergence Research, Research Institute of Halal Industrialization Technology, Chung-Ang University, Seoul, Republic of Korea

Melek Karaca
Department of Pharmaceutical Technology, Faculty of Pharmacy, Istanbul University, Istanbul, Turkey

Sukhjinder Kaur
Rayat-Bahra Institute of Pharmacy, Rayat-Bahra University, Hoshiarpur, Punjab, India

Mansoor A. Khan
Irma Lerma Rangel College of Pharmacy, Texas A&M Health Science Center, College Station, TX, United States

Sepideh Khoee
Polymer Laboratory, Chemistry Department, School of Science, University of Tehran, Tehran, Iran

Janardhan Reddy Koduru
Department of Environmental Engineering and, Kwangwoon University, Seoul, Republic of Korea

Katarína Kráľová
Institute of Chemistry, Faculty of Natural Sciences, Comenius University, Bratislava, Slovakia

Berrin Küçüktürkmen
Department of Pharmaceutical Technology, Ankara University Faculty of Pharmacy, Ankara, Turkey

Rasiah Ladchumananandasivam
Post Graduate Programme in Mechanical Engineering, Centre of Technology, Campus Universitário, Natal, Brazil

Afsaneh Lavasanifar
Faculty of Pharmacy and Pharmaceutical Sciences, University of Alberta, Edmonton, AB, Canada; Department of Chemical and Material Engineering, University of Alberta, Edmonton, AB, Canada

E. Manikandan
Department of Physics, TUCAS Campus, Thiruvalluvar University, Vellore, India; UNESCO UNISA Africa Chair in Nanosciences & Nanotechnology, College of Graduate Studies, University of South Africa, Pretoria, South Africa

Amaldoss M J Newton
School of Materials Science and Engineering, UNSW Sydney, NSW, Australia; Lowy Cancer Research Center, UNSW Sydney, NSW, Australia

Karen A. Ordaz-Martinez
School of Engineering and Sciences, Tecnologico de Monterrey, Monterrey, Mexico

Yıldız Özsoy
Faculty of Pharmacy, Department of Pharmaceutical Technology, Istanbul University, Istanbul, Turkey

Sharad P. Pandey
Department of Pharmacy, SGSITS, Indore, India

V.N. Pandey
Experimental Botany and Nutraceutical Laboratory, Department of Botany, DDU Gorakhpur University, Gorakhpur, India

V.S. Pandey
Experimental Botany and Nutraceutical Laboratory, Department of Botany, DDU Gorakhpur University, Gorakhpur, India

Karunamoorthy Panjakumar
Industrial Hygiene & Toxicology Division, ICMR-Regional Occupational Health Centre (Southern), Indian Council of Medical Research, Bangalore, India

Tae-Jung Park
Department of Chemistry, Institute of Interdisciplinary Convergence Research, Research Institute of Halal Industrialization Technology, Chung-Ang University, Seoul, Republic of Korea

Shahnaz Rahimi
Polymer Laboratory, Chemistry Department, School of Science, University of Tehran, Tehran, Iran

Ziyaur Rahman
Irma Lerma Rangel College of Pharmacy, Texas A&M Health Science Center, College Station, TX, United States

Anita Rao
Experimental Botany and Nutraceutical Laboratory, Department of Botany, DDU Gorakhpur University, Gorakhpur, India

Beerappa Ravichandran
Industrial Hygiene & Toxicology Division, ICMR-Regional Occupational Health Centre (Southern), Indian Council of Medical Research, Bangalore, India

Somnath Sen
Industrial Hygiene & Toxicology Division, ICMR-Regional Occupational Health Centre (Southern), Indian Council of Medical Research, Bangalore, India

Tripti Shukla
School of Pharmacy and Research, Peoples University, Bhopal, India

Bhupinder Singh
UGC Centre of Excellence in Nano Applications, University Institute of Pharmaceutical Sciences, Panjab University, Chandigarh, India

Gabriel Socol
Lasers Department, National Institute for Lasers, Plasma and Radiation Physics, Magurele, Romania

M.S. Sudheesh
VNS Institute of Pharmacy, Bhopal, India

Neha Tiwari
Experimental Botany and Nutraceutical Laboratory, Department of Botany, DDU Gorakhpur University, Gorakhpur, India

Neeraj Upmanyu
School of Pharmacy and Research, Peoples University, Bhopal, India

Matias Vazquez-Piñon
School of Engineering and Sciences, Tecnologico de Monterrey, Monterrey, Mexico

Ali Akbar Velayati
Mycobacteriology Research Centre (MRC), National Research Institute of Tuberculosis and Lung Disease (NRITLD), Shahid Beheshti University of Medical Sciences, Tehran, Iran

Rajesh Vijayvergiya
Department of Cardiology, Post Graduate Institute of Medical Education and Research, Chandigarh, India

Foreword

Nanoarchitectonics is a relatively new science aiming to arrange molecules and nanosized materials in a desired configuration in order to provide a new structure. This manipulation of materials at a nanoscale has allowed for the design of new materials and surfaces with wide applications in industry, biomedicine, pharmacy, engineering, chemistry, and others. Engineered nanomaterials have been intensively investigated in the biomedical field.

The purpose of this book is to reveal the latest progress and recently developed applications of nanoarchitectonics in biomedicine. Their applications in nanomedicine include, but are not limited to, drug delivery, targeting systems, detection approaches, controlled release agents, drug design, and bioavailability. Such nanoscale, engineered structures can be obtained by various methods and approaches and their synthesis parameters are essential for the biological effects. This book is a valuable resource describing the most innovative tools developed for the design of nanosized systems with diverse biomedical applications. This volume is an essential reference for researchers and engineers interested to investigate new alternatives for prophylaxis, diagnosis, and therapy based on nanoarchitectured solutions.

Alina Maria Holban

Faculty of Biology, University of Bucharest, Bucharest, Romania

Preface

The aim of this book is to offer new insight on the applications of nanoarchitectonics, starting with the diversity of the currently used methods employed for the synthesis and characterization of nanostructured materials, but also discusses more recent biomedical applications. This resourceful book contains 19 chapters focused on different applications of nanoarchitectonics in biomedicine.

Chapter 1, entitled *Basics in nanoarchitectonics*, prepared by Cristina Chircov et al., gives an up-to-date overview on nanoarchitectonics. The paradigm shift from nanotechnology to nanoarchitectonics is necessary for the fabrication of novel materials with enhanced properties. The concept is based on the mutual interactions between constituent nanostructures and their arrangements. The correlation between nanoarchitectonics key components, namely self-organization, chemical manipulations, new atom/molecule manipulation, field-induced interactions, and theory and simulations is necessary for further development of biomedical applications.

Chapter 2, entitled *Nanoarchitectonics prepared by laser processing and their biomedicinal applications*, prepared by Oana Gherasim et al., points out the multifunctional, biomedical potential related to different nanostructured coatings obtained by using pulsed laser deposition and matrix assisted pulsed laser evaporation techniques, as distinctive representatives of the ultraviolet laser processing of biomaterials.

Chapter 3, entitled *Microfluidics for drug delivery systems*, prepared by Hyundoo Hwang et al., presents the latest advances in microfluidic platforms for the fabrication of drug carriers and the development of direct drug delivery devices. The authors discuss challenges and opportunities of advanced drug delivery systems and the potential roles of nano- or microfluidic technologies toward developing smart drug delivery systems.

Chapter 4, entitled *Polymeric micelles as a novel carrier for ocular drug delivery*, prepared by Yildiz Özsoy et al., highlights the recent research findings and applications of polymeric micelles as ocular drug delivery systems for the treatment of eye diseases. Micelles can increase the aqueous solubility of hydrophobic or lipophilic compounds. Their small size also provides enhanced permeability and retention, thus targeting inflamed areas. Polymeric micelles have received much attention due to their potential, such as enhancement of topical bioavailability and improvement of ocular permeability of drugs.

Chapter 5, entitled *Reversible core/shell crosslinked micelles for controlled release of bioactive agents*, prepared by Sepideh Khoee et al., discusses various types of reversible core/shell crosslinked micelles with the main focus on the design and crosslinking strategy, improving biocompatibility and biodegradability and recent advances in their applications as drug delivery systems.

Chapter 6, entitled *Nanoemulsion preparation, characterization, and application in the field of biomedicine*, prepared by Cristina Chircov et al., focuses on the preparation, characterization techniques, and various applications of nanoemulsions in the field of biomedicine.

Chapter 7, entitled *Exosome nanocarriers: a natural, novel, and perspective approach in drug delivery system*, prepared by Jasvinder Singh Bhatti et al., gives an up-to-date overview on the formation, evaluation, drug targeting, and the mechanistic functioning of exosomes along with supporting evidence for the tasks performed by exosomes in many diseases.

Chapter 8, entitled *Recent advances in lipid nanocarriers applicable in the fight against cancer*, prepared by Josef Jampílek et al., focuses on the application of lipid-based nanocarriers representing a class of biocompatible and biodegradable vehicles that provide advanced delivery of therapeutic and imaging agents, improving its pharmacokinetic profile and safety in cancer therapy and diagnostics. Also, this chapter give an up-to-date overview on lipid-based delivery systems.

Chapter 9, entitled *Solid lipid nanoparticles for skin and drug delivery: methods of preparation and characterization and applications*, prepared by A.M.J. Newton et al., focuses on the recent potential of solid lipid nanoparticles in cosmetics, method of preparation, characterization, in vitro studies, stability, toxicity, and applications on the skin and in other drug delivery systems.

Chapter 10, entitled *A new approach for drug targeting to the central nervous system: lipid nanoparticles*, prepared by Berrin Küçüktürkmen et al., discusses the limitations in central nervous system drug delivery and the potential use of lipid nanoparticles for drug targeting to the brain. Lipid nanoparticles include advantages of both polymeric nanoparticles and liposomes and also have superior properties, such as the possibility of large-scale production. The methods currently in use for lipid nanoparticle production, physicochemical characterization, and formulation development for drug delivery and targeting to the brain are also discussed.

Chapter 11, entitled *Nanoparticles for improvement in oral bioavailability*, prepared by Naseem Ahmad Charoo et al., gives an up-to-date overview on basic concepts relevant to nanoparticles and their impact on bioavailability, as well as the effect of particle size and surface characteristics on biodistribution and cellular uptake of nanoparticles and biomedical application of nanoparticles.

Chapter 12, entitled *Targeted delivery systems for treatment of pancreatic cancer*, prepared by Melek Karaca et al., discusses the current progress on targeted drug delivery approaches in pharmaceutical technology for pancreatic cancer treatment.

Chapter 13, entitled *Targeted drug delivery and gene therapy through natural biodegradable nanostructures in pharmaceuticals*, prepared by V.N. Pandey et al., deals with insight into the use of plants and animals as diversified, sustainable, environment-benign, biodegradable resources, and a renewable platform for the production of useful nanostructures with promising pharmaceutical applications in ocular, brain, vaccine, oral, pulmonary, colon, dermal, and transdermal deliveries.

Chapter 14, entitled *Site-specific drug delivery, targeting and gene therapy*, prepared by Sharad P. Pandey et al., discusses the various concepts and systems for targeting and site-specific delivery, which has come into existence in the past few decades, with special concern to gene therapy. They addressed two different types of carriers and their physiochemical and biological characteristics which are needed to develop a site-specific drug delivery or for targeting. Attention is given to the various carriers and their mechanisms, including viral and nonviral vectors with special mention of polymeric systems used to deliver the gene to specific sites.

Chapter 15, entitled *Effects of block copolymer micelles on the pharmacokinetics of encapsulated drugs*, prepared by Ziyad Binkhathlan et al., provides an overview of the effects of polymeric micelles on the pharmacokinetics of encapsulated drugs, including those currently in clinical trials.

Chapter 16, entitled *Nano design of extracellular matrix for tissue engineering*, prepared by Jalaledin Ghanavi et al., offers an overview regarding applications of nanomatrices and their impact on tissue engineering. This chapter is very useful for researchers and clinicians.

Chapter 17, entitled *Polymer-based calcium phosphate scaffolds for tissue engineering applications*, prepared by K. Thanigai Arul et al., gives an up-to-date overview on polymeric-based calcium phosphate biomaterials employed as replacements for bone and tissue engineering highlighting studies on different polymer-based calcium phosphates that do not exhibit toxicity and reveal the generation of extracellular matrices by different cell types.

Chapter 18, entitled *Metal nanoparticles-based colorimetric methods for drug analyses*, prepared by Suresh Kumar Kailasa et al., summarizes the recent advances in functionalized gold (Au) and silver nanoparticles (Ag NPs) considered as good candidates for colorimetric assays of various drugs in various samples. This chapter aims to explore the advantages of organic molecules as ligands for tuning their interactions selectively with drugs, and to highlight the future perspectives of Au and Ag NPs in assaying drugs in pharmaceutical and biological samples.

Chapter 19, entitled *Source, effect, and risk assessment of nanoparticles with special reference to occupational exposure*, prepared by V. Dhananjayan et al., provides an overview on nanoparticle monitoring and risk assessment methods for the development and implementation of effective risk-management guidance.

Alexandru Mihai Grumezescu
University Politehnica of Bucharest, Bucharest, Romania

CHAPTER

Basics in nanoarchitectonics

1

Cristina Chircov[1] **and Alexandru Mihai Grumezescu**[2,3]

[1]*Faculty of Medical Engineering, University Politehnica of Bucharest, Bucharest, Romania*
[2]*Faculty of Applied Chemistry and Materials Science, University Politehnica of Bucharest, Bucharest, Romania* [3]*ICUB—Research Institute of University of Bucharest, University of Bucharest, Bucharest, Romania*

1.1 INTRODUCTION

The fabrication of nanomaterials has attracted a great deal of interest in a wide variety of research fields including medicine, electronics, cosmetics, food, and environment industries. Nanomaterials are materials with at least one dimension measuring 100 nm or less. In addition, nanomaterials possess unique physical and chemical properties due to their high surface area and nanoscale size that are different from the properties of the same materials in larger forms. Nanomaterials' fabrication encompasses top-down approaches using macroscopic initial structures that can be externally controlled, and bottom-up approaches including the miniaturization of materials' components up to atomic level and followed by further self-assembly processes.

The concept of nanoarchitectonics was first introduced by Masakazu Aono in 2000 in Tsukuba, Japan at the 1st International Symposium of Nanoarchitectonics Using Suprainteractions (Ariga et al., 2015). The term describes a novel technology system for the arrangement of nanostructural units in a predefined configuration. The purpose of this technology is the development of a way to control and modulate interactions within nanostructures. The concept of nanoarchitectonics can be applied to three different fields: nanomaterials' production, nanostructures' fabrication, and their application in different research areas, such as biomedicine. The need for material innovation is essential for the development of new technologies and applications, thus making the creation of nanomaterials through nanoarchitectonics crucial, not only by the creation and function elucidation of nanostructures, but also by the fabrication of macroscopic materials through the profound understanding of mutual interactions between constituent nanostructures and their arrangements (Ariga et al., 2011). The potential of nanoarchitectonics for technology development lies in its product examples (which are categorized as atomic/molecular manipulation), materials creation (which is the fundament for materials nanoarchitectonics), and advanced device materials (Ariga et al., 2015).

Nanoarchitectonics in Biomedicine. DOI: https://doi.org/10.1016/B978-0-12-816200-2.00001-3

In contrast with common nanofabrication strategies, nanoarchitectonics is an approach to design advanced materials through the harmonization of atomic/molecular-level control, chemical nanofabrication, self- and field-controlled organization, and theoretical modeling (Ariga, 2017; Ariga and Li, 2016), which are the key processes that lead to unexpected higher functions (Ariga et al., 2016). The applicability of nanoarchitectonics in a wide variety of research areas can be correlated to the applicability to all categories of materials, including organic and inorganic, as well as bio-based materials. The use of nanoarchitectonics has been reported for nanomaterials and nano-objects, such as graphene, carbon nanotubes, nanowires, nanomotors, nanoparticles, and colloidal nanocrystals, and fundamental chemistry and physics, including dynamic functions from atom to macro, mechanoresponsive luminescence, chiral twists and helices, and oriented self-assembly. Furthermore, nanoarchitectonics has been applied in biological studies and biomedical applications, such as cellulose-rich nanofibers, organelle mimics of soft matter, biomimetic light-harvesting, drug carriers, hemoglobin-based nanoarchitectures, cancer treatment and diagnosis, and nonviral delivery of nucleic acids (Ariga and Li, 2016).

1.2 FROM NANOTECHNOLOGY TO NANOARCHITECTONICS

Nanotechnology is currently one of the key domains in science and technology and its development is highly related to the development of observation technology which includes advanced microscopy techniques. The possibility to observe and manipulate individual atoms and molecules leads to the emergence of new opportunities in nanoscience (Ariga, 2015). Nanotechnology and nanoscience are generally based on their operation and functional size, which are defined in the nanometer scale. The misconception that nanoscience is an extension of the micrometer scale has led to a general belief that nanotechnology is an advanced version of microtechnology. Nonetheless, a reduction in size leads to unexpected phenomena due to many factors including uncertainties, thermal and/or statistical fluctuations, and quantum effects. Thus a simple transition from micro- to nanotechnology is not an appropriate solution since material structure and function cannot be determined (Ariga et al., 2016). Considering this, the improvement of nanoscale precision for materials' fabrication is highly important since it might result in significant advances of new functions with improved accuracy, specificity, and efficiency. The production of functional materials and structures through nanofabrication necessitates a careful combination of individual techniques. Still, despite the need for the development of novel and sophisticated materials and structures, nanostructured materials have only been fabricated by using simple tools (Ariga et al., 2014).

Thus a paradigm shift from nanotechnology to nanoarchitectonics is necessary (Ariga and Aono, 2018). Nevertheless, the nanoarchitectonics concept has been gradually introduced in nanoscience (Ariga et al., 2015) and is correlated to the shift from using single tool–based manufacturing to the total-construction-of-architecture (Ariga, 2015). Therefore its rapid progress is highly anticipated since the knowledge of nanotechnology can be applied in the development of nanoarchitectonics (Ariga et al., 2015). Since nanoarchitectonics is a general concept, it can be applied in a variety of fields, from materials science to biomedical applications (Ariga, 2015). Considering the bio-related field, nanoarchitectonics is of key importance due to the capacity to fuse artificial systems obtained through nanotechnology with biological systems (Ariga et al., 2015).

1.3 MATERIALS NANOARCHITECTONICS

Materials nanoarchitectonics is based on its product examples, which can be categorized as atomic/molecular manipulation, materials creation, and advanced device materials (Ariga et al., 2015). Manipulation of individual atoms and molecules proves the ultimate operation of miniaturized devices at the level of a single atom or molecule. The scanning tunneling microscope (STM) apparatus is of key importance in the development of technology at nanoscale. Atomic and molecular manipulation has been widely used for innovation in nanoelectronics and molecular nanomachines as well as for designing novel functionalized materials. There are two types of manipulation methods in STM: (1) manipulation by direct contact, which implies bringing the tip into close proximity with the adsorbate and inducing modifications by moving the tip, and (2) manipulation by inelastic electron tunneling, especially resonant inelastic electron tunneling, which will result in local vibrational excitation that can be used to desorb, move single atoms and molecules, or to dissociate molecules and isomerize or change the configuration of single molecules (Mayne and Dujardin, 2011). Materials creation through nanoarchitectonics is based on the synthesis involving the agglomeration of two or more existing materials as a strategy. The formation of layered structures is a versatile method that uses techniques such as the Langmuir–Blodgett and layer-by-layer assembly (Ariga et al., 2015). The Langmuir–Blodgett technique is based on the self-assembly of molecules to form a monolayer film at an interface and the subsequent transfer onto a solid substrate, which has the advantage of increased homogeneity with molecular-level precision (Wales and Kitchen, 2016). The layer-by-layer assembly method involves the coating of substrates by sequential adsorption of oppositely charged materials (such as polymers, colloids, biomolecules, and cells) offering increased control and versatility (Richardson et al., 2016). Advanced device materials in nanoarchitectonics are mostly referred to as bioinspired systems since there is a strong similarity to the operation of biological system through the cooperative action of the assembled components. Thus biological systems can be considered as fundamental nanoarchitectures based on the self-organization of functional units (Ariga et al., 2015).

1.3.1 ZERO-DIMENSIONAL NANOARCHITECTURES

The zero-dimensional category (which includes nanoparticles, nanocrystals, and quantum dots) is one of the most studied nanostructures among nanomaterials used in a broad range of applications. These nanostructures represent the simplest building blocks used for the design and construction of complex nanomaterials and nanodevices (Cao, 2017). Nanoparticles are one of the most important components of nanotechnology (Ealias and Saravanakumar, 2017). Depending on the material, nanoparticles can be based on polymers, molecular assemblies, and inorganic materials, metals, or their hybrids. Synthesis of nanoparticles through nanoarchitectonics brings the possibility of new properties for new applications. Nanoarchitectonics of nanoparticles implies the harmonization of different techniques and phenomena, such as structural control induced by physical stimuli, self-assembly and self-organization (Ji et al., 2018).

Self-assembly is a method to synthetize new materials with different properties based on intermolecular interactions, without the implication of covalent bonding. It is a method that ensures increased control of the properties due to its reversibility and response to external stimuli (Percebom et al., 2018). Examples of self-assembled nanoparticles are dendrimers, liposomes, micelles, fullerenes, and gold nanoparticles (Hormozi, 2015).

Dendrimers are a class of nanostructures formed by the periodical branching of molecules that radiate from a central unit, known as the core, and are characterized by the generation number. Dendrimer nanoparticles are functional nanomaterials with a distinguished structural symmetry, density gradient, and a defined number of terminal groups which exhibit exceptional electronic, optical, optoelectronic, magnetic, chemical, and biological properties (Singh, 2016). Dendrimers have proved their potential for the self-assembly of hierarchically organized structures, chirality mediation, and gluing supramolecular structures (Selin et al., 2016).

Since the mention of phospholipid self-assembly into closed bilayer vesicles in aqueous media, liposomes have gained significant scientific interest, especially as a drug delivery system, but also in various other fields, such as clinical medicine (Beziere et al., 2015). Due to their amphiphilic structure during the synthesis process, hydrophilic compounds can be encapsulated in the aqueous core and be attached within the bilayer by dissolution in the formulation solution (Kulkarni and Shaw, 2016).

An increased interest for polymeric micelles over conventional surfactant micelles lies in the advantages of the former for drug delivery and cancer therapy (Zhang et al., 2017). Core–shell micelle structures can be formed in an aqueous solution in the presence of a precise hydrophilic/hydrophobic balanced amphiphilic block copolymer. The use of different copolymer structures, including diblock, triblock, graft, star, or hyperbranched copolymers, can modify micelle architecture (Kotsuchibashi et al., 2016).

The concept of materials nanoarchitectonics has been used for the synthesis of nanoparticles and nanocrystals through the self-assembly of ideal zero-dimensional building block fullerenes (C_{60}). Fullerenes have long been used as building blocks for molecular engineering, supramolecular chemistry, novel materials fabrication, and medicinal chemistry. Water-soluble, functionalized fullerenes have also been shown to self-assemble to form nanostructures (Shrestha et al., 2013).

Self-assembly of small metallic nanoparticles, especially gold nanoparticles to form gold nanospheres, is another example of materials nanoarchitectonics. These nanostructures can be modified to be responsive to external stimuli, including changes in pH and temperature (Feng et al., 2017).

Synthesis of highly hierarchical nanoparticles from inorganic and hybrid materials mostly relies on bottom-up techniques, including template assembly and layer-by-layer assembly (Ji et al., 2018). Layer-by-layer assembly is a method to deposit molecular thin films on substrates. It is a highly versatile technique offering the possibility to engineer nanostructures from a wide range of materials with different architectures and functions, addressing physical, chemical, and biomedical applications. Layer-by-layer assembly is also useful for the nanoengineering of particle surfaces to obtain core–shell particles and hollow capsules by the subsequent immersion of the substrate into the layering and washing solutions. A commonly used template for the assembly of porous nanoparticles is mesoporous silica nanoparticles, which come in different dimensions, shapes, pore sizes, and morphologies. These nanoparticles, also called polymer replica particles, are synthetized through the infiltration of the polymers into the pores of the mesoporous silica, followed by crosslinking. After the formation of the polymer network, the silica template is dissolved. The architecture of the nanoparticles can be modified by changing template parameters or by combining different polymers and crosslinking strategies (Björnmalm et al., 2017).

1.3.2 ONE-DIMENSIONAL NANOARCHITECTURES

One-dimensional nanomaterials mostly consist of nanowires, nanotubes, nanofibers, and nanorods, but other morphologies are also possible. This category is of key importance for the field of energy technology (Wei et al., 2017), especially as a platform for charge transport and photo-generated charge carrier separation (Dalui et al., 2018). One-dimensional nanostructures have been extensively studied as platforms to investigate chemical and physical properties in dependence to dimensionality and size reduction, also known as quantum confinement (Peng et al., 2015). Ever since the discovery of carbon nanotubes in 1991, these nanostructures are the most promising building blocks for next-generation nanoelectronic and nano optoelectronic devices of the future (Frontmatter, 2013).

Nanoscale self-assembly has also led to the development of new applications in materials science and engineering through nanowire growth, especially due to the possibility to control their composition, shape, and structure. Semiconductor nanowires are one of the most promising candidates for bandgap engineering templating. Due to their dimensionality, denoting a small cross section compared to their length, nanowires allow high performances of operation when easily integrated in circuits and devices. Nanowires architecture can be modulated by different approaches, with the self-assembly of different quantum structures in a nanowire template being one of these. Depending on their dimensionality, the quantum structures used for nanowires assembly can be two-dimensional (such as quantum wells), one-dimensional (such as quantum wires), and zero-dimensional, including quantum dots. In order to fully understand these complex structures, the use of advanced electron microscopy tools is necessary (Arbiol et al., 2013).

The assembly of nanowires on specific substrates is usually performed in two steps, namely the aligned growth of nanowires followed by the transfer of the aligned nanowire to the substrates. Nonetheless, bottom-up approaches for nanowires growth usually result in random alignment and orientation, thus needing subsequent assembly and organization into integrated arrays for the development of devices. Nanowire device assembly methods include Langmuir–Blodgett assembly, electric/magnetic field-directed assembly, and chemically driven assembly. Studies demonstrate the potential of these methods for the organization of nanowires into aligned and hierarchical structures. Thus nanowires prove to be building blocks which exhibit unique properties for the design of nanodevices, but also for the nanoarchitectonics concept (Peng et al., 2015).

Carbon nanotubes are one-dimensional nanomaterials widely studied because of their unique structures and properties. The applications of carbon nanotubes mainly depend on the synthesis method. Many of the current bulk composite materials and thin films use unorganized carbon nanotubes which limits their properties. In contrast, organized carbon nanotubes architectures offer improved properties and, consequently, new functions and applications. Vertically aligned carbon nanotubes arrays and horizontally aligned carbon nanotubes arrays are examples of these organized architectures. The vertically aligned carbon nanotubes arrays are characterized by a forest-like structure, given by the perpendicular orientation and increased alignment of carbon nanotubes to the surface. Due to their structures, vertically aligned carbon nanotubes arrays have been extensively used in applications including energy-absorbing, thermal management, electromagnetic shielding coatings, strong fibers, nanocomposites, membranes, and high-performance electrodes (Zhang et al., 2016). The layer-by-layer technique is a technique that allows the formation of highly organized nanostructures, leading to enhanced material properties. For example, this technique has been used to prepare carbon nanotubes electrodes by producing films with controlled architecture. The layer-by-layer assembly offers the advantage of simplicity of the experimental apparatus that can deposit layers on surfaces of various shapes or sizes. It is often viewed as an extension of the self-assembly methods that use chemical interactions, covalent bonds mostly, between the deposited layers, in contrast to the physical interactions of oppositely charged layers (Oliveira et al., 2017).

Another example of one-dimensional nanomaterials is the naturally occurring nanosized tubular halloysite. Halloysite is a mineral from the class of kaolin, described as a diocthaedral 1:1 clay. The multilayer tubular structure forms due to the wrapping of the 1:1 clay mineral layer driven by a mismatch in the oxygen-sharing tetrahedral and octahedral sheets in the 1:1 layer (Yuan et al., 2015). Their tubular structure makes them an ideal candidate for the fabrication of nanoarchitectonic materials since it allows the incorporation of drugs in all its loading sites, including the lumen, the interwall spaces in multiwalled tubes or walls, and the smart modification of the tubes' endings and functionalization of the walls (Darrat et al., 2018).

1.3.3 TWO-DIMENSIONAL NANOARCHITECTURES

Two-dimensional nanomaterials are intensively studied in materials science with a major focus on synthesis, characterization, and application. The class of two-dimensional nanomaterials is extremely versatile since these sheet-like solids present a wide variety of chemical compositions, crystal phases, and physical forms. Their potential includes, but is not limited to, electronics, sensors, energy engineering, coatings, and barriers (Wang et al., 2016). Two-dimensional nanomaterials ultrathin nanomaterials are characterized by a high degree of anisotropy and chemical functionality (Chimene et al., 2015). Within two-dimensional spaces, the motion of components are highly restricted, thus system design and functional development are seriously affected by material packing and motional behavior (Ariga et al., 2018). Due to their unique structures, morphologies, and physicochemical properties, layer nanomaterials (such as graphene), transition metal dichalcogenides, layered metal oxides, black phosphorus, and hexagonal boron nitride have gained great interest (Kenry, 2017).

Ultrathin films are characterized by a layer thickness of less than 10 nm and they are of particular importance in nanotechnology (Cui and Li, 2018). Ultrathin films have been widely constructed through supramolecular assembly, which consists of supramolecular interactions, including ionic, hydrophobic, Van der Waals, hydrogen, and coordination bonds. This method leads to a certain intrinsic mobility that results in the formation of organized nanostructures through the equilibrium of aggregated and nonaggregated states (Busseron et al., 2013). Supramolecular assembly comes with the advantages of energy and time saving, and environmental-friendliness. The main strategies for ultrathin films through supramolecular assembly are the self-assembled monolayer method, Langmuir–Blodgett technique, and layer-by-layer assembly. Considering these approaches, nanoarchitectonics for ultrathin films involves a wide choice and combination of building blocks (Cui and Li, 2018).

Graphene is representative of a two-dimensional nanomaterial by having unique characteristics, namely its one-atom thickness and the hexagonal lattice consisting of sp^2 hybridized carbon atoms with a carbon–carbon distance of approximately 0.142 nm. Graphene is the basic building block of important allotropes, including three-dimensional graphite formed through graphene stacking, one-dimensional carbon nanotubes through graphene rolling, and zero-dimensional fullerenes formed through graphene wrapping. Its extraordinary thermal, electrical, and mechanical properties which make graphene a major focus in many research fields are a result of the long-range π conjugation. Other properties worth mentioning are its impermeability to gas and liquid and its high current density (Ghuge et al., 2017). Graphene has been studied for the fabrication of smart materials, some of the most common including the mechanically exfoliated perfect graphene, chemical vapor deposited high-quality graphene, chemically modified graphene (such as graphene oxide), and reduced graphene oxide, and the macroscopic materials and composites formed by their assemblies. A highly important characteristic of these materials is their sensitivity to various stimuli, such as pH value, electrical field, mechanical strain, thermal or optical excitation, and gas and biological molecules (Yu et al., 2017). However, it is a challenge to maintain the properties of nanoscale graphene at macroscale. Thus it is of key importance to develop methods for two-dimensional graphene nanosheets assembly into organized hierarchical architectures. Based on these nanoarchitectonics techniques, several methods have been developed to assemble graphene nanosheets to form macroscale structures, namely flow-directed assembly, layer-by-layer assembly, and the self-template strategy (Pan et al., 2015).

Despite the fact that two-dimensional nanomaterials that contain rigid components have attracted much attention, there is a need to focus on two-dimensional soft materials since the scientific views are not fully established (Ariga et al., 2018). In addition, two-dimensional soft nanomaterials, including two-dimensional polymers, covalent organic frameworks, and two-dimensional supramolecular organic nanostructures have the advantages of structural control and flexibility as well as the variety of fabrication methods (Zhuang et al., 2015). Two-dimensional soft materials can be categorized into well-packed and oriented states, mutually linked networks with certain motional flexibilities, and molecular systems with sufficient motional freedoms. Furthermore, these categories could also be defined as well-packed and oriented organic two-dimensional materials with rational design of component molecules and device applications, well-defined assemblies with two-dimensional porous structures as two-dimensional network materials, and two-dimensional control of molecular machines and receptors based on considerable motional freedom confined in two dimensions, respectively (Ariga et al., 2018).

1.3.4 THREE-DIMENSIONAL NANOARCHITECTURES

When compared to zero-, one-, or two-dimensional nanostructures, the morphology and properties of three-dimensional nanostructures offer new possibilities for the design and production of novel architectures. Their performance is improved by the hierarchical building of porous three-dimensional materials through the simultaneous enhancement of surface and diffusion capacity (Song et al., 2018). Self-assembly of zero-dimensional nanostructures, such as fullerenes, one-dimensional nanostructures such as carbon nanotubes, or two-dimensional nanostructures such as graphene, leads to the formation of three-dimensional nanomaterials with enhanced properties for specific applications.

The assembly of the zero-dimensional, 1-nm fullerene only produces fragile three-dimensional nanoarchitectures, which may limit their applicability. Studies have showed the fabrication of a three-dimensional scaffold through the vacuum vapor deposition method using a mask. Thus fullerenes were deposited on the microscopic glass coverslips under vacuum by using metallic masks with rectangular openings of 100 μm. Another commonly used method is the coating of fullerene onto a previously synthetized scaffold made from polymers or other materials (Nakanishi et al., 2014).

The practical use of carbon nanotubes usually requires the assembly of these nanomaterials into macroscopic architectures. Although the synthesis of long fibers or two-dimensional films using carbon nanotubes has been widely studied in the fields of energy storage, catalysis, electronics, and bioengineering, the need for developing three-dimensional carbon nanotubes architectures could broaden their range of application. These architectures with specific shapes and volumes are necessary for spatial requirements satisfaction and increasing the amount of active material. One representative three-dimensional architecture is the carbon nanotubes sponge, which has attracted great attention in many research fields. Its potential is based on the unique properties and its porous and hierarchical structure consisting of macropores, beneficial for mass transfer, and meso-, and micropores, which provide high surface areas and abundant reactive centers (Luo et al., 2017). Another example for three-dimensional carbon nanotubes architecture is the carbon nanotube field-effect transistors. Their assembly is based on the strategies of nanomanipulation and the electron beam–induced deposition technique into a scanning electron microscope. These methods are employed for building the structure through repetitive basic processes consisting of pickup, placement, fixing, and cutting of carbon nanotubes (Ning et al., 2017).

Three-dimensional graphene assembly advances show promising possibilities to fabricate solid porous materials lighter than air. In addition, the use of a single two-dimensional graphene structures is inefficient for mechanical functions. Thus the preassembly of graphene into three-dimensional scaffolds has the benefits of high stiffness and mechanical strength. The use of graphene as a building block has made progress for the synthesis of graphene-based porous materials which are applicable in many fields, including materials science, energy, and environmental innovations. The mechanical properties of three-dimensional graphene assemblies are very different from those of conventional polymers (Qin et al., 2017).

The concept of nanoarchitectonics is also applied when designing intelligent materials such as shape-memory materials. These materials exhibit the unique capacity to undergo chemo-responsive, electrically activated, light activated, microwave heating triggered, thermally triggered, and magnetically sensitive shape changes (Wang et al., 2016). Reports show the fabrication of various types of shape-memory materials, including alloys, ceramics, polymers, and supramolecular systems. Shape-memory alloys and soft shape-memory materials, including polymer and supramolecular systems, have been extensively researched in the field of biomedical devices, and automotive and robotics industries, due to the structural diversity and variation which allow for parameter tuning (Uto, 2018). The possibility to use functional fillers has helped to overcome the mechanical properties limitations. The functionalization with mesoporous, bioactive nanoparticles, nanotubes, and nanocomposite membranes is a way to prepare novel architectures with improved mechanical properties (Wang et al., 2016).

1.4 NANOARCHITECTONICS FOR BIOMEDICAL APPLICATIONS

Living organisms are created through the hierarchical organization of biological molecules, cells, and tissue, which proves the dependence between the biological response to a material and the nanoscale structural properties of the material. Thus it is safe to consider biological systems as the greatest prototype of nanoarchitectonics, which emphasizes the importance of nanoarchitectonics development for the design of biomaterials that can overcome current limitations and challenges (Ebara, 2016). The elaboration and future implementation of novel nanomaterials involve the correlation between the five components of nanoarchitectonics (self-organization, chemical manipulations, new atom/molecule manipulation, field-induced interactions, and theory and simulations through computer simulations and molecular modeling) and the effects generated through manipulation. It is necessary to examine the interdependent implementation of nanoarchitectonics-based technology that promotes nanotechnology in the biomedical field (Kujawa and Winnik, 2013).

1.4.1 DRUG DELIVERY SYSTEMS

The limitations of the delivery of therapeutic compounds to targeted sites affects the treatment of many diseases mostly due to the limited effectiveness, poor distribution, and lack of selectivity (Kesrevani and Sharma, 2016). Thus the development of advanced drug delivery systems is highly important to overcome these limitations. Nanoarchitectonics-based technologies can be applied to elaborate such systems, including dendrimers, liposomes, carbon nanotubes, and so on.

The unique properties of dendrimers given by the nanoscale globular shape, the peripheral functional groups, the hydrophilic/hydrophobic character of the internal cavities, and the reduced polydispersity make them ideal candidates as nanocarriers in drug delivery systems. In addition, they possess the capacity to cross cell barriers through both paracellular and transcellular pathways and to modify the number of functional surface groups in order to influence biodistribution, targeting, dosage, or release the profile of the drug from within the dendrimer which is extremely important for the overcoming of conventional drug delivery systems limitations (Parajapati et al., 2016). One specific example of dendrimer application is the antimicrobial activity of polypropylenimine dendrimers that have quaternary ammonium salts as functional groups. The surface-functionalized dendrimers exhibit an improved antimicrobial activity compared to common drug molecules due to the disrupting of bacterial cell walls which leads to the inhibition of bacterial growth (Sk, 2017).

The usage of liposomes as drug delivery systems lies in their capacity to stabilize therapeutic compounds to overcome the obstacles of cellular and tissue uptake and to improve biodistribution, which enables the effective delivery of the encapsulated compounds to the targeted sites with minimal systemic toxicity. Easy liposome manipulation is possible due to their flexible physicochemical properties (Sercombe et al., 2015). In addition, the synthetic nature and its production through self-assembly offer the possibility to encapsulate different and relatively large types of drugs. The functionalization of liposomes using polymers (such as polyethylene glycol as targeting ligands) improves homogeneity and circulatory properties, creating the so-called long-circulating liposomes (Johnsen et al., 2018).

Other ideal candidates as carriers for drug delivery are carbon nanotubes due to their high specific area which allows for an increased loading capacity in comparison to conventional nanomaterials. Drug encapsulation induces the formation of $\pi-\pi$ interactions between the drug and the surface of the carbon nanotube support (Roldo, 2016). Still, the modification of carbon nanotubes by functionalization or purification has the consequence of a certain toxicity. Thus it is necessary to establish the mechanisms for cellular internalization for improved efficiency and targeted delivery (Rakesh et al., 2014).

1.4.2 GENE DELIVERY

One way to provide lasting therapies or cures for previously untreatable, suboptimal, or temporary treatable diseases is through therapeutic gene transfer. Gene therapy consists of disease treatment based on transferring genes to the patient's cells or gene editing (Kumar et al., 2016). Still, the main challenge encountered in gene therapy research is the development of safe, nontoxic, and efficient vectors for nucleic acid delivery. Currently, the two approaches for gene delivery are based on either viral vectors or nonviral and synthetic vectors (Molla and Levkin, 2016). Different vectors have been proposed throughout the years which led to the development of gene therapy (Kumar et al., 2016). Nanoarchitectonics has played an important role for the elaboration of nonviral gene delivery vehicles by self-assembly of cationic lipids and polymers. Examples of nanostructures as gene delivery vehicles are nanoparticles, liposomes, and polymersomes (Molla and Levkin, 2016).

Many types of nanoparticles have been studied for gene delivery applications (including lipid-based, polymer-based, and inorganic nanoparticles) in order to overcome commonly associated challenges such as encapsulation and delivery efficiency, stability, degradation, endocytosis and endosomal escape, and toxicity

(Chen et al., 2016). The use of nanoparticles for gene delivery to regulate angiogenesis has been studied. Poly(ethyleneimine) and poly(L-lysine) are two of the most commonly investigated polymers that allow the nucleic acid binding to the support due to the presence of primary and secondary amines that impart a positive charge at physiological pH (Kim et al., 2017). Polymer-based nanoparticles also include liposomes, dendrimers, polymersomes, and so on, which have been used as gene delivery vehicles.

Inorganic nanoparticles, which include semiconductor-based nanoparticles consisting of nanoscale crystals, have high surface areas and remarkable optical and magnetic properties that offer the possibility to design numerous applications as gene delivery vectors. Quantum dots are commonly used inorganic nanoparticles, characterized by an increased flexibility through size control by adjusting temperature and duration of synthesis, but also the type of ligand molecules used for functionalization. The functionalization allows for covalently and electrostatically nucleic acid conjugation (Soleimani et al., 2016).

1.4.3 TISSUE ENGINEERING AND REGENERATIVE MEDICINE

The development of nanoarchitectonic-based materials could lead to great advances in regenerative medicine. The need for improvement in the field of regenerative medicine and tissue engineering is related to the growth of the aging population in developed countries. The goal is to create and implement novel methods for cell transplantation that could lead to improved repair in diseased and injured tissues through biological replacements (Fekrazad et al., 2016).

In contrast to conventional tissue engineering and regenerative medicine techniques, the scaffold-free technology is based on the cells assembly into three-dimensional tissues that about the brings the fundamental advantages of creating connections between cells and avoiding negative host response (Kobayashi and Okano, 2016). The cell-sheet technology relies on the use of poly(*N*-isopropylacrylamide) as a support for cell culturing. The extraordinary property of thermoresponsivity of this polymer offers the possibility to create a noninvasive harvest of cultured cell as an intact cell sheet with deposited extracellular matrix. The principle is based on the lower critical solution temperature at around 32°C of the polymer, which results in normal proliferation under normal conditions of 37°C until confluence since it is in a dehydrated compacted structure. When the temperature is decreased, the polymer will hydrate and expand resulting in the spontaneous detachment of the cells as a monolithic tissue-like cell sheet (Kanai et al., 2014). The cell-sheet technology has a wide range of applications, including cardiac tissue fabrication (Masuda and Shimizu, 2016), tendon regeneration (Gonçalves et al., 2017; Zhang et al., 2018), and reepithelialization and neovascularization of skin wounds (Cerqueira et al., 2014) and oral mucosal cell sheet for wound repair (Roh et al., 2017).

Nanoarchitectonics for tissue engineering and regenerative medicine also entails the component of cell manipulation. This method uses dynamic substrates, characterized by the possibility to control physicochemical surface aspects by extracellular stimulus, including head, light, and voltage. Dynamic substrates enable the mimic of dynamic changes in extracellular niches and the elucidation of phenotype change mechanism. Cell manipulations consist of coculturing heterotypic cell types, cell migration, and processes extension which offer new possibilities in drug testing and tissue engineering (Nakanishi, 2016).

1.4.4 THERANOSTICS

Theranostics is a promising combination of therapeutics and diagnostics. It usually consists of diagnosis followed by therapy to hierarchize patients with likely responses to treatment, therapy followed by diagnosis to monitor and predict treatment efficiency, or it could be the codeveloping of both therapeutics and diagnosis. One theranostic strategy is the design of nanoplatforms that could deliver both imaging and therapeutic agents (Chen and Wong, 2014). Nanotheranostics include a variety of multidisciplinary fields (including chemistry, physics, material science, nanotechnology, drug delivery, and pharmacology) that could revolutionize personalized medicine through the use of biomarkers, imaging and chemotherapeutic agents, and specific targeting ligands (Bai et al., 2015). The focus of nanoarchitectonics research revolves around the fabrication of various nanostructures that have the capacity to be used for cancer diagnosis and treatment. These nanoscale structures are being exploited for cancer therapeutics for both therapy and imaging due to their unique physical and optical properties (Pandey et al., 2014).

1.4.4.1 Cancer therapy

Cancer theranostics represents the combined diagnosis and therapeutic approaches for cancer with the purpose of reducing delays in treatment and ease patient care. It is the essential component of personalized cancer treatment (Chen and Wong, 2014). There is a general belief that if cancer growth could be restricted during the diagnosis procedure, it could ease the subsequent treatment since cancer growth is retarded, or the cancer burden is reduced (Lim et al., 2015). Nanomedicines, that is, nanoparticle-based therapeutics composed of various organic or inorganic nanomaterials, are commonly used for the treatment, diagnosis, monitoring, and control of biological systems. These nanostructures possess anticancer therapeutic abilities, such as drug encapsulation, surface-bound molecules or cancer cells targeting, and monitoring the response to treatment for further drug efficacy and safety modulation (Lim et al., 2015). Nanostructures can target cancer cells through two main approaches, namely passive targeting (in abnormal tumor cells, the defective physiology and structure results in the enhanced permeability and retention effect, which improves nanostructures permeation and accumulation in tumor cells) and active targeting, based on molecular recognition mechanisms such as ligand receptor, antigen–antibody, and other forms (Pandey et al., 2014).

The design and production of effective targeted theranostic nanoparticles for cancer therapy must fulfill several mail requirements. Firstly, the identification of a suitable biomarker highly expressed in tumor cells and relatively absent in normal cells is of key importance. Following this, it is necessary that a corresponding targeting ligand will bind to the receptor with high affinity. In addition, the production of large quantities for the applications is needed. The nanoparticles used for the development of theranostic agents must be stable, biocompatible, biodegradable, and generate low systemic toxicity. Also, after entering tumor tissues or cancer cells, they should have the capacity to release the therapeutic agent. There are many strategies for cancer therapy, the most commonly used include photothermal therapy that uses nanoparticles made of gold or carbon nanotubes to generate heat that destroys the surrounding tumor cells, and photodynamic therapy which uses nanoparticles conjugated with photosensitizers that are activated by specific wavelengths of light which will, subsequently, generate reactive oxygen species that destroy tumor cells (Zhu et al., 2017), and assist chemotherapy, gene therapy, and radiation (Wang et al., 2017).

1.4.4.2 Imagining diagnosis

Therapeutic strategies are usually integrated with one or more diagnostic imaging techniques, including computed tomography, magnetic resonance imaging, position emission tomography, single photon emission computed tomography, near-infrared fluorescence imaging, and ultrasonography. In these ways, molecular diagnostic tests and targeted therapeutics in clinics could be developed, resulting in the advance of personalized medicine (Wang et al., 2017). These imaging techniques have been widely applied in combination with probes for improving the contrast between healthy and cancerous tissues. The development of biocompatible polymeric nanoassemblies showed an improvement in precision for the detection of cancerous cells through the enhanced selectivity and sensitivity of the imaging material (Mi et al., 2017).

Nanoparticle-based contrast agents improves imaging quality, and property control, including size, could influence biological systems, through cellular uptake, tissue biodistribution, and tumor uptake (Dreifuss et al., 2015). A generalization of single type of material for theranostic applications is difficult, commonly used nanoscale biomaterials including magnetic nanoparticles, quantum dots, upconversion nanoparticles, mesoporous silica nanoparticles, carbon-based nanoparticles, and organic dye-based nanoparticles (Huang and Lovell, 2017).

1.4.5 MECHANOBIOLOGY

Mechanobiology is the emerging field that encompasses multidisciplinary areas, namely cell and developmental biology, bioengineering, and biophysics. The concept of mechanobiology consists of cells' active sensing and processing of mechanical information provided by the extracellular niche which results in the modification of growth, motility, and differentiation mechanisms. Deregulation of mechanical properties of the extracellular matrix is corelated to many diseases, thus making the understanding of these mechanisms crucial. In addition, these mechanisms could be used for programming stem cell differentiation for further applications in tissue engineering and regenerative medicine (Jansen et al., 2015).

Studying global mechanical interactions across large architectures, including cytoskeletons, organelles, cells, tissues, and organs is necessary for a complete understanding. Studies have shown that the application of compressive stress to cancer cells monolayers have resulted in invasive phenotypes. As cells are considered tensional integrity architectures, local distortion is transmitted to global architectural changes, and vice versa (Nakanishi, 2018). The concept of mechano-nanoarchitectonics has been introduced to include the methodology of nanoarchitectonics for the formation of functional structures and modulation of properties using mechanical processes. Thus an ideal medium to connect macroscopic mechanical actions and nanoscale functions is an interfacial, two-dimensional environment (Ariga, 2016).

One elegant approach to fabricate cell culture substrates for investigating cell mechanobiology are shape-memory polymers. The control of nanoarchitecture of the polymer networks is important for the further creation of shape-memory surfaces with tunable topography at a nanoscale. The required characteristic for these substrates is the actuation under biological conditions, especially in sharply narrow temperature ranges (Ebara, 2015).

1.4.6 DNA ASSEMBLY

Another application of nanoarchitectonics involves the use of DNA as molecular building blocks for the design of nanostructures. The design and fabrication using smaller nucleic acid strands leads to the formation of nanostructures with different shapes suitable for interfacial layers. DNA nanoarchitectonics is a promising research concept with studies showing the use of DNA nanostructures for homogeneous substrate coatings and defined nanopatterns and nanoarrays (Howorka, 2013).

Recent advances in DNA assembly involve the synthesis of particles characterized by controlled interactions that are chemically specific and directional in nature and involve the use of self-assembly approaches, but also of computational studies (Porter and Crocker, 2017). In addition, DNA molecules assembly provides a route for the synthesis of one-, two-, and three-dimensional hierarchical periodic lattices. The nucleic acid strands provide site-specific attachment, thus providing spatially positioned arrays of nanoparticles or macromolecules with nanoscale precision (Chandrasekaran and Zhuo, 2016).

Self-assembled DNA-based nanoarchitectures with controlled sizes and shapes offer great capacities of drug loading, cell internalization, stability, and biocompatibility. The fabrication of nanofibers, micelles, and vesicles in aqueous solutions through self-assembly of DNA-lock copolymers and DNA-dendron hybrids have been reported (Zhiyong et al., 2017).

1.5 CONCLUSIONS

The nanoarchitectonics concept can be applied for the synthesis of novel zero-, one-, two-, and three-dimensional nanoarchitectures that can be further used in various ranges of fields. The difference in properties at nanoscale is particularly important for the further development of nanomaterials. Besides electronics, sensors, energy engineering, coatings, and barriers, nanoarchitectonics is widely applied in the biomedical field with specific applications regarding drug and gene delivery, tissue engineering and regenerative medicine, theranostics, mechanobiology, and DNA assembly. Focus on nanoarchitectonics-related properties could further increase the potential of nanomaterials.

ACKNOWELEDGMENT

This work was supported by a grant from the Romanian National Authority for Scientific Research and Innovation, UEFISCDI, project number 45PCCDI/2018-PN-III-P1-1.2-PCCDI-2017-0749-Nanostructuri bioactive pentru strategii terapeutice inovatoare.

REFERENCES

Arbiol, J., de la Mata, M., Eickhoff, M., Morral, A.F.I., 2013. Bandgap engineering in a nanowire: self-assembled 0, 1 and 2D quantum structures. Mater. Today. 16 (6), 213–219.

Ariga, K., 2015. From nanotechnology to nanoarchitectonics. J. Inorg. Organomet. Polym. Mater. 25 (2), 177–178.

Ariga, K., 2016. Mechano-nanoarchitectonics for bio-functions at interfaces. Anal. Sci. 32 (11), 1141–1149.

Ariga, K., 2017. Nanoarchitectonics: a navigator from materials to life. Mater. Chem. Front. 1 (2), 208–211.

Ariga, K., Aono, M., 2018. Change thinking toward nanoarchitectonics. Mater. Nanoarchitectonics.

Ariga, K., Li, J., 2016. Nanoarchitectonics for advanced materials: strategy beyond nanotechnology. Adv. Mater. 28 (6), 987–988.

Ariga, K., Li, M., Richards, G.J., Hill, J.P., 2011. Nanoarchitectonics: a conceptual paradigm for design and synthesis of dimension-controlled functional nanomaterials. J. Nanosci. Nanotechnol. 11 (1), 1–13.

Ariga, K., Kawakami, K., Ebara, M., Kotsuchibashi, Y., Ji, Q., Hill, J.P., 2014. Bioinspired nanoarchitectonics as emerging drug delivery systems. New J. Chem. 38 (11), 5149–5163.

Ariga, K., Ji, Q., Nakanishi, W., Hill, J.P., Aono, M., 2015. Nanoarchitectonics: a new materials horizon for nanotechnology. Mater. Horiz. 2 (4), 406–413.

Ariga, K., Minami, K., Ebara, M., Nakanishi, J., 2016. What are the emerging concepts and challenges in NANO? Nanoarchitectonics, hand-operating nanotechnology and mechanobiology. Polym. J. 48, 371.

Ariga, K., Watanabe, S., Mori, T., Takeya, J., 2018. Soft 2D nanoarchitectonics. NPG Asia Mater.

Bai, R.G., Muthoosamy, K., Manickam, S., 2015. Chapter 12—Nanomedicine in theranostics A2—Thomas, Sabu. In: Grohens, Y., Ninan, N. (Eds.), Nanotechnology Applications for Tissue Engineering. William Andrew Publishing, Oxford, pp. 195−213.

Beziere, N., Lozano, N., Nunes, A., Salichs, J., Queiros, D., Kostarelos, K., et al., 2015. Dynamic imaging of PEGylated indocyanine green (ICG) liposomes within the tumor microenvironment using multi-spectral optoacoustic tomography (MSOT). Biomaterials. 37, 415−424.

Björnmalm, M., Cui, J., Bertleff-Zieschang, N., Song, D., Faria, M., Rahim, M.A., et al., 2017. Nanoengineering particles through template assembly. Chem. Mater. 29 (1), 289−306.

Busseron, E., Ruff, Y., Moulin, E., Giuseppone, N., 2013. Supramolecular self-assemblies as functional nanomaterials. Nanoscale. 5 (16), 7098−7140.

Cao, H., 2017. Synthesis, characterization, and applications of zero-dimensional (0D) nanostructures. In: Synthesis and Applications of Inorganic Nanostructures, Wiley.

Cerqueira, M.T., Pirraco, R.P., Martins, A.R., Santos, T.C., Reis, R.L., Marques, A.P., 2014. Cell sheet technology-driven re-epithelialization and neovascularization of skin wounds. Acta Biomater. 10 (7), 3145−3155.

Chandrasekaran, A.R., Zhuo, R.A., 2016. 'Tile' tale: hierarchical self-assembly of DNA lattices. Appl. Mater. Today. 2, 7−16.

Chen, J., Guo, Z., Tian, H., Chen, X., 2016. Production and clinical development of nanoparticles for gene delivery. Mol. Ther. Methods Clin. Dev. 3, 16023.

Chen, X., Wong, S.T.C., 2014. Chapter 1—Cancer theranostics: an introduction. Cancer Theranostics. Academic Press, Oxford, pp. 3−8.

Chimene, D., Alge, D.L., Gaharwar, A.K., 2015. Two-dimensional nanomaterials for biomedical applications: emerging trends and future prospects. Adv. Mater. 27 (45), 7261−7284.

Cui, W., Li, J., 2018. Nanoarchitectonics of multilayer shells toward biomedical application. Mater. Nanoarchitectonics.

Dalui, A., Khan, A.H., Pradhan, B., Ghosh, S., Acharya, S., 2018. Aspects of one-dimensional nanostructures: synthesis, characterization, and applications. Mater. Nanoarchitectonics .

Darrat, Y., Naumenko, E., Cavallaro, G., Lazzara, G., Lvov, Y., Fakhrullin, R., 2018. Tubular nanocontainers for drug delivery. Mater. Nanoarchitectonics.

Dreifuss, T., Betzer, O., Shilo, M., Popovtzer, A., Motiei, M., Popovtzer, R., 2015. A challenge for theranostics: is the optimal particle for therapy also optimal for diagnostics? Nanoscale. 7 (37), 15175−15184.

Ealias, A.M., Saravanakumar, M.P., 2017. A review on the classification, characterisation, synthesis of nanoparticles and their application. IOP Conf. Ser.: Mater. Sci. Eng. 263 (3), 032019.

Ebara, M., 2015. Shape-memory surfaces for cell mechanobiology. Sci. Technol. Adv. Mater. 16 (1), 014804.

Ebara, M., 2016. 1—Introductory guide to nanoarchitectonics. Biomaterials Nanoarchitectonics. William Andrew Publishing, pp. 1–4.

Fekrazad, R., Nokhbatolfoghahaei, H., Nikhalat, M., 2016. Advances in lasers and nanoparticles in treatment and targeting of epithelial originated cancers. In: Holban, A.M., Grumezescu, A.M. (Eds.), Nanoarchitectonics for Smart Delivery and Drug Targeting. William Andrew Publishing, pp. 613–645.

Feng, Q., Shen, Y., Fu, Y., Muroski, M.E., Zhang, P., Wang, Q., et al., 2017. Self-assembly of gold nanoparticles shows microenvironment-mediated dynamic switching and enhanced brain tumor targeting. Theranostics. 7 (7), 1875–1889.

Frontmatter, 2013. In: Zhai, T., Yao, J. (Eds.), One-Dimensional Nanostructures. Springer, Berlin, Heidelberg.

Ghuge, A.D., Shirode, A.R., Kadam, V.J., 2017. Graphene: a comprehensive review. Curr. Drug Targets. 18 (6), 724–733.

Gonçalves, A.I., Rodrigues, M.T., Gomes, M.E., 2017. Tissue-engineered magnetic cell sheet patches for advanced strategies in tendon regeneration. Acta Biomater. 63, 110–122.

Hormozi, B., 2015. Application of nanoparticles in cancer treatment. Adv. Theranostic Mater.

Howorka, S., 2013. DNA nanoarchitectonics: assembled DNA at interfaces. Langmuir. 29 (24), 7344–7353.

Huang, H., Lovell, J.F., 2017. Advanced functional nanomaterials for theranostics. Adv. Funct. Mater. 27 (2), 1603524.

Jansen, K.A., Donato, D.M., Balcioglu, H.E., Schmidt, T., Danen, E.H.J., Koenderink, G. H., 2015. A guide to mechanobiology: where biology and physics meet. Biochim. Biophys. Acta Mol. Cell Res. 1853 (11, Part B), 3043–3052.

Ji, Q., Liu, X., Yin, K., 2018. Architectonics in nanoparticles. Mater. Nanoarchitectonics.

Johnsen, K.B., Gudbergsson, J.M., Duroux, M., Moos, T., Andresen, T.L., Simonsen, J.B., 2018. On the use of liposome controls in studies investigating the clinical potential of extracellular vesicle-based drug delivery systems—a commentary. J. Controlled Release. 269, 10–14.

Kanai, N., Yamato, M., Okano, T., 2014. Chapter 5—Principles of cell sheet technology A2—Orlando, Giuseppe. In: Lerut, J., Soker, S., Stratta, R.J. (Eds.), Regenerative Medicine Applications in Organ Transplantation. Academic Press, Boston, MA, pp. 57–66.

Kenry, Lim C.T., 2017. Biocompatibility and nanotoxicity of layered two-dimensional nanomaterials. ChemNanoMat. 3 (1), 5–16.

Kesrevani, R.K., Sharma, A.K., 2016. 2—Nanoarchitectured biomaterials: present status and future prospects in drug delivery A2—Holban, Alina Maria. In: Grumezescu, A.M. (Ed.), Nanoarchitectonics for Smart Delivery and Drug Targeting. William Andrew Publishing, pp. 35–66.

Kim, J., Mirando, A.C., Popel, A.S., Green, J.J., 2017. Gene delivery nanoparticles to modulate angiogenesis. Adv. Drug Delivery Rev. 119, 20–43.

Kobayashi, J., Okano, T., 2016. 3.2—Cell sheet technologies A2—Ebara, Mitsuhiro. Biomaterials Nanoarchitectonics. William Andrew Publishing, pp. 97–113.

Kotsuchibashi, Y., Nakagawa, Y., Ebara, M., 2016. 2.1—Nanoparticles. Biomaterials Nanoarchitectonics. William Andrew Publishing, pp. 7–23.

Kujawa, P., Winnik, F.M., 2013. Innovation in nanomedicine through materials nanoarchitectonics. Langmuir. 29 (24), 7354–7361.

Kulkarni, V.S., Shaw, C., 2016. Chapter 4—Formulating creams, gels, lotions, and suspensions. Essential Chemistry for Formulators of Semisolid and Liquid Dosages. Academic Press, Boston, MA, pp. 29–41.

Kumar, S.R.P., Markusic, D.M., Biswas, M., High, K.A., Herzog, R.W., 2016. Clinical development of gene therapy: results and lessons from recent successes. Mol. Therapy—Methods Clin. Dev. 3.

Lim, E.-K., Kim, T., Paik, S., Haam, S., Huh, Y.-M., Lee, K., 2015. Nanomaterials for theranostics: recent advances and future challenges. Chem. Rev. 115 (1), 327–394.

Luo, S., Luo, Y., Wu, H., Li, M., Yan, L., Jiang, K., et al., 2017. Self-assembly of 3D carbon nanotube sponges: a simple and controllable way to build macroscopic and ultralight porous architectures. Adv. Mater. 29 (1), 1603549.

Masuda, S., Shimizu, T., 2016. Three-dimensional cardiac tissue fabrication based on cell sheet technology. Adv. Drug Delivery Rev. 96, 103–109.

Mayne, A.J., Dujardin, G., 2011. Chapter 1 - Introduction. In: Mayne, A.J., Dujardin, G. (Eds.), Frontiers of Nanoscience, vol. 2. Elsevier, pp. 1–14.

Mi, P., Wang, F., Nishiyama, N., Cabral, H., 2017. Molecular cancer imaging with polymeric nanoassemblies: from tumor detection to theranostics. Macromol. Biosci. 17 (1), 1600305.

Molla, M.R., Levkin, P.A., 2016. Combinatorial approach to nanoarchitectonics for nonviral delivery of nucleic acids. Adv. Mater. (Deerfield Beach, FA). 28 (6), 1159–1175.

Nakanishi, J., 2016. 3.3—Cell manipulation technologies A2—Ebara, Mitsuhiro. Biomaterials Nanoarchitectonics. William Andrew Publishing, pp. 115–135.

Nakanishi, J., 2018. Mechanobiology. Mater. Nanoarchitectonics.

Nakanishi, W., Minami, K., Shrestha, L.K., Ji, Q., Hill, J.P., Ariga, K., 2014. Bioactive nanocarbon assemblies: nanoarchitectonics and applications. Nano Today. 9 (3), 378–394.

Ning, Y., Qing, S., Masahiro, N., Huaping, W., Zhan, Y., Lining, S., et al., 2017. 3D assembly of carbon nanotubes for fabrication of field-effect transistors through nanomanipulation and electron-beam-induced deposition. J. Micromech. Microeng. 27 (10), 105007.

Oliveira, D.A., Oliveira, O.N., Siqueira, J.R., 2017. Amperometric Sensors Based on Carbon Nanotubes in Layer-by-Layer Films. Springer, Berlin, Heidelberg, pp. 1–21.

Pan, H., Zhu, S., Mao, L., 2015. Graphene nanoarchitectonics: approaching the excellent properties of graphene from microscale to macroscale. J. Inorg. Organomet. Polym Mater. 25 (2), 179–188.

Pandey, A.P., Girase, N.M., Patil, M.D., Patil, P.O., Patil, D.A., Deshmukh, P.K., 2014. Nanoarchitectonics in cancer therapy and imaging diagnosis. J. Nanosci. Nanotechnol. 14 (1), 828–840.

Parajapati, S.K., Maurya, S.D., Das, M.K., Tilak, V.K., Verma, K.K., Dhakar, R.C., 2016. Potential application of dendrimers in drug delivery: a concise review and update. J. Drug Deliv. Ther. 6 (2), 71–88.

Peng, L., Shi, Y., Yu, G., 2015. 5—Self-assembly and organization of nanowires. Semiconductor Nanowires. Woodhead Publishing, pp. 149–171.

Percebom, A.M., Towesend, V.J., de Paula Silva de Andrade Pereira, M., Gramatges, A.P., 2018. Sustainable self-assembly strategies for emerging nanomaterials. Curr. Opin. Green Sustain. Chem.

Porter, C.L., Crocker, J.C., 2017. Directed assembly of particles using directional DNA interactions. Curr. Opin. Colloid Interface Sci. 30, 34–44.

Qin, Z., Jung, G.S., Kang, M.J., Buehler, M.J., 2017. The mechanics and design of a lightweight three-dimensional graphene assembly. Sci. Adv. 3 (1).

Rakesh, K., Meenakshi, D., Sudhir, K., Brahma, N.S., Jayant, K.P., Vivek, R.S., 2014. Carbon nanotubes: a potential concept for drug delivery applications. Recent Patents Drug Deliv. Formulation. 8 (1), 12–26.

Richardson, J.J., Cui, J., Björnmalm, M., Braunger, J.A., Ejima, H., Caruso, F., 2016. Innovation in layer-by-layer assembly. Chem. Rev. 116 (23), 14828–14867.

Roh, J.-L., Lee, J., Kim, E.H., Shin, D., 2017. Plasticity of oral mucosal cell sheets for accelerated and scarless skin wound healing. Oral Oncol. 75, 81–88.

Roldo, M., 2016. Carbon nanotubes in drug delivery: just a carrier? Ther. Deliv. 7 (2), 55–57.

Selin, M., Peltonen, L., Hirvonen, J., Bimbo, L.M., 2016. Dendrimers and their supramolecular nanostructures for biomedical applications. J. Drug Deliv. Sci. Technol. 34, 10–20.

Sercombe, L., Veerati, T., Moheimani, F., Wu, S.Y., Sood, A.K., Hua, S., 2015. Advances and challenges of liposome assisted drug delivery. Front. Pharmacol. 6, 286.

Shrestha, L.K., Shrestha, R.G., Hill, J.P., Ariga, K., 2013. Self-assembled fullerene nanostructures. J. Oleo Sci. 62 (8), 541–553.

Singh, A.K., 2016. Chapter 2—Structure, synthesis, and application of nanoparticles. Engineered Nanoparticles. Academic Press, Boston, MA, pp. 19–76.

Sk, U.H., 2017. Chapter 12—Nanosize dendrimers: potential use as carriers and antimicrobials A2—Grumezescu, Alexandru Mihai. Antimicrobial Nanoarchitectonics. Elsevier, pp. 323–355.

Soleimani, M., Majid Al Zaabi, A., Merheb, M., Matar, R., 2016. Nanoparticles in gene therapy. Int. J. Integr. Biol. 17, 7–16.

Song, Y., Liu, T., Qian, F., Zhu, C., Yao, B., Duoss, E., et al., 2018. Three-dimensional carbon architectures for electrochemical capacitors. J. Colloid Interface Sci. 509, 529–545.

Uto, K., 2018. Shape-memory materials. Mater. Nanoarchitectonics .

Wales, D.J., Kitchen, J.A., 2016. Surface-based molecular self-assembly: Langmuir–Blodgett films of amphiphilic Ln(III) complexes. Chem. Cent. J. 10, 72.

Wang, F., Zhu, L., Yang, K., Park, D., 2017. Theranostic probes for cancer imaging. Contrast Media Mol. Imaging. 2017, 2.

Wang, W., Liu, Y., Leng, J., 2016. Recent developments in shape memory polymer nanocomposites: actuation methods and mechanisms. Coord. Chem. Rev. 320–321, 38–52.

Wang, Z., Zhu, W., Qiu, Y., Yi, X., von dem Bussche, A., Kane, A., et al., 2016. Biological and environmental interactions of emerging two-dimensional nanomaterials. Chem. Soc. Rev. 45 (6), 1750–1780.

Wei, Q., Xiong, F., Tan, S., Huang, L., Lan, E.H., Dunn, B., et al., 2017. Porous one-dimensional nanomaterials: design, fabrication and applications in electrochemical energy storage. Adv. Mater. (Deerfield Beach, FA). 29 (20).

Yu, X., Cheng, H., Zhang, M., Zhao, Y., Qu, L., Shi, G., 2017. Graphene-based smart materials. Nat. Rev. Mater. 2, 17046.

Yuan, P., Tan, D., Annabi-Bergaya, F., 2015. Properties and applications of halloysite nanotubes: recent research advances and future prospects. Appl. Clay Sci. 112-113, 75–93.

Zhang, C., Zhang, E., Yang, L., Tu, W., Lin, J., Yuan, C., et al., 2018. Histone deacetylase inhibitor treated cell sheet from mouse tendon stem/progenitor cells promotes tendon repair. Biomaterials.

Zhang, H., Xu, J., Xing, L., Ji, J., Yu, A., Zhai, G., 2017. Self-assembled micelles based on chondroitin sulfate/poly (D,L-lactide-*co*-glycolide) block copolymers for doxorubicin delivery. J. Colloid Interface Sci. 492, 101–111.

Zhang, S., Peng, D., Xie, H., Zheng, Q., Zhang, Y., 2016. Investigation on the formation mechanism of double-layer vertically aligned carbon nanotube arrays via single-step chemical vapour deposition. Nano-Micro Lett. 9 (1), 12.

Zhiyong, Z., Feng, L., Simin, L., 2017. Self-assembly of DNA-based nanomaterials and potential application in drug delivery. Curr. Top. Med. Chem. 17 (16), 1829–1842.

Zhu, L., Yang, L., Zhou, Z., 2017. Nanomaterials in cancer theranostics. In: Yan, B., Zhou, H., Gardea-Torresdey, J.L. (Eds.), Bioactivity of Engineered Nanoparticles. Springer Singapore, Singapore, pp. 173–206.

Zhuang, X., Mai, Y., Wu, D., Zhang, F., Feng, X., 2015. Two-dimensional soft nanomaterials: a fascinating world of materials. Adv. Mater. 27 (3), 403–427.

CHAPTER

2 Nanoarchitectonics prepared by laser processing and their biomedicinal applications

Oana Gherasim[1,2], Valentina Grumezescu[1,2], Gabriel Socol[2] and Anton Ficai[1]

[1]Department of Science and Engineering of Oxide Materials and Nanomaterials, Faculty of Applied Chemistry and Materials Science, Politehnica University of Bucharest, Bucharest, Romania [2]Lasers Department, National Institute for Lasers, Plasma and Radiation Physics, Magurele, Romania

2.1 INTRODUCTION

The beneficial conjunction between formidable technological evolution and innovative materials science and engineering methodologies enables unseen possibilities for exceeding the limits and drawbacks related to conventional medical practice. Being guided by the fundamental goal represented by the patient's quality of life, impressive knowledge and substantial multidisciplinary efforts have been oriented by worldwide healthcare practitioners and scientific researchers toward the design, development, and functional evaluation of novel biomaterials, medical devices, and unconventional treatments. In the particular case of common implantable medical devices (IMDs), some essential aspects which must be considered regarding their restricted multifunctional potential are the surface-related physicochemical features and biological characteristics (Rupp et al., 2018; Skoog et al., 2018; Woeppel et al., 2017).

It is very likely that the most handy and versatile approach adopted with respect to the performance enhancement of currently available medical devices consists in coating them with thin protective materials with the aim of inducing them novel functionality. Depending on their specific use in different biomedical fields, several limitations and shortcomings were related to conventional IMDs with a few relevant examples, including improper mechanical behavior (Jamil et al., 2017; Massil, 2017), susceptibility to corrosive processes (Asri et al., 2017; Manam et al., 2017), or increased tendency for pathogenic contamination (Elbourne et al., 2017; Preethanath et al., 2017; Silkey et al., 2017). In order to

Nanoarchitectonics in Biomedicine. DOI: https://doi.org/10.1016/B978-0-12-816200-2.00018-9

provide a genuine and practical solution to these particular issues, the nanotechnology-derived fabrication and auspicious biofunctionality related to thin protective coatings were demonstrated to be suitable choices.

Laser processing techniques provide an attractive, advantageous and challenging approach to fabricate thin coatings for biomedical materials and devices. The impressive versatility of laser-assisted modification of IMDs' surface is strongly related to the multiple possibilities implied during selecting the laser wavelength (Cristescu et al., 2016; Ge et al., 2014), the operating regime (Gomes et al., 2017; Tangermann-Gerk et al., 2016), and the pulse duration (Trtica et al., 2018; Ohtsu et al., 2017).

For laser-processed nanostructured materials considered for the applicative extension of commercial medical-related devices enhanced results were reported by using different laser-assisted methods, such as laser additive manufacturing (Berretta et al., 2018; Liu et al., 2017), direct laser writing (Daskalova et al., 2016; Xiang et al., 2016), laser cladding (Xue et al., 2017; Yang et al., 2018), laser engineered net shaping (Buciumeanu et al., 2018a; Li et al., 2017c), laser-induced forward transfer (Di Pietrantonio et al., 2015; Palla-Papavlu et al., 2014), laser texturing (Lee et al., 2018; Stango et al., 2018), laser welding (Chan, 2017; Dong et al., 2018), selective laser melting (Buciumeanu et al., 2018b; Hassanin et al., 2018), selective laser sintering (Du et al., 2017; Szustakiewicz et al., 2018), and two-photon polymerization (Kufelt et al., 2015; Brigo et al., 2017). For particular biomedical-related applications, protective coatings obtained under pulsed laser beams irradiation are the most attractive and promising candidates.

By far, the pulsed laser deposition (PLD) and matrix assisted pulsed laser evaporation (MAPLE) techniques represent the most explored laser processing methods with respect to the genuine functional coating of various IMDs. The versatile tunable parameters related to laser beams (wavelength, energy, fluence, and pulse duration) and the easily adjustable deposition features (solid target preparation, selective target irradiation, number of pulses, nature and pressure of deposition atmosphere, target-to-substrate distance, and substrate temperature) outline the indisputable advantages of both PLD and MAPLE methods.

High-quality thin coatings with complex stoichiometry and improved functionality were successfully fabricated by means of PLD. Briefly, during the synthesis process, a high-energy laser beam emitted in short pulses is focused onto the solid target. For a laser energy density that exceeds a certain ablation threshold value, the incident photons induce the fast evaporation of the target's surface in the form of a plasma plume that contains photons, atoms, ions, and molecules. The subsequent nucleation of the ablated species onto the concerned substrates enables the growth of thin-layered materials which composition, crystallographic structure, morphology, thickness, and functional properties can be easily modified by the proper adjustment of the deposition parameters (such as substrate temperature, pressure and nature of the ambient gas, separation distance between target and substrate, laser fluence, and repetition rate). Due to the physical mechanisms involved during the laser evaporation of the material, the PLD process is only appropriate for inorganic materials which constituent strong chemical bonds to withstand the degradation of the target material. Based on these premises, the PLD method was indicated as an inadequate processing approach for organic materials since their interactions with energetic laser radiation resulted in significant thermal and photochemical degradation of the organic phase.

In this respect, the MAPLE technique—representing a versatile and gentle extension of conventional PLD—proved its efficiency with respect to the processing of organic materials. The main difference to the PLD technique is related to the preparation of the target and the laser regime. Thus, in the case of MAPLE processing, the direct interaction of the laser beam with the organic phase is reduced by using a solvent which mainly absorbs the incident laser radiation. The solid target required for MAPLE deposition implies freezing the solution resulted by suspending or dissolving low amounts of organic compound (below 10 wt.%) into an appropriate solvent. In order to preserve the integrity of the evaporated species, the value of the laser energy density (named laser fluence) has to be under the ablation threshold. After laser irradiation, the solvent and organic molecules are ejected together from the target's surface. Further, the solvent is pumped away using a vacuum system while the organic molecules condensate on the substrate which is placed parallel to the target's surface.

With respect to biomedical applications of thin coatings obtained by PLD and MAPLE, the reported results recommend them as promising candidates for anti-infective therapy (Grumezescu et al., 2015a; Písařík et al., 2017; Visan et al., 2017), biomolecular detection (Califano et al., 2015; Dinca et al., 2018; Verrastro et al., 2016), and tissue functional restoration or regeneration (Major et al., 2016; Luculescu et al., 2018; Popescu-Pelin et al., 2017).

In this chapter, we aim to gather the latest results reported in thin coating synthesis by means of PLD and MAPLE methodologies. In particular, we directed our attention toward the studies performed by using ultraviolet (UV) pulsed laser beams. The compositional, microstructural, and functional features of the considered studies will be discussed as well as the biological behavior. Regarding the biomedical applications of both selected laser processing strategies, we took a thorough look at the anti-infective, bioactive, and multifunctional peculiarities related to these materials.

2.2 LASER PROCESSED ANTIMICROBIAL NANOSTRUCTURED COATINGS

A particular case of laser-processed nanostructured materials with specific biomedical applications is represented by the antimicrobial coatings. Tremendous interest, multidisciplinary knowledge, and substantial financial support have been directed toward this particular research field. Given the alarming data regarding the increasing number of pathogenic microorganisms which exhibit resistance to conventional pharmaceutical treatments (Frieri et al., 2017; Morio et al., 2017; Wang et al., 2017a), but also the concernings regarding the therapeutic inefficiency or severe side effects related to classic antimicrobial drugs (Campagna, 2016; Vitali et al., 2017; Wijdeven et al., 2016), there is a worldwide demand toward the experimental development and clinical implementation of novel and enhanced antimicrobials.

With respect to the laser-processed nanomaterials for antimicrobial applications, an attractive strategy embraced by the scientific community is to modify the clinically used medical devices by means of thin protective coatings embedded with commercial drugs or naturally derived compounds which possess acknowledged antimicrobial activity. In this case, the surface modification of IMDs by thin coatings is performed in order to induce them active potential against pathogenic microorganism contamination and colonization. Such particular coatings are designed as unharmful systems for normal cells and possess insignificant or absent effects on their physiology and functionality, acting only as active carriers or enhancers of the therapeutic agent and, thus, providing a local targeted or/and controlled anti-infective strategy. Promising results regarding the synthesis of protective coatings with enhanced antimicrobial efficiency were reported when considering the incorporation of bactericidal drugs (Bakhsheshi-Rad et al., 2017; García et al., 2017; Ionita et al., 2017; Li et al., 2017a), bacteriostatic drugs (Geißler et al., 2016; Gonzalez et al., 2018; Jabeen et al., 2017; Radda'a et al., 2017), antifungal drugs (Robles et al., 2016; Sharifi et al., 2016), and plant-derived phytochemical substances with acknowledged antimicrobial efficiency (Chan et al., 2016; Cristescu et al., 2016; Paris et al., 2017; Wang et al., 2017b).

As it was reported in a recent study performed by Grumezescu et al. (2017), the MAPLE method was successfully used to obtain coatings based on lysozyme-loaded polymeric microspheres for titanium-based materials. The incorporation of the naturally derived enzyme (Lys) within the poly(3-hydroxybutyrate-*co*-3-hydroxyvalerate) copolymer [P(3HB−3HV)] and P(3HB−3HV)/PEG (polyethylene glycol) composite was achieved by employing the solvent evaporation method followed by lyophilisation procedure. In terms of compositional integrity and preserved stoichiometry and morphology, the best results were obtained by using the 400 mJ/cm^2 laser fluence. This optimized laser processing strategy resulted in the synthesis of uniform thin coatings with sphere-like particulate morphology (the mean particle size being estimated around 2 μm) and particular nanoporous structure in the case of P(3HB−3HV)/PEG/Lys coatings. For the biological evaluation, the authors used human-derived SaOs2 osteoblast (OB)-like cells and EAhy926 endothelial cells. In comparison with the uncoated samples, the grade 4 titanium disks coated with P(3HB−3HV)/Lys and P(3HB−3HV)/PEG/Lys materials resulted in cellular proliferation improvement after 72 hours of treatment. For both groups of lysozyme-modified titanium materials, an enhanced biological behavior was noticed since the collected fluorescence and scanning electron micrographs evidenced the beneficial cellular attachment, spreading, morphology, and metabolism of the OB-like and endothelial cells. For all titanium-coated samples a prolonged efficiency against *Staphylococcus aureus* and *Pseudomonas aeruginosa* bacterial biofilm formation was evident for up to 72 hours. However, a particular enhanced and sustained antimicrobial activity was identified in the case of the PEG-containing composites where the intrinsic porous microstructure provided a faster and more efficient release of the lysozyme (Grumezescu et al., 2017).

The acknowledged pronounced intrinsic efficiency of lichen-synthesized usnic acid (UA) against Gram-positive pathogens directed various research studies toward the nanotechnology-assisted revaluation of this natural dibenzofuran derivative efficiency in order to extend its antimicrobial applications. The antibiofilm potential of composite coatings based on γ-cyclodextrin (γ-CD) loaded with UA for unconventional antibacterial applications was investigated by Iordache et al. (2015) who experimentally synthesized γ-CD/UA thin films using the MAPLE technique. In terms of chemical composition integrity and functional groups distribution, the best results were reported for composite γ-CD/UA materials processed by using the 500 mJ/cm^2 laser fluence which enabled the formation of continuous and uniform nanosized coatings with smooth topography, the estimated thickness being below 100 nm. When assessed in the presence of EAhy926 endothelial cultures, the γ-CD/UA-coated samples proved to be suitable materials for normal adhesion, migration, and proliferation of human-derived healthy eukaryote cells. The microbiology data, obtained after different treatment periods in the presence of *S. aureus* strain showed significant inhibitory effects against bacterial contamination and biofilm development. Moreover, the authors reported a prolonged antibiofilm activity, thus, emphasizing the potential revaluation of UA-loaded coatings to improve the resistance of medical devices against staphylococcal colonization (Iordache et al., 2015).

The same laser processing strategy was employed in a study performed by Grumezescu et al. (2014a) to modify the surface of titanium substrates with thin coatings based on polylactide (PLA) and polyvinyl alcohol (PVA) microspheres loaded with natural-derived UA. For synthesized PLA–PVA–UA composite materials, extensive infrared studies indicated the 300 mJ/cm^2 value as the optimal laser fluence for the stoichiometric transfer of the composite microspheres which resulted in the homogenous distribution of sphere-like microsized particles. In terms of cellular viability, cellular attachment, and particular cellular morphology, no significant alterations were reported after 24 hours of treatment of mesenchymal stem cells (MSCs) in the presence of titanium substrates coated with PLA–PVA–UA materials. The biocompatibility results emphasized the potential use of the obtained PLA–PVA–UA composite coatings as suitable materials for the normal development of human-derived cells. Moreover, the UA-loaded coatings significantly impaired the colonization of the samples by the *S. aureus* strain, being effective both during early and late stages of microbial biofilm development. It is worth mentioning that the PLA–PVA–UA coatings mostly inhibited the formation of bacterial biofilm after 72 hours of evaluation, possibly thanks to the prolonged release of the active substance from the composite microspheres (Grumezescu et al., 2014a).

In order to experimentally induce antimicrobial properties to available medical devices another attractive approach of the worldwide scientific community consists in the revaluation of metal-derived nanosystems with intrinsic efficiency against pathogenic microorganisms. In terms of enhanced antimicrobial efficiency of protective coatings (pristine ceramic or polymeric, and composite or hybrid materials), promising experimental data were reported when considering the embedding of metallic nanoparticles such as copper (Gotzmann et al., 2018; Hadidi et al., 2017; Wojcieszak et al., 2017) and silver (Nie et al., 2017; Písařík et al., 2017; Siegel et al., 2017), but also the incorporation of metallic oxide nanostructures, such as copper oxide (El Saeed et al., 2016; He et al., 2017; Sonia et al., 2016), titanium oxide (Esparza et al., 2017; Pessoa et al., 2017; Rodriguez et al., 2017), and zinc oxide (Alves et al., 2017a,b; Karbowniczek et al., 2017; Roknian et al., 2018). Such nanosystem-containing materials represent a confined compromise between the potential cytotoxicity manifested against normal cells and the antimicrobial effects exhibited against pathogenic organisms, whereas they can be experimentally tuned in terms of composition, microstructure and biofunctionality.

The potential use of porous composite coatings based on poly(methyl methacrylate) (PMMA) and silver nanoparticles (AgNPs) as antimicrobial materials for implantable devices was recently studied by Petrochenko et al. (2017). The authors experimentally used the PLD strategy and reported obtaining continuous PMMA/Ag layers with particulate and porous structures and with uniformly distributed spherical AgNPs (the particle size ranging between 10 and 200 nm). By increasing the applied number of pulses (10k, 14k, and 20k), the authors estimated the increasing tendency of the composite coatings microsized thickness, but also the enhanced accumulation and subsequent release of metallic silver from the composite porous materials under culture medium conditions (the metallic silver release values being estimated as 0.76, 1.05, and 1.67 μg/mL, respectively). For what concerns the quantitative biological assessment of the PMMA/Ag coatings in the presence of human bone marrow stromal cells for 24 hours, the collected data evidenced an indirect dependence between the cellular proliferation results and the amount of the incorporated AgNPs. Further qualitative flow cytometry assays revealed that both apoptotic and necrotic cellular processes occurred when increasing the Ag content within the composite coatings. In the particular case of the thinnest composite (10k PMMA/Ag), no necrotic processes were evidenced, while the reported apoptotic events (quantified as 12%) were statistically insignificant. A bacterial biofilm formation test performed under dynamic conditions against *Escherichia coli* strain after 24 hours of incubation evidenced the pronounced efficiency of all experimental coatings against bacterial contamination and colonization. The reported data evidenced the potential use of PLD technique for the synthesis of thin PMMA/Ag composite coatings with suitable compatibility for human-derived cells and enhanced antibacterial activity (Petrochenko et al., 2017).

AgNPs functionalized with sodium stearate with exclusive core/shell spherical morphology and 7.95 ± 2.55 nm average particle size resulted in no preferential affinity toward the most vital organs of albino mice, as evidenced both after 7 and 14 days of experimental treatment. Instead, significant histological alterations were reported at the splenic level since the white pulp hypertrophy and the dark-brown aggregate structures identified within the red pulp of the spleen were reported (although inversely related to the inoculation period). According to the infrared studies performed on the MAPLE-processed samples, the best data with respect to the chemical composition and functional groups distribution were obtained in the particular case of the 600 mJ/cm^2 laser fluence. The as-selected laser processing enabled the formation of uniform coatings characterized by particulate nanostructured morphology, rough topography, and $\sim$500 nm mean thickness. When assessed with regard to the antimicrobial activity, the AgNPs-coated polyvinyl chloride catheter sections significantly inhibited the bacterial adhesion and mature biofilm development of *E. coli* and *S. aureus* strains exhibiting a prolonged efficiency against biofilm formation. In comparison, a reduced (but still sustained) inhibitory activity against biofilm formation was reported in the case of *P. aeruginosa*, thus, suggesting the extensive potential use of AgNPs-based thin coatings as a versatile strategy to fight microbial contamination, colonization, and associated infection (Fufă et al., 2015).

The research group of Huang et al. (2016) used the MAPLE approach to obtain composite coatings based on zinc oxide (ZnO) nanoparticles stabilized with PEG onto the surface of silicone hydrogels. Prior to the MAPLE process, ZnO–PEG rod-shaped particles with 10 ± 4 nm average size were synthesized using the sol–gel method. The selected laser processing (performed by using the 532 nm laser beam of a Nd:YAG source) induced a preferential spherical morphology and lower dimension (9 ± 3 nm) of the ZnO–PEG nanoparticles, but the stoichiometric transfer of the composite nanostructures and the formation of a homogenous thin film with granular morphology onto the surface of photo-polymerized hydrogel were confirmed. The ZnO–PEG-coated hydrogels exhibited improved mechanical properties since higher Yong's modulus values were reported (0.815 ± 0.03 and 0.708 ± 0.03 MPa corresponding to the as-modified and pristine hydrogels, respectively) and the photoluminescent features of the metallic oxide were preserved. A significant inhibition of bovine serum albumin adsorption (up to 54%) was noticed in the case of ZnO–PEG-coated hydrogels, as assessed in comparison with the unmodified samples under physiologically simulated conditions. The biological tests performed on NIH/3T3 fibroblasts (derived from Swiss mouse embryo from National Institutes of Health) showed the enhanced cellular compatibility of the coated silicone hydrogels on eukaryote cells. Furthermore, it was reported that pristine silicone hydrogels possessed a substantial time-related affinity for bacterial contamination while the ZnO–PEG-coated materials resulted in the contrary time-related and significant efficiency against *E. coli* and *S. aureus* adhesion and biofilm development (Huang et al., 2016).

However, a particular interest was directed toward the incorporation of versatile metallic oxides within novel coatings which are designed to induce antimicrobial effects to the clinically used medical devices. Except for their acknowledged intrinsic activity against pathogenic microorganisms (Halbus et al., 2017; Kumar et al., 2017; Mehta, 2017; Raghunath & Perumal, 2017), the size-related features of nanostructured metallic oxides (including physical, chemical, biological, and functional characteristics) extended their preclinical use in anti-infective therapy. Therefore, the versatile tunability related to metallic oxide nanostructures reveals tremendous attractive strategies toward the development of novel platforms intended to provide an enhanced and even personalized antimicrobial therapy.

The research group of Rădulescu et al. (2016a) successfully used the solvent evaporation procedure to load the broad-spectrum antibiotic Zinforo (ZiN) within high purity spherical mesoporous silica nanoparticles which were obtained with uniform dimensional distribution (in the range 200–400 nm) and mean pore size around 2.3 nm by using the templating synthesis strategy. The as-obtained SiO_2/ZiN nanostructured systems proved favorable biological behavior when evaluated on a mouse animal model since only the splenic tissue revealed some histological modifications after the performed treatment. The presence of dark-brown inorganic aggregates within the splenic red pulp and the hypertrophy of splenic white pulp were evidenced both after 7 and 14 days of SiO_2/ZiN treatment. Such events were quite predictable, if we consider the acknowledged role of spleen within the innate immune system. Subsequently, the antibiotic-loaded silica nanostructures were comparatively processed by using the MAPLE method, the optimal results—in terms of improved deposition rate and chemical structure preservation—being assigned to the samples obtained at 500 mJ/cm^2 laser fluence. In this particular case, the performed infrared mapping and scanning electron microscopy studies confirmed the stoichiometric transfer of the SiO_2/ZiN nanosystems by MAPLE in the form of homogenous particulate coatings with a thickness below 400 nm. When compared to pristine commercial pure titanium, the metallic substrates covered with the SiO_2/ZiN laser-processed coatings resulted in enhanced cellular viability and unaltered metabolism of EAhy926 endothelial cells for up to 3 days of evaluation. Moreover, the SiO_2/ZiN-coated titanium samples also proved to significantly affect the development of *E. coli* biofilm, being effective especially during the early stage of bacterial contamination (the corresponding CFU/mL values were reduced with more than two orders of magnitude). The reported data suggest the potential use of SiO_2/ZiN-based materials for the fabrication of implantable devices with improved biological activity and effective antimicrobial action (Rădulescu et al., 2016a).

A modified coprecipitation method was implied in a study performed by Grumezescu et al. (2015b) in order to experimentally load small amounts

(~4 wt.%) of Kanamycin aminoglycoside onto the surface of 5 nm sized crystalline magnetite particles. In terms of preserved stoichiometry, the best results regarding the MAPLE processing of the obtained Fe_3O_4@KAN nanosystems were reported for the 500 mJ/cm^2 laser fluence value. Therefore, nanostructured thin coatings with particulate and porous morphology (due to the pronounced aggregation of functionalized nanosized magnetite) and rough topography (with 6.1 nm average roughness) were obtained. The biological assays performed on human-derived EAhy926 cells for up to 5 days showed the enhanced and sustained biocompatible behavior of the Fe_3O_4@KAN-coated substrates regardless the incubation time. Moreover, the biodistribution tests—performed on a BALBc mouse animal model for the collected composite coatings—emphasized that no histological modifications, nor physiological alterations were identified within the cerebral, hepatic, and renal tissues. Instead, the agglomeration of the Fe_3O_4@KAN nanosystems was reported at the pulmonary level, suggesting the potential use of such particular platforms for the targeted treatment of respiratory conditions. During the 72 hours of antimicrobial assessment, the aminoglycoside-loaded coatings resulted in prolonged and significant impairment of the monospecific biofilms associated to *E. coli* and *S. aureus* pathogens, with a particular remark against the initial contamination stages of biofilm development. Thereby, thin coatings based on antibiotic-functionalized magnetite nanoparticles were suggested as a promising unconventional strategy against the microbial contamination and colonization of medical devices (Grumezescu et al., 2015b).

The solvent evaporation process was reported by Stan et al. as an efficient approach to load the natural-derived UA onto the surface of 5 nm sized wurtzite hexagonal zinc oxide (ZnO) nanoparticles synthesized in the presence of sodium stearate (C_{18}). By further using the 248 nm excimer laser beam and the associated comparative infrared investigations, the optimal laser fluence (400 mJ/cm^2) for the stoichiometric MAPLE-assisted transfer of the synthesized nanosystems was identified. Continuous and uniform ZnO-based coatings (loaded or not with UA) with particulate morphology, irregular surface, and variable thickness (ranging between 350 and 600 nm) were, thus, obtained. However, in the particular case of the ZnO@C_{18}–UA coatings, a reduced tendency of aggregate structures formation was identified (possibly thanks to the beneficial spacer role of the natural compound). After performing qualitative (phase contrast microscopy) and quantitative (cellular viability and inflammatory NO release) biological assays in the presence of human-derived CCD-1070SK fibroblast cell cultures, the authors observed that both ZnO-coated series resulted in significant inhibition of eukaryotic cells adhesion and migration, but there were no clear signs of cytotoxicity. Moreover, the long-term evaluation of the ZnO@C_{18}-collected nanosystems on BALBc mice (performed for up to 10 days), evidenced the nonpreferential affinity for most vital organs (possessing or not possessing nonspecific immune systems) except for the spleen. When assessed in the presence of a *Salmonella enterica* strain, the

$ZnO@C_{18}$-coated materials resulted in the prominent inhibition of initial bacterial contamination, while the $ZnO@C_{18}$–UA-modified substrates significantly impaired both early pathogenic contamination and subsequent mature biofilm formation (Stan et al., 2016).

A complementary study performed by the same research group strengthened the purpose of nanostructured coatings loaded with UA as unconventional and effective antimicrobials. Prior to this, the authors synthesized myristic acid-functionalized magnetite nanoparticles ($Fe_3O_4@C_{14}$) with spherical morphology, physical dimensions ranging between 5 and 20 nm and core/shell structure. Subsequently, they loaded the UA and the as-modified magnetic nanosystems within composite particles based on poly(lactic-*co*-glycolic) acid (PLGA) and PVA with an exclusive sphere-like shape and microsized dimensions. The minimum inhibitory concentration of the as-synthesized composite microparticles against planktonic *S. aureus* pathogens was experimentally assessed as only 3.9 μg/mL. By applying adequate MAPLE processing at 300 mJ/cm^2 laser fluence (as identified after a comparative laser fluence study), PLGA–PVA–$Fe_3O_4@C_{14}$–UA composite coatings were obtained on commercially pure titanium substrates. The biological assays performed on MSCs derived from human bone marrow showed beneficial attachment and spreading of cells, but also superior cellular viability (the reported quantitative results being similar to those of control samples). Moreover, the UA-loaded composite coatings resulted in the significant impairment of staphylococcal biofilm, being specifically effective during the mature stages of biofilm formation (Grumezescu et al., 2014b). The promising results reported with respect to the enhanced antibacterial activity of UA-loaded materials processed by MAPLE confirms the extended antimicrobial efficiency and the potential clinical use of novel antimicrobials based on the naturally derived UA.

In similar research, the synthesis of composite coatings based on polymeric microspheres consisting in P(3HB–3HV) and PVA loaded with eugenol-functionalized Fe_3O_4 nanoparticles was reported. The naturally derived phenylpropene was firstly superficially immobilized onto spherically shaped magnetite nanoparticles (with mean size below 10 nm and narrow dimensional distribution), followed by the successful encapsulation of the resulted $Fe_3O_4@E$ nanosystems within composite P(3HV–3HV)–PVA matrix. The infrared studies performed on the complex P(3HV–3HV)–PVA–$Fe_3O_4@E$ materials processed by MAPLE means indicated the 300 mJ/cm^2 laser fluence as the optimal value for chemical and functional preservation of the initial composite microspheres. The as-performed experimental processing enabled the synthesis of homogenous and uniform complex nanostructured coatings, which were assessed as suitable materials for the beneficial adhesion, spreading, and proliferation of EAhy926 human endothelial cells for up to 5 days. The microbiology results proved that the P(3HV–3HV)–PVA–$Fe_3O_4@E$-coated samples significantly impaired the early stages of *P. aeruginosa* biofilm formation while the long-term inhibitory effects were considerably affected. Instead, when assessed in the presence of a *S. aureus* strain, the experimentally modified substrates resulted in the strong and durable inhibition of bacterial biofilm development. Therefore, the authors outlined the promising potential of magnetite-embedded materials for potentiating the intrinsic antimicrobial effects of various biosubstances (Grumezescu et al., 2014c).

2.3 LASER PROCESSED BIOACTIVE NANOSTRUCTURED COATINGS

Even if tremendous progress with respect to the development and clinical implementation of a new generation of biomaterials was reported, there are still various biological and medical applications in which evolution has been decelerated due to the use of biologically inert materials. For particular applications that require the utilization of biologically active biomaterials (especially tissue restoration, replacement, and regeneration), an attractive and promising approach to exceed the inertness of commercially available and clinically used medical devices is to induce them superficial bioactivity. In this respect, the experimental design and fabrication of thin protective coatings with enhanced biological activity represent a promising, versatile, and practical strategy toward the successful accomplishment of this particular desideratum.

By performing an extensive study, the research group of Li et al. (2017b) proposed a laser-assisted superficial modification of commercial polyethylene terephthalate (PET) grafts used for anterior cruciate ligament replacement with bioactive silicate glass (BG) embedded with copper (Cu–BG). The PLD process, performed under oxygen-enriched atmosphere, enabled the synthesis of amorphous uniform coatings consisting of spherical nanostructured aggregates with a narrow size distribution and a pronounced hydrophilic behavior (the water contact angles of both bioglass-coated series were $\sim 30^\circ$, while for the pristine PET specimens a mean value of $\sim 142^\circ$ was reported). The concentration of Ca, Si, and P ions was similar for up to 7 days of evaluation under culture medium conditions regardless the surface modification of PET samples. However, the Cu^{2+} release from Cu–BG/PET materials recorded a maximum value of 0.224 ppm after 3 days of experiments, followed by a decrease to 0.1479 ppm at the end of the immersion. Regarding the biological behavior of the polymeric grafts modified by PLD (BG/PET and Cu–BG/PET) after 7 days of treatment in the presence of MSCs isolated from rat bone marrow, favorable cellular attachment, enhanced cellular viability, increased Ca^{2+} levels, and improved expression of osteogenic and angiogenic genes and proteins were evident. Since the reported data were by far enhanced in the particular case of the PET grafts coated with Cu–BG materials, the beneficial presence of Cu^{2+} on exceeding the bioinertness of PET materials was established. The authors also reported opposite biological results after the addition of small interfering RNA encoding the gene of 1α hypoxia-inducible factor (HIF-1α), thus, emphasizing the beneficial expression of HIF-1α gene in manifold cellular processes, including proliferation, osteogenesis, and angiogenesis. After 12 weeks of healing assays performed on a goat animal model, superior bone regeneration and excellent neovascularization processes, but also enhanced pull-out load results, were identified in the particular case of the Cu–BG-coated PET grafts compared to both pristine PET and BG-coated PET devices. The copper-containing BG coatings synthesized by PLD method proved to represent an efficient and attractive strategy to enhance the physicochemical and biofunctional features of PET artificial grafts used in ligament reconstruction applications (Li et al., 2017b).

A multi-stage emulsion-diffusion-evaporation protocol was used for the successful microencapsulation of naturally derived quercetin (Q) within polymeric structures of PLGA. By using the MAPLE technique, thin PLGA/Q coatings were synthesized. The selected laser processing resulted in the physicochemical preservation of the initial composite materials and also the formation of uniform coatings with preferential grainy-like aspects and nanosized topography (experimentally evaluated as ~500 nm). The biological behavior of the PLGA/Q-coated samples was assessed on mesenchymal cells derived from mouse calvaria (MC3T3) for different periods. When compared to uncoated and pristine PLGA-coated samples (which encouraged similar cellular proliferation), the PLGA/Q-coated specimens resulted in slightly reduced cellular viability after 11 days of incubation. However, a particular multi-component behavior of the pre-OB cells was evidenced, the reported data rather suggesting the cellular uptake of flavonoid during the first week followed by a favorable activity related to quercetin on the remaining proliferation process. In terms of cellular attachment and proliferation, complementary fluorescence data confirmed the controversial activity of the PLGA/Q coatings during the first week of evaluation, but evidenced the preferential OB-like differentiation after 3 days of treatment. Moreover, during the second week of biological assessment, superior cellular density and preserved OB-like phenotypes were noticed, thus, providing evidence of the long-term biocompatible behavior of quercetin-embedded polymeric coatings (Fufă et al., 2017).

A particular interest has been directed toward the laser-assisted development of bioactive coatings for metallic implants used in orthopedic and orthodontic applications. Thanks to their intrinsic mechanical properties, physicochemical features, and native biocompatible behavior, pure titanium and its alloys represent the current *gold standard* in hard tissue restorative and replacing interventions. However, the acknowledged biological inertness of titanium-based materials, in conjunction with the recently reported data regarding the occurrence of related healthcare conditions (Behzadi et al., 2017; Manam et al., 2017; Revathi et al., 2017) encouraged worldwide healthcare and scientific communities to turn their attention toward nanotechnology-derived biomaterials. Concerning such particular medical applications, specific material-related features must be considered, such as osteoconduction, osteoinduction, and osseointegration (Chen et al., 2018; Martin & Bettencourt, 2018; Sethu et al., 2017; Su et al., 2018).

Given their specific similarity with natural bone apatite, but also thanks to their tunable compositional and microstructural features and versatile biofunctionality, calcium phosphate (CaP) ceramics are suitable materials for biomedical applications related to the modern treatment of osseous tissue diseases. Among CaP materials, the most-studied representatives for specific orthopedic and orthodontic applications are synthetic hydroxyapatite (HA) and tricalcium phosphate (TCP) (Fernandez-Yague et al., 2015; Ridi et al., 2017; Szcześ et al., 2017; Wang & Yeung, 2017). In this respect, tremendous attention and experimental research activity have been directed toward the utilization of these particular CaPs (both as pristine and composite materials), with promising results being reported regarding the development and clinical implementation of thin bioactive coatings based on HA (Domínguez-Trujillo et al., 2018; Shamray et al., 2017; Yu et al., 2017a) and TCP (Choy et al., 2017; Pillai et al., 2018; Prosolov et al., 2017).

A relatively unexplored, but attractive, representative of calcium phosphate materials with potential applications toward an improved osseous integration of metallic implants is octacalcium phosphate (OCP). The structural surface modification of pure titanium substrates by OCP coatings was recently reported. Smirnov et al. (2017) used the 532 nm Nd:YAG laser beam for the PLD of adherent and uniform microsized calcium carbonate (CC) layers, which have been used as precursors for the biomineralization of OCP materials. The synthesized sole calcite coatings were intermediary transformed into dicalcium phosphate dihydrate after 7 days of incubation in a calcium nitrate solution at 40°C under acidic conditions, followed by the entire transformation into OCP materials after a similar chemical treatment in sodium acetate solution performed under alkaline conditions. The as-obtained uniform and continuous CaP-based coatings, with their particular mesh-like morphology and thickness of hundreds of nanometers, proved superior mechanical adhesion onto the metallic substrates. When assessed in the presence of human-derived myofibroblast cells for up to 3 days and compared to pristine titanium and CC-coated samples, the metallic substrates modified with OCP coatings resulted in superior cellular viability and proliferative activity (assessed quantitatively and qualitatively by estimating the mitotic index and by determining the cellular attachment and spreading, respectively). Moreover, for both human myofibroblasts and bone marrow mesenchymal stromal cells, superior proliferation results were reported in the case of the OCP-coated titanium for up to 72 hours, thus, predicting the beneficial interfacial interactions between the bone tissue and the OCP-coated implants (Smirnov et al., 2017).

An extensive study performed by Boanini et al. (2015) reported the MAPLE synthesis of composite coatings based on calcium alendronate monohydrate ($CaAL{\bullet}H_2O$) and OCP for enhanced bone integration of titanium implants. In this respect, pristine OCP and OCP-based materials (obtained by prior reaction with solutions containing 8 and 20 mM of alendronate, respectively) were processed by using a KrF* excimer laser beam. For what concerns the composition and microstructure of the resulted coatings (denoted by the authors as cOCP, cAL8, and cAL20, respectively), the selected laser processing method enabled the complete preservation of the initial materials' stoichiometry and morphology. The composite nature of the synthesized coatings was confirmed only in the particular case of the cAL8 materials (corresponding to 48 wt.% CaAL•H_2O and 52 wt.% OCP), whereas for the cAL20 samples a complete transformation into CaAL•H_2O was reported. However, similar thickness ($\sim$100 nm) and average roughness (estimated around 0.2 μm) were evidenced, regardless the composition of the coatings. According to the degradation studies performed under physiologically simulated conditions, the highest amount of calcium was released from the cAL20-coated samples, reached a maximum value after 2 days, and stayed constant for up to 7 days. The biological assays performed for up to 2 weeks on co-cultures of stem cells-derived OBs and monocytes-derived osteoclasts (OCs) isolated from osteoporotic patients indicated the beneficial adhesion, spreading, and proliferation of the OB (the results being comparable for control, reference, and as-modified samples), but also the pronounced anti-osteoclastogenesis activity (the cellular viability and OC differentiation process being significantly affected by the cAL8 and cAL20 materials). Moreover, by monitoring different molecules, the authors emphasized a different metabolic activity of OBs and OCs after 14 days of single culture and co-culture assays and provided comprehensive insight regarding the anti-osteoporotic potential of OCP-based coatings (Boanini et al., 2015).

Even if the chemically assisted approach is preferred by most research groups to synthesize OCP coatings, the laser processing of this particular calcium phosphate represents a versatile and tunable strategy to synthesize OCP-based materials. The promising results reported with respect to the superior bioactivity of OCP nanostructured coatings obtained by PLD and MAPLE methods (Smirnov et al., 2017; Boanini et al., 2015) gathered with similar data regarding the use of OCP materials in bone-related restoring or regenerating applications (Forte et al., 2018; Jiang et al., 2015; Zorin et al., 2014), outline the idea that the increased amount of Ca^{2+} within OCP coatings seems to represent a beneficial strategy to improve the osseointegration of metallic implants for orthopedic and orthodontic applications.

Besides calcium phosphate materials, various ceramics have been studied as protective coatings with enhanced biological functionality and proved to be promising candidates for inducing bioactivity to metallic implants. Relevant examples of ceramic-based bioactive coatings include alumina (Wang et al., 2016; Zykova et al., 2016), zirconia (Liu et al., 2018; Wang et al., 2017c), titania (Bait et al., 2017; Jemat et al., 2018; Nasirpouri et al., 2017), and different glassy-containing ceramics (Bellucci et al., 2017a; Yu et al., 2017b). However, superior results (in terms of enabling, enhancing, or promoting the physiological growth of bone tissue) were reported when considering inorganic composites (Alves et al., 2017; Bellucci et al., 2017b; Ghorbel et al., 2017; Wang et al., 2017d) and biological-mimicking inorganic-organic composite materials (Al-Rashidy et al., 2018; Jugowiec et at., 2017a,b; Ren et al., 2018; Wang et al., 2017e).

According to the research study performed by Wang et al. (2018), PLD successfully enabled the synthesis of composite coatings based on HA and 45S5 BG. The experiments performed under an argon-enriched atmosphere at 200°C substrate temperature resulted in the formation of preponderant grainy compact films consisting in spherical nanostructures with a particularly smooth aspect. The thermal treatment applied to the HA/BG-coated samples at 800°C for 1 hour encouraged the formation of irregularly shaped nanostructured particles, while the same thermal treatment applied for 2 hours resulted in preferential acicular morphology and increased amount of HA crystalline phase. The post-deposition annealing protocol also resulted in improved superficial adhesion of the HA/BG coatings onto the metallic substrates since the corresponding scratch tests indicated an increasing tendency of the mean critical load from 15.03 N (corresponding to unmodified titanium substrates) to 17.17 and 16.83 N (corresponding to the composite coated samples thermally treated for 1 and 2 hours, respectively). The bioactive potential of the PLD-modified titanium implants was assessed on shin bones of a rabbit animal model. After 1 month of evaluation, reduced interactions between bone tissue and implant surface were evidenced in the case of initial HA/BG-coated samples, as well as no new bone formation. On the other hand, enhanced osteoinductive potential was observed for the titanium implants modified with both thermally treated HA/BG coatings. However, superior data (in terms of new bone formation and strong superficial interactions between implant and bone tissue) were assigned to the modified implants subjected to a 2-hour thermal protocol. Significant new bone formation was observed after 3 months post-intervention for all implants, but quite perfect osteointegration was reported for the titanium implants coated with thermally treated HA/BG materials (Wang et al., 2018).

In a previous study, the same research group proved the enhanced osteointegration of Ti6Al4V materials modified with composite coatings consisting in HA and silicate BG obtained by PLD, in this respect using an excimer laser source. The depositions were performed under oxygen atmosphere while applying different thermal treatments to the metallic substrates during experiments (200°C, 400°C, and 600°C, respectively). Thus, compact and uniform nanostructured HA/BG coatings with predominant particulate morphology were obtained and a temperature-related increasing tendency of Ca/P atomic ratio was identified (the corresponding values being 2.46, 2.55, and 2.6, respectively). Also, when increasing the substrate temperature, an increase of the HA crystalline phase was observed. One month after implantation in the rabbit shin bone, the HA/BG-coated titanium materials resulted in significant new bone tissue formation regardless the thermal treatment conditions. However, a particular remark was done with respect to the samples deposited at 600°C substrate temperature which resulted in the best osteogenic potential and excellent osteointegration of Ti6Al4V materials (Wang et al., 2017f).

Titanium-based binary alloys with various amounts of zirconium (Zr–Ti) were successfully modified with coatings consisting in HA, zirconium oxide, and AgNPs by means of PLD. The ultrasound-assisted multi-step synthesis process of the initial composite materials enabled the formation of spherical-shaped composite particles with 70 nm mean size, while the laser deposition method resulted in the formation of HA–ZrO_2–Ag continuous films with particular aggregate-like irregular morphology and uniform nanosized thickness onto the surface of Zr–Ti substrates. The performed compositional studies showed the stoichiometric laser-assisted transfer of the composite material. The electrochemical evaluation, performed in Ringer's solution containing albumin under a physiologically simulated environment revealed that the HA–ZrO_2–Ag-coated Zr–Ti samples exhibited improved stability and resistance to corrosive processes. Moreover, a prolonged time of electrochemical evaluation, accompanied by surface microstructural investigation, indicated that significantly enhanced stability and anti-corrosive behavior were particularly assigned to the composite-coated Zr–Ti alloy containing the highest amount of titanium. However, it is worth mentioning that all HA–ZrO_2–Ag-coated samples, overall, induced an improved long-term anti-corrosive behavior under physiologically simulated conditions (regardless the composition of the alloy substrate), possibly thanks to the beneficial adsorption of the considered human-derived protein (Mareci et al., 2016).

In order to develop vitroceramic-based thin coatings for Ti–Zr implantable alloys, Busuioc et al. (2017) used a PLD methodology to process materials pertaining to the SiO_2–CaO–P_2O_5 and SiO_2–CaO–P_2O_5–CaF_2 systems. The vitroceramics were processed under oxygen atmosphere in thermal conditions (at 300°C and 400°C). Regardless of the target composition, the PLD process resulted in the synthesis of uniform coatings with 350–400 nm mean thickness and granular aspect (related to the increased aggregation tendency of the constituent spherical grains with narrow size distribution, ranging between 10 and 25 nm). When compared to the pristine alloy, the vitroceramic-coated Ti–Zr samples were distinguished by a pronounced hydrophilic behavior and an increased surface energy which represents the essential aspects for beneficial cellular interactions. The biological behavior of the Ti–Zr coated and uncoated samples was assessed in the presence of amniotic fluid-derived MSCs. As the viability results showed, the coated alloy samples resulted in similar, but slightly reduced, cellular proliferation after 24 hours of treatment, when compared to the pristine Ti–Zr materials. Instead, after 96 hours of incubation, the cellular viability was significantly increased, although still similar to the control samples except for the alloy samples coated with CaF_2-free material at 300°C. In this particular case, the count of viable cells exceeded the other samples (either coated or uncoated) by far. When compared to the uncoated samples, the vitroceramic-coated Ti–Zr materials also resulted in significantly reduced antioxidant activity (as was evidenced by the total glutathione assay), which is a clear indicator of favorable long-term interactions with the cells. Moreover, the performed fluorescence microscopy studies did not show any signs of morphological and metabolic cellular impairment, confirming that the as-coated Ti–Zr materials are suitable substrates for the beneficial adhesion, proliferation, and normal growth of the progenitor cells (Busuioc et al., 2017).

In a similar study, vitroceramics were employed by the same research group in order to obtain thin coatings for medically graded titanium materials by means of PLD. The selected laser processing (performed under an oxygen environment at 400°C substrate temperature) enabled the formation of a majority of wollastonite-based materials with particular orthorhombic and tetragonal crystalline structures corresponding to the SiO_2–CaO–P_2O_5 and SiO_2–CaO–P_2O_5–CaF_2 vitroceramics, respectively. The morphological investigations revealed the synthesis of a particulate layer with thickness around 50 nm consisting in homogenously distributed quasi-spherical grains with a 25 nm mean size. In addition, the compositional analysis confirmed the stoichiometric laser transfer of the vitroceramics onto the metallic substrates while the contact angle measurements emphasized the increased hydrophilicity of the vitroceramic-coated titanium materials. The quantitative and qualitative biological assays performed on human-derived EAhy926 endothelial cell cultures after 24 hours of treatment showed the beneficial contribution of the vitroceramic-coated samples with respect to the enhanced cellular viability and normal cell growth. Similarly to the previous research study, significantly increased cellular adhesion and proliferation were reported in the case of the SiO_2–CaO–P_2O_5 vitroceramic-coated titanium, particularly assigned to the presence of orthorhombic calcium silicate within the coating composition (Voicu et al., 2016).

Composite coatings based on poly(3-hydroxybutyrate-*co*-hydroxyvalerate) (PHBV) copolymer and calcium phosphates (HA and β-TCP) were synthesized and evaluated by Raşoga et al. (2017) as bioactive coatings for enhanced osseous integration of titanium-based materials. The performed infrared spectroscopy and X-ray diffraction studies emphasized the successful use of the MAPLE technique for the stoichiometric transfer of both organic and inorganic materials regardless the laser beam parameters. However, depending on the selected calcium phosphate, the microstructure of the synthesized materials was significantly impacted since the PHBV/HA coatings possessed uniform surfaces with specific spherical particulate structures distributed within the polymer matrix while the surface of the PHBV/β-TCP coatings mainly consisted in randomly distributed irregular inorganic aggregates. As a general remark related to the distinctive morphological and topographical features of the synthesized coatings, the addition of both calcium phosphates within the copolymer matrix resulted in pronounced hydrophilic behavior, thus, indicating the potential beneficial interactions of such materials with eukaryotic cells. In the particular case of the PHBV/HA material, the long-term biocompatibility assay performed after 28 days evidenced an increased cellular attachment and proliferation of mesenchyme-derived progenitor cells (MSCs), thus, indicating that the obtained composite coatings may represent suitable substrates for prolonged favorable interactions with bone cells (Raşoga et al., 2017).

2.4 LASER PROCESSED MULTIFUNCTIONAL NANOSTRUCTURED COATINGS

By using the PLD technique, Hidalgo-Robatto et al. (2018) recently obtained thin coatings based on HA doped with zinc and copper (Zn–HA and Cu–HA, respectively) for Ti6Al4V and stainless steel materials. Thin films consisting in globular-shaped aggregates were thus synthesized, the thickness corresponding to the doped coatings (500 ± 100 nm) being significantly reduced compared to pristine HA materials (700 ± 100 nm). All the coatings adopted the irregular substrate surface morphology, but a particular interconnected structure was noticed in the case of Zn–HA materials, while for the Cu–HA coatings a rather heterogeneous structure was identified. Such a particular microstructure related to the doped HA coatings is a consequence of the increased incorporation of copper within the HA material which resulted in smaller crystallite size (38 nm in the case of Cu–HA, 50 and 55 nm in the case of Zn–HA and pristine HA, respectively), reduced amount of carbonate groups, and increased size of the coating inorganic aggregates (with dimensions ranging between 200 and 400 nm in the case of Zn–HA and HA and between 200 nm and 1 μm in the case of Cu–HA). The increased entrapment of Cu within the HA composition was also confirmed by X-ray photoelectron spectroscopy studies which revealed the incorporation of both metallic species in oxide forms (ZnO and CuO_2) and estimated the PLD transfer efficiency of both Zn and Cu up to 30%. The biocompatibility of the coatings was assessed in the presence of pre-OB MC3TC-E1 cell cultures. The confocal laser scanning microscopy and scanning electron microscopy studies (performed after 11 and 21 days of treatment, respectively) evidenced that Zn–HA and Cu–HA were suitable materials for the cellular attachment, spreading, proliferation, and normal development of OB-like cells. The quantitative biocompatibility results evidenced that Zn–HA and Cu–HA encouraged significant cellular proliferation after 7 and 21 days of treatment, the reported data being similar with those corresponding to the pristine HA-coated samples. Moreover, the osteogenic potential of the doped HA coatings was emphasized by monitoring the activity of alkaline phosphatase, which levels were significantly increased after the prolonged treatment for all experimental samples. However, a slightly reduced enzyme activity was identified for Zn–HA and Cu–HA when compared to HA-coated specimens. Furthermore, in comparison to pristine HA-coated samples, both doped materials significantly inhibited the biofilm development of *E. coli* and *S. aureus* after 24 hours (Hidalgo-Robatto et al., 2018).

By using a Nd:YAG laser source, the research group of Tirca et al. (2017) synthesized composite coatings based on PEG and PVA embedded with natural curcumin (CM) by MAPLE means as an unconventional alternative for local antitumor treatment of osteosarcoma. For comparative reasons, the authors used different solvents during the frozen target preparation, but maintained constant laser processing parameters. The performed infrared studies confirmed the stoichiometric transfer of the composite materials (loaded or not with curcumin) onto the substrates, while complementary UV-visible investigations evidenced the successful embedding of the unaltered natural polyphenol within the composite matrix regardless the solvent type. Instead, the nature of the solvent was evidenced to significantly impact the microstructure of the curcumin-embedded composite coatings (PEG-PVA + CM), since preponderant porous materials were obtained when using chloroform and dimethyl sulfoxide (DMSO) while preferential interconnected and insular morphologies were observed when using benzene and water or a water-ethanol mixture, respectively. Exclusive nanosized roughness was reported for all experimental samples, with mean values lower than 100 nm. According to the degradation and release studies performed for up to 48 hours, the composite coatings resulting from using chloroform and benzene during target preparation exhibited initial swelling and subsequent erosion processes accompanied by the gradual release of curcumin of 49% and 60%, respectively. In the case of the other obtained composite materials, a faster and increased release of CM was evidenced, with the maximal release being reported for the thin films obtained by using DMSO. The biological behavior of the synthesized composite coatings was quantitatively and qualitatively assessed in the presence of MG-63 cells (human OB-like cells derived from osteosarcoma). The curcumin-free PEG-PVA materials proved to be highly cytocompatible, while the CM-embedded coatings resulted in significant impairment of MG-63 cells. The cytotoxic and anti-proliferative effects of PEG-PVA + CM coatings were directly related to the curcumin release profile and the copolymer degradation kinetics (Tirca et al., 2017).

By using the same UV laser processing experimental setup, composite coatings based on block copolymer consisting in PEG, poly(ε-caprolactone) (PCL) and methyl ether (ME) embedded with spherical HA nanoparticles and lactoferrin (Lf) were obtained by Rusen et al. (2017) and proposed as multifunctional platforms for bone tissue applications. The MAPLE-resulted PEG–PCL–ME–HA–Lf materials (with particular 1.58 Ca/P ratio) preserved the stoichiometry of initial materials and possessed hydrophilic features. Also, the composites uniformly coated the substrates by adopting a preferential grainy morphology (resulting from the uniform distribution of HA within the copolymer matrix) and possessed nanosized topography (the evaluated roughness being significantly decreased by the addition of HA and Lf when compared to pristine polymer samples). The nanostructured composite coatings showed increased stability and gradual degradation upon various periods of evaluation under physiologically simulated conditions. When assessed in the presence of MC3TC-E1 cell cultures, the PEG–PCL–ME–HA–Lf-coated samples proved beneficial impact on cellular adhesion, as evidenced from the initial hours. In comparison to the samples coated with individual or dual composite materials, the ternary composites also resulted in best cellular proliferation results, as revealed after 5 days of assessment. By performing long-term biological assays (up to 6 weeks), the authors revealed the beneficial role of composite materials on the mineralization process. Taking into account the acknowledged intrinsic therapeutic effects related to the selected glycoprotein, the present study outlines the promising potential of Lf-embedded composites toward the development of new highly biocompatible and multifunctional coatings for osseous restorative and regenerative strategies (Rusen et al., 2017).

By using the PLD approach, Li et al. (2016) reported the successful synthesis of thin nanosized amorphous coatings ($\sim$40 nm thickness) based on copper-containing silicate bioglass (Cu–BG) onto an eggshell membrane (ESM). The selected laser processing resulted in the synthesis of homogenous particulate inorganic coatings which maintained the preferential fibrillar structure of pristine ESM. When compared to a pristine collagenous membrane, the Cu–BG-coated ESM materials exhibited totally hydrophilic behavior and improved surface hardness. The metallic ions release studies performed in a modified culture medium for up to 7 days showed that Cu^{2+} reached a maximum concentration after 24 hours of evaluation and then followed a sustained release. As it was assessed in the presence of human umbilical vein endothelial cell (HUVEC) cultures after 1 week of incubation, the Cu–BG/ESM materials resulted in superior biological behavior in terms of attachment, migration, and proliferation of the human-derived cells. Moreover, when compared to pristine ESM and BG-coated materials, the ESM samples modified with composite Cu–BG nanostructured coatings resulted in a significantly improved expression of angiogenic genes and proteins in a manner dependent on the amount of copper embedded within the silicate glass ceramics. Also, the assessment of wound-healing potential on albino mice revealed that the Cu–BG-coated collagenous membranes resulted in improved angiogenesis and enhanced formation of new epidermis during the initial 2 post-operative weeks (in comparison with pristine ESM and sole BG-coated ESM specimens). Thanks to the native occurring proteins (ovotransferrin) and enzymes (lysozyme and β-*N*-acetyl-glucosaminidase) within the ESM, excellent intrinsic antibacterial efficiency of pristine materials against *E. coli* was reported in this study. Still, the Cu–BG-coated samples resulted in significantly enhanced activity against bacterial contamination (the survival rate being decreased with more than 10% compared to pristine ESM), the beneficial addition of copper ions on the impairment of pathogenic microorganisms being, thus, evidenced (Li et al., 2016).

According to the research study performed by Rădulescu et al. (2016b), the MAPLE technique (performed by using a KrF* excimer laser beam) proved to represent an efficient strategy to synthesize composite coatings based on HA and PLGA copolymer (HAp/PLGA) embedded with ceftriaxone (CFX) and cefuroxime (CFR) antibiotics for the prevention of microbial contamination related to commercial pure titanium. In terms of chemical composition and stoichiometry, the best results were experimentally established for the 500 mJ/cm^2 laser fluence. Complementary investigation methods confirmed the efficient entrapment of both antibiotics within the composite materials which uniformly covered the substrates and possessed a particular grainy morphology (as a consequence of the nanosize-related aggregation of rod-shaped calcium phosphate particles). When compared to the antibiotic-free HAp/PLGA materials, where the thickness varied between 200 and 300 nm, the addition of CFX and CFR resulted in an increased thickness of the composite samples in the range 400–600 nm. After 24 hours of treatment in the presence of 1.19 human fetal osteoblastic cells the antibiotic-embedded composites proved to inhibit the cellular attachment and to diminish the number of viable cells, the data being proportional with increasing the laser fluence. However, all coatings resulted in a slight decrease of inflammatory NO release regardless the laser fluence used during laser processing, with a particular weakly increase of nitric oxide levels in the case of HAp/PLGA/CFX coatings. Nevertheless, all the synthesized composite coatings were concluded to be suitable substrates for bone cell development and to represent promising materials for bone tissue applications. Both HAp/PLGA/CFX and HAp/PLGA/CFR proved their strong efficiency against the adhesion and proliferation of pathogenic *E. coli* after 24 hours of evaluation, with a significant impact on the CFU/mL values and on the bacterial cells' morphology. Even if the effectiveness of these materials against the development of bacterial biofilm was significantly reduced after 48 and 72 hours of experiments, the CFX-containing samples were more efficient against the colonization of Gram-negative pathogen. If we consider the critical first hours after surgery, the proposed antibiotic-embedded HAp/PLGA materials may represent promising coatings for titanium-based implants against short-term bacterial contamination (Rădulescu et al., 2016b).

2.5 CONCLUSIONS

Extensive endeavors have been directed toward exceeding the restrictions related to conventional medical devices, especially susceptibility to microbial contamination and colonization, intrinsic predisposition for corrosion occurrence, biological inertness, and limited biofunctionalily. In this respect, the surface modification by implementing the challenging and versatile synthesis of thin coatings by laser-processing methods represent an appropriate approach for enhancing the performances of commercial medical devices. Particularly, PLD and MAPLE were experimentally proved to represent suitable choices for the synthesis of thin and nanostructured coatings with improved physicochemical features and biofunctional activity. The available data shows the tremendous potential of the as-processed coatings in targeted anti-infective treatment, tissue healing, and regeneration applications, so we can predict the beneficial impact of the PLD and MAPLE coatings on the progress of current healthcare practice.

ACKNOWLEDGMENT

The authors acknowledge the project no. 13PCCDI/2018 (PN-III-P1-1.2-PCCDI-2017-0697)–Intelligent therapies for noncommunicable diseases based on controlled release of pharmacological compounds from encapsulated engineered cells and targeted bionanoparticles–INTERA, funded by the Romanian Executive Unit for Financing Higher Education, Research, Development and Innovation (UEFISCDI).

REFERENCES

Al-Rashidy, Z.M., Farag, M.M., Abdel Ghany, N.A., Ibrahim, A.M., Abdel-Fattah, W.I., 2018. Orthopaedic bioactive glass/chitosan composites coated 316L stainless steel by green electrophoretic co-deposition. Surf. Coat. Technol. 334, 479–490.

Alves, M.M., Prosek, T., Santos, C.F., Montemor, M.F., 2017a. In vitro degradation of ZnO flowered coated Zn–Mg alloys in simulated physiological conditions. Mater. Sci. Eng., C 70 (Part1), 112–120.

Alves, S.A., Patel, S.B., Sukotjo, C., Mathew, M.T., Filho, P.N., Celis, J.P., et al., 2017b. Synthesis of calcium-phosphorous doped TiO_2 nanotubes by anodization and reverse polarization: a promising strategy for an efficient biofunctional implant surface. Appl. Surf. Sci. 399, 682–701.

Asri, R.I.M., Harun, W.S.W., Samykano, M., Lah, N.A.C., Ghani, S.A.C., Tarlochan, F., et al., 2017. Corrosion and surface modification on biocompatible metals: a review. Mater. Sci. Eng., C 77, 1261–1274.

Bait, L., Azzouz, L., Madaoui, N., Saoula, N., 2017. Influence of substrate bias voltage on the properties of TiO_2 deposited by radio-frequency magnetron sputtering on 304L for biomaterials applications. Appl. Surf. Sci. 395, 72–77.

Bakhsheshi-Rad, H.R., Hadisi, Z., Hamzah, E., Ismail, A.F., Aziz, M., Kashefian, M., 2017. Drug delivery and cytocompatibility of ciprofloxacin loaded gelatin nanofibers-coated Mg alloy. Mater. Lett. 207, 179–182.

Behzadi, S., Luther, G.A., Harris, M.B., Farokhzad, O.C., Mahmoudi, M., 2017. Nanomedicine for safe healing of bone trauma: opportunities and challenges. Biomaterials 146, 168–182.

Bellucci, D., Bianchi, M., Graziani, G., Gambardella, A., Berni, M., Russo, A., et al., 2017a. Pulsed electron deposition of nanostructured bioactive glass coatings for biomedical applications. Ceram. Int. 43 (17), 15862–15867.

Bellucci, D., Anesi, A., Salvatori, R., Chiarini, L., Cannillo, V., 2017b. A comparative in vivo evaluation of bioactive glasses and bioactive glass-based composites for bone tissue repair. Mater. Sci. Eng., C 79, 286–295.

Berretta, S., Evans, K., Ghita, O., 2018. Additive manufacture of PEEK cranial implants: manufacturing considerations versus accuracy and mechanical performance. Mater. Des. 139, 141–152.

Boanini, E., Torricelli, P., Forte, L., Pagani, S., Mihailescu, N., Ristoscu, C., et al., 2015. Antiresorption implant coatings based on calcium alendronate and octacalcium phosphate deposited by matrix assisted pulsed laser evaporation. Colloids Surf., B: Biointerfaces 136, 449–456.

Brigo, L., Urciuolo, A., Giulitti, S., Della Giustina, G., Tromayer, M., Liska, R., et al., 2017. 3D high-resolution two-photon crosslinked hydrogel structures for biological studies. Acta Biomater. 55, 373–384.

Buciumeanu, M., Bagheri, A., Shamsaei, N., Thompson, S.M., Silva, F.S., Henriques, B., 2018a. Tribocorrosion behavior of additive manufactured Ti–6Al–4V biomedical alloy. Tribol. Int. 119, 381–388.

Buciumeanu, M., Almeida, S., Bartolomeu, F., Costa, M.M., Alves, N., Silva, F.S., et al., 2018b. Ti6Al4V cellular structures impregnated with biomedical PEEK – new material design for improved tribological behavior. Tribol. Int. 119, 157–164.

Busuioc, C., Voicu, G., Zuzu, I.D., Miu, D., Sima, C., Iordache, F., et al., 2017. Vitroceramic coatings deposited by laser ablation on Ti–Zr substrates for implantable medical applications with improved biocompatibility. Ceram. Int. 43, 5498–5504.

Califano, V., Ausanio, G., Bloisi, F., Aronne, A., Vicari, L.R.M., Nasti, L., 2015. m-DOPA addition in MAPLE immobilization of lipase for biosensor applications. Sens. Bio-Sens. Res. 6, 103–108.

Campagna, M.C., 2016. Chemotherapy complications. Hosp. Med. Clin. 5 (3), 400–412.

Chan, C.W., 2017. A study on the corrosion fatigue behaviour of laser-welded shape memory NiTi wires in a simulated body fluid. Surf. Coat. Technol. 320, 574–578.

Chan, Y.W., Siow, K.S., Ng, P.Y., Gires, U., Majlis, B.Y., 2016. Plasma polymerized carvone as an antibacterial and biocompatible coating. Mater. Sci. Eng., C 68, 861–871.

Chen, S., Guo, Y., Liu, R., Wu, S., Fang, J., Huang, B., et al., 2018. Tuning surface properties of bone biomaterials to manipulate osteoblastic cell adhesion and the signaling pathways for the enhancement of early osseointegration. Colloids Surf., B: Biointerfaces 164, 58–69.

Choy, M.T., Yeung, K.W., Chen, L., Tang, C.Y., Tsui, G.C.P., Law, W.C., 2017. In situ synthesis of osteoconductive biphasic ceramic coatings on Ti6Al4V substrate by laser-microwave hybridization. Surf. Coat. Technol. 330, 92–101.

Cristescu, R., Visan, A., Socol, G., Surdu, A.V., Oprea, A.E., Grumezescu, A.M., et al., 2016. Antimicrobial activity of biopolymeric thin films containing flavonoid natural compounds and silver nanoparticles fabricated by MAPLE: a comparative study. Appl. Surf. Sci. 374, 290–296.

Daskalova, A., Nathala, C.S.R., Kavatzikidou, P., Ranella, A., Szoszkiewicz, R., Husinsky, W., et al., 2016. FS laser processing of bio-polymer thin films for studying cell-to-substrate specific response. Appl. Surf. Sci. 382, 178–191.

Di Pietrantonio, F., Benetti, M., Cannatà, D., Verona, E., Palla-Papavlu, A., Fernández-Pradas, J.M., et al., 2015. A surface acoustic wave bio-electronic nose for detection of volatile odorant molecules. Biosens. Bioelectron. 67, 516–523.

Dinca, V., Zaharie-Butucel, D., Stanica, L., Brajnicov, S., Marascua, V., Bonciu, A., Cristocea, A., et al., 2018. Functional *Micrococcus lysodeikticus* layers deposited by laser technique for the optical sensing of lysozyme. Colloids Surf., B: Biointerfaces 162, 98–107.

Domínguez-Trujillo, C., Peón, E., Chicardi, E., Pérez, H., Rodríguez-Ortiz, J.A., Pavón, J. J., et al., 2018. Sol–gel deposition of hydroxyapatite coatings on porous titanium for biomedical applications. Surf. Coat. Technol. 333, 158–162.

Dong, P., Li, H., Wang, Q., Zhou, J., 2018. Microstructural characterization of laser micro-welded Nitinol wires. Mater. Charact. 135, 40–45.

Du, Y., Liu, H., Yang, Q., Wang, S., Wang, J., Ma, J., et al., 2017. Selective laser sintering scaffold with hierarchical architecture and gradient composition for osteochondral repair in rabbits. Biomaterials 137, 37–48.

El Saeed, E.M., El-Fattah, M.A., Azzam, A.M., Dardir, M.M., Bader, M.M., 2016. Synthesis of cuprous oxide epoxy nanocomposite as an environmentally antimicrobial coating. Int. J. Biol. Macromol. 89, 190–197.

Elbourne, A., Crawford, R.J., Ivanova, E.P., 2017. Nano-structured antimicrobial surfaces: from nature to synthetic analogues. J. Colloid Interface Sci. 508, 603–616.

Esparza, J., García Fuentes, G., Bueno, R., Rodríguez, R., García, J.A., Vitas, A.I., et al., 2017. Antibacterial response of titanium oxide coatings doped by nitrogen plasma immersion ion implantation. Surf. Coat. Technol. 314, 67–71.

Fernandez-Yague, M.A., Abbah, S.A., McNamara, L., Zeugolis, D.I., Pandit, A., Biggs, M. J., 2015. Biomimetic approaches in bone tissue engineering: integrating biological and physicomechanical strategies. Adv. Drug Delivery Rev. 84, 1–29.

Forte, L., Torricelli, P., Bonvicini, F., Boanini, E., Gentilomi, G.A., Lusvardi, G., et al., 2018. Biomimetic fabrication of antibacterial calcium phosphates mediated by polydopamine. J. Inorg. Biochem. 178, 43–53.

Frieri, M., Kumar, K., Boutin, A., 2017. Antibiotic resistance. J. Infect. Public Health 10 (4), 369–378.

Fufă, O., Andronescu, E., Grumezescu, V., Holban, A.M., Mogoantă, L., Mogoşanu, G.D., et al., 2015. Silver nanostructurated surfaces prepared by MAPLE for biofilm prevention. Biointerface Res. Appl. Chem. 5 (6), 1011–1017.

Fufă, O., Socol, M., Preda, N., Grigorescu, S., Croitoru, S., Socol, G., 2017. MAPLE deposition of PLGA microspheres for medical applications. Dig. J. Nanomater. Biostruct. 12 (1), 73–80.

García, M.C., Aldana, A.A., Tártara, L.I., Alovero, F., Strumia, M.C., Manzo, M.H., et al., 2017. Bioadhesive and biocompatible films as wound dressing materials based on a novel dendronized chitosan loaded with ciprofloxacin. Carbohydr. Polym. 175, 75–86.

Ge, W., Yu, Q., López, G.P., Stiff-Roberts, A.D., 2014. Antimicrobial oligo(*p*-phenylene-ethynylene) film deposited by resonant infrared matrix-assisted pulsed laser evaporation. Colloids Surf., B: Biointerfaces 116, 786–792.

Geißler, S., Tiainen, H., Haugen, H.J., 2016. Effect of cathodic polarization on coating doxycycline on titanium surfaces. Mater. Sci. Eng., C 63, 359–366.

Ghorbel, H.F., Guidara, A., Danlos, Y., Bouaziz, J., Coddet, C., 2017. Alumina-fluorapatite composite coating deposited by atmospheric plasma spraying: an agent of cohesion between bone and prostheses. Mater. Sci. Eng., C 71, 1090–1098.

Gomes, G.C., Borghi, F.F., Ospina, R.O., López, E.O., Borges, F.O., Mello, A., 2017. Nd: YAG (532 nm) pulsed laser deposition produces crystalline hydroxyapatite thin coatings at room temperature. Surf. Coat. Technol. 329, 174–183.

Gonzalez, G., Sagarzazu, A., Cordova, A., Gomes, M.E., Salas, J., Contreras, L., et al., 2018. Comparative study of two silica mesoporous materials (SBA-16 and SBA-15) modified with a hydroxyapatite layer for clindamycin controlled delivery. Microporous Mesoporous Mater. 256, 251–265.

Gotzmann, G., Jorsch, C., Wetzel, C., Funk, H.W.R., 2018. Antimicrobial effects and dissolution properties of silver copper mixed layers. Surf. Coat. Technol. 336, 22–28.

Grumezescu, V., Socol, G., Grumezescu, A.M., Holban, A.M., Ficai, A., Truşcă, R., et al., 2014a. Functionalized antibiofilm thin coatings based on PLA–PVA microspheres loaded with usnic acid natural compounds fabricated by MAPLE. Appl. Surf. Sci. 302, 262–267.

Grumezescu, V., Holban, A.M., Grumezescu, A.M., Socol, G., Ficai, A., Vasile, B.S., et al., 2014b. Usnic acid-loaded biocompatible magnetic PLGA–PVA microsphere thin films fabricated by MAPLE with increased resistance to staphylococcal colonization. Biofabrication 6, 035002.

Grumezescu, V., Holban, A.M., Iordache, F., Socol, G., Mogoşanu, G.D., Grumezescu, A. M., et al., 2014c. MAPLE fabricated magnetite@eugenol and (3-hidroxybutyric acid-*co*-3-hidroxyvaleric acid)–polyvinyl alcohol microspheres coated surfaces with antimicrobial properties. Appl. Surf. Sci. 306, 16–22.

Grumezescu, V., Andronescu, E., Holban, A.M., Socol, G., Grumezescu, A.M., Ficai, A., et al., 2015a. Fabrication and characterization of functionalized surfaces with 3-amino propyltrimethoxysilane films for anti-infective therapy applications. Appl. Surf. Sci. 336, 401–406.

Grumezescu, V., Andronescu, E., Holban, A.M., Mogoantă, L., Mogoşanu, G.D., Grumezescu, A.M., et al., 2015b. MAPLE fabrication of thin films based on kanamycin functionalizedmagnetite nanoparticles with anti-pathogenic properties. Appl. Surf. Sci. 336, 188–195.

Grumezescu, V., Holban, A.M., Sima, L.E., Chiritoiu, M.B., Chiritoiu, G.N., Grumezescu, A.M., et al., 2017. Laser deposition of poly(3-hydroxybutyric acid-*co*-3-hydroxyvaleric acid) – lysozyme microspheres based coatings with anti-microbial properties. Int. J. Pharm. 521, 184–195.

Hadidi, M., Bigham, A., Saebnoori, E., Hassanzadeh-Tabrizi, S.A., Rahmati, S., Alizadeh, Z. M., et al., 2017. Electrophoretic-deposited hydroxyapatite-copper nanocomposite as an antibacterial coating for biomedical applications. Surf. Coat. Technol. 321, 171–179.

Halbus, A.F., Horozov, T.S., Paunov, V.N., 2017. Colloid particle formulations for antimicrobial application. Adv. Colloid Interface Sci. 249, 134–148.

Hassanin, H., Finet, L., Cox, S.C., Jamshidi, P., Grover, L.M., Shepherd, D.E.T., et al., 2018. Tailoring selective laser melting process for titanium drug-delivering implants with releasing micro-channels. Addit. Manuf. 20, 144–155.

He, X., Zhang, G., Wang, X., Hang, R., Huang, X., Qin, L., et al., 2017. Biocompatibility, corrosion resistance and antibacterial activity of TiO_2/CuO coating on titanium. Ceram. Int. 43 (18), 16185–16195.

Hidalgo-Robatto, B.M., López-Álvarez, M., Azevedo, A.Z., Dorado, J., Serra, J., Azevedo, N.F., et al., 2018. Pulsed laser deposition of copper and zinc doped hydroxyapatite coatings for biomedical applications. Surf. Coat. Technol. 333, 168–177.

Huang, G., Chen, Y., Zhang, J., 2016. Nanocomposited coatings produced by laser-assisted process to prevent silicone hydrogels from protein fouling and bacterial contamination. Appl. Surf. Sci. 360, 383–388.

Ionita, D., Bajenaru-Georgescu, D., Tote, G., Mazare, A., Schmuki, P., Demetrescu, I., 2017. Activity of vancomycin release from bioinspired coatings of hydroxyapatite or TiO_2 nanotubes. Int. J. Pharm. 517 (1-2), 296–302.

Iordache, F., Grumezescu, V., Grumezescu, A.M., Curuţiu, C., Diţu, L.M., Socol, G., et al., 2015. Gamma-cyclodextrin/usnic acid thin film fabricated by MAPLE for improving the resistance of medical surfaces to *Staphylococcus aureus* colonization. Appl. Surf. Sci. 336, 407–412.

Jabeen, S., Islam, A., Ghaffar, A., Gull, N., Hameed, A., Bashir, A., et al., 2017. Development of a novel pH sensitive silane crosslinked injectable hydrogel for controlled release of neomycin sulfate. Int. J. Biol. Macromol. 97, 218–227.

Jamil, T., Ansari, U., Najabat Ali, M., Mir, M., 2017. A review on biomechanical and treatment aspects associated with anterior cruciate ligament. Innov. Res. BioMed. Eng. 38 (1), 13–25.

Jemat, A., Ghazali, M.J., Razali, M., Otsuka, Y., Rajabi, A., 2018. Effects of TiO_2 on microstructural, mechanical properties and in-vitro bioactivity of plasma sprayed yttria stabilised zirconia coatings for dental application. Ceram. Int. 44 (4), 4271–4281.

Jiang, P., Liang, J., Song, R., Zhang, Y., Ren, L., Zhang, L., et al., 2015. Effect of octacalcium-phosphate-modified micro/nanostructured titania surfaces on osteoblast response. ACS Appl. Mater. Interfaces 7 (26), 14384–14396.

Jugowiec, D., Łukaszczyk, A., Cieniek, Ł., Kot, M., Reczyńska, K., Cholewa-Kowalska, K., et al., 2017a. Electrophoretic deposition and characterization of composite chitosan-based coatings incorporating bioglass and sol–gel glass particles on the Ti–13Nb–13Zr alloy. Surf. Coat. Technol. 319, 33–46.

Jugowiec, D., Łukaszczy, A., Cieniek, Ł., Kowalski, K., Rumian, Ł., Pietryga, K., et al., 2017b. Influence of the electrophoretic deposition route on the microstructure and properties of nano-hydroxyapatite/chitosan coatings on the Ti–13Nb–13Zr alloy. Surf. Coat. Technol. 324, 64–79.

Karbowniczek, J., Cordero-Arias, L., Virtanen, S., Misra, S.K., Valsami-Jones, E., Tuchscherr, L., et al., 2017. Electrophoretic deposition of organic/inorganic composite coatings containing ZnO nanoparticles exhibiting antibacterial properties. Mater. Sci. Eng., C 77, 780–789.

Kufelt, O., El-Tamer, A., Sehring, C., Meißner, M., Schlie-Wolter, S., Chichkov, B.N., 2015. Water-soluble photopolymerizable chitosan hydrogels for biofabrication via two-photon polymerization. Acta Biomater. 18, 186–195.

Kumar, R., Umar, A., Kumar, G., Nalwa, H.S., 2017. Antimicrobial properties of ZnO nanomaterials: a review. Ceram. Int. 43 (5), 3940–3961.

Lee, B.E.J., Exir, H., Weck, A., Grandfield, K., 2018. Characterization and evaluation of femtosecond laser-induced sub-micron periodic structures generated on titanium to improve osseointegration of implants. Appl. Surf. Sci. 441, 1034–1042.

Li, J., Zhai, D., Lv, F., Yu, Q., Ma, H., Yin, J., et al., 2016. Preparation of copper-containing bioactive glass/eggshell membrane nanocomposites for improving angiogenesis, antibacterial activity and wound healing. Acta Biomater. 36, 254–266.

Li, D., Gong, Y., Chen, X., Zhang, B., Zhang, H., Jin, P., et al., 2017a. Room-temperature deposition of hydroxyapatite/antibiotic composite coatings by vacuum cold spraying for antibacterial applications. Surf. Coat. Technol. 330, 87–91.

Li, H., Li, J., Jiang, J., Lv, F., Chang, J., Chen, S., et al., 2017b. An osteogenesis/angiogenesis-stimulation artificial ligament for anterior cruciate ligament reconstruction. Acta Biomater. 54, 399–410.

Li, Y., Hu, Y., Cong, W., Zhi, L., Guo, Z., 2017c. Additive manufacturing of alumina using laser engineered net shaping: effects of deposition variables. Ceram. Int. 43 (10), 7768–7775.

Liu, C., Zhang, M., Chen, C., 2017. Effect of laser processing parameters on porosity, microstructure and mechanical properties of porous Mg–Ca alloys produced by laser additive manufacturing. Mater. Sci. Eng., A 703, 359–371.

Liu, X., Yang, Q., Li, Z., Yuan, W., Zheng, Y., Cui, Z., et al., 2018. A combined coating strategy based on atomic layer deposition for enhancement of corrosion resistance of AZ31 magnesium alloy. Appl. Surf. Sci. 434, 1101–1111.

Luculescu, C.R., Acasandrei, A.M., Mustaciosu, C.C., Zamfirescu, M., Dinescu, M., Calin, B.S., et al., 2018. Electrically responsive microstructured polypyrrole–polyurethane composites for stimulated osteogenesis. Appl. Surf. Sci. 433, 166–176.

Major, R., Trembecka-Wójciga, K., Kot, M., Lackner, J.M., Wilczek, P., Major, B., 2016. In vitro hemocompatibility on thin ceramic and hydrogel films deposited on polymer substrate performed in arterial flow conditions. Mater. Sci. Eng., C 61, 15–22.

Manam, N.S., Harun, W.S.W., Shri, D.N.A., Ghani, S.A.C., Kurniawan, T., Ismail, M.H., et al., 2017. Study of corrosion in biocompatible metals for implants: a review. J. Alloys Compd. 701, 698–715.

Mareci, D., Trincă, L.C., Căilean, D., Souto, R.M., 2016. Corrosion resistance of ZrTi alloys with hydroxyapatite–zirconia–silver layer in simulated physiological solution containing proteins for biomaterial applications. Appl. Surf. Sci. 389, 1069–1075.

Martin, V., Bettencourt, A., 2018. Bone regeneration: biomaterials as local delivery systems with improved osteoinductive properties. Mater. Sci. Eng., C 82, 363–371.

Mehta, R.V., 2017. Synthesis of magnetic nanoparticles and their dispersions with special reference to applications in biomedicine and biotechnology. Mater. Sci. Eng., C 79, 901–916.

Morio, F., Jensen, R.H., Le Pape, P., Cavling Arendrup, M., 2017. Molecular basis of antifungal drug resistance in yeasts. Int. J. Antimicrob. Agents 50 (5), 599–606.

Massil, P., 2017. How does total knee replacement technique influence polyethylene wear? Orthop. Traumatol. Surg. Res. 103 (1), S21–S27.

Nasirpouri, F., Yousefi, I., Moslehifard, E., Khalil-Allafi, J., 2017. Tuning surface morphology and crystallinity of anodic TiO_2 nanotubes and their response to biomimetic bone growth for implant applications. Surf. Coat. Technol. 315, 163–171.

Nie, C., Yang, Y., Cheng, C., Ma, L., Deng, J., Wang, L., et al., 2017. Bioinspired and biocompatible carbon nanotube-Ag nanohybrid coatings for robust antibacterial applications. Acta Biomater. 51, 479–494.

Ohtsu, N., Kozuka, T., Yamane, M., Arai, H., 2017. Surface chemistry and osteoblast-like cell response on a titanium surface modified by a focused Nd:YAG laser. Surf. Coat. Technol. 309, 220–226.

Palla-Papavlu, A., Patrascioiu, A., Di Pietrantonio, F., Fernández-Pradas, J.M., Cannatà, D., Benetti, M., et al., 2014. Preparation of surface acoustic wave odor sensors by laser-induced forward transfer. Sens. Actuators, B: Chem. 192, 369–377.

Paris, J.P., Seyer, D., Jouenne, T., Thébault, P., 2017. Elaboration of antibacterial plastic surfaces by a combination of antiadhesive and biocidal coatings of natural products. Colloids Surf., B: Biointerfaces 156, 186–193.

Pessoa, R.S., dos Santos, V.P., Cardoso, S.B., Doria, A.C.O.C., Figueira, F.R., Rodrigues, B.M.V., et al., 2017. TiO_2 coatings via atomic layer deposition on polyurethane and polydimethylsiloxane substrates: properties and effects on *C. albicans* growth and inactivation process. Appl. Surf. Sci. 422, 73–84.

Petrochenko, P.E., Zheng, J., Casey, B.J., Bayati, M.R., Narayan, R.J., Goering, P.L., 2017. Nanosilver-PMMA composite coating optimized to provide robust antibacterial efficacy while minimizing human bone marrow stromal cell toxicity. Toxicol. in Vitro 44, 248–255.

Pillai, R.S., Frasnelli, M., Sglavo, V.M., 2018. HA/β-TCP plasma sprayed coatings on Ti substrate for biomedical applications. Ceram. Int. 44 (2), 1328–1333.

Písařík, P., Jelínek, M., Remsa, J., Mikšovský, J., Zemek, J., Jurek, K., et al., 2017. Antibacterial, mechanical and surface properties of Ag–DLC films prepared by dual PLD for medical applications. Mater. Sci. Eng., C 77, 955–962.

Popescu-Pelin, G., Sima, F., Sima, L.E., Mihailescu, C.N., Luculescu, C., Iordache, I., et al., 2017. Hydroxyapatite thin films grown by pulsed laser deposition and matrix assisted pulsed laser evaporation: comparative study. Appl. Surf. Sci. 418 (Part B), 580–588.

Preethanath, R.S., AlNahas, N.W., Bin Huraib, S.M., Al-Balbeesi, H.O., Almalik, N.K., Dalati, M.H.N., et al., 2017. Microbiome of dental implants and its clinical aspect. Microb. Pathog. 106, 20–24.

Prosolov, K.A., Popova, K.S., Belyavskaya, O.A., Rau, J.V., Gross, K.A., Ubelis, A., et al., 2017. RF magnetron-sputtered coatings deposited from biphasic calcium phosphate targets for biomedical implant applications. Bioact. Mater. 2 (3), 170–176.

Radda'a, N.S., Goldmann, W.H., Detsch, R., Roether, J.A., Cordero-Arias, L., Virtanen, S., et al., 2017. Electrophoretic deposition of tetracycline hydrochloride loaded halloysite nanotubes chitosan/bioactive glass composite coatings for orthopedic implants. Surf. Coat. Technol. 327, 146–157.

Raghunath, A., Perumal, E., 2017. Metal oxide nanoparticles as antimicrobial agents: a promise for the future. Int. J. Antimicrob. Agents 49 (2), 137–152.

Raşoga, O., Sima, L.E., Chiriţoiu, M., Popescu-Pelin, G., Fufă, O., Grumezescu, V., et al., 2017. Biocomposite coatings based on poly(3-hydroxybutyrate-*co*-3-hydroxyvalerate)/calcium phosphates obtained by MAPLE for bone tissue engineering. Appl. Surf. Sci. 417, 204–212.

Ren, Y., Babaie, E., Bhaduri, S.B., 2018. Nanostructured amorphous magnesium phosphate/poly(lactic acid) composite coating for enhanced corrosion resistance and bioactivity of biodegradable AZ31 magnesium alloy. Prog. Org. Coat. 118, 1–8.

Revathi, A., Dalmau Borrás, A., Igual Muñoz, A., Richard, C., Manivasagam, G., 2017. Degradation mechanisms and future challenges of titanium and its alloys for dental implant applications in oral environment. Mater. Sci. Eng., C 76, 1354–1368.

Ridi, F., Meazzini, I., Castroflorio, B., Bonini, M., Berti, D., Baglioni, P., 2017. Functional calcium phosphate composites in nanomedicine. Adv. Colloid Interface Sci. 244, 281–295.

Robles, E., Salaberria, A.M., Herrera, R., Fernandes, S.C.M., Labidi, J., 2016. Self-bonded composite films based on cellulose nanofibers and chitin nanocrystals as antifungal materials. Carbohydr. Polym. 144, 41–49.

Rodriguez, O., Stone, W., Schemitsch, E.H., Zalzal, P., Waldman, S., Papini, M., et al., 2017. Titanium addition influences antibacterial activity of bioactive glass coatings on metallic implants. Heliyon 3 (10), e00420.

Roknian, M., Fattah-alhosseini, A., Gashti, S.O., Keshavarz, M.K., 2018. Study of the effect of ZnO nanoparticles addition to PEO coatings on pure titanium substrate: microstructural analysis, antibacterial effect and corrosion behavior of coatings in Ringer's physiological solution. J. Alloys Compd. 740, 330–345.

Rupp, F., Liang, L., Geis-Gerstorfer, J., Scheideler, L., Hüttig, F., 2018. Surface characteristics of dental implants: a review. Dent. Mater. 34 (1), 40–57.

Rusen, L., Brajnicov, S., Neacsu, P., Marascu, V., Bonciu, A., Dinescu, M., et al., 2017. Surf. Coat. Technol. 325, 397–409.

Rădulescu, D., Voicu, G., Oprea, A.E., Andronescu, E., Grumezescu, V., Holban, A.M., et al., 2016a. Mesoporous silica coatings for cephalosporin active release at the bone-implant interface. Appl. Surf. Sci. 374, 165–171.

Rădulescu, D., Grumezescu, V., Andronescu, E., Holban, A.M., Grumezescu, A.M., Socol, G., et al., 2016b. Biocompatible cephalosporin–hydroxyapatite–poly(lactic-*co*-glycolic acid)-coatings fabricated by MAPLE technique for the prevention of bone implant associated infections. Appl. Surf. Sci. 374, 387–396.

Sethu, S.N., Namashivayam, S., Devendran, S., Nagarajan, S., Tsai, W.B., Narashiman, S., et al., 2017. Nanoceramics on osteoblast proliferation and differentiation in bone tissue engineering. Int. J. Biol. Macromol. 98, 67–74.

Shamray, V.F., Sirotinkin, V.P., Smirnov, I.V., Kalita, V.I., Fedotov, A.Yu, et al., 2017. Structure of the hydroxyapatite plasma-sprayed coatings deposited on pre-heated titanium substrates. Ceram. Int. 43 (12), 9105–91019.

Sharifi, H., Aliofkhazraei, M., Bratai Darband, G., Sabour Rouhaghdam, A., 2016. Tribological properties of PEO nanocomposite coatings on titanium formed in electrolyte containing ketoconazole. Tribol. Int. 102, 463–471.

Siegel, J., Lyutakov, O., Polívková, M., Staszek, M., Hubáček, T., Švorčík, V., 2017. Laser-assisted immobilization of colloid silver nanoparticles on polyethyleneterephthalate. Appl. Surf. Sci. 420, 661–668.

Silkey, J.R., Ludtke, S.L., Acharya, K., 2017. Orthopedic infections. Physician Assist. Clin. 2 (2), 261–276.

Skoog, S.A., Kumar, G., Narayan, R.J., Goering, P.L., 2018. Biological responses to immobilized microscale and nanoscale surface topographies. Pharmacol. Ther. 182, 33–55.

Smirnov, I.V., Rau, J.V., Fosca, M., De Bonis, A., Latini, A., Teghil, R., et al., 2017. Structural modification of titanium surface by octacalcium phosphate via pulsed laser deposition and chemical treatment. Bioact. Mater. 2, 101–107.

Sonia, S., Jayasudha, R., Jayram, N.D., Kumar, P.S., Mangalaraj, D., Prabagaran, S.R., 2016. Synthesis of hierarchical CuO nanostructures: biocompatible antibacterial agents for Gram-positive and Gram-negative bacteria. Curr. Appl Phys. 16 (8), 914–921.

Stan, M.S., Constanda, S., Grumezescu, V., Andronescu, E., Ene, A.M., Holban, A.M., et al., 2016. Thin coatings based on ZnO@C18-usnic acid nanoparticles prepared by MAPLE inhibit the development of *Salmonella enterica* early biofilm growth. Appl. Surf. Sci. 374, 318–325.

Stango, A.R.X., Karthick, D., Swaroop, S., Mudali, U.K., Vijayalakshmi, U., 2018. Development of hydroxyapatite coatings on laser textured 316 LSS and Ti–6Al–4V and its electrochemical behavior in SBF solution for orthopedic applications. Ceram. Int. 44 (3), 3149–3160.

Su, Y., Luo, C., Zhang, Z., Hermawan, H., Zhu, D., Huang, J., et al., 2018. Bioinspired surface functionalization of metallic biomaterials. J. Mech. Behav. Biomed. Mater. 77, 90–105.

Szcześ, A., Hołysz, L., Chibowski, E., 2017. Synthesis of hydroxyapatite for biomedical applications. Adv. Colloid Interface Sci. 249, 321–330.

Szustakiewicz, K., Stępak, B., Antończak, A.J., Maj, M., Gazińska, M., Kryszak, B., et al., 2018. Femtosecond laser-induced modification of PLLA/hydroxypatite composite. Polym. Degrad. Stab. 149, 152–161.

Tangermann-Gerk, K., Barros, G., Weisenseel, B., Greil, P., Fey, T., Schmidt, M., et al., 2016. Laser pyrolysis of an organosilazane-based glass/ZrO_2 composite coating system. Mater. Des. 109, 644–651.

Tirca, I., Mitran, V., Marascu, V., Brajnicov, S., Ion, V., Stokker-Cheregi, F., et al., 2017. In vitro testing of curcumin based composites coatings as antitumoral systems against osteosarcoma cells. Appl. Surf. Sci. 425, 1040–1051.

Trtica, M., Stasic, J., Batani, B., Benocci, R., Narayanan, V., Ciganovic, J., 2018. Laser-assisted surface modification of Ti-implant in air and water environment. Appl. Surf. Sci. 428, 669–675.

Verrastro, M., Cicco, N., Crispo, F., Morone, A., Dinescu, M., Dumitru, M., et al., 2016. Amperometric biosensor based on laccase immobilized onto a screen-printed electrode by matrix assisted pulsed laser evaporation. Talanta 154, 438–445.

Visan, A., Cristescu, R., Stefan, N., Miroiu, M., Nita, C., Socol, M., et al., 2017. Antimicrobial polycaprolactone/polyethylene glycol embedded lysozyme coatings of Ti implants for osteoblast functional properties in tissue engineering. Appl. Surf. Sci. 417, 234–243.

Vitali, M., Ripamonti, C.I., Roila, F., Proto, C., Signorelli, D., Imbimbo, M., et al., 2017. Cognitive impairment and chemotherapy: a brief overview. Crit. Rev. Oncol. Hematol. 118, 7–14.

Voicu, G., Miu, D., Dogaru, I., Jinga, S.I., Busuioc, C., 2016. Vitroceramic interface deposited on titanium substrate by pulsed laser deposition method. Int. J. Pharm. 510, 449–456.

Wang, W., Yeung, K.W.K., 2017. Bone grafts and biomaterials substitutes for bone defect repair: a review. Bioact. Mater. 2 (4), 224–247.

Wang, X., Liu, F., Song, Y., Liu, Z., Qin, D., 2016. Structure and properties of Al_2O_3 coatings formed on NiTi alloy by cathodic plasma electrolytic deposition. Surf. Coat. Technol. 285, 128–133.

Wang, H.P., Zhang, H.J., Liu, J., Dong, Q., Duan, S., Ge, J.Q., et al., 2017a. Antimicrobial resistance of 3 types of Gram-negative bacteria isolated from hospital surfaces and the hands of health care workers. Am. J. Infect. Control 45 (11), e143–e147.

Wang, X., Jing, S., Liu, Y., Liu, S., Tan, Y., 2017b. Diblock copolymer containing bioinspired borneol and dopamine moieties: synthesis and antibacterial coating applications. Polymer 116, 314–323.

Wang, H., Sun, T., Chang, L., Liu, F., Liu, B., Zhao, C., et al., 2017c. Preparation of Ca doping ZrO_2 coating on NiTi shape memory alloy by cathodic plasma electrolytic deposition and its structure, in-vitro bioactivity and biocompatibility analysis. Surf. Coat. Technol. 325, 136–144.

Wang, W., He, X., Gao, Y., Zhang, X., Yao, X., Tang, B., 2017d. Antimicrobial property, cytocompatibility and corrosion resistance of Zn-doped ZrO_2/TiO_2 coatings on Ti6Al4V implants. Mater. Sci. Eng., C 75, 7–15.

Wang, T., Weng, Z., Liu, X., Yeung, K.W.K., Pan, H., Wu, S., 2017e. Controlled release and biocompatibility of polymer/titania nanotube array system on titanium implants. Bioact. Mater. 2 (1), 44–50.

Wang, D.G., Zhang, W.L., Li, H.J., Zhang, J.H., Chen, C.Z., 2017f. HA/BG composite films deposited by pulse laser under O_2 atmosphere. Ceram. Int. 43, 672–676.

Wang, D.G., Ming, X.C., Zhang, W.L., Zhang, J.H., Chen, C.Z., 2018. Effect of heat treatment on the properties of HA/BG composite films. Ceram. Int. 44, 7228–7233.

Wijdeven, R.H., Pang, B., Assaraf, Y.G., Neefjes, J., 2016. Old drugs, novel ways out: drug resistance toward cytotoxic chemotherapeutics. Drug Resist. Updat. 28, 65–81.

Woeppel, K., Yang, Q., Cui, X.T., 2017. Recent advances in neural electrode–tissue interfaces. Curr. Opin. Biomed. Eng. 4, 21–31.

Wojcieszak, D., Mazur, M., Kalisz, M., Grobelny, M., 2017. Influence of Cu, Au and Ag on structural and surface properties of bioactive coatings based on titanium. Mater. Sci. Eng., C 71, 1115–1121.

Xiang, Y.C., Martínez-Martínez, R.M., Torres-Costa, V., Agulló-Rueda, F., García-Ruiz, J. P., Manso-Silván, M., 2016. Direct laser writing of nanorough cell microbarriers on anatase/Si and graphite/Si. Mater. Sci. Eng., C 66, 8–15.

Xue, B., Guo, L., Chen, X., Fan, Y., Ren, X., Li, B., et al., 2017. Electrophoretic deposition and laser cladding of bioglass coating on Ti. J. Alloys Compd. 710, 663–669.

Yang, L., Li, Z., Zhang, Y., Wei, S., Liu, F., 2018. Al–TiC in situ composite coating fabricated by low power pulsed laser cladding on AZ91D magnesium alloy. Appl. Surf. Sci. 435, 1187–1198.

Yu, N., Cai, S., Wang, F., Zhang, F., Ling, R., Li, Y., et al., 2017a. Microwave assisted deposition of strontium doped hydroxyapatite coating on AZ31 magnesium alloy with enhanced mineralization ability and corrosion resistance. Ceram. Int. 43 (12), 2495–2503.

Yu, C., Zhuang, J., Dong, L., Cheng, K., Weng, Q., 2017b. Effect of hierarchical pore structure on ALP expression of MC3T3-E1 cells on bioglass films. Colloids Surf., B: Biointerfaces 156, 213–220.

Zorin, V.L., Komlev, V.S., Zorina, A.I., Khromova, N.V., Solovieva, E.V., Fedotov, A.Y., et al., 2014. Octacalcium phosphate ceramics combined with gingiva-derived stromal cells for engineered functional bone grafts. Biomed. Mater. 9 (5), 055005.

Zykova, A., Safonov, V., Goltsev, A., Dubrava, T., Rossokha, I., Smolik, J., et al., 2016. The effect of surface treatment of ceramic oxide coatings deposited by magnetron sputtering method on the adhesive and proliferative activity of mesenchymal stem cells. Surf. Coat. Technol. 301, 114–120.

CHAPTER

Microfluidics for drug delivery systems

3

Lucero G. Aceves-Serrano[1], Karen A. Ordaz-Martinez[1], Matias Vazquez-Piñon[1] and Hyundoo Hwang[2]

[1]School of Engineering and Sciences, Tecnologico de Monterrey, Monterrey, Mexico [2]BBB Inc., Seoul, Republic of Korea

3.1 INTRODUCTION

The development of novel drugs and delivery techniques have been an important goal for many researchers in recent years to improve the efficiency of treatments for patients. Drug delivery is the method or process to administer a pharmaceutical compound to a patient to achieve a therapeutic effect on a determined disease (Gao and Li, 2013). In most traditional drug delivery techniques, the pharmaceutical agents are evenly distributed throughout the body and is not possible to selectively reach the pathological tissues of interest. In this case, to reach a target area, drugs must pass through various biological barriers which decrease their efficacy and cause undesirable effects on untargeted organs or tissues (Torchilin, 2000).

Recent advances in microfluidics and microtechnologies allow the development of miniaturized systems that can process small amounts of samples with high efficiency and throughput. These new technologies have been successfully used to miniaturize otherwise bulky devices. Microfluidic techniques not only take advantage of the most evident characteristic, which is the small dimension scale, capable of processing even single particle/cells, but also of an unconventional transport phenomena in the microscale, such as laminar flow. The main asset of microfluidics is that it fundamentally offers new capabilities for the control of concentrations of molecules in space and time (Whitesides, 2006).

Microfluidic devices make use of a variety of basic elements to control the fluids, for instance, micropumps, microneedles, and reservoirs. Furthermore, complex designs can be assembled from the basic elements into a single, integrated system (Gao and Li, 2013). For this reason, the Lab-on-a-Chip devices have currently been under development for a wide variety of applications in biotechnology and biomedical engineering.

Several fabrication techniques have been developed for microfluidic systems, including micromilling (Guckenberger et al., 2015), micromachining (Ziaie et al., 2004), photolithography (Lin et al., 2002), stereolithography (Hwang and Cho, 2012), and the assembly of microcapillaries (Benson et al., 2013), among others.

Nanoarchitectonics in Biomedicine. DOI: https://doi.org/10.1016/B978-0-12-816200-2.00002-5

Moreover, fabrication materials can significantly affect many aspects of the overall system, including fabrication methodology, cost, and assay and detection procedures. The most traditionally used materials for the fabrication of microfluidic devices include silicon, glass, polydimethylsiloxane (PDMS), and, more recently, paper (Dou et al., 2015).

Microfluidics has become a powerful tool for the design of drug carriers because it allows the precise handling and manipulation of nanoliter and picoliter sample volumes. Compared to conventional drug carrier preparation techniques which are based on reaction vessels, microtiter plates,or test tubes, microfluidic approaches require less volume of materials, and also provides higher throughput. The application of microfluidics on drug carrier fabrication enhances accuracy and efficiency on sample preparation, as well as dramatically reduces fabrication and operational costs.

For a reliable drug carrier design, it is essential the active control of drug release rates, and switchable on–off mode capabilities, especially when unexpected changes in the physiological conditions of the patient require corresponding adjustments in dosage (Pirmoradi et al., 2013). The popularity on the use of microfluidics for the synthesis of drug carriers is due to the fact that it provides a tool for controlled release profiles with a reduced dosage amount. Furthermore, microfluidics allow the production of well-controlled drug carriers that improve mass transfer between miscible or immiscible phases and give a precise flow control using small reagent volumes in a continuous flow operation (Khan et al., 2013). Although bulky tools can be used to produce micro- and nanoscale drug carriers, challenges such as wide droplet size and shape distribution, uncontrolled drug encapsulation efficiency complicate the development of precise systems. Since microfluidics allow control over the drug carrier's physical and chemical characteristics, it has become a powerful candidate to address these challenges.

In this chapter, we summarize the state-of-the-art on microfluidics approaches on drug delivery systems. We present the latest advances in microfluidics approaches for the fabrication of drug carriers and direct drug delivery devices. We also discuss challenges and opportunities of advanced drug delivery systems, as well as the potential roles of micro-technologies toward developing novel drug delivery systems.

3.2 MICROFLUIDICS FOR DRUG CARRIER FABRICATION

The design of effective drug carriers is focused on finding ways to control the release rate, improve bioavailability, and reduce the side effects on patients. Moreover, the administration method has to be considered during the design process to effectively reach the location of interest in the body. For example, oral delivery carriers are required to withstand the low stomach pH and be small enough, or covered with the proper markers, to pass through the intestinal mucosal barrier.

The control and reliability of drug release profiles are key factors for an efficient drug delivery system. These factors depend on the size, shape, uniformity, and composition of the drug carriers (Champion et al., 2007). Other important parameters include surficial properties due to the interaction with particles, cells, and tissues, thus efforts have also been focused on improving biocompatibility on the surface of these systems. Drug carrier fabrication techniques such as self-assembly, single and multiple emulsion systems, Janus droplets, and flow lithography have been used to develop systems with tunable characteristics. These techniques are covered in detail next.

3.2.1 SELF-ASSEMBLED DRUG CARRIERS

Nano/microparticles can be synthesized in bulk solutions; however, this approach offers poor control over size, morphology, compactness, and surface functionality compared to the microfluidics approach. Large-scale techniques usually result in poorly reproducible systems with insufficient control over particle characteristics. To overcome this issue, self-assembly has been used to fabricate nano- to micro-sized vesicles, which allows the formation of polymeric systems that feature an increased bioavailability, enhanced permeation and retention, targeted delivery, and improved stability of labile agents from degradation (Dashtimoghadam et al., 2013).

Particle self-assembly can be achieved using microfluidic channels through hydrodynamic flow focusing. In this approach, two miscible phases are employed: one continuous and one discrete, and microchannels drive the flow, focusing the continuous phase into the discrete phase. This technique has been demonstrated to be suitable to produce nanoparticles due to the rapid and tunable mixing properties provided by microchannels (Jahn et al., 2007). Moreover, different microfluidic approaches such as flow lamination (Bessoth et al., 1999), fluid folding (Johnson et al., 2002), and chaotic mixing (Chen et al., 2012) have been used for the production of self-assembled drug carriers. The use of microfluidic channels to produce self-assembled particles has proven to provide precise control over particle size distribution, composition, and morphology.

Niosomes—synthetic membrane vesicles—allow the encapsulation of hydrophilic molecules in their inner aqueous core, thus reducing systemic toxicity by minimizing clearance of such agents from the body through slow drug release rate. Niosomes are not only employed as carriers of treatment agents due to improved bioavailability, but also as carriers of contrast agents for targeted clinical imaging applications (Luciani et al., 2004). Niosomes can be produced using microfluidic hydrodynamic flow focusing for the controlled self-assembly of monodisperse nanoparticles formed from nonionic surfactants and provides a reliable technique for drug carrier assembly.

Microfluidic hydrodynamic flow focusing has also been used for the production of other drug carriers, including polymeric complexes (Khan et al., 2015a), liposomes-hydrogel hybrid particles (Hong et al., 2010), block copolymer micelles (Wang et al., 2011), and homogeneous nanoparticles (Tran et al., 2012), among others. Furthermore, it can be used for the development and optimization of biomimetic colloidal systems for nanomedicine applications (Lo et al., 2010).

3.2.2 DROPLET-BASED CARRIERS

The main difference between droplet formation and self-assembly techniques is that the latter produces smaller particles and miscible fluids are mixed, whereas in the former, immiscible fluids are used. The droplet-based method is one of the most popular carrier synthesis techniques and has been used to generate reproducible and homogenous drug-loaded particles, microcapsules, microbubbles, and microgels (Riahi et al., 2015). Droplet size generated via microfluidics is highly uniform and devices have been produced with a size variation in the dripping regime of less than 3% (Silva et al., 2016).

When using microchannels for the droplet formation, wettability is an important factor to be considered in the droplet production process. The continuous phase should wet the channel surface favorably, so the dispersed phase can be disfavored by the channel walls. For this reason, material selection and surface modification processes play an important role on the design of microfluidic devices for droplet formation. These devices have been satisfactorily fabricated and tested using materials such as PDMS (Huang et al., 2007), silicon (Pollack et al., 2002), perfluoropolyether (Tran et al., 2012), polymethylmethacrylate (Xu et al., 2012), polyurethane (Nie et al., 2006), polycarbonate (Ogończyk et al., 2011), and stainless steel (Xie et al., 2012), among others.

In planar flow focusing geometries, four different regimes have been identified: squeezing, dipping, jetting, and tip streaming. Due to the large number of variables, there is still a lack of scaling laws to determine the transition between regimes (Gu et al., 2011); however, it has been proven that squeezing and dripping regimes depend on the upstream geometry and flow field, whereas tip streaming depends on the downstream channel and flow field (Lee et al., 2009). It has also been reported that the squeezing regime is dependent on the increase of pressure difference that causes partial and temporal blocking of the orifice by the stream (Funfschilling et al., 2009).

Passive droplet formation is the most widely used method for creating microdroplets using the microfluidic devices. In the passive methods, the interface between two immiscible fluids is deformed by their flow field, causing interfacial instability (Gu et al., 2011). The most common strategies for passive droplet formation in planar devices are T-junctions (Fig. 3.1A) and flow focusing (Fig. 3.1B). Nonplanar geometries—such as concentric capillaries (Liu et al., 2013a)—are also used for passive droplet formation. These designs bring the two phases to meet at a junction, where local flow field is deformed. Here, droplet formation occurs due to the competition between viscous shear stresses acting to deform the liquid interface, while capillary pressure resists this deformation.

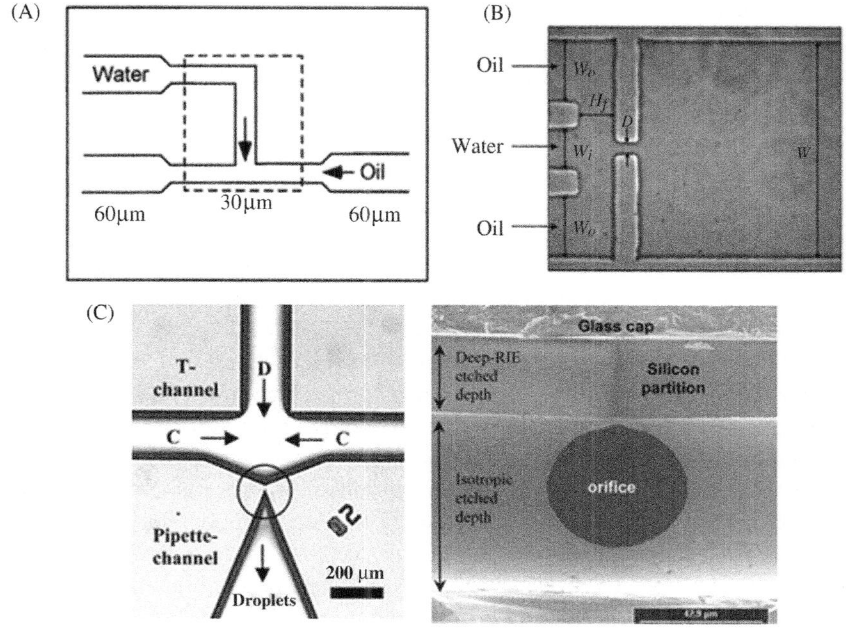

FIGURE 3.1

Planar microfluidic devices for droplet formation. (A) T-junction device (Thorsen et al., 2001). (B) Flow focusing method. (C) Axisymmetric flow focusing method.

(A) Reprinted figure with permission from Thorsen, L. et al., 2001. Dynamic pattern formation in a vesicle-generating microfluidic device. Phys. Rev. Lett 86, 4163–4166, Copyright (2001) by the American Physical Society. (B) Reproduced from Yobas, L., et al., 2006. High-performance flow-focusing geometry for spontaneous generation of monodispersed droplets. Lab. Chip. 6 (8), 1073–1079 with permission of The Royal Society of Chemistry. (C) Reproduced from Anna, S.L., Bontoux, N., Stone, H.A., 2003. Formation of dispersions using "flow focusing" in microchannels. Appl. Phys. Lett. 82 (3), 364–366 with permission from AIP Publishing.

The T-junction configuration was first reported in 2001 (Thorsen et al., 2001). Further experiments led to the conclusion that droplet size depended not only on the flow rate of the two phases, but also on the microchannel dimensions, relative viscosity of the two phases (Tice et al., 2004), and type and concentration of surfactants (Baret et al., 2009).

The method based on flow focusing geometries was first proposed in 2003 (Anna et al., 2003) with a design that consists of three inlet channels merging into a main channel through a narrow orifice. The disperse phase, flowing through the middle channel, is restrained by the shear forces of the continuous phase on the lateral channels. Both phases pass through the small orifice as the disperse flow becomes narrow, breaking into a droplet. The diameter and production ratio of droplets depend upon the flow rate and viscosity of both phases, as well as on the channel geometry (Garstecki et al., 2006).

Axisymmetric flow focusing (Fig. 3.1C) has also been used for the formation of smaller monodisperse droplets. This microchannel geometry allows the flow of the disperse phase to pass through an orifice and the pinch-off is achieved by the combination of shear stress and wetting upon contact with the inner channel surface.

Active droplet formation is utilized to increase the flexibility on control of parameters such as pressure, flow rate, viscosity of phases, and others, resulting in higher controllability in the droplet formation process. A wide variety of techniques, including electrowetting (EW) (Gu et al., 2008), membrane valves (Guo et al., 2010; Sun et al., 2013), pulsating inlet pressure (Moon et al., 2015), magnetic control of ferrofluid droplets (Wu et al., 2014), temperature control (Murshed et al., 2009), and surface acoustic waves (Schmid and Franke, 2013) have been demonstrated for the active droplet formation.

With these techniques, emulsions with tunable size and surface characteristics can be produced, making microfluidics an exceptional tool to produce drug delivery carriers based on droplet formation. For instance, oil-in-water (O/W) emulsification can be used to produce droplets loaded with active lipophilic molecules to increase their bioavailability. Droplets can also be used as gene carriers by forming water droplets stabilized by a cationic lipid surfactant (dimethyldioctadecylammonium bromide) to allow DNA bonding to the droplet by electrostatic attractions, which is achieved by mixing the droplet with plasmid DNA (Wu et al., 2009).

Droplets can also be formed as water-in-water (W/W) emulsions by using a two-phase aqueous system. This allows the replacement of the conventional water organic system, thus avoiding the use of harmful solvents. Successful emulsification of poly(ethylene glycol) (PEG)-based droplets has been produced by mechanically perturbing the stable W/W jet and inducing gelatination at the W/W interface (Cheung Shum et al., 2012). The low interfacial tension between phases in the two-phase system is mechanically broken by modifying the pressure used to inject the dispersed phase into a glass microcapillary, yielding droplet formation. Such an aqueous environment is useful to encapsulate drugs that are not stable in organic solvents.

Due to the wide variety of approaches to produce droplets and the control they provide over emulsion characteristics, drug delivery carriers can also be designed to provide targeting characteristics. As an example, superparamagnetic particles have been encapsulated using microfluidics droplet formation. This brings about a passive-targeting drug delivery system since the devices can be controlled and manipulated using an external magnetic field (Velev and Bhatt, 2006).

Moreover, different polymers and other materials have been used to produce porous microparticles to control the release and encapsulation profiles. Microfluidics allows for the formation of monodisperse porous microspheres with well-defined pore characteristics which, in turn, enables encapsulation of cargo inside the particle and pores. It is also possible to change the release kinetics of the active agents (Fontana et al., 2016). Porous materials such as porous silicon have been used to produce single emulsions with two or more drugs with different physicochemical properties, and even as a stimuli-responsive drug carrier. These types of carriers have attractive properties such as remarkable stability and biocompatibility, and a high drug-loading capacity due to the large volume and surface area available in the pores (Liu et al., 2013a). Imaging and therapeutic agents are two examples that have been successfully loaded into this type of carrier. Polymeric materials are commonly used for the production of drug carriers, and poly(lactic-*co*-glycolic acid) (PLGA) is one of the most successfully developed biodegradable polymers. Material benefits such as biodegradability, biocompatibility, ability to protect drug degradation, sustained release, modifiable surface, improved interaction with biological materials, and the possibility to target nanoparticles to specific organs or cells, make PLGA a very attractive material for drug carriers. Moreover, PLGA is approved by the US Food and Drug Administration and European Medicines Agency for drug delivery systems for parenteral administration and has a well-described formulation and production method adapted to various types of drugs (Danhier et al., 2012).

3.2.3 MULTIPLE-EMULSION DRUG CARRIERS

When encapsulating lipophilic drugs, O/W emulsion systems have shown to successfully retain the drugs of interest; nevertheless, poor encapsulation efficiency is observed in hydrophilic drugs. This issue is solved by using double or multiple emulsion techniques. With this, a relatively simple technique makes possible to encapsulate highly soluble compounds. These emulsions can be classified as water-in-oil-in-water (W/O/W) or as oil-in-water-in-oil (O/W/O), according to their composition sequence. This nomenclature refers to "water" as any aqueous or polar phase and to "oil" as any hydrophobic phase. Since there are sets of oils and aqueous phases that are immiscible in each other, O/O and W/W emulsion can also be designed. A double emulsion allows greater control and flexibility for encapsulation and release, and adding an intermediate phase brings up an extra barrier, thus enhancing control over cargo release.

Microfluidics play an important role in the precise fabrication of multiphase emulsions by allowing the production of controlled complex capsules for sequential or simultaneous drug release. Several types of multiphase emulsification configurations have been reported, ranging from multiple concentric shell drops (Kim et al., 2011) to higher-order emulsions with a controller number of inner and middle drops (Chu et al., 2007), and also including combinations and variations of the two configurations.

Core shell drops can be produced in planar and non-planar devices. In one hand, planar devices take advantage of microchannel junctions for droplet production using different geometries, such as T-junctions connected in series. These intersections allow precise control over droplet size and shell thickness. To allow the formation of O/W and W/O emulsifications, wettability must also be varied by employing two microfluidic devices with different wall properties (Okushima et al., 2004). For W/O emulsifications, a hydrophobic T-junction is often connected to a second chip with a hydrophilic T-junction, resulting in an aqueous core emulsion enclosed by an oil drop. However, this process can be also attained in a single device, making the process simpler and more reproducible.

The high reproducibility of these microfluidic devices is achieved by using three-dimensional (3D)-printed microfluidic elements that later undergo a hydrophobic and hydrophilic-initiated chemical vapor deposition coating process to change surface wettability (Thompson et al., 2016). Furthermore, this coating technique can also be used in conventional PDMS-based microchannels where vapor-phase reactants can be diffused into the channels to completely coat them.

To overcome the limitations of planar microchannels, 3D microfluidic channels can be designed using soft-lithography. These devices avoid contact of the discontinuous phase with the channel walls, thus removing the need of specific coatings in each step of the multiphase drop formation. Such devices have been fabricated with PDMS using asymmetric flow focusing, where a cylindrical microchannel was achieved by using insulated optical fiber as the mold to be removed after polymer curing. One section of the optical fiber was exposed, meaning that part of the insulator was removed to form the channel central region which served as the flow focusing orifice (Takeuchi et al., 2005).

Non-planar microfluidics, such as capillary devices, are made using coaxial assemblies of glass capillaries attached onto the surface of a glass microscope slide (Vladisavljević, 2016). Core shell drops are made using three capillaries: two round capillaries are inserted into the opposite side of a squared capillary. The fluids are injected using hypodermic needles and plastic hubs attached to the capillaries. The inner phase is pumped in through one of the round capillaries, and the remaining is used to collect the formed drops, while the middle phase is parallelly pumped into the inner phase between the space of the squared and round capillaries. The outer phase is injected in the opposite direction of the middle phase through the interstitial space between the round and square capillaries.

With this approach is possible to reuse materials to fabricate other microfluidic systems and it allows changing the distance between capillaries, thus changing the characteristics of the produced droplets. The shell thickness obtained with these devices usually ranges from several to hundreds of micrometers. Formation of thinner shells is limited by the ability of the pumps to provide lower flow rates in the middle phase; however, to form thinner shells, the same configuration can be used (i.e., one squared and two round capillaries) with the addition of a third, round capillary into the inner phase capillary. The inner phase is then injected in the smaller capillary while the middle phase flows in the larger round capillary, and the outer phase is pumped in the squared capillary. The three fluids flow into the orifice of the collection capillary and form the core–shell drops with ultrathin shells.

The shell–core droplet technology can be used to produce microcapsules with tunable size and morphology, allowing control over the drug release kinetics (Fig. 3.2).

(A)

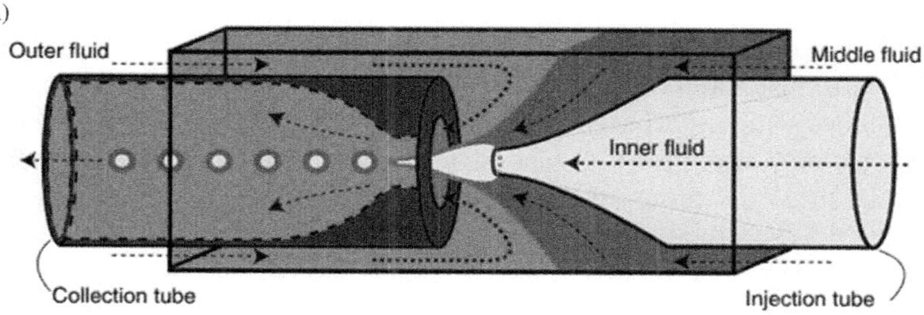

(B)

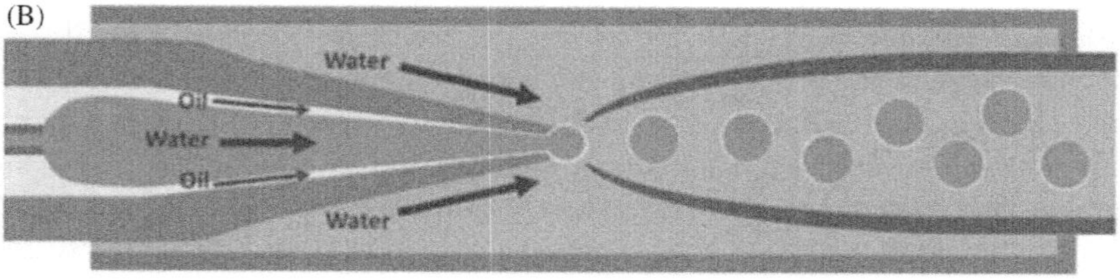

FIGURE 3.2

Microcapillary device for double emulsion production. (A) Coflow focusing method. (B) Small central capillary-based method.

(A) Reproduced from Kim, S.-H., et al., 2011. Double-emulsion drops with ultra-thin shells for capsule templates. Lab Chip 11 (18), 3162–3166 with permission of The Royal Chemical Society. (B) Utada, A.S., et al. Dripping, jetting, drops, and wetting: the magic of microfluidics. MRS Bull. 32, 9, 702–708, reproduced with permission.

O/W and W/O emulsions can be made using polymers and biopolymers. Such particles can then be obtained by evaporation, extraction, and other techniques. During these processes, tuning its characteristics—varying the rate of solvent evaporation, for example—allows adding surface texture to such particles. Polymers like PLGA and poly(lactic acid) (PLA), which are biocompatible and biodegradable, can be used to produce O/W emulsions, as they dissolve in the oil phase. Biopolymers such as alginate, chitosan, pectin, and cellulose are also used for drug carrier production due to their water solubility, thus the formation of microcapsules is achieved through W/O emulsions. Furthermore, the core–shell technique allows for encapsulation of quantum dots, magnetic nanoparticles (Huang et al., 2011), and gold nanoparticles (Huang et al., 2006) when used in the disperse phase.

The design of capsules using microfluidics allows the precise fabrication of monodisperse stimuli-responsive carriers. Glucose-responsive microcapsules have been produced using the double emulsion technique with poly(*N*-isopropylacrylamide-*co*-3-aminophenylboronic acid-*co*-acrylic acid) [P(NIPAM-*co*-AAPBA-*co*-AAc)] hydrogel as microcapsule shell. In this case, PNIPAM thermo-responsive network and the glucose-responsive AAPBA were used for actuation and glucose responsiveness. With these microcapsules, the fabrication technique was shown to be feasible for the development of sensors and self-regulated delivery systems for diabetes and cancer therapies (Zhang et al., 2013).

Cationic pH-responsive microcapsules can be prepared by the O/W/O approach by using poly(*N,N*-dimethylaminoethyl methacrylate) (PDM), a cationic pH-responsive material with good biocompatibility. The obtained PDM microcapsules exhibited cationic pH-sensitivity, and it was shown that preparation conditions significantly affect the pH-responsive swelling ratios of such microcapsules (Wei et al., 2011). pH-responsive microcapsules can also be fabricated using chitosan, a highly versatile cationic polysaccharide. Moreover, in acidic media with pH values ranging from 1.5 to 4.7, microcapsules showed an acid-triggered membrane decomposition due to the acidic hydrolysis of the crosslinked chitosan networks. The decomposition rate of networked chitosan was shown to be directly dependent on the environmental pH value (Liu et al., 2011). This property is a key factor for the design of oral drug delivery carriers, since chitosan not only exhibits excellent biological activity, good biocompatibility, and biodegradability, but it also possesses antiulcer and antimicrobial properties which make it a matrix capable of gastric drug delivery (Hejazi and Amiji, 2003).

PLGA–alginate core–shell microspheres show a sustained drug release profile. For this reason, alginate has been used as shell material to modulate the drug release profile. It was found that a change on the microcapsule size consequently changes the release profile. Furthermore, these particles promote a sustained release of the model drug for over a month (Wu et al., 2013). Further examples of core–shell drug carriers and their microfluidic platform are listed in Table 3.1.

Table 3.1 Examples of Core–Shell Microfluidic Design of Drug Delivery Carriers

Droplet Material	Encapsulated Cargo	Device	Reference
PEG-4MAL microgel	VEGF	PDMS-based microfluidic channels	Foster et al. (2017)
Alginate	Fluorescein isothiocyanate (FITC)-labeled mouse IgG antibodies	PDMS-based microfluidic channels	Mazutis et al. (2015)
Ranitidine HCl loaded poly (acrylamide)	Ketoprofen loaded poly(methyl acrylate)	Co-axial capillaries	Khan et al. (2015b)
Sodium alginate microparticles	BSA and FITC	Capillaries	Liu et al. (2013b)
Crosslinked chitosan	LR300	Capillaries	Liu et al. (2011)
PEG-b-PLA polymerosomes	Hydrophilic maghemite $\gamma-Fe_2O_3$ magnetic nanoparticles and hydrophobic CdSe/ZnS Quantum Dots	Capillaries	Seth et al. (2012)

PEG-4MAL, *4-Arm poly(ethylene glycol)-maleimide;* VEGF, *vascular endothelial growth factor;* PDMS, *polydimethylsiloxane;* BSA, *bovine serum albumin;* LR300, *Lumogen F Red 300.*

Structures such as lipospheres can be used to encapsulate and deliver different types of drugs such as antiinflammatory compounds, antibiotics, and vaccines. Gas can also be encapsulated for enhanced ultrasound-added particle location and delivery. These agents can selectively be concentrated and disrupted at regions of ultrasound interest. Standard liposomes cannot behave this way since their density and compressibility are similar to the surrounding blood, thus encapsulation of gases allows the control of the delivery location. For higher specificity, targeting ligands can be added to the shell surface (Hettiarachchi et al., 2009).

Multiple emulsions with two or more inner cores are produced using capillary devices with sequential coflow drop generators. Here, the number of inner droplets is controlled by varying the flow rate of each phase. These structures are produced by sequentially aligning glass capillaries forming serial coflow structures. Particles with multiple cores allow a distinctive kinetic release profile, and microfluidic tools allow precise design and control over them.

Multicore emulsions with inner drops made up from different materials can be designed using planar (Okushima et al., 2004) and capillary microfluidics (Chu et al., 2007). Coflow of two different disperse phases using an injection tube with two separate internal channels in a capillary device (Fig. 3.3A) allows the

(A)

(B)

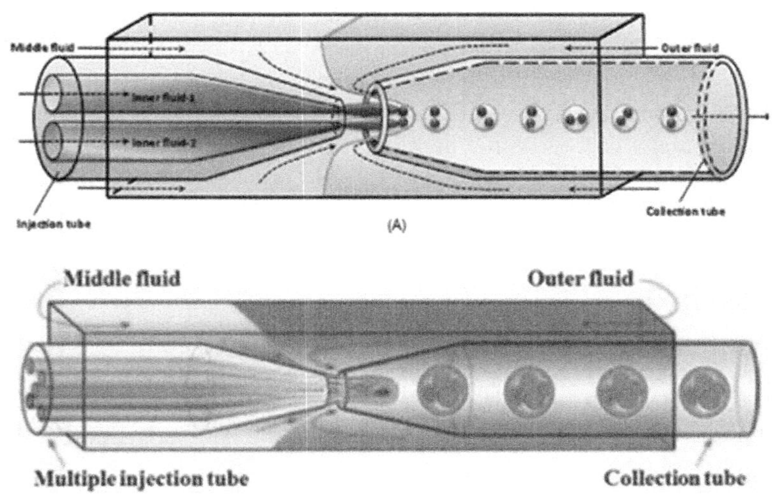

FIGURE 3.3

Capillary microfluidic devices for multicore particles. (A) Capillary microfluidic device for generation of two-bore double emulsions using an injection tube with two internal channels. (B) Capillary device for multicore particles using an injection tube with several inner microchannels.

(A) Reproduced from Sun, B.J., et al., 2010. Microfluidic melt emulsification for encapsulation and release of actives. ACS Appl. Mater. Interfaces 2 (12), 3411–3416 with permission from the American Chemical Society. (B) Reprinted by permission from Macmillan Publishers Ltd: [NPG ASIA MATERIALS] Zhao, Y., et al. Multifunctional photonic crystal barcodes from microfluidics 4(9), e25, copyright (2012).

formation of two-core droplets (Sun et al., 2010). These techniques constitute a promising solution for the concurrent delivery of multiple drugs, as well as allowing tuning of the outer shell properties to control the release profile of the actives. Multiple core, double emulsion particles can be designed using multiple injection tubes to form the most inner part of the particle (Fig. 3.3B), which is achieved by fabricating a multicore capillary by tapering to achieve a specific orifice diameter (Zhao et al., 2012).

Capillaries for multiple emulsion formation have been used to encapsulate nanotubes inside microspheres. For instance, this technique was used to deliver colon cancer-preventing and cancer-inhibiting drugs through the harsh conditions of the gastrointestinal tract (atorvastatin and celecoxib). The nanotube-in-microsphere technology employed microfluidics to encapsulate halloysite nanotubes in a pH-responsive hydroxypropyl methylcellulose acetate succinate polymer. Results showed not only fast release and enhanced drug permeability, but also the inhibition of colon cancer cell proliferation at pH 7.4 (Li et al., 2016).

3.2.4 JANUS DROPLETS

Janus droplets are particles formed from two separate halves with different chemical and physical compositions, and microfluidics has demonstrated to be a powerful tool to assemble these structures. Different emulsions have been produced in microfluidic devices by injecting dispersed phases to form parallel streams that flow as a continuous phase. On one hand, planar microfluidics allows the use of a photolithography process with a high resolution and replication rate. On the other hand, capillary devices are easier to assemble and the wetting effect is smaller, hence a higher production rate can be achieved.

Janus droplets are used for four main purposes: (1) encapsulation of active pharmaceutical ingredients with opposite solubility; (2) dual-phase release kinetics; (3) drug targeting; and (4) theranostics (Tran et al., 2014). As an example, Janus droplets were used to overcome the limitation of polymeric particles which are not able to load agents with different solubilities (Xie et al., 2012). This design consisted of Janus particles made from the biodegradable polymer PLGA containing two anticancer agents, demonstrating the release kinetics of hydrophobic and hydrophilic agents.

Janus droplets are also used to improve targeted delivery of agents to tumors. Since the combination of two or more therapeutic strategies with different mechanisms can cooperatively inhibit cancer development, Janus particles are used to carry and deliver chemotherapeutic active pharmaceutical ingredients and siRNA. It has been demonstrated that siRNA and paclitaxel can be simultaneously delivered to the same tumor cells, either in vivo or in vitro (Sun et al., 2011).

Moreover, dual-phase release kinetics is also achieved using Janus particles. The bi-compartmental structure allows designing particles that combine physical and chemical interactions to create two separated degradation kinetics. Poly (acrylamide-*co*-acrylic acid), either mixed with poly(ethylene oxide) into an interpenetrating polymer network on one hand, or crosslinked to dextran on the other hand, showed two different pH-dependent degradations. This allows a very slow release from the more stable compartment and a faster release from the more unstable compartment (Hwang and Lahann, 2012). Finally, implementation of theranostics using Janus particles allows loading of both diagnostic and therapeutic agents in the same carrier. Although conventional droplet technologies can be used in the design theranostics systems, the bi-compartmental characteristic of the Janus particles reduces the risk of exclusion of one agent by the other (Tran et al., 2014).

The design of Janus droplet makers highly depends on the interfacial tension between the employed phases. The morphology of the resulting droplet will be highly influenced by surface tension; thus the final structure can be a shell–core droplet, a Janus droplet, or two separated droplets. Miscibility is another important factor in the sense that if the two dispersed phases are miscible, a cross-junction configuration can be used where the two dispersed phases are injected at the same rate in the two facing channels, while the continuous phase flows perpendicular to them with a higher rate. T-junctions can be used with miscible phases, and in this case, both phases flow parallel to each other at the same rate, and the stress force produced at the junction due to the flow of the continuous phase breaks the stream, forming Janus droplets. For these two methods, streams of the three flows of the phases need to be balanced and their viscosity should be similar, otherwise swirls could be formed within the droplets. Moreover, convective mixing can also occur due to the interaction between droplets and the microchannel walls. However, one of the techniques used to prevent this phenomenon is photopolymerization right after the droplet is formed to stabilize the particle.

Designing has less limitations in the case of immiscible phases. Swirling of the two disperse phases is minimum, which allows different design configurations and simplifies the microfluidic setup. Janus droplets can even be formed from core–shell droplets if the phases have the proper interfacial condition. Incorporation of components such as dendrimers into the polymer network allows a prolonged drug release and it can be tuned by dendrimer-drug conjugation (Acton et al., 2013). The use of ligands also improves the capabilities of Janus drug carriers since they allow a superior targeting efficiency compared to classical particles by having homogeneous ligand distribution (Djohari and Dormidontova, 2009).

PLGA–alginate core–shell microspheres have a sustained drug release profile. As an example, a single PLGA core was used due to its biocompatibility and alginate was used as the shell material to modulate the drug release profile. In this study, it was found that a change in the size of the microcapsules changes the release profile. It was also determined that the particles promoted a sustained release of the model drug for over 1 month (Wu et al., 2013). Janus droplets were also used for triggered and controlled drug release. Interfacial tension can dynamically change due to a certain stimulus such as a response to temperature, light, or pH which is possible by using stimuli responsive and cleavable surfactants

(Zarzar et al., 2015). These drug carriers are very useful for controlled release of drugs in vivo.

Janus droplets can be produced from two separated particles that later come in contact with each other to form polymers with two compartments to encapsulate two different active (W/O/W) emulsions, where, after solvent evaporation, the emulsification changes its shape and transforms into a Janus droplet (Shum et al., 2011).

Multiple-core emulsions with inner drops made from different materials can also be designed using planar (Nisisako et al., 2005) and capillary microfluidic devices (Zhao et al., 2012). Co-axial flow of two different disperse phases using an injection tube with two separate internal channels in capillary devices allows the formation of two-core droplets (Sun et al., 2010). This technique is a promising method to deliver multiple drugs at the same time, and it also allows tuning the outer shell characteristics to manipulate the release profile of the actives and for further control of their release.

3.2.5 FLOW LITHOGRAPHY

Flow lithography takes advantage of the photocrosslinkability property of certain polymers. Here, photosensitive phases are pumped into a microchannel where the particles are polymerized by directing a light beam through a photomask to achieve the desired feature shape. This technique allows the formation of either spherical or nonspherical particles, to improve the control over the release kinetics of the drug carrier. Furthermore, the particle shape also affects the in vivo biodistribution, uptake mechanism, and circulation in the bloodstream (He et al., 2010; Geng et al., 2007). Finally, the development of nonspherical particles through flow lithography allows shaping drug carriers to mimic natural entities and controlling the surface-to-volume ratio, which affects the attachment to cells membrane.

Flow lithography can be classified in two main categories: continuous flow and stop flow. Continuous flow lithography refers to the technique that allows a constant flow of phases, and the light source is controlled to polymerize particles. For the stop flow lithography, once the channels have been filled up, the flow is stopped and a constant light radiation is provided to crosslink the particles. With these techniques it is not only possible to synthesize micrometer-sized microparticles in a variety of 2D microchannels, but also allows tuning the chemical anisotropy of the formed structures (Jang et al., 2007).

Continuous flow lithography was proposed in 2006 using sequential ultraviolet light pulses through a photomask to a continuous flow stream (Dendukuri et al., 2006). The main drawback of this technique was a low resolution due to the flowing phases. The flow stream creates blurred particles due to the moving phases during polymerization. To overcome this issue, stop flow lithography was created, and the production and resolution were improved by incrementing flow rates to flush particles and optimizing the patterning process.

Both conventional flow lithography techniques can only produce particles with uniform height (2D structure). To fabricate 3D structures, alternative techniques have been created, such as in situ photopolymerization, which is set as a vertical continuous flow lithography configuration. This method was developed using a digital micromirror as a dynamic photomask and a vertical flow stream of PEG diacrylate in a microchannel (Habasaki et al., 2013). This technique allows the formation of long fibers with tunable diameter and shape. Additionally, the two-photon polymerization method has also been used to produce 3D particles. In this case, it is possible to change the focal plane across the depth of the channel, but has the drawback of being a low throughput technique.

Compared to droplet-based microfluidics, flow lithography offers improved control over particle formation and simpler operational setup. However, it also has limitations and the use for the development of drug delivery carriers is reserved for nonlight-sensitive drugs. Regarding the fabrication process, the smallest particle size that can be fabricated is restricted by the resolution of the illumination source (Riahi et al., 2015).

3.3 DIRECT DRUG DELIVERY DEVICES

Microfluidic systems have also been used to directly deliver active molecules to specific locations in the body. This increases the drug availability to the targeted tissue or organ on the precise dose, hence minimizing detrimental effects of the administered drugs on the rest of the body.

The delivery mechanism is the core feature of a drug delivery device as it must provide the driving force to dispense the stored drug from the reservoir toward the targeted site (Jian et al., 2012). Microfluidic systems have allowed the precise targeting of drugs on a highly controlled basis, which provides a significant advantage over more traditional drug release approaches (McGinty and Pontrelli, 2015). Table 3.2 summarizes the most relevant parameters considered in microfluidic drug delivery systems.

3.3.1 LOCALIZED DRUG DELIVERY

In a drug delivery system, it is important to precisely control the dose of active substances: drug concentrations above the prescribed dose can be toxic, while lower concentrations can diminish the benefits or have no therapeutic effect. Moreover, the delivery mechanism depends on the working mechanisms of molecules; for instance, insulin and other hormones require a pulsatile release pattern to be more effective (Santini et al., 2000).

Implantable drug delivery devices provide effective delivery of drugs into localized regions. Such devices can be designed to release drugs with different dosages and for intermittent and continuous delivery. Such devices are designed

Table 3.2 Relevant Fabrication Parameters of the Most Widely Used Drug Delivery Mechanisms

Direct Drug Delivery Mechanism	Relevant Parameters	References
Micropumps	The most widely used micropumps: 1. Mechanical. Displacement pumps that apply pressure differentials to fluid 2. Electro and magneto-kinetic. Provide continuous energy to the fluid by means of an external electric signal applied to the actuators The pressure generation requirements for implantable micropumps can be as high as 25 kPa	Zhou et al. (2016), Laser and Santiago (2004)
Microneedles	Typical geometries range from 25 to 2500 μm in height, 50 to 250 μm in base width, and between 1 and 25 μm in tip diameter	Yung et al. (2012), McAllister et al. (2003)
Microfluidic system	• Usually comprises a pump or actuator, valve, reservoir, and membrane to control release rate • Usually are PDMS-based due to the physical and chemical properties of the material • These devices are able to process nano- and picoliter volumes	Riahi et al. (2015)

PDMS, *Polydimethylsiloxane.*

to operate for short periods (days) or for extended periods (~1 year), and some can be refilled while in use (Lavan et al., 2003). However, the most relevant challenge in microfluidics for the successful development of implantable devices is precise delivery control (Kim et al., 2016). Controlled drug release can be triggered by different stimuli, such as temperature, pH, and magnetic and electric fields, among others (Singh et al., 2013).

In most implantable devices, drug release is passive and dosing cannot be stopped after implantation, except by surgical removal. Thus, it is essential to have active control of the release rates and switchable on–off mode capabilities, particularly for unexpected changes in the physiological conditions of the patient that might require a corresponding adjustment of the dose. Moreover, implantable devices must have an optimized drug reservoir size, relative size, extended operative lifetime, controllability, and biocompatibility. An ideal implantable device not only has to include a large enough drug reservoir to minimize refilling, but must also have a small size relative to the region of implantation to reduce interference with surrounding tissues and to avoid discomfort after implantation (Jian et al., 2012).

Micropumps are essential components of drug delivery devices. Pumping methods of microfluidic devices can be categorized as mechanical or nonmechanical micropumps (Nisar et al., 2008). Nonmechanical approaches include mechanisms such as electroosmosis, electrochemical, EW, and flexural plate wave (FPW). Such mechanisms are easy to be implemented since they require no moving components. On the other hand, mechanical pumps require moving components for operation. For this category of micropumps, a pressure differential is applied to drive the samples out of the reservoir. Some of the most common mechanical pumps utilize electrostatic, electromagnetic, piezoelectric, and electrothermal principles to induce the movement of pumping elements. Other types of mechanical and nonmechanical micropumps are described elsewhere (see, e.g., Nisar et al., 2008; Tetteh et al., 2014).

Despite the operational differences, the common characteristic of most of the reported micropumps is the requirement of a power source (usually a battery) to operate, which increases the size and complexity of the system (Mousoulis et al., 2011). Moreover, other aspect to be considered in the development of localized delivery and implantable devices is maintenance; particularly, the battery recharging. Recommended designs are those where batteries can be recharged during the same session as reservoir refilling.

Electrostatic micropumps are devices that drive the fluids via the mechanical movement of a membrane. These systems work due to the Coulombic attraction force between two charged plates, where the membrane of the pump deflects in a certain direction, depending on the amplitude of the applied voltage. Because electrostatic actuators require small currents, fabrication process is simpler, compared to other mechanical systems, and also have faster time response, which makes this mechanism one of the most popular actuation methods for micropumps.

Electromagnetic pumps use ferromagnetic materials usually deposited on a cantilever beam. These devices use an induction coil under the cantilever to generate a magnetic field that causes the cantilever to bend. The use of this approach is mainly limited by a complex fabrication process and high power consumption. However, magnetic actuators are evolving as promising candidates for localized drug delivery in implantable devices. The working principle relies on the fabrication of an elastic magnetic PDMS membrane with a micro aperture that seals the device reservoir. During actuation, a magnetic field causes the membrane to deform, thus, increasing the hydrodynamic pressure inside the reservoir that pushes the active out. Volume and rate of release can be controlled by tuning the magnetic actuator parameters—number of actuators and strength of the external magnetic field (Pirmoradi et al., 2011). This type of micropump has been used for the delivery of antiepileptic drugs and was found to have low-power dissipation. For this application, the proposed device quickly injects the medication at the seizure onset. The delivery system is composed of two parts: (1) a seizure detector; and (2) an embedded drug delivery system (Hamie et al., 2013).

An alternative system makes use of piezoelectric actuators that base their working principle on the use of materials that deform as a response to an applied electric field. This method provides a better linearity compared to the two methods discussed, and it also has a fast switching time. Finally, electrothermal micropumps use temperature variations as a mechanism to change an object's length or thickness. It commonly uses a membrane or two dissimilar strips with different thermal coefficients, therefore, when temperature changes, the membrane or strips deflect, pushing the actives out of the reservoir. Such actuators are simple to fabricate, but have high power consumption and a slow time response. Wireless-controlled microfluidic pumps have been developed using this type of actuator and were designed to administrate controlled drug dosages to the brain. Here, a soft microfluidic neural probe was created with a minimal invasive effect and to allow the direct in vivo infusion of pharmacological agents into the brain with reduced formation of glial scars. The delivery mechanism relies on a thermally-actuated micropump formed by four drug reservoirs covered by copper membranes, a thermally expandable layer, four microheaters, and a substrate. The temperature is increased by applying an electrical current that causes the thermal layers to expand, thus increasing the pressure in the reservoirs, leading to the rupture of the copper membrane and the release of the actives through the microchannels. The process is controlled by a wireless transmitter and receiver switching the heaters on and off (Jeong et al., 2015). Reservoirs covered with silicon nitride membranes can also be used for the same purpose (Elman et al., 2009). This approach works by heating the liquid inside the actuation layer, generating bubbles in the membrane that later burst, jetting the reservoir's contained solution. With this, a fast drug delivery without significant degradation is achieved. Such devices are designed to work under emergency treatments where rapid and reliable drug delivery is crucial, rather than constant and continuous administration of the drugs.

Non-mechanical micropumps use the conversion of non-mechanical energy to kinetic energy, providing momentum to the fluid to be supplied. Such systems do not require mechanical pumps, moving parts or valves, thus fabrication is simple. Electroosmotic actuators, for instance, take advantage of the electrokinetic phenomenon to move fluids—an ionic solution moves toward a charged surface when an electric field is externally applied. A high-pressure pump has been designed using this method and provided a maximum flow rate of 0.47 L/min with a pressure of 72 kPa. The device was made using silica nanospheres confined within the channels with the objective of creating an efficient flow. It was assumed that the porous structure of such nanospheres would help in the formation of an electrical double layer between them (Takemori et al., 2005).

In electrochemical micropumps, the driving force is provided via electrolysis of water. The structure is basically composed by electrodes for electric stimulation, an electrolysis chamber, microchannels, and reservoirs. Proof of concept has shown reliability for periodical drug administration and as an automatic refilling system. The devices showed a drug release rate of 2.2 μg ± 0.3 μg per pulse. Their solid-drug reservoir system provides long-term therapeutic application (Yi et al., 2015).

In EW, surface tension is used to move the fluid. Here, an electric potential is applied to a dielectric layer, causing a decrease in the interfacial energy of the solid–liquid surface, thus forcing the fluid to move. EW actuation can be used in capillary microfluidic devices since the capillary flow is controlled by using a hydrophobic monolayer within the microchannel. Then, the electric potential is applied, making it hydrophilic, thus, allowing a continuous flow (He and Nugen, 2014).

FPW is another pumping principle, which enables the development of an active microchannel and overcomes the high fluidic resistance encountered during delivery of fluids through channels. Using this approach, a device consisting of a flow sensor and FPW micropump was developed (Nguyen et al., 2000). This device was the first complex microfluidic system with an FPW pump.

Implantable devices are useful to administrate actives to systems that block drugs when administrated through oral or intravascular routes, such as the cochlear barrier. Delivery in the inner ear is a major challenge in drug-based treatments for auditory disorders since anatomical barriers make difficult the administration of multiple doses. In such a case, implantable devices can be an alternative to the common approaches such as gel placement in the middle ear or intratympanic injections, which results in unpredictable drug kinetics. To address these challenges, a particular system that infuses microliter scale doses periodically was designed using a cannula to deliver the active, and it was implanted in the base of the scala tympani. This specific system uses a reciprocating microfluidic pump that infuses actives and then withdraws an identical volume of a less-concentrated fluid, resulting in a mass transfer of the active with zero net volume displacement of fluid (Kim et al., 2014). In this approach, drug concentration is maintained using a reservoir made by a long microchannel and a loading pump that delivers fresh solutions prior each infuse–withdraw cycle.

Microfluidic platforms have been developed to overcome some of the challenges that traditional drug delivery systems have not been able to solve. For example, the eye, one of the most fragile components of the human body, can be damaged using common methods for drug delivery that affect the optic nerve or healthy surrounding tissue. In this case, non-invasive drug delivery systems, such as an ocular iontophoretic device, can be used to tackle this problem. Ionophoretic methods enhance penetration of ionized drugs into tissues in a non-invasive basis, and the integration this technique with a biocompatible planar Poly(3,4-ethylenedioxythiophene) (PEDOT) electrode (Zhang et al., 2016) allows the devices to be placed in a reduced location of the eyeball, reducing the possible damage caused by ions, since delivery of drugs over a large area could lead to damage of the cornea or sclera. Biocompatibility among materials is of great importance to obtain high effectiveness of the actives into the eye. In vivo experiments on rabbit eyes demonstrated the possibility of performing ocular iontophoresis effectively, simply, and conveniently (Fig. 3.4).

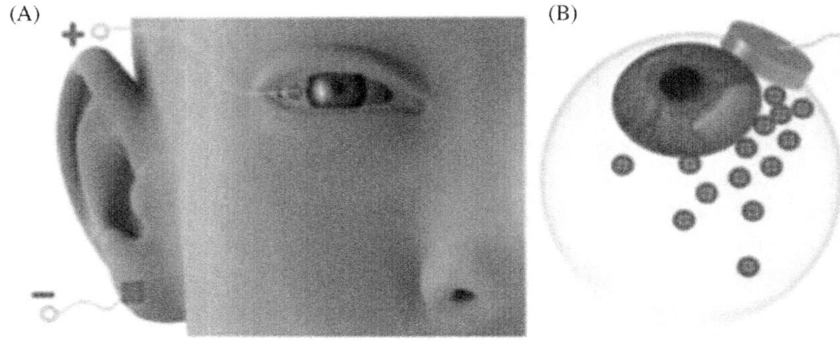

FIGURE 3.4

Schematic diagram of an iontophoretic device. (A) A simple device based on PDMS with planar PEDOT electrode, can be placed under eyelid. (B) Deliver of ions through a small area by iontophoresis.

Reprinted from Sens. Actuators, B: Chem., 237, Zhang Y. et al., An ocular iontophoretic device using PEDOT electrode for local drug delivery, 1007–1014, Copyright (2016), with permission from Elsevier.

Given that microfluidic devices can deliver small and precise amounts of bioactive compounds into localized areas, they have also been used for treatment of neural tissue. For instance, a multichannel silicon probe was developed to deliver compounds into neural tissue while simultaneously recording electrical signals from neurons in vivo. To achieve this task, it was crucial to determine the optimal pressure needed to deliver a specific amount of fluid (Mcallister et al., 2000). Moreover, a 3D multifunctional probe was developed to combine chemical, optical, and electrical stimulations along a recording system to monitor the effects of different actives administrated into the brain. In this specific design, Tetodotoxin was administrated in vitro through a PDMS microchannel bonded to the neural probe (Shin et al., 2016). Integrated implantable devices provide better control over drug kinetics since they allow monitoring the effects of said actives, and could be used in the future to develop a close-loop drug delivery system able to automatically determine the precise dosage while recording its effects in situ.

One of the objectives of microfluidic systems for drug delivery is the implementation of a screening system to assess the dynamics of the active uptakes and effects on the organism. For this reason, a real-time cell monitoring platform was presented to provide statistically-relevant information about the dynamics of intracellular processes underlying the interaction of cells with different nanomaterial formulations. In this case, a microfluidic device consisting of three inlets, one outlet and one trap array chamber was developed. This platform provided relevant information from a cell population comparable to that obtained with flow cytometry, as well as allowing the additional assessment of cell/nanoparticle interaction dynamics at the same time (Cunha-Matos et al., 2016).

Implantable microfluidic devices must cover five basic requirements to achieve effectiveness: (1) drug compatibility; (2) actuation safety; (3) controllable flow rate; (4) small size; and (5) low-power consumption. In addition to this, since such systems will directly interact with the organism, they must be fabricated using biocompatible materials. One of the most widely used substrate materials is PDMS. This elastic polymer has extensively been used in biomedical applications because of the moderate cost, ease of preparation (usually made by soft-lithography), relatively good biocompatibility, and clearness for optical inspection; however, PDMS often requires complex immobilization procedures for surface functionalization (Dou et al., 2015). Substrate materials for the fabrication microfluidic devices have advantages and limitations, and are chosen on the basis of many parameters including type of drugs, delivery mechanisms, implantation periods, and so on.

3.3.2 TRANSDERMAL DRUG DELIVERY

Transdermal delivery refers to the direct drug delivery through the skin. The most commonly used technique is via hypodermic needles, which bring discomfort to patients, such as pain, needle phobia, and potential transmission of infectious diseases through the needle. Microneedles are an alternative method to conventual injections. This method overcomes the drawback of gastrointestinal drug degradation and eradicates the pain and discomfort of intravenous and intramuscular injections (Razzacki et al., 2004). Since microneedles can penetrate the skin across the stratum corneum, avoiding nerve fibers and blood vessels, it provides a painless method for drug delivery as well as preventing bleeding.

There are four traditional microneedle designs: (1) solid; (2) coated; (3) dissolving; and (4) hollow (Tuan-Mahmood et al., 2013). For the solid microneedle, delivery is achieved by applying the drug through the skin by the microneedles after being removed. In this method, the drug moves by passive ionic diffusion and it can be applied in a transdermal patch or directly applying the active in form of a gel or cream.

Coated microneedles are covered by the drug which stays inserted after the needles are removed. After application, the drug dissolves in the skin. Dissolving microneedles rely on a similar strategy, but instead of coating the microneedles with the drug, they are made with the drug itself and, upon contact with the skin, the needles dissolve. Based on this principle, biodegradable microneedles were fabricated by a molding process, and the drugs were encapsulated within biodegradable polymers. With this approach, the microneedles release the drugs only after skin penetration (Imaeda et al., 2015). For this particular design, tip-separable microneedles allow the administration of actives through sharp tips containing the drug after penetration. With this method, dissolution time of the biodegradable needles within the skin is reduced, compared to previous mechanisms with similar administration techniques. Previously developed biodegradable microneedles have a relatively long dissolution time within the skin, which means that the patient must maintain them in their skin for prolonged times. These microneedles, shaped as eight-sided and rhombic pyramids, were developed by photolithography and anisotropic wet etching.

A different strategy using hollow microneedles allows controlled and continuous drug administration through either diffusion or pressure-driven flows, where the fluid can be injected through the hollow microneedles. This specific approach can be integrated with sensing and actuation systems to achieve preprogrammed drug delivery. For the optimal effectiveness of such devices, microneedle architecture must be taken into consideration. In a 2010 study, it was proven that in pressure-assisted flow in hollow needles, the increase of flow rate is proportional to the inner diameter of the microneedle and inversely proportional to the length, providing a better understanding on how the design can affect the release profile of actives (Bodhale et al., 2010). Drug delivery is no longer limited by the drug amount that can be loaded into the microneedles themselves, so delivery of greater doses over longer time periods is also possible. For instance, for an alternative application, microneedles can be used to extract skin interstitial fluid for diagnostic or therapeutic monitoring purposes (Singh et al., 2013).

Micropumps can be integrated into transdermal systems to allow programmed drug administration. A microfabricated transdermal dispenser or pump using a low-boiling-point phase-changing liquid was developed as a low-cost and batteryless pumping option. In this disposable transdermal device, the flow of a stored drug is initiated by the temperature rise due to heat conduction when the device is placed in contact with the skin. Only a soft touch on the skin is required to trigger the pumping action so there is no need to apply any manual force. The prototype device consists of four stacked layers of PDMS on a silicon substrate which provides the interface for heating and cooling of the working liquid, causing evaporation and condensation, respectively (Mousoulis et al., 2011).

Microneedles could be used as part of an integrated platform for drug delivery, but challenges such as prolonged adhesion, clogging, possible bacterial infections, and fibrosis must be taken into account at the design stage.

3.4 CONCLUSIONS

In this chapter, we have discussed different microfluidic platforms for drug delivery. Related applications including implantable and localized drug delivery devices were analyzed. We also highlighted the relevance of delivering precise doses to the body to avoid unnecessary damage to untargeted organs and tissues. Additionally, benefits and innovative improvements of microfluidic systems for special diseases such as cancer and diabetes were also discussed. Microfluidic approaches for drug delivery seem to be the future of medicine by testing and delivering compounds or biological samples in a short time. Microfluidic devices will also be important for medical healthcare in terms of accessibility and speed of results. Most implantable microfluidic devices provide site-specific administration of drug formulations to the affected area where the drugs have optimized effects in treating the disease.

Research in microfluidic devices has enabled more accurate drug delivery carrier designs. Nevertheless, these systems still have some technologic challenges that limit their widespread usage. These systems have a low throughput because the flow rates and volumes used are small. A single chip does not produce milliliters per minute, as could be required in some applications. Finding solutions for these challenges is of great importance for the wide production of drug delivery microfluidic platforms. Current efforts in microfluidics are been focused on the development of systems that allow performing different steps in the development of carriers. Without a doubt, microfluidic techniques will continue helping researchers in the development of reproducible, innovative, and efficient drug delivery systems.

REFERENCES

Acton, A.L., et al., 2013. Janus PEG-based dendrimers for use in combination therapy: controlled multi-drug loading and sequential release. Biomacromolecules 14 (2), 564–574.

Anna, S.L., Bontoux, N., Stone, H.A., 2003. Formation of dispersions using "flow focusing" in microchannels. Appl. Phys. Lett. 82 (3), 364–366.

Baret, J.C., et al., 2009. Kinetic aspects of emulsion stabilization by surfactants: a microfluidic analysis. Langmuir 25 (11), 6088–6093.

Benson, B., Stone, H.A., Prudhomme, R.K., 2013. An "off-the-shelf" capillary microfluidic device that enables tuning of the droplet breakup regime at constant flow rates. Lab Chip 13 (c), 4507–4511.

Bessoth, F.G., DeMello, A.J., Manz, A., 1999. Microstructure for efficient continuous flow mixing. Anal. Commun. 36 (6), 213–215.

Bodhale, D.W., Nisar, A., Afzulpurkar, N., 2010. Structural and microfluidic analysis of hollow side-open polymeric microneedles for transdermal drug delivery applications. Microfluid. Nanofluid. 8 (3), 373–392.

Champion, J.A., Katare, Y.K., Mitragotri, S., 2007. Particle shape: a new design parameter for micro- and nanoscale drug delivery carriers. J. Controlled Release 121 (1–2), 3–9.

Chen, D., et al., 2012. Rapid discovery of potent siRNA-containing lipid nanoparticles enabled by controlled microfluidic formulation. J. Am. Chem. Soc. 134 (16), 6948–6951.

Cheung Shum, H., Varnell, J., Weitz, D.A., 2012. Microfluidic fabrication of water-in-water (w/w) jets and emulsions. Biomicrofluidics 6 (1), 012808–012808-9.

Chu, L.Y., et al., 2007. Controllable monodisperse multiple emulsions. Angew. Chem. Int. Ed. Engl. 46 (47), 8970–8974.

Cunha-Matos, C.A., et al., 2016. Real-time assessment of nanoparticle-mediated antigen delivery and cell response. Lab Chip 16, 3374–3381.

Danhier, F., et al., 2012. PLGA-based nanoparticles: an overview of biomedical applications. J. Controlled Release 161 (2), 505–522. Available at: https://doi.org/10.1016/j.jconrel.2012.01.043.

Dashtimoghadam, E., et al., 2013. Microfluidic self-assembly of polymeric nanoparticles with tunable compactness for controlled drug delivery. Polymer 54 (18), 4972–4979.

Dendukuri, D., et al., 2006. Continuous-flow lithography for high-throughput microparticle synthesis. Nat. Mater. 5 (5), 365–369.

Djohari, H., Dormidontova, E.E., 2009. Kinetics of nanoparticle targeting by dissipative particle dynamics simulations. Biomacromolecules 10 (11), 3089–3097.

Dou, M., et al., 2015. A versatile PDMS/paper hybrid microfluidic platform for sensitive infectious disease diagnosis. Anal. Chem. 86, 7978–7986.

Elman, N.M., Ho Duc, H.L., Cima, M.J., 2009. An implantable MEMS drug delivery device for rapid delivery in ambulatory emergency care. Biomed. Microdevices 11 (3), 625–631.

Fontana, F., et al., 2016. Microfluidics as a cutting-edge technique for drug delivery applications. J. Drug Deliv. Sci. Technol. 34, 76–87.

Foster, G.A., et al., 2017. Protease-degradable microgels for protein delivery for vascularization. Biomaterials 113, 170–175.

Funfschilling, D., et al., 2009. Flow-field dynamics during droplet formation by dripping in hydrodynamic-focusing microfluidics. Phys. Rev. E: Stat. Nonlinear Soft Matter Phys. 80 (1), 2–5.

Gao, N., Li, X.J., 2013. Controlled drug delivery using microfluidic devices. Microfluid. Devices Biomed. Appl. 185e, 167–184.

Garstecki, P., et al., 2006. Formation of droplets and bubbles in a microfluidic T-junction-scaling and mechanism of break-up. Lab Chip 6 (3), 437–446.

Geng, Y., et al., 2007. Shape effects of filaments versus spherical particles in flow and drug delivery. Nat. Nanotechnol. 2 (4), 249–255.

Gu, H., et al., 2008. Electrowetting-enhanced microfluidic device for drop generation. Appl. Phys. Lett. 93 (18), 183507.

Gu, H., Duits, M.H.G., Mugele, F., 2011. Droplets formation and merging in two-phase flow microfluidics. Int. J. Mol. Sci. 12 (4), 2572–2597.

Guckenberger, D.J., et al., 2015. Micromilling: a method for ultra-rapid prototyping of plastic microfluidic devices. Lab Chip 15 (11), 2364–2378.

Guo, F., et al., 2010. Valve-based microfluidic device for droplet on-demand operation and static assay. Appl. Phys. Lett. 97 (23).

Habasaki, S., et al., 2013. Vertical continuous flow lithography for fabricating long 3D structures. Proceedings of the IEEE International Conference on Micro Electro Mechanical Systems (MEMS), pp. 369–372.

Hamie, A., Ghafar-Zadeh, E., Sawan, M., 2013. An implantable micropump prototype for focal drug delivery. 2013 IEEE International Symposium on Medical Measurements and Applications (MeMeA), IEEE, pp. 278–281.

He, C., et al., 2010. Effects of particle size and surface charge on cellular uptake and biodistribution of polymeric nanoparticles. Biomaterials 31 (13), 3657–3666.

He, F., Nugen, S.R., 2014. Automating fluid delivery in a capillary microfluidic device using low-voltage electrowetting valves. Microfluid. Nanofluid. 16 (5), 879–886.

Hejazi, R., Amiji, M., 2003. Chitosan-based gastrointestinal delivery systems. J. Controlled Release 89 (2), 151–165.

Hettiarachchi, K., et al., 2009. Controllable microfluidic synthesis of multiphase drug-carrying lipospheres for site-targeted therapy. Biotechnol. Progr. 25 (4), 938–945.

Hong, J.S., et al., 2010. Microfluidic directed self-assembly of liposome-hydrogel hybrid nanoparticles. Langmuir 26 (13), 11581–11588.

Huang, C.-J., et al., 2007. Integrated microfluidic systems for automatic glucose sensing and insulin injection. Sens. Actuators, B: Chem. 122 (2), 461–468.

Huang, K.-S., Lai, T.-H., Lin, Y.-C., 2006. Manipulating the generation of Ca-alginate microspheres using microfluidic channels as a carrier of gold nanoparticles. Lab Chip 6 (7), 954–957.

Huang, K.-S., et al., 2011. In situ synthesis of twin monodispersed alginate microparticles. Soft Matter 7 (14), 6713.

Hwang, H., Cho, Y.-K., 2012. Stereolithography. In: Bhushan, B. (Ed.), Encyclopedia of Nanotechnology. Springer Netherlands, Dordrecht, pp. 2503–2509.

Hwang, S., Lahann, J., 2012. Differentially degradable janus particles for controlled release applications. Macromol. Rapid Commun. 33 (14), 1178–1183.

Imaeda, K., Bessho, K., Shikida, M., 2015. Sharp tip-separable microneedle device for trans-dermal drug delivery systems. 2015 18th International Conference on Solid-State Sensors, Actuators and Microsystems (TRANSDUCERS), pp. 1715–1718.

Jahn, A., et al., 2007. Microfluidic directed formation of liposomes of controlled size. Langmuir 23 (11), 6289–6293.

Jang, J.H., et al., 2007. A route to three-dimensional structures in a microfluidic device: stop-flow interference lithography. Angew. Chem. Int. Ed. Engl. 46 (47), 9027–9031.

Jeong, J.W., et al., 2015. Soft microfluidic neural probes for wireless drug delivery in freely behaving mice. 2015 Transducers—2015 18th International Conference on Solid-State Sensors, Actuators and Microsystems, TRANSDUCERS 2015, (C), pp. 2264–2267.

Jian, D., et al., 2012. Approaches and challenges of engineering implantable microelectro-mechanical systems (MEMS) drug delivery systems for in vitro and in vivo applications. Micromachines 3, 615–631.

Johnson, T.J., Ross, D., Locascio, L.E., 2002. Rapid microfluidic mixing. Anal. Chem. 74 (1), 45–51.

Khan, I.U., et al., 2013. Microfluidics: a focus on improved cancer targeted drug delivery systems. J. Controlled Release 172 (3), 1065–1074.

Khan, I.U., et al., 2015a. Production of nanoparticle drug delivery systems with microfluidics tools. Expert Opin. Drug Deliv. 12, 547–562.

Khan, I.U., et al., 2015b. Microfluidic conceived pH sensitive core-shell particles for dual drug delivery. Int. J. Pharm. 478 (1), 78–87.

Kim, B., et al., 2016. A smart multi-pipette for hand-held operation of microfluidic devices. Analyst 141, 5753–5758.

Kim, E.S., et al., 2014. A microfluidic reciprocating intracochlear drug delivery system with reservoir and active dose control. Lab Chip 14 (4), 710–721.

Kim, S.-H., et al., 2011. Double-emulsion drops with ultra-thin shells for capsule templates. Lab Chip 11 (18), 3162–3166.

Laser, D.J., Santiago, J.G., 2004. A review of micropumps. J. Micromech. Microeng. 14 (6), R35–R64.

Lavan, D.A., Mcguire, T., Langer, R., 2003. Small-scale systems for in vivo drug delivery. Nat. Biotechnol. 21 (10), 1184–1191.

Lee, W., Walker, L.M., Anna, S.L., 2009. Role of geometry and fluid properties in droplet and thread formation processes in planar flow focusing. Phys. Fluids 21 (3).

Li, W., et al., 2016. Microfluidic assembly of a nano-in-micro dual drug delivery platform composed of halloysite nanotubes and a pH-responsive polymer for colon cancer therapy. Acta Biomater. 48, 238–246.

Lin, C.H., et al., 2002. A new fabrication process for ultra-thick microfluidic microstructures utilizing SU-8 photoresis. J. Micromech. Microeng. 7, 2146–2150.

Liu, D., et al., 2013a. Microfluidic templated mesoporous silicon–solid lipid microcomposites for sustained drug delivery. ACS Appl. Mater. Interfaces 5 (22), 12127–12134.

Liu, L., et al., 2011. Monodisperse core-shell chitosan microcapsules for pH-responsive burst release of hydrophobic drugs. Soft Matter 7, 4821–4827.

Liu, L., et al., 2013b. Preparation of monodisperse calcium alginate microcapsules via internal gelation in microfluidic-generated double emulsions. J. Colloid Interface Sci. 404, 85–90.

Lo, C.T., et al., 2010. Controlled self-assembly of monodisperse niosomes by microfluidic hydrodynamic focusing. Langmuir 26 (11), 8559–8566.

Luciani, A., et al., 2004. Glucose-receptor MR imaging of tumors: study in mice with PEGylated paramagnetic niosomes. Radiology 231 (1), 135–142.

Mazutis, L., Vasiliauskas, R., Weitz, D.A., 2015. Microfluidic production of alginate hydrogel particles for antibody encapsulation and release. Macromol. Biosci. 15 (12), 1641–1646.

Mcallister, D.V., Allen, M.G., Mark, R., 2000. Microfabricated microneedles for gene and drug delivery. Annu. Rev. Biomed. Eng. 2 (1-754), 289–313.

McAllister, D.V., et al., 2003. Microfabricated needles for transdermal delivery of macromolecules and nanoparticles: fabrication methods and transport studies. Proc. Natl. Acad. Sci. U.S.A. 100 (24), 13755–13760.

McGinty, S., Pontrelli, G., 2015. A general model of coupled drug release and tissue absorption for drug delivery devices. J. Controlled Release 217, 327–336.

Moon, B.-U., et al., 2015. Microfluidic generation of aqueous two-phase system (ATPS) droplets by controlled pulsating inlet pressures. Lab Chip 15 (11), 2437–2444.

Mousoulis, C., et al., 2011. A skin-contact-actuated micropump for transdermal drug delivery. IEEE Trans. Biomed. Eng. 58, 1492–1498.

Murshed, S.M.S., et al., 2009. Microdroplet formation of water and nanofluids in heat-induced microfluidic T-junction. Microfluid. Nanofluid. 6 (2), 253–259.

Nguyen, N., et al., 2000. Integrated flow sensor for in situ measurement and control of acoustic streaming in flexural plate wave micropumps. Sens. Actuators, A: Phys. 79 (2), 115–121.

Nie, Z., et al., 2006. Janus and ternary particles generated by microfluidic synthesis: design, synthesis, and self-assembly. J. Am. Chem. Soc. 128 (29), 9408–9412.

Nisar, A., et al., 2008. MEMS-based micropumps in drug delivery and biomedical applications. Sens. Actuators, B: Chem. 130 (2), 917–942.

Nisisako, T., Okushima, S., Torii, T., 2005. Controlled formulation of monodisperse double emulsions in a multiple-phase microfluidic system. Soft Matter 1 (1), 23. Available at: http://xlink.rsc.org/?DOI = b501972a.

Ogończyk, D., Siek, M., Garstecki, P., 2011. Microfluidic formulation of pectin microbeads for encapsulation and controlled release of nanoparticles. Biomicrofluidics 5 (1), 13405.

Okushima, S., et al., 2004. Controlled production of monodisperse double emulsions by two-step droplet breakup in microfluidic devices. Langmuir 20 (23), 9905–9908.

Pirmoradi, F.N., et al., 2011. Delivery of an anti-cancer drug from a magnetically controlled MEMS device show cytotoxicity in PC3 and HUVEC cells. 2011 16th International Solid-State Sensors, Actuators and Microsystems Conference, TRANSDUCERS'11, pp. 2831–2834.

Pirmoradi, F.N., et al., 2013. Controlled delivery of antiangiogenic drug to human eye tissue using a MEMS device. Proceedings of the IEEE International Conference on Micro Electro Mechanical Systems (MEMS), pp. 1–4.

Pollack, M.G., Shenderov, A.D., Fair, R.B., 2002. Electrowetting-based actuation of droplets for integrated microfluidicsElectronic supplementary information. Lab Chip 2 (2), 96.

Razzacki, S.Z., et al., 2004. Integrated microsystems for controlled drug delivery. Adv. Drug Delivery Rev. 56 (2), 185–198.

Riahi, R., et al., 2015. Microfluidics for advanced drug delivery systems. Curr. Opin. Chem. Eng. 7, 101–112.

Santini, J.T., et al., 2000. Microchips as controlled drug-delivery devices. Angew. Chem. Int. Ed. Engl. 39 (14), 2396–2407.

Schmid, L., Franke, T., 2013. SAW-controlled drop size for flow focusing. Lab Chip 13 (9), 1691–1694.

Seth, A., et al., 2012. Design of vesicles using capillary microfluidic devices: from magnetic to multifunctional vesicles. Adv. Mater. 24 (26), 3544–3548.

Shin, H., et al., 2016. 3D multi-functional neural probe array for mapping functional connectivities in a 3D neuron chip. Proceedings of the IEEE International Conference on Micro Electro Mechanical Systems (MEMS), 2016 January, pp. 321–324.

Shum, H.C., et al., 2011. Multicompartment polymersomes from double emulsions. Angew. Chem. Int. Ed. Engl. 50 (7), 1648–1651.

Silva, B.F.B., Rodriguez-Abreu, C., Vilanova, N., 2016. Recent advances in multiple emulsions and their application as templates. Curr. Opin. Colloid Interface Sci. 25, 98–108.

Singh, T.R.R., et al., 2013. Microneedles for Drug Delivery and Monitoring. Woodhead Publishing Limited.

Sun, B.J., et al., 2010. Microfluidic melt emulsification for encapsulation and release of actives. ACS Appl. Mater. Interfaces 2 (12), 3411–3416.

Sun, T.M., et al., 2011. Simultaneous delivery of siRNA and paclitaxel via a "two-in-one" micelleplex promotes synergistic tumor suppression. ACS Nano 5 (2), 1483–1494.

Sun, X., et al., 2013. Controlled dispensing and mixing of pico- to nanoliter volumes using on-demand droplet-based microfluidics. Microfluid. Nanofluid. 15 (1), 117–126.

Takemori, Y., et al., 2005. High pressure electroosmotic pump packed with uniform silica nanospheres. Digest of Technical Papers—International Conference on Solid State Sensors and Actuators and Microsystems, TRANSDUCERS'05, 2, pp. 1573–1576.

Takeuchi, S., et al., 2005. An axisymmetric flow-focusing microfluidic device. Adv. Mater. 17 (8), 1067–1072.

Tetteh, E.A., Boatemaa, M.A., Martinson, E.O., 2014. A review of various actuation methods in micropumps for drug delivery appliactions. Proceedings of the 11th International Conference on Electronics, Computer and Computation, ICECCO 2014.

Thompson, B., et al., 2016. Engineered hydrophobicity of discrete microfluidic elements for double emulsion generation. Microfluid. Nanofluid. 20 (5), 1–5.

Thorsen, T., et al., 2001. Dynamic pattern formation in a vesicle-generating microfluidic device. Phys. Rev. Lett. 86 (18), 4163–4166.

Tice, J.D., Lyon, A.D., Ismagilov, R.F., 2004. Effects of viscosity on droplet formation and mixing in microfluidic channels. Anal. Chim. Acta 507 (1), 73–77.

Torchilin, V.P., 2000. Drug targeting. Eur. J. Pharm. Sci. 11, S81–S91.

Tran, L.-T.-C., Lesieur, S., Faivre, V., 2014. Janus nanoparticles: materials, preparation and recent advances in drug delivery. Expert Opin. Drug Deliv. 5247 (September), 1–14.

Tran, T.H., et al., 2012. Microfluidic approach for highly efficient synthesis of heparin-based bioconjugates for drug delivery. Lab Chip 12, 589.

Tuan-Mahmood, T.M., et al., 2013. Microneedles for intradermal and transdermal drug delivery. Eur. J. Pharm. Sci. 50 (5), 623–637.

Utada, A.S., et al., 2007. Dripping, jetting, drops, and wetting: the magic of microfluidics. MRS Bull. 32 (9), 702–708.

Velev, O.D., Bhatt, K.H., 2006. On-chip micromanipulation and assembly of colloidal particles by electric fields. Soft Matter 2 (9), 738.

Vladisavljević, G.T., 2016. Recent advances in the production of controllable multiple emulsions using microfabricated devices. Particuology 24, 1–17.

Wang, C.W., Sinton, D., Moffitt, M.G., 2011. Flow-directed block copolymer micelle morphologies via microfluidic self-assembly. J. Am. Chem. Soc. 133 (46), 18853–18864.

Wei, J., et al., 2011. Novel cationic pH-responsive poly(*N,N*-dimethylaminoethyl methacrylate) microcapsules prepared by a microfluidic technique. J. Colloid Interface Sci. 357 (1), 101–108. Available at: https://doi.org/10.1016/j.jcis.2011.01.105.

Whitesides, G.M., 2006. The origins and the future of microfluidics. Nature 442 (7101), 368–373.

Wu, H.W., et al., 2009. Exploitation of a microfluidic device capable of generating size-tunable droplets for gene delivery. Microfluid. Nanofluid. 7 (1), 45–56.

Wu, J., et al., 2013. Fabrication and characterization of monodisperse PLGA-alginate core-shell microspheres with monodisperse size and homogeneous shells for controlled drug release. Acta Biomater. 9 (7), 7410–7419.

Wu, Y., et al., 2014. Active control of ferrofluid droplet breakup dynamics in a microfluidic T-junction. Microfluid. Nanofluid. 18 (1), 19–27.

Xie, H., et al., 2012. One-step fabrication of polymeric Janus nanoparticles for drug delivery. Langmuir 28 (9), 4459–4463.

Xu, J.H., et al., 2012. A novel microfl uidic approach for monodispersed chitosan microspheres with controllable structures. Adv. Healthcare Mater. 1 (1), 106–111.

Yi, Y., et al., 2015. Electromagnetically powered electrolytic pump and thermo-responsive valve for drug delivery. 2015 IEEE 10th International Conference on Nano/Micro Engineered and Molecular Systems, NEMS 2015, pp. 5–8.

Yobas, L., et al., 2006. High-performance flow-focusing geometry for spontaneous generation of monodispersed droplets. Lab Chip 6 (8), 1073–1079.

Yung, K.L., et al., 2012. Sharp tipped plastic hollow microneedle array by microinjection moulding. J. Micromech. Microeng. 22 (1), 15016.

Zarzar, L.D., et al., 2015. Dynamically reconfigurable complex emulsions via tunable interfacial tensions. Nature 518 (7540), 520–524.

Zhang, M.J., et al., 2013. Microfluidic fabrication of monodisperse microcapsules for glucose-response at physiological temperature. Soft Matter 9 (16), 4150–4159.

Zhang, Y., et al., 2015. An ocular iontophoretic device for local drug delivery using PEDOT electrode. 2015 Transducers—2015 18th International Conference on Solid-State Sensors, Actuators and Microsystems (TRANSDUCERS), IEEE, pp. 1041–1044.

Zhang, Y., et al., 2016. An ocular iontophoretic device using PEDOT electrode for local drug delivery. Sens. Actuators, B: Chem. 237, 1007–1014.

Zhao, Y., et al., 2012. Multifunctional photonic crystal barcodes from microfluidics. NPG Asia Mater. 4 (9), e25.

Zhou, C., et al., 2016. Chemistry pumps: a review of chemically powered micropumps. Lab Chip 16 (10), 1797–1811.

Ziaie, B., et al., 2004. Hard and soft micromachining for BioMEMS: review of techniques and examples of applications in microfluidics and drug delivery. Adv. Drug Delivery Rev. 56 (2), 145–172.

FURTHER READING

Astolfi, M., et al., 2015. Micro-dissected tumor tissues on chip: an ex vivo method for drug testing and personalized therapy. Lab Chip 16, 312–325.

CHAPTER

Polymeric micelles as a novel carrier for ocular drug delivery

4

Yıldız Özsoy, Sevgi Güngör, Emine Kahraman and M. Ezgi Durgun

Faculty of Pharmacy, Department of Pharmaceutical Technology, Istanbul University, Istanbul, Turkey

4.1 INTRODUCTION

Topical ocular drug delivery is used to treat ophthalmic diseases in a localized manner. It is grouped as topical administration of the formulation onto the ocular surface and injections directly ocular tissues (i.e., intraocular). However, the eye is a unique organ due to its anatomical and physiological characteristics to protect itself from external dangerous and toxic substances. Both anatomical structure and physiological functions of the eye restricts the effective passage of drugs across ocular barriers and the entry of drugs to the target sites of the eye. Thus effective ocular drug delivery is one of the most challenging issues to overcome the barriers of the eye and to achieve therapeutically effective drug concentrations in the targeted tissues of the eye (Patel et al., 2013; Yellepeddi and Palakurthi, 2016).

In terms of therapeutic approach, the eye is formed by anterior and posterior segments. Both anterior and posterior segments of the eye are affected by sight-threatening diseases. Glaucoma, allergic conjunctivitis, anterior uveitis, and cataracts are disorders affecting the anterior segment. The most serious diseases affecting the posterior segment of the eye are age-related macular degeneration (AMD), diabetic retinopathy (DR), and posterior uveitis (Willoughby et al., 2010; Cholkar et al., 2013).

Drugs are mostly administered to the anterior segment of the eye via conventional dosage forms including eye drops, suspensions, or ointments due to the ease of administration and low cost. However, topical drug delivery is limited due to the short contact period of the drugs applied via conventional dosage forms on the ocular surface. For instance, when the targeted site of drugs is the aqueous humor, the need to pass through the corneal barriers. However, the topical ocular bioavailability of drugs is very low due to certain physiological features of the eye such as naso-lachrymal drainage, blinking, tear turnover, and protein binding. Besides these, static ocular barriers such as the conjunctiva and cornea also

Nanoarchitectonics in Biomedicine. DOI: https://doi.org/10.1016/B978-0-12-816200-2.00005-0

diminish drug absorption through intraocular tissues and play a major role in the limitation of drug absorption across ocular tissues. Thus less than 5% of the administered dose of the drug can penetrate through ocular tissues. As a result, conventional topical dosage forms result in low ocular drug bioavailability, necessitating frequent dosing which often leads to low patient compliance (Patel et al., 2013; Yellepeddi and Palakurthi, 2016).

For the treatment of posterior segment diseases, drugs should reach the posterior of the eye usually by transscleral route around the conjunctiva, through the sclera, choroid, and retina. However, treatment of posterior segment diseases is limited due to anatomical and physiological barriers of the eye. The vitreous body does not have cellular components which reduces the passage of drug molecules to the posterior segment (Fischer et al., 2011; Delplace et al., 2015; Barar et al., 2016).

Alternative drug delivery strategies, namely intravitreal or periocular injections, have also been used to enhance the ocular bioavailability of drugs. However, the intravitreal injection option is very invasive and has serious side effects including intravitreal hemorrhage, retinal detachment, and increased risk of cataract development. Periocular drug delivery is an alternative method to achieve high drug concentrations in the vitreous cavity and is less invasive compared to intravitreal injections. Periocular drug administration also has side effects such as high intraocular pressure, cataract development, and corneal decompensation. In addition, permeation of drugs via intravitreal or periocular injections is also restricted by ocular static barriers including the sclera, choroid, and retinal pigment epithelium (RPE). Dynamic barriers include the blood flow in the conjunctiva and choroid, and lymphatic flow in the conjunctiva and episclera (Patel et al., 2013; Gaudana et al., 2010).

Systemic delivery is another option for delivering drugs to the posterior segment of the eye; however, high drug doses are administered to achieve sufficient therapeutic amounts of the drugs in the retina via systemic drug delivery. However, either application of high drug doses or frequent drug administration lead to poor patient compliance and increased risk of systemic side effects. The anterior and posterior barriers which separate the retina and the vitreous humor from the systemic circulation in the eye also limit the efficacy of the drugs following the systemic delivery of drugs. It has been reported that only 2% of systemically administered drugs reaches the vitreous cavity (Gaudana et al., 2010).

In this context, the treatment of both anterior and posterior segment diseases is limited due to the effective barrier characteristics of the eye, thus it is still major challenging issue for pharmaceutical scientists. The residence time of the drugs on barriers such as cornea or conjunctiva could be increased for the effective drug delivery to anterior and posterior segments of the eye, afterwards it results in the prolonged contact period of the drugs in the eye and achievement of sustained drug release. The development of novel carriers to deliver drugs to targeted eye segments would satisfy these unmet medical needs (Yavuz and Kompella, 2016). For this application nanotechnology-based formulations have been widely investigated over the past few decades (Barar et al., 2016; Liu et al., 2012; Reimondez-Troitiño et al., 2015). Numerous drug delivery carriers have been developed nano-sized polymeric micellar carriers as well as other polymeric and lipid-based drug carriers such as nanoparticles (Mahaling and Katti, 2016), liposomes (Lajunen et al., 2016; Tan et al., 2017), dendrimers (Yavuz et al., 2016; Bravo-Osuna et al., 2016), nanosuspensions (Luschmann et al., 2013b), and microemulsions (Kumar and Sinha, 2014; Lidich et al., 2016) to overcome the limitations of topical ocular drug delivery and to improve the ocular bioavailability of the drugs.

Among these drug carriers, polymeric micelles have the potential to deliver drugs to targeted areas (Cholkar et al., 2012; Pepic et al., 2012; Vaishya et al., 2014a; Mandal et al., 2017). Polymeric micelles are nano-sized aggregates generated from synthetic block copolymers or graft copolymers. Their superior properties include having nano-size, relatively low toxicity for biological membranes, solubilizing of drugs in aqueous media, and targeting ability to the target sites. The hydrophobic core in the inner part of micelles dissolves hydrophobic or lipophilic drugs while the hydrophilic shell in the outer part of micelles interacts with cell membranes and provides effective steric protection of the micellar structure (Torchilin, 2007; Güngör et al., 2015).

Polymeric micelles have been widely examined for their ability to increase the solubility of hydrophobic drugs and to enhance drug absorption across biological barriers, including the ocular topical bioavailability of drugs. As a drug delivery system, polymeric micelles have proved to be potential carriers to enhance hydrophobic drug and protein delivery through biological membranes, to increase the chemical stability of unstable compounds, and to control the release of drugs. They might have the ability to bind with cellular barriers and may improve penetration of drugs through the cornea since they are in a similar size range as membrane proteins, scleral pores (cornea), and other biomolecules (Vadlapudi et al., 2014). Their small size also provides enhanced permeability and retention (EPR), thus targeting inflamed areas (Kanoujia et al., 2014). The superiority and potential applications of micellar carriers in ocular drug delivery have been reviewed (Cholkar et al., 2012; Pepic et al., 2012; Vaishya et al., 2014a). The recent advances of micelles as effective carriers for ocular administration and patented inventions have been reviewed by Mandal et al. (2017).

This chapter will provide an overview of: the basic anatomical structure of the eye; potential dynamic and static barriers of the eye; information on eye disorders which need more effective treatments; the obstacles of topical ocular drug delivery; and the micellar carrier systems which have been developed to deliver drugs to targeted ocular tissues for the treatment of ocular disorders.

4.2 ANATOMY OF THE EYE

The eye is a complex organ consisting of mainly two segments, namely the anterior and posterior segments. The anterior segment includes the cornea, iris, ciliary body, aqueous body, crystalline lens, and conjunctiva, and anterior chamber is filled with aqueous humor. The posterior segment of the eye consists of the retina, the choroid, the sclera (as thin layers), and optic nerves, and the posterior chamber is filled with vitreous humor (Weng et al., 2016).

The eyeball (anatomically called the *Bulbus oculi*) is settled in the orbit and protected by orbital bones consisting of three layers: (from outside to inside) *tunica fibrosa bulbi*, *tunica vasculosa bulbi*, and *tunica interna bulbi* (Di Fiore, 2001).

Tunica fibrosa bulbi is the fibrous layer of the eyeball and is also known as a cornea-scleral layer (Di Fiore, 2001). The cornea is the thicker and more transparent part located in the front section of the eyeball (Murphy and Pollock, 1993). This layer is where light rays penetrate the eye following the basic structure of light refraction (Williams et al., 1989; April, 1997). Connective tissue constitutes the biggest part of this layer (April, 1997). The sclera is the tough, white, outer surface layer surrounding the outside of the left side cornea and provides mechanical strength and protection (April, 1997).

Tunica vasculosa bulbi, located between the retina and the sclera, is a thick and vascular layer (Murphy and Pollock, 1993). The choroid, corpus siliare, and iris are the other parts of this vascular layer. The iris consists of pigmented epithelial cells (Di Fiore, 2001). This circular structure is responsible for controlling the diameter and size of the pupil and, thus determines the amount of light reaching the retina (Keeton et al., 1993).

The *Tunica interna bulbi* is the neural, sensory, and most light-sensitive layer located in the innermost area of the eyeball (Murphy and Pollock, 1993).

4.3 BARRIERS OF THE EYE

The delivery of drugs to both anterior and posterior ocular segments is restricted due to the precorneal, dynamic and static ocular barriers. The barriers limiting ocular bioavailability in both anterior and posterior segment diseases are shown in Fig. 4.1.

4.3.1 TEAR

Tear film is one of the precorneal barriers. Tear turnover dilutes the concentration of the effective drug following its administration. The rate of drug elimination from the tear fluid is 0.5–2.2 μL/min. Normally, the tear film has a total volume of 7–10 μL. However, 20–50 μL volume is administrated in a dosage. The excess volume exits through the nasolacrimal duct (Mishima et al., 1966; Kuno and Fujii, 2011).

4.3.2 CORNEA

The cornea is the major barrier in the anterior chamber of the eye for conventional topical drugs (Dey et al., 2003). Corneal *epithelium*, *stroma*, and *endothelium* are

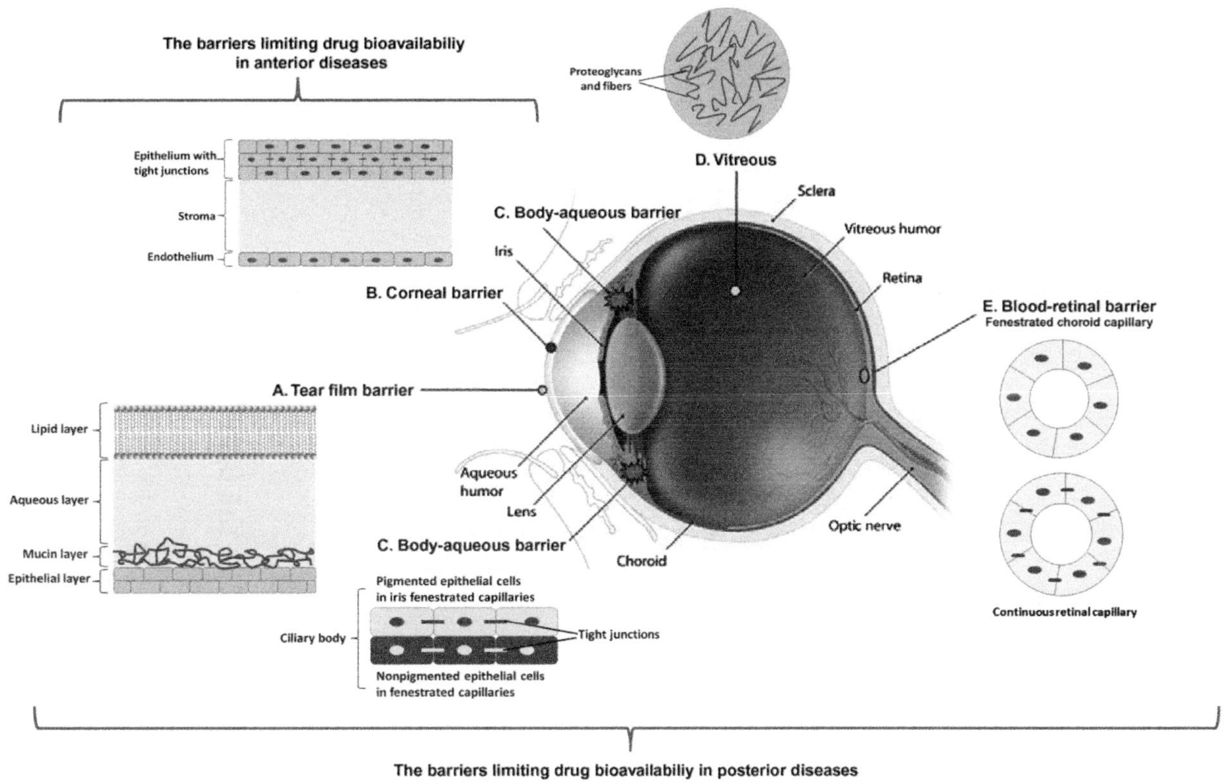

FIGURE 4.1

The barriers limiting ocular bioavailability in anterior and posterior segment diseases.

the main layers of the cornea and have different polarities. These layers play a role in the transcorneal permeation of drugs. *Epithelium*-bearing squamous, nonkeratinized, and stratified layer has a lipophilic character and restricts *paracellular* drug permeation from the tear film. The *stroma* is highly hydrated due to consisting of hydrophilic collagen fibrils. Its hydrophilic character is a major barrier for lipophilic drugs. *Endothelium*, the innermost part of cornea, has tight junctions. Large molecules up to 70 kDa in size pass through these junctions easily (Kuno and Fujii, 2011; Mitra, 2003; Cox Gad, 2008). Contrastly, another research group have also suggested that the endothelium does not have the tight junctions, thus could not act as a barrier for permeant molecules (Vadlapudi et al., 2015).

4.3.3 CONJUNCTIVA

Conjunctiva is a thin and transparent mucosal membrane which has many capillaries and lymphatics (Cox Gad, 2008) lining the inside of eyelids (Mitra, 2003). Because of these microvessels, conjunctiva is a part of the eye, which the administrated drugs are diluted by the blood circulation and lymph. Conjunctiva has two to three layers of epithelial cells (Kuno and Fujii, 2011). These conjunctival epithelium have tight junctions, although not as tight as corneal epithelium. The macromolecules up to 20 kDa may pass conjunctival epithelium, while corneal tight junctions might be permeable for compounds to 5 kDa (Huang et al., 1989). Drug molecules can be transported from these tight junctions to the vessels by pinocytosis and/or active transport mechanism (Raviola, 1983).

4.3.4 SCLERA

The sclera is hydrated because of that collagen fibrils (Olsen et al., 1998). Therefore, the polarity of drug molecules is the most effective parameter for their transport through the sclera. Lipophilic drugs present lower permeability (Cruysberg et al., 2002). The positive charge of the drug molecules reduces their permeability across the sclera (Dunlevy and Rada, 2004).

4.3.5 IRIS–CILIARY BODY (THE BLOOD–AQUEOUS HUMOR BARRIER)

The iris and the ciliary body exist in the vascular layer (*Tunica Vasculosa Bulbi*). The iris has pigmented epithelial cells. The ciliary body produces aqueous humor which provides for the avascular ocular tissues and maintains the intraocular pressure (Di Fiore, 2001; Murphy and Pollock, 1993). The ciliary body is highly vascularized by the choroidal capillaries which are fenestrated and leaky. The capillaries contain tight junctions (Coca-Prados, 2014; Lee and Pelis, 2016). These tight junctions, epithelial tissues, and endothelium cells compose the blood–aqueous humor barrier (BAB). The solute exchange between the blood and the intraocular fluid is regulated by the BAB. Whereas small and lipophilic drug molecules can enter the uveal blood circulation, hydrophilic molecules are eliminated by the aqueous humor turnover (Cox Gad, 2008).

4.3.6 BLOOD–RETINAL BARRIER

The retina, the innermost tissue of the eye, has photoreceptors, bipolar cells, ganglion cells, and capillary vessels (Andrade et al., 2006). Retinal capillary endothelial cells have tight junctions which restrict drug transport from blood into the retina. The blood–retinal barrier (BRB) is subdivided into the inner and outer BRBs (Tomi and Hosoya, 2010; Kim et al., 2014). The drug is eliminated from the eye by BRB and, consequently its therapeutic dose and the potential toxic side effect of drugs is decreased (Cunha-Vaz, 1997; Duvvuri et al., 2003).

4.4 OCULAR DISORDERS

Different types of ocular disorders can affect the eye and, consequently, vision. Besides the posterior segment diseases such as AMD, DR, posterior uveitis, and retinitis which mostly threatened the seeing, the ocular surface diseases which affect the anterior segment of the eye including *conjunctivitis*, dry eye syndrome, glaucoma, keratitis, and other infections, also affect the sight of the eye.

4.4.1 OCULAR SURFACE DISEASES

Conjunctivitis: This is inflammation of the conjunctiva (Rathore and Nema, 2009). Infection is the most common reason and allergy is another common cause of conjunctivitis (Sowka et al., 2015). Topical ophthalmic solutions, ophthalmic suspensions and soft contact lens are used to treatment of conjunctivitis (Barar et al., 2016).

Dry Eye Syndrome: This is one of frequently observed eye diseases (Kymionis et al., 2008). The National Eye Institute workshops, diagnosed dry eye as *a disorder of the tear film due to tear deficiency or excessive evaporation that causes damage to the inter-palpebral ocular surface and is associated with symptoms of discomfort* (Lemp, 1995). Anti-inflammatory therapy is the first-line therapy. The other treatment option is artificial tear which is used to relieve burning, irritation, and discomfort caused by dry eyes used (Kymionis et al., 2008). Ophthalmic suspension, ophthalmic emulsion and ophthalmic solution are the dosage forms of therapy (Barar et al., 2016).

Glaucoma: Glaucoma is a group of eye diseases which result in damage to the optic nerve and, ultimately, vision loss. Glaucoma is divided into two types as the open and closed angle glaucoma. Pressure in the eye is increased by degenerated drainage of the aqueous humor in open-angle glaucoma. Topical ophthalmic solution as an eye drops is used to treatment of glaucoma (Barar et al., 2016).

Keratitis: Inflammation of the cornea. Virus, bacteria, fungus, and amoebic are causes of the keratitis (Rathore and Nema, 2009). Ophthalmic suspension, ophthalmic emulsion, and ophthalmic solution are the dosage forms of therapy (Barar et al., 2016).

Scleritis: An inflammation of sclera may occur with any contact by an infecting agent or due to autoimmune disorders (Ramenaden and Raiji, 2013). This case which has known as *scleritis* may be observed at six subtypes. In general, *scleritis* can be treated with systemic or topical steroids, oral non-steroidal antienflamatuars, corticosteroids, anti-metabolites (Beardsley et al., 2013).

4.4.2 POSTERIOR SEGMENT DISEASES

AMD: AMD is a complex disease in which genetic, environmental, and inflammatory factors influence its onset and progression, resulting in blurred or vision loss among people aged 55 and older. The disease causes damage to the macula of the retina occurs via two processes: (1) through exudative neovascularizations (called wet AMD) which often lead to sudden vision loss; and (2) by nonexudative geographic atrophy (referred to as dry AMD) which progresses more slowly than wet AMD (Hennig and Goepferich, 2015). There is no cure that improves the sight, but mostly antivascular endothelial growth factor (VEGF) is intravitreally administered to slow down the course of the disease (National, 2017).

DR: DR is the most prevalent retinal microvascular complication of diabetic patients and may lead to blindness. In diabetic patients, hyperglycemia causes to the vascular injuries and the transcription factors are activated. Then, the structure and function of the retina are deteriorated due to alteration in the expression of VEGF (Mitra et al., 2016). Recent studies have indicated that multiple doses of intravitreal injections of anti-VEGF drugs had good results (O'Malley, 2012).

Posterior uveitis: Posterior uveitis, which is an intraocular inflammatory disease, affects intermediate and/or posterior segments of the eye such as the choroid and retina. It leads to 10%–15% of blindness cases in developed countries (Durrani et al., 2004). In the treatment of the posterior uveitis, steroids have been used (Sudharshan et al., 2010) either systemically or topically. While topical dosage forms are well tolerated, drug levels systemically achieved in intermediate and posterior ocular segments are not well tolerated by the patients.

4.4.3 CORNEAL TRANSPLANTATION

Corneal transplantation, also known as corneal grafting, is the most common form of solid tissue transplant in humans (Faraj et al., 2012). The graft rejection is one of the major postoperative complications. In order to overcome rejection problems, topical and systemic steroids, topical corticosteroids, and cyclosporine are used after operations as the medical treatment agents (Kumar and Kumar, 2014).

4.5 OCULAR DRUG DELIVERY AND ITS OBSTACLES

Ocular drug delivery encounters some major challenges due to unique anatomical and physiological features of the eye which is designed to protect itself from external dangerous and toxic substances (Weiner and Gilger, 2010). But, this also limits delivery of drugs to the targeted sites of action at therapeutically effective concentrations. From a point of drug delivery, the *precorneal*, dynamic, and static ocular barriers are the main obstacles in the therapy of eye diseases. Moreover, therapeutic drug concentrations in target tissues of the eye cannot be prolonged, which is important to decrease the repetition of application in the treatment and, consequently, improve patient compliance. Thus the main challenge of ocular drug delivery is to overcome these protective barriers in order to provide effective drug concentrations and to achieve sustained drug levels in the targeted ocular tissues (Patel et al., 2013; Gaudana et al., 2010; Yavuz and Kompella, 2016).

In terms of therapeutic approaches, ocular diseases could be classified as anterior and posterior segment (back of the eye) diseases. Both anterior and posterior segments of the eye are affected by sight-threatening diseases. Eye diseases are mainly treated by the application of conventional, small, chemical compounds and new therapeutic agents applied topically or systemically (Cholkar et al., 2013).

Topical ocular drug delivery: Topical drug delivery is considered to be the first option and it is generally preferred in the treatment of anterior segment diseases (Vadlapudi et al., 2015; Faraj et al., 2012). Low-cost, patient compliance, and ease of administration are the main reasons for choosing the topical route (Vadlapudi et al., 2015; Kim et al., 2014; Duvvuri et al., 2003). Conventional delivery systems such as eye drops, suspensions, and ointments are mostly used for the treatment of ocular diseases affecting anterior segment of the eye. The most common ophthalmic drugs are mostly administered topically as eye drops to the corneal surface. However, anatomical and physiological barriers of the eye prevent the efficiency of drugs to entry into the posterior segment. Thus topical ocular delivery mostly results in low ocular bioavailability.

Ointment, eye drops, and gel are the pharmaceutical dosage forms administrated by the topical route (Weiner and Gilger, 2010). The most preferred dosage form is eye drops which are made up of solution, emulsion, and suspension. Basically, the short-contact time of drugs with the eye surface following application of conventional topical dosage forms, the permeability problems of drugs across anatomical barriers, and the elimination of drugs by the physiological barriers are general limitations of the topical route. Therefore topical drug delivery to the eye has low drug bioavailability (Kim et al., 2014; Duvvuri et al., 2003). Some 90% of the applied dose is lost due to tear turnover, tear dilution, and nasolacrimal drainage. Less than 5% of the applied dose can reach the aqueous humor (Vadlapudi et al., 2015). Moreover, its anatomical barriers such as conjunctiva and cornea play major roles in the limitation of drug absorption across ocular tissues (Kumar and Kumar, 2014). Routes and limitations for penetration of topically administered drugs are shown in Fig. 4.2. The drug molecules which may

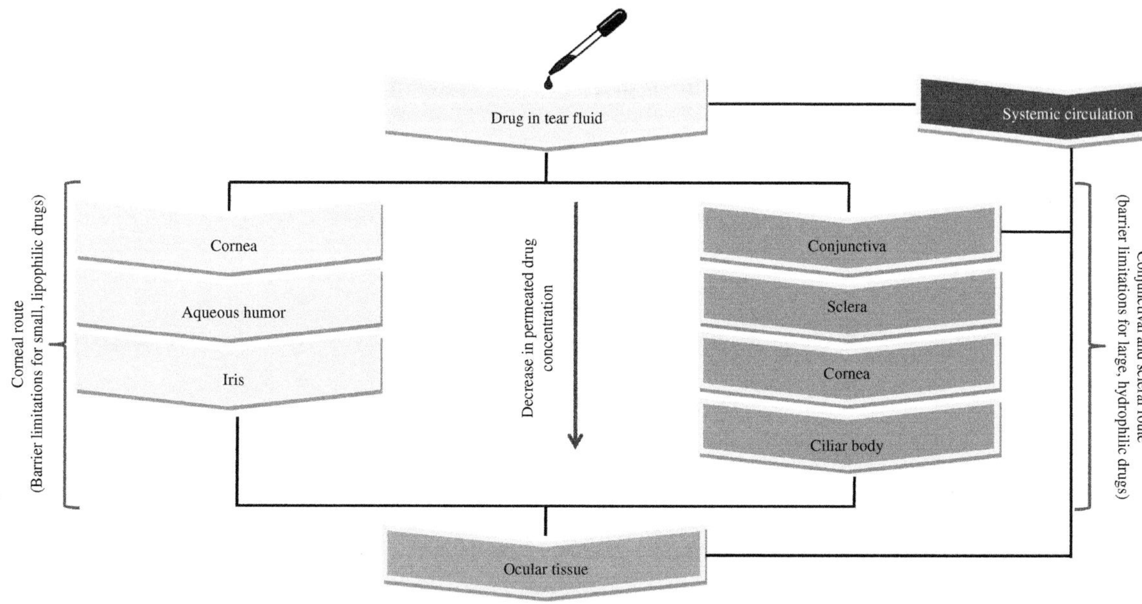

FIGURE 4.2

Administration routes and limitations for penetration of topically administered drugs.

permeate these barrier is also diluted by the blood-ocular barriers (BAB, BRB) and goes the systemic circulation.

In addition, either intravitreal or periocular injections have been applied to maintain effective drug concentration in the targeted sites of the eye, thereby improving the ocular bioavailability of the drugs. Generally the intravitreal injection option is the most widely used drug delivery strategy to treat posterior segment diseases. However, it has severe side effects such as *endophthalmitis*, hemorrhage, retinal detachment due to repeated intravitreal injections. Moreover, poor patient compliance is a crucial problem for the intravitreal injection administration. Periocular drug administration, which results in transscleral drug delivery is another alternative drug delivery strategy for the treatment of posterior ocular diseases. But, even if transscleral delivery is a relatively less invasive, easier, and better tolerated strategy for patients, ocular static barriers such as sclera, choroid and RPE, and dynamic barriers include lymphatic flow in the conjunctiva and episclera, and the blood flow in conjunctiva and choroid are still obstacle for the efficacy of the ocular therapy (Gaudana et al., 2010).

Systemic ocular drug delivery: Systemic drug delivery is another option for the treatment of ocular disorders and is mostly preferred for the treatment of posterior segment diseases. However, it has disadvantages such as serious systemic side effects, blood–ocular barriers limit the efficacy of the drug particularly to the posterior segments. Because of the physiological and anatomical barriers, the drug concentration is very low in the aqueous humor after the systemic administration (Kansara et al., 2007). It was reported that less than 2% of systemically administered drugs reaches the vitreous cavity. This concentration is not enough for therapeutic efficacy. Thus, high drug doses are required with the systemic route for the treatment of ocular diseases because of the inner and outer BRB that separate the retina and the vitreous humor from the systemic circulation. However, the administration of high doses or frequent application to achieve effective drug levels in the ocular sites often results in poor patient compliance and increased risk of systemic side effects. In particular, nonspecific binding causing the systemic side effect is the most important disadvantage of this route (Vadlapudi et al., 2015).

The development of alternative novel drug carriers for the treatment of ocular diseases is one of the most interesting and significant research studies in the pharmaceutical area (Gaudana et al., 2009).

4.6 ENHANCEMENT STRATEGIES IN OCULAR TOPICAL DRUG DELIVERY

Ideally, an ocular drug delivery system should have numerous characteristics in terms of the efficacy of the treatment and patient compliance (Liu et al., 2012):

1. In order to reduce application frequency of the drug and to maintain therapeutic drug levels for a prolonged period of time, the drug delivery system should provide a controlled and sustained release profile.

2. In order to improve therapeutic efficacy and minimize or limit side effects, specific targeting and prolonged retention of drugs in the targeted ocular tissues should be provided.
3. In order to improve the patient compliance, side effects directly based on the administration methods should be minimized or eliminated.

Therefore numerous efforts have been performed to develop novel drug delivery systems to improve efficacy of topical ocular drug delivery with high patient compliance. Nanotechnology-based drug carriers (Barar et al., 2016; Liu et al., 2012; Reimondez-Troitiño et al., 2015) and other ocular drug delivery strategies including in situ gels, implants, contact lenses, iontophoretic systems, and microneedles have been investigated (Patel et al., 2013; Delplace et al., 2015; Gaudana et al., 2010; Yavuz and Kompella, 2016). Among these approaches, nanosized ocular drug delivery systems appear to be one of the most promising strategies to be an ideal technique (Cholkar et al., 2012; Pepic et al., 2012; Vaishya et al., 2014a; Mandal et al., 2017).

Due to their small sizes, nano drug carriers are considered to have the potential to enhance drug diffusion through biological membranes including the corneal epithelium, which results in the improved corneal permeability of drugs. Nanosized drug carriers can also have high interaction capability with the mucous membrane of the corneal surface because of their high surface-area-to-volume ratio, thus the retention of the drug administered topically might be prolonged. Nanotechnology-based drug delivery system carriers also have a potential to deliver drugs to the posterior segment such as the retina or choroid via noninvasive delivery approaches (Barar et al., 2016; Liu et al., 2012; Reimondez-Troitiño et al., 2015).

Over the past few decades, numerous types of nano-sized carriers include colloidal drug carriers and polymeric delivery systems have been developed to provide therapeutically effective and desired drug concentrations in the sites of the eye. Nanoparticles (Mahaling and Katti, 2016), liposomes (Lajunen et al., 2016; Tan et al., 2017), dendrimers (Yavuz et al., 2016; Bravo-Osuna et al., 2016), nanosuspensions (Luschmann et al., 2013b), microemulsions (Kumar and Sinha, 2014; Lidich et al., 2016), and micelles (Di Tommaso et al., 2012; Ribeiro et al., 2013) have shown promising results in enhancing the corneal and scleral penetration of drugs.

4.7 MICELLES IN OCULAR DRUG DELIVERY

4.7.1 POLYMERIC MICELLES

Polymeric micelles are self-assembling carriers generally between 10 and 100 nm in size and core–shell structures formed from amphiphiles (Fig. 4.3). The hydrophobic core in the inner part of micelles causes to dissolve hydrophobic or

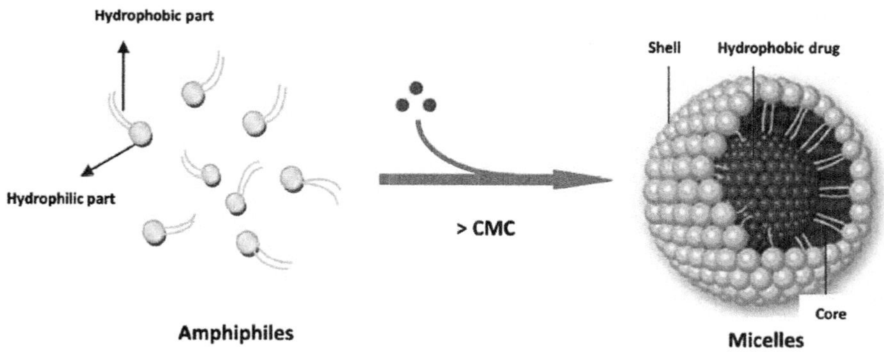

FIGURE 4.3

The formation of drug-loaded micelles.

lipophilic drugs while the hydrophilic shell in the outer part of micelles acts for interaction with cell membranes, providing effective steric protection of micellar structure (Torchilin, 2007; Güngör et al., 2015).

Polymeric micelles are mostly fabricated using amphiphilic copolymers which enable sustained release, in vivo stability, and targeting (Mandal et al., 2017). Their main structures are di-block, tri-block, and graft copolymers which consist of hydrophilic and hydrophobic units in specific sequence (Fig. 4.4). Polyethylene glycol (PEG) and polyethylene oxide (PEO) are the most commonly used hydrophilic copolymers. PEG with high mobility, high aqueous solubility, large exclusion volume, low toxicity, immunogenicity, which forms shell provides steric stabilization of the micelles and resists phagocytosis by macrophages (Pepic et al., 2012).

Polymethacrylic acid derivates, *N*-isopropylacrylamide, and vinyl pyrrolidone might also be used as the hydrophilic block for ocular administration (Rafie et al., 2010; Prosperi-Porta et al., 2016). Polyester [polycaprolactone, poly(D,L-lactide)], polyether (polypropylene oxide), or poly aminoacid [poly(β-benzoyl-L-aspartate)] derivatives are mostly preferred as the core part of these copolymers which can be synthesized or tailor-made to meet specific requirements for ocular administration (Di Tommaso et al., 2012; Ribeiro et al., 2013; Civiale et al., 2009; Li et al., 2015). Polyoxyethylated amphiphilic surfactants have been widely studied for ophthalmic micelle formulations. The some researchers have argued that they might prolong the drug's time on the precorneal area, improve drug solubility, and increase cornea permeability and bioavailability compared with other liquid formulations (Ribeiro et al., 2013). Pluronics or poloxamers consisting of PEG and polypropylene oxide are well-known examples (Kanoujia et al., 2014). However, Pluronic micelles with low mucoadhesive activity (like other nonmucoadhesive copolymers) do not strongly resist the barriers of the eyes (Lin and Chang, 2013). On the other hand, phenylboronic acid (PBA)-based polymeric micelles (Prosperi-Porta et al., 2016), Pluronic F127 micelles + chitosan

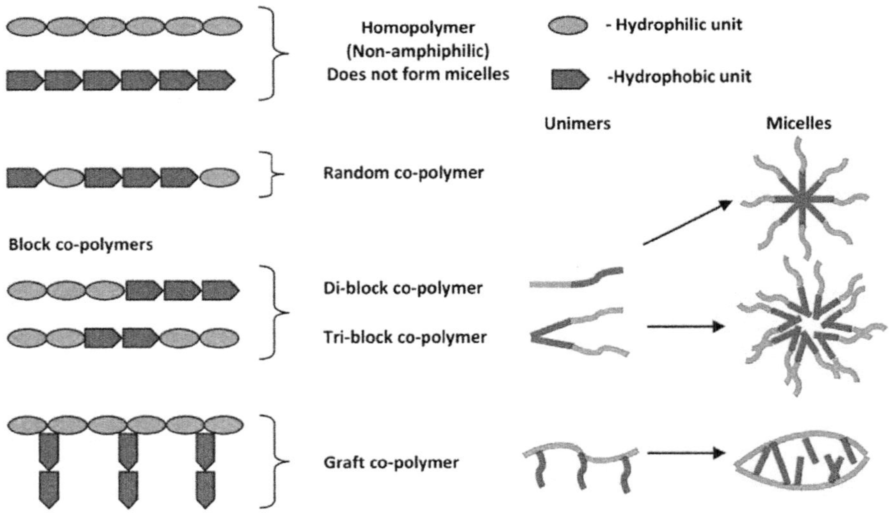

FIGURE 4.4

Main structural types of copolymers and micelles formed from amphiphilic copolymers.

(Lin and Chang, 2013), thiolated polyethylene glycol-b-poly D,L-lactic acid (PEG-PDLLA) micelles (Yang et al., 2014), and polyethylene glycol-b-poly ε-caprolactone (PEG-PCL) micelles in hydrogels (Zhang et al., 2016; Yingfang et al., 2016) were studied to overcome this issue.

In aqueous media, the copolymer unimers spontaneously generate the micelles to decrease surface tension of the system above/at the concentration that is called as the critical micelle concentration. Intermolecular forces such as hydrophobic, hydrogen bonding, steric, electrostatic and van der Waals interactions critically act in this micellization process. The hydrophobic units mainly attract with hydrophobic interactions, drift/aggregate to escape from the aqueous medium while steric and electrostatic interactions occur between hydrophilic units. The micelles generate after the hydrogen bond network reestablishes in the aqueous medium (Cholkar et al., 2012).

4.7.2 ADVANTAGES OF MICELLES FOR OCULAR ADMINISTRATION

Polymeric micelles have recently attracted attention due to capability of solubilizing the poor soluble drugs, thus increase ocular bioavailability and provide more effective treatment with less drug, minimal/no side effects. These carriers might have the ability to bind with cellular barriers and improve penetration of drugs through the cornea since they have a similar size range as membrane proteins, scleral pores (cornea), and other biomolecules (Vadlapudi et al., 2014). The small

Table 4.1 Advantages and Limitations of Polymeric Micelles

Advantages	Limitations
• Small size • Solubilizing the poorly soluble drugs • Controllable drug releasing properties • Increased drug stability • Minimized cytotoxicity • Injectable • Ease of sterilization • Versatility in monomer species • Well-defined polymer structure • Surface modification • Multi-functionality • Targeted delivery • Applicable to protein and gene delivery • Relatively easy preparation method	• Low drug loading • Low colloidal stability • Concentration-dependent properties due to critical micelle concentration at in vivo conditions

size also provides EPR, thus targeting inflamed areas (Kanoujia et al., 2014). Moreover, ophthalmic micelle formulations offer unique advantages including less toxicity, increased drug stability in the core (Ribeiro et al., 2013), easy preparation method, relatively simple sterile production (Pepic et al., 2012), not interfering with the vision of patient due to its aqueous character (Vadlapudi et al., 2014), sustained drug release as the core acts as a drug depot (Cholkar et al., 2012; Ribeiro et al., 2013), and reduced frequency of drug administration (Cholkar et al., 2012). These carrier systems have especially attracted attention due to their ability of solubilizing the lipophilic drugs and nano-sizes in recent years. They have many advantages including increased drug stability, surface modification, and others, and only a few disadvantages (Table 4.1) (Torchilin, 2001; Lee et al., 2007b; Cha et al., 2009).

4.7.3 CHARACTERISTIC FEATURES OF MICELLES IN OCULAR DELIVERY

The retention time of drugs in the anterior part of the eye or the effective transport of the drug to the posterior part of eye via micelles depends on various micelle characteristics including size, encapsulation efficiency, surface charge, release, and stability (Yang et al., 2014). Additionally, the osmolality and pH of the formulation are important for ophthalmic micelle formulations, as with other ocular preparations (Cholkar et al., 2014).

Size and Size distribution: Polymeric micelles usually range in size between 10 and 100 nm (Torchilin, 2007; Güngör et al., 2015; Torchilin, 2001) and the transport of nanocarriers through the static and dynamic barriers depends on their size (Duan et al., 2015). The scleral channels or pore diameter is in range of 20–80 nm, which enables the diffusion of hydrophilic molecules (Cholkar et al., 2014). Therefore these nanosized micelles in aqueous media permeate through the aqueous scleral pores (Duan et al., 2015). Additionally, the size is an important factor for determining endocytosis efficiency, the mechanism of internalization of the micelles in the cell, and targeting of the micelles in inflamed areas by EPR. Moreover, their size and size distribution affect the physical and biological stability of polymeric micelles. Considering in terms of patient compliance, the nano size is required for ophthalmic formulations as larger sizes occur foreign body sensation and blurring vision (Pepic et al., 2012).

Encapsulation efficiency: The encapsulation efficiency is affected by polymer-related factors (copolymer concentration, crystallinity, length, and hydrophobicity of the blocks forming the core, structure, and length of the block forming the shell, etc.), interactions between the drug and the micelle core, polarity/hydrophobicity and charge of the drug, structure of the organic used solvent and solvent ratio, and polymer-drug miscibility (Torchilin, 2001; Lee et al., 2007a; Zhang and Ma, 2009; Guo et al., 2015). Polymeric micelles present a high encapsulation efficiency ($>80\%$) in ocular studies (Vadlapudi et al., 2014; Lin and Chang, 2013; Yingfang et al., 2016) and encapsulation efficiency is a crucial factor for physical stability and access of therapeutic dose into the eye.

Surface charge: The surface charge of polymeric micelles is a crucial factor in terms of their biological and physicochemical properties (Pepic et al., 2012). Their surface charge may affect the retention time of micelles on the anterior cornea. The positively charged micelles are attracted to the negatively charged anterior cornea, so that the removal of the micelles from the eye is prevented by their adhesion to the anterior cornea (Lin and Chang, 2013). In the case of posterior targeting, positively charged micelles are contrastly bound to the negatively charged scleral matrix and result in poor permeability (Pepic et al., 2012). Also, the surface charge of polymeric micelles might improve the physical stability of the micelles via electrostatic forces (Heurtault et al., 2003).

Release characteristics of drugs from micelles: The release of drugs physically encapsulated in polymeric micelles is controlled by the rate of drug diffusion in the micellar core or break up of the micelles. The drug diffusion rate of polymeric micelles are affected by drug localization in core–shell structure, micellar size, stability, molecular volume of the drug, drug loading, rigidity of core, and polymer properties (Pepic et al., 2012; Yingfang et al., 2016; Torchilin, 2001; Lee et al., 2007a; Lee et al., 2007b). Fast drug release should be preferred in the case of formulation applications targeted at the anterior segment due to rapid drainage. However, drug delivery into the back of the eye is required for sustained drug release without burst effect (Pepic et al., 2012). The release of drugs covalently conjugated into polymeric micelles occurs with the breakage of the bond which exists between drug and polymer via stimulant (Talelli et al., 2010). Drug release is dependent on the rate of micellar dissociation.

Stability: The stability of micelles can be examined via thermodynamic and kinetic stability tests. Thermodynamic stability implies on the micelle formation and reaching equilibrium. In order to determine the thermodynamic stability is used CMC values and an increase in thermodynamic stability is explained with a decrease in CMC. The core–drug interaction and drug encapsulated in the micelles may increase the thermodynamic stability decreasing CMC (Owen et al., 2012). Kinetic stability elucidates the behavior of the system over time and micelle dissociation. The high kinetic stability in micelles allows to continue their integrity, even if slow dissociation upon dilution and interactions of the micelles with components of eye fluids proteins, mucin, glycosaminoglycans, etc. occurs. Kinetic stability is important for sustained drug release into the posterior segment (Owen et al., 2012). Also, micelles increase the chemical stability of drugs (Ribeiro et al., 2013; Wichit et al., 2012).

Osmolality and pH: In order to avoid side effects on the eye, ocular formulations should exhibit optimal osmolality and pH. These values are approximately 300 mOsm/kg and in range of 6.5–7.6, respectively, which is the same as tear values (Cholkar et al., 2014).

4.7.4 OCULAR TRANSPORT OF MICELLAR CARRIERS

Drug-loaded polymeric micelles administered via the topical route that possibly permeate into the eye across the cornea would first enter the aqueous humor and then be spread out to the surrounding tissues (Pepic et al., 2012). Corneal barriers are rate-determining steps if the action site of polymeric micelles in aqueous humor. The permeation through the sclera, choroid and retina is most probably route of polymeric micelles which are used for treatment of posterior segment (Mitra et al., 2010). However, polymeric micelles have limitations in successfully delivering drugs to the desired site of eye segments following topical administration (Pepic et al., 2012).

Polymeric micelles should have a sustained drug release profile to get a more efficient drug permeation on the cornea or conjunctiva. The release of the drug micellar carriers is one of the crucial parameters for the effective delivery of drugs to the eye. The rate of drug diffusion from the micelles, micelle stability, and the rate of biodegradation of the copolymer have an effect on the drug's release from polymeric micelles. The strength of the interaction between the drug and the core-forming block are the most influential parameters on the release rate when the micelle is stable and the rate of biodegradation of the copolymer is slow (Allen et al., 1999).

The other important requirement is to increase the carriers' contact period on the ocular surface to overcome the dynamic barriers' effects. The negative charge of the corneal and conjunctival surface is created by mucin in hydrophilic tear forms which covers the ocular surface. Thus to prolong the residence time of the carriers at the ocular surface, micelles should contact with the negatively charged mucins at the corneal surface. It is assumed that the hydrophilic micelle shell which has a positive charge would interact with the negatively charged mucin (de la Fuente et al., 2010). On the other hand, the micelles would be eliminated from the precorneal site almost as quickly as the aqueous solution when the bioadhesive character of the carrier is weak. Thus polymeric micelles prepared with mucoadhesive polymers like chitosan show higher bioavailability (Pepić et al., 2010).

It has been reported that polymeric micelles could increase the permeability of drugs via transcellular and/or paracellular pathways. Amphiphilic copolymers (unimers), which change the physical properties of the cell membranes, can be incorporated into the lipid bilayer (Kim et al., 2010). These polymers affect the physical structure of the cell membranes, resulting in forming polar defects. Following the saturation of the lipid bilayer, the mixed micelles are formed and then the phospholipids of the cell membranes are removed which leads to the increase of transcellular permeability (Jiao, 2008). Also, intact polymeric micelles may enter cells via the endocytosis mechanism (Savic et al., 2003); however, this mechanism has not been defined exactly yet. It has also been suggested that particles with sizes <100nm could permeate across the corneal membrane (Nagarwal et al., 2009).

Paracellular transport of hydrophilic drugs across the epithelia occurs through tight junction. In normal circumstances, 3–5 nm paracellular pores will be excluded from this uptake route because the solutes have a molecular radius exceeding 1.5 nm. On the other hand, permeation enhancers can overcome the barrier features of the cornea. For instance, chitosan and poly-L-arginine increase the transepithelial absorption by opening the tight junctions limiting paracellular permeation of drugs into the ocular epithelia (de la Fuente et al., 2010). Thus both the elastic structure and permeation enhancement capability of micelles lead to the opening of tight junctions and extension of the pore dimensions on the epithelia. It is also suggested that the transport of intact micelles through membranes occurred by opening the hydrophilic paracellular pores (Nagarwal et al., 2009).

It has also been suggested that most polyoxyethylated nonionic surfactants might behave as P-glycoprotein efflux pump inhibitors that can increase drug absorption. Although this mechanism is not clear, nonionic surfactant systems have been used [e.g., dexamethasone (DEX), timolol] to improve drug efficacy (Jiao, 2008).

It has also been suggested that the transscleral transport of micelles increases the permeability of drugs through the trans-scleral route due to their small sizes and hydrophilic surface. Polymeric micelles could pass through the hydrophilic pores of the sclera. In addition, hydrophilic chains of the micelle corona may escape from wash-out by the conjunctival or choroidal blood vessels and lymphatics leading the micelles to access the retina (Mitra et al., 2010). The elimination of polymeric surfactants and drug absorbed from the posterior eye tissues may occurs with either choroidal blood flow or, lightly possibility, diffusing across the vitreous to posterior chamber and, then via aqueous turnover and uveal blood flow (Kuno and Fujii, 2011).

4.8 APPLICATION OF MICELLAR CARRIERS IN THE TREATMENT OF EYE DISEASES

Hydrophobic and high-molecular weight compounds have been widely studied to form nanomicellar carriers for ocular drug delivery. The chemical structures and physicochemical properties of drugs prepared as ocular micellar carriers are given in Table 4.2.

4.8.1 CYCLOSPORINE

Cyclosporine A (Cyc A) (cyclosporine) is a potent immunosuppressant drug. It is widely used to inhibit T-cell activation which leads to associated eye diseases (KCS, primary Sjögren's syndrome, secondary Sjögren's syndrome, corneal transplantation, Behçet's disease, Peripheral corneal melting syndrome, etc.) as well as many autoimmune diseases (Bacman et al., 2001; Sullivan et al., 2002). Currently, it has been used topically with oily ophthalmic solutions or eye drop emulsions (Restasis). However, cyclosporine is a hydrophobic and highly lipophilic drug, thus it has very poor aqueous solubility. Its commercial preparations have some side effects such as eye burning, redness, discharge, pain, itching, stinging and may be poorly tolerated by patients. Cyclosporine can also be administered orally or intravenously to treat ocular diseases affecting the posterior segments of eye. However, systemic cyclocsporine administration has serious side effects. Therefore ophthalmic formulation of cyclosporine is still challenging to overcome its issues for the researchers (Agarwal and Rupenthal, 2016). Several drug delivery strategies have been examined to develop novel ocular formulations of cyclosporine based on cyclodextrin inclusion complexes and colloidal carriers (liposomes, nanoparticles, nanoemulsions, nanomicelles, and solid lipid nanoparticles) (Di Tommaso et al., 2012; Aksungur et al., 2012; Luschmann et al., 2014). However, none of these carriers have been commercialized despite their efficiency and tolerability.

Over the past decade, researchers have focused on the development of micellar carriers as an attractive strategy for improving the corneal and conjunctival penetration of cyclosporine. In 2002, to avoid the various side effects associated with systemic delivery, Kuwano et al. (2002) developed cyclosporine micelles of the nonionic surfactant, polyoxyl 40 stearate at a concentration of 2% w/v for topical administration. A rabbit eye model was used for the in vivo studies and the micelles had 60-fold higher concentration in the cornea than the 0.1% w/v cyclosporine castor oil solution (Kuwano et al., 2002; Lallemand et al., 2003). After that, MPEG-hexylsubstituted poly(lactides) (hexPLA) micelle formulations of

Table 4.2 Chemical Structures and Physicochemical Properties of Drugs Prepared as Ocular Micellar Carriers

Drugs	Structure	Molar Mass (g/mol)	Log P	Water Solubility (mg/mL)
Cyclosporine $C_{62}H_{111}N_{11}O_{12}$ *CAS number: 59865-13-3*	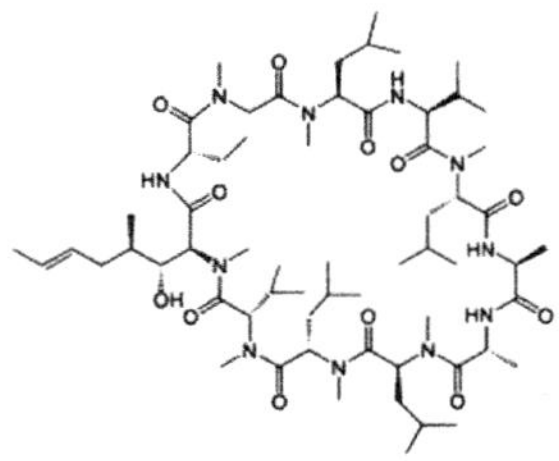	1202.63	3.64	0.027
Rapamycin $C_{51}H_{79}NO_{13}$ *CAS number:* 53123-88-9	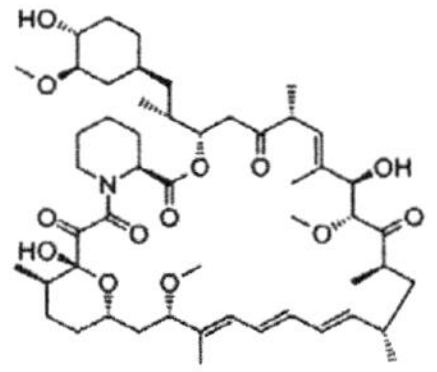	914.17	4.3	0.002
Pimecrolimus $C_{43}H_{68}ClNO_{11}$ *CAS number:* 137071-32-0	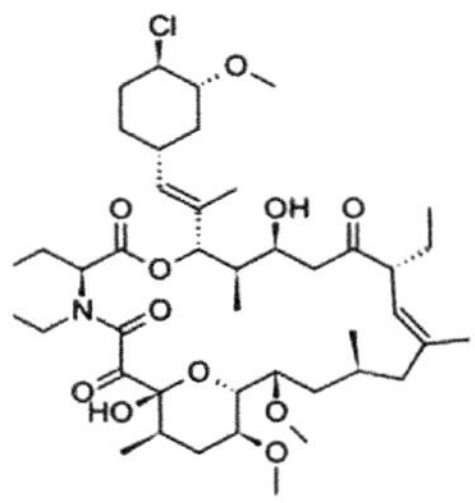	810.46	4.4	0.001
Dexamethasone $C_{22}H_{29}FO_5$ *CAS number*: 50-02-2	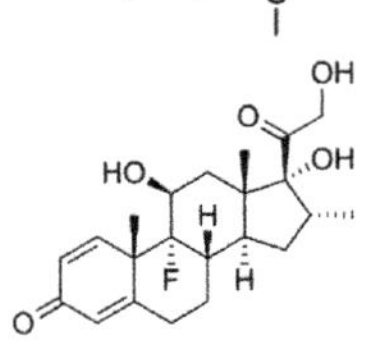	392.46	1.93	0.051
Acyclovir $C_8H_{11}N_5O_3$ *CAS number:* 59277-89-3	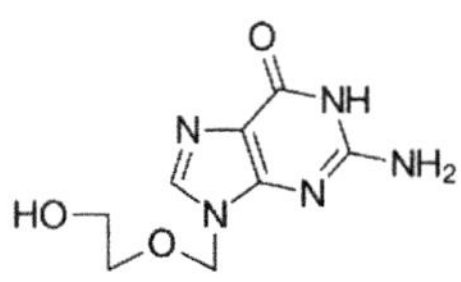	225.20	−0.95	9.08
Alpha-tocopherol $C_{29}H_{50}O_2$ *CAS number:* 59-02-9	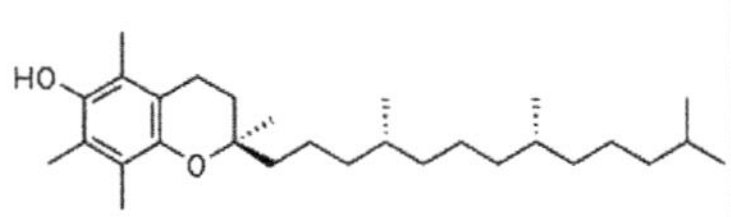	430.70	8.84	None
Ethoxzolamide $C_9H_{10}N_2O_3S_2$ *CAS number:* 452-35-7	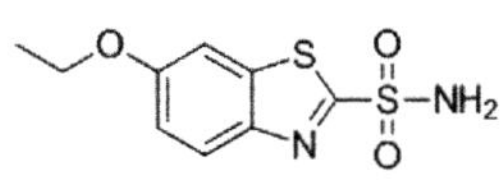	258.31	1.87	0.688

cyclosporine were prepared by Di Tommaso et al. (2012). The analyses showed that micelles were produced which were solvan-free, noncytotoxic, and stable at +4°C storage condition for more than a year. MPEG-hexPLA micelles had therapeutic cyclosporine concentration in ocular tissue to prevent the corneal graft rejection (Di Tommaso et al., 2011).

Guo et al. (2015) developed nanomicelles of cyclosporine as topical ophthalmic formulations with an amphiphilic polymer, polyvinyl caprolactam-polyvinyl acetate PEG graft copolymer (PVCL-PVA-PEG, Soluplus) in order to improve its tolerance and efficacy. They indicated that the preparation procedure of the ophthalmic nanomicelle solution of cyclosporine was simple and the formulation was stable, noncytotoxic, and its ocular tolerance in rabbit eyes was excellent.

Following treatment of labeled nanomicelles to the human corneal epithelial cells (HCECs), significant high fluorescence intensity has been observed compared to the cells treated with free Cou-6. Significant enhancement in cellular uptake of the drug was obtained. The cellular uptake pathway of nanomicelles was the intracellular endocytosis that involved early endosomes, lately endosomes, lysosomes, and ER. The high level of cyclosporine has been existed in the cornea, when compared to its oil-based 10 mg/mL ophthalmic solution. Excellent in vivo ocular tolerability of nanomicelles were shown in rabbit eyes. The authors concluded that nanomicelles prepared with PVCL-PVA-PEG could be a promising topical delivery system for ocular administration of cyclosporine and these micellar carriers delivered high levels of cyclosporine into the cornea and may provide an effective local treatment for immune-mediated corneal disease.

Over the past decade Gurny's research group widely investigated the preparation of polymeric micelles based on PEG-hexPLA block copolymers (Trimaille et al., 2006). They showed that these micelles had high drug-loading capacity for cyclosporine and had good stability in aqueous formulations (Mondon et al., 2011). They also demonstrated that the lyophilization of MPEG-hexPLA micelle solutions could improve the stability of the formulation (Di Tommaso et al., 2010). They then examined the ocular biocompatibility of micellar carriers in order to avoid any disruption of the corneal tissues. They have shown the nontoxic features of well-characterized cyclosporine/MPEG-hexPLA micelle formulation on HCE cells via MTT and clonogenic tests. In addition, no activation of apoptosis mechanisms in the cells has been shown by immunohistochemistry analyses. Moreover, the CLSO studies demonstrated that these micelles exhibited excellent topical ocular tolerance in similar to 0.9 % of NaCI solution. Based on these data, the researchers proposed that MPEG-hexPLA micelles of cyclosporine may be used as eye drop formulations for the treatment of specific diseases such as dry eye syndrome and autoimmune uveitis, as well as for the prevention of corneal graft rejection (Di Tommaso et al., 2011).

Then, these group have examined in vivo the efficiency of cyclosporine loaded polymeric micelles for the prevention of cornea graft rejection with a keratoplasty procedure. Following the administration of micelles labeled fluorescent, the penetration of micelles into all corneal layers have been viewed via confocal microscopy analysis images. The authors indicated that MPEG-hexPLA micelle carriers could be considered as effective drug carriers which overcome corneal barriers delivering effective cyclosporine concentrations therapeutically in ocular tissues in the prevention of corneal graft rejection. They proposed that novel biocompatible topical formulation of cyclosporine could be an alternative to its systemic preparations used in the prevention of graft immune rejection (Di Tommaso et al., 2012).

Luschmann et al. (2013a) developed novel ophthalmic cyclosporine delivery systems using a simple micellar solution containing nonionic surfactants. The authors showed that these carriers had a potential to improve cyclosporine resorption in vitro. In this study, the authors demonstrated both cornea tolerability and uptake of cyclosporine loading micelles in rabbit corneal tissue in vivo. They observed no irritation at the anterior segment of the eye after administration of the micellar solutions. The efficacy of this formulation was shown by very high Cyc A levels in corneal tissue.

One of the most frequent approaches to improve ocular permeability of drugs into ocular tissues is the use of mucoadhesive polymers in formulations. This strategy has been applied to the formulation of micellar carriers. Prosperi-Porta et al. (2016) prepared and characterized micelles prepared with poly(L-lactide)-*b*-poly(methacrylic acid-*co*-3-acrylamido-PBA) block copolymer as a mucoadhesive vehicle. PBA is a synthetic molecule and boronic acid has affinity to diol groups, this feature has been utilized in mucoadhesive delivery systems (Springsteen and Wang, 2002). Poly(L-lactide)-*b*-poly(methacrylic acid-*co*-PBA) copolymer micelles composed of different amounts of PBA has been prepared. They showed that the physicochemical and mucoadhesive features of micelles were affected by their PBA content in the formulation. It was presented that the micelles have increased mucoadhesion in comparison to chitosan, a well-known mucoadhesive polymer. In this study, enhancement of Cyc A delivery was also demonstrated. In addition, the low in vitro cytotoxicity of micelles against HCECs was shown, which was then confirmed with in vivo ocular irritation tests on rat models. The authors concluded that mucoadhesive micelles provides a significant potential to treat anterior segment eye diseases, increasing the ocular bioavailability. Pluronic is an amphiphilic block copolymer and is one of the most widely used polymers to formulate micelles; however, Pluronic has low mucoadhesive characteristics. Lin and Chang (2013) prepared Cyc A loaded Pluronic micelles, they have modified the nanomicelle, using chitosan solution to improve its ocular bioadhesiveness. Chitosan has strong bioadhesive features due to the electrostatic attraction between its positive charge and the negative charge of the cornea. Chitosan can also enhance the cellular permeability with the possibility of additional intracellular pathways. Pluronic micelles in which the surfaces were modified by chitosan have been successfully prepared. Based on data obtained from the in vitro and in vivo studies, the micellar carriers have been proposed as a potential in decreasing frequency of administration and improving patient compliance for ocular drug delivery.

4.8.2 RAPAMYCIN

Rapamycin, known as sirolimus, has an immunosuppressant effect in humans. It is used as an antifungal, immunosuppressive, cytostatic and antiangiogenic agent in anterior and posterior ocular diseases (Phillips and Wroblewski, 2010). Rapamycin is poorly soluble in water (Simamora et al., 2001) with a log P value of 5.8 (Shin et al., 2011) and is, therefore, not suitable for topical ocular administration. Buech et al. (2007) developed liposomes, microemulsions, and cyclodextrin solutions of rapamycin. However, these formulations showed poor ocular permeation. Particle size is an important parameter to increase the permeation through the retina. In the studies, it seems that under 20 nm particles are more permeable (Amrite and Kompella, 2007; Kim et al., 2009). Elsaid et al. (2014) used a copolymer composed of cholesterol and PEG. The sizes of micelles with 1 and 5 kDa weight copolymers were 11.7 and 13.8 nm. The contact time of both copolymer micelles with sclera was significantly longer than the rapamycin solution. Cholkar et al. (2014) developed a clear aqueous rapamycin-loaded nanomicellar formulation for the posterior segment of the eye. 0.2% rapamycin micelles are prepared with a copolymer of vitamin E tocopherol PEG succinate and octoxynol-40. In this way rapamycin was successfully delivered to the retina–choroid. In a rat experimental autoimmune uveitis model, the therapeutic efficacy of rapamycin-conjugated MPEG-PCL micelles was evaluated by Wu et al. (2016) over 14 days following intravitreal injection, and concentration of rapamycin conjugated in micelles in the retinal tissue was significantly higher than unconjugated rapamycin over this period.

4.8.3 PIMECROLIMUS

Ascomycin macrolactam derivative, pimecrolimus, can inhibit the transcription and release of proinflammatory cytokines in T cells at nanomolar concentrations. This molecule is an immunomodulating agent used in the treatment of atopic dermatitis (Yingfang et al., 2016; Allen et al., 2003). Pimecrolimus has a high lipophilic character (Billich et al., 2004). Ofri et al. (2009) revealed that pimecrolimus oil-based eye drops has been more efficient than Cyc A ointment to alleviate the keratoconjunctivitis in dogs. In order to treat keratoconjunctivitis sicca, Yingfang et al. (2016) prepared nanomicelles and thermosensitive hydrogel of pimecrolimus, using mPEG-PCL and PCEC copolymers, respectively. Pimecrolimus nanomicelles formulation had the best therapeutic effect and sustained-release profile.

4.8.4 DEXAMETHASONE

DEX is a poorly water soluble corticosteroid that prevents the release of substances in the body that cause inflammation (Li et al., 2014). DEX is used to treat many different inflammatory conditions such as allergic disorders, skin conditions, ulcerative colitis, arthritis, lupus, psoriasis, and breathing disorders. DEX and the other corticosteroids are applied via the topical route to prevent or suppress postoperative inflammation (Chopra et al., 2012). Civiale et al. (2009) prepared DEX micelles by using three different copolymers made from polyhydroxyethylaspartamide (PHEA), PEG, and hexadecylamine (C16). The polymeric micelles based on PHEA-PEG, PHEA-PEG-C16, and PHEA-C16 copolymers were studied on primary cultured rabbit conjunctival and corneal epithelia cells for in vitro permeability. DEX-loaded PHEA-C16 and PHEAPEG-C16 micelles provided greater drug permeation across the ocular epithelia than simple drug solutions or suspensions. According to the results, PHEA-PEG-C16 micelles were chosen to study in vivo bioavailability. DEX-loaded PHEA-PEG-C16 micelles had great bioavailability in comparison to that obtained with a water suspension of the same drug after ocular administration to rabbit eyes.

Sodium taurocholate (TA) or combination it with egg lecithin (LE) was used as drug carrier system for DEX in another study. In order to study tissue acrossing, Chopra et al. (2012) used passive and iontophoretic transport with excised human sclera in vitro. The drug release of the micelles from the sclera was found to be significantly higher compared to the control after passive and iontophoretic delivery. Vaishya et al. (2014b) reported that they had developed DEX nanomicelles for the treatment of uveitis. They studied a copolymer of ε-caprolatone and mPEG. Nanomicelles exhibited average sizes in the range of 25–30 nm with a low polydispersity of 0.125. Comparing with DEX suspension, the nanomicelles increased DEX permeability by 2 and 2.5 times through conjunctival cell line and excised rabbit sclera, respectively.

4.8.5 ACYCLOVIR

Acyclovir has an antiviral effect and is primarily used for the treatment of herpes simplex virus infections, chickenpox, and shingles. It is also a medical option to prevent cytomegalovirus infections following transplant and severe complications of Epstein-Barr virus infection (Blake et al., 2012; Weller et al., 2015). The micellar formulation of Biotin-12-Hydroxystearic acid-acyclovir (B-12HS-ACV) was prepared as a novel carrier for topical ocular delivery of biotinylated lipid prodrugs and evaluated for the treatment of herpetic keratitis by Vadlapudi et al. (2014). The nanomicelles comprised of vitamin E TPGS and octoxynol-40 loaded with 0.1% B-12HS-ACV were small enough, spherical, and homogenous, without any aggregates, and perfectly transparent (similar to pure water).

4.8.6 α-TOCOPHEROL

α-Tocopherol is commonly used as a stabilizer in drug formulations. TOC has antiproliferative and protein kinase C-suppressing effects which makes it useful for the prevention of glaucoma, cataracts, and posterior capsule proliferation (Ribeiro et al., 2013). Ribeiro et al. (2013) used Tetronic to prepare TOC polymeric micelles. Tetronic is a copolymer of PPO and PEO. The micelles had a sustained-release profile in lachrymal fluid.

4.8.7 OTHERS

Curcumin and ethoxzolamide (ETOX) are other active ingredients that have been studied for the optimization of micellar ocular formulations.

ETOX is used in the treatment of glaucoma and duodenal ulcers as well as a diuretic. Ribeiro et al. (2012) used a combination of Tetronic (T908, T1107, T1307) to prepare ETOX polymeric micelles for the treatment of glaucoma.

Curcumin, a natural polyphenol, has an antiinflammatory effect and is used to treat ocular diseases. Because of its poorly soluble in water, curcumin cannot easily pass through the corneal barriers. Duan et al. (2015) developed curcumin micelles in situ gels using P123/TPGS as a carrier system. The micelles were dispersed in gellan gum solution (0.2%, w/w) at the ratio of 3:1 and 1:1 (v/v), respectively. They found that both ratios had sustained-release profiles.

4.9 CONCLUSION

Topical ocular drug delivery is limited due to the effective barrier characteristics of the eye, thus it presents a major challenging task for pharmaceutical scientists. Delivery of drugs to the ocular tissues to treat both anterior and posterior segment diseases remains an obstacle to be solved. With this aim, the development of novel carriers to deliver drugs to targeted ocular tissues is crucial to satisfy this unmet medical need. The nanotechnology-based formulations have been widely investigated in last decades. Different types of nanocarriers have been developed to overcome the limitations of topical drug delivery and to improve the ocular bioavailability of drugs. Over the past few decades, nano-sized micellar drug delivery systems have been widely investigated to overcome the natural barrier of the eye, which limits the entrance of the drugs into targeted ocular tissues.

Polymeric micelles have shown potential in delivering drugs to the eye due to their features such as having nano-size, relatively low toxicity for biological membranes, solubilizing of drugs in aqueous media, and targeting ability. It has been suggested that the hydrophobic core of micelles incerases the solubility of lipophilic drugs while the hydrophilic shell of micelles acts for interaction with cell membranes, providing effective steric protection of micellar structure. Nano-sized micelles have been shown to be a highly promising strategy for delivering drugs into the posterior segment of the eye. A considerable number of research studies have been performed to develop micellar carriers to improve the ocular delivery of the drugs. Numerous polymeric micellar formulations have been patented to date, indicating the possibility of commercialization of novel ophthalmic micellar formulations in the near future. Highly hydrophobic drugs and relatively high-molecular weight compounds such as cyclosporine, rapamycine, and pimecrolimus could be efficiently delivered by nano-sized micellar carriers to anterior and posterior ocular tissues. Based on the preclinical data obtained from these studies polymeric micelles were found to have great potential to improve drug stability, prolong drug release, and to enhance the therapeutic activity of the drugs. Micellar carriers also seem to be promising carriers in terms of increasing the permeability of the drugs across the ocular barriers. In addition, it has been proposed that the EPR effect is established because of the small size of the micelles, resulting in their localization in the targeted tissues with a minimum amount of (or no) drug accumulation in normal tissues. One of the most important superiority of polymeric micelles is their easy-scale up process and low cost of production in terms of industrial perspective.

In conclusion, nano-sized micellar carriers to deliver drugs as a no-invasive topical administration strategy has gained great interest due to the characteristic features of micelles drugs might be delivered therapeutic drug levels into targeted ocular tissues. Based on the research results, it might be expected that nano-sized micellar carriers might come onto the market in the near future to be used as an effective treatment for anterior and posterior eye diseases and replace the current invasive drug delivery options such as periocular and intravitreal injections.

REFERENCES

Agarwal, P., Rupenthal, I.D., 2016. Modern approaches to the ocular delivery of cyclosporine A. Drug Discov. Today 21 (6), 977–988.

Aksungur, P., Demirbilek, M., Denkbas, E.B., Unlu, N., 2012. Comparative evaluation of cyclosporine A/HPbetaCD-incorporated PLGA nanoparticles for development of effective ocular preparations. J. Microencapsul. 29 (6), 605–613.

Allen, C., Maysinger, D., Eisenberg, A., 1999. Nano-engineering block copolymer aggregates for drug delivery. Colloids Surf., B: Biointerfaces 16, 3–27.

Allen, B.R., Lakhanpaul, M., Morris, A., Lateo, S., Davies, T., Scott, G., et al., 2003. Systemic exposure, tolerability, and efficacy of pimecrolimus cream 1% in atopic dermatitis patients. Arch Dis. Child. 88 (11), 969–973.

Amrite, A.C., Kompella, U.B., 2007. Nanoparticles and microparticles: particle engineering, cell uptake, in vivo disposition & efficacy. Drug Deliv. Technol. 7, 52–56.

Andrade, F.A., Fiorot, S.H., Benchimol, E.I., Provenzano, J., 2006. The autoimmune diseases of the eyes. Autoimmun. Rev. 15, 258–271.

April, W.E., 1997. Clinical Anatomy., third ed. Williams&Wilkins A Waverly Company, Egypt.

Bacman, S., Berra, A., Sterin-Borda, L., Borda, E., 2001. Muscarinic acetylcholine receptor antibodies as a new marker of dry eye Sjögren syndrome. Invest. Ophthalmol. Vis. Sci. 42 (2), 321–327.

Barar, J., Aghanejad, A., Fathi, M., Omidi, Y., 2016. Advanced drug delivery and targeting technologies for the ocular diseases. Bioimpacts 6 (1), 49–67.

Beardsley, R.M., Suhler, E.B., Rosenbaum, J.T., Lin, P., 2013. Pharmacotherapy of scleritis: current paradigms and future directions. Expert Opin. Pharmacother. 14 (4), 411–424.

Billich, A., Aschauer, H., Aszódi, A., Stuetz, A., 2004. Percutaneous absorption of drugs used in atopic eczema: pimecrolimus permeates less through skin than corticosteroids and tacrolimus. Int. J. Pharm. 269, 29–35.

Blake, J.A., Bareiss, B., Jimenez, L., Griffith, M., Scaiano, J.C., 2012. Design of xanthone propionate photolabile protecting group releasing acyclovir for the treatment of ocular herpes simplex virus. Photochem. Photobiol. Sci. 11, 539–547.

Bravo-Osuna, I., Vicario-de-la-Torre, M., Andres-Guerrero, V., Sanchez-Nieves, J., Guzman-Navarro, M., de la Mata, F.J., et al., 2016. Novel water-soluble mucoadhesive carbosilane dendrimers for ocular administration. Mol. Pharm. 13 (9), 2966–2976.

Buech, G., Bertelmann, E., Pleyer, U., Siebenbrodt, I., Borchert, H., 2007. Formulation of sirolimus eye drops and corneal permeation studies. J. Ocul. Pharmacol. Ther. 23, 292–303.

Cha, E.J., Kim, J.E., Ahn, C.H., 2009. Stabilized polymeric micelles by electrostatic interactions for drug delivery system. Eur. J. Pharm. Sci. 38 (4), 341–346.

Cholkar, K., Patel, A., Dutt Vadlapudi, A., Mitra, K., Novel Nanomicellar, A., 2012. Formulation approaches for anterior and posterior segment ocular drug delivery. Recent Pat. Nanomed. 2 (2), 82–95.

Cholkar, K., Patel, S.P., Vadlapudi, A.D., Mitra, A.K., 2013. Novel strategies for anterior segment ocular drug delivery. J. Ocul. Pharmacol. Ther. 29 (2), 106–123.

Cholkar, K., Gunda, S., Earla, R., Pal, D., Mitra, A.K., 2014. Nanomicellar topical aqueous drop formulation of rapamycin for back-of-the-eye delivery. AAPS PharmSciTech. 16 (3), 610–622.

Chopra, P., Hao, J., Li, S.K., 2012. Sustained release micellar carrier systems for iontophoretic transport of dexamethasone across human sclera. J. Control. Release 160 (1), 96–104.

Civiale, C., Licciardi, M., Cavallaro, G., Giammona, G., Mazzone, M.G., 2009. Polyhydroxyethylaspartamide-based micelles for ocular drug delivery. Int. J Pharm. 378 (1-2), 177–186.

Coca-Prados, M., 2014. The blood-aqueous barrier in health and disease. J Glaucoma 23 (8 Suppl 1), S36–S38.

Cox Gad S., 2008. Ocular Drug Delivery—Pharmaceutical Manufacturing Handbook: Production and Processes. John Wiley & Sons, Inc., Hoboken, New Jersey.

Cruysberg, L., Nuijts, R., Geroski, D., Koole, L., 2002. In vitro human scleral permeability of fluorescein, dexamethasone-fluorescein, methotrexate-fluorescein and rhodamine 6g and the use of a coated coil as a new drug delivery system. J. Ocul. Pharmacol. Ther. 18 (6), 559–569.

Cunha-Vaz, J.G., 1997. The blood-ocular barriers: past, present, and future. Doc. Ophthalmol. 93 (1-2), 149–157.

de la Fuente, M., Ravina, M., Paolicelli, P., Sanchez, A., Seijo, B., Alonso, M.J., 2010. Chitosan-based nanostructures: a delivery platform for ocular therapeutics. Adv. Drug Deliv. Rev. 62 (1), 100–117.

Delplace, V., Payne, S., Shoichet, M., 2015. Delivery strategies for treatment of age-related ocular diseases: from a biological understanding to biomaterial solutions. J. Control. Release 219, 652–668.

Dey, S., Anand, B.S., Patel, J., Mitra, A.K., 2003. Transporters/receptors in the anterior chamber: pathways to explore ocular drug delivery strategies. Expert Opin. Biol. Ther. 3 (1), 23–44.

Di Fiore, M.S.H., Cartographer, 2001. Atlas of Human Histology. Lea and Febiger, Philadelphia, PA.
Di Tommaso, C., Como, C., Gurny, R., Moller, M., 2010. Investigations on the lyophilisation of MPEG-hexPLA micelle based pharmaceutical formulations. Eur. J. Pharm. Sci. 40 (1), 38–47.
Di Tommaso, C., Torriglia, A., Furrer, P., Behar-Cohen, F., Gurny, R., Möller, M., 2011. Ocular biocompatibility of novel cyclosporin A formulations based on methoxy poly (ethylene glycol)-hexylsubstituted poly(lactide) micelle carriers. Int. J. Pharm. 416 (2), 515–524.
Di Tommaso, C., Bourges, J.-L., Valamanesh, F., Trubitsyn, G., Torriglia, A., Jeanny, J.-C., et al., 2012. Novel micelle carriers for cyclosporin A topical ocular delivery: in vivo cornea penetration, ocular distribution and efficacy studies. Eur. J. Pharm. Biopharm. 81 (2), 257–264.
Duan, Y., Cai, X., Du, H., Zhai, G., 2015. Novel in situ gel systems based on P123/TPGS mixed micelles and gellan gum for ophthalmic delivery of curcumin. Colloids Surf., B: Biointerfaces 128, 322–330.
Dunlevy, J.R., Rada, J.A., 2004. Interaction of lumican with aggrecan in the aging human sclera. Invest. Ophthalmol. Vis. Sci. 45 (11), 3849–3856.
Duvvuri, S., Majumdar, S., Mitra, A.K., 2003. Drug delivery to the retina: challenges and opportunities. Expert Opin. Biol. Ther. 3 (1), 45–56.
Durrani, O.M., Tehrani, N.N., Marr, J.E., Moradi, P., Stavrou, P., Murray, P.I., 2004. Degree, duration, and causes of visual loss in uveitis. Br. J. Ophthalmol. 88 (9), 1159–1162.
Elsaid, N., Somavarapu, S., Jackson, T.L., 2014. Cholesterol-poly(ethylene) glycol nanocarriers for the transscleral delivery of sirolimus. Exp. Eye Res. 121, 121–129.
Faraj, L.A., Hashmani, K., Khatib, T., Al-Aqaba, M., Dua, H.S., 2012. The changing face of corneal graft rejection. Br. J. Ophthalmol. 96 (8), 1049–1050.
Fischer, N., Narayanan, R., Loewenstein, A., Kuppermann, B.D., 2011. Drug delivery to the posterior segment of the eye. Eur. J. Ophthalmol. 21 (6), 20–26.
Gaudana, R., Jwala, J., Boddu, S.H., Mitra, A.K., 2009. Recent perspectives in ocular drug delivery. Pharm. Res. 26 (5), 1197–1216.
Gaudana, R., Ananthula, H.K., Parenky, A., Mitra, A.K., 2010. Ocular drug delivery. AAPS J. 12 (3), 348–360.
Güngör, S., Kahraman, E., Özsoy, Y., 2015. Polymeric micelles for cutaneous drug delivery. In: Naik, J. (Ed.), Nano Based Drug Delivery. IAPC Publishing, Croatia, pp. 367–388.
Guo, C., Zhang, Y., Yang, Z., Li, M., Li, F., Cui, F., et al., 2015. Nanomicelle formulation for topical delivery of cyclosporine A into the cornea—in vitro mechanism and in vivo permeation evaluation. Sci. Rep. 5 (12968). Available from: https://doi.org/10.1038/srep12968.
Hennig, R., Goepferich, A., 2015. Nanoparticles for the treatment of ocular neovascularizations. Eur. J. Pharm. Biopharm. 95, 294–306.
Heurtault, B., Saulnier, P., Pech, B., Proust, J.E., Benoit, J.P., 2003. Physico-chemical stability of colloidal lipid particles. Biomaterials 24 (23), 4283–4300.
Huang, A.J.W., Tseng, S.C.G., Kenyont, K.R., 1989. Paracellular permeability of corneal and conjunctival epithelia. Invest. Ophthalmol. Vis. Sci. 30 (4), 684–689.
Jiao, J., 2008. Polyoxyethylated nonionic surfactants and their applications in topical ocular drug delivery. Adv. Drug Deliv. Rev. 60 (15), 1663–1673.

Kanoujia, J., Kushwaha, P.S., Saraf, S.A., 2014. Evaluation of gatifloxacin pluronic micelles and development of its formulation for ocular delivery. Drug Deliv. Transl. Res 4 (4), 334–343.

Kansara, V., Budda, B., Mitra, A.K., 2007. Advancements in ocular drug delivery. In: Mashkevich, B.O. (Ed.), Drug Delivery Research Advances. Nova Science Publishers, New York.

Keeton, W.T., Gould, J.L., Grant Gould, C., 1993. Biological Science, third ed. W.W. Norton&Company, New York.

Kim, J.H., Kim, J.H., Kim, K.-W., Kim, M.H., Yu, Y.S., 2009. Intravenously administered gold nanoparticles pass through the blood–retinal barrier depending on the particle size, and induce no retinal toxicity. Nanotechnology 20 (50), 505101.

Kim, S., Shi, Y., Kim, J.Y., Park, K., Cheng, J.X., 2010. Overcoming the barriers in micellar drug delivery: loading efficiency, in vivo stability, and micelle-cell interaction. Expert Opin. Drug Deliv. 7, 49–62.

Kim, Y.C., Chiang, B., Wu, X., Prausnitz, M.R., 2014. Ocular delivery of macromolecules. J. Control. Release 190, 172–181.

Kumar, R., Sinha, V.R., 2014. Preparation and optimization of voriconazole microemulsion for ocular delivery. Colloids Surf., B: Biointerfaces 117, 82–88.

Kumar, V., Kumar, A., 2014. Immunological aspects of corneal transplant. Immunol. Invest. 43 (8), 888–901.

Kuno, N., Fujii, S., 2011. Recent advances in ocular drug delivery systems. Polymers 3 (4), 193–221.

Kuwano, M., Ibuki, H., Morikawa, N., Ota, A., Kawashima, Y., 2002. Cyclosporine A formulation affects its ocular distribution in rabbits. Pharm. Res. 19, 108–111.

Kymionis, G.D., Bouzoukis, D.I., Diakonis, V.F., Siganos, C., 2008. Treatment of chronic dry eye: focus on cyclosporine. Clin. Ophthalmol. 2 (4), 829–836.

Lajunen, T., Nurmi, R., Kontturi, L., Viitala, L., Yliperttula, M., Murtomäki, L., et al., 2016. Light activated liposomes: functionality and prospects in ocular drug delivery. J. Control. Release 244, 157–166.

Lallemand, F., Felt-Baeyens, O., Besseghir, K., Behar-Cohen, F., Gurny, R., 2003. Cyclosporine A delivery to the eye: a pharmaceutical challenge. Eur. J. Pharm Biopharm. 56 (3), 307–318.

Lee, J., Pelis, R.M., 2016. Drug transport by the blood–aqueous humor barrier of the eye. Drug Metab. Dispos. 44 (10), 1675–1681.

Lee, H., Soo, P.L., Liu, J., Butler, M., Allen, C., 2007a. Polymeric micelles for formulation of anti-cancer drugs. In: Amiji, M.M. (Ed.), Nanotechnology for Cancer Therapy. CRC Press Taylor & Francis Group, pp. 317–356.

Lee, S.C., Huh, K.M., Ooya, T., Park, K., 2007b. Hydrotropic polymer micelles for cancer therapeutics. In: Amiji, M. (Ed.), Nanotechnology for Cancer Therapeutics. CRC Press, pp. 385–408.

Lemp, M.A., 1995. Report of the National Eye Institute/Industry workshop on clinical trials in dry eyes. CLAO J. 21 (4), 221–232.

Li, N.N., Lin, J., Gao, D., Zhang, L.M., 2014. A macromolecular prodrug strategy for combinatorial drug delivery. J. Colloid Interface Sci. 417, 301–309.

Li, J., Li, Z., Zhou, T., Zhang, J., Xia, H., Li, H., et al., 2015. Positively charged micelles based on a triblock copolymer demonstrate enhanced corneal penetration. Int. J. Nanomed. 10, 6027–6037.

Lidich, N., Wachtel, E.J., Aserin, A., Garti, N., 2016. Water-dilutable microemulsions for transepithelial ocular delivery of riboflavin phosphate. J. Colloid Interface Sci. 463, 342–348.
Lin, H.-R., Chang, P.-C., 2013. Novel pluronic-chitosan micelle as an ocular delivery system. J. Biomed. Mater. Res. B: Appl. Biomater. 101B (5), 689–699.
Liu, S., Jones, L., Gu, F.X., 2012. Nanomaterials for ocular drug delivery. Macromol. Biosci. 12 (5), 608–620.
Luschmann, C., Herrmann, W., Strauß, O., Luschmann, K., Goepferich, A., 2013a. Ocular delivery systems for poorly soluble drugs: an in-vivo evaluation. Int. J Pharm. 455 (1-2), 331–337.
Luschmann, C., Tessmar, J., Schoeberl, S., Strauss, O., Frammed, C., Luschmanne, K., et al., 2013b. Developing an in situ nanosuspension: a novel approach towards the efficient administration of poorly soluble drugs at the anterior eye. Eur. J. Pharm. Sci. 50, 385–392.
Luschmann, C., Tessmar, J., Schoeberl, S., Strauss, O., Luschmann, K., Goepferich, A., 2014. Self-assembling colloidal system for the ocular administration of cyclosporine A. Cornea 33 (1), 77–81.
Mahaling, B., Katti, D.S., 2016. Understanding the influence of surface properties of nanoparticles and penetration enhancers for improving bioavailability in eye tissues in vivo. Int. J. Pharm. 501, 1–9.
Mandal, A., Bisht, R., Rupenthal, I.D., Mitra, A.K., 2017. Polymeric micelles for ocular drug delivery: from structural frameworks to recent preclinical studies. J. Control. Release 248, 96–116.
Mishima, S., Gasset, A., Klyce, S.D., Baum, J.L., 1966. Determination of tear volume and tear flow. Invest. Ophthalmol. 5 (3), 264–276.
Mitra A.K., 2003. Ophtalmic Drug Delivery System. Marcel Dekker, Inc. New York.
Mitra A.K., Velagaleti P.R., Grau U.M., 2010. Topical drug delivery for ophthalmic use. WO2010144194.
Mitra, R.N., Nichols, C.A., Guo, J., Makkia, R., Cooper, M.J., Naash, M.I., et al., 2016. Nanoparticle-mediatedmiR200-b delivery for the treatment of diabetic retinopathy. J. Control. Release 236, 31–37.
Mondon, K., Zeisser-Labouèbe, M., Gurny, R., Möller, M., 2011. Novel cyclosporin A formulations using MPEG-hexyl-substituted polylactide micelles: a suitability study. Eur. J. Pharm. Biopharm. 77 (1), 56–65.
Murphy, C.J., Pollock, R.V.S., 1993. Anatomy of the Dog. The Eye. WB Sanders Co, Philadelphia, PA.
Nagarwal, R.C., Kant, S., Singh, P.N., Maiti, P., Pandit, J.K., 2009. Polymeric nanoparticulate system: a potential approach for ocular drug delivery. J. Control. Release 136 (1), 2–13.
National Eye Institute, 2017. Facts About Age-Related Macular Degeneration. Last accessed: 26.01.17.
O'Malley, P.G., 2012. Comparative effectiveness of anti–growth factor therapies for diabetic macular edema. Arch. Intern. Med. 172 (13), 1014–1015.
Ofri, R., Lambrou, G.N., Allgoewer, I., Graenitz, U., Pena, T.M., Spiess, B.M., et al., 2009. Clinical evaluation of pimecrolimus eye drops for treatment of canine keratoconjunctivitis sicca: a comparison with cyclosporine A. Vet. J. 179 (1), 70–77.
Olsen, T.W., Aaberg, S.Y., Geroski, D., Fedelhauser, H.F., 1998. Human sclera: thickness and surface area. Am. J. Ophthalmol. 125 (2), 237–241.

Owen, S.C., Chan, D.P.Y., Shoichet, M.S., 2012. Polymeric micelle stability. Nano Today 7 (1), 53–65.

Patel, A., Cholkar, K., Agrahari, V., Mitra, A.K., 2013. Ocular drug delivery systems: an overview. World J. Pharmacol. 2 (2), 47–64.

Pepić, I., Hafner, A., Lovrić, J., Pirkić, B., Filipović-Grčić, J., 2010. A nonionic surfactant/chitosan micelle system in an innovative eye drop formulation. J. Pharm. Sci. 99, 4317–4325.

Pepic, I., Lovric, J., Filipovic-Grcic, J., 2012. Polymeric micelles in ocular drug delivery: rationale, strategies and challenges. Chem. Biochem. Eng. Q. 26 (4), 365–377.

Phillips, B.N., Wroblewski, K.J., 2010. A retrospective review of oral low-dose sirolimus (rapamycin) for the treatment of active uveitis. J. Ophthalmic. Inflamm. Infect. 1 (1), 29–34.

Prosperi-Porta, G., Kedzior, S., Muirhead, B., Sheardown, H., 2016. Phenylboronic-acid-based polymeric micelles for mucoadhesive anterior segment ocular drug delivery. Biomacromolecules 17 (4), 1449–1457.

Rafie, F., Javadzadeh, Y., Javadzadeh, A.R., Ghavidel, L.A., Jafari, B., Moogooee, M., et al., 2010. In vivo evaluation of novel nanoparticles containing dexamethasone for ocular drug delivery on rabbit eye. Curr. Eye Res. 35 (12), 1081–1089.

Ramenaden, E.R., Raiji, V.R., 2013. Clinical characteristics and visual outcomes in infectious scleritis: a review. Clin. Ophthalmol. 7, 2113–2122.

Rathore, K.S., Nema, R.K., 2009. An insight into ophthalmic drug delivery. Int. J. Pharm. Sci. 1 (1), 1–5.

Raviola, G., 1983. Conjunctival and episcleral blood vessels ore permeable to blood-borne horseradish peroxidase. Invest. Ophthal. Vis. Sci. 24, 725–736.

Reimondez-Troitiño, S., Csaba, N., Alonso, M.J., de la Fuente, M., 2015. Nanotherapies for the treatment of ocular diseases. Eur. J. Pharm. Biopharm. 95, 279–293.

Ribeiro, A., Sosnik, A., Chiappetta, D.A., Veiga, F., Concheiro, A., Alvarez-Lorenzo, C., 2012. Single and mixed poloxamine micelles as nanocarriers for solubilization and sustained release of ethoxzolamide for topical glaucoma therapy. J. R. Soc. Interface 9 (74), 2059–2069.

Ribeiro, A., Sandez-Macho, I., Casas, M., Alvarez-Pérez, S., Alvarez-Lorenzo, C., Concheiro, A., 2013. Poloxamine micellar solubilization of α-tocopherol for topical ocular treatment. Colloids Surf., B: Biointerfaces 103, 550–557.

Savic, R., Luo, L., Eisenberg, A., Maysinger, D., 2003. Micellar nanocontainers distribute to defined cytoplasmic organelles. Science 300 (5619), 615–618.

Shin, H.C., Alani, A.W.G., Cho, H., Bae, Y., Kolesar, J.M., Kwon, G.S., 2011. A 3-in-1 polymeric micelle nanocontainer for poorly water-soluble drugs. Mol. Pharm. 8 (4), 1257–1265.

Simamora, P., Alvarez, J.M., Yalkowsky, S.H., 2001. Solubilization of rapamycin. Int. J. Pharm. 213, 25–29.

Sowka, J.W., Gurwood, A.S., Kabat, A.G., 2015. The handbook of ocular disorders management. Rev. Optometry 15, 1A–82A.

Springsteen, G., Wang, B., 2002. A detailed examination of boronic acid-diol complexation. Tetrahedron 58 (26), 5291–5300.

Sudharshan, S., Ganesh, S.K., Biswas, J., 2010. Current approach in the diagnosis and management of posterior uveitis. Indian J. Ophthalmol. 58 (1), 29–43.

Sullivan, B.D., Evans, J.E., Dana, M.R., Sullivan, D.A., 2002. Impact of androgen deficiency on the lipid profiles in human meibomian gland secretions. Adv. Exp. Med. Biol. 506 (Pt A), 449–458.

Talelli, M., Iman, M., Varkouhi, A.K., Rijcken, C.J., Schiffelers, R.M., Etrych, T., et al., 2010. Core-crosslinked polymeric micelles with controlled release of covalently entrapped doxorubicin. Biomaterials 31 (30), 7797–7804.

Tan, G., Yu, S., Pan, H., Li, J., Liu, D., Yuan, K., et al., 2017. Bioadhesive chitosan-loaded liposomes: a more efficient and higher permeable ocular delivery platform for timolol maleate. Int. J. of Biol. Macromol. 94, 355–363.

Tomi, M., Hosoya, K., 2010. The role of blood-ocular barrier transporters in retinal drug disposition: an overview. Expert Opin. Drug Metab. Toxicol. 6 (9), 1111–1124.

Torchilin, V.P., 2001. Structure and design of polymeric surfactant-based drug delivery systems. J. Control. Release 73 (2-3), 137–172.

Torchilin, V.P., 2007. Micellar nanocarriers: pharmaceutical perspectives. Pharm. Res. 24 (1), 1–16.

Trimaille, T., Mondon, K., Gurny, R., Möller, M., 2006. Novel polymeric micelles for hydrophobic drug delivery based on biodegradable poly(hexyl-substituted lactides). Int. J. Pharm. 319 (1-2), 147–154.

Vadlapudi, A.D., Cholkar, K., Vadlapatla, R.K., Mitra, A.K., 2014. Aqueous nanomicellar formulation for topical delivery of biotinylated lipid prodrug of acyclovir: formulation development and ocular biocompatibility. J. Ocul. Pharmacol. Ther. 30 (1), 49–58.

Vadlapudi, A.D., Cholkar, K., Dasari, S.R., Mitra, A.K., 2015. Ocular drug delivery. In: Mitra, A.K., Kwatra, D., Vadlapudi, A.D. (Eds.), Drug Delivery. Jones and Bartlett Learning, NJ.

Vaishya, R.D., Khurana, V., Patel, S., Mitra, A.K., 2014a. Controlled ocular drug delivery with nanomicelles. Wiley Interdiscip. Rev. Nanomed. Nanobiotechnol. 6 (5), 422–437.

Vaishya, R.D., Gokulgandhi, M., Patel, S., Minocha, M., Mitra, A.K., 2014b. Novel dexamethasone-loaded nanomicelles for the intermediate and posterior segment uveitis. AAPS PharmSciTech. 15 (5), 1238–1251.

Weiner, A.L., Gilger, B.C., 2010. Advancements in ocular drug delivery. Vet. Ophthalmol. 13 (6), 395–406.

Weller, J.M., Bergua, A., Mardin, C.Y., 2015. Retinopathy in a patient with acute Epstein-Barr virus infection: follow-up analysis using spectral domain optical coherence tomography. Retin. Cases Brief Rep. 9 (1), 72–77.

Weng, Y., Liu, J., Jin, S., Guo, W., Liang, X., Hu, Z., 2016. Nanotechnology-based strategies for treatment of ocular disease. Acta Pharm. Sin. B 7, 281–291.

Wichit, A., Tangsumranjit, A., Pitaksuteepong, T., Waranuch, N., 2012. Polymeric micelles of PEG-PE as carriers of all-trans retinoic acid for stability improvement. AAPS PharmSciTech. 13 (1), 336–343.

Williams, P.L., Warwick, R., Dyson, M., Bannister, L.H., 1989. Gray's Anatomy, 37th ed. Churchill Livingstone, New York.

Willoughby, C.E., Ponzin, D., Ferrari, S., Lobo, A., Landau, K., Omidi, Y., 2010. Anatomy and physiology of the human eye: effects of mucopolysaccharidoses disease on structure and function—a review. Clin. Experiment. Ophthalmol. 38, 2–11.

Wu, W., He, Z., Zhang, Z., Yu, X., Song, Z., Li, X., 2016. Intravitreal injection of rapamycin-loaded polymeric micelles for inhibition of ocular inflammation in rat model. Int. J. Pharm. 513 (1-2), 238–246.

Yang, J., Yan, J., Zhou, Z., Amsden, B.G., 2014. Dithiol-PEG-PDLLA micelles: preparation and evaluation as potential topical ocular delivery vehicle. Biomacromolecules 15 (4), 1346–1354.

Yavuz, B., Kompella, U.B., 2016. Ocular drug delivery. Handb. Exp. Pharmacol. 22 (9), 1135–1146.

Yavuz, B., Bozdag Pehlivan, S., Sumer Bolu, B., Nomak Sanyal, R., Vural, I., Unlu, N., 2016. Dexamethasone— PAMAM dendrimer conjugates for retinal delivery: preparation, characterization and in vivo evaluation. J. Pharm. Pharmacol. 68 (8), 1010–1020.

Yellepeddi, V.K., Palakurthi, S., 2016. Recent advances in topical ocular drug delivery. J. Ocul. Pharmacol. Ther. 32 (2), 67–82.

Yingfang, F., Zhuang, B., Wang, C., Xu, X., Xu, W., Lv, Z., 2016. Pimecrolimus micelle exhibits excellent therapeutic effect for keratoconjunctivitis sicca. Colloids Surf., B: Biointerfaces 140, 1–10.

Zhang, J., Ma, P.X., 2009. Polymeric core–shell assemblies mediated by host–guest interactions: versatile nanocarriers for drug delivery. Angew. Chem. Int. Ed. Engl. 48, 964–968.

Zhang, Z., He, Z., Liang, R., Ma, Y., Huang, W., Jiang, R., et al., 2016. Fabrication of a micellar supramolecular hydrogel for ocular drug delivery. Biomacromolecules 17 (3), 798–807.

FURTHER READING

Biswas, S., Vaze, O.S., Movassaghian, S., Torchilin, V.P., 2013. Polymeric micelles for the delivery of poorly soluble drugs. In: Douroumis, D.F.A. (Ed.), Drug Delivery Strategies for Poorly Water-Soluble Drugs. John Wiley & Sons Ltd, New York.

CHAPTER

5 Reversible core–shell crosslinked micelles for controlled release of bioactive agents

Sepideh Khoee and Shahnaz Rahimi

Polymer Laboratory, Chemistry Department, School of Science, University of Tehran, Tehran, Iran

5.1 INTRODUCTION

Polymeric micelles are generally nanosized spheres (10–100 nm) having core–shell structures which are mainly prepared by self-assembly of amphiphilic block copolymers in an aqueous solution. In the past few decades, micellar nanoparticles have attracted growing attention as nanocarrier systems for tumor-targeting delivery of anticancer drugs (Lee et al., 2007; Oh et al., 2007; Mikhail and Allen, 2010), genetic drugs (Convertine et al., 2010; Kim et al., 2008), and imaging agents (Shiraishi et al., 2009). These nanoparticles exhibit several advantages for the drug molecules, such as improved stability in blood, preferential agglomeration at the tumor site via the enhanced permeation and retention (EPR) effect and the enhanced permeability, prolonged in vivo circulation time, better pharmacological profiles compared to the conventional clinical techniques, and release rate control of the entrapped drugs (Nie et al., 2007; Davis et al., 2008). However, despite the recent progress in the field of micellar nanoparticles, there are certain limitations which preclude their use in practical applications as drug nanocarriers which need to be solved. One fundamental limitation is their low structural stability since the thermodynamically self-assembled micellar carriers are always in equilibrium between the micelles and unimers in an aqueous environment. Thus, after being injected into the blood stream, conventional self-assembled (SA) polymeric micelles are easily dissociated into unimers on dilution or changes in external conditions, because the polymer concentration falls below the critical micelle concentration (O'Reilly et al., 2006; Bronich et al., 2005). In addition, in blood circulation, the drug-loaded micelles, unless crosslinked covalently, will disrupt upon interaction with blood proteins and lipoprotein particles

Nanoarchitectonics in Biomedicine. DOI: https://doi.org/10.1016/B978-0-12-816200-2.00016-5

such as high density lipoprotein, low density lipoprotein, very low density lipoprotein, and chylomicronmay (Li et al., 2007). This process can result in rapid dissociation or aggregation of micelles upon intravenous administration and are excreted from the body via the kidneys, leads to premature drug release at normal tissues or organs and slow drug release at the pathological site, thus, compromising the therapeutic effects. In order to solve the stability issue and circumvent dilution effects of amphiphilic block copolymer micelles for both in vitro storing and in vivo circulation, different approaches of crosslinking the block copolymer micelles, including core crosslinked (CCL) micelles (CCLM) (Guo et al., 1996; Tao et al., 1997; Kakizawa et al., 1999), shell crosslinked (SCL) micelles (SCLM) (Thurmond et al., 1996; Liu and Armes, 2001; Jiang et al., 2006), and core–shell interface crosslinked micelles (Koo et al., 2008) through photo crosslink, chemical crosslink, or polymerization after the micelle construction has been proposed and developed. However, in some cases, excessively covalently crosslinked micelles may prohibit the release of cancer drugs to tumor tissue or inside tumor cell, thus, decreasing the therapeutic efficacy (Li et al., 2009; Huang et al., 2012). In order to render the crosslinking reversible and improve intracellular drug release, several environmentally responsive crosslinkers, such as reducible disulfide (DS) bonds (Miyata et al., 2004; Talelli et al., 2010) and pH cleavable (Chan et al., 2008), have been employed to design stimuli-responsive crosslinked micelles (SCMs) that respond to external stimuli such as pH values, redox, or ultraviolet-light conditions to afford a change in the structure, morphology, or controlled release event. They represent excellent stability in blood circulation and can further preserve the drug contents. The special micelles that are adequately stable under extracellular conditions, but release drugs quickly and exclusively in the target site, are often called "smart" or "intelligent" micellar nanoparticles. In addition, shell crosslinking of the block copolymer micelles has been considered as a promising approach to enhance the structural stability of micelles and induce particular intracellular drug release.

The first paper on SCL micelles were published in 1996 by Wooley et al., who oligomerized pendant styrene groups within the shells of micellar assemblies by using free radical chemistry (Thurmond et al., 1996). Armes et al. also reported SCL micelles using ABC triblock copolymers which were composed from hydrophobic cores, crosslinkable inner shells (i.e., the B block), and poly (ethylene glycol) (PEG) outer coronas that act as a steric stabilizer (Butun et al., 2000). Moreover, the SCL micelles periphery can be subsequently functionalized with reactive components for the conjugation with targeting agents (Qi et al., 2004). To date, numerous approaches have been employed in the literature to prepare core/shell crosslinking of micelles, including reacting with bifunctional crosslinking reagents, such as linking diamines to carboxylic acid groups with carbodiimide coupling chemistry as reported by Wooley's group (Huang et al., 1997; Zhang et al., 2000), Liu et al. (2002) utilized the reaction of hydroxyl groups with divinyl sulfone to successfully crosslink hydroxylated segments, quaternization of amino groups, for instance the use of 1,2-bis(2-iodoethoxy)ethane to crosslink 2-(dimethylamino)ethyl methacrylate residues, which have been employed by Armes et al. (Butun et al., 1998; Pilon et al., 2005), UV-induced coupling of cinnamoyl groups (Ding and Liu, 1998; Jiang et al., 2006) and other crosslinking approaches such as click chemistry (Joralemon et al., 2005), ionically induced polyelectrolyte complexation (Weaver et al., 2004), and multivalent Au ligation through in situ reduction (Li et al., 2007; Smith et al., 2009) have also been reported. Although advantageous, it is recognized that most techniques used for core/shell crosslinking of the polymeric micelles were limited by reagent insolubility, multi-step procedures requiring organic solvents and protecting group chemistry, low reaction efficiency, and extensive methods of purification to remove small molecule by-products. Thus, because of low delivery efficiency within target sites these crosslinked micelles are usually not desirable for drug delivery systems. Hence degradable linkages should be used for the facilitated release of loaded drugs to target cells. In this chapter, we classify the reversible CCLM/SCL micelles into three main categories: (1) pH-reversible CCLM/SCL micelles; (2) photo-reversible CCLM/SCL micelles; and (3) reduction-reversible CCLM/SCL micelles, that most of the reported work on reversible SCL and CCL nanovehicles relies on the reductive cleavage of micelles in cancerous cells or in the endocytic compartments (Rijcken et al., 2007; Dai et al., 2011). This chapter summarizes recent advances in reversible crosslinked micellar nanovehicles with the main focus being on the design, crosslink strategy, improving their biocompatibility and biodegradability, and preclinical evaluations.

5.2 pH-REVERSIBLE CROSSLINKED MICELLES

In recent years, pH-reversible crosslinked micelles have revealed several specifications that make them promising candidates as a great category nanovehicle for drug delivery systems based on the smart pH changes in the human body. For instance, the pH in the stomach is 1–2, that of intestine varies from 5 to 8, while the plasma pH is 7.4. The intracellular release of encapsulated chemotherapeutic drugs could be increased using designed pH and reduction-sensitive biodegradable polymeric micelles. In particular, the pH-triggered polymeric micelles gained much attention due to the existence of the different pH environments at the internal and external tumor cells—e.g., the pH range of extracellular in the tumor site is slightly acidic (6.5–7.2), while the pH value in blood and normal tissue is about (7.4)—which will hydrolyze the acid-liable groups and accelerate the dissociation of micelles in the tumor cells, resulting in the quick release of loaded drugs due to the pH difference between healthy and cancerous organs (Kellum et al., 2010; Xu et al., 2009; Dai et al., 2011). Also, the pH-triggered control of release, which is potentially site-selective, results in minimizing the toxicity of therapeutic and diagnostic agents. In addition, the incorporation of pH-reversible core/shell crosslinks into polymer micelles may enable the formation of useful intracellular nanovehicles.

To date, different categories of pH-reversible crosslinked micelles have been investigated (Jiang et al., 2006, 2007b). The most common pH-reversible CCLM/SCL micelles include ionically induced polyelectrolyte complexation, hydrazones, imines, acetals/ketals, and others.

5.2.1 pH-REVERSIBLE IMINE CROSSLINKED MICELLES

Imines as a pH-reversible crosslinker reagent that have been employed at polymeric nanostructures are formed from the reaction between simple amines (e.g., lysine) and aldehydes or ketones that are commonly reversible in aqueous solutions. The pH-triggered imine crosslinker can be utilized in the reversible fabrication of CCLM/SCL micelles. For example, Xu et al. (2011) reported reversible SCL micelles with pH-reversible imine (Schiff base) crosslinks, which can be reversibly cleaved by readily regulating the solution pH. At basic condition (pH > 7.0), these reversible SCL micelles are stable, while at acidic conditions (pH < 6.5), in which the imine linkages are cleaved, they are disrupted into unimers. This system is based on a thermoresponsive ABC triblock copolymer α-methoxypoly(ethylene oxide)-*b*-poly(*N*-(3-aminopropyl)methacrylamide)-*b*-poly(*N*-isopropylacrylamide) (mPEO-PAPMA-PNIPAM), synthesized using aqueous reversible addition-fragmentation chain-transfer (aRAFT) polymerization and composed of a hydrophilic mPEO block as outer coronas stabilize the micelles, the PAPMA block containing primary amine groups as crosslinkable middle shells, and the PNIPAM block as thermoresponsive cores (Fig. 5.1). Micellar nanoparticles are formed from the self-assembly of this thermoresponsive triblock copolymer in aqueous media at above the lower critical solution temperature (LCST) of the PNIPAM block. Subsequently, the reversible SCL micelles were successfully prepared by incorporation of the terephthaldicarboxaldehyde (TDA) as a dialdehyde crosslinker into a micellar solution, which can react with the primary amine groups of the triblock copolymer. A model hydrophobic drug, prednisolone acetate (PA) was loaded into the reversible SCL micelle core to investigate the release behavior at pH and temperature values that simulate that of the human body. The pH-triggered burst release of PA was observed within 2 hours at a lower pH (i.e., pH 5.5, 37°C, endosomes have pH values from 5.0 to 6.5). In contrast, only about of 40%–50% of PA was released after 10 hours under normal physiological conditions (i.e., pH 7.4, 37°C). Burst release of PA at endosomal conditions is attributed to hydrolysis of the imine linkages and micelle dissociation.

In addition, biodegradable and biocompatible core-crosslinked polymeric micelles based on two water soluble methoxy-poly(ethylene glycol)-block-poly(aspartamide) (mPEG-PA) copolymers using two amine and aldehyde modified mPEG-PA precursors have been developed (Wang et al., 2014a), which can be reversibly crosslinked by imine bonds (Fig. 5.2). The process of crosslinking of

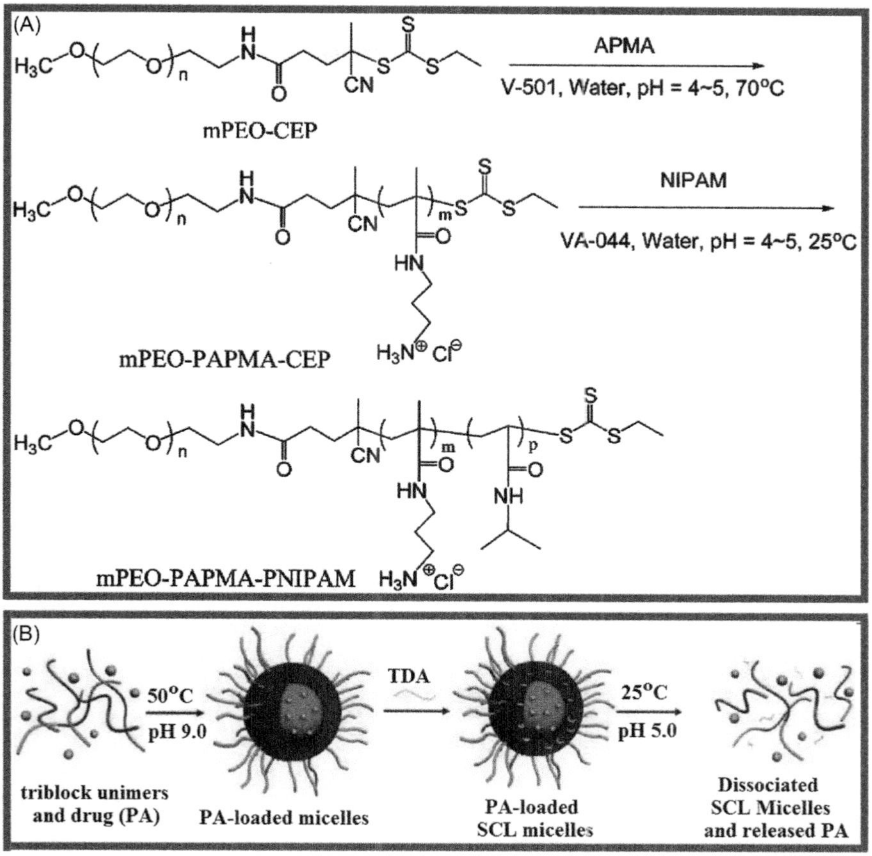

FIGURE 5.1

(A) Synthetic route for preparation of temperature-responsive mPEO-PAPMA-PNIPAM triblock copolymer via aqueous RAFT polymerization. (B) Schematic image of the formation of reversible shell cross-linked (SCL) micelles from the triblock copolymer, mPEO113-PAPMA12-PNIPAM136 (Xu et al., 2011).

the core via imine bonds is the driving force to construct the core–shell micelles. The hydrophobic anticancer drug doxorubicin (DOX) was encapsulated into mPEG-PA micelles and their performance as a smart drug delivery system was investigated in terms of the stabilization under physiological conditions and in vitro drug release under pH stimulus. The nanocarriers possessed an excellent pH-responsive drug release behavior. Under physiological conditions, mPEG-PA core-crosslinked micelles demonstrated strong stability and a small encapsulated drug release. Quick release of the encapsulated DOX can be obtained under mildly acidic conditions due to cleavage of the imine bonds and induce the dissociation of the core-crosslinked micelles.

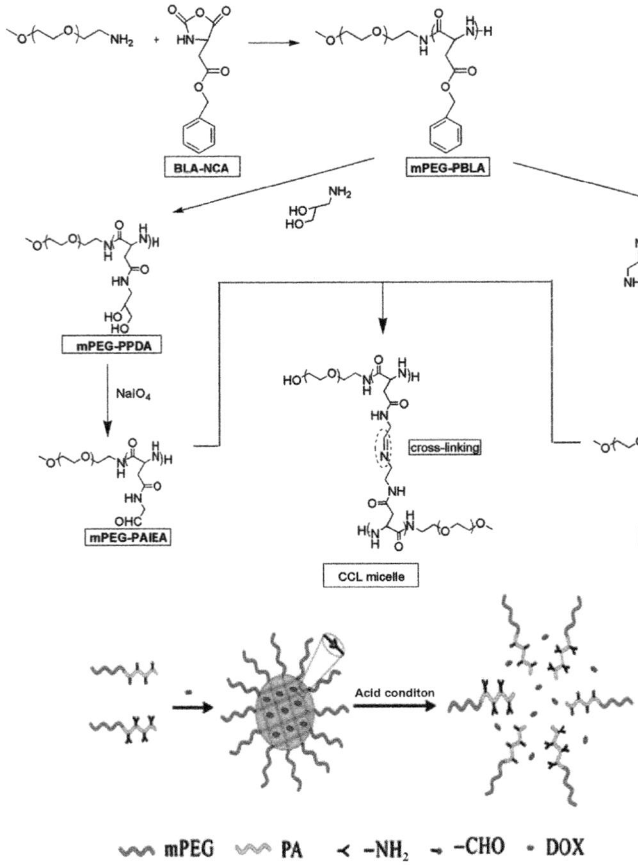

FIGURE 5.2

Synthesis of mPEG-PAmEA, mPEG-PAIEA and schematic image of core-crosslinked mPEG-PA micelles (Wang et al., 2014a).

5.2.2 pH-REVERSIBLE HYDRAZONE CROSSLINKED MICELLES

Compared to reversible imine linkages, hydrazone bonds (as the most famous pH-reversible cross-linker agents) are another promising type of dynamic covalent bond and possess several characteristics that make them powerful candidates for use in biomedicine. The acid-sensitive hydrazone bonds can be formed from condensation of an aldehyde or ketone with a hydrazine derivative under very mild conditions, including aqueous media, that is more particular attractive as a method due to their thermodynamic stability, extensive structural variability, and applications in both drug delivery and materials science (Ono et al., 2005, 2007; Nguyen and Huc, 2003). At neutral or basic conditions ($pH > 7.0$), the hydrazone bond is relatively stable, while at mildly acidic conditions (e.g., endosomal and lysosomal pH ~ 5–6), it can be rapidly dissociated or hydrolyzed. This particular feature has been widely employed to construct new drug delivery systems. The first example of utilizing hydrazone linkages in drug–vehicle conjugates was reported by Hurwitz et al. in 1980, who demonstrated the activity of daunorubicin linked to polypeptide (polyglutamic acid) or dextran polymers (West and Otto, 2005). The hydrazone bonds were prepared by carbodiimide coupling of carboxylic acids to hydrazine or via reacting carboxylic ester precursors with hydrazine hydrate. In vivo studies on the hydrazone conjugates demonstrated considerable cytotoxic activity, while analogues containing noncleavable linkers were overall inactive. Kataoka et al. prepared intracellular pH-reversible polymeric micelles by grafting DOX to a PEG-polyaspartate block copolymer, which were synthesized by utilizing a hydrophobic block containing pH-triggered hydrazone linkers (Bae et al., 2003). It can stably retain drugs at basic conditions (pH 7.4) and release them in the cell nuclei at mildly acidic pH (e.g., endosomes and lysosomes) after cellular uptake.

Wang et al. (2014b) prepared well-defined hydrazide-containing copolymer scaffolds poly(poly(ethylene glycol) methacrylate-*co*-methacryoyl hydrazide) P(PEGMA-*co*-MAH) for bioconjuagtion and were synthesized through RAFT polymerization. The bioconjugation of the copolymer with glucose and the aldehyde-functionalized biotin group generated glycoconjugated copolymer and biotinylated copolymer via a hydrazone bond for protein recognition. Subsequently, the core-crosslinked nanoparticles were generated through reversible acylhydrazone bonds by the reaction of hydrazide-containing copolymers and TDA (Fig. 5.3). The crosslinked nanoparticles demonstrated reversible pH-dependent formation and disintegration and were further adorned via the reaction of the residual hydrazide functions with an aldehyde-modified biotin derivative and subsequently with fluorescein isothiocyanate. In vitro cytotoxicity studies confirmed that the CCL nanoparticles showed good biocompatibility against HeLa cells. These nanoparticles can encapsulate with hydrophobic dye Nile red as a model drug, which demonstrate pH-triggered release behavior around the acidic tumoral environment, implying that through hydrazone linkages these nanoparticles may have various potential applications in biological applications and drug delivery systems.

5.2.3 pH-REVERSIBLE ACETAL/KETAL CROSSLINKED MICELLES

Among different reactions for the chemical derivation approach applied so far, reactions of diols with aldehydes or ketones to generate corresponding cyclic acetals or ketals have been most commonly employed because: (1) various compounds utilized as drugs and fragrances contain these reactive functional groups; and (2) acetal or ketal are stable at basic environment, while at low pH they can be rapidly dissociated or hydrolyzed and lead to controlled release of the encapsulated chemotherapeutic drugs by regulating the pH of the environments.

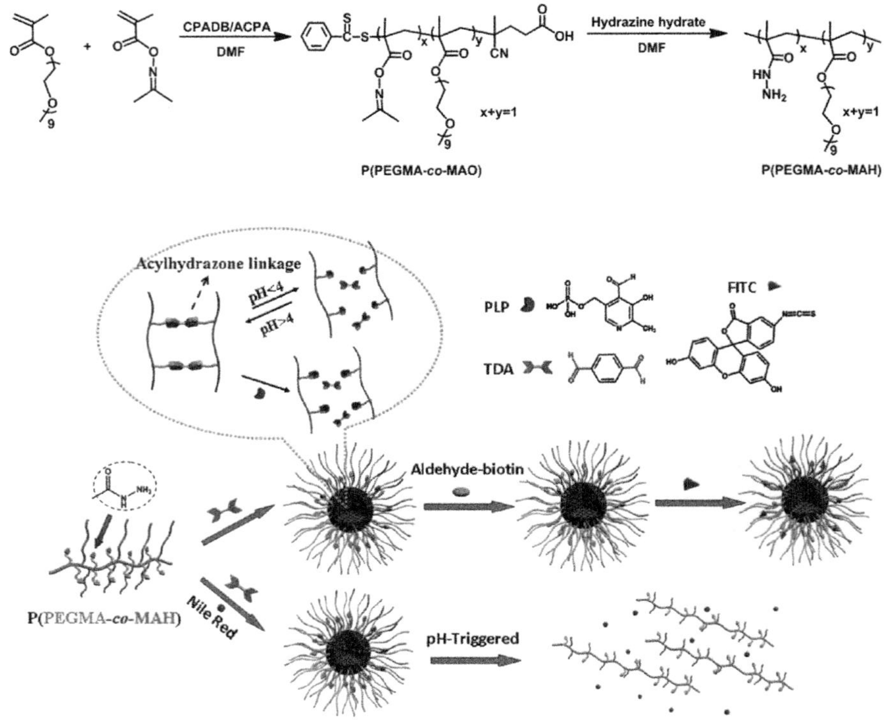

FIGURE 5.3

Outline for the conjugation of biomolecules to P(PEGMA-*co*-MAH) copolymer scaffold. Schematic representation of core-cross-linked polymeric nanoparticles via reversible acylhydrazone linkages and subsequent functionalizations with biotin and FITC (Wang et al., 2014b).

In addition, acetal/ketal-based delivery systems have various aspects such as hydrophobicity, quick degradation upon endo/lysosomal pH conditions, lack of acidic degradation products, and versatile preparation and applications. Frechet et al. have used acetals as acid-labile linkers in the design of pH-responsive polymers, nanoparticles, micelles, and hydrogels (Gillies et al., 2004; Gillies and Frechet, 2005; Jain et al., 2007). In particular, they found that trimethoxybenzylidene acetals associated either to the side chain of poly(aspartic acid) (PAA) segment of PEG-PAA block copolymers or to the dendron surface of PEG dendritic block copolymers indicated unique sensitivity in mildly acidic pH. The hydrolysis of acetals/ketals effectively disassembled the polymeric micelles and rapidly released encapsulated Nile red or DOX in mildly acidic pH (Gillies et al., 2004).

A type of acid-labile crosslinked micelle system was also described using acetal bisacrylate as a degradable crosslinker by Bulmus et al. (2007), which has been utilized as a drug vehicle controlling the pH-triggered release of an antitumor drug.

A robust, biocompatible polymer micelle containing pH-hydrolyzable shell crosslinks using a crosslinking agent bearing ketal groups for efficient intracellular delivery of DOX in response to endosomal pH has been developed (Lee et al., 2011). The ketal crosslinked micelles (KCLMs) are prepared from a model triblock copolymer, poly(ethylene glycol)-*b*-poly(L-aspartic acid)-*b*-poly(L-phenylalanine) (PEG-PAsp-PPhe), which SA to constitute core-shell-corona polymer micelles with three well-defined, separate domains of the PEG outer corona, the PAsp middle shell, and the PPhe inner core. Subsequently, these micellar nanoparticles are crosslinked by the reaction of the bifunctional ketal crosslinkers with Asp groups within the middle shell, yielding a crosslinked shell without considerably affecting the original micelle size and the spherical morphology (Fig. 5.4). Compared to normal physiological pH (7.4), ketal crosslinked micelles demonstrated the quick entrapped DOX release at intracellular endosomal compartments. Confocal laser scanning microscopy (CLSM) studies revealed that KCLMs were taken up by the cancer cells through endocytosis and transferred into endosomes to hydrolyze the crosslinks by lowered pH and, thus, facilitate the DOX release to inhibit proliferation of cancer cells.

In addition, a facile strategy for the development of folate-conjugated core-crosslinked polyphosphoester micelles with acid-cleavable acetal groups (ACCL-FA) has been utilized (Hu et al., 2015), where these ACCL-FA micelles are prepared by crosslinking of an acid-cleavable agent azide-modified tetraethylene glycol (N_3-*a*-TEG-*a*-N_3) and folate-conjugated amphiphilic block copolymers comprising biodegradable polyphosphoesters (PBYP-*b*-PEEP or PBYP-*b*-PEEP-FA) via a Cu(I)-catalyzed azide-alkyne cycloaddition (CuAAC) "Click" reaction (Fig. 5.5). These polyphosphoester-based crosslinked micelles could encapsulate DOX and also exhibited good biodegradability, low cytotoxicity, and much smaller size than those of the corresponding non-crosslinked micelles (NCLM). The in vitro release studies confirmed that the DOX-loaded ACCL micelles exhibited enhanced stability with minimal drug release in extracellular environments, while they could be quick dissociated in the presence of acid or phosphodiesterase I due to the acid-cleavable and enzymatic hydrolysis of polymeric vehicle, and then released DOX in the environment of cancer cells.

5.2.4 pH-REVERSIBLE INTERPOLYELECTROLYTE CROSSLINKED MICELLES

Inter-polyelectrolyte complexation of oppositely charged polymers is another approach to generate pH-reversible crosslinked micelles, which has been utilized in DNA condensation (Rungsardthong et al., 2001). In addition, Kabanov et al. described the formation of polymeric micelles by adding either a cationic

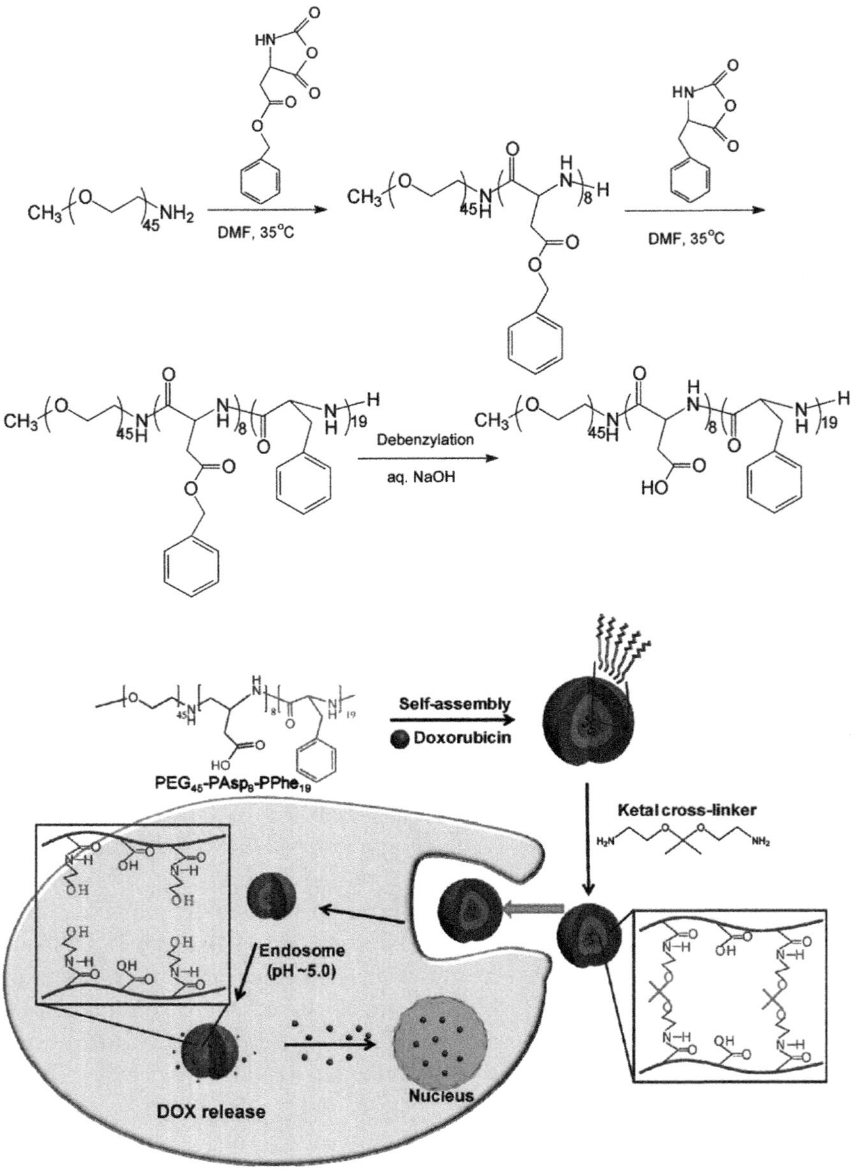

FIGURE 5.4

Synthetic route for the PEG45-PAsp8-PPhe19 triblock copolymer. Illustration of shell cross-linking of DOX-loaded polymer micelles with a pH-labile ketal cross-linker and intracellular release of DOX triggered by endosomal pH (Lee et al., 2011).

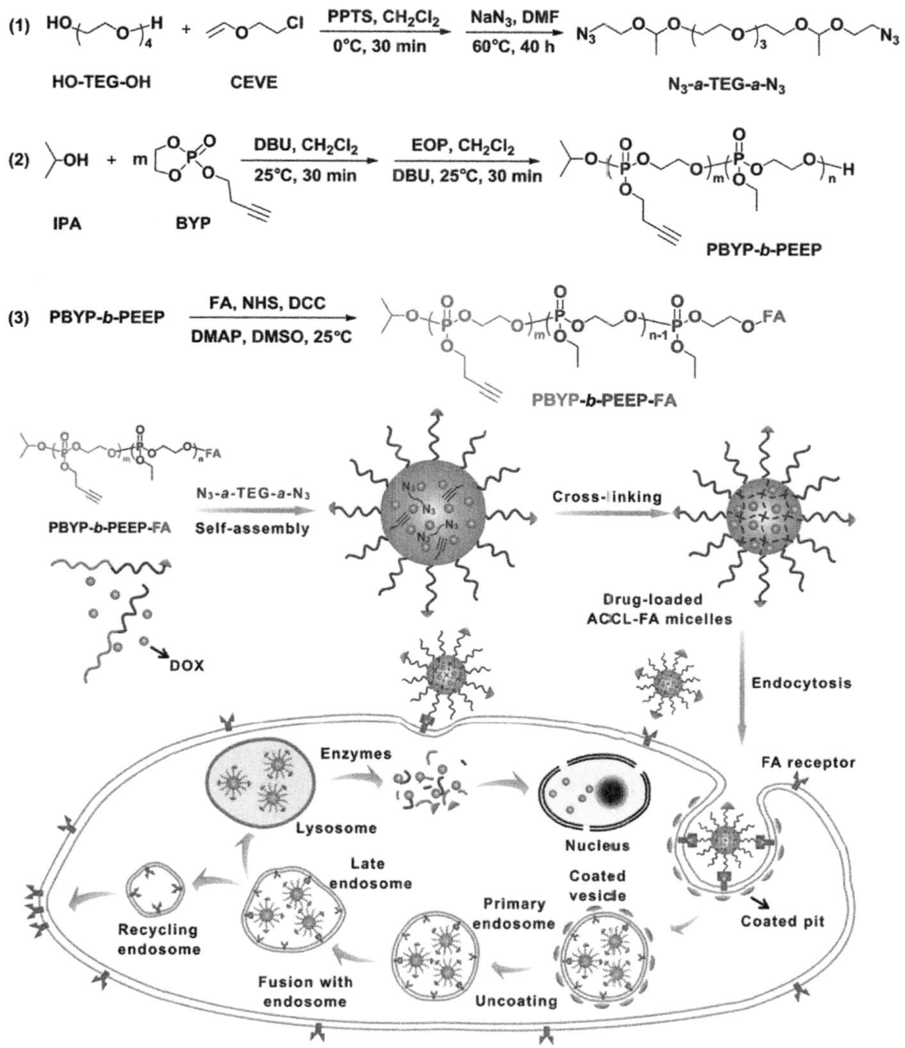

FIGURE 5.5

Synthesis routes of ACCL-FA micelles via a combination of ROP and CuAAC "Click" reaction. Illustration of folate-conjugated and core cross-linked polyphosphoester micelles for efficient intracellular release of hydrophobic anticancer drugs triggered by enzymes and acidic microenvironment inside the tumor tissue (Hu et al., 2015).

homopolyelectrolyte (quaternized poly(4-vinylpyridine)) or a cationic surfactant (hexadecyltrimethylammonium bromide) to an aqueous media of a molecularly dissolved neutral–anionic AB diblock copolymer, for instance, poly(ethylene oxide-block-sodium methacrylate) (Weaver et al., 2004).

In principle, polyelectrolyte complexation provides many advantages over conventional small-molecule crosslinking reactions: (1) lack of by-products, thus, purification is straightforward; (2) polyelectrolytes are usually regarded as possess minimal toxicity; (3) physical crosslinking by means that there is no possibility of undesired chemical modification of the active hydrophobic agent to be encapsulated; (4) the use of aqueous solution; and (5) ionic crosslinking that lead to the readily reversibility of the nanoparticles in the presence of sufficient added salts.

Both temperature- and pH-responsive systems capable of fabricating these inter-polyelectrolyte complex (IPEC) micelles have been reported by McCormick's group, who showed their reversibility by the introduction of electrolytes (Lokitz et al., 2007). They also designed the synthesis of pH-reversible shell-crosslinked micelles based on a doubly hydrophilic polyanionic diblock copolymer and inter-polyelectrolyte complex P(AMPS-*b*-AAL) (Kellum et al., 2010). These diblock copolymers are prepared through a RAFT polymerization technique and are composed of poly(sodium 2-acrylamido-2-methyl-1-propanesulfonate) segment acted as an anionically charged corona, and pH-responsive poly (*N*-acryloyl-L-alanine) (PAAL) blocks have an insoluble protonated core with varying block lengths that undergo a reversible unimer-to-micelle transition with changes in pH. Reversible SCL micelles were obtained through interpolyelectrolyte complicated by the addition of the cationic homopolymer of poly(*N*,*N*-dimethylaminoethyl methacrylate) (PDMAEMA) ($pKa = 7.3$) as a crosslinker. Subsequently, these crosslinked micelles swell by deprotonation of the PAAL segment, but remain stable. They showed reversibility of the crosslinking which suggests the possible future application of IPEC micelles as drug delivery carriers.

Polysaccharides-based polyelectrolyte complex nanoparticles (PCNs) have been designed (Guo et al., 2013) as a novel drug delivery system and were successfully synthesized by the self-assembly nanoparticles deoxycholic acid hydrophobically modified chitosan (CS-DCA) core to generate nanoparticles in water, followed by the surface of positively charged CS-DCA nanoparticles being rapidly coated with the polyanionic sodium alginate (ALG) shell to enhance the stability structural of PCNs (Fig. 5.6). The model drug, bovine serum albumin, could be readily incorporated into the PCNs with EE up to 69% and keep a sustained release manner for one month without burst release when exposed to PBS (pH 7.4) at 37°C.

5.2.5 pH/CIS-DIOL-REVERSIBLE CROSSLINKED MICELLES

The reaction of boronic acids with diols to boronate ester is fully reversible and display quick dual sensitivity to external pH value and competing diols, which are used as structural blocks to design carbohydrate sensors, drug delivery systems, nanoreactors, and self-healing materials (Qin et al., 2005; Li et al., 2012). Between deals, cuticles are considered as a superior reactant for the fabrication process of boronate esters. Based on this principle, synthesis of a novel type of dual pH- and boronate-responsive SCL micelles (BCMs) with particle size of

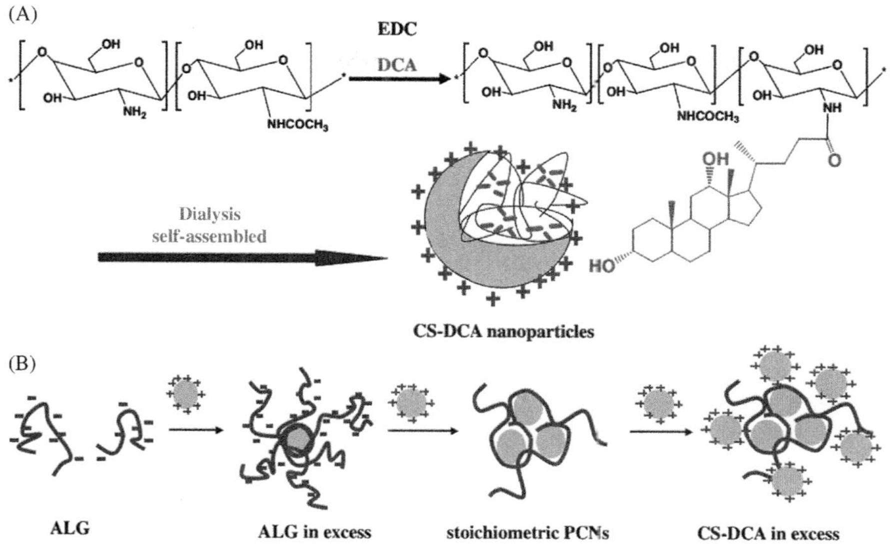

FIGURE 5.6

Synthesis route of chitosan-deoxycholic acid (CS-DCA) and self-assembly of polyelectrolyte complex nanoparticles (PCNs) (Guo et al., 2013).

20–30 nm has recently been developed for anticancer drug delivery (Fig. 5.7) (Li et al., 2012). These reversible boronate crosslinked micelles preserve the encapsulated paclitaxel (PTX) at physiological glucose level (2–10 mm) and/or under physiological conditions in blood circulation in vivo. Whereas, rapid release of the payload in response to the acidic pH value of the tumor microenvironment or when mannitol is introduced as an on-demand triggering agent into the solution, indicate that the cleavage of a critical percentage of boronate occurred. In addition, the combination of 100 mM mannitol and pH 5.0 could greatly accelerate the encapsulated PTX release of the BCM4 micelles.

5.3 PHOTO-REVERSIBLE CROSSLINKED MICELLES

Among the different crosslinking procedures, the photo-induced crosslinking method is a good choice to tune the stability of polymeric micelles that demonstrate very interesting advantages, such as no chemical reagents being required and no unwanted by-products being constructed during the procedure. Additionally, the crosslinking density, can be readily controlled by tuning the light wavelength, intensity, and irradiation time. The UV light-triggered photodimerization of cinnamic acids, anthracenes, chalcones, coumarins, and thymines contain double bonds to form cyclobutane rings by intermolecular $[2\pi + 2\pi]$ cycloaddition reactions has been well discussed (Long et al., 2016). Light-triggered crosslinking of

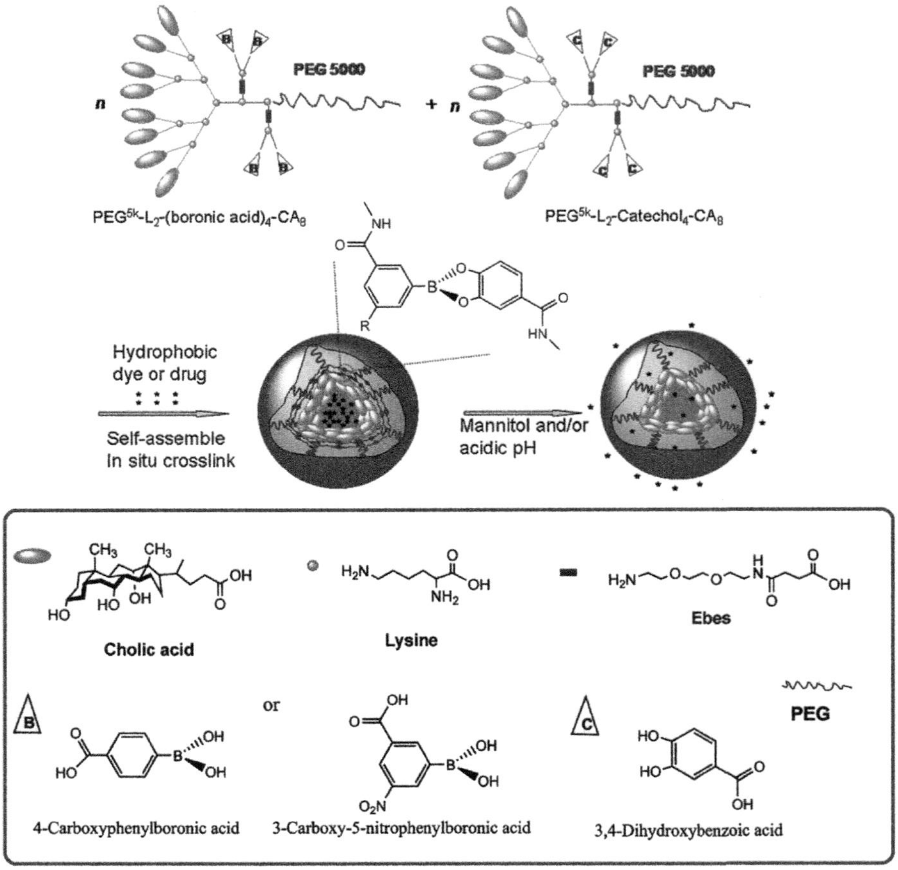

FIGURE 5.7

Schematic representation of the telodendrimer pair [PEG5k-(boronic acid/catechol)4-CA8] and the resulting boronate crosslinked micelles (BCMs) in response to mannitol and/or acidic pH (Li et al., 2012).

micelles was firstly used in the $[2\pi + 2\pi]$ photocycloaddition of cinnamic esters for photo crosslinking by the cinnamic acid–truxillic acid reversible photochemical $[2\pi + 2\pi]$ cycloaddition reaction by Liu et al. (Ding and Liu, 1998).

Coumarin is another interesting candidate of photo-triggered crosslinking groups. It was first known and synthesized in 1868 and has been used extensively in the pharmaceutical industry (Murray et al., 1982). An attractive feature of the use of coumarin groups is that they provide a reversible photo-crosslinking reaction by the photodimerization of coumarin moieties, which has been well-known since the beginning of the 20th century. Under UV light at $\lambda > 365$ nm, two coumarin groups will form stable dimerization through the photochemical $[2\pi + 2\pi]$

cycloaddition reaction and construction of cyclobutane rings. The cyclobutane bridges resulting from dimerization can also be photo-cleaved by exposure to UV light at shorter wavelengths of $\lambda < 260$ nm. Coumarin has been exploited to form reversibly photo-crosslinked nanoparticles, hydrogels, and micelles. For example, in recent years, Zhao et al. introduced coumarin derivatives in novel amphiphilic diblock copolymers and showed that it is possible to reversibly control the crosslinking of micelles by UV irradiation at wavelengths above 310 nm and decrosslink the process upon irradiation at wavelengths below 260 nm (Jiang et al., 2007a).

Core or shell reversibly photo-crosslinked micelles with dual (temperature and pH) responsive behaviors were designed and synthesized by Jin et al. (2010). This system is based on poly(butanedioic acid, 1-[3-[(2-methyl-1-oxo-2-propen-1-yl) oxy]propyl] ester)-*b*-poly(methoxydi(ethylene glycol) methacrylate-*co*-4-methyl-[7-(methacryloyl)oxyethyloxy] coumarin) (PSPMA-*b*-P(DEGMMA-*co*-CMA)) double-hydrophilic block copolymer, which SA into P(DEGMMA-*co*-CMA)-core micelles at pH 10 and 35°C (i.e., above the LCST) and PSPMA-core micelles at pH 3 and 20°C. After the two types of micelles were irradiated with 365 nm UV light, reversibly SCL and CCL micelles were readily generated, while the decrosslinking process of coumarin groups can be accomplished upon exposure to UV light at 254 nm. These results clearly indicate that the crosslinking density can be readily regulated by tuning the irradiation time and the intensity and wavelength of irradiation light. On the other hand, by transforming the hydrophobic core CCL and SCL micelles into hydrophilic through regulating the pH and temperature, reversibly photo-crosslinked block copolymer nanogels were fabricated. Therefore, the introduction of reversibly photo-crosslinkable groups into the double responsive block copolymers could lead to more sophisticated, controllable, and smarter nanovehicles that might have excellent potential in biomedical applications.

The use of surface-initiated atom transfer radical polymerization (ATRP) to prepare SCL reverse micelles (SCRM) via combining block copolymer self-assembly and postfunctionalization has also been developed (Babin et al., 2008). This system is prepared from an amphiphilic diblock copolymer of poly(dimethylaminoethyl methacrylate)$_{54}$-*block*-poly(methyl methacrylate$_{55}$-*random*-coumarin methacrylate$_{19}$) (PDMAEMA$_{54}$-*b*-P(MMA$_{55}$-*co*-CMA$_{19}$)), in which reverse polymer micelles with a dense ionic inner core of PDMAEMA$_{54}$ can be stabilized via crosslinking of the hydrophobic shell of P(MMA$_{55}$-*co*-CMA$_{19}$) using the photoinduced dimerization of the coumarin side groups upon UV irradiation at $\lambda > 310$ nm (Fig. 5.8A). In addition, the halide groups of SCRM initiated the grafting of styrene and PDMAEMA chains onto SCRM via ATRP renders the nanoparticles soluble in water in response to changes in pH and temperature. The cyclobutane bridges can be photo-cleaved upon UV irradiation ($\lambda < 260$ nm), which results in photo-de-crosslinking of the hydrophobic shell and, thus, the disintegration of SCRM (Fig. 5.8B).

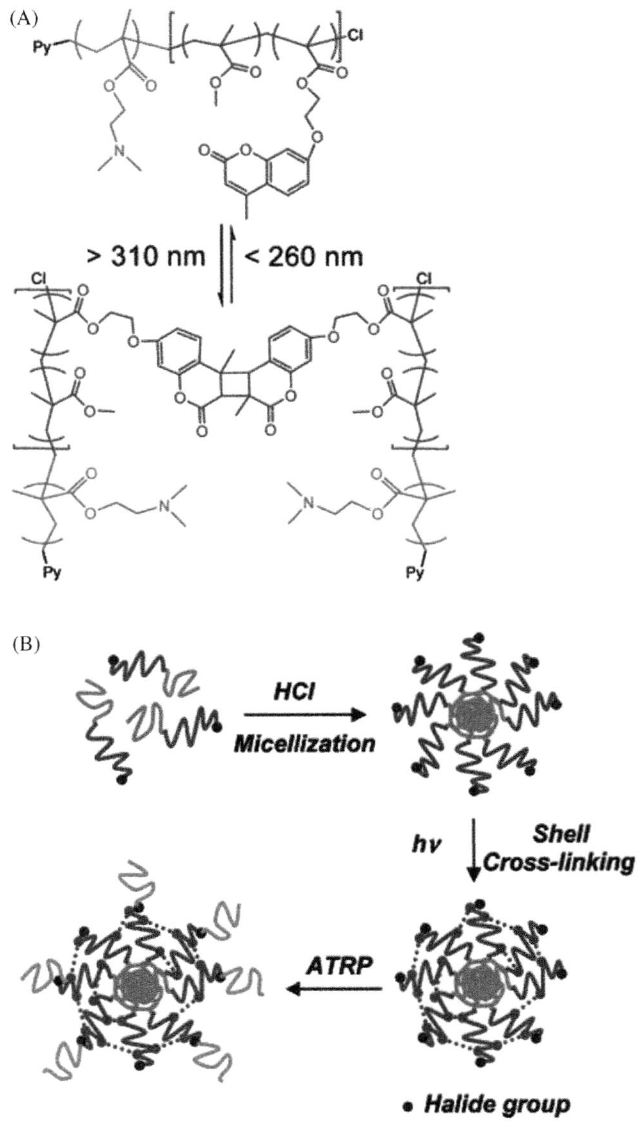

FIGURE 5.8

(A) UV photodimerization and photocleavage of coumarin side groups in PDMAEMA54-b-P(MMA55-co-CMA19) block copolymer. (B) Illustration of the approach used to obtain and "decorate" shell crosslinked reverse micelles (SCRM) formed by the diblock copolymer (Babin et al., 2008).

Moreover, recently, a novel ultraviolet- and pH-responsive micellar system based on polyurethane (PU) with photo-responsive coumarin derivative i.e., 4-methyl-7-((4-oxopentyl)oxy)-2H-chromen-2-one (OMC) and pH-responsive hydrazone (hydra) groups for targeted cancer chemotherapy has been designed (Long et al., 2016). This system is based on PU-hydra–OMC/PU-hydra–OMC–FA copolymer, in which PU formed the backbone of the copolymer and consists of hydrophobic, nontoxic L-lysine ethyl ester diisocyanate, and hydrophilic PEG. Then, OMC groups, as a crosslinker, were attached to the polymer chains via a hydrazone bond and subsequently, for targeted therapy, folic acid (FA), as a model targeting molecule, was introduced into the PU via an amidation reaction (Fig. 5.9). The copolymers could be easily SA into micelles in an aqueous solution and after UV light irradiation at 365 nm, crosslinked micelles were fabricated and, thus, the stability of micelles was improved. The in vitro experiments of DOX-loaded crosslinked micelles against cancer cells (L929 cells and HeLa cells) exhibited an accelerated DOX release due to the partially dissociation at endosomal pH and an improve in cancer cellular uptake due to the existence of FA in crosslinked micelles.

It is remarkable that light not only affords the stability of polymer micellar aggregates by photo-crosslinking; however, it also permits the release of encapsulated guest molecules via light-induced disruption of the micelles, but it turned out to be incomplete and cannot be used to totally disrupt the micellar aggregates. Moreover, none of the photo-crosslinked micelles have found any application so far, for instance, in the field of nanoscale carriers as a potential platform for drug delivery. In addition, the use of these for controlled release in vivo is limited because ultraviolet light radiations are harmful to healthy normal tissues and their penetration depth into the living body is rather hard. Thus, the use of other reversible linkages is more suitable, which will be discussed next.

5.4 REDUCTION-REVERSIBLE CROSSLINKED MICELLES

Given the inimitable features of the redox condition in the human body, reducible crosslinkers have been highly utilized to design an intelligent micellar nanoparticles system. The formation of DS bonds (–S–S–) is one of the first molecules covalent linkage proposed, which is generated as a result of the versatile oxidation two thiol (SH) groups of cysteines or other SH-containing material. DS linkage is thought to provide relative stability under normal physiological conditions such as in the blood stream, but it can quickly be cleaved into free thiols (–SH) in response to the presence of glutathione (GSH) or different reducing agents such as dithiothreitol (DTT) and *b*-mercaprelativelytheretoethanol. In addition, there are other techniques of reducing DS bonds to thiols, such as zinc dust and UV light (Overman et al., 1974; Kolano et al., 2007). Therefore, DS bond formation and cleavage are reversible processes and this matter frequently plays an

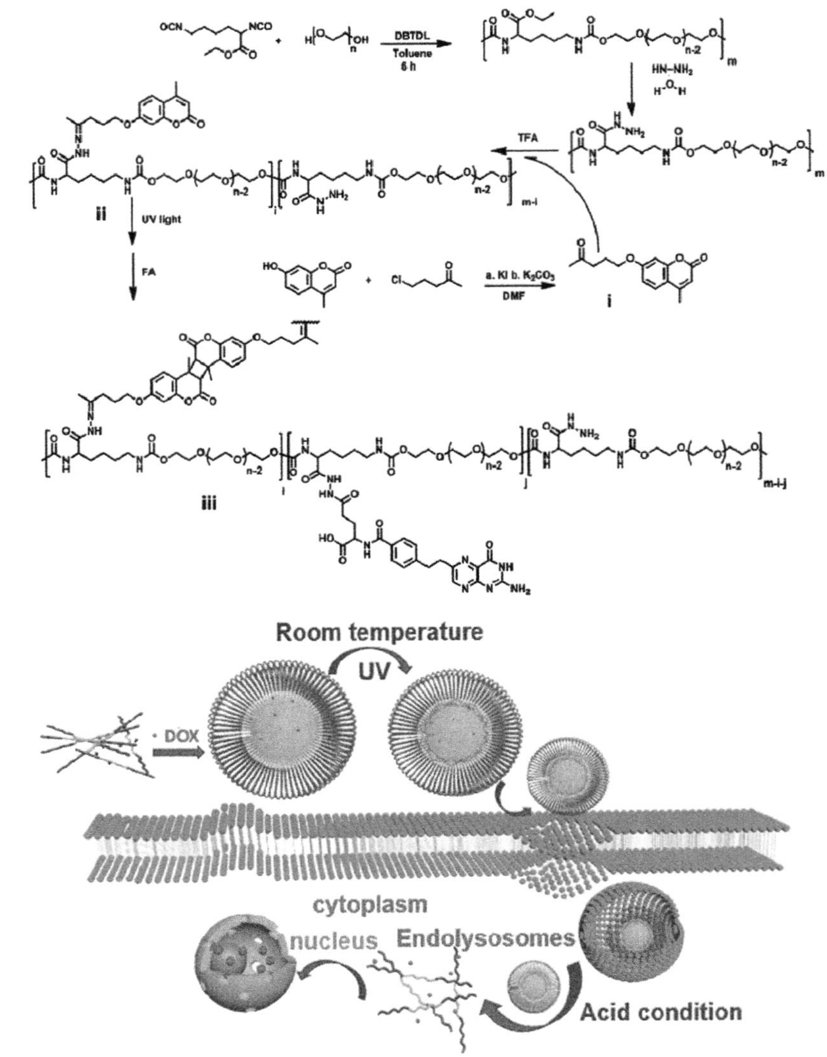

FIGURE 5.9

Synthesis process of OMC (i), PU-hydra–OMC (ii) and PU-hydra–OMC–FA (iii). Illustration of core cross-linked polyurethane micelles for efficient intracellular release of hydrophobic anticancer drugs triggered by acidic microenvironment inside the tumor tissue (Long et al., 2016).

important role in the biological utility of its diverse processes such as enzyme catalysis, protection against oxidative damage, the stabilization of extracellular proteins, and the regulation of biological activity. In principle, the ability to construct and cleave a particular DS bond under appropriate biological conditions depends on four factors: (1) the stability of DS bond; (2) the nature of the oxidant or reductant; (3) the kinetics of the forward and reverse reactions; and (4) the nature and redox state of the environment in which the reaction occurs.

The thiol–DS exchange reaction plays a significant role in the metabolic activity of living cells. Glutathione tripeptide (γ-glutamyl-cysteinyl-glycine; GSH) is an antioxidant regarding its reducing properties and it can trap reactive hydroxyl free radicals, other oxygen-centered free radicals, radical centers on DNA as well as electrophilic and oxidant species. GSH also plays a significant role in the reduction of peroxides (hydrogen and lipid peroxides) and producing glutathione disulfide (GSSG) by assistant of glutathione peroxidase. In turn, GSSG is reduced by dihydronicotinamide-adenine dinucleotide phosphate (NADPH) and glutathione disulfide reductase to 2 GSHs.

It is well-known that in body fluids and normal extracellular matrices, the concentration of GSH is extremely low (approximately 2–20 μM). Meanwhile, in the cytoplasm, the concentration of GSH is significantly higher (0.5–10 mM). It should further be noted that a high reducing potential has also been found in the endosomal compartment in which the redox potential is modulated by a specific reducing enzyme *g*-interferoninducible lysosomal thiol reductase in the copresence of a reducing agent such as cysteine (but not GSH). Moreover, the cytosolic concentration of GSH in some particular tumor tissues has been observed to be at least four times higher than normal tissues (Kuppusamy et al., 2002; Black and Wolf, 1991).

Thus, because of the dramatic variation in the GSH level and its reversibility and relative stability in plasma, stimuli-responsive polymers with DS linkages have attracted a great amount of attention. The reductive degradation of DS crosslinked polymers also have been utilized to design vehicles that rapidly react in a reducing environment in diverse biomedical applications, including gene delivery, diagnostic imaging, and intracellular drug delivery due to the existence of numerous differences in the redox potential between the mildly oxidizing extracellular environment and the reducing of the intracellular environment.

To date, researchers have also employed the reduction sensitive crosslinked intelligent micelles to deliver DOX, PTX, vincristine, methotrexate (MTX), DNA, and small interfering RNA (siRNA) to the target site following the pioneer attempts of the Kataoka group (Sun et al., 2009; Li et al., 2014).

In addition, in principle, CCLM/SCL micelles containing GSH-cleavable DS bonds have been extensively utilized to encapsulate different drugs for cancer chemotherapy and would be comparatively stable in the extracellular environment and may prevent the release of inserted drugs (e.g., during circulation). However, such promising drug delivery vehicles with a bioreductive DS crosslinked shell could readily show controlled release of the incorporated drugs intracellularly at the target site through reversible cleavage of the DS bonds under a reductive environment, e.g., in the cytosol of cells.

In this chapter, considering the differences of crosslinking strategies, we regrouped the reversible CCLM/SCL micelles containing DS linkages into three main types: (1) reduction-reversible CLMs with bifunctional DS-containing crosslinkers; (2) reduction-reversible CLMs with self-crosslinking technique; and (3) reduction-reversible CLMs with lipoic acid (LA).

5.4.1 REDUCTION-REVERSIBLE CLMS WITH BIFUNCTIONAL DISULFIDE-CONTAINING CROSSLINKERS

Among the different approaches for core/shell crosslinking of micellar nanoparticles, bifunctional redox-sensitive crosslinkers are most generally used, such as 3,3′-dithiodipropionic acid (DTA) and its derivatives, bis(acryloyloxyethyl) DS (BAEDS), and cystamine.

5.4.1.1 Reduction-reversible CLMs with 3,3′-dithiodipropionic acid and its derivatives

DTA and its derivatives, including dimethyl 3,3′-dithiobispropionimidate (DTBP), dithiolbis(propanoic dihydrazide) (DTP), and 3,3′-dithiobis(sulfosuccinimidylpropionate) (DTSSP) are crosslinkers containing a bioreductive DS bond, which can be cleaved in the presence of a redox reagent and have been utilized to design reversible crosslinked micelles.

For example, a series of redox-responsive SCL micelles based on amphiphilic copolymers poly(γ-benzyl-L-glutamate)-*block*-dextran (PBLG-*b*-dextran) with good biocompatibility have been designed (Zhang et al., 2013b). These block copolymers are synthesized using click chemistry and, followed by the SCL redox-responsive micelles, are obtained by crosslinking of dextran with DTA (Fig. 5.10). The hydrodynamic radius of DS SCL micelles (SCMs) increases gently in PBS because of the dissociation of the DS bond in the micellar shell caused by the presence of a reductive environment (GSH). The in vitro DOX release profiles from the micelles revealed that up to about 85% of loaded DOX can be rapidly released under intracellular-mimicking reductive environments (0.010 M GSH), while only a small amount of loaded DOX was released at 59.5 hours without GSH. MTT and flow cytometry approaches revealed that DOX-loaded SCMs exhibit a faster drug release behavior against GSH-OEt pretreated HeLa cells than that in the BSO pretreated and unpretreated cells.

In addition, Cao et al. (2016) prepared a facile drug release system of prednisolone 21-acetate (PA) based on dual responsive (pH- and reduction-responsive) CCL polymeric micelles with click chemistry. The block copolymers of poly(ethylene oxide)-*b*-poly(glycidyl methacrylate) (PEO-*b*-PGMA) comprise a PEO hydrophilic shell and a PGMA hydrophobic core, were prepared by RAFT polymerization of glycidyl methacrylate using a PEO-based macroinitiator, and the PGMA segments were then functionalized with azido groups by the epoxide ring-opening process. Afterwards, the Huisgens cycloaddition reaction simultaneously took place between azide groups of the PGMA-N_3 core and the alkyne-functionalized hydrazone-containing PA to form covalently entrapped PA and

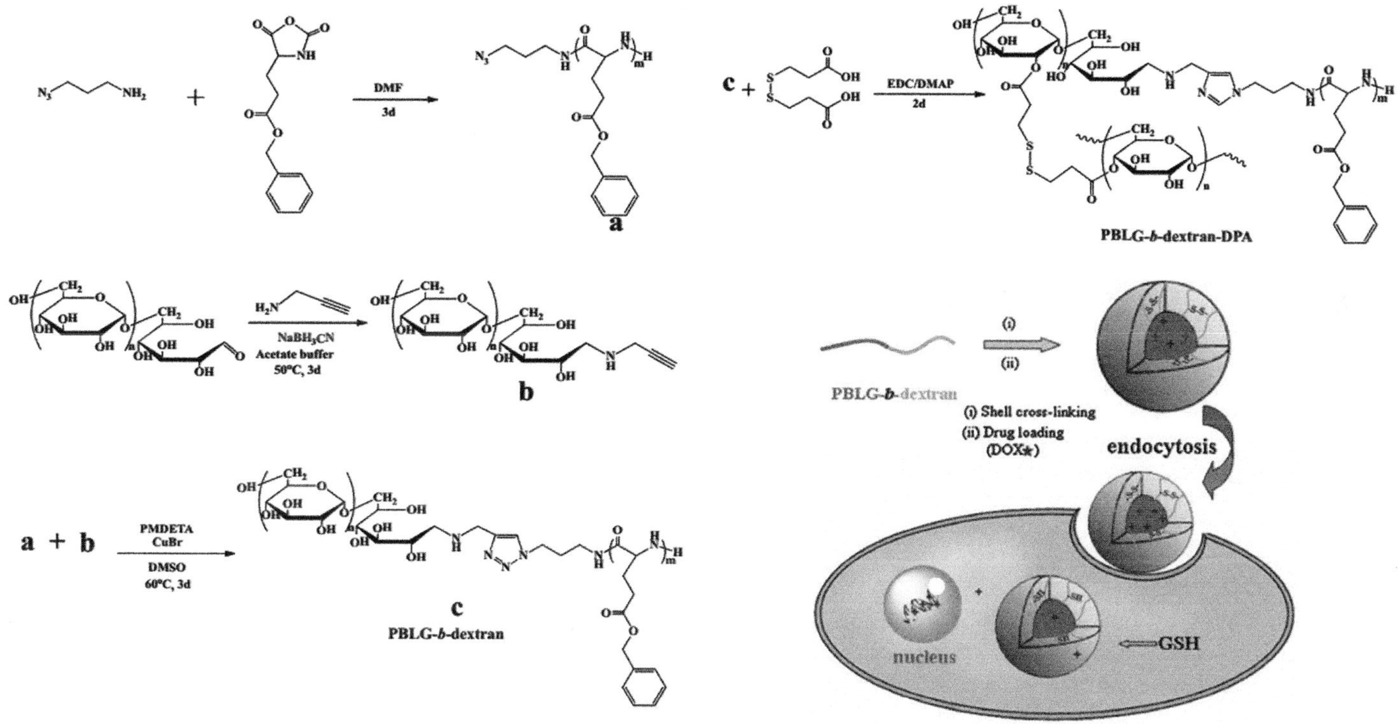

FIGURE 5.10

Synthetic routes of PBLG-b-dextran-DAP copolymers. Schematic illustration of the formation of shell crosslinked micelles and intracellular drug release (Zhang et al., 2013b).

dialkyne crosslinkers of dipropargyl adipate and/or dipropargyl 3,3′-dithiodipropionate that are synthesized from the reaction between propargyl alcohol and adipic acid and/or DTA to achieve CCL micelles in the presence of $CuSO_4{\cdot}5H_2O$ and sodium ascorbate (Fig. 5.11). The CCL micelles exhibited a high PA loading efficiency up to 83% and the excellent structural stability compared to physically loaded CCL micelles under physiological condition, while the de-crosslinking

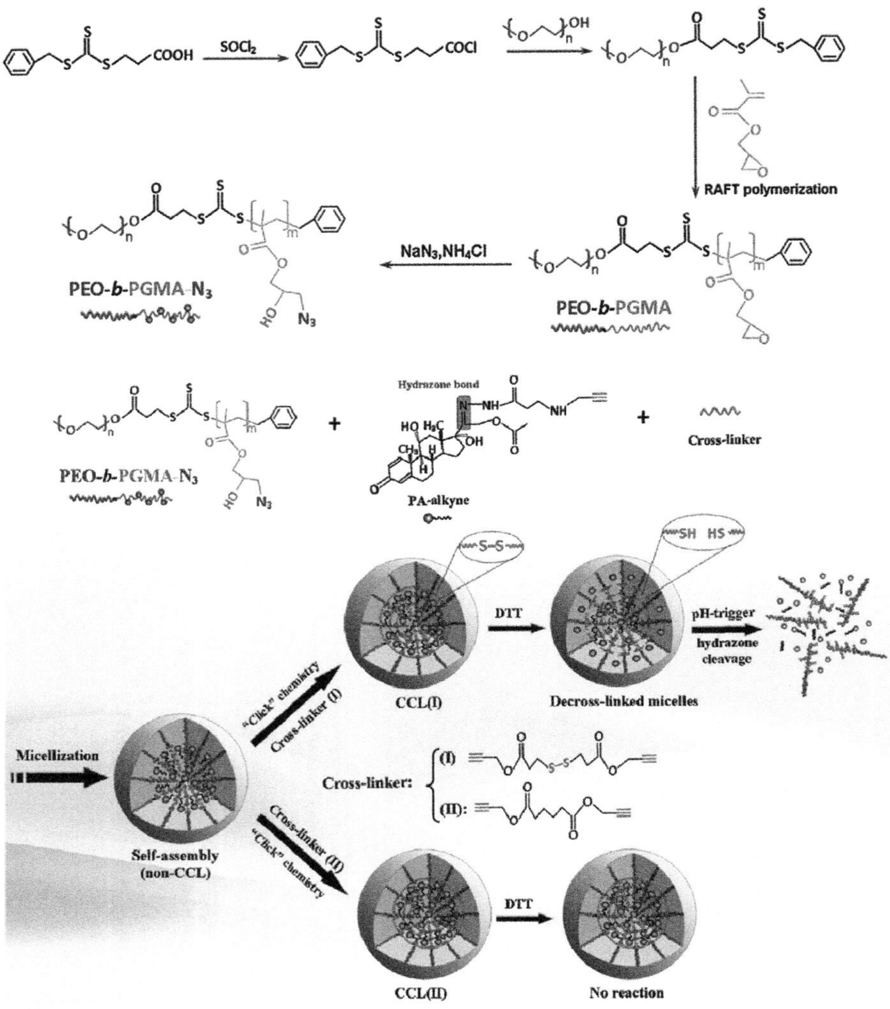

FIGURE 5.11

Synthesis of the PEO-b-PGMA-N3 block copolymer by RAFT polymerization followed by azidation process. CCL micelles of PEO-b-PGMA-N3 block copolymers, followed by degradation and drug release in the presence of DTT (Cao et al., 2016).

process via the cleavage of DS groups took place quickly in a reducing environment such as DTT. Hydrolysis of the hydrazone groups in loaded PA CCL micelles showed burst release behavior at pH 5 and in the presence of 10 mM DTT.

In another example, a series of SCL micelles based on DS-containing triblock copolymers, poly(ethylene glycol)-*b*-poly(L-lysine)-*b*-poly(rac-leucine) (PEG-SS-PLys-PLeu), have been synthesized, in which the PEG block has been attached to the PLys block through a DS linkage to achieve the redox-sensitive nanostructure (Wang et al., 2013). In an aqueous solution, the copolymers could self-assemble to generate core–shell-corona micelles with the PEG corona, PLys shell, and PLeu core. The PLys middle shell could be crosslinked using DTA to improve structural stability. In the presence of GSH, the micelles underwent destruction of the crosslinked shell with detachment of the PEG corona due to the cleavage of DS bonds, followed by the collapse of the micelles (Fig. 5.12). Camptothecin (CPT) as a hydrophobic model drug was loaded in the SCL micelles. The in vitro studies revealed that the CPT-loaded SCL micelles exhibited reduced drug loss in extracellular environments and accelerated drug release at the cytoplasmic GSH level, leading to improved growth inhibition of cancer cells.

Regarding the use of DTSSP as a DS-containing crosslinker to generate reversible crosslinked micelles, a biocompatible core–shell–corona polypeptide micellar system was designed (Koo et al., 2008). Amphiphilic ABC triblock copolymer of poly(ethylene glycol)-*b*-poly(L-lysine)-*b*-poly(L-phenylalanine) (PEG-*b*-PLys-*b*-PPha), SA in water to form micelles with three different functional domains composed of the PPha segments that formed the hydrophobic inner core for drug loading, the PLys middle shell containing reactive primary amines for DS crosslinking, and the hydrophilic PEG outer corona to prolong circulation, which can preferentially release entrapped anticancer drugs at the cytoplasm in response to an intracellular glutathione level. Crosslinking of PLys middle shells are performed using DTSSP (Fig. 5.13). The in vitro GSH-mediated drug release studies confirmed that shell crosslinking inhibited the release of entrapped MTX, an antineoplastic agent for osteosarcoma and lung and breast cancers, compared to non-crosslinked micelles (NCLM). The cytotoxicity experiments of SCL micelles containing MTX showed a clear relevance with the intracellular GSH levels in A549 cells. The in vivo studies of the DTX-loaded core–shell–corona PEG-*b*-PLys-*b*-PPhe micelles with DS shell cross-links (DTX-SSCLM) in MDAMB-231 tumor-bearing mice showed enhanced tumor accumulation because of the prolonged stable circulation in the blood stream via the PEG outer corona (minimized uptake by RES) due to the stable DS crosslinks, the EPR effect, the facilitated release of the drug in target cells by breaking DS bonds, and more effective antitumor therapeutic efficacy in comparison with the DTX-loaded non-crosslinked micelles (DTX-NCLM).

Dimethyl DTBP is a bifunctional, reversibly cleavable crosslinker that contains an amine-reactive imidoester at each end of an eight-atom DS-containing spacer arm and have been exploited previously to study the spatial arrangement

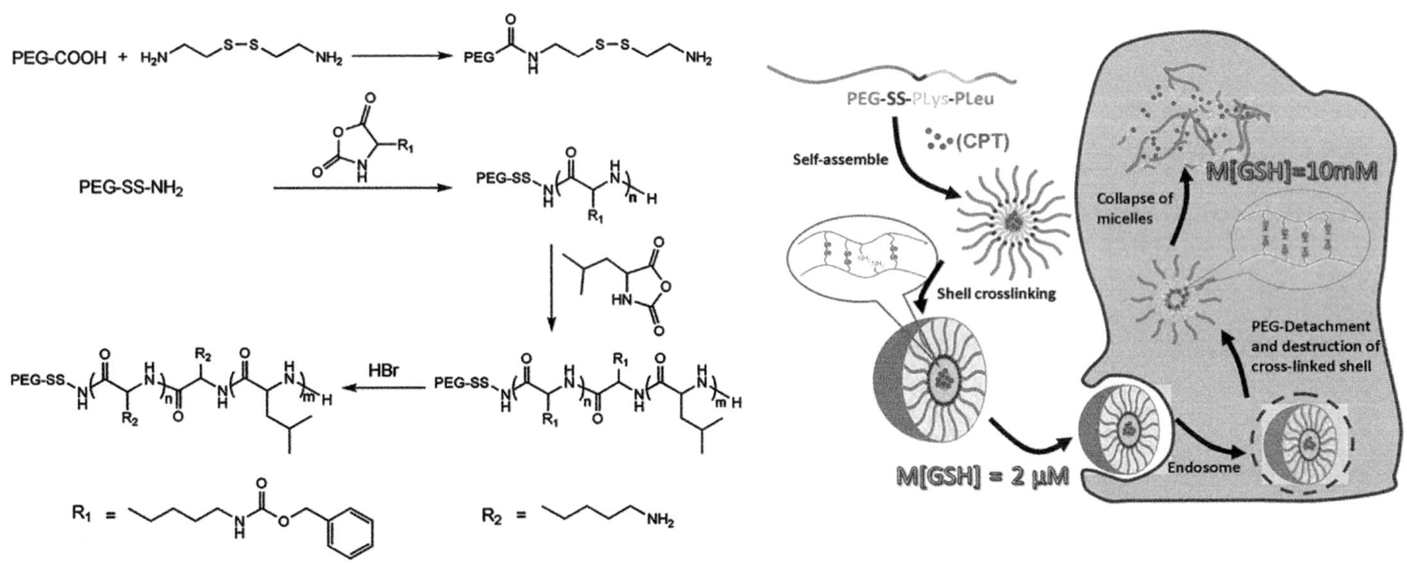

FIGURE 5.12

Synthesis of the PEG-SS-PLys-PLeu copolymer. Schematic illustration of the formation of the PEG-SS-PLys-PLeu shell cross-linked (SCL) micelle and intracellular drug release (Wang et al., 2013).

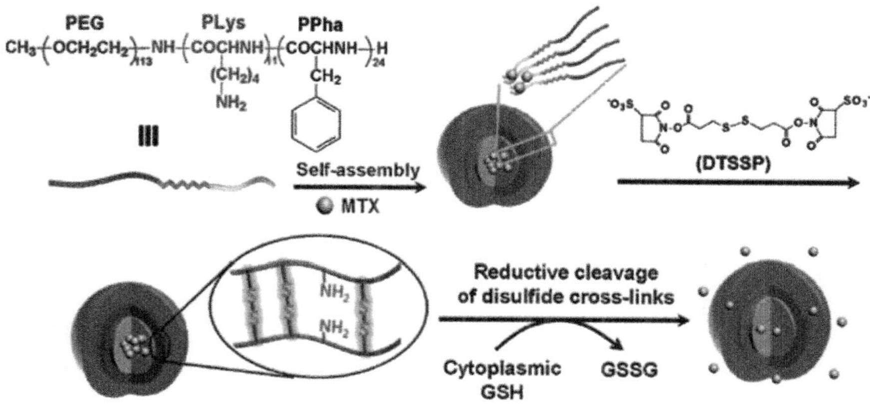

FIGURE 5.13

Illustration of shell cross-linking in drug-loaded polymer micelles and facilitated drug release in response to cellular GSH (Koo et al., 2008).

of proteins in erythrocyte membranes, nucleosomes, and ribosomes (Charulatha and Rajaram, 2001).

In this respect, reversible SCL micelles based on a pH-responsive ABC triblock copolymer, α-methoxy poly(ethylene oxide)-*b*-poly(*N*-(3-aminopropyl) methacrylamide)-*b*-poly(2-(diisopropylamino)ethyl methacrylate) (mPEO-PAPMA-PDPAEMA) have been prepared using a simple one-pot procedure through aRAFT polymerization undergoes pH-induced micellization in water (Xu et al., 2009). Above pH 6.0, this triblock copolymer SA into micelles composed of a pH-responsive core (PDPAEMA), an amine- containing shell (PAPMA) as well as a permanently hydrophilic corona (MPEO). Crosslinking of shell micelles are obtained at room temperature in water by mixing the triblock copolymers and DTBP. These SCL micelles are readily cleaved through a thiol exchange reaction of the resulting DS crosslinks and the reducing agent, DTT. The re-formation and re-crosslinking of these DTBP crosslinked micelles is accomplished simply by increasing the solution pH and oxidation of thiol functionality in the shell segment by exposure to air.

Another DTA derivative, DTP, a water-soluble and bifunctional DS-containing linkage which has been used to design pH- and thiol-sensitive SCL micelles containing aldehyde functionalized in the hydrophilic block using an amphiphilic diblock copolymer prepared from poly(ε-caprolactone) (PCL) and poly(oligo(ethylene glycol) monomethyl ether methacrylate) (POEGMA)/poly (p-(methacryloxyethoxy)benzaldehyde) (PMAEBA) (PCL-*b*-P(OEGMA-*co*-MAEBA)) (Hu et al., 2013a). The block copolymer is synthesized through ring-opening polymerization (ROP) of ε-caprolactone (CL) and ATRP of oligo(ethylene glycol) monomethyl ether methacrylate (OEGMA) and *p*-(methacryloxyethoxy)benzaldehyde (MAEBA) co-monomers (Fig. 5.14). Spherical micelles were prepared from the self-assembly

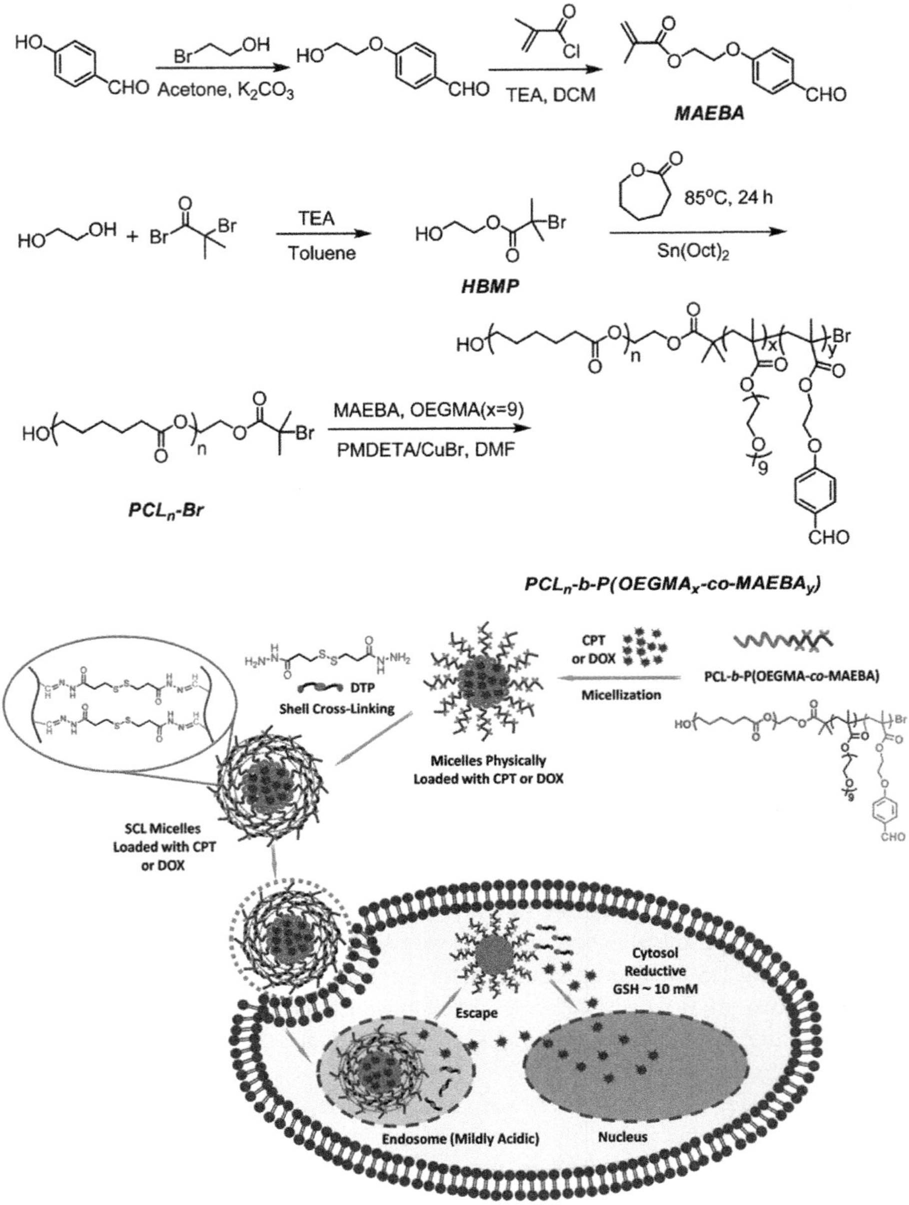

FIGURE 5.14

Synthetic routes employed for the preparation of p-(methacryloxyethoxy)benzaldehyde (MAEBA) monomer and well-defined PCL-b-P(OEGMA-co-MAEBA) diblock copolymer. Schematic illustration for the fabrication of shell cross-linked (SCL) micelles of PCL-b-P (OEGMA-co-MAEBA) diblock copolymers physically loaded with camptothecin (CPT) or doxorubicin (DOX) within hydrophobic PCL cores (Hu et al., 2013a).

of this diblock copolymer in an aqueous solution composed of hydrophobic PCL cores and hydrophilic P(OEGMA-*co*-MAEBA) coronas covalently incorporated with aldehyde groups. The subsequent reversible SCL micelles were obtained at pH 6.2 by the addition of the dihydrazide crosslinker DTP which can react with pendent aldehyde groups to produce acylhydrazone linkages within the hydrophilic shell. The obtained dynamic covalent SCL micelles exhibit thiol and pH dual responsive features, which have been utilized as nanocarriers for controlled release of CPT drug into cancer cells leading to prominent cell nucleic drug agglomeration and effective cell apoptosis. Compared to CPT-loaded non-crosslinked micelles, CPT-loaded SCL micelles can efficiently minimize drug leakage under normal physiological conditions, whereas accelerated drug release can be achieved under mildly acidic pH and/or thiol-rich reductive microenvironments. Cell cytotoxicity studies confirmed that the blank SCL micelles were almost nontoxic, while CPT-loaded SCL micelles can efficiently deliver CPT into cancer cells and lead to prominent cell nucleic drug agglomeration at an extended incubation duration.

They also prepared (Hu et al., 2013b) FA-functionalized dually responsive SCL miclles based on two types of amphiphilic diblock copolymers, consisting of P(CL-*g*-CPT)-*b*-P(OEGMA-*co*-MAEBA) amphiphilic diblock copolymer conjugated with photocaged CPT prodrug in hydrophobic inner core and PCL-*b*-P(OEGMA-*co*-MAEBA-*co*-FA) amphiphilic diblock copolymer functionalized with FA moieties and aldehyde groups in the hydrophilic outer corona for subsequent shell crosslinking and cancer cell targeting. Upon comicellization, hydrophilic shell is crosslinked at pH 6.2 by the addition of DTP. The obtained FA-decorated SCL micelles, contain DS and acylhydrazone linkages in the crosslinking motif, thus, exhibited de-crosslinking feature to two biologically relevant environment, mildly acidic pH and reductive conditions, that is, endosomal/lysosomal pH or high GSH level in the cytosol. In vitro studies revealed that the reported cancer cell-targeting and multi-responsive SCL micelles possess advantages such as controlled drug release profiles, more prominent A549 cellular uptake that leading to superior cytotoxicity against A549 cells, higher endosomal escape capability, and demonstrating at least ~9.7-fold enhanced cytotoxicity upon light irradiation.

5.4.1.2 Reduction-reversible CLMs with bis(acryloyloxyethyl) disulfide

BAEDS is a type of small molecule crosslinker containing bioreductive DS bond which can be cleaved at the presence of a redox-reagent, such as DTT. Thus, if this type of crosslinker can be added into the target nanoparticles, it would be useful to regulate the release rate of the drugs.

Based on this principle, An et al. (2015) designed a PEGylated and multifunctional core-crosslinked nanovehicle with pH- and reduction-responsive and fluorescence imaging functionality for tumor-targeting drug delivery system. This CCL polymeric nanovehicle is prepared by the lactobionic acid (LA) modified poly(1-*O*-acryloyloxyethyl-[2-deoxy-2-amino]-D-glucopyranoside)/poly(oligo[ethylene glycol])-*b*-poly(2-[5,5-dimethyl-1,3-dioxan-2-yloxy] ethyl acrylate-*co*-bis[2-acryloyloxyethyl] DS) copolymer, and obtained LA-pDAGEA/pPEGA-*b*-p(DMDEA-*co*-BADS) multifunctional copolymers using RAFT polymerization. The RAFT chain transfer agent was modified with boron–dipyrromethene dye (BODIPY) as a fluorescence dye, to exhibit the cellular internalization process. In addition, to improve the cancer cellular uptake, galactose was chemically linked to the LA's gluconic acid via ether linkage, The presence of galactose residues on the surface of nanovehicle would encourage it to selectively bind to asialoglycoprotein receptors (ASGP-R) overexpressed on the surface of the HepG2 cell, a liver tumor. Finally, the drug-loaded nanovehicles displayed efficient anticancer activity in vitro and have a potential use for in vivo applications.

Recently, Bafkary and Khoee (2016) designed novel carbon nanotube-based dual-responsive (thermo- and reduction-responsive) nanovehicles, poly(NIPAM-HEMA-SS)/MN-MWCNT, for efficient intracellular anticancer drug delivery. The PNIPAM-*co*-PHEMA copolymer is thermo-responsive due to the presence of a PNIPAM segment and reduction-responsive owing to the existence of DS bonds for intracellular delivery of DOX into the nuclei of cancer cells. The nanovehicle was synthesized by traditional free radical polymerization of NIPAM, HEMA monomers, and cystine-containing DS crosslinker on the surface of acrylated magnetite MWCNTs in an aqueous media (Fig. 5.15). Under normal physiological conditions (temperature 37°C and absence of DTT), DOX-poly(NIPAM-HEMA-SS)/MN-MWCNT nanoparticles showed strong stability and small encapsulated drug release, while rapid release of the encapsulated DOX can be obtained with the increase of temperature and in the presence of 20 mM DTT due to the cleavage of the DS bonds (–S–S–) to thiol functional groups (–SH HS–). The in vitro and in vivo experiments confirmed that the Poly(NIPAM-HEMA-SS)/MN-MWCNT nanoparticles exhibited low cytotoxicity, high biocompatibility, and no toxicity in the liver and kidney of mice.

Biocompatible star polymers with PEG arms and cores containing cationic and degradable functionalities have been synthesized (Cho et al., 2011) by ATRP copolymerization of PEG methyl ether methacrylate macromonomer (PEGMA), 2-(dimethylamino)ethyl methacrylate (DMAEMA), and BAEDS as a degradable DS crosslinker (SS) via an "arm-first" method for the delivery of siRNA and cellular internalization of star polymers complexed with siRNA (Fig. 5.16). The star polymers were degraded into individual polymeric chains in an aqueous solution under redox conditions in the presence of GSH (0.2–10.0 mM) due to the cleavage of the DS bond. To determine biocompatibility and cellular internalization, the cationic star polymers were cultured with mouse calvarial preosteoblast-like cells, embryonic day 1, subclone 4 (MC3T3-E1.4). MTS assay results revealed that the synthesized polymers were biocompatible, with ≥80% cell proliferation after 48 hours of incubation even at a higher concentration of star polymers (100–800 μg/mL).

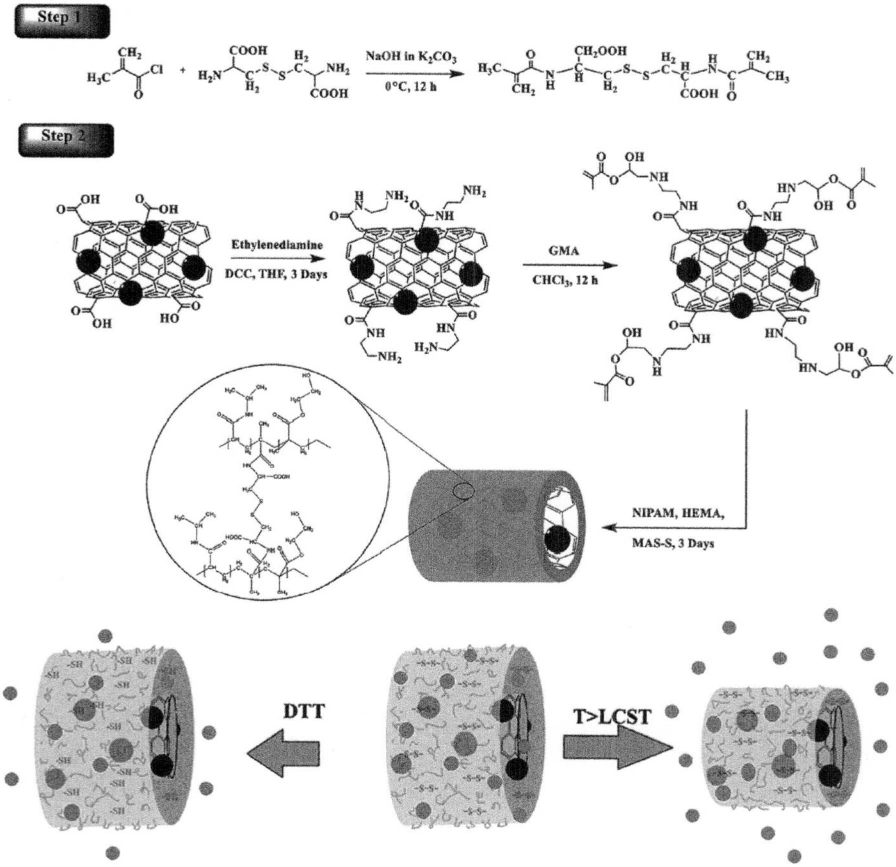

FIGURE 5.15

Synthetic route for the preparation of poly(NIPAM- HEMA-SS)/MN-MWCNT (Bafkary et al., 2016).

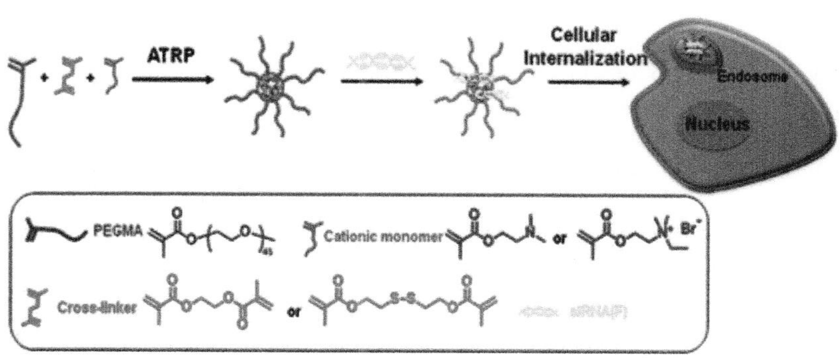

FIGURE 5.16

Synthesis of star polymers with PEG arms and with cationic and degradable core by ATRP via an "Arm-First" method and cell internalization of star polymers complexed with siRNA (Cho et al., 2011).

5.4.1.3 Reduction-reversible CLMs with cystamine

Cystamine is a linear aliphatic diamine reversible crosslinking agent in the development of polymers and a water-soluble functional group in nanoparticles developed for siRNA and DNA delivery composed of a central, cleavable DS bridge, produced from the oxidation of two cysteamine residues that can react with carboxyl groups on the pendant of a polymer chain. In addition, it has been discussed as a potential radioprotective agent shown to protect the liver against acetaminophen poisoning through their increasing of the antioxidant glutathione (GSH) system. A main feature of the use of a bioreducible crosslinker such as cystamine at the polymeric micelles is that, in origin, the block copolymer chain is conveniently eliminated from the human body via the kidneys after in situ micelle degradation.

According to this, a facile and effective synthetic strategy by using cystamine small molecules to form reduction-responsive core–shell–corona micelles based on self-assembly of poly(ethylene glycol)-*b*-poly(acrylic acid-*co*-*tert*-butyl acrylate)-poly(ε-caprolactone) (PEG_{43}-*b*-P(AA_{30}-*co*-tBA_{18})-*b*-PCL_{53}) triblock copolymer has been developed (Zhao and Liu, 2015). DOX-loaded shell-crosslinked micelles obtained with crosslinking PAA shells by using cystamine showed much higher in vitro stability in comparison with the non-crosslinked micelles. In addition, this reduction-responsive SCL micelles demonstrated slow sustained release in normal physiological conditions and a quick release upon exposing to the simulated intracellular reductive conditions. The release amount and release rate of DOX could be tuned by the crosslinked PAA shells. In addition, CLSM and HepG2 cell growth inhibition methods revealed their remarkable biocompatibility and similar great anticancer activity as the free DOX.

In another work, a novel reduction-responsive controlled drug delivery system composed of a mesoporous silica nanoparticle (MSN) core as the drug vehicle and poly(acrylic acid) (PAA) SCL through DS linkages as a drug release switcher has been developed (Fig. 5.17) (Li et al., 2013). To preserve the pore structure of MSN intact for high-loading efficiency of the drug, PAA chains were covalently grafted to the exterior surface of MSN before removing the structure-template through radical polymerization. After removing the structure-template and loading DOX, the channels of MSN were sealed by crosslinking the PAA shell using a cystamine crosslinker through an amidation reaction. Because the crosslinking network of PAA was dissociated with the action of glutathione, the release behavior of DOX could be controlled by varying the concentration of reductant. The release rate of DOX in a simulated environment of tumor cells after 24 hours with the existence of 2 mM glutathione was about three times higher than that without adding glutathione in normal human cells. The in vitro cell assays showed that MSN–PAA was highly biocompatible and suitable to use as a drug carrier while the DOX/MSN–PAA-crosslink demonstrated an excellent inhibition to the proliferation of HeLa cells (human cancer cells) and relatively lower cytotoxicity to 293 cells (human normal cells).

In addition, reduction-responsive shell-crosslinked micelles have been constructed from biodegradable Y-shaped amphiphilic block copolymer, mPEG-*b*-PLG-*b*-$(PLA)_2$ by ROP of γ-Benzyl-L-glutamate *N*-carboxyanhydride

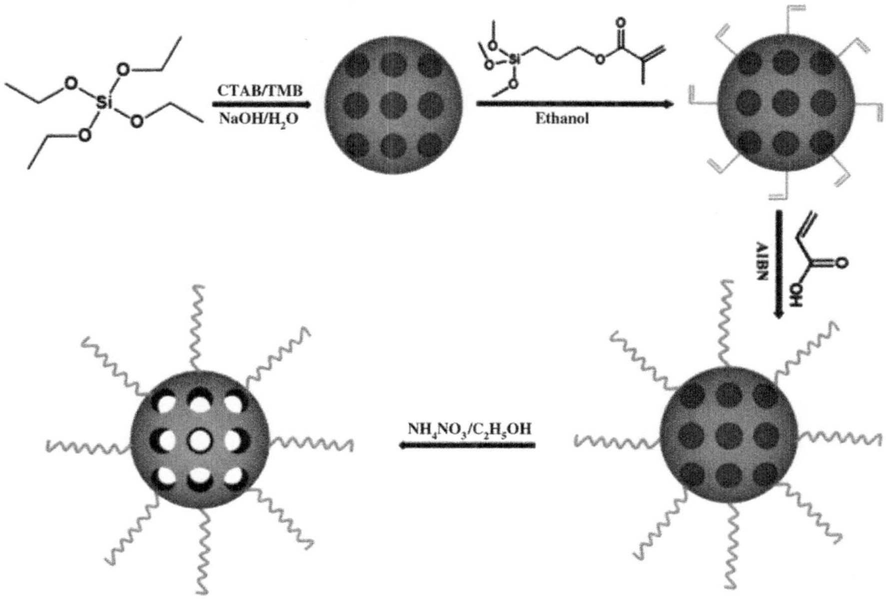

FIGURE 5.17

Schematic preparation process of PAA–MSN (Li et al., 2013).

(BLG-NCA) and L-Lactide (LA) sequentially, where the hydrophilic mPEG-*b*-PLG segment forms the shell of the micelles and two PLA segments form the hydrophobic core of the micelles (Fig. 5.18) (Yue et al., 2012). DOX-loaded micelles are prepared from the coassembly of DOX and the amphiphilic block copolymer in selective solvent prior to the crosslinking reactions. The carboxyl groups of the PLG segment as the inner shell of the micelles are crosslinked via a simple carbodiimide coupling strategy with cystamine as a reduction-responsive diamine crosslinker. In vitro experiments showed that the drug loading and entrapment efficiency, reduction-responsive release of DOX, and the stability in vitro for the crosslinked micelles (CLMs) were much better than non-crosslinked micelles (NCLMs). The total amount of DOX released increased owing to the increased solubility of DOX upon down-regulating the pH value of the release medium. The in vivo studies demonstrated that DOX loaded in the crosslinked micelles with PEG5K as the outer shell exhibited reduction-responsive release of DOX and the longest blood circulation life in SCLMs after intravenous injection in comparison to non-crosslinked micelles.

PTX-loaded dual-responsive SCL polymer nanoparticles have also been obtained (Samarajeewa et al., 2013) in aqueous media by the coassembly of PTX and a block copolymer micellar system composed of a enzymatically and hydrolytically degradable poly(D,L-lactic acid) (PDLLA) core that can encapsulate PTX and a hydrophilic shell consisting of poly(oligo(ethylene glycol) methyl ether

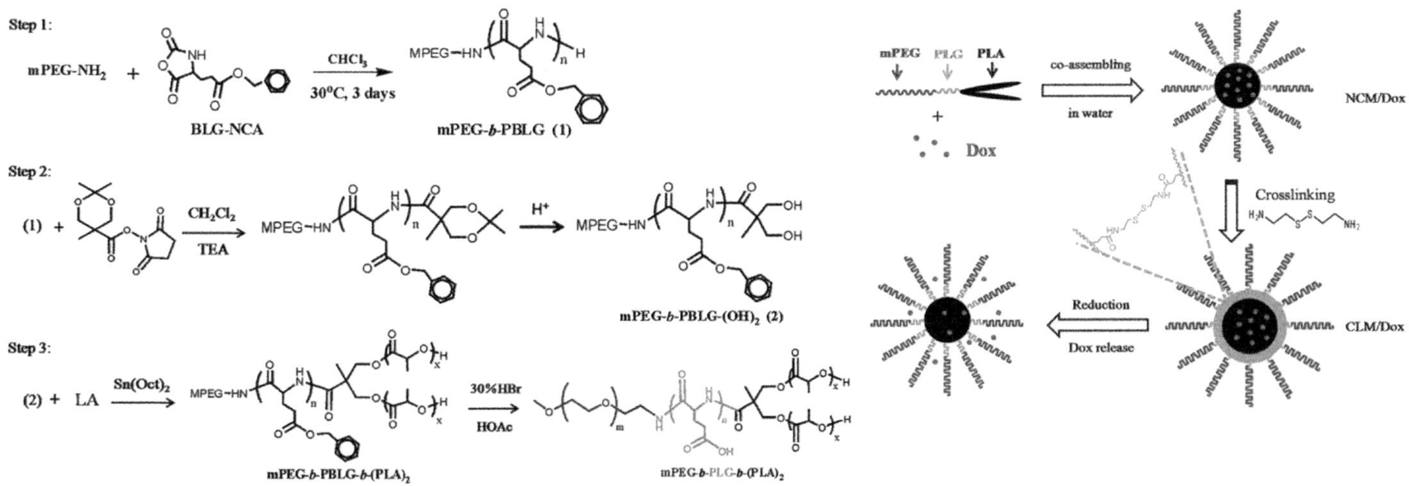

FIGURE 5.18

Synthesis of Y-shaped block copolymer mPEG-b-PLG-b-(PLA)2. Preparation of Dox-loaded, inner-shell crosslinked micelles and their reduction-response (Yue et al., 2012).

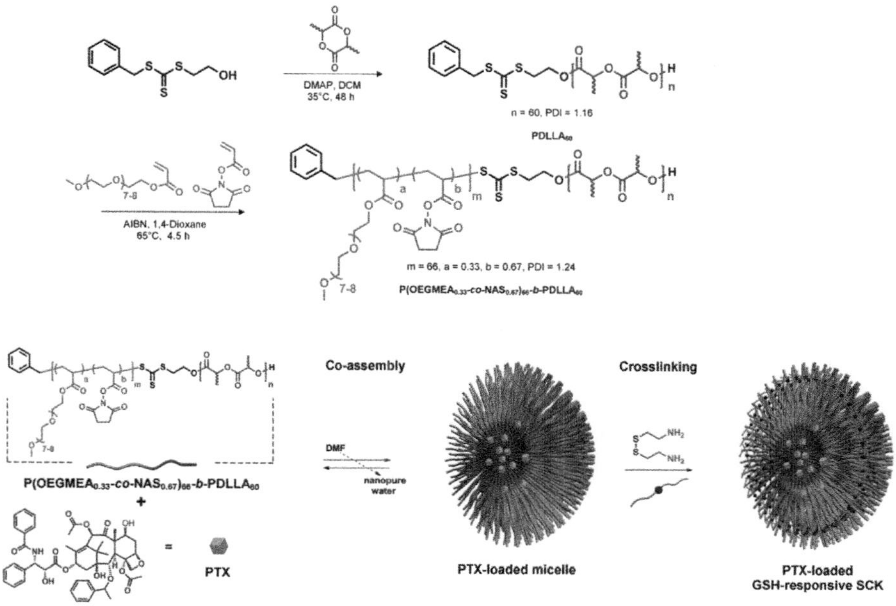

FIGURE 5.19

Syntheses of PDLLA60 homopolymer and P(OEGMEA0.33-co-NAS0.67)66-b-PDLLA60 diblock copolymer by sequential ROP and RAFT polymerization. b) Preparation of a SCK nanoparticle having glutathione-responsive shell cross-links and a hydrolytically degradable core domain by the co-assembly of amphiphilic diblock copolymers P (OEGMEA0.33-co-NAS0.67)66-b-PDLLA60 with PTX followed by covalent shell crosslinking with cystamine small molecules (Samarajeewa et al., 2013).

acrylate-copolymer-*N*-acryloyloxysuccinimide) (P(OEGMEA-*co*-NAS)), are subsequently crosslinked by the addition of cystamine small molecules (Fig. 5.19). PTX release from the SCK nanoparticles in a PBS buffer was accelerated in the presence of GSH at both pH 5.5 and pH 7.4, reaching c.65% cumulative drug release, while just c.50% and 35% cumulative release of PTX were observed in the absence of GSH after 8 days at pH 5.5 and pH 7.4, respectively. Enzyme-catalyzed hydrolysis of the nanoparticle core resulted in the degradation of c.30% of the poly(lactic acid) core to lactic acid within 12 hours, with c.37% PTX release as opposed to only c.17% release from the uncatalyzed SCK-PTX nanoparticles at pH 7.4. The degradability of the crosslinked shell and the PDLLA core, together with the higher cancer cell-killing efficacy of the PTX-loaded SCKs (in comparison to the free drug and the Taxol-mimicking formulation) can be promising for potential in vivo applications.

Finally, Han et al. (2014) fabricated a redox-responsive CCL micelle containing cypate moieties and cisplatin Pt(IV) complex prodrugs within the cores for combined photothermal and chemotherapy of cancer. The amphiphilic diblock copolymer, poly[(2-(2-methoxyethoxy)ethyl methacrylate)-*co*-(*N*-methacryloxy succinimide)]-block-poly(*N*-(2-hydroxypropyl) methacrylamide) (P(MEO_2MA-*co*-MASI)-*b*-PHPMA) was prepared through successive (RAFT) polymerization. Subsequently, partial amidation reaction of succinimidyl moieties with 3-azidopropylamine, the alkynyl-functionalized cypate and Pt(IV) complex were conjugated through click reaction. The stable CCL micelles were obtained by core cross-linking at 37°C in an aqueous solution by using cystamine as the reduction-responsive crosslinker. In aqueous solution, these CCL micelles demonstrated reduction-responsive crosslinker cleavage and cisplatin drug release in the presence of DTT. The cypate molecules in the cores of CCL micelles showed photothermal temperature increase and reactive oxygen species generation effects under 805 nm near infrared laser irradiation, which not only generated cytotoxicity against cisplatin-resistant human lung cancer cells A549R, but also enhanced the therapeutic effect of the cisplatin through drug diffusion increase and endo/lysosomal disruption.

5.4.2 REDUCTION-REVERSIBLE CLMs WITH A SELF-CROSSLINKING TECHNIQUE

Among the various crosslinking approaches to design reversible SCL micelles, crosslinking via oxidizing free mercapto groups in the prepolymer (or a self-crosslinking process) demonstrated many advantages over traditional crosslinking reactions. Firstly, no additional crosslinking agents are required and no by-products are formed during the process, so the purification procedure is straightforward in comparison with traditional crosslinking strategies that need an extensive purification procedure to remove unreacted reagents. Secondly, this method is low cost and can potentially be used in commercial applications as well as in agreement with green chemistry. Thirdly, the crosslinking conditions are mild to avoid association of polymeric micelles caused via the unwanted inter-micellar crosslinking.

According to this, reversibly self-crosslinked core–shell–corona polymeric micelles based on a biocompatible and biodegradable triblock copolymer (PCL-*b*-PPE-*b*-PEG) have been developed (Wang et al., 2010) composed of poly(ε-CL) segments formed the hydrophobic inner core, polyphosphoester formed the middle shell, and the hydrophilic outer corona of PEG bearing thiols as the side-chain groups on the middle block by consecutive ROP. After deprotection of the polyphosphoester, core–shell–corona micelles can be observed from the self-assembly of this amphiphilic triblock copolymer in an aqueous solution. Selective reversibly self-crosslinking of the micellar shell via DS bonds stabilized the micellar structure against dilution. Subsequently, this system showed the preferential release of an encapsulated anticancer drug (DOX) in response to intracellular GSH and also the enhanced cytotoxicity in A549 cells (a kind of human lung cancer). Hence, selective crosslinking of the micelles retarded the drug release in a nonreductive milieu, but the intracellularly breakage of DS bonds resulted in enhanced drug release and cytotoxicity.

In eukaryotic cells, cysteines are mainly bridged in the endoplasmic reticulum through the DS bond and this indicates that poly(L-cysteine) (PLC) is a superior choice to construct DS-containing self-crosslinked micelles. As a polypeptide, PLC has a significant number of advantages in biodegradability and biocompatibility, and show the feasibility of simulation of the structure and biological performance of natural proteins and peptides (Deming, 2005).

SCL micelles from poly(ethylene glycol)-*b*-poly(L-cysteine)-*b*-poly(L-phenylalanine) (PEG-PCys-PPhe) triblock copolymers have been designed and fabricated as PEG−polypeptide hybrid drug carriers (Wang et al., 2012). The core−shell−corona micelles are prepared from the self-assembly of these triblock copolymers in aqueous solutions. The PCys blocks constructed the middle shell of the micelles, which has the ability to self-crosslink by the oxidation of the −SH groups. The crosslinked shell could keep the shape of spherical micelles and also could enhance the structural stability even in severe environments. HeLa and 293 T cells were employed for evaluate the cytotoxicity of DOX-loaded SCL micelles in the intracellular environment by using glutathione monoester (GSH-OEt) (a glycine esterified form of GSH) (Fig. 5.20). The in vitro drug release experiments demonstrated that the DOX-loaded SCL micelles could reduce drug loss in the extracellular environment and quickly release drugs in response to the intracellular GSH level. Therefore, these GSH-responsive SCL micelles might have numerous potential applications in intracellular drug delivery systems.

The self-crosslinked SCL micelles are constructed by oxidizing the thiol groups into DS bonds. However, free thiol groups are radical transfer agents, thermally unstable, and quickly oxidized. Another disadvantage of the use of thiol groups to fabricate DS bonds is the protection and deprotection of the thiol groups during the synthetic procedure which is hard and unendurable work. Thus, the synthesis of amphiphilic polymers containing pendant thiol groups, without requiring protection and deprotection processes, is a promising strategy for fabrication reductive SCL micelles and a more straightforward method.

In this respect, Gao et al. (2014) reported a one-pot method for the preparation of biodegradable and biocompatible SCL micelles with reduction-responsive DS bonds for controlled drug release (Fig. 5.21). This system was formed from an amphiphilic triblock copolymer, PCL-*b*-POEGTMA-*b*-PCL, containing multiple pendant thiol groups in the hydrophilic block which was synthesized by a composition of polycondensation and ROP in one pot without any protection or deprotection procedures. SCL micelles were obtained in a facile procedure through the self-assembly of this amphiphilic copolymer in aqueous media and the poly(oligo(ethylene glycol)thiomalicate) (POEGTMA) as the micellar shell was promoted by adding diluted H_2O_2. In comparison with the bifunctional crosslinkers containing DS bonds, H_2O_2 is a greener and less toxic agent and is more easily removed from the micellar environment. These SCL micelles were employed for release of PTX, one of the most successful anticancer drugs, in the blood circulation. The

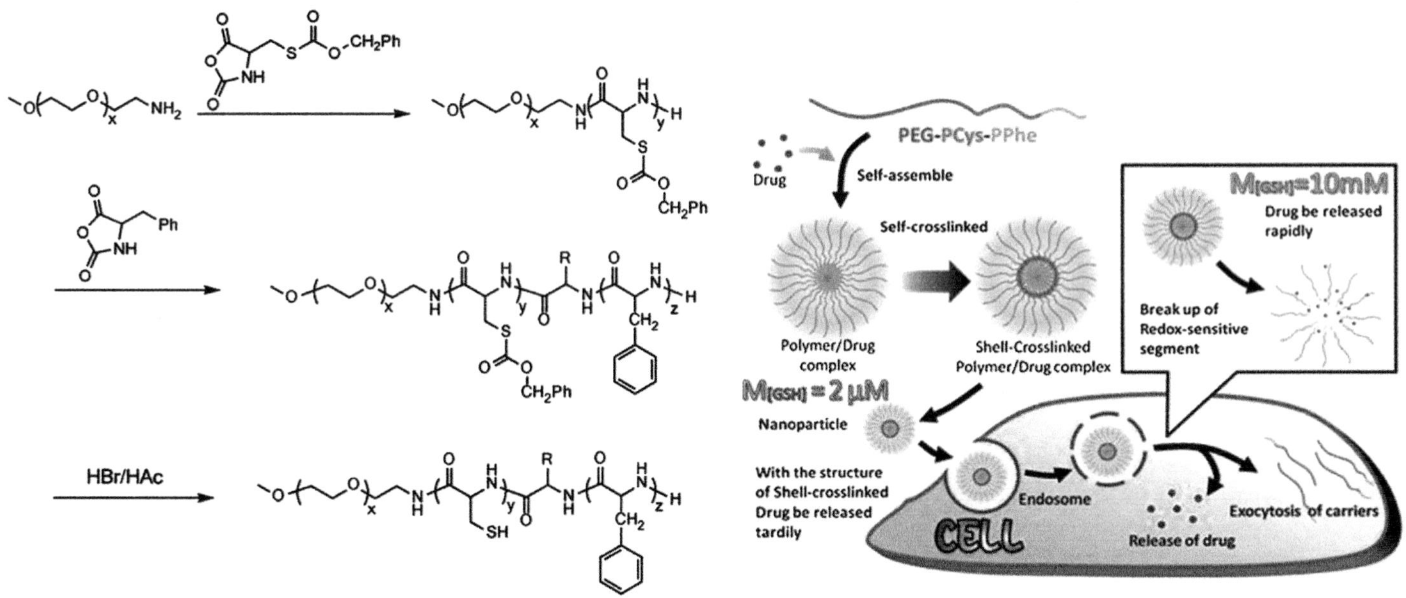

FIGURE 5.20

Synthetic route to the PEG-PCys-PPhe copolymer. Schematic illustration of the formation of shell cross-linked micelles and intracellular drug release (Wang et al., 2012).

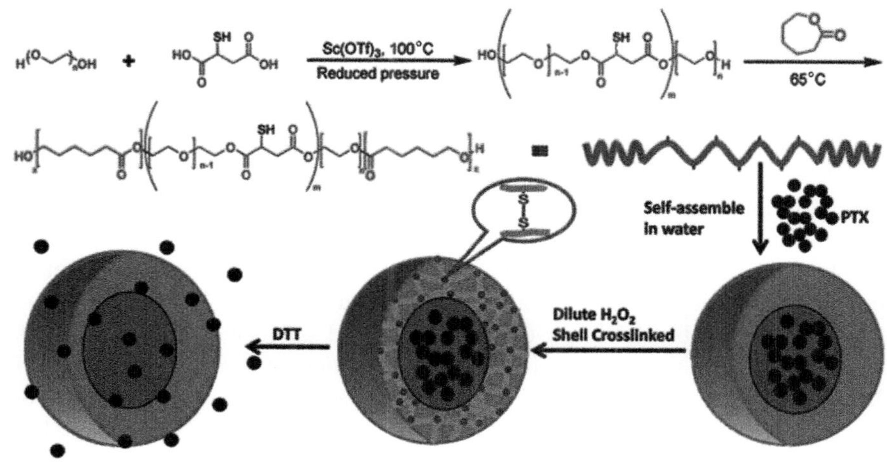

FIGURE 5.21

Preparation of SCL micelles for redox-responsive drug release (Gao et al., 2014).

in vitro PTX release profile from SCL micelles demonstrates the quick release of drug molecules in response to the bioreductive environment.

Lee et al. (2013) reported that DS bonded micelles possess high blood-stability and tumor-adaptability for the design of drug carriers composed of methoxypoly(ethylene glycol)-(Cysteine)$_4$-poly(D,L-lactic acid) (mPEG-(Cys)$_4$-PDLLA) and DS bonds were introduced at the interface between the hydrophobic core and hydrophilic shell. In vitro studies revealed that the DS-bonded micelles delivered DOX to the cancer cell, with around 7-fold increase in DOX delivery to tumor and 1.9-fold decrease in the heart, compared to SA, non-crosslinked mPEG-PDLLA micelles. Moreover, in vivo experiments of the DS-bonded micelles demonstrated enhanced antitumor activity.

Han et al. (2015) reported the synthesis of HA nanoparticles based on a biocompatible and biodegradable HA-*b*-poly(caprolactone) (HA-*b*-PCL) diblock copolymer to develop robust, tumor-targeted drug delivery in vivo. These nanoparticles can be crosslinked through intraparticular bioreducible DS bonds, endowing the nanovehicles with robust blood stability in the extracellular environment and high targetability to CD44-overexpressing cancer cells.

5.4.3 REDUCTION-REVERSIBLE COPOLYMER MICELLES WITH LIPOIC ACID

LA is another DS-containing crosslinker used to design reversible crosslinked micelles. LA, a naturally occurring substance produced by the human body, shows no cytotoxicity even at high concentrations and has been characterized as an efficient antioxidant drug in medicine, which is a potential therapeutic agent for the treatment or prevention of varying diseases such as diabetes, polyneuropathy, HIV/AIDS, and Alzheimer's disease (Ziegler et al., 1999; Packer et al., 1995). LA acts as an optically active compound that is composed of a carboxyl group and a five-membered ring containing a DS bond and functions as a cofactor in multi-enzyme complexes. The DS bond is unstable in the LA ring and is degraded through light or heat stress and, thus, a linear poly DS is formed by ROP in the polymeric micelles containing lipoyl groups, which can be conveniently crosslinked in the presence of a catalytic amount of DTT in an aqueous environment.

A type of reversibly DS crosslinked dextran nanoparticle for the controlled intracellular delivery of DOX have been developed (Li et al., 2009) which were conveniently assembled through dextran-lipoic acid conjugates (Dex-LAs) and, subsequently, were simply crosslinked by the addition of a catalytic amount of DTT. It was found that these Dex-LA conjugates with average sizes of 85.3–142.5 nm before crosslinking shrunk by 20–35 nm after crosslinking. The reported dextran nanoparticles are stable under extracellular conditions but quickly de-crosslinked under a high concentration of glutathione (GSH) tripeptides and triggered the rapid and efficient release of DOX into cancer cells. Notably, these crosslinked dextran nanoparticles (C-DNPs) showed a high drug loading efficiency in contrast with the non-crosslinked dextran nanoparticles. The in vitro release studies using both HeLa and K562 cells demonstrated that the release of DOX-loaded C-DNP decreased to minimal at about 10% even under extensive dilution, but in the presence of 10 mm DTT over 90% of the DOX was released in 11 hours which mimics those of the intracellular reductive environment, such as the cytoplasm and the cell nucleus.

DS CCLM based on amphiphilic starch-graft-poly(ethylene glycol) (starch-*g*-PEG) have also been readily prepared (Zhang et al., 2013a) through grafting carboxyl group terminated PEG to the starch backbone, followed by conjugation with lipoic acid to produce DS crosslinking micelles and used for controlled delivery of DOX (Fig. 5.22). The in vitro studies revealed that the DS CCL micelles show high structural stability under normal physiological conditions, while they can de-crosslink under a reductive intracellular environment (e.g., GSH) through reversible cleavage into free thiols. In addition, CLSM and flow cytometry methods confirmed that the DOX-loaded crosslinked micelles exhibited the faster DOX release behavior and higher cellular proliferation inhibition efficacy in GSH-OEt pretreated HeLa and HepG2 cells compared to the nonpretreated and BSO pretreated cells.

A highly nonfouling micelle stabilized by reversible cross-linkage as a robust and smart nanovehicle for the controlled release of the anticancer drug DOX in response to an intracellular reductive environment based on a simple amphiphilic random copolymer, poly(CBMA-*co*-MAEL), has been formed (Lin et al., 2014) using traditional radical copolymerization, with nonfouling zwitterionic carboxybetaine methacrylate (CBMA) acting as a hydrophilic block and DS containing 2-(methacryloyloxy)ethyl lipoate (MAEL) acting as a hydrophobic and crosslinked block (Fig. 5.23). This nanovehicle system has the characteristics of the controlled

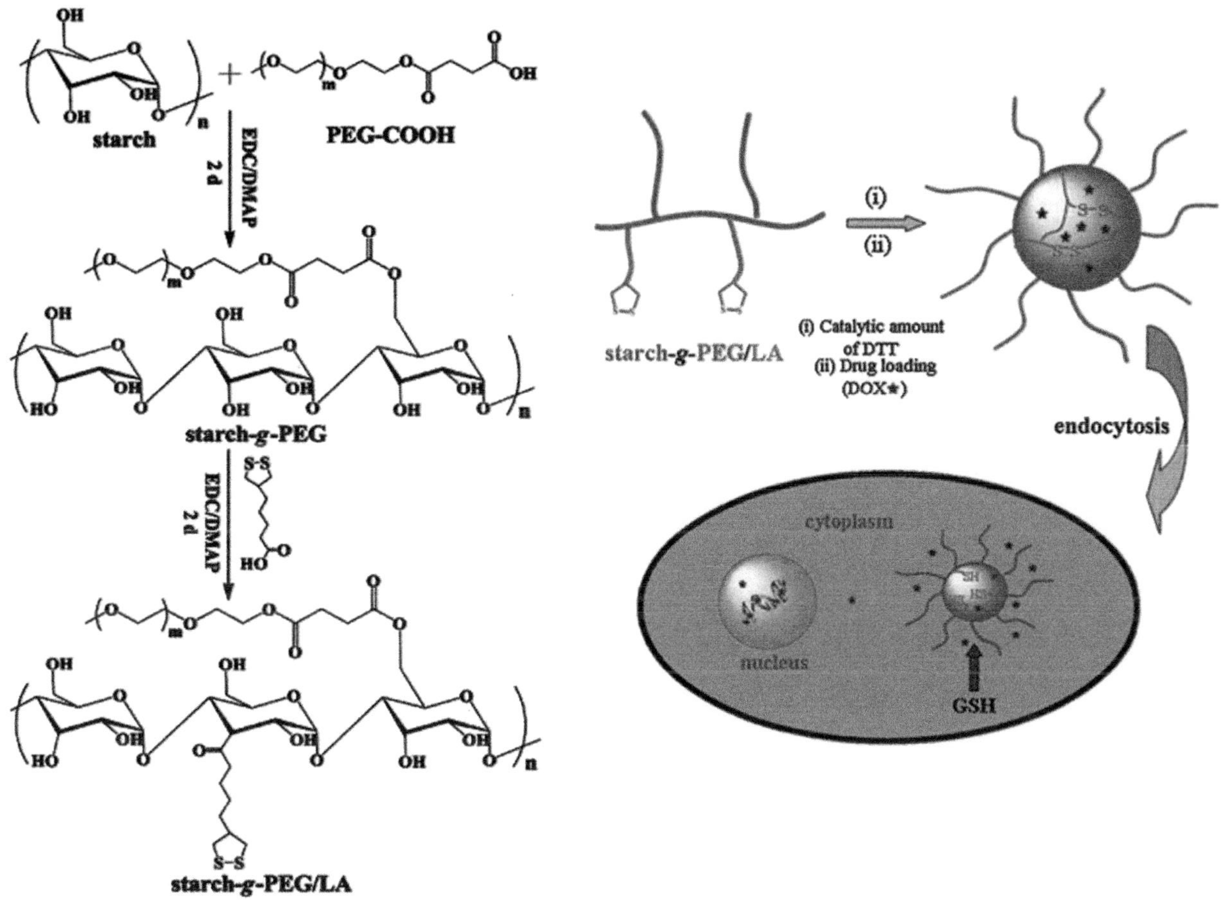

FIGURE 5.22

Synthetic pathway of starch-g-PEG–LA graft copolymers. Illustration of reversible disulfide crosslinked starch-g-PEG micelles for intracellular drug release triggered by GSH (Zhang et al., 2013a).

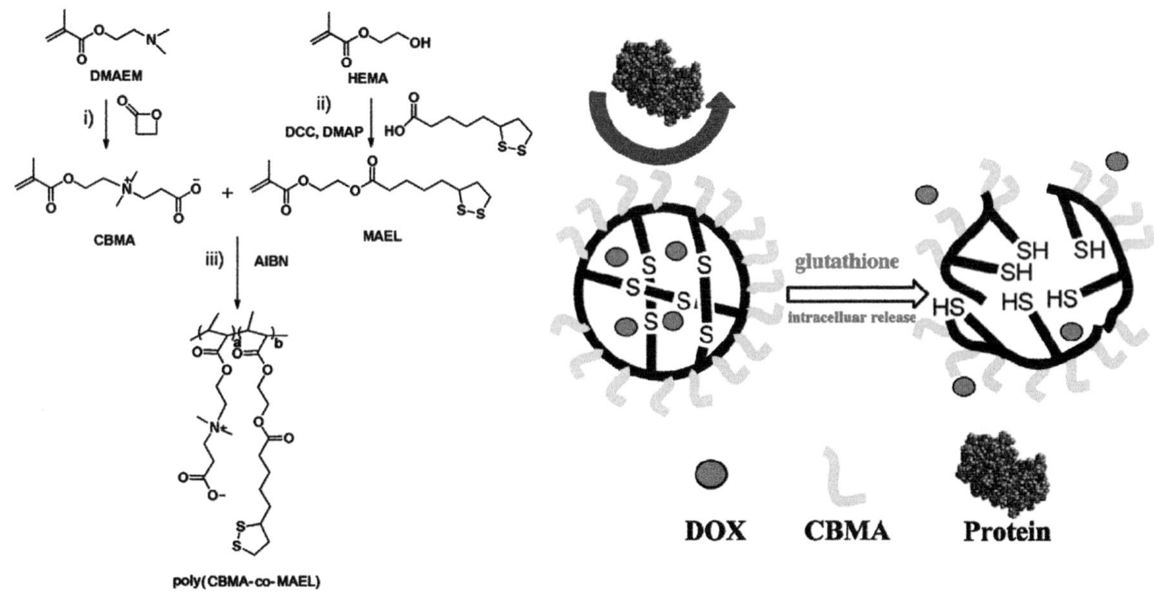

FIGURE 5.23

Synthesis of poly(CBMA-co-MAEL). Schematic representation of hemocompatible reversible cross-linkedmicelles for intracellular release of DOX (Lin et al., 2014).

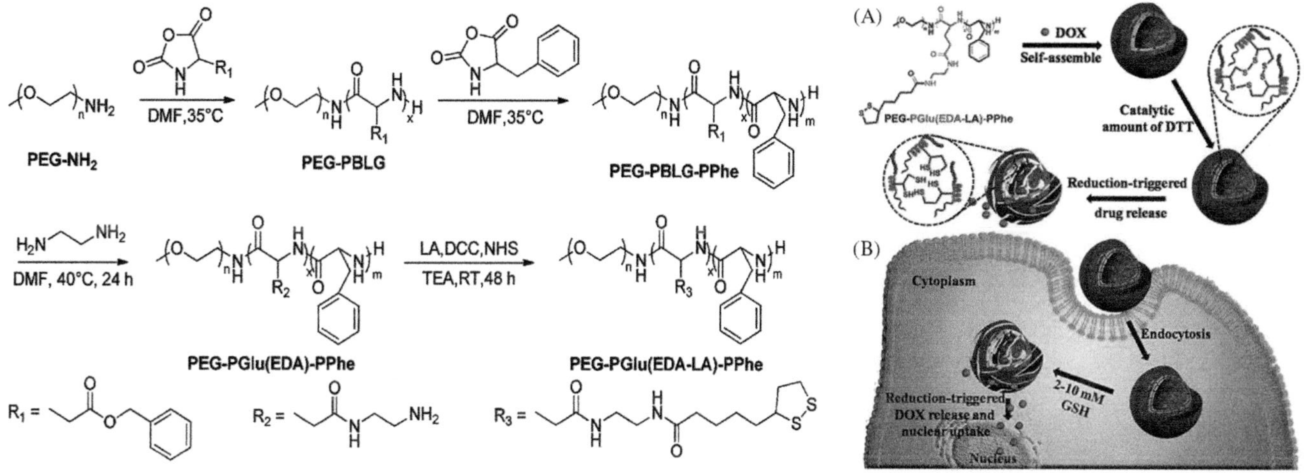

FIGURE 5.24

Synthetic route for the preparation of PEG–PGlu(EDA–LA)–PPhe copolymers. Schematic illustration of the preparation of DOX-loaded PEG–PGlu(EDA–LA)–PPhe crosslinked micelles for reduction responsive intracellular drug release (Li et al., 2015).

release of DOX in a reducing environment, excellent micellar stability in fibrinogen (1 mg/mL) PBS solution, and 50% fetal bovine serum (FBS) as well as very low cytotoxicity. The high hemocompatibility is attributed to the nonfouling pCBMA on the nanocarrier periphery and the stable crosslinked shell of the micelles. In addition, with the increasing percentage of reductive GSH concentration, the hemocompatible DOX-loaded micelles exhibited high drug-loading content, excellent loading efficiency of DOX, enhanced intracellular drug release, and increased toxicity to target cancer cells.

In another work, Li et al. (2015) reported on the preparation of a type of biocompatible and degradable reversibly SCL poly(amino acid)s micelles based on a LA decorated poly(ethylene glycol)-*b*-poly(*g*-benzyl-L-glutamate)-*b*-poly(L-phenylalanine) (PEG–PGlu(EDA–LA)–PPhe) amphiphilic triblock copolymers for smart loading and efficient reduction responsive intracellular delivery of DOX into cancer cells, resulting in excellent antitumor activity (Fig. 5.24). The amphiphilic triblock copolymer was synthesized via consecutive ROP of γbenzyl-L-glutamate *N*-carboxyanhydride (BLG-NCA) and L-phenylalanine *N*-carboxyanhydride (Phe-NCA) by PEG–NH_2 as macroinitiator, subsequently, conjugation with LA to form reversible shell cross-linking. This amphiphilic copolymer was self assembled in aqueous solution into core shell corona micelles in which the hydrophilic PEG outer is corona, the functional PBLG is middle shell, and the PPhe inner core acts as a reservoir of hydrophobic anticancer drugs. Afterwards, these smart triblock polymeric micelles containing DS linkages were crosslinked based on the thiol-DS exchange reaction in the presence of a catalytic amount of DTT in order to increase structural stability in physiological conditions (pH 7.4, 37°C). However, the SCLM obtained from LA conjugates was quickly dissociated into unimers and DOX was delivered and released in reductive environments mimicking those of the cytoplasm and the nuclei of cancer cells.

5.5 CONCLUSION

Reversible crosslinked micelles are various substances that have become a fascinating new category of organic, functional nanoparticles in cancer research to enhance drug-therapeutic efficacies and decrease drug-adverse toxicities. In this chapter, we succinctly introduced some main knowledge of reversible CCLM/SCL micelles, including the definition, the design, the classification, cross-linking strategies such as DS/thiol chemistry, polyelectrolyte complexation, self-crosslinking process, UV-induced coupling, and their recent applications in drug delivery systems. In recent years, the research on intelligent RCL micelles triggered by enhanced stability and a stimuli responsive factor (such as pH-responsive, reduction-responsive, and photo-responsive CCLM/SCL micelles) has been extensive. It is clear from the few in vivo studies published in the literature that reversibly crosslinked micelles can reduce the problem of precocious drug release and, thus, substantially enhance their therapeutic efficacies.

Future research must focus on improving the biocompatibility and biodegradability of the selected functionalized reversible crosslinked micelles for their use as nanocarriers for controllable biopharmaceutical delivery. Novel, less toxic, and more commercially crosslinking chemistries are still extremely favorable. Quick crosslinking is exclusively important if polymeric micelles are to be physically loaded with hydrophobic drugs, since the remarkably diffusional loss of such drugs can happen if crosslinking proceeds too gently. The design of multi-responsive reversible CCLM/SCL micelles is another field for further development. Considerable work in the field of these micelles should also be directed toward more efficient, less expensive, and time-consuming fabrication of these substances. In addition, new characterization methods that will enable the real degree of crosslinking to be determined would be highly effective for the progression of this fascinating sub-field. It is a promising and efficacious approach to exclusively reduce precocious drug release from their carriers during blood circulation and deliver high concentrations of drugs to tumor cells on demand. It demonstrates excellent candidate in formulating many existing and new drugs for decreasing their side effects and increasing of their efficacy against cancers and other diseases.

REFERENCES

An, J., Dai, X., Zhao, Y., Guo, Q., Wu, Z., Zhang, X., et al., 2015. A biodegradable and fluorescent nanovehicle with enhanced selective uptake by tumor cells. Polym. Chem. 6, 6529–6542.

Babin, J., Lepage, M., Zhao, Y., 2008. "Decoration" of shell cross-linked reverse polymer micelles using ATRP: a new route to stimuli-responsive nanoparticles. Macromolecules 41, 1246–1253.

Bae, Y., Fukushima, S., Harada, A., Kataoka, K., 2003. Design of environment-sensitive supramolecular assemblies for intracellular drug delivery: polymeric micelles that are responsive to intracellular pH change. Angew. Chem., Int. Ed. 42, 4640–4643.

Bafkary, R., Khoee, S., 2016. Carbon nanotube-based stimuli-responsive nanocarriers for drug delivery. RSC Adv. 6, 82553–82565.

Black, S.M., Wolf, C.R., 1991. The role of glutathione-dependent enzymes in drug resistance. Pharmacol. Ther. 51, 139–154.

Bronich, T.K., Keifer, P.A., Shlyakhtenko, L.S., Kabanov, A.V., 2005. Polymer micelle with cross-linked ionic core. J. Am. Chem. Soc. 127, 8236–8237.

Bulmus, V., Chan, Y., Nguyen, Q., Tran, H.L., 2007. Synthesis and characterization of degradable P(HEMA) microgels: use of acid-labile crosslinkers. Macromol. Biosci. 7, 446–455.

Butun, V., Billingham, N.C., Armes, S.P., 1998. Synthesis of shell cross-linked micelles with tunable hydrophilic/hydrophobic cores. J. Am. Chem. Soc. 120, 12135–12136.

Butun, V., Wang, X.S., Banez, M.V.D., Robinson, K.L., Billingham, N.C., Armes, S.P., et al., 2000. Synthesis of shell cross-linked micelles at high solids in aqueous media. Macromolecules 33, 1–3.

Cao, X.T., Kim, Y.H., Park, J.M., Lim, K.T., 2016. One-pot syntheses of dual-responsive core cross-linked polymeric micelles and covalently entrapped drug by click chemistry. Eur. Polym. J. 78, 264–273.

Chan, Y., Wong, T., Byrne, F., Kavallaris, M., Bulmus, V., 2008. Acid-labile core cross-linked micelles for pH-triggered release of antitumor drugs. Biomacromolecules 9, 1826–1836.

Charulatha, V., Rajaram, A., 2001. Dimethyl 3,3′-dithiobispropionimidate: a novel cross-linking reagent for collagen. J. Biomed. Mater. Res. 54, 122–128.

Cho, H.Y., Srinivasan, A., Hong, J., Hsu, E., Liu, S., Shrivats, A., 2011. Synthesis of biocompatible PEG-based star polymers with cationic and degradable core for siRNA delivery. Biomacromolecules 12, 3478–3486.

Convertine, A.J., Diab, C., Prieve, M., Paschal, A., Hoffman, A.S., Johnson, P.H., et al., 2010. pH-responsive polymeric micelle carriers for siRNA drugs. Biomacromolecules 11, 2904–2911.

Dai, J., Lin, S., Cheng, D., et al., 2011. Interlayer-crosslinked micelle with partially hydrated core showing reduction and pH dual sensitivity for pinpointed intracellular drug release. Angew. Chem. Int. Ed. Engl. 50, 9404–9408.

Davis, M.E., Chen, Z.G., Shin, D.M., 2008. Nanoparticle therapeutics: an emerging treatment modality for cancer. Nat. Rev. Drug Discov. 7, 771–782.

Deming, T.J., 2005. Polypeptide hydrogels via a unique assembly mechanism. Soft Matter 1, 28–35.

Ding, J., Liu, G., 1998. Polystyrene-*b* lock-poly(2-cinnamoylethyl methacrylate) nanospheres with cross-linked shells. Macromolecules 31, 6554–6558.

Gao, L., Luo, Q., Wang, Y., Du, H., Li, X., Shena, Z., et al., 2014. Facile preparation of shell crosslinked micelles for redox-responsive anticancer drug release. RSC Adv. 4, 4177–4180.

Gillies, E.R., Frechet, J.M.J., 2005. pH-Responsive copolymer assemblies for controlled release of doxorubicin. Bioconjugate Chem. 16, 361–368.

Gillies, E.R., Jonsson, T.B., Frechet, J.M.J., 2004. Stimuli-responsive supramolecular assemblies of linear-dendritic copolymers. J. Am. Chem. Soc. 126, 11936–11943.

Guo, A., Liu, G.J., Tao, J., 1996. Star polymers and nanospheres from cross-linkable diblock copolymers. Macromolecules 29, 2487–2490.

Guo, R., Chen, L., Cai, S., Liu, Z., Zhu, Y., Xue, W., et al., 2013. Novel alginate coated hydrophobically modified chitosan polyelectrolyte complex for the delivery of BSA. Mater. Med. 24, 2093–2100.

Han, Y., Li, J., Zan, M., Luo, S., Ge, Z., Liu, S., 2014. Redox-responsive core cross-linked micelles based on cypate and cisplatin prodrugs-conjugated block copolymers for synergistic photothermal–chemotherapy of cancer. Polym. Chem. 5, 3707–3718.

Han, H.S., Thambi, T., Choi, K.Y., et al., 2015. Bioreducible shell-cross-linked hyaluronic acid nanoparticles for tumor-targeted drug delivery. Biomacromolecules 16, 447–456.

Hu, X., Li, H., Luo, S., Liu, T., Jianga, Y., Liu, S., 2013a. Thiol and pH dual-responsive dynamic covalent shell cross-linked micelles for triggered release of chemotherapeutic drugs. Polym. Chem. 4, 695–706.

Hu, X., Tian, J., Liu, T., Zhang, G., Liu, S., 2013b. Photo-triggered release of caged camptothecin prodrugs from dually responsive shell cross-linked micelles. Macromolecules 46, 6243–6256.

Hu, J., He, J., Cao, D., Zhang, M., Ni, P., 2015. Core cross-linked polyphosphoester micelles with folate-targeted and acid-cleavable features for pH-triggered drug delivery. Polym. Chem. 6, 3205–3216.

Huang, H., Kowalewski, T., Remsen, E.E., Gertzmann, R., Wooley, K.L., 1997. Hydrogel-coated glassy nanospheres: a novel method for the synthesis of shell cross-linked knedels. J. Am. Chem. Soc. 119, 11653–11659.

Huang, W.C., Yang, Y.S., Lai, T.Y., Jan, J.S., 2012. Lysine-block-tyrosine block copolypeptides: self-assembly, cross-linking, and conjugation of targeted ligand for drug encapsulation. Polymer 53, 913–922.

Jain, R., Standley, S.M., Frechet, J.M.J., 2007. Synthesis and degradation of pH-sensitive linear poly(amidoamine)s. Macromolecules 40, 452–457.

Jiang, X.Z., Luo, S.Z., Armes, S.P., Shi, W.F., Liu, S.Y., 2006. UV irradiation-induced shell cross-linked micelles with pH-responsive cores using ABC triblock copolymers. Macromolecules 39, 5987–5994.

Jiang, J., Qi, B., Lepage, M., Zhao, Y., 2007a. Polymer micelles stabilization on demand through reversible photo-cross-linking. Macromolecules 40, 790–792.

Jiang, X.Z., Ge, Z.S., Xu, J., Liu, H., Liu, S.Y., 2007b. Fabrication of multiresponsive shell cross-linked micelles possessing pH-controllable core swellability and thermo-tunable corona permeability. Biomacromolecules 8, 3184–3192.

Jin, Q., Liu, X., Liu, G., Ji, J., 2010. Fabrication of core or shell reversibly photo cross-linked micelles and nanogels from double responsive water-soluble block copolymers. Polymer 51, 1311–1319.

Joralemon, M.J., O'Reilly, R.K., Hawker, C.J., Wooley, K.L., 2005. Shell click-crosslinked (SCC) nanoparticles: a new methodology for synthesis and orthogonal functionalization. J. Am. Chem. Soc. 127, 16892–16899.

Kakizawa, Y., Harada, A., Kataoka, K., 1999. Environment-sensitive stabilization of core-shell structured polyion complex micelle by reversible cross-linking of the core through disulfide bond. J. Am. Chem. Soc. 121, 11247–11249.

Kellum, M.G., Smith, A.E., York, S.K., McCormick, C.L., 2010. Reversible interpolyelectrolyte shell cross-linked micelles from pH/salt-responsive diblock copolymers synthesized via RAFT in aqueous solution. Macromolecules 43, 7033–7040.

Kim, S.H., Jeong, J.H., Lee, S.H., Kim, S.W., Park, T.G.J., 2008. Local and systemic delivery of VEGF siRNA using polyelectrolyte complex micelles for effective treatment of cancer. J. Controlled Release 129, 107–116.

Kolano, C., Helbing, J., Bucher, G., Sander, W., Hamm, P., 2007. Intramolecular disulfide bridges as a phototrigger to monitor the dynamics of small cyclic peptides. J. Phys. Chem. B 111, 11297–11302.

Koo, A.N., Lee, H.J., Kim, S.E., Chang, J.H., Park, C., Kim, C., et al., 2008. Disulfidecross-linked PEG-poly(amino acid)s copolymer micelles for glutathione-mediated intracellular drug delivery. Chem. Commun. 6570–6572.

Kuppusamy, P., Li, H., Ilangovan, G., et al., 2002. Noninvasive imaging of tumor redox status and its modification by tissue glutathione levels. Cancer Res 62, 307–312.

Lee, Y., Fukushima, S., Bae, Y., Hiki, S., Ishii, T., Kataoka, K.J., 2007. A protein nanocarrier from charge-conversion polymer in response to endosomal pH. Am. Chem. Soc. 129, 5362–5363.

Lee, S.J., Min, K.H., Lee, H.J., et al., 2011. Ketal cross-linked poly(ethylene glycol)-poly (amino acid)s copolymer micelles for efficient intracellular delivery of doxorubicin. Biomacromolecules 12, 1224–1233.

Lee, S.Y., Kim, S., Tyler, J.Y., Park, K., Cheng, J.X., 2013. Blood-stable, tumor-adaptable disulfide bonded mPEG-(Cys)4-PDLLA micelles for chemotherapy. Biomaterials 34, 552–561.

Li, Y., Smith, A.E., Lokitz, B.S., McCormick, C.L., 2007. In situ formation of gold-"decorated" vesicles from a RAFT-synthesized, thermally responsive block copolymer. Macromolecules 40, 8524–8526.

Li, Y.L., Zhu, L., Liu, Z.Z., Cheng, R., Meng, F.H., Cui, J.H., et al., 2009. Reversibly stabilized multifunctional dextran nanoparticles efficiently deliver doxorubicin into the nuclei of cancer cells. Angew. Chem., Int. Ed. 48, 9914–9918.

Li, Y., Xiao, W., Xiao, K., Berti, L., Luo, J., Tseng, H.P., et al., 2012. Well-defined, reversible boronate crosslinked nanocarriers for targeted drug delivery in response to acidic pH values and cis-diols. Angew. Chem., Int. Ed. Engl. 51, 2864–2869.

Li, H., Zhang, J.Z., Tang, Q., Du, M., Hu, J., Yang, D., 2013. Reduction-responsive drug delivery based on mesoporous silica nanoparticle core with crosslinked poly(acrylic acid) shell. Mater. Sci. Eng. C 33, 3426–3431.

Li, Y., Xiao, K., Zhu, W., Denga, W., Lama, K.S., 2014. Stimuli-responsive cross-linked micelles for on-demand drug delivery against cancers. See Comment in PubMed Commons Below Adv Drug Deliv Rev 66, 58–73.

Li, Y., Wang, S., Zhu, D., Shen, Y., Du, B., Liu, X., et al., 2015. Reversibly cross-linked poly(ethylene glycol)–poly(amino acid)s copolymer micelles: a promising approach to overcome the extracellular stability versus intracellular drug release challenge. RSC Adv. 5, 20025–20034.

Lin, W., He, Y., Zhang, J., Wang, L., Wang, Z., Ji, F., et al., 2014. Highly hemocompatible zwitterionic micelles stabilized by reversiblecross-linkage for anti-cancer drug delivery. Colloids Surf., B: Biointerfaces 115, 384–390.

Liu, S.Y., Armes, S.P., 2001. The facile one-pot synthesis of shell cross-linked micelles in aqueous solution at high solids. J. Am. Chem. Soc. 123, 9910–9911.

Liu, S., Weaver, J.V.M., Save, M., Armes, S.P., 2002. Synthesis of pH-responsive shell cross-linked micelles and their use as nanoreactors for the preparation of gold nanoparticles. Langmuir 18, 8350–8357.

Lokitz, B.S., York, A.W., Stempka, J.E., Treat, N.D., Li, Y., Jarrett, W.L., et al., 2007. Aqueous RAFT synthesis of micelle-forming amphiphilic block copolymers containing *N*-acryloylvaline. Dual mode, temperature/pH responsiveness, and "locking" of micelle structure through interpolyelectrolyte complexation. Macromolecules 40, 6473–6480.

Long, Y.B., Gu, W.X., Pang, C., Ma, J., Gao, H., 2016. Construction of coumarin-based cross-linked micelles with pH responsive hydrazone bond and tumor targeting moiety. J. Mater. Chem. B. 4, 1480–1488.

Miyata, K., Kakizawa, Y., Nishiyama, N., Harada, A., Yamasaki, Y., Koyama, H., et al., 2004. Block catiomer polyplexes with regulated densities of charge and disulfide cross-linking directed to enhance gene expression. J. Am. Chem. Soc. 126, 2355–2361.

Mikhail, A.S., Allen, C., 2010. Poly(ethylene glycol)-*b*-poly(ε-caprolactone) micelles containing chemically conjugated and physically entrapped docetaxel: synthesis, characterization, and the influence of the drug on micelle morphology. Biomacromolecules 11, 1273–1280.

Murray, R.D.H., Mendez, J., Brown, S.A., 1982. The Natural Coumarins: Occurrence, Chemistry, and Biochemistry. John Wiley & Sons, New York.

Nguyen, R., Huc, I., 2003. Optimizing the reversibility of hydrazone formation for dynamic combinatorial chemistry. Chem. Commun. 8, 942–943.

Nie, S., Xing, Y., Kim, G.J., Simons, J.W., 2007. Nanotechnology applications in cancer. Annu. Rev. Biomed. Eng. 9, 257–288.

Oh, K.T., Yin, H., Lee, E.S., Bae, Y.H.J., 2007. Polymeric nanovehicles for anticancer drugs with triggering release mechanisms. Mater. Chem. 17, 3987–4001.

Ono, T., Nobori, T., Lehn, J.M., 2005. Dynamic polymer blends component recombination between neat dynamic covalent polymers at room temperature. Chem. Commun. 12, 1522–1524.

Ono, T., Fujii, S., Nobori, T., Lehn, J.M., 2007. Soft-to-hard transformation of the mechanical properties of dynamic covalent polymers through component incorporation. Chem. Commun. 1, 46–48.

O'Reilly, R.K., Hawker, C.J., Wooley, K.L., 2006. Cross-linked block copolymer micelles: functional nanostructures of great potential and versatility. Chem. Soc. Rev. 35, 1068–1083.

Overman, L.E., Smoot, J., Overman, J.D., 1974. The reduction of aryl disulfides with triphenylphosphine and water. Synthesis 1, 59–60.

Packer, L., Witt, E.H., Tritschler, H.J., 1995. Alpha-lipoic acid as a biological antioxidant. Free Radical Biol. Med. 19, 227–250.

Pilon, L.N., Armes, S.P., Findlay, P., Rannard, S.P., 2005. Synthesis and characterization of shell cross-linked micelles with hydroxy-functional coronas: a pragmatic alternative to dendrimers? Langmuir 21, 3808–3813.

Qi, K., Ma, Q.G., Remsen, E.E., Clark, C.G., Wooley, K.L., 2004. Determination of the bioavailability of biotin conjugated onto shell cross-linked (SCK) nanoparticles. J. Am. Chem. Soc. 126, 6599–6607.

Qin, Y., Sukul, V., Pagakos, D., Cui, C.Z., Jakle, F., 2005. Preparation of organoboron block copolymers via ATRP of silicon and boron-functionalized monomers. Macromolecules 38, 8987–8990.

Rijcken, C.J., Snel, C.J., Schiffelers, R.M., et al., 2007. Hydrolysable core-crosslinked thermosensitive polymeric micelles: synthesis, characterisation and in vivo studies. Biomaterials 28, 5581–5593.

Rungsardthong, U., Deshpande, M., Bailey, L., Vamvakaki, M., Armes, S.P., Garnett, M. C., et al., 2001. Copolymers of amine methacrylate with poly(ethylene glycol) as vectors for gene therapy. J. Controlled Release 73, 359–380.

Samarajeewa, S., Shrestha, R., Elsabahy, M., Karwa, A., Li, A., Zentay, R.P., et al., 2013. In vitro efficacy of paclitaxel-loaded dual-responsive shell cross-linked polymer nanoparticles having orthogonally degradable disulfide cross-linked corona and polyester core domains. Mol. Pharmaceutics 10, 1092–1099.

Shiraishi, K., Kawano, K., Minowa, T., Maitani, Y., Yokoyama, M.J., 2009. Preparation and in vivo imaging of PEG-poly(L-lysine)-based polymeric micelle MRI contrast agents. J. Controlled Release 136, 14–20.

Smith, A.E., Xu, X., Abell, T.U., Kirkland, S.E., Hensarling, R.M., McCormick, C.L., 2009. Tuning nanostructure morphology and gold nanoparticle "locking" of multi-responsive amphiphilic diblock copolymers. Macromolecules 42, 2958–2964.

Sun, H., Guo, B., Cheng, R., Meng, F., Liu, H., Zhong, Z., 2009. Biodegradable micelles with sheddable poly(ethylene glycol) shells for triggered intracellular release of doxorubicin. Biomaterials 30, 6358–6366.

Talelli, M., Iman, M., Varkouhi, A.K., Rijcken, C.J., Schiffelers, R.M., Etrych, T., et al., 2010. Core-crosslinked polymeric micelles with controlled release of covalently entrapped doxorubicin. Biomaterials 31, 7797–7804.

Tao, J., Liu, G.J., Ding, J.F., Yang, M.L., 1997. Cross-linked nanospheres of poly(2-cinnamoylethyl methacrylate) with immediately attached surface functional groups. Macromolecules 30, 4084–4087.

Thurmond, K.B., Kowalewski, T., Wooley, K.L., 1996. Water-soluble knedel-like structures: the preparation of shell-cross-linked small particles. J. Am. Chem. Soc. 118, 7239–7240.

Wang, Y.C., Li, Y., Sun, T.M., Xiong, M.H., Wu, J., Yang, Y.Y., et al., 2010. Core–shell–corona micelle stabilized by reversible cross-linkage for intracellular drug delivery. Macromol. Rapid Commun. 31, 1201–1206.

Wang, K., Luo, G.F., Liu, Y., Li, C., Cheng, S.X., Zhuo, R.X., et al., 2012. Redox-sensitive shell cross-linked PEG–polypeptide hybrid micelles for controlled drug release. Polym. Chem. 3, 1084–1090.

Wang, K., Liu, Y., Yi, W.J., Li, C., Li, Y.Y., Zhuo, R.X., et al., 2013. Novel shell-cross-linked micelles with detachable PEG corona for glutathione-mediated intracellular drug delivery. Soft Matter 9, 692–699.

Wang, J., Zhao, D., Wang, Y., Wu, G., 2014a. Imine bond cross-linked poly(ethylene glycol)-block-poly(aspartamide) complex micelle as a carrier to deliver anticancer drugs. RSC Adv. 4, 11244–11250.

Wang, X., Wang, L., Yang, S., Zhang, M., Xiong, Q., Zhao, H., et al., 2014b. Construction of multifunctionalizable, core-cross-linked polymeric nanoparticles via dynamic covalent bond. Macromolecules 47, 1999–2009.

Weaver, J.V.M., Tang, Y.Q., Liu, S., Iddon, P.D., Grigg, R., Armes, S.P., et al., 2004. Preparation of shell cross-linked micelles by polyelectrolyte complexation. Angew. Chem. 43, 1389–1392.

West, K.R., Otto, S., 2005. Reversible covalent chemistry in drug delivery. Curr. Drug Discov. Technol. 2, 123–160.

Xu, X.W., Smith, A.E., McCormick, C.L., 2009. Facile'one-pot' preparation of reversible, disulfidecontaining shell cross-linkedmicelles from a RAFT-synthesized, pH-responsive triblock copolymer in water at room temperature. Aust. J. Chem. 62, 1520–1527.

Xu, X.W., Flores, J.D., McCormick, C.L., 2011. Reversible imine shell cross-linked micelles from aqueous RAFT-synthesized thermoresponsive triblock copolymers as potential nanocarriers for "pH-Triggered" drug release. Macromolecules 44, 1327–1334.

Yue, J., Wang, R., Liu, S., Wu, S., Xie, Z., Huanga, Y., et al., 2012. Reduction-responsive shell-crosslinked micelles prepared from Y-shaped amphiphilic block copolymers as a drug carrier. Soft Matter 8, 7426–7435.

Zhang, Q., Remsen, E.E., Wooley, K.L., 2000. Shell cross-linked nanoparticles containing hydrolytically degradable, crystalline core domains. J. Am. Chem. Soc. 122, 3642–3651.

Zhang, A., Zhang, Z., Shi, F., Xiao, C., et al., 2013a. Disulfide crosslinked PEGylated starch micelles as efficient intracellular drug delivery platforms. Soft Matter 9, 2224–2233.

Zhang, A., Zhang, Z., Shi, F., Xiao, C., et al., 2013b. Redox-sensitive shell-crosslinked polypeptide-block-polysaccharide micelles for efficient intracellular anticancer drug delivery. Macromol. Biosci. 13, 1249–1258.

Zhao, X., Liu, P., 2015. Reduction-responsive core-shell-corona micelles based on triblock copolymer: novel synthetic strategy, characterization, and application as tumor microenvironment-responsive drug delivery system. ACS Appl. Mater. Interfaces 7, 166–174.

Ziegler, D., Nowak, H., Kempler, P., Vargha, P., Low, P.A., 1999. Treatment of symptomatic diabetic polyneuropathy with the antioxidant alpha-lipoic acid: a meta-analysis. Diabet Med 22, 114–121.

CHAPTER

Nanoemulsion preparation, characterization, and application in the field of biomedicine

6

Cristina Chircov[1] and Alexandru Mihai Grumezescu[2,3]

[1]*Faculty of Medical Engineering, University Politehnica of Bucharest, Bucharest, Romania*
[2]*Faculty of Applied Chemistry and Materials Science, University Politehnica of Bucharest, Bucharest, Romania* [3]*ICUB—Research Institute of University of Bucharest, University of Bucharest, Bucharest, Romania*

6.1 INTRODUCTION

The main characteristics of an ideal drug delivery system are based on the objective of maximizing the therapeutic effect by minimizing the toxicity. The progress of science and technology has allowed drug delivery systems to evolve from simple mixtures and pills to extremely sophisticated systems (Chime et al., 2014).

Emulsions are soft materials composed of droplets of one liquid dispersed in a different immiscible liquid that are typically obtained through an emulsification process (Kim and Mason, 2017). Droplet size of classical emulsions (macroemulsions) varies in the range 1–100 μm, while for nanoemulsions—also called miniemulsions, ultrafine emulsions, and submicron emulsions (Đekić and Primorac, 2017)—and microemulsions the size should be less than 500 and 100 nm, respectively (Gupta et al., 2016). Nanoemulsions act as solid sphere carriers with an amorphous, lipophilic, and negatively charged surface (Jaiswal et al., 2015). Depending on the component types, ratios, and preparation method the physicochemical properties of nanoemulsions are highly variable, being more suitable for some applications than others. Major applications of nanoemulsions involve drug delivery, especially for poorly soluble drugs, for the treatment of chronic diseases, cancer therapy, involving treatment and imaging applications, and gene therapy, as nonviral vectors (Scheme 6.1).

6.2 NANOEMULSIONS: CONCEPTS, TYPES, AND PROPERTIES

Nanoemulsions are kinetically stable (no apparent flocculation or coalescence over long periods of time), isotropic, and heterogenous systems (Rai et al., 2018)

Nanoarchitectonics in Biomedicine. DOI: https://doi.org/10.1016/B978-0-12-816200-2.00019-0

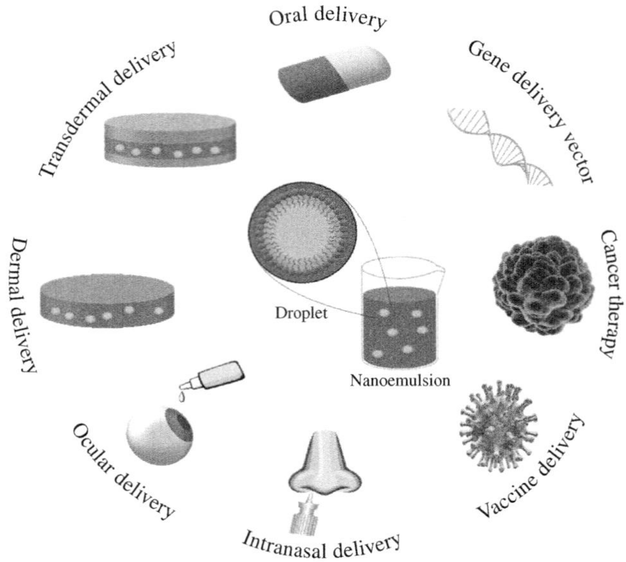

SCHEME 6.1

The main applications of nanoemulsions.

characterized by a transparent or translucent appearance and obtained by droplet dispersion of one liquid into another (Patravale et al., 2012).

Nanoemulsions are mainly composed of oil, emulsifying agents, and aqueous phases (Jaiswal et al., 2015). Considering the application and the different parameters such as oil toxicity, drug solubility, and route of administration, there are various types of oils—from medium- and long-chain fatty acids to essential oils—used for the formulation of nanoemulsions (Franklyne et al., 2016). A high solubility of the drug will minimize the volume of the formulation necessary to deliver the therapeutic agent (Talegaonkar and Negi, 2015). Oils used in nanoemulsion formulations are Captex 355, Captex 8000, Witepsol, Myritol 318, Capryol 90, Sefsol-218, triacetin, isopropyl myristate, castor oil, and olive oil (Chime et al., 2014). The emulsifying agent is the material through which the stability of the mixture of lipid and aqueous phase is achieved (Sharma et al., 2017). The behavior of surfactants available differs due to the variety of molecular structure, physicochemical properties, and functional attributes (Bai and McClements, 2016b). There are four types of surfactants based on the nature of the polar group, cationic, anionic, zwitterionic, and nonionic (Talegaonkar and Negi, 2015), and they may include Capryol 90, Gelucire 44/14, 50/13, Cremophor RH 40, Imwitor 191, 742, 780 k, 928, 988, Labrafil CS, M, 2125 CS, Lauroglycol 90, PEG MW $>$ 4000, Plurol Oleique CC 497, Poloxamer 124 and 188, Softigen 701, 767, Labrasol, Cremophor EL, Tween 20, Tween 60, and Tween 80 (Chime et al., 2014). A surfactant should have the following properties: it should reduce the surface tension to below 10 dyn/cm; it should be adsorbed rapidly around dispersed phase globule to form a complete and coherent film to prevent coalescence; it should optimally stabilize the system by building up an adequate zeta potential and viscosity; and it should be effective in a low concentration (Jaiswal et al., 2015). Due to the small droplet size and the large oil/water interface, stability of nanoemulsions can rarely be achieved by using only a surfactant, so the addition of a cosurfactant may be necessary (Talegaonkar and Negi, 2015). Cosurfactants usually used in the preparation of nanoemulsions are transcutol P, glycerin, ethylene glycol, ethanol, propanol, isopropyl alcohol, *n*-butanol, PEG 400, Carbitol, and propylene glycol (Chime et al., 2014).

Depending on the oil/water ratio, there are three types of nanoemulsions that can form: oil-in-water (O/W), water-in-oil (W/O), and a bi-continuous nanoemulsion (also known as multiple emulsion) which are all stabilized by the use of a mixture of surfactant and/or cosurfactant (Sharma et al., 2013). O/W nanoemulsions implicate the dispersion of oil phases in continuous aqueous phases (Rai et al., 2018) favored by the water solubility of the surfactant (Singh et al., 2017) in contrast to W/O nanoemulsions where water droplets are dispersed in continuous oil phases (Rai et al., 2018) which are favored by the lipid solubility of the surfactant (Singh et al., 2017). A bi-continuous nanoemulsion usually forms when the volumes of oil and water are approximately equal, the phases then being inter-dispersed through the system (Rai et al., 2018; Talegaonkar and Negi, 2015).

Nanoemulsions are systems with unique properties (small droplet size, transparence, stability, and tunable rheology) which make them desirable candidates for a variety of applications in medicine and the pharmaceutical, food, and cosmetic industries. Nanoemulsions become transparent with the decrease in droplet size. Naturally, the system will be more viscous with an increase of the fraction of the internal phase (Koroleva and Evgenii, 2012). Nanoemulsions tend to separate into different phases over time due to several destabilization mechanisms such as flocculation, coalescence, Ostwald ripening, and creaming/sedimentation. On the contrary, these systems are more stable when it comes to temperature and pH changes or dilution (Gupta et al., 2016). The interfacial tension at the limits should be in the range 1–10 mN/m (Koroleva and Evgenii, 2012).

Rheology tunability can be achieved with the control of the dispersed phase volume fraction and droplet size. Studies show that by changing the number of passes in the high-pressure homogenization process, nanoemulsions can show a slowly relaxing fluid or gel-like behavior. The same effect can be obtained by adding salt or depletion agents (Gupta et al., 2016; Helgeson, 2016). The possibility of rheology tuning is mostly important in the cosmetic industry as it often dictates the consumer's perception of the product (Gupta et al., 2016).

6.3 NANOEMULSION PREPARATION

Nanoemulsions are thermodynamically unstable systems, thus, their production requires the use of energy. The formation and stability of nanoemulsions is determined by the choice of preparation method. There are two main types of preparation methods: high energy methods and low energy methods (Jasmina et al., 2017).

6.3.1 HIGH ENERGY METHODS

High energy methods are based on the use of mechanical devices (microfluidizers, homogenizers, or ultrasonic devices) that are capable of generating high amounts of energy that form small oil droplets (Jasmina et al., 2017). Thus, as the amount of energy input increases, the droplet size decreases. This may constitute as an disadvantage since the high level of energy required to obtain nanoemulsions is cost-inefficient, especially since only around 0.1% of the energy produced is used for the actual emulsification (Solans and Solé, 2012).

6.3.1.1 Microfluidization

Microfluidization is one of the most effective high energy methods when it comes to the production of small droplets with uniform particle size distributions. The first step of microfluidization involves the blending of oil, water, and surfactant/cosurfactant in a high shear mixer. Using a pneumatic chamber, the O/W emulsion is forced in a homogenizer through an interaction chamber. There it is divided through two narrow channels at high velocities. The design of the channels causes the collision between the two, generating disruptive forces (turbulence, shear and cavitation) that form small droplets (Bai and McClements, 2016a).

6.3.1.2 High pressure homogenization

High pressure homogenization is one of the high energy methods to prepare O/W nanoemulsions that is mostly used for the continuous production of finely dispersed emulsions (Dumay et al., 2013). This method involves the use of pressure to force a liquid through a specifically designed homogenization valve to form suspended particles with a uniform size distribution. Modern homogenizers can apply pressures ranging between 20 and 100 MPa, reducing the particle size and increasing nanoemulsion stability by avoiding destabilization mechanisms such as creaming and coalescence. These homogenizers are largely used in the pharmaceutical, food, and cosmetic industries (Patrignani and Lanciotti, 2016).

6.3.1.3 Ultrasonication

Ultrasonication is widely used since it is one of the most cost-effective and easy-to-use high energy methods (Sugumar et al., 2016). This technique involves high-intensity ultrasound waves that generate shear forces, high pressure, and temperature through cavitation resulting in a change in the liquid's characteristics (Hashtjin and Abbasi, 2015). Through ultrasonication, the oil and aqueous phases are disrupted and blended into small dispersed oil droplets (Rodríguez-Burneo et al., 2017).

6.3.2 LOW ENERGY METHODS

Low energy methods are based on the system's properties and its complex interfacial hydrodynamic mechanisms (Jasmina et al., 2017). Low energy techniques are classified into either thermal or isothermal methods. The emulsion formation due to temperature-dependent changes in surfactant properties is typical for thermal methods while emulsion formation due to changes at constant temperature is typical for isothermal methods. Among isothermal methods it is worth mentioning spontaneous emulsification (self-emulsification) and emulsion phase inversion, whereas the phase inversion temperature method is an example of the thermal method (Komaiko and McClements, 2014, 2015). Spontaneous emulsification involves the addition of an oil–surfactant mixture to water, while for the emulsion phase inversion the emulsion forms when water is added to the oil–surfactant mixture (Komaiko and McClements, 2014). The addition of water to the mixture produces the formation of a W/O emulsion, then a multiple emulsion is formed after which there is an O/W emulsion (Komaiko and McClements, 2015). An emulsion is formed through the phase inversion temperature method when the temperature of the mixture of oil, water, and surfactant is rapidly dropped below the phase inversion temperature under continuous mixing (Komaiko and McClements, 2014). These methods have the advantage of producing small droplets without the necessity of using specialized equipment, thus, they are cost-effective and easy to use (Bai and McClements, 2016a).

6.4 CHARACTERIZATION AND EVALUATION OF NANOEMULSIONS

6.4.1 DROPLET SIZE, Z-POTENTIAL

The most common particle size analysis currently is dynamic light scattering, which can also be integrated with the Z-potential analyzer to gain information of surface charge. Nanoemulsion droplets can also be measured by using the disc centrifuge analyzer which can determine particle sizes ranging between 5 nm and 75 μm (Dasgupta and Ranjan, 2018).

6.4.2 STABILITY AND pH ANALYSIS

Stability of nanoemulsions can be determined by the naked eye by observing the occurrence of destabilization mechanisms (flocculation, coalescence, creaming, and sedimentation). Similarly, pH analysis is a key measurement because of the interaction of nanoemulsions with biological tissues and can be performed with a pH meter or a pH pen (Dasgupta and Ranjan, 2018).

6.4.3 VISCOSITY, CONDUCTIVITY

Viscosity and conductivity are parameters which provide information at the macroscopic level. Viscosity measurements can reveal the aspect of the existent micelles, while conductivity measurements indicate which of the phases is continuous as well as the mechanism of phase inversion (Chime et al., 2014).

6.4.4 MORPHOLOGY

Imaging techniques such as transmission electron microscopy and scanning electron microscopy can provide information regarding the morphology of nanoemulsions: size, distribution, and shape (Chime et al., 2014; Dasgupta and Ranjan, 2018). Another imaging technique that can be used is atomic force microscopy since it gives the same information and can serve as a confirmation of the others (Dasgupta and Ranjan, 2018; Sugumar et al., 2015).

6.5 APPLICATIONS OF NANOEMULSIONS

6.5.1 DRUG DELIVERY

6.5.1.1 Oral delivery of poorly soluble drugs

Since oral administration of pharmaceuticals is highly preferred, there is a special need to enhance the extremely low bioavailability of many hydrophobic drugs. Over the past few years, nanoemulsions have been studied as potential carriers for poorly soluble drugs, showing promising results for increasing bioavailability and modifying drug release. Since drugs are incorporated in the oil phase, its lipid solubility is highly important because it will determine the features of the formulation, such as the ability to maintain the drug solubilized and the volume of the formulation (Khani et al., 2016). Nanoemulsions as systems for oral delivery of poorly soluble drugs have numerous applications that are dependent on the primary drug effects.

Many studies of improved oral bioavailability of hydrophobic drugs have been reported. Studies performed on cardiovascular drugs have showed that incorporating mebudipine (Khani et al., 2016), a calcium blocker with relaxant effects on vascular and atrial smooth muscle, while cilostazol (Mahour et al., 2015), a therapeutic agent used for antithrombotic activity, in nanoemulsion systems can improve their bioavailability compared to suspensions, oily solutions, or micellar solutions. Similar results have been reported on hepatoprotective drugs like silymarin that also showed an enhanced antioxidant activity of the drug (Nagi et al., 2017), and on the anti-obesity drug *N*-oleoylethanolamine, a lipid mediator that acts as a satiety factor which showed no adverse effects or toxicity after continuous oral administration of the nanoemulsions (Wulff-Perez et al., 2014). Research performed on pyridoclax, an original oligopyridine used for the treatment of chemoresistant cancers, showed that entrapment in a nanoemulsion drastically increases the bioavailability of the drug, even when compared to di-hydrochloride salt synthesis where precipitation was observed (Groo et al., 2017).

Besides the obvious improvement of bioavailability, incorporation of hydrophobic drugs into nanoemulsion systems can also reduce the administered dose for therapeutic effects.

6.5.1.2 Transdermal drug delivery

The skin is an appealing site for drug delivery systems, being accessible especially due to its large surface being around 1.7 m^2. Drug administration through the skin is of high interest and there are continuously developing new and innovative approaches (Zaid Alkilani et al., 2015). The skin constitutes a variety of primary barriers which need to be overcome in order to improve drug targeting and release. Due to their small size, nanoemulsion particles can easily be absorbed through skin pores into the systemic circulation (Chime et al., 2014). Because nanoemulsions are sometimes insufficiently viscous and, therefore, can be removed from the skin, they need to be converted into gels (Wais et al., 2013; Yadav et al., 2015; Aqil et al., 2016; Khurana et al., 2013).

Transdermal drug delivery is a promising administration route for drugs. Compared to the more conventional drug delivery systems like oral or intravenous methods, transdermal administration has the advantages of painless drug delivery (Lee et al., 2017) and a special design that allows drug release at a predetermined and controlled rate, enhancing the bioavailability of the drug (Ahad et al., 2015). Since the release profiles are more uniform, the risk of toxic side effects is reduced to a minimum and patient compliance is improved due to reduced dosing frequencies. Transdermal drug delivery is mostly suitable for unconscious patients or those who rely on self-administration (Zaid Alkilani et al., 2015). Transdermal drug delivery is of great interest for chronic diseases that require long-term self-administration of drugs. Most studies are performed on chronic diseases such as hypertension (Yadav et al., 2015; Aqil et al., 2016), diabetes (Wais et al., 2013; Mostafa et al., 2015a), and rheumatoid arthritis (Khurana et al., 2013).

Hypertension is a cardiovascular disease of great interest because of its high prevalence (there are one billion affected individuals worldwide) and being a leading risk of stroke, ischemic heart disease, kidney disease, hardening of arteries, and eye damage if left untreated (Aqil et al., 2016). One study demonstrated that nanoemulsion formulation is a potential carrier for transdermal delivery. The selected drug is olmesartan, a lipophilic anti-hypertensive drug. Nanoemulsion formulations were obtained through spontaneous emulsification and clove oil was used as the oil phase due to the high lipid solubility of olmesartan. Results showed good pharmacokinetic features with increased bioavailability compared to oral administration (Aqil et al., 2016). Another anti-hypertensive drug used is nimodipine, which belongs to the class of calcium channel blockers. Similarly, results showed an improved performance, especially since the half-life of nimodipine is around 1−2 hours, which allows increases in dosing frequency (Yadav et al., 2015).

Diabetes is a severe disease associated with low insulin-levels or the improper cell response to insulin. Glibenclamide is a sulfonylurea which stimulates insulin release from pancreatic beta cells, but can be associated with the risk of hypoglycemia. The glibenclamide loaded O/W nanoemulsion system showed increased efficacy with improved bioavailability (Wais et al., 2013). One study focused on the potential of transdermal delivery of *Foeniculum vulgare* (fennel), an essential oil with antidiabetic and antioxidant activities. Nanoemulsion systems were prepared through ultrasonication method by incorporating the essential oil in the oil phase. Results showed an increased and prolonged antidiabetic effect because of the administration route (Mostafa et al., 2015a).

Transdermal drug delivery can also be used for the delivery of anti-inflammatories for the treatment of rheumatoid arthritis, an autoimmune disease which primarily affects the joints. The use of meloxicam, a nonsteroidal anti-inflammatory in a nanoemulsion system has been studied due to its favorable properties for transdermal administration. Caprylic acid was used as the oil phase and the system showed good attributes for transdermal delivery such as the release profile (Khurana et al., 2013).

Depending on the effect of the drug, applications of transdermal drug delivery can also involve cancer therapy, antioxidants, and hepatoprotective drugs, etc. (Shakeel and Ramadan, 2010; Gao et al., 2014; Mostafa et al., 2015b). Cancer therapy studies are based on the use of transdermally applied caffeine (1,3,7-trimethylhanthine) in a W/O nanoemulsion system that is suitable for the treatment of skin cancer. Although results have not shown improved anticancer efficacy, they showed a potential for using W/O nanoemulsions for transdermal delivery of caffeine (Shakeel and Ramadan, 2010). Another study showed good results for 10,11-methylenedioxycamptothecin loaded hyaluronic acid nanoemulsion system as a transdermal delivery system for keloid therapy (Gao et al., 2014). Cumin is an herb with significant antioxidant capacity. The administration of cumin essential oil through a transdermal nanoemulsion system showed increased antioxidant activity as well as hepatoprotective effects for more than 7 days. Applied once weekly the system could prevent several chronic diseases (Mostafa et al., 2015b).

6.5.1.3 Dermal drug delivery

Nanoemulsions are used as dermal drug delivery systems due to their distinct advantages: increased stability, mild nature, and low quantities of surfactants as well as its formulation through simple methods. Unlike other colloidal drug delivery systems, nanoemulsions do not present the limitation of low drug loading or irritating side effects. Compared to lipid nanoparticles, nanoemulsions achieve a higher rate of skin permeation and drug accumulation (Klang et al., 2015).

Dermal drug delivery through nanoemulsions is also suitable for chronic and autoimmune diseases, such as psoriasis and vitiligo (Barradas et al., 2017; Kaur et al., 2017). The use of clobitasol propionate, a corticosteroid with anti-inflammatory activity, in combination with calcipotriol, a vitamin D analog, has been reported in the literature and it proved the inhibition of hyperproliferation of keratinocytes in epidermis. The use of nanoemulsions prepared through self-emulsification improves efficiency due to the permeation through psoriasis plakes and the deliverability of drugs due to hydration mechanisms. Besides a good healing process, the system reduced skin irritation compared to the free drug in gel (Kaur et al., 2017). Another type of treatment involves the use of 8-methoxypsoralen, a naturally occurring tricyclic furocoumarins, in association with ultraviolet (UV) light. The use of nanoemulsions is necessary to reduce severe side effects caused by the combination of the drug and UV light. Since nanoemulsions can have low viscosities, they need to be thickened with hydrogels formed by polymers like chitosan (Barradas et al., 2017).

The use of antiviral drugs for dermal delivery through nanoemulsions has been studied for a variety of applications, especially for the treatment of herpes. The use of incorporated aciclovir in W/O/W nanoemulsions prepared by the phase inversion temperature method showed good skin penetration of the drug (Schwarz et al., 2012). However, new studies show that the use of aciclovir and penciclovir determined the apparition of drug-resistant strains. A new approach has been suggested; the use of gallates, *n*-alkyl esters of gallic acid with antioxidant, antitumoral, antibacterial, antifungal, and antiviral activity. The self-emulsifying nanoemulsions with incorporated pentyl gallate showed good skin penetration and antiherpetic activity (Kelmann et al., 2016).

Anti-inflammatory activity of dermally delivered drugs using nanoemulsion has been studied (Alvarado et al., 2015; Isailović et al., 2016). Aceclofenac, a nonsteroidal anti-inflammatory drug, has been administered through a nanoemulsion drug delivery system and showed good skin absorption without cutaneous adverse effects. The nanoemulsion was prepared through the high pressure homogenization method (Isailović et al., 2016). Another study used oleanane acid and ursolic acid, anti-inflammatory and anti-carcinogenic compounds belonging to the class of pentacyclic triterpenes. The systems showed adequate physical properties with high penetration capacity and greater anti-inflammatory effect than in dermal delivery systems (Alvarado et al., 2015).

Other dermal drug delivery applications of nanoemulsions involved the treatment of candidiasis and aspergillosis with amphotericin B antifungal (Sosa et al., 2017) and wound healing by using a curcumin (diferuloyl methane) loaded chitosan-based nanoemulsion gel (Thomas et al., 2017).

6.5.1.4 Ocular drug delivery

Ocular application is one of the most common drug administration routes for the treatment of various eye diseases. Most liquid ophthalmic formulations have a low bioavailability (typically between 1% and 10%) caused by continuous lacrimal drainage. Recently, O/W nanoemulsions have gained great interest as vehicles for improved ocular bioavailability of drugs (Gallarate et al., 2013).

Among the common applications, glaucoma is of high significance in research since it is one of the leading causes of irreversible blindness worldwide (Morsi et al., 2017). Timolol maleate, a β-adrenergic-blocking agent, is frequently used to lower the intraocular pressure. The incorporation of timolol maleate in the O/W nanoemulsions lead to insignificant changes in physical properties. The presence of thickening and mucoadhesive polymers, along with chitosan, not only increased viscosity and, therefore, prolonged residence time on the eye surface, but it also improved drug permeation. Final results demonstrated the use of nanoemulsions as a promising drug delivery system for ocular application (Gallarate et al., 2013). Another drug used for reducing intraocular pressure is acetazolamide, a carbonic anhydrase inhibitor. The study focused on the drug release using a system which converted the nanoemulsion into a gel formed in situ. The effect of the system applied directly on the ocular surface showed slow release with superior lowering of intraocular tension compared to tablets or common eye drops (Morsi et al., 2017).

Nanoemulsions were also studied as carriers for lutein, an effective medication in age-related macular degeneration, due to the increased bioavailability. Physicochemical properties of nanoemulsions were not affected by the use of lutein and drug release properties were significantly enhanced (Lim et al., 2016). The delivery of antisense oligonucleotides can reduce neovascularization complicating age-related macular degeneration because they can interfere with gene expression in targeted cells. Cationic nanoemulsions can protect the degradation of antisense oligonucleotides. Thus, cationic nanoemulsions for delivery of antisense oligonucleotides have a great potential in the prevention of ocular neovascularization (Hagigit et al., 2012). The use of immunosuppressants such as cyclosporin A onto the eye surface is necessary in cases like cornea transplant rejection or keratoconjunctivitis sicca (dry eye disease). Because of the poor aqueous solubility of the drug it must be encapsulated in nanoemulsions. The use of chitosan solution improves mucoadhesive strength of nanoemulsions. The system showed high ocular retention and improved pharmacokinetics and biodistribution of cyclosporin A (Akhter et al., 2016).

6.5.1.5 Intranasal drug delivery

Intranasal drug delivery has recently become a novel noninvasive technique used for drug targeting to the brain. One of the main advantages is the presence of a unique connection between the olfactory epithelium and central nervous system provided by the olfactory and trigeminal nerves. This route helps drug delivery directly into the brain by bypassing the blood–brain barrier. In addition, extensive hepatic and intestinal metabolism is avoided and there is better patient compliance, reducing systemic side effects caused by increased exposure of the drug (Abdou et al., 2017; Haider et al., 2018). However, intranasal delivery is affected by mucociliary clearance and reduced nasal permeability of drugs. To overcome these problems, nanoemulsions can be used as colloidal drug delivery systems (Sood et al., 2014).

Such systems have been used as treatments for Alzheimer's disease, a neurodegenerative disorder that causes the death of brain cells. Rivastigmine hydrochloride, a choline esterase inhibitor, is a suitable drug candidate, but due to its hydrophilicity it shows low concentration in the brain and a bioavailability of 36%, thus, necessitating frequent dosing. Permeation studies across ophthalmic mucosa showed great potential of rivastigmine hydrochloride-loaded nanoemulsions as intranasal delivery systems reducing the side effects of frequent dosing (Haider et al., 2018). Curcumin has also showed potential for treating Alzheimer's disease. The addition of chitosan enhances the mucoadhesive property of nanoemulsions. Incorporated curcumin chitosan-based nanoemulsion showed improved nasal mucosa permeation, demonstrating potential in intranasal drug delivery (Sood et al., 2014). Other studies involved the use intranasal cationic nanoemulsions encapsulating an anti-tumor necrosis factor (TNFα) small interfering ribonucleic acid (siRNA) for potential anti-inflammatory therapy in neurodegenerative disorders. The use of siRNA nanoemulsions showed site-specific down regulation of TNFα, a cytokine factor toxic to neurons, demonstrating the potential to prevent neuroinflammation diseases by enhanced drug delivery to the brain due to the use of nanoemulsions (Yadav et al., 2016). Zolmitriptan is a drug used in migraine treatment that shows reduced bioavailability caused by the low concentration that reaches the brain. Zolmitriptan-loaded nanoemulsions were prepared by the ultrasonication method and chitosan was added as a mucoadhesive agent. Results showed increased bioavailability and treatment efficacy with no abnormalities caused in nasal mucosa (Abdou et al., 2017).

6.5.1.6 Vaccine delivery

The most common vaccine administration route is by inoculation with injections despite the disadvantage of the high risk of transmission of pathogens like an immunodeficiency virus, hepatitis B virus, or hepatitis C virus. Different strategies have been proposed to improve vaccine delivery, one being direct vaccination through the skin since it contains important immune system components such as Langerhans and dendritic cells. Nanoemulsions offer two main advantages as a vaccine delivery system: the possibility to include substances to improve transdermal absorption of lipophilic compounds and the presence of penetration enhancers like surfactants (Tamayo et al., 2017).

Reports in the literature discussed the administration of different vaccines using nanoemulsion delivery systems. Vaccines administered to prevent infection with pulmonary pathogens using interleukin-17 have been studied. The use of interleukin-17 nanoemulsion system has improved the efficacy of protection against the hypervirulent *Mycobacterium tuberculosis* and the limitation of tuberculosis (Ahmed et al., 2017). Enhanced memory cell formation and virus protection using a crystalline imiquimid solid nanoemulsion for transcutaneous vaccination have been demonstrated (Lopez et al., 2017). Nanoemulsion systems have been evaluated for their potential to incorporate vesicles from *Salmonella enterica* to facilitate penetration through the skin. This formulation enhanced antibody responses, demonstrating the possibility of using nanoemulsions for topical vaccination (Tamayo et al., 2017).

6.5.1.7 Parenteral drug delivery

Parenteral nanoemulsions incorporating therapeutic agents are widely used as drug delivery systems due to their many advantages, such as extended bioavailability, improved effectiveness, and the safety of drugs. Used in intravenous systems, the droplet size of nanoemulsions is critical because they may cause embolism (Hörmann and Zimmer, 2016).

Nanoemulsions were synthetized through an ultrasonication method and the incorporation of valproic acid, a therapeutic drug used for epilepsy, schizoaffective, and bipolar disorder, and it did not affect physical properties. The necessity to use nanoemulsions as carriers is based on the occurrence of severe side effects caused by the high dosage of valproic acid due to its high plasma protein binding that lowers its concentration in the brain (Tan et al., 2016). O/W nanoemulsions prepared by the high pressure homogenization method were studied as systems for parenteral nutrition (Mirković et al., 2017).

6.5.2 CANCER THERAPY

Cancer therapy involves a new theranostic approach, photodynamic therapy, which can be noninvasively applied to malignant and nonmalignant diseases and to visualize cancer cells through fluorescent imaging (Leandro et al., 2017). Photodynamic therapy relies on the combination of red light (600–800 nm), molecular oxygen, and a photosensitizing compound. The light excites the photosensitizer and it transfers its energy to oxygen, turning it into a reactive oxygen species (de Menezes Furtado et al., 2017). The photodynamic process will result in preferential absorption of the photosensitizer drug by the diseased tissue, causing irreversible photodamage to a specific region (tumor tissues and blood vessels) and activation of antitumor, immunity, and inflammatory responses (Monge-Fuentes et al., 2017; Castilho-Fernandes et al., 2017).

Since near-infrared region photons are preferred because they penetrate the most and they are the least harmful to healthy tissue, an intense infrared absorption is desirable for cancer imaging and photodynamic therapy (Leandro et al., 2017). Unfortunately, none of the photosensitizing compounds used so far are ideal for these applications (Monge-Fuentes et al., 2017).

Photosensitizer drugs are hydrophobic compounds and the incorporation into specific systems to increase their bioavailability are necessary. O/W nanoemulsions represent promising drug delivery systems for photodynamic therapy and cancer imaging (Castilho-Fernandes et al., 2017).

The use of chloro-aluminum phthalocyanine has been extensively studied as a potential theranostic agent (Leandro et al., 2017; Castilho-Fernandes et al., 2017; Franchi et al., 2016). Chloro-aluminum phthalocyanine is a second-generation photosensitizer, with photophysical properties suitable for photodynamic therapy and cancer imaging, such as absorption of far and near infrared light, high quantum yields and having long lifetime (Leandro et al., 2017; Castilho-Fernandes et al., 2017). The use of chloro-aluminum phthalocyanine-loaded nanoemulsion systems has been studied as a therapy for glioblastoma (Castilho-Fernandes et al., 2017), prostate cancer (Leandro et al., 2017), lung adenocarcinoma, and cervical adenocarcinoma (Franchi et al., 2016). The systems demonstrated efficient suppression of glioblastoma cells, inducing apoptosis or necrosis. Even though the study was performed in vitro, nanoemulsions proved to be suitable delivery systems for chloro-aluminum phthalocyanine to glioma cells (Castilho-Fernandes et al., 2017). A study on prostate cancer proved the better theranostic behavior of chloro-aluminum phthalocyanine loaded nanocapsules than that through chloro-aluminum phthalocyanine loaded nanoemulsions (Leandro et al., 2017). A comparison between lung and cervical adenocarcinoma showed an improved effect on cervical cancer cells, where increased cytotoxicity and apoptosis were observed (Franchi et al., 2016). Acai oil and extracts from *Tectona grandis* leaves were studied as new potential photosensitizer agents. The use of incorporated natural compounds in nanoemulsions showed efficiency in the treatment of melanoma, with significant cell death and tumor reduction (de Menezes Furtado et al., 2017; Monge-Fuentes et al., 2017).

Another approach for cancer therapy is based on the hyperthermia effect induced by applying a radiofrequency alternating magnetic field in nanoemulsions containing magnetic nanoparticles. The observed effect demonstrates the potential of this system for multimodal therapy due to the possibility of loading anticancer drugs or photodynamic agents (Lahiri et al., 2017, 2018).

Breast cancer treatment using camptothecin-loaded nanoemulsions stabilized with chitosan (Natesan et al., 2017) or Poloxamer (Sugumaran et al., 2017) has demonstrated the potential of nanoemulsions as carriers for cancer therapy. Loading of Paclitaxel in nanoemulsions for cancer treatment was observed. High tumor-accumulation and effectiveness demonstrates its potential as a promising candidate drug system for clinical applications (Chen et al., 2017). Additionally, coating of nanoemulsions with hyaluronic acid provides better tumor targeting and excellent stability (Kim and Park, 2017). Other studies proved the efficiency of incorporation of different anticancer drugs into nanoemulsions (Abu-Fayyad and Nazzal, 2017; Ahmad et al., 2017; Chen et al., 2018). Conjugation of gemcitabine to the γ-tocotrienol isomer of vitamin E and entrapping into nanoemulsions were studied for pancreatic cancer treatment, showing a significantly enhanced anticancer activity (Abu-Fayyad and Nazzal, 2017). The use of new-generation taxoid encapsulated into nanoemulsions showed better anticancer activity than peclitaxel systems (Ahmad et al., 2017). Another study showed better anticancer activity of lycobetaine-loaded liposomes than nanoemulsions (Chen et al., 2018).

An ultrasonic nanoemulsion based on the use of *Nigella sativa* essential oil was investigated for the treatment of breast cancer. The use of traditional medicinal systems induced cancer cell apoptosis, proving its potential in therapeutic applications for cancer (Periasamy et al., 2016).

6.5.3 GENE DELIVERY VECTOR

Gene therapy has gained a lot of interest as a promising strategy for a variety of hereditary and acquired diseases (Silva et al., 2016). Gene therapy is based on the delivery of nucleic acids through viral or nonviral vectors, because of their poor cell uptake and biological instability. Considering their advantages, such as low immune response, ease of production, cost-effectiveness, and safety, nonviral vectors have been extensively preferred (Teixeira et al., 2017). Liposomes, nanoparticles, nanoemulsions, and dendrimers, among others, are nonviral potential vectors which have been considered to replace viruses as an alternative to delivery nucleic acids (Silva et al., 2016). O/W cationic nanoemulsions were proposed as delivery systems for single- and double-stranded nucleic acids. Their use has the potential to overcome problems like particle aggregation. Plenty of research has been done to investigate the best production conditions of cationic nanoemulsion for applications in the gene therapy domain (Teixeira et al., 2017).

Cationic nanoemulsions have great potential to overcome most of the barriers that nucleic acids delivery pose. The formulation of an ideal system requires the consideration of oil–water–surfactant interactions, toxicity profile, and transfection efficiency (Teixeira et al., 2017).

Reports in the literature proved high efficiency of a nanoemulsion prepared by high pressure homogenization method and stabilized with different lipids as an intravenous gene transfer system. Transfection efficiency was observed for the treatment of mucopolysaccharidosis type I disorder and is measured as α-L-iduronidase enzyme activity (Fraga et al., 2015).

The ability to load plasmid DNA into a nanoemulsion system was studied. The addition of searylamine is mandator, since gene delivery requires a positive charge to the system. The ultrasonic stearylamine-containing nanoemulsion can be used for gene therapy due to its suitable stability, but further studies on transfection efficiency are necessary (Silva et al., 2016). The same study was performed on ultrasonic nanoemulsions stabilized with arginine, lysine, and histidine. The highest transfection efficiency was observed for lysine and arginine, followed by histidine, proving that positively charged cationic nanoemulsions have improved effects (Liu and Yu, 2010).

6.6 CONCLUSIONS

Nanoemulsions are extensively used for the delivery of therapeutic agents. These systems offer several advantages over conventional drug delivery systems. The main application of nanoemulsions in the field of biomedicine is the delivery of poor soluble drugs to the specific site. The main applications of nanoemulsions in the field of biomedicine include the delivery of poorly soluble drugs and active pharmaceutical compounds to the specific site through various administration routes, cancer therapy, as carriers for photosensitizer agents, hyperthermia effect, or anticancer drugs, and gene delivery, as nonviral vectors. Their main characteristics are the improvement of the bioavailability of drugs, being considered as a safe and effective delivery system. Further research and development must be carried out in the future regarding nanoemulsion systems.

REFERENCES

Abdou, E.M., Kandil, S.M., Miniawy, H.M.F.E., 2017. Brain targeting efficiency of antimigrain drug loaded mucoadhesive intranasal nanoemulsion. Int. J. Pharm. 529 (1), 667–677.

Abu-Fayyad, A., Nazzal, S., 2017. Gemcitabine-vitamin E conjugates: synthesis, characterization, entrapment into nanoemulsions, and in-vitro deamination and antitumor activity. Int. J. Pharm. 528 (1), 463–470.

Ahad, A., Al-Saleh, A.A., Akhtar, N., Al-Mohizea, A.M., Al-Jenoobi, F.I., 2015. Transdermal delivery of antidiabetic drugs: formulation and delivery strategies. Drug Discovery Today 20 (10), 1217–1227.

Ahmad, G., El Sadda, R., Botchkina, G., Ojima, I., Egan, J., Amiji, M., 2017. Nanoemulsion formulation of a novel taxoid DHA-SBT-1214 inhibits prostate cancer stem cell-induced tumor growth. Cancer Lett. 406, 71–80.

Ahmed, M., Smith, D.M., Hamouda, T., Rangel-Moreno, J., Fattom, A., Khader, S.A., 2017. A novel nanoemulsion vaccine induces mucosal interleukin-17 responses and confers protection upon *Mycobacterium tuberculosis* challenge in mice. Vaccine 35 (37), 4983–4989.

Akhter, S., Anwar, M., Siddiqui, M.A., Ahmad, I., Ahmad, J., Ahmad, M.Z., et al., 2016. Improving the topical ocular pharmacokinetics of an immunosuppressant agent with mucoadhesive nanoemulsions: formulation development, in-vitro and in-vivo studies. Colloids Surf., B: Biointerfaces 148, 19–29.

Alvarado, H.L., Abrego, G., Souto, E.B., Garduño-Ramirez, M.L., Clares, B., García, M. L., et al., 2015. Nanoemulsions for dermal controlled release of oleanolic and ursolic acids: in vitro, ex vivo and in vivo characterization. Colloids Surf., B: Biointerfaces 130, 40–47.

Aqil, M., Kamran, M., Ahad, A., Imam, S.S., 2016. Development of clove oil based nanoemulsion of olmesartan for transdermal delivery: Box–Behnken design optimization and pharmacokinetic evaluation. J. Mol. Liq. 214, 238–248.

Bai, L., McClements, D.J., 2016a. Development of microfluidization methods for efficient production of concentrated nanoemulsions: comparison of single- and dual-channel microfluidizers. J. Colloid Interface Sci. 466, 206–212.

Bai, L., McClements, D.J., 2016b. Formation and stabilization of nanoemulsions using biosurfactants: rhamnolipids. J. Colloid Interface Sci. 479, 71–79.

Barradas, T.N., Senna, J.P., Cardoso, S.A., Nicoli, S., Padula, C., Santi, P., et al., 2017. Hydrogel-thickened nanoemulsions based on essential oils for topical delivery of psoralen: permeation and stability studies. Eur. J. Pharm. Biopharm. 116, 38–50.

Castilho-Fernandes, A., Lopes, T.G., Primo, F.L., Pinto, M.R., Tedesco, A.C., 2017. Photodynamic process induced by chloro-aluminum phthalocyanine nanoemulsion in glioblastoma. Photodiagn. Photodyn. Ther. 19, 221–228.

Chen, L., Chen, B., Deng, L., Gao, B., Zhang, Y., Wu, C., et al., 2017. An optimized two-vial formulation lipid nanoemulsion of paclitaxel for targeted delivery to tumor. Int. J. Pharm. 534 (1), 308–315.

Chen, T., Gong, T., Zhao, T., Fu, Y., Zhang, Z., Gong, T., 2018. A comparison study between lycobetaine-loaded nanoemulsion and liposome using nRGD as therapeutic adjuvant for lung cancer therapy. Eur. J. Pharm. Sci. 111, 293–302.

Chime, S.A., Kenechukwu, F.C., Attama, A.A., 2014. Nanoemulsions—advances in formulation, characterization and applications in drug delivery. In: Sezer, A.D. (Ed.), Application of Nanotechnology in Drug Delivery. InTech, Rijeka, p. Ch. 03.

Dasgupta, N., Ranjan, S., 2018. Research Updates on Different Vitamins Based Nanoemulsions and Characterization of Nanoemulsions. An Introduction to Food Grade Nanoemulsions. Springer Singapore, Singapore, pp. 105–122.

de Menezes Furtado, C., de Faria, F.S.E.D.V., Azevedo, R.B., Py-Daniel, K., dos Santos Camara, A.L., da Silva, J.R., et al., 2017. *Tectona grandis* leaf extract, free and associated with nanoemulsions, as a possible photosensitizer of mouse melanoma B16 cell. J. Photochem. Photobiol., B 167, 242–248.

Đekić, L., Primorac, M., 2017. Chapter 3—Microemulsions and nanoemulsions as carriers for delivery of NSAIDsA2—Čalija, Bojan. Microsized and Nanosized Carriers for Nonsteroidal Anti-Inflammatory Drugs. Academic Press, Boston, MA, pp. 69–94.

Dumay, E., Chevalier-Lucia, D., Picart-Palmade, L., Benzaria, A., Gràcia-Julià, A., Blayo, C., 2013. Technological aspects and potential applications of (ultra) high-pressure homogenisation. Trends Food Sci. Technol. 31 (1), 13–26.

Fraga, M., de Carvalho, T.G., Diel Dda, S., Kretzmann Filho, N.A., Teixeira, H.F., Matte, U., 2015. Cationic nanoemulsions as a gene delivery system: proof of concept in the mucopolysaccharidosis I murine model. J. Nanosci. Nanotechnol. 15 (1), 810–816.

Franchi, L.P., Amantino, C.F., Melo, M.T., de Lima Montaldi, A.P., Primo, F.L., Tedesco, A.C., 2016. In vitro effects of photodynamic therapy induced by chloroaluminum phthalocyanine nanoemulsion. Photodiagn. Photodyn. Ther. 16, 100–105.

Franklyne, J.S., Mukherjee, A., Chandrasekaran, N., 2016. Essential oil micro- and nanoemulsions: promising roles in antimicrobial therapy targeting human pathogens. Lett. Appl. Microbiol. 63 (5), 322–334.

Gallarate, M., Chirio, D., Bussano, R., Peira, E., Battaglia, L., Baratta, F., et al., 2013. Development of O/W nanoemulsions for ophthalmic administration of timolol. Int. J. Pharm. 440 (2), 126–134.

Gao, Y., Cheng, X., Wang, Z., Wang, J., Gao, T., Li, P., et al., 2014. Transdermal delivery of 10,11-methylenedioxycamptothecin by hyaluronic acid based nanoemulsion for inhibition of keloid fibroblast. Carbohydr. Polym. 112, 376–386.

Groo, A.C., De Pascale, M., Voisin-Chiret, A.S., Corvaisier, S., Since, M., Malzert-Fréon, A., 2017. Comparison of 2 strategies to enhance pyridoclax solubility: nanoemulsion delivery system versus salt synthesis. Eur. J. Pharm. Sci. 97, 218–226.

Gupta, A., Eral, H.B., Hatton, T.A., Doyle, P.S., 2016. Nanoemulsions: formation, properties and applications. Soft Matter 12 (11), 2826–2841.

Hagigit, T., Abdulrazik, M., Valamanesh, F., Behar-Cohen, F., Benita, S., 2012. Ocular antisense oligonucleotide delivery by cationic nanoemulsion for improved treatment of ocular neovascularization: an in-vivo study in rats and mice. J. Controlled Release 160 (2), 225–231.

Haider, M.F., Khan, S., Gaba, B., Alam, T., Baboota, S., Ali, J., et al., 2018. Optimization of rivastigmine nanoemulsion for enhanced brain delivery: in-vivo and toxicity evaluation. J. Mol. Liq. 255, 384–396.

Hashtjin, A.M., Abbasi, S., 2015. Optimization of ultrasonic emulsification conditions for the production of orange peel essential oil nanoemulsions. J. Food Sci. Technol. 52 (5), 2679–2689.

Helgeson, M.E., 2016. Colloidal behavior of nanoemulsions: interactions, structure, and rheology. Curr. Opin. Colloid Interface Sci. 25, 39–50.

Hörmann, K., Zimmer, A., 2016. Drug delivery and drug targeting with parenteral lipid nanoemulsions—a review. J. Controlled Release 223, 85–98.

Isailović, T., Đorđević, S., Marković, B., Ranđelović, D., Cekić, N., Lukić, M., et al., 2016. Biocompatible nanoemulsions for improved aceclofenac skin delivery: formulation approach using combined mixture-process experimental design. J. Pharm. Sci. 105 (1), 308–323.

Jaiswal, M., Dudhe, R., Sharma, P.K., 2015. Nanoemulsion: an advanced mode of drug delivery system. 3 Biotech. 5 (2), 123–127.

Jasmina, H., Džana, O., Alisa, E., Edina, V., Ognjenka, R. (Eds.), 2017. Preparation of Nanoemulsions by High-Energy and Lowenergyemulsification Methods. Springer Singapore, Singapore.

Kaur, A., Katiyar, S.S., Kushwah, V., Jain, S., 2017. Nanoemulsion loaded gel for topical co-delivery of clobitasol propionate and calcipotriol in psoriasis. Nanomed. Nanotechnol. Biol. Med. 13 (4), 1473–1482.

Kelmann, R.G., Colombo, M., De Araújo Lopes, S.C., Nunes, R.J., Pistore, M., Dall Agnol, D., et al., 2016. Pentyl gallate nanoemulsions as potential topical treatment of herpes labialis. Thai J. Pharm. Sci. 105 (7), 2194–2203.

Khani, S., Keyhanfar, F., Amani, A., 2016. Design and evaluation of oral nanoemulsion drug delivery system of mebudipine. Drug Delivery 23 (6), 2035–2043.

Khurana, S., Jain, N.K., Bedi, P.M.S., 2013. Nanoemulsion based gel for transdermal delivery of meloxicam: physico-chemical, mechanistic investigation. Life Sci. 92 (6), 383–392.

Kim, H.S., Mason, T.G., 2017. Advances and challenges in the rheology of concentrated emulsions and nanoemulsions. Adv. Colloid Interface Sci. 247, 397–412.

Kim, J.-E., Park, Y.-J., 2017. High paclitaxel-loaded and tumor cell-targeting hyaluronan-coated nanoemulsions. Colloids Surf., B: Biointerfaces 150, 362–372.

Klang, V., Schwarz, J.C., Valenta, C., 2015. Nanoemulsions in dermal drug delivery. In: Dragicevic, N., Maibach, H.I. (Eds.), Percutaneous Penetration Enhancers Chemical Methods in Penetration Enhancement: Drug Manipulation Strategies and Vehicle Effects. Springer Berlin Heidelberg, Berlin, Heidelberg, pp. 255–266.

Komaiko, J., McClements, D.J., 2014. Optimization of isothermal low-energy nanoemulsion formation: hydrocarbon oil, non-ionic surfactant, and water systems. J. Colloid Interface Sci. 425, 59–66.

Komaiko, J., McClements, D.J., 2015. Low-energy formation of edible nanoemulsions by spontaneous emulsification: factors influencing particle size. J. Food Eng. 146, 122–128.

Koroleva, M.Y., Evgenii, V.Y., 2012. Nanoemulsions: the properties, methods of preparation and promising applications. Russ. Chem. Rev. 81 (1), 21.

Lahiri, B.B., Ranoo, S., Philip, J., 2018. Effect of orientational ordering of magnetic nanoemulsions immobilized in agar gel on magnetic hyperthermia. J. Magn. Magn. Mater. 451, 254–268.

Lahiri, B.B., Ranoo, S., Zaibudeen, A.W., Philip, J., 2017. Magnetic hyperthermia in magnetic nanoemulsions: effects of polydispersity, particle concentration and medium viscosity. J. Magn. Magn. Mater. 441, 310–327.

Leandro, F.Z., Martins, J., Fontes, A.M., Tedesco, A.C., 2017. Evaluation of theranostic nanocarriers for near-infrared imaging and photodynamic therapy on human prostate cancer cells. Colloids Surf., B: Biointerfaces 154, 341–349.

Lee, H., Song, C., Baik, S., Kim, D., Hyeon, T., Kim, D.-H., 2018. Device-assisted transdermal drug delivery. Adv. Drug Delivery Rev. 127, 35–45.

Lim, C., Kim, D.-W., Sim, T., Hoang, N.H., Lee, J.W., Lee, E.S., et al., 2016. Preparation and characterization of a lutein loading nanoemulsion system for ophthalmic eye drops. J. Drug Delivery Sci. Technol. 36, 168–174.

Liu, C.-H., Yu, S.-Y., 2010. Cationic nanoemulsions as non-viral vectors for plasmid DNA delivery. Colloids Surf., B: Biointerfaces 79 (2), 509–515.

Lopez, P.A., Denny, M., Hartmann, A.-K., Alflen, A., Probst, H.C., von Stebut, E., et al., 2017. Transcutaneous immunization with a novel imiquimod nanoemulsion induces superior T cell responses and virus protection. J. Dermatol. Sci. 87 (3), 252–259.

Mahour, R., Sahni, J.K., Sharma, S., Kumar, S., Ali, J., Baboota, S., 2015. Nanoemulsion as a tool for improvement of cilostazol oral bioavailability. J. Mol. Liq. 212, 792–798.

Mirković, D., Ibrić, S., Balanč, B., Knez, Ž., Bugarski, B., 2017. Evaluation of the impact of critical quality attributes and critical process parameters on quality and stability of parenteral nutrition nanoemulsions. J. Drug Delivery Sci. Technol. 39, 341–347.

Monge-Fuentes, V., Muehlmann, L.A., Longo, J.P.F., Silva, J.R., Fascineli, M.L., de Souza, P., et al., 2017. Photodynamic therapy mediated by acai oil (*Euterpe oleracea* Martius) in nanoemulsion: a potential treatment for melanoma. J. Photochem. Photobiol., B 166, 301–310.

Morsi, N., Ibrahim, M., Refai, H., El Sorogy, H., 2017. Nanoemulsion-based electrolyte triggered in situ gel for ocular delivery of acetazolamide. Eur. J. Pharm. Sci. 104, 302–314.

Mostafa, D.M., Abd El-Alim, S.H., Asfour, M.H., Al-Okbi, S.Y., Mohamed, D.A., Awad, G., 2015a. Transdermal nanoemulsions of *Foeniculum vulgare* Mill. essential oil: preparation, characterization and evaluation of antidiabetic potential. J. Drug Delivery Sci. Technol. 29, 99–106.

Mostafa, D.M., Kassem, A.A., Asfour, M.H., Al Okbi, S.Y., Mohamed, D.A., Hamed, T. E.-S., 2015b. Transdermal cumin essential oil nanoemulsions with potent antioxidant and hepatoprotective activities: in-vitro and in-vivo evaluation. J. Mol. Liq. 212, 6–15.

Nagi, A., Iqbal, B., Kumar, S., Sharma, S., Ali, J., Baboota, S., 2017. Quality by design based silymarin nanoemulsion for enhancement of oral bioavailability. J. Drug Delivery Sci. Technol. 40, 35–44.

Natesan, S., Sugumaran, A., Ponnusamy, C., Thiagarajan, V., Palanichamy, R., Kandasamy, R., 2017. Chitosan stabilized camptothecin nanoemulsions: development, evaluation and biodistribution in preclinical breast cancer animal mode. Int. J. Biol. Macromol. 104, 1846–1852.

Patravale, V., Dandekar, P., Jain, R., 2012. 1—Nanoparticulate systems as drug carriers: the need. Nanoparticulate Drug Delivery. Woodhead Publishing, pp. 1–28.

Patrignani, F., Lanciotti, R., 2016. Applications of high and ultra high pressure homogenization for food safety. Front. Microbiol. 7, 1132.

Periasamy, V.S., Athinarayanan, J., Alshatwi, A.A., 2016. Anticancer activity of an ultrasonic nanoemulsion formulation of *Nigella sativa* L. essential oil on human breast cancer cells. Ultrason. Sonochem. 31, 449–455.

Rai, V.K., Mishra, N., Yadav, K.S., Yadav, N.P., 2018. Nanoemulsion as pharmaceutical carrier for dermal and transdermal drug delivery: formulation development, stability issues, basic considerations and applications. J. Controlled Release 270, 203–225.

Rodríguez-Burneo, N., Busquets, M.A., Estelrich, J., 2017. Magnetic nanoemulsions: comparison between nanoemulsions formed by ultrasonication and by spontaneous emulsification. Nanomaterials 7 (7), 190.

Schwarz, J.C., Klang, V., Karall, S., Mahrhauser, D., Resch, G.P., Valenta, C., 2012. Optimisation of multiple W/O/W nanoemulsions for dermal delivery of aciclovir. Int. J. Pharm. 435 (1), 69–75.

Shakeel, F., Ramadan, W., 2010. Transdermal delivery of anticancer drug caffeine from water-in-oil nanoemulsions. Colloids Surf., B: Biointerfaces 75 (1), 356–362.

Sharma, S., Hegde, M., Sadananda, V., Matthews, B., 2017. In vitro study to evaluate laser fluorescence device for monitoring the effect of aluminum gallium arsenide laser on noncavitated enamel lesions. J. Dent. Lasers 11, 2–6.

Sharma, N., Mishra, S., Sharma, S., Deshpande, R., Sharma, R., 2013. Preparation and optimization of nanoemulsions for targeting drug delivery. Int. J. Drug Dev. Res. 5 (4), 37–48.

Silva, A.L., Marcelino, H.R., Verissimo, L.M., Araujo, I.B., Agnez-Lima, L.F., do Egito, E.S., 2016. Stearylamine-containing cationic nanoemulsion as a promising carrier for gene delivery. J. Nanosci. Nanotechnol. 16 (2), 1339–1345.

Singh, Y., Meher, J.G., Raval, K., Khan, F.A., Chaurasia, M., Jain, N.K., et al., 2017. Nanoemulsion: concepts, development and applications in drug delivery. J. Controlled Release 252, 28–49.

Solans, C., Solé, I., 2012. Nano-emulsions: formation by low-energy methods. Curr. Opin. Colloid Interface Sci. 17 (5), 246–254.

Sood, S., Jain, K., Gowthamarajan, K., 2014. Optimization of curcumin nanoemulsion for intranasal delivery using design of experiment and its toxicity assessment. Colloids Surf., B: Biointerfaces 113, 330–337.

Sosa, L., Clares, B., Alvarado, H.L., Bozal, N., Domenech, O., Calpena, A.C., 2017. Amphotericin B releasing topical nanoemulsion for the treatment of candidiasis and aspergillosis. Nanomed. Nanotechnol. Biol. Med. 13 (7), 2303–2312.

Sugumar, S., Mukherjee, A., Chandrasekaran, N., 2015. Nanoemulsion formation and characterization by spontaneous emulsification: investigation of its antibacterial effects on *Listeria monocytogenes*. Asian J. Pharm. 9 (1), 23.

Sugumar, S., Singh, S., Mukherjee, A., Chandrasekaran, N., 2016. Nanoemulsion of orange oil with non ionic surfactant produced emulsion using ultrasonication technique: evaluating against food spoilage yeast. Appl. Nanosci. 6 (1), 113–120.

Sugumaran, A., Ponnusamy, C., Kandasamy, P., Krishnaswami, V., Palanichamy, R., Kandasamy, R., et al., 2018. Development and evaluation of camptothecin loaded polymer stabilized nanoemulsion: targeting potential in 4T1-breast tumour xenograft model. Eur. J. Pharm. Sci. 116, 15–25.

Talegaonkar, S., Negi, L.M., 2015. Nanoemulsion in drug targeting. In: Devarajan, P.V., Jain, S. (Eds.), Targeted Drug Delivery: Concepts and Design. Springer International Publishing, Cham, pp. 433–459.

Tamayo, I., Gamazo, C., de Souza Rebouças, J., Irache, J.M., 2017. Topical immunization using a nanoemulsion containing bacterial membrane antigens. J. Drug Delivery Sci. Technol. 42, 207–214.

Tan, S.F., Masoumi, H.R.F., Karjiban, R.A., Stanslas, J., Kirby, B.P., Basri, M., et al., 2016. Ultrasonic emulsification of parenteral valproic acid-loaded nanoemulsion with response surface methodology and evaluation of its stability. Ultrason. Sonochem. 29, 299–308.

Teixeira, H.F., Bruxel, F., Fraga, M., Schuh, R.S., Zorzi, G.K., Matte, U., et al., 2017. Cationic nanoemulsions as nucleic acids delivery systems. Int. J. Pharm. 534 (1), 356–367.

Thomas, L., Zakir, F., Mirza, M.A., Anwer, M.K., Ahmad, F.J., Iqbal, Z., 2017. Development of curcumin loaded chitosan polymer based nanoemulsion gel: in vitro, ex vivo evaluation and in vivo wound healing studies. Int. J. Biol. Macromol. 101, 569–579.

Wais, M., Samad, A., Nazish, I., Khale, A., Aqil, M., Khan, M., 2013. Formulation development ex-vivo and in-vivo evaluation of nanoemulsion for transdermal delivery of glibenclamide. Int. J. Pharm. Pharm. Sci. 5 (4), 747–754.

Wulff-Perez, M., Pavon, F.J., Martin-Rodriguez, A., de Vicente, J., Alen, F., de Fonseca, F.R., et al., 2014. Preparation, characterization and in vivo evaluation of nanoemulsions for the controlled delivery of the antiobesity agent *N*-oleoylethanolamine. Nanomedicine 9 (18), 2761–2772.

Yadav, S., Gandham, S.K., Panicucci, R., Amiji, M.M., 2016. Intranasal brain delivery of cationic nanoemulsion-encapsulated TNFα siRNA in prevention of experimental neuroinflammation. Nanomed. Nanotechnol. Biol. Med. 12 (4), 987–1002.

Yadav, S.A., Singh, D.K., Poddar, S.S., 2015. Formulation, in-vitro and in-vivo evaluation of nanoemulsion gel for transdermal drug delivery of nimodipine. Asian J. Pharm Clin. Res. 8, 119–124.

Zaid Alkilani, A., McCrudden, M.T.C., Donnelly, R.F., 2015. Transdermal drug delivery: innovative pharmaceutical developments based on disruption of the barrier properties of the stratum corneum. Pharmaceutics 7 (4), 438–470.

CHAPTER

7

Exosome nanocarriers: a natural, novel, and perspective approach in drug delivery system

Jasvinder Singh Bhatti[1], Rajesh Vijayvergiya[2], Bhupinder Singh[3] and Gurjit Kaur Bhatti[3,4]

[1]Department of Biotechnology, Sri Guru Gobind Singh College, Chandigarh, India [2]Department of Cardiology, Post Graduate Institute of Medical Education and Research, Chandigarh, India [3]UGC Centre of Excellence in Nano Applications, University Institute of Pharmaceutical Sciences, Panjab University, Chandigarh, India [4]University Institute of Pharmacy Sciences, Chandigarh University, Mohali, Punjab, India

ABBREVIATIONS

miRNA	micro RNA
ILVs	intraluminal vesicles
MVE	multivesicular endosomes
EVs	extracellular vesicles
RLPs	retrovirus-like particles
MHC	major histocompatibility molecules
mRNA	messenger RNA
ESCRT	endosomal sorting complexes required for transport
IL	interleukins
DCs	dendritic cells
APCs	antigen presenting cells
CD	cluster of differentiation
NF-κB	nuclear factor kappa-light-chain-enhancer of activated B-cells
TNF-α	tumor necrosis factor-α
PD	Parkinson's disease
AD	Alzheimer's disease
DNA	deoxyribonucleic acid
RNA	ribonucleic acid
T2DM	type 2 diabetes mellitus
BBB	blood–brain barrier
siRNA	small interfering ribonucleic acid
EFGR	epidermal growth factor receptor

Nanoarchitectonics in Biomedicine. DOI: https://doi.org/10.1016/B978-0-12-816200-2.00008-6

7.1 INTRODUCTION

The extracellular vesicles (EVs) are naturally present in many biological fluidsand are classified by their functionality, morphological properties, and origin/biogenesis (Street et al., 2012; Bobrie et al., 2011; Buzas et al., 2014). EVs are mainly classified into four broad types according to their biogenesis as exosomes, microvesicles (MVs), apoptotic bodies, and retrovirus-like particles (RLPs) (EL Andaloussi et al., 2013; van der Pol et al., 2012). The origin of MVs, apoptotic bodies, and RLPs occurs by direct budding from the plasma membrane. However, exosomes originate by the process of exocytosis of the multivesicular bodies (MVBs). Recent studies demonstrated that MVs are bigger than exosomes (Akers et al., 2013).

MVs bear different nucleic acids and proteins such as actin, tubulin, β1 integrins, and vesicle-associated membrane protein 3 (Buzas et al., 2014). The main feature of all these vesicles is the endosomal biogenesis as intraluminal vesicles (ILVs) within multivesicular endosomes (MVEs), which is different from the origin of other vesicles shedding from the plasma membrane (Kalra et al., 2012; Raposo and Stoorvogel, 2013). Inward invagination of the membrane results in the continuous addition of vesicles in the lumen of MVE into the extracellular space and was first observed in rat reticulocytes (Harding et al., 2013). It has been recognized that after the cells undergo apoptosis they form small vesicles known as apoptotic bodies (Emanueli et al., 2015). In fact, these small vesicles were identified as debris and considered as a means of wrapping the leftovers of the dead cells following an approach that there should not be any collateral damage to the cells in surrounding area. On the other hand, evidence has already materialized the fact that the EVs, in the size range of 30 nm to 1 μm, are not released during cell death but necessarily have their own a biological function (Emanueli et al., 2015).

Exosomes are tiny-sized (ranging in 30–100 nm in diameter) membranous vesicles secreted by a variety of cell types in the body through the endosomal pathway. They were first reported in 1981 as exfoliation from monolayer cultures and subsequently demonstrated by Johnstone and his team (Johnstone et al., 1991; Trams et al., 1981). They are known to be associated with diverse pathological conditions and represent a variety of functions like delivering biomolecules and genetic material to recipient cells.

Exosomes are enriched with a bundle of specific molecules which are different from proteins normally located in the plasma membrane including tetraspanins (CD9, CD63, CD81, and CD82), heat-shock proteins (HSP60, HSP70, HSPA5, CCT2, and HSP90), major histocompatibility molecules (MHC-I and MHC-II), cytoskeleton proteins (actin, tubulin, cofilin, myosin, vimentin, fibronectin, advillin, talin, CAP1, keratins, and meosin), microRNA (miRNA), messenger RNA (mRNA), Annexins (I, II, IV, V, VI, VII, and X1), apoptotic bodies, and many more (Akers et al., 2013; Keller et al., 2006; Hristov et al., 2004).

Exosomes also contain a domain of lipids including cholesterol, sphingomyelin, ceramide, and ganglioside called membrane rafts (Fig. 7.1). There is another class of cytosolic proteins found in the exosomes includes Rabs, a family of GTPases involved in vesicular transport and protein complexes that regulate membrane fusion and docking. About 40 Rab proteins have been identified in the exosomes RAB27A, RAB27B, RAB11, and RAB35 (Bobrie et al., 2011; Ostrowski et al., 2010). Exosomes are special in their way as they possess particular cell surface markers as a legacy from the cell of their origin. These cell surface

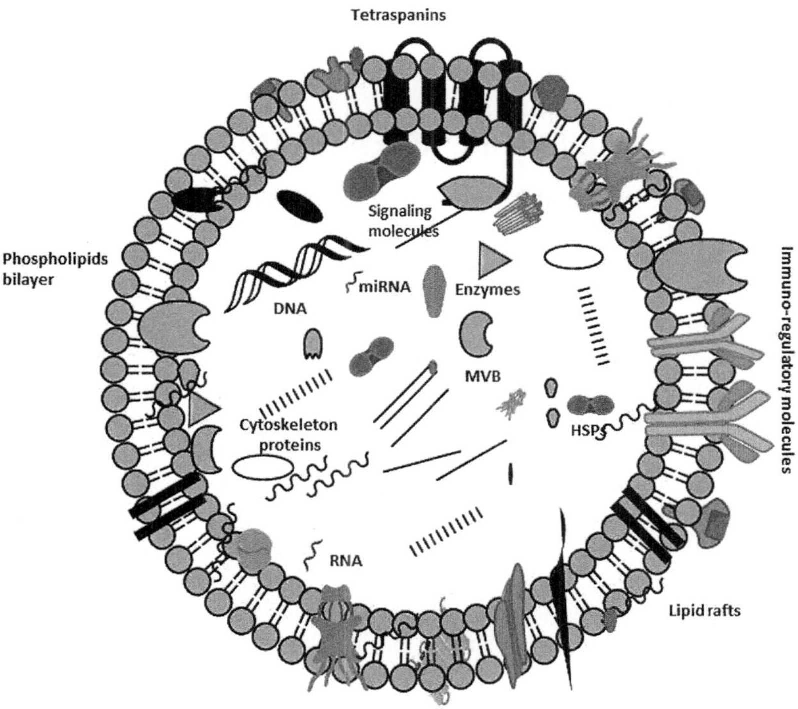

FIGURE 7.1

Structural components of the exosome. Exosomes are composed of a bundle of specific molecules including tetraspanins (CD9, CD63, CD81, CD82), HSPs (HSP60, HSP70, HSPA5, CCT2, and HSP90), MHCs (MHC-I and -II), cytoskeleton proteins (actin, tubulin, cofilin, myosin, vimentin, fibronectin, advillin, talin, CAP1, keratins, meosin), microRNA and mRNA, Annexins I, II, IV, V, VI, VII and X1, and apoptotic bodies. Exosomes also have a domain of lipids including cholesterol, sphingomyelin, ceramide, and ganglioside called membrane rafts. There is another class of cytosolic proteins found in exosomes including Rabs, a family of GTPases involved in vesicular transport and protein complexes that regulate membrane fusion and docking. *HSP*, Heat-shock protein; *MHC*, major histocompatibility molecule; *mRNA*, messenger RNA.

markers provide the exosomes the means of interaction with neighboring and distant cells (Skokos et al., 2003). Possession of these qualities has made exosomes a valuable therapeutic tool with inherent qualities in the biomedicine research area for targeted drug delivery to counter deadly diseases. In addition to lipids, proteins, and mRNA, exosomes also transfer miRNA to recipient cells which can alter the pathologic genetic expression (Valadi et al., 2007). It has been a long journey to come to understand the actual morphology and functions of exosomes concerning the dynamic role in intercellular communication by transporting a broad range of bioactive molecules between different cells and tissues (Tkach and Thery, 2016). This function allows exosomes to regulate many physiological activities, including the immune response.

7.2 EXOSOME BIOGENESIS, RELEASE, AND UPTAKE

The secretion of exosomes in the cells occurs by fusion of the MVB and the cell membrane's most vibrant compartments which are actively involved in internalization of extracellular ligands, post membranous invagination, and release of various kinds of endocytic vesicles. During these processes, some membrane proteins get transferred to these vesicles. Exosomes are formed by the inward budding of the endocytic vesicles and contain cytoplasmic components. Continuous formation of exosomes leads to the development of MVBs. The membrane of MVBs is semipermeable and ensures that cargoes are selectively loaded into the exosomes. Exosomes can have similar kinds of membrane constituents even if they have originated from different types of cells (Bartel, 2004). Exosomes keep the basic topology of the membrane intact; phosphatidylserine on the exterior while the cytosolic side of lipid bilayer remains inside. This arrangement causes the isolation of signal domains of the membrane receptors, thereby restricting their functionality of dampening the signal transduction. On the other hand, the luminal part remains exposed. Fig. 7.2 shows the exosome biogenesis and release of exosomes from the endocytic compartment of live cells.

The preliminary steps involved in sorting the proteins undergoing endocytosis are the inward budding processes on the cell surface, which can be clathrin-dependent and independent ways. The endocytotic vesicles are transferred to endosomes which have an internal acidic environment. The acidic pH gets reduced by some modification in the protein composition in the late endosomes which shed into the endosomal lumen to form MVBs and further leads to the formation of ILVs. MVBs are abundant in cholesterol which can have dire consequences for cell survival, so they get fused with the plasma membrane and release exosomes but MVBs with a low amount of cholesterol fuse with the lysosomes and are involved in proteolysis. There are mainly two mechanisms involved in the formation of exosomes; ESCRT (endosomal sorting complexes required for transport) dependent (which comprises four proteins O, I, II, III, and lipid-dependent)

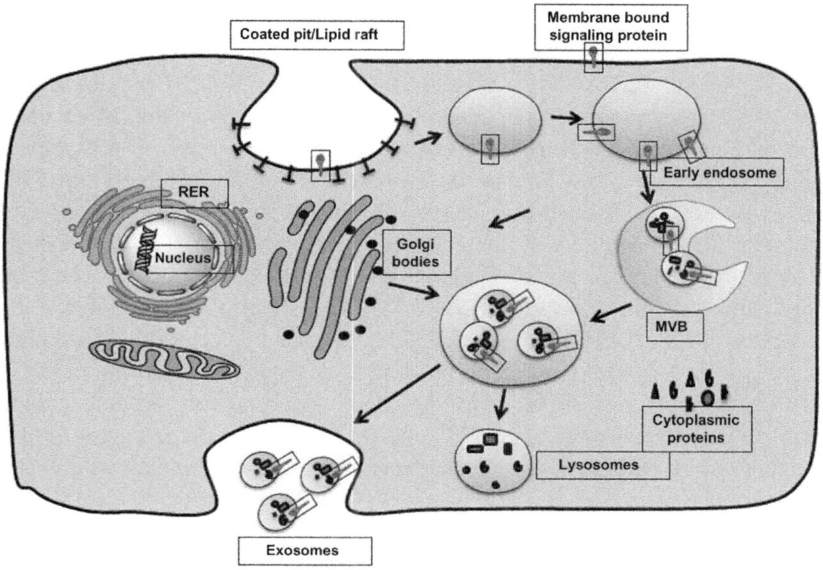

FIGURE 7.2 Mechanism of exosome biogenesis.

Early endosomes arose through clathrin-mediated endocytosis at the plasma membrane and converted into late endosomes by changing the protein content and acidification. Then, MVBs are derived from late endosomes budding into their endosomal lumen, and cholesterol-rich MVBs interact with the plasma membrane to secrete exosomes. Exosomes are formed by invagination MVBs that fuse with the plasma membrane. *MVB*, Multivesicular body.

and ESCRT independent. ESCRT-O separates ubiquitinated proteins, I and II complexes help in budding, and III completes the process and leads to the formation of ILVs in the MVBs rich population.

Exosomes firstly originate as ILVs then get matured to MVEs (Huotari and Helenius, 2011; Hanson and Cashikar, 2012). Tetraspanins such as CD81, CD9, and CD63 play a fundamental role in the composition of ESCRT-independent loading into exosomes by organizing the membrane into tetraspanin-enriched domains. The ESCRT complex mechanism, both dependent and independent, for packing of protein cargo into exosomes is a sign of the existence of different MVB and exosome populations. Different types of cancerous cells secrete different types of exosomes which vary in morphology and miRNA composition (Palma et al., 2012; Penfornis et al., 2016), but some secret exosomes which get differentiated by surface antigens having different protein content with various functions. The exosomes released from apical and basolateral cell membranes may have the heterogeneous population of vesicles, but the existence of controlled and specialized mechanisms control the selective, specific sorting of protein and miRNA cargo into these vesicles. While numerous researchers explained the biogenesis of exosomes, most of these mechanisms are still unclear.

7.3 EXOSOMES IN BIOLOGICAL FLUIDS

Exosomes play a vital role in intercellular communication or as disease biomarkers. An attribute that is independently measured and evaluated as an indicator of a biologic, pathogenic process or pharmacologic response to a therapeutic intervention is referred to a biomarker (Biomarkers Definitions Working, 2001). Indeed, exosomes are secreted by a variety of biological fluids and cells including extracellular fluids, amniotic fluid, cultured cells, cerebrospinal fluid, synovial fluid, bronchoalveolar lavage fluid, blood, breast milk, saliva, urine, semen, bile, and nasal mucus (Johnstone et al., 1987; Michael et al., 2010; Admyre et al., 2007; Keller et al., 2011; Vella et al., 2008; Skriner et al., 2006). Particularly, exosomes released from serum or plasma fluids are more accurate for the development of biomarkers. These blood-based biomarkers can predict patients' biochemical parameters more accurately and further predict the response to therapeutic intervention helping in earlier diagnosis and inhibiting the development of severe complications.

Exosomes from various tissues of the body released into the blood stream can be distributed to distance tissues, and in the specified target tissue they regulate gene transcription, posttranslational modification, and signal transduction. Assorted cellular cargoes released by exosomes become a source of blood-based biomarkers and these biomarkers point toward the pathophysiological mechanisms which could be extremely useful in identifying cardiovascular complications associated with diabetes. That is the reason that plasma- and serum-derived exosomes act as a specific molecular marker for various tissues and determine the pathophysiological condition if any. Exosomes released from the other fluids like milk, saliva, and cerebrospinal fluid can be tissue specific. Although the exosomes exist in almost every biological body fluid, their existence in the serum and plasma is known to be of particular importance and can be exploited as biomarkers. The most important step in the isolation of exosomes and their development into biomarkers is to make sure of the exosomes' enrichment in the extracellular milieu, which is quite complicated. Plasma and serum are the biological fluids which represent the extracellular environment of almost every organ, thereby; the efficacy of exosomes is entirely dependent on the matrix of the sample. The procedure of isolation of exosomes from biological fluids is a tedious method. Therefore many methods for isolation of exosomes have been standardized but a single optimized method has not been established till date. Out of available methods, the most widely used is differential centrifugation. Different sizes of vesicles get separated after the subsequent rounds of centrifugation, that is, large vesicles (2000 g, 10 minutes), MVs greater than 150 nm (10,000 g, 30 minutes), and exosomes (100,000 g, 70 minutes) (Schulte et al., 2016). However, this method also requires further optimization for the duration of each centrifugation according to the rotor type used. Sedimentation efficiency and density of EVs of varying sizes are dependent on the sedimentation path length and the duration of centrifugation, which should be calculated for every individual rotor to minimize the presence of nonexosomal vesicles and MVs (Livshits et al., 2016; Cocucci et al., 2007; Eken et al., 2008). The size-based characterization of the isolated exosomes is done with at least three exosomal markers, for example, tetraspanins, TSG101, and Alix. Nonexosomal markers including Grp94, GM130, and cytochrome C are used for confirming the absence/low levels of contamination (Lotvall et al., 2014). Moreover, morphological characters are also analyzed by electron microscopy. During ultracentrifugation, the membrane of the exosomes may get disrupted and lead to the formation of aggregates of different shapes and sizes which may pose difficulties in the interpretation of the biomarkers (Linares et al., 2015). Size exclusion chromatography is also one of the methods used to isolate exosomes by size and chances of getting vesicles of regular and homogeneous shape with negligible damage are more when using this method. Currently, standard kits are commercially available in the market to isolate exosomes from the lesser volume of biological fluids.

7.4 ROLE OF EXOSOMES IN THE PATHOGENESIS, DIAGNOSTICS, AND THERAPEUTICS

Exosomes display a wide variety of functions in the cell including cell–cell communication, diagnostic markers, and potential roles in therapeutics. Furthermore, recent studies demonstrated the possible role of exosomes as a biomarker for early diagnosis of diseases (Simpson et al., 2009; Pisitkun et al., 2006). In the past few years more attention has been focused on the role of exosomes in the pathophysiology of a variety of diseases including cancer, metabolic disorders, cardiovascular diseases, immune diseases, and neurodegenerative diseases.

7.4.1 EXOSOMES IN INFLAMMATORY DISEASES

Exosomes are known to play a vital role in treating various inflammatory reactions and autoimmune diseases as potential therapeutic agents. On the one hand, exosomes released from infected cells demonstrate a variety of functions. Further, the exosomes released from pathogen-stimulated brain cells have miRNAs which could interfere with the gene expression of various other cells (Gupta and Pulliam, 2014). In asthmatic patients, the release of exosomes from the bronchial epithelial cells in the lungs is enhanced and might be responsible for the increase in intercellular signaling. Exosomes released from the epithelial cells treated with interleukins 13 (IL-13) stimulate the inflammation along with the cell proliferation, but some specific drugs that result in the inhibition of exosomes production can lead to diminishing the inflammation (Kulshreshtha et al., 2013).

Exosomes from *Mycobacterium tuberculosis* infected macrophages have mycobacteria-derived antigens and cause the induction of inflammatory immune responses while inhibiting IFN-γ dependent activation of macrophages (Bhatnagar et al., 2007; Singh et al., 2011). In a similar way, exosomes released from *Mycobacterium smegmatis* or *Mycobacterium avium* infected macrophages will have great Hsp70 expression; however, the stimulation of the nuclear factor kappa-light-chain-enhancer of activated B-cells (NF-κB) pathway helps in the production of high levels of tumor necrosis factor-α (TNF-α). Exosomes with Hsp70 expression can produce a proinflammatory response with the release of TNF-κB by macrophages and promotion of NK cell activity (Gastpar et al., 2005). Curcumin loaded exosomes have been used to activate myeloid cells, which are involved in control and/or development of autoimmune diseases associated with inflammation. These hydrophobic compounds loaded onto exosomes were readily internalized by mouse lymphoma cells (EL-4) (Sun et al., 2010). Exosomal curcumin helps in decreasing the number of CD11b + Gr-1 + cells which are responsible for causing acute lung inflammation by increasing the solubility, stability, and bioavailability of curcumin after the encapsulation into the lipid bilayer of exosomes. Nevertheless, it becomes easier to deliver the encapsulated curcumin to the brain through intranasal administration. Recently, an in vivo study explained the efficacy and effectiveness of curcumin-loaded exosomes in the treatment of neuroinflammatory diseases (Zhuang et al., 2011).

Exosomes released from the dendritic cells (DCs) have received much attention lately. Pretreated DCs with adenovirus express IL-10 and exhibiting an anti-inflammatory response. Conversely, exosomes released from DCs with the expression of IL-10 were established as immunosuppressive and help in suppressing T-cell proliferation and the delayed type of hypersensitivity response in collagen-induced arthritis (Kim et al., 2005). Exosomes released from the DCs produced much improved therapeutic results in comparison to the parental DCs (Kim et al., 2007). Intravenous administration of 10 μg of TGF-β/exosomes may have prevented weight loss with reduced intestinal bleeding in mice with inflammatory bowled disease. The suppression of the symptoms might be due to stable TGF-β in exosomes following intravenous administration and, hence, increased activity (Cai et al., 2012b). These results indicated that exosomes were effective nanocarriers for the delivery of biological agents for treating inflammatory diseases.

Previous *in vitro* studies demonstrated that the exosomes released from the Epstein–Barr virus cause some changes in the B-cells, and present MHC class II molecules lying on the membrane to CD4 + T-cells (Raposo et al., 1996; Admyre et al., 2006; Liu et al., 2006). Other reports demonstrated that exosomes released from DCs stimulate CD8 + T-cell-dependent antitumor immune response in vivo experiments (Zitvogel et al., 1998). Some research reports indicate that exosomes released from the antigen presenting cells (APCs) have the ability to stimulate the T-cells, but other reports suggest that APCs need exosomes beforehand in the case of T-cell stimulation (Admyre et al., 2006). However, it was revealed that MHC molecules fostered with exosomes have the capability of coming together with the T-cell receptors and can initiate the production of CD4 + and CD8 + cells. Some reports came forward with the fact that MHC peptides loaded with exosomes, when transferred to DC, can stimulate naïve T-cells (Thery et al., 2002a). T-cell stimulation by means of exosomes is recognized to be dependent on the physiological conditions of the origin cell, but matured DC-derived exosomes trigger T-cell activation in a much more efficient manner compared to immature ones (Segura et al., 2005; Montecalvo et al., 2008). Exosomes also carry antigenic properties so they can be degraded in that way too. Exosomes derived from pathogen-infected cells need to carry some specific antigenic properties which may be inherited from that particular pathogen and can induce CD4 + and CD8 + T-cell proliferation (Montecalvo et al., 2008; Bhatnagar and Schorey, 2007).

Previous studies have revealed that exosomes secreted from cancerous cells release certain cytokines into the microenvironment which are attributed with triggering the immune response against tumorigenic tissue (Wolfers et al., 2001; Dai et al., 2006). On the contrary, exosomes derived from tumor tissue demonstrate exactly the opposite immune response in a manner of bearing different immunosuppressive molecules. This happens because of either decline in CD4 +, CD8 +, and natural killer cell stimulation or amplification in the differentiation of the immunosuppressive cells (Bobrie et al., 2011). Injecting exosomes derived from tumor cells supported metastasis by decreasing natural killer cells activation, but increase the differentiation of myeloid cells and lead to the development of melanoma and carcinoma (Liu et al., 2006). Till date, there is no relevant explanation for these controversial results reported by different scientists, but it could be speculated to be due to the heterogeneous population of exosomes.

7.4.2 EXOSOMES IN CANCER

Cancer is the leading cause of mortality worldwide. Current treatment methods include surgery, chemotherapy, and radiation which further raises toxicity concerns. Since these methods are unable to cure metastasis of the cancerous cells there is a great need to develop more effective and less toxic therapies. Immunotherapy is a much-explored way of treatment for many types of cancers. Antibodies have exhibited a clinical accomplishment for cancer immunotherapy as well (Scott et al., 2012). Recent research reports show the potential role of exosomes in cancer treatment as well as diagnosis and prognosis (Salido-Guadarrama et al., 2014; Mahaweni et al., 2013; Chaudhuri et al., 2011; Gross et al., 2012; Lai et al., 2013; Zhang et al., 2016; Chaput et al., 2003; Basu and Ludlow, 2016; Viaud et al., 2008; Cho et al., 2012; Soekmadji et al., 2013). The proteins and RNA content from tumor-derived exosomes are involved in disseminating malignancy to recipient cells. This may provide diagnostic information and aid in therapeutic selection for cancer patients. The most important fact about exosomes is their ability to stimulate the immune response that can be exploited for the development of cancer vaccines (Tan et al., 2010).

Many studies have demonstrated the therapeutic potential of DC-derived exosomes in cancer physiology (Chaput et al., 2006; Munich et al., 2012; Pitt et al., 2014, 2016; Viaud et al., 2010; Delcayre et al., 2005; Romagnoli et al., 2014; Zitvogel et al., 1999). Currently, much attention is being given toward exosomes and how to use them as gene delivery systems to specifically target tumor cells (Shtam et al., 2013). Exosomes also have the ability to act as a shuttle for antigen presentation and could be used as cancer vaccines in the near future (Natasha et al., 2014). Exosomes released by immune-suppressive DCs can suppress inflammatory responses and this potential could be used in treating autoimmune diseases (Kim et al., 2005; Yang and Robbins, 2012). Exosomes released from DCs carrying tumor-specific peptides can provoke cytotoxic T lymphocytes to respond and, consequently, inhibit tumor growth (Wolfers et al., 2001). Exosomes released from APCs contain tumor antigens which have been shown to encourage CD4 + and CD8 + T-cells and finally results in suppressing tumor growth (Zitvogel et al., 1998). Exosomes can initiate apoptosis by the activation of PTEN and GSK-3β that leads to the inactivation of PI3K/AKT which, in turn, plays a major role in tumor growth (Ristorcelli et al., 2008).

Cancer cells are efficient in modulating the normal functioning of immune cells of the body's defense mechanism and become malignant. The macrophages uptake of exosomes released by cancerous cells leads to the activation of NF-κB and, as a result, there are upregulation inflammatory cytokines (IL-6, TNF-α, GCSF, and CCL2) in the macrophages. Toll-like receptor 2 expressions on the surface of macrophages of these exosomes regulates the immune regulation which is mediated by the presence of palmitoylated protein ligands on the surface of tumor-derived exosomes (Chow et al., 2014).

7.4.3 EXOSOMES IN NEURODEGENERATIVE DISEASES

Exosomes are involved in a variety of neurological disorders including Parkinson's disease (PD) and Alzheimer's disease (AD) (Haney et al., 2015; Wu et al., 2016; Russo et al., 2012; Rajendran et al., 2006; Malm et al., 2016; Vella et al., 2016). They are released by numerous cell types, whether normal or pathogenic, into the extracellular environment. Particularly, exosomes released by tumor tissues initiate the metastasis and highlight the fact that exosomes have the ability to spread diseases within the body (Costa-Silva et al., 2015; Hoshino et al., 2015). This ability of exosomes to spread the diseases is considered to be responsible for neurodegenerative disorders. The cause of neurodegenerative diseases is mainly the deposition of certain misfolded and aggregated forms of protein molecules in brain tissue. These misfolded proteins further spread to different parts of the body as the disease progresses (Braak et al., 2003). Biochemical and molecular analysis of cerebral spinal fluid and blood of patients suffering from several neurodegenerative diseases illustrated the presence of many misfolded and aggregated proteins which are carried by exosomes. Exosomes contain many of the proteins which are prone to aggregate and are responsible for causing PD, AD,

Creutzfeldt–Jakob disease, and Amyotrophic lateral sclerosis (Brettschneider et al., 2015). Proteins like Aβ, tau, and α-synuclein have been reported to show prion-like activity in causing the pathological conversion of normal conformers of these proteins to those associated with disease. The aggregated nature and stable conformation of these proteins are resistant to proteolytic digestion. Many studies using cells and rat models came forward with evidence that the proteins causing neurodegeneration could be transmissible (Prusiner et al., 2015; Aguzzi and Rajendran, 2009). Although the accurate mechanism of transmission is yet to be discovered, various studies have revealed that it could be either by cell to cell communication or via tunneling nanotubes (Kanu et al., 2002; Gousset et al., 2009). A number of research studies have illustrated the involvement of exosomes in the transmission as neurodegenerative proteins were found to be secreted from the cell in association with exosomes (Fevrier et al., 2004). Due to their small size and relative stability, exosomes can cross the blood–brain barrier (BBB) and, therefore, reveal bright perspectives toward diagnosis and as therapeutic tools for future neuro medicine (Aryani and Denecke, 2016; Jarmalaviciute and Pivoriunas, 2016; Howitt and Hill, 2016; Vella et al., 2016; Tsilioni et al., 2014; Wu et al., 2016; Lai and Breakefield, 2012). An exosomal biomarker signature has also recently been established to detect AD (Lugli et al., 2015; Cheng et al., 2015).

7.4.4 EXOSOMES IN METABOLIC DISORDERS

The burden of metabolic disorders including type 2 diabetes, dyslipidemia, insulin resistance, and cardiovascular disease is exponentially increasing worldwide (Raj et al., 2014; Bhatti et al., 2016). Recent in vivo and in vitro studies demonstrated that exosomes secreted by a variety of cells play an important role in the regulation of the metabolic functions and may be responsible for the progression of some diseases (Lee et al., 2016). Exosomes are found to be rich in various biomolecules like proteins, mRNA, miRNA, noncoding RNA, and lipids. Recent studies have shown that exosomes isolated from biological fluids can be implemented as biomarkers for early diagnosis of metabolic diseases and can give an idea about the progression of the disease. Exosomes are capable of transferring DNA, RNA, and protein molecules which can regulate the metabolic functioning by interacting directly by transmembrane contact with the cell surface receptors of the nearby cells in remote areas (Aoki et al., 2007; Conde-Vancells et al., 2008; EL Andaloussi et al., 2013). Exosomes have the ability to perform vital functions on various platforms, including having a significant role in tumor progression (Rak and Guha, 2012), tissue regeneration (Timmers et al., 2011), enhancing the immune escape way and immune responses (Cai et al., 2012a; Simhadri et al., 2008; Chivet et al., 2012), transmission of neurological proteins (Bellingham et al., 2012; Emmanouilidou et al., 2010), and metabolic dysfunctions (Aswad et al., 2014).

Amplification in the number of exosomes can lead to obesity and inflammatory complications (Feng et al., 2010; Kim et al., 2012). Exosomes released from the surface molecules of skeletal muscle, platelets, and T-cells contribute in metabolic dysfunctions and atherogenic tendencies (Aswad et al., 2014; Diamant et al., 2002). Previous studies linked the functioning of the exosomes with their number that was assessed to be higher in the plasma of type 2 diabetes mellitus (T2DM) patients having insulin resistance. Exosomes have the ability to organize insulin signaling which has become an emerging concept in the exosome-mediated regulation of metabolism in T2DM (Aswad et al., 2014; Safdar et al., 2016; Jalabert et al., 2016; Lee et al., 2015). Diverse complications associated with the T2DM represent miscellaneous patterns of exosomal biomolecules signifying exosomes that may contribute to the pathophysiology of type 2 diabetes and its complications (Lawson et al., 2016; Lee et al., 2016). A study reported that Exosomes-like vesicles released by adipose tissue can act as a mode of communication between adipose tissues and macrophages and subsequent development of insulin resistance in a mouse model (Deng et al., 2009). Moreover, visceral adipose tissue (VAT)–derived exosomes are also functional in the intercellular transmission of signals as they possess adipokines (Lee et al., 2015). Skeletal muscle regulates the metabolism of the whole body and the exosomes released from it can facilitate the amendments in muscle homeostasis (Aswad et al., 2014). Keeping in view the merits of exosomes, efforts are being made to flourish them as novel prognostic, and even diagnostic, biomarkers, especially for the exosomes released from biological fluids. The presence of miRNA in blood, specific for the disease and condition/stage of illness, can act as a biomarker for prediction and diagnosing the complications associated with that particular disease (Jansen et al., 2013; Wang et al., 2014; Schulte et al., 2016; O'Neill et al., 2016).

Exosomes secreted by various cells and tissues emerged as a new therapeutic strategy in cardiovascular diseases (Sahoo and Losordo, 2014; Davidson et al., 2016; Ibrahim et al., 2014; Ong and Wu, 2015; Cervio et al., 2015; Huber and Holvoet, 2015; Yellon and Davidson, 2014; Emanueli et al., 2015; Zhang et al., 2017; Zhao et al., 2015; Huang et al., 2015; Lawson et al., 2016; Suzuki et al., 2016; Xitong and Xiaorong, 2016). It has been reported that after myocardial infarction, the cardiac tissue discharges many soluble chemokines, cytokines, and growth factors; induce inflammatory responses; and recruit stem and progenitor cells to accelerate the repair process (Liu et al., 2016). Exosomes are reported to be involved in cardiac repair and regeneration by communication at intercellular and tissue levels (Sahoo and Losordo, 2014; Yuan et al., 2016). The exosomes from cardiomyocyte can transfect endothelial cells, stem cells, fibroblasts, and smooth muscle cells to induce cellular changes. Exosomes derived from DCs improve cardiac function via activation of CD4+ T lymphocytes after myocardial infarction (Liu et al., 2016). A recent study demonstrated exosomes as potential alternatives to stem cell therapy in mediating cardiac regeneration (Ong and Wu, 2015).

7.5 EXOSOMES AS NANOCARRIERS

Nanoparticle drug delivery systems offer many advantages like their ability to load high quantities (Natasha et al., 2014), sustained release of the loaded drug, ability to load both drug along with the adjuvant for the induction of costimulation within a particular nanocarrier (Amoozgar and Goldberg, 2015; Hamdy et al., 2011). Although there has been rigorous and thorough experimentation, partial accomplishments have been developed in clinical trials owing to deficiency in safety and efficiency of the target drug to the desired cells. Different types of nanocarriers have been investigated at length for the targeted drug delivery of drugs/genes to a specific site of cancerous tissue. Nanocarriers are most-exploited in this field because they are more permeable along with the ability to accumulate into tumors passively as well as having the capability of longer retention, for example, DCs. Macrophages and the primary cells are the natural phagocytotic involved in antitumor immunity and inflammatory response. Nano-drug delivery systems are promising platforms to deliver immunomodulatory compounds to these cells (Amoozgar and Goldberg, 2015). Exosomes, also called biological nanocarriers, are becoming visible as a promising alternative to this new concept of drug delivery (Ha et al., 2016; Aryani and Denecke, 2016). Exosomes are secreted by many types of cells, under both normal and pathological conditions. The matter of great interest is that exosomes are secreted by tumor cells and are much more in number compared with the normal cells (Thery et al., 2002b). RNA carried by exosomes could be exploited as drug and gene delivery systems; however, donor cells should not release exosomes with any immunomodulation ability and should be stable enough for the sufficient period to carry the cargo.

7.6 EXOSOMES IN DRUG AND GENE DELIVERY

About 98% of drugs cannot cross the BBB. Even the nano-drugs that can solve this problem to some extent have associated issues of nanotoxicity and rapid drug clearance by the mononuclear phagocytic system (MPS) (Pardridge, 2012). Polyethylene glycol is used to solve this problem, but decreases the drug distribution to the brain and the interaction between target cells (Peng et al., 2013). These types of complications can be handled by using exosomes as a drug delivery system as it is a body's natural product and can be personalized to cross the BBB and increasing the drug distribution to the brain by decreasing the MPS drug clearance (Yoshida et al., 1992). The morphology of the exosomes can be exploited in a manner to facilitate drug delivery as they can be used as vehicles because of their many advantages over other drug delivery systems in practice like liposomal and nano-particulate drug delivery systems (Ferguson and Nguyen, 2016). The phenomenon of communication between cells is an integral part of the multicellular organisms for the maintenance of homeostasis. Gap junctions connect the nearby cells and distant intercellular communication can be facilitated via hormones by sending signals through circulatory systems. EVs can provide many advantages in distant cell communication as EVs can carry a cellular cargo of lipids, proteins, receptors, and active molecules to recipient cells (Lai et al., 2013).

7.6.1 DRUG DELIVERY VEHICLES FOR SMALL MOLECULES

Liposomal and the polymeric nanoparticle systems are preferred systems for drug delivery (Sercombe et al., 2015). Liposomes, a synthetic vesicle of the phospholipidic membrane, have the capacity of self-assembling in various sizes and shapes in an aqueous medium whereas polymeric nanoparticles can entrap, encapsulate, or attach drug molecules (Agrawal et al., 2014; Raemdonck et al., 2014). Both systems have been used for delivering anticancer, antifungal, and analgesic drugs. The main advantage of liposomes used as the delivery system is their stability in circulation for a longer duration without causing toxicity and, moreover, their capability of evading the immune system (Li et al., 2015; Aryani and Denecke, 2016). Polymeric nanoparticles have better stability compared to liposomes, but biocompatibility and toxicity issues are of major concern. Fig. 7.3 shows the role of exosomes as nanocarriers for the delivery of small molecules including miRNA, small interference RNA (siRNA), peptides, and drugs.

Above all, exosomes come with all the features of longer duration in the circulatory system, the ability to target the specific tissue, enhanced biocompatibility

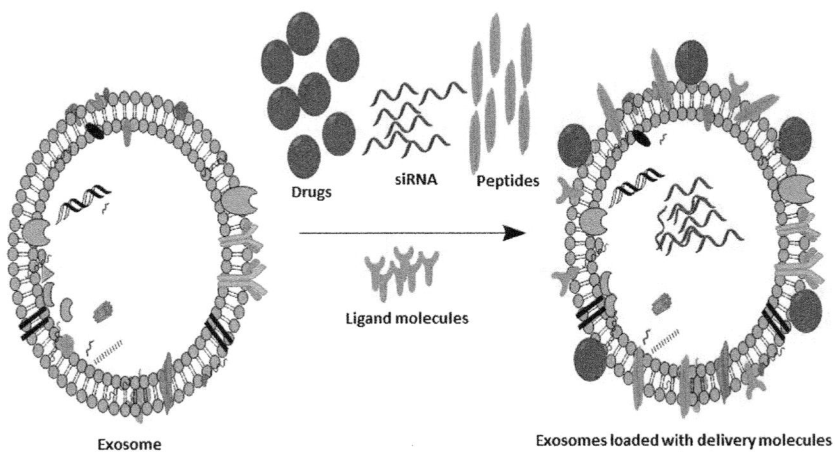

FIGURE 7.3 Exosomes as nanocarriers for drugs, siRNA, or peptide delivery.

Exosomes are nanosized and stable molecules capable of targeted delivery of nucleic acids, proteins, and drugs and, therefore, can be exploited for gene therapy and drug delivery systems. *siRNA*, Small interference RNA.

along with the minimal inherent toxicity issues. Hence exosomes are preferred over previously used drug delivery systems due to their unbeatable characteristics (Turturici et al., 2014; Rani and Ritter, 2016). Extensive research has been done using exosomes as vehicles for therapeutic drug delivery. Exosome complexes with curcumin enhance efficacy and safety when administered in cancer patients (Dhillon et al., 2008). Experiments on exosomes carrying curcumin demonstrate the increased solubility, stability, bioavailability, and antiinflammatory activity both in vitro and in vivo. In in vitro experiments, when macrophages were treated with exosomal curcumin, the production of the TNF and IL-6 is less in comparison to those treated with curcumin alone, proving the antiinflammatory capability of exosomes. In lipopolysaccharide (LPS) induced a septic shock model in which CD11bþGr-1þ cells increases response to LPS in the lungs and leads to acute lung inflammation. Exosomal curcumin showed significantly longer survival compared with mice treated with curcumin alone, again verifying the fact that exosomes enhance the antiinflammatory characteristics of curcumin. Exosomes illustrate the potential of brain drug delivery across the BBB by carrying paclitaxel and doxorubicin drugs and also demonstrate the transport mechanism in a zebrafish model (Yang et al., 2015). Exosomes were isolated from the glioblastoma astrocytoma U-87 MG, endothelial bEND.3, neuroectodermal tumor PFSK-1, and glioblastoma A-172 loaded with rhodamine 123, paclitaxel, and doxorubicin. Brain tissue was examined for the presence of the rhodamine 123 fluorescence and drug distribution observed in the brain region of the zebrafish embryos suggested the ability of exosomes to deliver the drug across the BBB. Further, in subsequent experiments, the primary brain cancer model was developed and anticancer drugs loaded with or without exosomes were applied. Drugs loaded with the exosomes showed the therapeutic efficacy of the exosomes better than the drug alone. Exosomes have the potential to deliver the drugs to the brain tissue by crossing the BBB and helps tremendously in curing brain cancers and other neurological disorders (Ha et al., 2016).

7.6.2 DRUG DELIVERY VEHICLES FOR PROTEINS

Exosomes can also have the ability to deliver molecules bigger in size. Mounting evidence shows that exosomes loaded with the antioxidant protein catalase was successfully delivered across the BBB resulting in an improved disease state in PD (Haney et al., 2015). Patients suffering from PD have lower levels of antioxidant enzymes like catalase and superoxide dismutase levels and cause oxidative stress along with neurodegeneration (Ambani et al., 1975; Riederer et al., 1989; Abraham et al., 2005). Catalase can neutralize one million free radicals per second per molecule in a single catalytic reaction cycle (Haney et al., 2015), but the delivery of catalase across the BBB is a big task. Exosomes have undertaken this tedious task and fulfilled it efficiently, so are a promising option in PD therapy.

7.6.3 DRUG DELIVERY VEHICLES FOR NUCLEIC ACIDS

Exosomes are exceptionally excellent delivery systems which can also carry DNA and RNA-like big and heavy molecules to the target cells, thereby instigating genetic modification in the biological processes. This strategy can be employed as a promising tool for delivering genetic material and altering the gene expression in certain diseases and help improve gene therapy.

7.6.4 SMALL INTERFERENCE RNA

siRNA is a molecule with low stability and it degrades quickly. It can be used to change gene expression and can be very useful in gene therapy to alter pathological symptoms. Exosomes can deliver siRNA to the cell of interest (Alvarez-Erviti et al., 2011; Wahlgren et al., 2016). A recent study explored the ability of exosomes to transfer nucleic acids to cells in a specific manner using elf exosomes loaded with chemically modified siRNAs and displaying specific targeting molecules (Alvarez-Erviti et al., 2011). The DCs were modified to express LAMP-2b fused to the central nervous system-specific RVG peptide that specifically binds to the acetylcholine receptor. Exosomes were loaded with exogenous siRNA by electroporation and, after in vitro experiments, were injected into mice together with a series of control exosome preparations. The injection of RVG exosomes resulted in a significant knockdown of GAPDH mRNA in the brain regions expressing the target of the RVG ligand-nicotinic acetylcholine receptors. These results not only showed the therapeutic potential of RVG exosome technology for new therapeutic approaches against neurodegenerative diseases, but also strongly supported the use of implemented and innovative exosome technologies for targeting therapies for a number of diseases. The siRNA delivery through exosomes has numeral benefits over typical viruses, lipid nanocarriers, and polycationic delivery mediators currently in practice (van den Boorn et al., 2011). Most importantly, exosomes deliver their load straight into the cytosol, evading the requirement for endosomal escape, while their inertness circumvents clearance in the extracellular environment (van den Boorn et al., 2013). Many studies reported exosomes as an efficient therapeutic vehicle for exogenous and to T-cells and monocytes (Wahlgren et al., 2012). There is no other vehicle available for delivering genetic material, but exosomes proved to be safe, efficient, and target specific. Electroporation was found to be the best method for successful introduction of siRNA (Wahlgren et al., 2016). Exosomes with siRNA in recipient cells showed a decrease in MAPK1 expression indicating downregulation of the specific gene and successful gene silencing. Exosomes delivered siRNA against RAD51, a eukaryotic gene protein which assists in DNA repair and stops the proliferation of cancer cells (Shtam et al., 2013). Recent studies indicated that exosomes could deliver the exogenous molecules to the cells with all the functions still intact (Banizs et al., 2014).

7.6.5 MICRORNA

miRNAs are noncoding RNAs with nonprotein nucleotides and are present in eukaryotic cells and considered as key functional elements. The exosomes are natural carriers of miRNA that could be exploited as an RNA drug delivery system. miRNA-containing exosomes are found in a variety of body fluids and cell regulate gene expression in the target cells (Kosaka et al., 2010; Thery, 2011). miRNA bind to the complementary sequences on the mRNA and function to control the posttranslational gene expression (Bartel, 2004, 2009). A recent study showed that exosomes can efficiently deliver miRNA to epidermal growth factor receptor (EGFR)-expressing breast cancer cells (Ohno et al., 2013). The epidermal growth factor and EGFR-specific peptide (GE11) were incorporated onto the surfaces of exosomes that carried let-7a in order to deliver let-7a to EGFR-expressing cancer tissue. Mounting evidence suggests that miRNA, like let-7a, functions as a tumor suppressor and prevents the development of cancer by reducing RAS and high-mobility group AT-hook protein (HMGA2) expression (Ohno et al., 2013). In this study, modified exosomes with GE11 peptide or EGF on their surfaces delivered miRNA to EGFR-expressing cancer tissues; intravenously injected exosomes targeting EGFR delivered let-7a specifically to xenograft breast cancer cells in RAG2 − / − mice. These data indicate that exosomes targeted to EGFR-expressing cells may provide a platform for miRNA replacement therapies in the treatment of various cancers. Numerous human tumors of epithelial origin display elevated EGFR expression, suggesting that EGFR could serve as a receptor target in cancer drug delivery systems (Gazdar, 2010; Harichand-Herdt and Ramalingam, 2008; Huang and Harari, 1999; Woodburn, 1999). Table 7.1 shows the therapeutic cargo molecules loaded into exosomes in cancer (Johnsen et al., 2014).

Recent studies also suggested exosomal miRNA as a biomarker of AD (Lugli et al., 2015; Cheng et al., 2015) and a variety of cancers including acute myeloid leukemia, colorectal cancer, and hepatocellular carcinoma (Hornick et al., 2015; Matsumura et al., 2015; Ogata-Kawata et al., 2014; Sugimachi et al., 2015). Among proteins, lipids, DNA, and RNA, the miRNA is found in higher concentrations (Goldie et al., 2014) and could be transferred between cells via exosomes (Valadi et al., 2007). Exosomes carrying miRNA released from the cells circulate and can transfer it to distant cells where the miRNA becomes functional in the recipient cells. Some studies done in both immune cells and other cell types demonstrated that transferred miRNAs inhibit target mRNAs in recipient cells (Montecalvo et al., 2012; Kosaka et al., 2010; Mittelbrunn et al., 2011; Katakowski et al., 2010; Alexander et al., 2015). A decade of research has identified many exosomal miRNAs which are involved in the diagnosis, therapeutics, and pathophysiology of many diseases including cancer, neurodegenerative diseases, metabolic disorders, cardiovascular diseases, and immune disorders.

Table 7.1 Various Therapeutic Cargo Molecules Loaded into Exosomes for Targeted Cancer Therapy

Study	Cargo
Interfering RNAs	
Shtam et al. (2013)	siRNA against RAD51 and RAD52
Wahlgren et al. (2012)	MAPK1 siRNA
Alvarez-Erviti et al. (2011)	GAPDH siRNA and BACE1 siRNA
Pan et al. (2012)	shNS5b, shCD81
Chen et al. (2014)	miR-214
Bryniarski et al. (2013)	miR-150
Zhang et al. (2010)	miR-150
Katakowski et al. (2013)	miR-146b
Kosaka et al. (2012)	miR-143
Pan et al. (2012)	miR-122
Xin et al. (2012)	miR-133b
Ohno et al. (2013)	Let-7a
Munoz et al. (2013)	Cy5-anti-miR-9
Other Types of Therapeutic Cargo	
Tian et al. (2014)	Doxorubicin
Jang et al. (2013)	Doxorubicin
Hood et al. (2014)	SPION5
Mizrak et al. (2013)	CD fused with UPRT and EGFP
Maguire et al. (2012)	Adeno-associated viral vector
Zhuang et al. (2011)	Curcumin and JSI-124
Sun et al. (2010)	Curcumin

CD, *Cytosine deaminase;* EGFP, *enhanced green fluorescence protein;* siRNA, *small interference RNA;* SPION, *superparamagnetic iron oxide nanoparticles;* UPRT, *uracil phosphoribosyltransferase.*

7.7 CONCLUDING REMARKS AND FUTURE PERSPECTIVES

Exosomes are nanosized vesicles secreted by a variety of cells and have ideal features as nanocarriers for drug and gene delivery. Exosomes are secreted by a variety of cells and body fluids including serum, plasma, urine, semen, breast milk, sweat, synovial fluids, etc. They are composed of a lipid bilayer, proteins, DNA, RNA, enzymes, and many other essential molecules. Due to their normal biological functioning in the cell, exosomes help cells to regulate other cells at a distance and communicate with them. Exosomes are known to be associated with diverse pathologies and represent a variety of functions like delivering biomolecules such as mRNAs, miRNAs, siRNA, proteins, and drugs to the recipient cells by communication at intercellular and tissue levels. Exosomes display a wide variety of functions in the cell including cell–cell communication, diagnostic markers, and their potential role in therapeutics. Some exosomes possess immunostimulatory or immunosuppressive effects which can be used as immunotherapies for cancer or autoimmune diseases. Primary studies have also reported the successful delivery of chemotherapeutic drug-loaded exosome mimetic MVs to tumor tissue in vivo and in vitro. Furthermore, exosomes are involved in various pathologic conditions including AD, PD, and metabolic disorders. Emerging evidence has revealed the promising therapeutic potential of exosomes as effective nano-vaccines and immunological treatment for inflammatory diseases. Exosomes are also known to play important role as diagnostic biomarkers of various diseases. In addition, they are considered in many clinical studies as a delivery system to transport RNA, protein, and drug molecules as they can easily cross the BBB. However, developing a strategy to produce exosomes with desirable components and less undesirable effects may have great potential for clinical application. More research is required to fully illucidate the use of exosomes in diagnostic and therapeutic measures.

CONFLICT OF INTEREST

None declared.

ACKNOWLEDGMENTS

JSB and GKB were financially supported by University Grant Commission, New Delhi, India under Raman Post-Doctoral Research Fellowship in United States [F.No. 5-82/2016 (IC)] and Rajiv Gandhi Post Doc Fellowship (PDFSS-2011-12-SC-CHA-3405), respectively.

REFERENCES

Abraham, S., Soundararajan, C.C., Vivekanandhan, S., Behari, M., 2005. Erythrocyte antioxidant enzymes in Parkinson's disease. Indian J. Med. Res. 121, 111–115.

Admyre, C., Johansson, S.M., Paulie, S., Gabrielsson, S., 2006. Direct exosome stimulation of peripheral human T cells detected by ELISPOT. Eur. J. Immunol. 36, 1772–1781.

Admyre, C., Johansson, S.M., Qazi, K.R., Filen, J.J., Lahesmaa, R., Norman, M., et al., 2007. Exosomes with immune modulatory features are present in human breast milk. J. Immunol. 179, 1969–1978.

Agrawal, U., Sharma, R., Gupta, M., Vyas, S.P., 2014. Is nanotechnology a boon for oral drug delivery? Drug Discov. Today 19, 1530–1546.

Aguzzi, A., Rajendran, L., 2009. The transcellular spread of cytosolic amyloids, prions, and prionoids. Neuron 64, 783–790.

Akers, J.C., Gonda, D., Kim, R., Carter, B.S., Chen, C.C., 2013. Biogenesis of extracellular vesicles (EV): exosomes, microvesicles, retrovirus-like vesicles, and apoptotic bodies. J. Neurooncol. 113, 1–11.

Alexander, M., Hu, R., Runtsch, M.C., Kagele, D.A., Mosbruger, T.L., Tolmachova, T., et al., 2015. Exosome-delivered microRNAs modulate the inflammatory response to endotoxin. Nat. Commun. 6, 7321.

Alvarez-Erviti, L., Seow, Y., Yin, H., Betts, C., Lakhal, S., Wood, M.J., 2011. Delivery of siRNA to the mouse brain by systemic injection of targeted exosomes. Nat. Biotechnol. 29, 341–345.

Ambani, L.M., Van woert, M.H., Murphy, S., 1975. Brain peroxidase and catalase in Parkinson disease. Arch. Neurol. 32, 114–118.

Amoozgar, Z., Goldberg, M.S., 2015. Targeting myeloid cells using nanoparticles to improve cancer immunotherapy. Adv. Drug Deliv. Rev. 91, 38–51.

Aoki, N., Jin-No, S., Nakagawa, Y., Asai, N., Arakawa, E., Tamura, N., et al., 2007. Identification and characterization of microvesicles secreted by 3T3-L1 adipocytes: redox- and hormone-dependent induction of milk fat globule-epidermal growth factor 8-associated microvesicles. Endocrinology 148, 3850–3862.

Aryani, A., Denecke, B., 2016. Exosomes as a nanodelivery system: a key to the future of neuromedicine? Mol. Neurobiol. 53, 818–834.

Aswad, H., Forterre, A., Wiklander, O.P., Vial, G., Danty-Berger, E., Jalabert, A., et al., 2014. Exosomes participate in the alteration of muscle homeostasis during lipid-induced insulin resistance in mice. Diabetologia 57, 2155–2164.

Banizs, A.B., Huang, T., Dryden, K., Berr, S.S., Stone, J.R., Nakamoto, R.K., et al., 2014. In vitro evaluation of endothelial exosomes as carriers for small interfering ribonucleic acid delivery. Int. J. Nanomed. 9, 4223–4230.

Bartel, D.P., 2004. MicroRNAs: genomics, biogenesis, mechanism, and function. Cell 116, 281–297.

Bartel, D.P., 2009. MicroRNAs: target recognition and regulatory functions. Cell 136, 215–233.

Basu, J., Ludlow, J.W., 2016. Exosomes for repair, regeneration and rejuvenation. Expert Opin. Biol. Ther. 16, 489–506.

Bellingham, S.A., Guo, B.B., Coleman, B.M., Hill, A.F., 2012. Exosomes: vehicles for the transfer of toxic proteins associated with neurodegenerative diseases? Front. Physiol. 3, 124.

Bhatnagar, S., Schorey, J.S., 2007. Exosomes released from infected macrophages contain *Mycobacterium avium* glycopeptidolipids and are proinflammatory. J. Biol. Chem. 282, 25779–25789.

Bhatnagar, S., Shinagawa, K., Castellino, F.J., Schorey, J.S., 2007. Exosomes released from macrophages infected with intracellular pathogens stimulate a proinflammatory response in vitro and in vivo. Blood 110, 3234–3244.

Bhatti, G.K., Bhadada, S.K., Vijayvergiya, R., Mastana, S.S., Bhatti, J.S., 2016. Metabolic syndrome and risk of major coronary events among the urban diabetic patients: North Indian Diabetes and Cardiovascular Disease Study-NIDCVD-2. J. Diabet. Complications 30, 72–78.

Biomarkers Definitions working, G, 2001. Biomarkers and surrogate endpoints: preferred definitions and conceptual framework. Clin. Pharmacol. Ther. 69, 89–95.

Bobrie, A., Colombo, M., Raposo, G., Thery, C., 2011. Exosome secretion: molecular mechanisms and roles in immune responses. Traffic 12, 1659–1668.

Braak, H., Rub, U., Gai, W.P., Del tredici, K., 2003. Idiopathic Parkinson's disease: possible routes by which vulnerable neuronal types may be subject to neuroinvasion by an unknown pathogen. J. Neural Transm. (Vienna) 110, 517–536.

Brettschneider, J., Del tredici, K., Lee, V.M., Trojanowski, J.Q., 2015. Spreading of pathology in neurodegenerative diseases: a focus on human studies. Nat. Rev. Neurosci. 16, 109–120.

Bryniarski, K., Ptak, W., Jayakumar, A., Pullmann, K., Caplan, M.J., Chairoungdua, A., et al., 2013. Antigen-specific, antibody-coated, exosome-like nanovesicles deliver suppressor T-cell microRNA-150 to effector T cells to inhibit contact sensitivity. J. Allergy Clin. Immunol. 132, 170–181.

Buzas, E.I., Gyorgy, B., Nagy, G., Falus, A., Gay, S., 2014. Emerging role of extracellular vesicles in inflammatory diseases. Nat. Rev. Rheumatol. 10, 356–364.

Cai, Z., Yang, F., Yu, L., Yu, Z., Jiang, L., Wang, Q., et al., 2012a. Activated T cell exosomes promote tumor invasion via Fas signaling pathway. J. Immunol. 188, 5954–5961.

Cai, Z., Zhang, W., Yang, F., Yu, L., Yu, Z., Pan, J., et al., 2012b. Immunosuppressive exosomes from TGF-beta1 gene-modified dendritic cells attenuate Th17-mediated inflammatory autoimmune disease by inducing regulatory T cells. Cell Res. 22, 607–610.

Cervio, E., Barile, L., Moccetti, T., Vassalli, G., 2015. Exosomes for intramyocardial intercellular communication. Stem Cells Int. 2015, 482171.

Chaput, N., Schartz, N.E., Andre, F., Zitvogel, L., 2003. Exosomes for immunotherapy of cancer. Adv. Exp. Med. Biol. 532, 215–221.

Chaput, N., Flament, C., Viaud, S., Taieb, J., Roux, S., Spatz, A., et al., 2006. Dendritic cell derived-exosomes: biology and clinical implementations. J. Leukoc. Biol. 80, 471–478.

Chaudhuri, A.A., So, A.Y., Sinha, N., Gibson, W.S., Taganov, K.D., O'connell, R.M., et al., 2011. MicroRNA-125b potentiates macrophage activation. J. Immunol. 187, 5062–5068.

Chen, L., Charrier, A., Zhou, Y., Chen, R., Yu, B., Agarwal, K., et al., 2014. Epigenetic regulation of connective tissue growth factor by microRNA-214 delivery in exosomes from mouse or human hepatic stellate cells. Hepatology 59, 1118–1129.

Cheng, L., Doecke, J.D., Sharples, R.A., Villemagne, V.L., Fowler, C.J., Rembach, A., et al., 2015. Prognostic serum miRNA biomarkers associated with Alzheimer's disease shows concordance with neuropsychological and neuroimaging assessment. Mol. Psychiatry 20, 1188–1196.

Chivet, M., Hemming, F., Pernet-Gallay, K., Fraboulet, S., Sadoul, R., 2012. Emerging role of neuronal exosomes in the central nervous system. Front. Physiol. 3, 145.

Cho, J.A., Park, H., Lim, E.H., Lee, K.W., 2012. Exosomes from breast cancer cells can convert adipose tissue-derived mesenchymal stem cells into myofibroblast-like cells. Int. J. Oncol. 40, 130–138.

Chow, A., Zhou, W., Liu, L., Fong, M.Y., Champer, J., Van haute, D., et al., 2014. Macrophage immunomodulation by breast cancer-derived exosomes requires Toll-like receptor 2-mediated activation of NF-kappaB. Sci. Rep. 4, 5750.

Cocucci, E., Racchetti, G., Podini, P., Meldolesi, J., 2007. Enlargeosome traffic: exocytosis triggered by various signals is followed by endocytosis, membrane shedding or both. Traffic 8, 742–757.

Conde-Vancells, J., Rodriguez-Suarez, E., Embade, N., Gil, D., Matthiesen, R., Valle, M., et al., 2008. Characterization and comprehensive proteome profiling of exosomes secreted by hepatocytes. J. Proteome Res. 7, 5157–5166.

Costa-Silva, B., Aiello, N.M., Ocean, A.J., Singh, S., Zhang, H., Thakur, B.K., et al., 2015. Pancreatic cancer exosomes initiate pre-metastatic niche formation in the liver. Nat. Cell Biol. 17, 816–826.

Dai, S., Zhou, X., Wang, B., Wang, Q., Fu, Y., Chen, T., et al., 2006. Enhanced induction of dendritic cell maturation and HLA-A*0201-restricted CEA-specific CD8(+) CTL response by exosomes derived from IL-18 gene-modified CEA-positive tumor cells. J. Mol. Med. (Berl.) 84, 1067–1076.

Davidson, S.M., Takov, K., Yellon, D.M., 2016. Exosomes and cardiovascular protection. Cardiovasc. Drugs Ther.

Delcayre, A., Shu, H., Le pecq, J.B., 2005. Dendritic cell-derived exosomes in cancer immunotherapy: exploiting nature's antigen delivery pathway. Expert Rev. Anticancer Ther. 5, 537–547.

Deng, Z.B., Poliakov, A., Hardy, R.W., Clements, R., Liu, C., Liu, Y., et al., 2009. Adipose tissue exosome-like vesicles mediate activation of macrophage-induced insulin resistance. Diabetes 58, 2498–2505.

Dhillon, N., Aggarwal, B.B., Newman, R.A., Wolff, R.A., Kunnumakkara, A.B., Abbruzzese, J.L., et al., 2008. Phase II trial of curcumin in patients with advanced pancreatic cancer. Clin. Cancer Res. 14, 4491–4499.

Diamant, M., Nieuwland, R., Pablo, R.F., Sturk, A., Smit, J.W., Radder, J.K., 2002. Elevated numbers of tissue-factor exposing microparticles correlate with components of the metabolic syndrome in uncomplicated type 2 diabetes mellitus. Circulation 106, 2442–2447.

Eken, C., Gasser, O., Zenhaeusern, G., Oehri, I., Hess, C., Schifferli, J.A., 2008. Polymorphonuclear neutrophil-derived ectosomes interfere with the maturation of monocyte-derived dendritic cells. J. Immunol. 180, 817–824.

EL Andaloussi, S., Mager, I., Breakefield, X.O., Wood, M.J., 2013. Extracellular vesicles: biology and emerging therapeutic opportunities. Nat. Rev. Drug Discov. 12, 347–357.

Emanueli, C., Shearn, A.I., Angelini, G.D., Sahoo, S., 2015. Exosomes and exosomal miRNAs in cardiovascular protection and repair. Vascul. Pharmacol. 71, 24–30.

Emmanouilidou, E., Melachroinou, K., Roumeliotis, T., Garbis, S.D., Ntzouni, M., Margaritis, L.H., et al., 2010. Cell-produced alpha-synuclein is secreted in a calcium-dependent manner by exosomes and impacts neuronal survival. J. Neurosci. 30, 6838–6851.

Feng, B., Chen, Y., Luo, Y., Chen, M., Li, X., Ni, Y., 2010. Circulating level of microparticles and their correlation with arterial elasticity and endothelium-dependent dilation in patients with type 2 diabetes mellitus. Atherosclerosis 208, 264–269.

Ferguson, S.W., Nguyen, J., 2016. Exosomes as therapeutics: the implications of molecular composition and exosomal heterogeneity. J. Controlled Release 228, 179–190.

Fevrier, B., Vilette, D., Archer, F., Loew, D., Faigle, W., Vidal, M., et al., 2004. Cells release prions in association with exosomes. Proc. Natl. Acad. Sci. U.S.A. 101, 9683–9688.

Gastpar, R., Gehrmann, M., Bausero, M.A., Asea, A., Gross, C., Schroeder, J.A., et al., 2005. Heat shock protein 70 surface-positive tumor exosomes stimulate migratory and cytolytic activity of natural killer cells. Cancer Res. 65, 5238–5247.

Gazdar, A.F., 2010. Epidermal growth factor receptor inhibition in lung cancer: the evolving role of individualized therapy. Cancer Metastasis Rev. 29, 37–48.

Goldie, B.J., Dun, M.D., Lin, M., Smith, N.D., Verrills, N.M., Dayas, C.V., et al., 2014. Activity-associated miRNA are packaged in Map1b-enriched exosomes released from depolarized neurons. Nucleic Acids Res. 42, 9195–9208.

Gousset, K., Schiff, E., Langevin, C., Marijanovic, Z., Caputo, A., Browman, D.T., et al., 2009. Prions hijack tunnelling nanotubes for intercellular spread. Nat. Cell Biol. 11, 328–336.

Gross, J.C., Chaudhary, V., Bartscherer, K., Boutros, M., 2012. Active Wnt proteins are secreted on exosomes. Nat. Cell Biol. 14, 1036–1045.

Gupta, A., Pulliam, L., 2014. Exosomes as mediators of neuroinflammation. J. Neuroinflammation 11, 68.

Ha, D., Yang, N., Nadithe, V., 2016. Exosomes as therapeutic drug carriers and delivery vehicles across biological membranes: current perspectives and future challenges. Acta Pharm. Sin. B 6, 287–296.

Hamdy, S., Haddadi, A., Hung, R.W., Lavasanifar, A., 2011. Targeting dendritic cells with nano-particulate PLGA cancer vaccine formulations. Adv. Drug Deliv. Rev. 63, 943–955.

Haney, M.J., Klyachko, N.L., Zhao, Y., Gupta, R., Plotnikova, E.G., He, Z., et al., 2015. Exosomes as drug delivery vehicles for Parkinson's disease therapy. J. Controlled Release 207, 18–30.

Hanson, P.I., Cashikar, A., 2012. Multivesicular body morphogenesis. Annu. Rev. Cell Dev. Biol. 28, 337–362.

Harding, C.V., Heuser, J.E., Stahl, P.D., 2013. Exosomes: looking back three decades and into the future. J. Cell Biol. 200, 367–371.

Harichand-Herdt, S., Ramalingam, S.S., 2008. Targeted therapy for the treatment of non-small cell lung cancer: focus on inhibition of epidermal growth factor receptor. Semin. Thorac. Cardiovasc. Surg. 20, 217–223.

Hood, J.L., Scott, M.J., Wickline, S.A., 2014. Maximizing exosome colloidal stability following electroporation. Anal. Biochem. 448, 41–49.

Hornick, N.I., Huan, J., Doron, B., Goloviznina, N.A., Lapidus, J., Chang, B.H., et al., 2015. Serum exosome microRNA as a minimally-invasive early biomarker of AML. Sci. Rep. 5, 11295.

Hoshino, A., Costa-Silva, B., Shen, T.L., Rodrigues, G., Hashimoto, A., Tesic mark, M., et al., 2015. Tumour exosome integrins determine organotropic metastasis. Nature 527, 329–335.

Howitt, J., Hill, A.F., 2016. Exosomes in the pathology of neurodegenerative diseases. J. Biol. Chem. 291, 26589–26597.

Hristov, M., Erl, W., Linder, S., Weber, P.C., 2004. Apoptotic bodies from endothelial cells enhance the number and initiate the differentiation of human endothelial progenitor cells in vitro. Blood 104, 2761–2766.

Huang, L., Ma, W., Ma, Y., Feng, D., Chen, H., Cai, B., 2015. Exosomes in mesenchymal stem cells, a new therapeutic strategy for cardiovascular diseases? Int. J. Biol. Sci. 11, 238–245.

Huang, S.M., Harari, P.M., 1999. Epidermal growth factor receptor inhibition in cancer therapy: biology, rationale and preliminary clinical results. Invest. New Drugs 17, 259–269.

Huber, H.J., Holvoet, P., 2015. Exosomes: emerging roles in communication between blood cells and vascular tissues during atherosclerosis. Curr. Opin. Lipidol. 26, 412–419.

Huotari, J., Helenius, A., 2011. Endosome maturation. EMBO J. 30, 3481–3500.

Ibrahim, A.G., Cheng, K., Marban, E., 2014. Exosomes as critical agents of cardiac regeneration triggered by cell therapy. Stem Cell Rep. 2, 606–619.

Jalabert, A., Vial, G., Guay, C., Wiklander, O.P., Nordin, J.Z., Aswad, H., et al., 2016. Exosome-like vesicles released from lipid-induced insulin-resistant muscles modulate gene expression and proliferation of beta recipient cells in mice. Diabetologia 59, 1049–1058.

Jang, S.C., Kim, O.Y., Yoon, C.M., Choi, D.S., Roh, T.Y., Park, J., et al., 2013. Bioinspired exosome-mimetic nanovesicles for targeted delivery of chemotherapeutics to malignant tumors. ACS Nano 7, 7698−7710.

Jansen, F., Yang, X., Hoelscher, M., Cattelan, A., Schmitz, T., Proebsting, S., et al., 2013. Endothelial microparticle-mediated transfer of microRNA-126 promotes vascular endothelial cell repair via SPRED1 and is abrogated in glucose-damaged endothelial microparticles. Circulation 128, 2026−2038.

Jarmalaviciute, A., Pivoriunas, A., 2016. Exosomes as a potential novel therapeutic tools against neurodegenerative diseases. Pharmacol. Res. 113, 816−822.

Johnsen, K.B., Gudbergsson, J.M., Skov, M.N., Pilgaard, L., Moos, T., Duroux, M., 2014. A comprehensive overview of exosomes as drug delivery vehicles—endogenous nanocarriers for targeted cancer therapy. Biochim. Biophys. Acta 1846, 75−87.

Johnstone, R.M., Adam, M., Hammond, J.R., Orr, L., Turbide, C., 1987. Vesicle formation during reticulocyte maturation. Association of plasma membrane activities with released vesicles (exosomes). J. Biol. Chem. 262, 9412−9420.

Johnstone, R.M., Mathew, A., Mason, A.B., Teng, K., 1991. Exosome formation during maturation of mammalian and avian reticulocytes: evidence that exosome release is a major route for externalization of obsolete membrane proteins. J. Cell Physiol. 147, 27−36.

Kalra, H., Simpson, R.J., Ji, Hong, Aikawa, E., Altevogt, P., Askenase, P., et al., 2012. Vesiclepedia: a compendium for extracellular vesicles with continuous community annotation. PLoS Biol. 10, e1001450.

Kanu, N., Imokawa, Y., Drechsel, D.N., Williamson, R.A., Birkett, C.R., Bostock, C.J., et al., 2002. Transfer of scrapie prion infectivity by cell contact in culture. Curr. Biol. 12, 523−530.

Katakowski, M., Buller, B., Wang, X., Rogers, T., Chopp, M., 2010. Functional microRNA is transferred between glioma cells. Cancer Res. 70, 8259−8263.

Katakowski, M., Buller, B., Zheng, X., Lu, Y., Rogers, T., Osobamiro, O., et al., 2013. Exosomes from marrow stromal cells expressing miR-146b inhibit glioma growth. Cancer Lett. 335, 201−204.

Keller, S., Sanderson, M.P., Stoeck, A., Altevogt, P., 2006. Exosomes: from biogenesis and secretion to biological function. Immunol. Lett. 107, 102−108.

Keller, S., Ridinger, J., Rupp, A.K., Janssen, J.W., Altevogt, P., 2011. Body fluid derived exosomes as a novel template for clinical diagnostics. J. Transl. Med. 9, 86.

Kim, S.H., Lechman, E.R., Bianco, N., Menon, R., Keravala, A., Nash, J., et al., 2005. Exosomes derived from IL-10-treated dendritic cells can suppress inflammation and collagen-induced arthritis. J. Immunol. 174, 6440−6448.

Kim, S.H., Bianco, N.R., Shufesky, W.J., Morelli, A.E., Robbins, P.D., 2007. Effective treatment of inflammatory disease models with exosomes derived from dendritic cells genetically modified to express IL-4. J. Immunol. 179, 2242−2249.

Kim, S.J., Moon, G.J., Cho, Y.H., Kang, H.Y., Hyung, N.K., Kim, D., et al., 2012. Circulating mesenchymal stem cells microparticles in patients with cerebrovascular disease. PLoS One 7, e37036.

Kosaka, N., Iguchi, H., Yoshioka, Y., Takeshita, F., Matsuki, Y., Ochiya, T., 2010. Secretory mechanisms and intercellular transfer of microRNAs in living cells. J. Biol. Chem. 285, 17442−17452.

Kosaka, N., Iguchi, H., Yoshioka, Y., Hagiwara, K., Takeshita, F., Ochiya, T., 2012. Competitive interactions of cancer cells and normal cells via secretory microRNAs. J. Biol. Chem. 287, 1397−1405.

Kulshreshtha, A., Ahmad, T., Agrawal, A., Ghosh, B., 2013. Proinflammatory role of epithelial cell-derived exosomes in allergic airway inflammation. J. Allergy Clin. Immunol. 131, 1194–1203. 1203 e1–14.

Lai, C.P., Breakefield, X.O., 2012. Role of exosomes/microvesicles in the nervous system and use in emerging therapies. Front. Physiol. 3, 228.

Lai, R.C., Yeo, R.W., Tan, K.H., Lim, S.K., 2013. Exosomes for drug delivery—a novel application for the mesenchymal stem cell. Biotechnol. Adv. 31, 543–551.

Lawson, C., Vicencio, J.M., Yellon, D.M., Davidson, S.M., 2016. Microvesicles and exosomes: new players in metabolic and cardiovascular disease. J. Endocrinol. 228, R57–R71.

Lee, J.E., Moon, P.G., Lee, I.K., Baek, M.C., 2015. Proteomic analysis of extracellular vesicles released by adipocytes of Otsuka Long-Evans Tokushima fatty (OLETF) rats. Protein J. 34, 220–235.

Lee, M.J., Park, D.H., Kang, J.H., 2016. Exosomes as the source of biomarkers of metabolic diseases. Ann. Pediatr. Endocrinol. Metab. 21, 119–125.

Li, C., Zhang, J., Zu, Y.J., Nie, S.F., Cao, J., Wang, Q., et al., 2015. Biocompatible and biodegradable nanoparticles for enhancement of anti-cancer activities of phytochemicals. Chin. J. Nat. Med. 13, 641–652.

Linares, R., Tan, S., Gounou, C., Arraud, N., Brisson, A.R., 2015. High-speed centrifugation induces aggregation of extracellular vesicles. J. Extracell. Vesicles 4, 29509.

Liu, C., Yu, S., Zinn, K., Wang, J., Zhang, L., Jia, Y., et al., 2006. Murine mammary carcinoma exosomes promote tumor growth by suppression of NK cell function. J. Immunol. 176, 1375–1385.

Liu, H., Gao, W., Yuan, J., Wu, C., Yao, K., Zhang, L., et al., 2016. Exosomes derived from dendritic cells improve cardiac function via activation of CD4(+) T lymphocytes after myocardial infarction. J. Mol. Cell. Cardiol. 91, 123–133.

Livshits, M.A., Khomyakova, E., Evtushenko, E.G., Lazarev, V.N., Kulemin, N.A., Semina, S.E., et al., 2016. Corrigendum: isolation of exosomes by differential centrifugation: theoretical analysis of a commonly used protocol. Sci. Rep. 6, 21447.

Lotvall, J., Hill, A.F., Hochberg, F., Buzas, E.I., Di vizio, D., Gardiner, C., et al., 2014. Minimal experimental requirements for definition of extracellular vesicles and their functions: a position statement from the International Society for Extracellular Vesicles. J. Extracell. Vesicles 3, 26913.

Lugli, G., Cohen, A.M., Bennett, D.A., Shah, R.C., Fields, C.J., Hernandez, A.G., et al., 2015. Plasma exosomal miRNAs in persons with and without Alzheimer disease: altered expression and prospects for biomarkers. PLoS One 10, e0139233.

Maguire, C.A., Balaj, L., Sivaraman, S., Crommentuijn, M.H., Ericsson, M., Mincheva-Nilsson, L., et al., 2012. Microvesicle-associated AAV vector as a novel gene delivery system. Mol. Ther. 20, 960–971.

Mahaweni, N.M., Kaijen-Lambers, M.E., Dekkers, J., Aerts, J.G., Hegmans, J.P., 2013. Tumour-derived exosomes as antigen delivery carriers in dendritic cell-based immunotherapy for malignant mesothelioma. J. Extracell. Vesicles 2.

Malm, T., Loppi, S., Kanninen, K.M., 2016. Exosomes in Alzheimer's disease. Neurochem. Int. 97, 193–199.

Matsumura, T., Sugimachi, K., Iinuma, H., Takahashi, Y., Kurashige, J., Sawada, G., et al., 2015. Exosomal microRNA in serum is a novel biomarker of recurrence in human colorectal cancer. Br. J. Cancer 113, 275–281.

Michael, A., Bajracharya, S.D., Yuen, P.S., Zhou, H., Star, R.A., Illei, G.G., et al., 2010. Exosomes from human saliva as a source of microRNA biomarkers. Oral Dis. 16, 34–38.

Mittelbrunn, M., Gutierrez-Vazquez, C., Villarroya-Beltri, C., Gonzalez, S., Sanchez-Cabo, F., Gonzalez, M.A., et al., 2011. Unidirectional transfer of microRNA-loaded exosomes from T cells to antigen-presenting cells. Nat. Commun. 2, 282.

Mizrak, A., Bolukbasi, M.F., Ozdener, G.B., Brenner, G.J., Madlener, S., Erkan, E.P., et al., 2013. Genetically engineered microvesicles carrying suicide mRNA/protein inhibit schwannoma tumor growth. Mol. Ther. 21, 101–108.

Montecalvo, A., Shufesky, W.J., Stolz, D.B., Sullivan, M.G., Wang, Z., Divito, S.J., et al., 2008. Exosomes as a short-range mechanism to spread alloantigen between dendritic cells during T cell allorecognition. J. Immunol. 180, 3081–3090.

Montecalvo, A., Larregina, A.T., Shufesky, W.J., Stolz, D.B., Sullivan, M.L., Karlsson, J. M., et al., 2012. Mechanism of transfer of functional microRNAs between mouse dendritic cells via exosomes. Blood 119, 756–766.

Munich, S., Sobo-Vujanovic, A., Buchser, W.J., Beer-Stolz, D., Vujanovic, N.L., 2012. Dendritic cell exosomes directly kill tumor cells and activate natural killer cells via TNF superfamily ligands. Oncoimmunology 1, 1074–1083.

Munoz, J.L., Bliss, S.A., Greco, S.J., Ramkissoon, S.H., Ligon, K.L., Rameshwar, P., 2013. Delivery of functional anti-miR-9 by mesenchymal stem cell-derived exosomes to glioblastoma multiforme cells conferred chemosensitivity. Mol. Ther. Nucleic Acids 2, e126.

Natasha, G., Gundogan, B., Tan, A., Farhatnia, Y., Wu, W., Rajadas, J., et al., 2014. Exosomes as immunotheranostic nanoparticles. Clin. Ther. 36, 820–829.

O'neill, S., Bohl, M., Gregersen, S., Hermansen, K., O'driscoll, L., 2016. Blood-based biomarkers for metabolic syndrome. Trends Endocrinol. Metab. 27, 363–374.

Ogata-Kawata, H., Izumiya, M., Kurioka, D., Honma, Y., Yamada, Y., Furuta, K., et al., 2014. Circulating exosomal microRNAs as biomarkers of colon cancer. PLoS One 9, e92921.

Ohno, S., Takanashi, M., Sudo, K., Ueda, S., Ishikawa, A., Matsuyama, N., et al., 2013. Systemically injected exosomes targeted to EGFR deliver antitumor microRNA to breast cancer cells. Mol. Ther. 21, 185–191.

Ong, S.G., Wu, J.C., 2015. Exosomes as potential alternatives to stem cell therapy in mediating cardiac regeneration. Circ. Res. 117, 7–9.

Ostrowski, M., Carmo, N.B., Krumeich, S., Fanget, I., Raposo, G., Savina, A., et al., 2010. Rab27a and Rab27b control different steps of the exosome secretion pathway. Nat. Cell Biol. 12, 19–30. Suppl. 1–13.

Palma, J., Yaddanapudi, S.C., Pigati, L., Havens, M.A., Jeong, S., Weiner, G.A., et al., 2012. MicroRNAs are exported from malignant cells in customized particles. Nucleic Acids Res. 40, 9125–9138.

Pan, Q., Ramakrishnaiah, V., Henry, S., Fouraschen, S., De ruiter, P.E., Kwekkeboom, J., et al., 2012. Hepatic cell-to-cell transmission of small silencing RNA can extend the therapeutic reach of RNA interference (RNAi). Gut 61, 1330–1339.

Pardridge, W.M., 2012. Drug transport across the blood–brain barrier. J. Cereb. Blood Flow Metab. 32, 1959–1972.

Penfornis, P., Vallabhaneni, K.C., Whitt, J., Pochampally, R., 2016. Extracellular vesicles as carriers of microRNA, proteins and lipids in tumor microenvironment. Int. J. Cancer 138, 14–21.

Peng, Q., Zhang, S., Yang, Q., Zhang, T., Wei, X.Q., Jiang, L., et al., 2013. Preformed albumin corona, a protective coating for nanoparticles based drug delivery system. Biomaterials 34, 8521–8530.

Pisitkun, T., Johnstone, R., Knepper, M.A., 2006. Discovery of urinary biomarkers. Mol. Cell Proteomics 5, 1760–1771.

Pitt, J.M., Charrier, M., Viaud, S., Andre, F., Besse, B., Chaput, N., et al., 2014. Dendritic cell-derived exosomes as immunotherapies in the fight against cancer. J. Immunol. 193, 1006–1011.

Pitt, J.M., Andre, F., Amigorena, S., Soria, J.C., Eggermont, A., Kroemer, G., et al., 2016. Dendritic cell-derived exosomes for cancer therapy. J. Clin. Invest. 126, 1224–1232.

Prusiner, S.B., Woerman, A.L., Mordes, D.A., Watts, J.C., Rampersaud, R., Berry, D.B., et al., 2015. Evidence for alpha-synuclein prions causing multiple system atrophy in humans with parkinsonism. Proc. Natl. Acad. Sci. U.S.A. 112, E5308–E5317.

Raemdonck, K., Braeckmans, K., Demeester, J., De smedt, S.C., 2014. Merging the best of both worlds: hybrid lipid-enveloped matrix nanocomposites in drug delivery. Chem. Soc. Rev. 43, 444–472.

Raj, R., Bhatti, J.S., Badada, S.K., Ramteke, P.W., 2014. Genetic basis of dyslipidemia in disease precipitation of coronary artery disease (CAD) associated type 2 diabetes mellitus (T2DM). Diabetes Metab. Res. Rev.

Rajendran, L., Honsho, M., Zahn, T.R., Keller, P., Geiger, K.D., Verkade, P., et al., 2006. Alzheimer's disease beta-amyloid peptides are released in association with exosomes. Proc. Natl. Acad. Sci. U.S.A. 103, 11172–11177.

Rak, J., Guha, A., 2012. Extracellular vesicles—vehicles that spread cancer genes. Bioessays 34, 489–497.

Rani, S., Ritter, T., 2016. The exosome—a naturally secreted nanoparticle and its application to wound healing. Adv. Mater. 28, 5542–5552.

Raposo, G., Stoorvogel, W., 2013. Extracellular vesicles: exosomes, microvesicles, and friends. J. Cell Biol. 200, 373–383.

Raposo, G., Nijman, H.W., Stoorvogel, W., Liejendekker, R., Harding, C.V., Melief, C.J., et al., 1996. B lymphocytes secrete antigen-presenting vesicles. J. Exp. Med. 183, 1161–1172.

Riederer, P., Sofic, E., Rausch, W.D., Schmidt, B., Reynolds, G.P., Jellinger, K., et al., 1989. Transition metals, ferritin, glutathione, and ascorbic acid in parkinsonian brains. J. Neurochem. 52, 515–520.

Ristorcelli, E., Beraud, E., Verrando, P., Villard, C., Lafitte, D., Sbarra, V., et al., 2008. Human tumor nanoparticles induce apoptosis of pancreatic cancer cells. FASEB J. 22, 3358–3369.

Romagnoli, G.G., Zelante, B.B., Toniolo, P.A., Migliori, I.K., Barbuto, J.A., 2014. Dendritic cell-derived exosomes may be a tool for cancer immunotherapy by converting tumor cells into immunogenic targets. Front. Immunol. 5, 692.

Russo, I., Bubacco, L., Greggio, E., 2012. Exosomes-associated neurodegeneration and progression of Parkinson's disease. Am. J. Neurodegener. Dis. 1, 217–225.

Safdar, A., Saleem, A., Tarnopolsky, M.A., 2016. The potential of endurance exercise-derived exosomes to treat metabolic diseases. Nat. Rev. Endocrinol. 12, 504–517.

Sahoo, S., Losordo, D.W., 2014. Exosomes and cardiac repair after myocardial infarction. Circ. Res. 114, 333–344.

Salido-Guadarrama, I., Romero-Cordoba, S., Peralta-Zaragoza, O., Hidalgo-Miranda, A., Rodriguez-Dorantes, M., 2014. MicroRNAs transported by exosomes in body fluids as mediators of intercellular communication in cancer. Onco Targets Ther. 7, 1327–1338.

Schulte, E.C., Altmaier, E., Berger, H.S., Do, K.T., Kastenmuller, G., Wahl, S., et al., 2016. Alterations in lipid and inositol metabolisms in two dopaminergic disorders. PLoS One 11, e0147129.

Scott, A.M., Wolchok, J.D., Old, L.J., 2012. Antibody therapy of cancer. Nat. Rev. Cancer 12, 278–287.

Segura, E., Nicco, C., Lombard, B., Veron, P., Raposo, G., Batteux, F., et al., 2005. ICAM-1 on exosomes from mature dendritic cells is critical for efficient naive T-cell priming. Blood 106, 216–223.

Sercombe, L., Veerati, T., Moheimani, F., Wu, S.Y., Sood, A.K., Hua, S., 2015. Advances and challenges of liposome assisted drug delivery. Front. Pharmacol. 6, 286.

Shtam, T.A., Kovalev, R.A., Varfolomeeva, E.Y., Makarov, E.M., Kil, Y.V., Filatov, M.V., 2013. Exosomes are natural carriers of exogenous siRNA to human cells in vitro. Cell Commun. Signal 11, 88.

Simhadri, V.R., Reiners, K.S., Hansen, H.P., Topolar, D., Simhadri, V.L., Nohroudi, K., et al., 2008. Dendritic cells release HLA-B-associated transcript-3 positive exosomes to regulate natural killer function. PLoS One 3, e3377.

Simpson, R.J., Lim, J.W., Moritz, R.L., Mathivanan, S., 2009. Exosomes: proteomic insights and diagnostic potential. Expert Rev. Proteomics 6, 267–283.

Singh, P.P., Lemaire, C., Tan, J.C., Zeng, E., Schorey, J.S., 2011. Exosomes released from *M. tuberculosis* infected cells can suppress IFN-gamma mediated activation of naive macrophages. PLoS One 6, e18564.

Skokos, D., Botros, H.G., Demeure, C., Morin, J., Peronet, R., Birkenmeier, G., et al., 2003. Mast cell-derived exosomes induce phenotypic and functional maturation of dendritic cells and elicit specific immune responses in vivo. J. Immunol. 170, 3037–3045.

Skriner, K., Adolph, K., Jungblut, P.R., Burmester, G.R., 2006. Association of citrullinated proteins with synovial exosomes. Arthritis Rheum. 54, 3809–3814.

Soekmadji, C., Russell, P.J., Nelson, C.C., 2013. Exosomes in prostate cancer: putting together the pieces of a puzzle. Cancers (Basel) 5, 1522–1544.

Street, J.M., Barran, P.E., Mackay, C.L., Weidt, S., Balmforth, C., Walsh, T.S., et al., 2012. Identification and proteomic profiling of exosomes in human cerebrospinal fluid. J. Transl. Med. 10, 5.

Sugimachi, K., Matsumura, T., Hirata, H., Uchi, R., Ueda, M., Ueo, H., et al., 2015. Identification of a bona fide microRNA biomarker in serum exosomes that predicts hepatocellular carcinoma recurrence after liver transplantation. Br. J. Cancer 112, 532–538.

Sun, D., Zhuang, X., Xiang, X., Liu, Y., Zhang, S., Liu, C., et al., 2010. A novel nanoparticle drug delivery system: the anti-inflammatory activity of curcumin is enhanced when encapsulated in exosomes. Mol. Ther. 18, 1606–1614.

Suzuki, E., Fujita, D., Takahashi, M., Oba, S., Nishimatsu, H., 2016. Stem cell-derived exosomes as a therapeutic tool for cardiovascular disease. World J. Stem Cells 8, 297–305.

Tan, A., De Lapena, H., Seifalian, A.M., 2010. The application of exosomes as a nanoscale cancer vaccine. Int. J. Nanomed. 5, 889–900.

Thery, C., 2011. Exosomes: secreted vesicles and intercellular communications. F1000 Biol. Rep. 3, 15.

Thery, C., Duban, L., Segura, E., Veron, P., Lantz, O., Amigorena, S., 2002a. Indirect activation of naive CD4+ T cells by dendritic cell-derived exosomes. Nat. Immunol. 3, 1156–1162.

Thery, C., Zitvogel, L., Amigorena, S., 2002b. Exosomes: composition, biogenesis and function. Nat. Rev. Immunol. 2, 569–579.

Tian, Y., Li, S., Song, J., Ji, T., Zhu, M., Anderson, G.J., et al., 2014. A doxorubicin delivery platform using engineered natural membrane vesicle exosomes for targeted tumor therapy. Biomaterials 35, 2383–2390.

Timmers, L., Lim, S.K., Hoefer, I.E., Arslan, F., Lai, R.C., Van oorschot, A.A., et al., 2011. Human mesenchymal stem cell-conditioned medium improves cardiac function following myocardial infarction. Stem Cell Res. 6, 206–214.

Tkach, M., Thery, C., 2016. Communication by extracellular vesicles: where we are and where we need to go. Cell 164, 1226–1232.

Trams, E.G., Lauter, C.J., Salem Jr., N., Heine, U., 1981. Exfoliation of membrane ecto-enzymes in the form of micro-vesicles. Biochim. Biophys. Acta 645, 63–70.

Tsilioni, I., Panagiotidou, S., Theoharides, T.C., 2014. Exosomes in neurologic and psychiatric disorders. Clin. Ther. 36, 882–888.

Turturici, G., Tinnirello, R., Sconzo, G., Geraci, F., 2014. Extracellular membrane vesicles as a mechanism of cell-to-cell communication: advantages and disadvantages. Am. J. Physiol. Cell Physiol. 306, C621–C633.

Valadi, H., Ekstrom, K., Bossios, A., Sjostrand, M., Lee, J.J., Lotvall, J.O., 2007. Exosome-mediated transfer of mRNAs and microRNAs is a novel mechanism of genetic exchange between cells. Nat. Cell Biol. 9, 654–659.

van den boorn, J.G., Schlee, M., Coch, C., Hartmann, G., 2011. SiRNA delivery with exosome nanoparticles. Nat. Biotechnol. 29, 325–326.

van der pol, E., Boing, A.N., Harrison, P., Sturk, A., Nieuwland, R., 2012. Classification, functions, and clinical relevance of extracellular vesicles. Pharmacol. Rev. 64, 676–705.

van den boorn, J.G., Dassler, J., Coch, C., Schlee, M., Hartmann, G., 2013. Exosomes as nucleic acid nanocarriers. Adv. Drug Deliv. Rev. 65, 331–335.

Vella, L.J., Greenwood, D.L., Cappai, R., Scheerlinck, J.P., Hill, A.F., 2008. Enrichment of prion protein in exosomes derived from ovine cerebral spinal fluid. Vet. Immunol. Immunopathol. 124, 385–393.

Vella, L.J., Hill, A.F., Cheng, L., 2016. Focus on extracellular vesicles: exosomes and their role in protein trafficking and biomarker potential in Alzheimer's and Parkinson's disease. Int. J. Mol. Sci. 17, 173.

Viaud, S., Thery, C., Ploix, S., Tursz, T., Lapierre, V., Lantz, O., et al., 2010. Dendritic cell-derived exosomes for cancer immunotherapy: what's next? Cancer Res. 70, 1281–1285.

Viaud, S., Ullrich, E., Zitvogel, L., Chaput, N., 2008. Exosomes for the treatment of human malignancies. Horm. Metab. Res. 40, 82–88.

Wahlgren, J., De, L.K.T., Brisslert, M., Vaziri sani, F., Telemo, E., Sunnerhagen, P., et al., 2012. Plasma exosomes can deliver exogenous short interfering RNA to monocytes and lymphocytes. Nucleic Acids Res. 40, e130.

Wahlgren, J., Statello, L., Skogberg, G., Telemo, E., Valadi, H., 2016. Delivery of small interfering RNAs to cells via exosomes. Methods Mol. Biol. 1364, 105–125.

Wang, X., Huang, W., Liu, G., Cai, W., Millard, R.W., Wang, Y., et al., 2014. Cardiomyocytes mediate anti-angiogenesis in type 2 diabetic rats through the exosomal transfer of miR-320 into endothelial cells, J. Mol. Cell. Cardiol., 74. pp. 139–150.

Wolfers, J., Lozier, A., Raposo, G., Regnault, A., Thery, C., Masurier, C., et al., 2001. Tumor-derived exosomes are a source of shared tumor rejection antigens for CTL cross-priming. Nat. Med. 7, 297–303.

Woodburn, J.R., 1999. The epidermal growth factor receptor and its inhibition in cancer therapy. Pharmacol. Ther. 82, 241–250.

Wu, X., Zheng, T., Zhang, B., 2016. Exosomes in Parkinson's disease. Neurosci. Bull.

Xin, H., Li, Y., Buller, B., Katakowski, M., Zhang, Y., Wang, X., et al., 2012. Exosome-mediated transfer of miR-133b from multipotent mesenchymal stromal cells to neural cells contributes to neurite outgrowth. Stem Cells 30, 1556–1564.

Xitong, D., Xiaorong, Z., 2016. Targeted therapeutic delivery using engineered exosomes and its applications in cardiovascular diseases. Gene 575, 377–384.

Yang, C., Robbins, P.D., 2012. Immunosuppressive exosomes: a new approach for treating arthritis. Int. J. Rheumatol. 2012, 573528.

Yang, T., Martin, P., Fogarty, B., Brown, A., Schurman, K., Phipps, R., et al., 2015. Exosome delivered anticancer drugs across the blood–brain barrier for brain cancer therapy in *Danio rerio*. Pharm. Res. 32, 2003–2014.

Yellon, D.M., Davidson, S.M., 2014. Exosomes: nanoparticles involved in cardioprotection? Circ. Res. 114, 325–332.

Yoshida, K., Burton, G.F., Mckinney, J.S., Young, H., Ellis, E.F., 1992. Brain and tissue distribution of polyethylene glycol-conjugated superoxide dismutase in rats. Stroke 23, 865–869.

Yuan, M.J., Maghsoudi, T., Wang, T., 2016. Exosomes mediate the intercellular communication after myocardial infarction. Int. J. Med. Sci. 13, 113–116.

Zhang, Y., Liu, D., Chen, X., Li, J., Li, L., Bian, Z., et al., 2010. Secreted monocytic miR-150 enhances targeted endothelial cell migration. Mol. Cell 39, 133–144.

Zhang, X., Pei, Z., Chen, J., Ji, C., Xu, J., Zhang, X., et al., 2016. Exosomes for immunoregulation and therapeutic intervention in cancer. J. Cancer 7, 1081–1087.

Zhang, Y., Hu, Y.W., Zheng, L., Wang, Q., 2017. Characteristics and roles of exosomes in cardiovascular disease. DNA Cell Biol.

Zhao, W., Zheng, X.L., Zhao, S.P., 2015. Exosome and its roles in cardiovascular diseases. Heart Fail Rev. 20, 337–348.

Zhuang, X., Xiang, X., Grizzle, W., Sun, D., Zhang, S., Axtell, R.C., et al., 2011. Treatment of brain inflammatory diseases by delivering exosome encapsulated anti-inflammatory drugs from the nasal region to the brain. Mol. Ther. 19, 1769–1779.

Zitvogel, L., Regnault, A., Lozier, A., Wolfers, J., Flament, C., Tenza, D., et al., 1998. Eradication of established murine tumors using a novel cell-free vaccine: dendritic cell-derived exosomes. Nat. Med. 4, 594–600.

Zitvogel, L., Fernandez, N., Lozier, A., Wolfers, J., Regnault, A., Raposo, G., et al., 1999. Dendritic cells or their exosomes are effective biotherapies of cancer. Eur. J. Cancer 35 (Suppl. 3), S36–S38.

CHAPTER

Recent advances in lipid nanocarriers applicable in the fight against cancer*

8

Josef Jampílek[1] and Katarína Kráľová[2]

[1]*Regional Centre of Advanced Technologies and Materials, Palacky University, Olomouc, Czech Republic* [2]*Institute of Chemistry, Faculty of Natural Sciences, Comenius University, Bratislava, Slovakia*

8.1 INTRODUCTION

Tumors of various etiologies are one of the leading causes of morbidity and mortality worldwide, with approximately 14 million new cases in 2012 (Ferlay et al., 2013), which indicates that cancer is the second leading cause of death globally, and was responsible for 8.8 million deaths in 2015. Globally, nearly 1 in 6 deaths is due to cancer. The most frequent causes of deaths are cancers of lung (1.69 million deaths), liver (788,000 deaths), colorectal (774,000 deaths), stomach (754,000 deaths), and breast (571,000 deaths). Moreover, the number of new cases is expected to rise by about 70% over the next two decades. Malignant degeneration of cells resulting in malignant growth can be caused by physical (e.g., ultraviolet (UV), ionizing radiation, etc.), chemical (e.g., asbestos, components of tobacco smoke, various food contaminants, various drinking water contaminants, etc.), and biological (viruses, bacteria, parasites) carcinogens. Although approximately 70% of deaths from cancer occur in low- and middle-income countries (due to economy reasons and absence of sufficient diagnosis and treatment), from statistics it is evident that incidence is in direct connection with the income of the region. High-income regions have had a manifold number of estimated malignancies compared to low-income countries (with some exceptions). The disease is connected with an unhealthy lifestyle, unhealthy diet, physical inactivity, stress, and pollution of environment, etc., but also with a higher frequency of routine checkups. However, the decreasing trend of mortality in high-income regions can be clearly connected to the discovery of new drugs to treat various neoplasms and early diagnosis that can reduce cancer mortality significantly (WHO, 2017).

*This chapter is sincerely dedicated to the memory of Prof. Ervin Wolfram (1923–85), the long-time head of the Department of Colloid Chemistry and Colloid Technology, the Faculty of Science, Eötvös Loránd University, Budapest, Hungary.

Nanoarchitectonics in Biomedicine. DOI: https://doi.org/10.1016/B978-0-12-816200-2.00009-8

Over the past few decades, several major groups of antineoplastics have been developed, such as different types of alkylating agents, antimetabolites, intercalating agents, topoisomerase inhibitors, and mitotic inhibitors. Other groups are different sensitizers used in phototherapy and radiotherapy. Developmental biology, physiology, and chemical science enabled the creation and synthesis of monoclonal antibodies. The latest group that, immediately after its discovery, has become one of the most important classes of anticancer agents are inhibitors of protein kinases. A number of various natural and synthetic compounds and their combinations have been under investigation (Steinhilber et al., 2010; Oktábec and Jampílek, 2013; Musioł et al., 2007; Podeszwa et al., 2007; Opletalová et al., 2008; Mrozek-Wilczkiewicz et al., 2010; Král et al., 2010, 2011; Kauerová et al., 2016; Pentak et al., 2016).

These compounds belong to various structural types but, in general, all these compounds show limited solubility otherwise they are recognized as substrates of P-glycoprotein (P-gp) which results in low bioavailability. In addition, many cancer cells are able to develop resistance to anticancer drugs. The rapid development of various nanosystems have offered new opportunities in designing novel treatment strategies for cancers, whether nano-based drug delivery systems (drug carriers with targeted biodistribution) or application of nanomaterials with increased therapeutic efficiency (killing effect of nanosize alone or various combinations with antineoplastics and supportive agents). Thus application of nanosystems is advantageous primarily in terms of the increased bioavailability of the drugs and the reduction of toxicity as well as due to controlled drug release (Jampílek and Králová, 2017, 2018).

This chapter is focused on the application of lipid-based nanocarriers representing a class of biocompatible and biodegradable vehicles that provide advanced delivery of therapeutic and imaging agents, improving pharmacokinetic profile and safety in cancer therapy and diagnostics. Discussed lipid-based delivery systems include micelles, liposomes, nanoemulsions (NEs), nanosuspensions (NSs), solid lipid nanoparticles (SLNPs), nanostructured lipid carriers (NLCs), and lipid–polymer hybrid nanoparticles (LPNPs). Enhanced targeting potential, imaging, and controlled release of anticancer drugs or therapeutic molecules could be possible through multifunctional nanocarriers. Attention is focused primarily on nanoformulations of anticancer drugs usually used in clinical practice for treatment of the most widespread types of tumors.

8.2 MICROEMULSIONS AND NANOEMULSIONS

According to the IUPAC definition, microemulsion (ME) is "dispersion made of water, oil, and surfactant(s) that is an isotropic and thermodynamically stable system with dispersed domain diameter varying approximately from 1 to 100 nm, usually 10–50 nm," whereby the term "oil" refers to any water-insoluble liquid (Slomkowski et al., 2011). Structurally, they are divided into oil-in-water (o/w), water-in-oil (w/o), and bicontinuous MEs. MEs are characterized with very low interfacial tension and large interfacial area, require minimum energy for formation, and are able to solubilize otherwise immiscible liquids with droplet sizes usually ranging from 10 to 100 nm (Singh et al., 2011). NEs that are formed of very small emulsion droplets, generally oil droplets in water, with sizes <300 nm occur from a thermodynamic point of view in a nonequilibrium state, although they could be considered as kinetically stable because they are characterized by slow kinetics of destabilization. In contrast to MEs, the NEs are generally prepared using the so-called high-energy methods that require an adequate supply of energy to increase the water/oil interfacial area for generating nanosized droplets, which is provided, for example, by ultrasound generators or high-pressure homogenizers (Anton and Vandamme, 2011).

Surface modification of nanocarriers is attractive for their enhanced functions on imaging, targeting, and delivery. Targeting ligands attached to the surface of nanocarriers could prolong circulation time, increase drug bioavailability, reduce undesirable side effects, and minimize nonspecific uptake, thus, allowing specific cancer targeting to certain target cells within the cancer sites or even intracellular localization to target organelles (Yu and Zhang, 2009). A bitargeted ME prepared using coix seed oil and coix seed polysaccharide (acting as anticancer components as well as functional excipients) and modified by tumor-targeting ligand galactose stearate showed spherical particles of 47 nm and zeta potential of -4.37 ± 1.76 mV and were effectively accumulated and retained in tumor tissue, whereby they exhibited improved inhibition of tumor growth, prolonged survival time, as well as more efficacious induction of tumor cell apoptosis compared to unmodified MEs (Qu et al., 2016). Incorporating the cationic lipid stearylamine together with indocyanine green (ICG) in the shell of NEs resulted in the change of the NEs surface charge from a negative to a positive value and improved physical stability. Subcutaneous injection of this nanoformulation led to an increased ICG concentration accumulated in popliteal lymph nodes without affecting the kinetics of lymph node uptake of NEs compared to the stearylamine-free NE (Lee et al., 2015). Multifunctional lipid NEs combine several advantages, for example, tissue and cell targeting, imaging analysis, barrier permeability enhancement, and therapeutic purposes. Shinde et al. (2011) highlighted engineering of functional MEs/NEs into multifunctional MEs/NEs as a strategy to further enhance brain targeting in a review paper, and in another review paper focused on the use of NEs in anticancer therapy, Souto et al. (2011) analyzed the technological aspects of pharmaceutical formulation of these carriers and exploited their advantages in small interfering RNA (siRNA) therapy. The current state of NEs in translational research and its role in targeted cancer therapy was summarized by Ganta et al. (2014).

Lipidic NEs concentrate in neoplastic cells after injection into the bloodstream and, thus, can be used as drug carriers to tumor sites. For example, a folate-based docetaxel (DTX) lipid NE was found to show tumor targeting ca. 5- and 2-fold over controls and its effect was also superior in tumor regression (Afzal et al., 2016). DTX NEs showing highly effective anticancer efficiency and low toxicity suitable for intravenous (i.v.) injection to animals were reported by Li et al. (2011). A DTX-loaded NE with particle size of 72.3 nm, the average zeta potential of −6.38 mV and 93.1% encapsulation efficiency (EE) exhibited similar antiproliferative effects on both human primary glioblastoma cells (U87) and bEnd.3 cells and considerably lower in vivo toxicity than Taxotere and it was demonstrated that it could target glioma sites and prolong the median survival time of mice with gliomas (Gaoe et al., 2012). An albumin-anchored DTX lipid NE showed higher in vitro cytotoxicity against MCF-7 and HeLa cells than the plain NE, exhibited 80.01% ± 2.74% inhibition of solid tumors induced in C57BL/6 mice compared to 55.62% ± 5.41% inhibition caused by administration of the plain NE and showed 3-fold higher tumor-targeting activity than the plain NE (Muzammil et al., 2016). NEs of DTX surface-functionalized with folate with particle size <150 nm showed a 270-fold drop in the concentration causing 50% inhibition (IC_{50}) in chemoresistant ovarian cancer cells SKOV3TR compared to DTX alone, and in SKOV3 tumor-bearing mice they delivered DTX by folate receptor (FR) mediated endocytosis showing enhanced cytotoxicity capable of overcoming ABC transporter mediated taxane resistance (Ganta et al., 2016). A NE of DTX prepared using soybean oil, lecithin, Pluronic F-68, polyethylene glycol (PEG) 4000, and ethanol with the droplet size of 233 nm, the negative zeta potential of −43.66 ± 1.9 mV and 96.76% ± 1.5% drug content showed ca. 3-fold higher cell uptake than the control and strong cytotoxic activity against MCF-7 cancer cells with IC_{50} = 13.6 μg/mL at 72 hours, and after administration of a dose of 20 mg/kg liver and kidney tissues of mice did not show toxicity or necrosis indicating that such NE might be used as an alternative DTX therapy (Verma et al., 2016).

A D-α-tocopherol PEG succinate (TPGS) incorporating paclitaxel (PTX) NE with droplets of the mean diameter of ca. 25 nm showing IC_{50} = 5.39 mg/mL in resistant MCF-7/ADR cells (estimated IC_{50} for the free drug was 101.45 mg/mL) efficiently reduced tumor volume in a resistant tumor xenograft model, which represented only 10.06% of that estimated with PTX solution application, whereby the tumor inhibitory rate of TPGS incorporating NE-PTX was 93.84%, indicating that such NE could be applied for circumventing PTX resistance in breast cancer (Bu et al., 2014). Vitamin E derivative-based NEs containing PTX considerably enhanced the antiproliferation effect on PTX-resistant human ovarian carcinoma A2780/Taxol cells compared to the free drug, decreased mitochondrial potential, increased intracellular accumulation of rhodamine 123, which is a P-gp substrate, and also reduced the B-cell lymphoma 2 (Bcl-2) protein level and increased the Bax protein level indicating that this nanoformulation would be suitable for modulating intrinsic and/or acquired multidrug resistance (MDR) (Zheng et al., 2016). Coadministration of PTX and ceramide in NE formulations with droplet size of ca. 200 nm in diameter exhibited significantly enhanced cytotoxicity against brain tumor cells U-118, and also enhancement in the apoptotic activity compared to the administration of individual agents was estimated (Desai et al., 2008). A ME containing the linoleic acid conjugated with PTX (CLA-PTX) with the average droplet size of 176 nm showed IC_{50} value of 1.61 μM for a C6 glioma cell line, which was similar to that of free PTX and CLA-PTX solutions, exhibited the antitumor activity against brain tumors in in vivo C6 glioma tumor-bearing nude mice as well as in a rat model; mice tolerated this ME at doses up to 200 mg/kg, while estimated LD_{50} of the CLA-PTX solution was 103.9 mg/kg (Li et al., 2012). NEs loaded with PTX and phosphatidylinositol-3-kinases/protein kinase B/mammalian target of rapamycin dual inhibitor BEZ235 were found to be more effective than the free drug, and combination treatment with 10 nM NE-BEZ235 and 0.6 and 1.78 nM NE-PTX killed 50% of human colon cancer cells HCT-116 and human colorectal adenocarcinoma cells (HT-29), respectively. While estimated IC_{50} values for HCT-116 and HT-29 cells were 127.1 and 145.0 nM, respectively, for BEZ235 and 9.7 and 9.5 nM, respectively, for PTX, the application of NE decreased these IC_{50} values to 52.6 and 55.6 nM (BEZ235) and 1.9 and 2.3 nM (PTX), respectively. Treatment with these NEs caused apoptotic cell death, which coincided with decreased expression of antiapoptotic protein Bcl-2, and the NEs could be used in effective colon cancer treatment (Zou et al., 2016). PTX-loaded (6 mg/mL) commercially available NE formulations, Clinoleic 20% and Intralipid 20%, with the droplet size of 264.7 and 294.6 nm and loading efficiencies 70.4%–80.2% and 44.2%–57.4%, respectively, decreased the viability of U87MG cells. Clinoleic loaded with PTX decreased the viability of grade IV glioma cells U87MG to 6.4% ± 2.3% (Clinoleic formulation), and this decrease for PTX-loaded Intralipid formulation represented 21.29% ± 3.82%. Both NEs showed less toxicity against the normal glial cells (SVGP12), in which the cell viability decreased to 25%–35% (Najlah et al., 2016). In breast and pancreatic cancer animal models, ultrasound-mediated therapy with PTX-loaded perfluoro-15-crown-5-ether NEs showed excellent therapeutic properties characterized by tumor regression and suppression of metastasis (Rapoport et al., 2011). A PTX-loaded PEG 400 mediated lipid NE showed the same cytotoxicity against HeLa cells as Taxol (while its LD_{50} was ca. 3-fold higher than that of Taxol), considerably reduced extraction of reticuloendothelial system organs, increased tumor uptake and showed superior antitumor potency on A2780 (human ovarian cancer cell line) and Bcap-37 (human breast cancer cell line) tumor-bearing nude mice compared to the conventional PTX-loaded lipid NE (Jing et al., 2014). Coadministration of PTX and curcumin (CUR) in NE formulations designed to overcome MDR in tumor cells was found to be very effective in enhancing cytotoxicity in wild-type SKOV3 and drug-resistant SKOV3 (TR) human ovarian adenocarcinoma cells by promoting the apoptotic response (Ganta and Amiji, 2009).

Simvastatin, which decreases low-density lipoprotein (LDL)-cholesterol, used as an adjuvant to PTX associated in lipid NEs showed significantly increased antitumor activity in B16F10 melanoma-bearing mice compared to the control and a PTX NE, which was reflected in 95% tumor growth inhibition, reduced metastasis and higher survival rates, whereby the tumors showed reduced cellular density and increased collagen fibers I and III and lower matrix metallopeptidase 9 expression (Kretzer et al., 2016).

Lipid NEs loaded with PTX + etoposide (ETP) administered via three intraperitoneal injections to the B16F10 melanoma-bearing mice caused a reduction in the number of animals bearing metastases, a reduction of cellular density, blood vessels, and an increase of collagen fibers in tumor tissues, a reduction of melanoma-related anemia and thrombocytosis and showed 65% arrest in the G_2/M-phase of the cell cycle as well as reduced toxicity compared to administration of both free drugs (Kretzer et al., 2012). The antiproliferative effect against A549 cells of a ETP-loaded coix seed oil ME with droplet size of 35 nm showed IC_{50} values 4- and 10-fold lower compared to the pure drug or the mixture of the drug and coix seed oil; ME formulation also improved the induction of cell apoptosis (Qu et al., 2015a).

The core-matched NEs functionalized by vitamin E and TPGS designed to codeliver PTX and 5-fluorouracil (5-FU) exhibited time-dependent cytotoxicity against PTX-sensitive human cervix carcinoma KB-3-1 cells and PTX-resistant KB-8-5 cervical cells as well as an obviously increased G_2/M-phase bloc, effectively induced the apoptosis of tumor cells via upregulation of tumor suppressor p53 and β-tubulin and by the significant inhibition of cell cycle progression caused dramatic inhibition of tumor growth, which could be connected with the synergism of PTX and 5-FU and the reverse of MDR by the inhibition of ATPase activity by vitamin E and TPGS (Ma et al., 2014).

An optimized cholesterol-rich NE of cholesteryl-succinyl-5-FU (Chol-S-5-FU) showed considerable cytotoxic effects in MDA-MB-231 breast cancer and LoVo colon cancer cells and about 2-fold higher bioavailability in male Wistar rats than ethanolic solution of Chol-S-5-FU, indicating that this NE could be used for chemoprevention of breast and colon cancers (Shakeel et al., 2015a). A cholesterol-rich NE for the drug delivery potential of cholesteryl-maleoyl-5-FU was designed by Alanazi et al. (2015). An optimized low hydrophilic–lipophilic balance surfactant NE of 5-FU showed enhancement ratio of 22.33 compared to the control in in vitro skin permeation testing through rat skin using Franz diffusion cells and also higher cytotoxicity against human skin malignant melanoma cells SK-MEL-5 than the free drug indicating that such NE could be used for topical chemoprevention of skin cancer (Shakeel et al., 2015b). Topical MEs of 5-FU prepared using Span 80, Span 85, Tween 80, and isopropyl alcohol with different oils (e.g., oleic acid, triacetin, and isopropyl myristate) with droplet size 100 nm showed improved drug release compared to marketed 5-FU cream and the safety of this nanoformulation for topical use confirmed in vivo tests with rats (Kumar and Sinha, 2016).

A doxorubicin (DOX)-loaded NE administered to Ehrlich ascites carcinoma-bearing Swiss albino mice reduced levels of serum enzymes and lowered damage to heart tissues, prolonged more than 2-fold the mean survival time, and the life-span was by ca. 5-fold higher compared to administration of DOX solution

(Alkhatib et al., 2016). A lipid NE with coencapsulated DOX and bromotetrandrine (W198) developed to reverse MDR in breast cancer showed considerably higher cytotoxicity in DOX-resistant human breast cancer cells (MCF-7/ADR) compared with DOX-loaded lipid NEs, exhibited enhanced intracellular uptake of DOX in MCF-7/ADR cells and, using a nude mice xenograft model, enhanced tumor uptake of DOX in animals treated with DOX/W198 lipid NE was estimated. In addition, also reduced cardiac toxicity and gastrointestinal injury in rats were observed (Cao et al., 2015). Investigation of the effect of DOX encapsulation in micelles or NEs on the ultrasound-mediated intracellular delivery and nuclear trafficking performed by Mohan and Rapoport (2010) showed that ultrasound triggers DOX trafficking into the cell nuclei, which is particularly noticeable in the presence of NEs that convert into microbubbles under ultrasound action, and due to the microbubble cavitation, transient permeabilization of both plasma and nuclear membranes occurs, allowing DOX penetration into the cell nuclei which is reflected in dramatically enhanced therapeutic efficacy of this nanoformulation.

N-oleyl-daunorubicin (DRC) in a cholesterol-rich NE formulation applied at dose 7.5 mol/kg reduced tumor growth in tumor-bearing mice by $59 \pm 2\%$ (while at administration of DRC this reduction was only $23\% \pm 2\%$), improved survival rates and reduced the number of mice with metastasis compared to those estimated in animals to which the free drug was administered (30% and 82%, respectively) (Contente et al., 2014).

Dacarbazine (DAC)-loaded NEs showed significantly reduced tumor size in epidermoid carcinoma xenograft mice, both with intramuscular injection and topical administration, compared to the corresponding administration of DAC suspensions (Kakumanu et al., 2011). Similar findings concerning the significantly increased anticancer efficacy of DAC loaded in NE preparations in a xenograft mouse melanoma model were described by Tagne et al. (2008).

A bimodal imaging contrast agent, a perfluorocarbon (PFC)/rhodamine NE was found to be successfully delivered into FR positive tumor xenograft models and showed significantly enhanced signal intensities of ^{19}F magnetic resonance imaging (MRI) and fluorescence imaging in the tumor area (Bae et al., 2013). Phase-shift NEs consisting of liquid PFC droplets that can be vaporized into microbubbles were found to accelerate the rate of high-intensity focused ultrasound (HIFU)-mediated heating and could dramatically improve the efficiency and clinical feasibility of magnetic resonance-guided HIFU, a noninvasive technology to treat solid tumors (Kopechek et al., 2013). Improvement of the efficiency of HIFU-mediated thermal ablation of solid tumors by phase-shift NEs was reported also by Zhang et al. (2013). The development of PFC NEs for theranostics of tumor and plaque angiogenesis was reviewed by Lanza et al. (2010). Acoustically active PFC NEs prepared using liquid PFCs and coconut oil as the cores of the inner phase and stabilized by phospholipids and/or Pluronic F-68 loaded with camptothecin (CPT) with the droplet diameter of 220–420 nm showed retarded drug release, and formulations with a lower oil concentration exhibited cytotoxicity against melanomas and ovarian cancer cells; however, increased drug release from the system with lower oil concentration under application of 1 MHz ultrasound was estimated (Fang et al., 2009). A review paper focused on the drug-loaded PFC nanodroplets for ultrasound-mediated drug delivery that have great potential for unloading encapsulated drugs at a desired time and location in the body in response to directed ultrasound was presented by Rapoport (2016). The CPT-loaded-MEs and magnetic MEs fabricated using benzyl alcohol:Captex 300 (3:1), TPGS:Tween 80 (2:1), and water were characterized by an acceptable level of hemolytic activity, considerable cytotoxicity against MCF-7 cancer cells and low DNA damage in lymphocytes, and introduction of external magnetic field resulted in higher concentrations of magnetic MEs at the target tissue of 4T1 breast cancer-induced BALB/c mice. At application of both types of MEs, the active targeting of 5067.56 ± 354.72 ng/g and passive targeting of 1677.58 ± 134.20 ng/g CPT to breast cancer were estimated (Natesan et al., 2014).

MEs containing tamoxifen (TAM) citrate prepared using sesame oil and Tween 80 as drug solvent oil and a surfactant, respectively, and edible glycerin as a cosurfactant with average droplet size ranging from 16.64 to 64.62 nm released 85.79% of drug at $pH = 7.4$ and 100% at $pH = 5.5$ in the first 24 hours and markedly reduced the tumor growth rate in cancerous tumors in BALB/c female mice in vivo, and the weight loss of animals was lower than after administration of commercial drug tablets (Dehghani et al., 2017).

Didodecyl methotrexate (ddMTX) associated with a lipid NE with the particle size of 60 nm showing 98% ddMTX incorporation released ca. 28% ddMTX after 48 hours of incubation with plasma and showed higher cytotoxicity against K562 and HL60 neoplastic cells with IC_{50} values of 1.6 and 0.2 mM, respectively, than methotrexate (MTX) (IC_{50}: 18.2 and 26 mM) and also 90-fold higher uptake in K562 cells and 75-fold higher uptake in HL60 cells than MTX (Moura et al., 2011).

The anticancer effect of tocotrienols entrapped in parenteral lipid NEs evaluated using highly malignant + stearic acid (SA) mouse mammary epithelial cells showed that stabilizing the NEs with poloxamer or sodium oleate resulted in enhanced activity with IC_{50} values of 3 and 6 μM, respectively, compared to tocotrienol/albumin conjugate ($IC_{50} = 10$ μM), and the antiproliferative activity of the NEs inversely correlated with particle size (Alayoubi et al., 2014). Parenteral lipid NEs for the concurrent delivery of tocotrienol rich fraction (TRF) of vitamin E and simvastatin with the droplet size of 200 nm and the surface potential of −45 mV exhibited antiproliferative activity against MCF-7 and MDA-MB-231 human mammary tumor cells with IC_{50} 10 and 4.8 μM, respectively, while the corresponding IC_{50} values estimated for TRF NE without simvastin were higher (14 and 7 μM, respectively) (Alayoubi et al., 2013).

Using castor oil and Cremophor ELP, an aluminum-phthalocyanine chloride (AlPc) NE with ca. 25 nm nanodroplets showing intense photodynamic activity was prepared by Muehlmann et al. (2015), whereby photoactivated AlPc exhibited strong cytotoxic effect on cell spheroids of human mammary adenocarcinoma MCF-7 and for application to MCF-7 cell monolayers, even IC_{50} of 6.0 nM, was estimated.

NEs loaded with prednisolone acetate valerate, iron oxide nanocrystals and the fluorescent dye Cy7 for near-infrared (NIR) fluorescence imagining were found to represent a flexible and unique theranostic NPs platform that can be applied for imaging guided therapy of cancer (Gianella et al., 2011).

The NE system for the transdermal delivery of granisetron hydrochloride, which may be as effective as oral or i.v. dosage forms and avoid some difficulties associated with these dosage forms, was designed by Zheng et al. (2010).

Folic acid (FA)-functionalized NEs loaded with carbon monoxide releasing molecule-2 (CORM-2) showed significant antitumor effect and an increased survival of BALB/c mice bearing subcutaneous A20 lymphoma tumors, and in vitro and in vivo experiments confirmed efficacy against Bcl cells (Loureiro et al., 2015).

The NE-based lipid nanosystems containing CUR prepared by loading w/o NEs into lipid nanosystems showed 2- to 4-fold higher absorption constants and 4- to 8-fold higher effective permeabilities in different gastrointestinal tracts than bulk CUR and relative bioavailability representing 734% of that of the free drug. In addition, they exhibited also stronger inhibitory effects on the viabilities of A549 cells than free CUR (Wan et al., 2016). Vecchione et al. (2016) reported that oil-in-water NEs coated with thiol modified chitosan (CS) (14%–15%) used for the codelivery of CUR and piperine (weight ratio 100:1) did not show any cytotoxic effect on normal fibroblasts, while they were able to promote death in colon cancer cells. After gavage administration of the curcuminoid-loaded ME prepared using labrafac lipophile WL 1349, Cremophor RH40 and glycerin as oil phase, a surfactant and a cosurfactant, respectively, to rats, approximately 10-fold enhancement of relative bioavailability was estimated compared to that of curcuminoid suspension (Xiao et al., 2013).

A ME consisting of sucrose oleate, ethanol, isopropyl myristate, and water excellently increased the amount of skin incorporation of natural antioxidant resveratrol which was rapidly incorporated into skin and situated predominantly in the dermis; and it was assumed that enhanced skin delivery process of the drug could be connected with the ME-skin interaction (Yutani et al., 2016).

A highly stable *Nigella sativa* L. essential oil NE formulated using ultrasonic emulsification, polysorbate 80 and water with droplet size of about 20–50 nm in diameter reduced the viability and induced apoptosis in MCF-7 cells (Periasamy et al., 2016). Eugenol (EU) loaded NEs emulsified with modified starch and containing canola oil induced apoptosis in human liver (HB8065) and colon (HTB37) cancer cells via reactive oxygen species (ROS) generation (Majeed et al., 2014). The application of a nanoscale formulation of argan oil and TPGS (40%–80% wt. of emulsifiers) resulted in five to 9-fold lower IC_{50} values compared to free TPGS and free argan-oil NEs in murine breast and colon carcinoma cells and demonstrated significant proapoptotic effect on both test cancer cell lines

(Jordan et al., 2012). Optimized NE formulations of carvone Schiff base of isoniazid (CSB-INH) with the droplet size of 19.4 nm and the optimal value of zeta potential of −30.8 mV releasing 94.5% of the drug after 24 hours showed approximately 9-fold higher cytotoxicity to human colon cancer cells than free CSB-INH and more rapid absorption in albino rats than CSB-INH suspension formulation (Bhat et al., 2015).

NEs loaded with piplartine, an alkaloid found in black pepper exhibiting anticancer activity showed a 1.5-fold increase in oral bioavailability as compared to free piplartine as well as noticeable antitumor activity in melanoma tumor-bearing mice at the administration of a dose of 10 mg/kg (Fofaria et al., 2016).

A cashew nut shell liquid NE was found to cause loss of viability and death of human breast cancer cells MCF-7, mostly through apoptosis and/or necrosis (Al-Hazzani et al., 2012).

A catechin NE composed of catechin extract, lecithin, Tween 80, and deionized water with the mean particle size of 11.45 nm and the zeta potential of −66.3 mV showing EE of 88.1% more effectively inhibited the growth of PC-3 tumor cells (IC_{50} = 8.5 μg/mL) than catechin extract (IC_{50} = 15.4 μg/mL) and induced apoptosis of PC-3 cells, and both caspase-dependent and caspase-independent pathways may be involved in apoptosis of these cells (Tsai and Chen, 2016).

Treatment with lycopene (LP)-nanogold (nanoAu) NE prepared using Tween 80 as an emulsifier and containing 0.16 ppm nanoAu + 0.4 μM LP showed a 5- and 15-fold rise in early apoptotic cells of HT-29, respectively, compared to the control, and the application of the combo (10 ppm nanoAu + 12 μM LP) significantly reduced the expressions of procaspases 8, 3, and 9 as well as poly(ADP-ribose) polymerase 1 and Bcl-2, and a 5-fold decline in the migration capability of HT-29 cells was observed for this NE when compared to the control. Because nanoAu and LP doses were found to be 250- and 120-fold lower, respectively, than in the combo treatment, this NE could have application in colon cancer therapy (Huang et al., 2015).

A mannosylated lycorine lipid NE formulation showing 82.7% ± 1.6% EE was preferably taken up by adenocarcinomic human alveolar basal epithelial cells (A549) compared to an uncoated lipid NE and exhibited also better growth inhibition activity on A549 cell lines than the lipid NE without coating (Guo et al., 2012).

An optimized NE composed of Miglyol 812N/Labrasol/Tween 80/Lipoid E80/water (10%/10%/2.5%/1.2%/76.3%) for incorporation of the natural flavonoid fisetin (3,3′,4′,7-tetrahydroxyflavone) with the oil droplet diameter of 153 nm and the negative zeta potential of −28.4 ± 0.6 mV when administered intraperitoneally showed a 24-fold increase in fisetin relative bioavailability compared to free fisetin and exhibited the antitumor activity in Lewis lung carcinoma (LLC) bearing mice at lower doses (36.6 mg/kg) than free fisetin (223 mg/kg) (Ragelle et al., 2012).

The i.v. administration of a PEGylated NE loaded with lycobetaine (LBT)-oleic acid complex to rats showed higher concentration of LBT in the lung and considerably lower LBT concentration in the heart, live, and kidney and exhibited also higher growth inhibitory effect and longer survival time in both heterotopic and lung metastatic tumor models than free LBT (Zhao et al., 2013a).

An emodin (EMO) NE containing Cremophor EL with spherical droplets of ca. 116 nm increased the apparent permeability of EMO 2-fold compared to the pure drug solution and enhanced the transcellular permeation of the drug into UGT1A1-overexpressing MDCKII cells (or MDCK1A1 cells), which was connected with the inhibition of uridine 5′-diphospho-glucuronosyltransferase (UGT) metabolism, indicating that such nanoformulation can improve the oral bioavailability of EMO (Zhang et al., 2016a). An EMO NE effectively improved the drug bioavailability and prolonged in vivo mean residence time (MRT) via oral administration to rats, and EMO amount in individual organs decreased in the following order: liver, lung, kidney, heart or spleen, and brain, but it is necessary to note that the MRT of EMO in the brain was 2-fold longer than that in the other tissues (Shi et al., 2015a).

Lactoferrin (Lf) from camel milk loaded on a perfluorooctyl bromide NE with particle size ca. 100 nm and the zeta potential of 21.61 mV was reported to cause the death of MCF-7 breast cancer cells, while the percentage of dead cells increased with increasing Lf concentration and it might be used to deliver cytolytic peptides and proteins (Keykanlu et al., 2016).

Epidermal growth factor receptor (EGFR) and FR functionalized NEs incorporating phosphatidylinositol 3-kinase inhibitor PIK_{75} and C_6-ceramide exhibited an increase in cytotoxicity against SKOV3 cells, which was reflected in 57% decrease in IC_{50} value of PIK75 following treatment with a EGFR targeted NE and 40% decrease following treatment with a FR targeted NE and could be connected with the increase in cellular apoptosis and the activity of highly ROS (Talekar et al., 2012).

A self-nanoemulsifying drug delivery system (SNEDDS) for coadministration of CUR and PTX showed ca. 3-fold enhancement in the dissolution rate of both drugs from SNEDDS compared with the pure drug suspension and ca. 3-fold reduction in the cytotoxicity on MCF-7 cells as well as 6-fold improvement in the oral bioavailability in vivo compared to the pure PTX suspension (Sandhu et al., 2015). An optimized SNEDDS prepared using capryol propylene glycol monocaprylate (oil phase), Tween-20 (surfactant), Carbitol (cosurfactant), and water (aqueous phase) with the droplet size of 28 nm and the zeta potential of −32.8 mV controlled the ibrutinib absorption compared with the drug suspension and enhanced the drug bioavailability about 3-fold in female Wistar rats (Shakeel et al., 2016). Also pomegranate seed oil was found to exert synergistic effects with *trans*-resveratrol in the SNEDDS (Lu et al., 2015). Shakeel et al. (2014) developed double w/o/w NEs for a SNEDDS of 5-FU for chemoprevention of colon cancer that were 5-fold more efficacious than free 5-FU on human colon cancer cell lines (HT-29).

8.3 NANOSUSPENSIONS

Nanosuspensions (NSs) are aqueous suspensions containing one or several submicron-sized drug substances and appropriate stabilizers used to prevent particle agglomeration or aggregation and crystal growth by steric hindrance and/or electrostatic repulsion. Steric hindrance could be secured by adsorbing polymers onto particles, while for electrostatic repulsion, ionic surfactants, or polymers could be used (Merisko-Liversidge et al., 2003; Yadollahi et al., 2015). To ensure pure electrostatic stabilization, zeta potential of least ± 30 mV is necessary, while at combined electrostatic and steric stabilization, the aggregation and precipitation could be prevented already with zeta potential of ± 20 mV (Jacobs and Müller, 2002). The NSs represent a promising approach in nanoparticulate drug delivery systems, including anticancer drugs (Rao et al., 2004). For satisfactory stabilization of a NS, the drug substance:stabilizer ratio ranging between 20:1 and 2:1 (w/w) is desirable. For preparation of solid dosage form using NSs, redispersants such as sucrose, lactose and mannitol are applied to guarantee complete redispersion of NPs into their primary, predrying state (Merisko-Liversidge et al., 2003).

NS technologies for delivery of poorly soluble drugs were summarized by Yadollahi et al. (2015) and different technologies and methods applied in drug nanocrystal preparation, suspensions, and stabilization with comparison of nano- and micro-sized drug crystals in pharmaceutical applications were overviewed by Chang et al. (2015). Correlation of the characteristics of NS formulations, such as drug property, particle size, crystallinity, stabilizer, and surface property with their transport, pharmacokinetics, bioactivity, and toxicity after delivery by different administration routes are presented in the review paper of Wang et al. (2014a). The differences between nanocrystals and NSs are described in Liu et al. (2012).

Sharma et al. (2015) reported that for fabrication of PTX nanocrystals using high-pressure homogenizer without application of any organic solvent, the high molecular weight polymers glycol CS and sodium alginate were found to be less efficient stabilizers than Tween 80 and sodium polystyrene sulfonate, and PTX nanocrystals were more effective against MCF-7 and MDA-MB breast cancer cells than the drug solution and showed almost a 9- to 10-fold increase in the area under the plasma-concentration time curve (AUC) in an in vivo study compared to plain PTX crystals. Yin et al. (2016) prepared a PEG modified NS based on human serum albumin (HSA) loaded with PTX showing prolonged blood circulation as well as enhanced antitumor efficacy in S180 mice compared to Taxol and a NS without PEG modification. Moreover, at i.v. administration the PEG modified NS showed also lower hemolytic activity, which was reflected in the increased LD_{50} value and the absence of local venous irritation. The prolonged circulation is connected with the incorporation of PEG showing the steric stabilization effect, while the enhanced tumor uptake and reduced side effects are related to HSA-mediated targeting resulting in active receptor-mediated endocytosis (Yin et al., 2015). The PEGylated PTX nanocrystals with the mean particle size of 330 nm showed higher stability than drug nanocrystals under both storage and physiological conditions and also exhibited improved anticancer effects after i.v. administration in breast cancer xenografted mice. In a model of lung tumor metastasis, their anticancer activity exceeded also the effect of Taxol achieving an 82% reduction at the end of the experiment (Zhang et al., 2015a). Monodisperse PTX NSs fabricated using Pluronic F-127 as a stabilizer showed cytotoxicity and maximum-tolerated dose comparable with those of Taxol; however, higher bioavailability and the application of this NS in hyperthermic intraperitoneal chemotherapy resulted in a significant reduction of tumor volume which also was comparable with the effect of Taxol, but the rats recovered faster (De Smet et al., 2012). Both a PTX solution and a PTX NS with TPGS stabilizer on the surface of nanocrystals showed cytotoxicity against sensitive human lung cancer cells H460; however, the NS exhibited significantly higher cytotoxicity against resistant H460/RT cells and 5-fold higher inhibition rate of tumor growth after i.v. administration compared to the application of combined PTX and TPGS solution, indicating that this NS could effectively reverse drug resistance of H460/RT cells (Gao et al., 2014).

DTX nanocrystals prepared through high-pressure homogenization technique using Pluronic F-127 as a stabilizer were found to be more effective against MCF-7 breast cancer cells, showed enhancement of DTX induced G2-M arrest and the most effective apoptosis after 24 hours incubation of the cells compared to free drug and Taxotere. Higher depolarization of mitochondria estimated for DTX nanocrystals than for free DTX and Taxol suggested that apoptosis induced by these nanocrystals was linked to altered mitochondrial membrane potential. Based on the in vivo studies in mice, the DTX nanocrystals also showed acceptable hemolytic potential and they could be considered as a promising nanoformulation for parenteral drug administration (Pawar et al., 2015).

The in vivo antitumor efficacy of lyophilized *N*-3-*O*-toluyl-FU-loaded lipid-based NSs with the average particle size of 180 nm and the zeta potential of -8.02 ± 1.43 mV was similar to that of 5-FU injection; however, they released lower drug concentration in the heart and kidney (Zhang et al., 2014).

A sorafenib-lipids NS prepared by the solvent displacement technique using soya lecithin and sorafenib dissolved in methanol as an organic phase and aqueous solution of Tween 80 as an aqueous phase and having the particle size of 164.5 nm and the zeta potential of -11.0 mV were found to exhibit higher in vitro cytotoxicity against human liver cancer cells HepG2 and Bel-7402 cells than the free drug, and their i.v. administration to H22-bearing liver cancer xenograft mice resulted in higher antitumor efficacy and reduced tumor volume compared with i.v. administration of the same dose of the pure drug or oral administration of the 2-fold dose of the drug lipid NS. Consequently, this nanoformulation could be suitable to treat hepatocellular carcinoma (Yang et al., 2016a).

The i.v. administration of a CPT NS prepared by antisolvent precipitation with TPGS as a stabilizer resulted in higher drug bioavailability, intratumor accumulation, and anticancer effect compared to the drug solution, and an enhanced cellular drug uptake was estimated in hypoxic and acid conditions (Tang et al., 2014).

Annonaceous acetogenins NSs designed using monomethoxyPEG 2000-poly (ε-caprolactone) (PCL) as a stabilizer with the particle size 123.2 nm and the zeta potential −20.17 mV were found to be more cytotoxic against 4T1, MCF-7 and HeLa cells than the free drug, and their administration even at 10-fold lower dose than that of usually used oil solution to 4T1 tumor-bearing mice resulted in much better therapeutic efficacy (Hong et al., 2016).

Spherical sucrose ester stabilized oleanolic acid NSs with particle size ca. 100 nm in diameter significantly increased oleanolic acid saturation solubility and showed a markedly higher in vitro dissolution rate and cytotoxicity and in vivo bioavailability than the pure drug (Li et al., 2014).

8.4 MICELLES

Micelles are self-assembling nanosized colloidal particles with a hydrophobic core and hydrophilic shell which can solubilize compounds insoluble in water or showing low aqueous solubility and, thus, they could be used as carriers of insoluble drugs. Beside surfactants, micelles could be formed also by amphiphilic copolymers, that is, polymers consisting of a hydrophobic block and a hydrophilic block or by conjugates of soluble copolymers with lipids (such as PEG-phosphatidyl ethanolamine (PE) conjugate) forming lipid-core micelles (Torchilin, 2007). The hydrophobic core of micelles of amphiphilic polymers functions as a drug reservoir and encapsulates hydrophobic drugs, whereby the hydrophilic shell prevents the aggregation of micelles and also prolongs their systemic circulation in vivo (Li et al., 2016). For example, phospholipid micelles can increase the accumulation of loaded drugs in tumor tissues by taking advantage of the enhanced permeability and retention effect and by ligand-mediated active targeting, and the intracellular delivery of the cargo could be enhanced by tailoring the composition of the micelles (Sawant and Torchilin, 2010). Also Gill et al. (2015) reviewed findings concerning PEG-lipid micelles as delivery carriers for poorly soluble drugs, their biological implications and recent developments in exploring their active-targeting potential. Wang et al. (2012a) reported that PEGylated phospholipid micelles induce endoplasmic reticulum-dependent apoptosis of cancer cells, but not normal cells, and they have the potential to alter the sensitivity of cancer cells to some chemotherapeutic agents by selectively activating unfolded protein response in cancer cells.

Investigation of the self-assembly of PE conjugate with water-soluble drugs (DOX hydrochloride, vinorelbine (VNB) tartrate, and vincristine sulfate) showed that electrostatic interaction and hydrophobic forces are responsible for the intercalation of drugs into PEG-PE micelles, and using NMR it was estimated that the anthracycline ring of DOX was inserted between PE phospholipids, and its amino sugar located in the outer shell of micelle between PEG chains (Wang et al., 2010). Oxidized phospholipid, 1-palmitoyl-2-azelaoyl-sn-glycero-3-phosphocholine (PazPC) micelles loaded with anthracycline anticancer drugs DOX and idarubicin (IDA) exhibited pH-sensitive behavior with accelerated release of DOX or IDA in an acidic endosome environment and the drug-loaded PazPC micelles enhanced drug uptake and exhibited higher cytotoxicity in both leukemia (P388 and its resistant subline P388/ADR) cell lines than free drugs, indicating that this micellar formulation can be useful in delivering anthracycline anticancer drugs, contributes to the increase of the therapeutic index and overcomes MDR for leukemia treatment (Wang et al., 2012b). Cell membrane-mimetic micelles of amphiphilic copolymers prepared by grafting hydrophobic SA onto a zwitterionic phosphorylcholine functionalized CS oligosaccharide showed excellent stability under extreme conditions including protein containing media, high salt content, and a wide pH range and the hydrophobic interaction between DOX and SA segments enabled successfully entrapping of the drug into polymeric micelles. These biomimetic polymeric micelles may be considered as promising drug delivery systems in cancer therapy due to internalization by cancer cells and following sustained drug release from micelles to the cytoplasm and nucleus (Liu et al., 2015a). Oleic acid-coated MnO NPs encapsulated in micelles composed of PEG2000, PE, cholesteryl 3β-*N*-(dimethylaminoethyl)carbamate hydrochloride (DC-cholesterol), and dioleoyl-phosphatidylethanolamine represent multifunctional lipid-micellar NPs, which can be taken up in vitro by human embryonic kidney (HEK293), LLC1, and A549 cells; they are not cytotoxic and are able to enhance T-1 MRI contrast. Moreover, these NPs loaded with plasmid DNA and/or DOX are efficiently taken up by HEK293 cells in vitro and by target cells in vivo and when administered to mice intranasally, they preferentially accumulate in the lungs, indicating their exploitability as a lung theranostic, especially for lung cancers (Howell et al., 2013). Sterically stabilized, biocompatible, and biodegradable phospholipid nanomicelles with particle size ca. 15 nm were found to incorporate VNB at high quantities within the interface between the core and palisade sections of the micelles, whereby the incorporation ratio of the drug within sterically stabilized micelles increased as the total amount of the drug in the system increased and the nanomicellar formulation of VNB was about 7-fold more potent on MCF-7 cell line than VNB dissolved in dimethyl sulfoxide (Bahadori et al., 2014).

The synthesis and optimization of novel polymeric micelles based on hyaluronic acid (HA) and phospholipids including 1,2-dimiristoyl phosphatidylethanolamine and 1,2-distearoyl phosphatidylethanolamine (DSPE) for PTX delivery and their in vitro and in vivo evaluation was reported by Saadat et al. (2014). The PTX-loaded micelle formulations were stable both physically and chemically for at least 6 months and in vivo real-time image analysis showed that the micellar system was mostly accumulated in the liver, spleen, and heart. Sterically stabilized mixed micelles composed of biocompatible and biodegradable phospholipids, the surface of which was grafted with a ligand, human vasoactive intestinal peptide, selectively interacting with its cognate receptors overexpressed on breast cancer cells, were tested for delivery of PTX to breast cancer in rats. Administration of clinically relevant PTX dose (5 mg/kg) using these micelles eradicated carcinogen-induced orthotopic breast cancer in rats, and a 5-fold lower dose (1 mg/kg) of drug in actively targeted micelles was associated with similar to 80% reduction in tumor size, while the response to PTX was significantly less (Dagar et al., 2012). PTX-loaded PEG-b-poly(lactic acid) (PLA) micelles increased the cellular accumulation of PTX in PTX-resistant human ovarian cell line A2780/T, which resulted in more apoptosis compared with free PTX and PEG-b-PLA micelles inhibited P-gp function and P-gp ATPase activity without affecting P-gp protein expression, induced cell membrane depolarization, and enhanced membrane microviscosity (Xiao et al., 2011). Micelles of self-assembling copolymer stearoyl-PEG-poly-sulfadimethoxine methacrylate showing responsive behavior in the physiopathologic pH range with the size of 13 nm in water at pH 7.4 and the critical micelle concentration of 36 μM with the PTX loading capacity 3.25% ± 0.25% (w/w) were found to be taken up at pH 6.5 by MCF-7 tumor cells and at pH 6.5 the PTX-loaded micelles had a higher cytotoxic effect than the micelles incubated at pH 7.4 (Ravazzolo et al., 2013).

Pluronic F-127 conjugated to cholesterol via succinyl linkage formed spherical micelles with the mean particle size of 248–279 nm and zeta potential ranging from −17.2 ± 8.7 to −28.4 ± 12.7 mV and DTX loading efficiency >98%. The in vitro release study showed the sustained release behavior of DTX within 144 hours, and DTX-loaded micelles exhibited higher cytotoxicity against B16F10 melanoma and HepG2 cell lines compared to the free drug (Varshosaz et al., 2015).

Amphiphilic statistical copolymers poly(cholesteryl acrylate-co-methoxypoly (ethylene glycol) methacrylate), poly[CHOLy-co-mPEG(n,x)] (for $n = 5$, $x = 110$ and $y = 15$, and for $n = 23$, $x = 22$ and $y = 3$), with copolymer composition (x:y) of 7:1 were found to form stable polymeric micelles in water above a relatively low critical concentration and were tested as a delivery system for water-insoluble anticancer agent, *S*-(+)-CPT. The copolymers were not only hemocompatible and nontoxic in the allowed concentration range, but also easily permeated into cancer cells (MCF-7 and HeLa) and in an in vitro experiment using cancer cells (HeLa), the CPT-loaded polymeric micelles showed very good chemotherapeutic activity in the biocompatible concentration range of the copolymers (Laskar et al., 2014).

Micelles of a novel folate-targeted copolymer synthesized using dextran stearate coupled to FA with the particle size of 106 nm and the zeta potential of −21.2 mV when loaded with ETP showed 20-fold higher cytotoxicity against CT26 colorectal carcinoma cell line compared to the pure drug (IC_{50}: 0.49 vs 9.41 μg/mL) and these drug-loading micelles are promising in reducing ETP resistance of colorectal cancer by boosting drug cellular uptake (Varshosaz et al., 2014).

By conjugating SA to gemcitabine (GEM) and subsequent encapsulating stearoyl GEM within PEG-DSPE/TPGS, mixed micelles pronouncedly prolonged circulation time of the drug and elevated its concentration in the tumor 3-fold, which resulted in superior antitumor efficacy in mice bearing human pancreatic cancer BxPC-3 xenografts (Wang et al., 2014b).

The photosensitizer zinc phthalocyanine (ZnPc)- and upconversion nanocrystals (UCNs)-loaded octadecyl-quaternized small-lipid-grafted poly(γ-glutamic acid)-PEG/arginyl-glycyl-aspartic acid (RGD)/transactivator of transcription (TAT) lipid micelles with the average size of 25 nm, are stable in water with low leakage of photosensitizer, and can be used in photodynamic therapy (PDT) of cancer. The UCNs acting as a nanotransducer convert deep-penetrating NIR light to visible light for activating the photosensitizer, and the visible fluorescence emitted from the UCNs upon excitation by the NIR laser at 980 nm can activate the photosensitizer to produce singlet oxygen for PDT, while the targeting RGD peptide and cell-penetrating TAT peptide on the surface help the NPs getting into cancer cells (Wang et al., 2014c).

Lanoline acid–modified PEG-b-PCL copolymer forming spherical micelles with the average diameter of 21 nm which could be used as a delivery system for CUR with the high drug-loading capacity of 12.80% and the EE of 98.53% was developed by Song et al. (2014). It was found that CUR existed in the polymeric matrices under π–π conjugation and hydrogen bond interaction with the copolymer and in vitro drug release from copolymer micelles was controlled with a lower release rate than that from the linolenic acid–unmodified copolymer micelles, showing higher cytotoxicity against HeLa and A549 cells. Linolenic acid–modified copolymer was found to be safe for i.v. injection and its micelles prolonged drug retention in blood circulation.

Micelles made of 1,2-distearoyl-sn-glycero-3-phosphoethanolamine-*N*-[methoxy (polyethylene glycol)-2000] mixed with TPGS in a 3:1 molar ratio increased berberine solubilization by 300% and drug-containing micelles significantly amplified apoptosis and overall cytotoxic effectiveness against monolayer and spheroid cultures of human prostate carcinomas, showing 16- to 18-fold lower IC_{50} values in PC-3 human prostate cancer cells and androgen-sensitive human prostate adenocarcinoma cells LNPaC compared to free berberine (Shen et al., 2016).

Mucoadhesive nanoscale buccal films containing phospholipid-bile salts-mixed micelles as an effective carrier for cucurbitacin B (Cu B) delivery were developed by Lv et al. (2015). The absorption of Cu B from carboxymethyl CS films containing Cu B loaded phospholipid-bile salts-mixed micelles resulted in a 3-fold increase in bioavailability as compared to conventional tablet formulation and 10-fold higher bioavailability as compared to the carboxymethyl CS films formulation.

8.5 LIPOSOMES

Liposomes are spherical vesicles having an aqueous solution core surrounded by a hydrophobic membrane in the form of a lipid bilayer composed most often of phospholipids, for example, phosphatidylcholine (PC), and eventually also of other lipids such as PE and they could occur as multilamellar vesicles with several lamellar phase lipid bilayers, small unilamellar vesicles with one lipid bilayer or large unilamellar vesicles. Because polar drug molecules could be encapsulated in the liposomal polar core and amphiphilic and lipophilic molecules could be solubilized within the phospholipid bilayer, liposomes represent widely used lipid carriers for drug delivery (Akbarzadeh et al., 2013; Xing et al., 2016).

Urbinati et al. (2012) published an overview of the past 40 years of liposome research until the last liposomal formulations commercially available or undergoing clinical trials and summarized the main concepts over liposome domain, which can count on encouraging results as target therapy associated with targeted delivery. Liposomes constitute one of the most popular nanocarriers for the delivery of cancer therapeutics due to better solubility, higher stability and slower release of drugs compared to free drug administration. Xing et al. (2016) summarized recent developments of liposomes as nanocarriers for theranostic applications focusing on diagnostic applications to detect proteins, DNA, and small molecule targets using fluorescence, magnetic resonance, ultrasound and nuclear imaging, therapeutic applications based on small molecule-based therapy, gene therapy and immunotherapy, and theranostic applications for simultaneous detection and treatment of heavy metal toxicity and cancers. Liposomes represent an attractive delivery system due to their flexible physicochemical properties which allow easy manipulation in order to address different delivery considerations (Hua and Wu, 2013; Deshpande et al., 2013). In general, at tumor-specific drug delivery using liposomes, the passive accumulation of liposomes that have an increased circulation half-life is observed which could be connected with the leakiness of tumor neovasculature as well as the active targeting of liposomes using surface-bound ligands (Lindner and Hossann, 2010). The uses of liposomes as a drug delivery system was summarized by Kim (2016) and Petersen et al. (2012) reviewed the major advances in the development of radiolabeled liposomes for imaging as a tool in personalized medicine.

Drug delivery to tumors using liposomal chemotherapy can be strongly enhanced by applying hyperthermia that increases vascular permeability in solid tumors resulting in increased levels of liposome accumulation and by the use of thermosensitive liposomes that can be triggered to release their contents upon hyperthermia (Koning et al., 2010). Also folate-targeted bimodal paramagnetic and fluorescent liposomes composed of both a paramagnetic and a fluorescent lipids and a PEG-lipid amphiphile for prolonged in vivo circulation showed rapid accumulation within the tumor tissue in a FR expressing tumor model compared to nontargeted liposomes, indicating their potential for dynamic real-time solid tumor MRI (Kamaly et al., 2009). A new drug delivery nanosystem combining liposomal and dendrimeric technology (liposomal locked-in dendrimers, LLDs) for cancer therapy was reported in which the interaction between liposomes and dendrimers could result in higher loading and slower drug release from the LLDs comparing to pure liposomes (Gardikis et al., 2010). Liposomal NPs were found to be effective in delivering siRNA into tumor tissues by improving stability and bioavailability; however, cationic lipids/liposomes are highly toxic and negatively charged liposomes are rapidly cleared from circulation. Therefore Ozpolat et al. (2014) developed highly safe and effective neutral lipid-based nanoliposomes that provided robust gene silencing in tumors following systemic (i.v.) administration and exhibited remarkable antitumor efficacy in various orthotopic human cancer models in animals.

Qin et al. (2011) developed partially polymerized liposomes composed of polydiacetylene lipids and saturated lipids in which cross-linking of the diacetylene lipids prevented the drug leakage even at 40°C for days and more than 70% of the encapsulated drug can be instantaneously released by a laser that matches the plasmon resonance of the tethered AuNPs on the liposomes, showing the therapeutic effect in cancer cells. Zhu et al. (2012) developed a matrix metalloprotease 2—responsive multifunctional liposomal nanocarrier for enhanced tumor targeting and improved cancer cell-specific delivery of loaded drugs. This system representing a surface-functionalized liposomal nanocarrier, for which two functional PEG-lipid conjugates were synthesized and the functionalized liposome was further modified with the tumor cell-specific antinucleosome monoclonal antibody (mAb 2C5), could enhance the targetability and internalization in cancer cells.

By encapsulation of phosphoethanolamine (PHO-S) into dioctadecyldimethylammonium chloride (DODAC) cationic liposomes, its cytotoxic effect and selectivity against tumor cells could increase. The PHO-S encapsulation efficacy was about 50% and the formulation showed the IC_{50} value of 0.8 mM for B16F10 cells and 0.2 mM for Hepa1c1c7 cells without significant effects on endothelial cells. The B16F10 cells and human umbilical vein endothelial cells (HUVECs) were less sensitive to DODAC/PHO-S formulation than Hepa1c1c7 cells and their application on B16F10 cells induced G2/M-phase cell cycle arrest with the proportion significantly greater than that estimated at treatment with PHO-S alone. Moreover, cationic liposomal formulation for PHO-S delivery promoted cytotoxicity more selectively and effectively against B16F10 and Hepa1c1c7 cells (Luna et al., 2016).

Mei et al. (2014) designed a multistage liposome system comodified with RGD, TAT, and cleavable PEG that combined the advantages of PEG, specific ligand, and penetrating peptide. The cleavable PEG could increase the stability and circulation time of liposomes during circulation and RGD specifically recognized integrins overexpressed on various malignant tumors and mediated efficient internalization in the synergistic effect of RGD and TAT. It was shown that in vitro the system could not only be selectively and efficiently taken up by cells overexpressing integrins, but also penetrate the tumor cells to reach the depths of the avascular tumor spheroids, while in vivo profoundly improved distribution within tumor tissues was observed and the RGD and TAT ligands on liposomes produced a strong synergistic effect that promoted the uptake of liposomes into cells after the systemic administration of L-cysteine.

Octa-arginine (R8)-modified PEGylated liposomal DOX (R8-PlDOX) designed for the treatment of non–small cell lung cancer and tested with non–small cell lung cancer cell line A549 exhibited significantly higher penetration into the hypoxic A549 tumor spheroids compared to PEGylated liposomal DOX, induced a higher level of apoptosis to A549 tumor xenograft and caused dramatic inhibition of tumor volume as well as reduction of tumor weight; the treatment of tumor lysate with R8-PlDOX resulted in an elevated caspase 3/7 expression compared to PEGylated liposomal DOX treatment (Biswas et al., 2013). Joo et al. (2013) prepared cross-linked multilamellar liposome (CML) which stably encapsulated DOX in the vesicle that exhibited a remarkable controlled rate of release compared to that of the unilamellar liposome (UL) with the same lipid composition or Doxil-like liposome (DLL). The CMLs were mainly internalized through a caveolin-dependent pathway and further trafficked through endosome–lysosome compartments. Moreover, this CML-DOX formulation reduced systemic toxicity and significantly improved therapeutic activity in inhibiting tumor growth compared to that of UL-DOX and DLL-DOX. Omacetaxine mepesuccinate (OMT)-loaded liposomes prepared by loading the drug in the lipid bilayers and then electrostatically complexed with DOX yielded double-loaded liposomes (DOLs), while DOX-loaded liposomes were formulated by electrostatic interaction with negatively charged empty liposomes. The combination of DOX and OMT showed ratio-dependent synergistic activities with very strong synergism observed at a molar ratio of 4:1 (DOX:OMT). In an in vitro experiment DOLs showed higher antitumor activity than OMT- and DOX-loaded liposomes or empty liposomes in cervical carcinoma HeLa cells, in vivo DOLs reduced the tumor growth by 98.6% and 97.3% relative to the untreated control on day 15 and 25 after the cessation of treatment, respectively, suggesting that the liposomal codelivery of DOX and OMT could synergistically potentiate antitumor effects (Shim et al., 2014).

A nanocarrier system with particle size of about 100 nm constructed by the modification of stealth liposomes with a metastatic cancer specific peptide using unmodified liposomes as the control was found to penetrate deeper into tumor mass and have a longer retention within the malignant tissue and greatly facilitated the efficacy of DOX loaded in terms of inhibiting xenograft tumor growth and inducing cancer cell apoptosis with only minor side effects. It could be mentioned that highly metastatic cancer cells (MDA-MB-435S and MDA-MB-231) were found to endocytose more metastatic cancer specific peptide modified liposomes in a faster way than liposomes through a receptor-mediated pathway proved by specific receptor inhibition (Wang et al., 2012c). Targeting liposomes encapsulating DOX to solid tumors by nonthermal diode laser was reported by Ghannam et al. (2016).

Dicheva et al. (2014) prepared DOX-loaded cationic thermosensitive liposomes (DOX-CTSLs) carrying targeting functions addressing tumor cells and tumor vasculature and having a heat-responsive lipid bilayer, which demonstrated superior uptake by and toxicity to different tumor cell lines and endothelial cells compared to nontargeted TSLs. Upon an external heat-trigger, the DOX-CTSLs delivered 3-fold higher DOX quantity to tumors than TSLs and caused significant tumor vessel damage which was not observed when nontargeted TSLs were administered. Compared to DOX-loaded nontargeted TSLs, hyperthermia applied prior to administration of DOX-CTSLs resulted in enhanced DOX delivery to B16BL6 murine melanoma causing tumor suppression and prolonged survival of animals and their efficacy was confirmed also in LLC tumor model (Dicheva et al., 2016). The i.v. administration of long-circulating phosphatidyldiglycerol-based TSLs with encapsulated DOX into a syngeneic soft tissue sarcoma (BN 175) located on both hind legs of male Brown Norway rats having tumor temperature >40°C that was maintained for 60 minutes confirmed more efficacious drug delivery into the tumor resulting in DOX concentration of 4.5–23.1 ng/mg as compared to nonheated tumors (0.5 ± 0.1 ng/mg), whereby a 13- and 15-fold higher DOX deposition was estimated in large tumors and small tumors, respectively, due to the use of the laser-based applicator (Willerding et al., 2016). Also, local hyperthermia combined with liposomal DOX was found to markedly increase antitumor efficacy in rabbit VX2 tumors and reduce systemic toxicity (Xiong et al., 2015).

DOX-loaded HA-modified pH-sensitive liposomes showing faster drug release at pH 5 compared to that at pH 7.4 and cytotoxicity on CD44 receptor expressing MCF-7 cells ($IC_{50} = 1.9\ \mu M$ after 48 hours of incubation) were more effective in cells expressing high levels of CD44 than unmodified liposomes and exhibited only minimal cardiotoxicity (Paliwal et al., 2016). Thermosensitive CS/β-glycerophosphate hydrogel containing liposomal DOX had better antineoplastic effect and less toxicity and showed higher $AUC_{(0-t)}$ and longer mean resistance time (MRT) in blood and tumor as well as lower toxicity to normal tissues compared to DOX in CS/β-glycerophosphate (Ren et al., 2016). DOX-loaded glycyrrhetinic acid (GA) conjugated liposomes showing both pH-sensitive features and active-targeting ability showed considerable inhibition of tumor growth and prolonged survival time and exhibited lower systematic toxicity and higher therapeutic effect (Chen et al., 2016a).

Rizzitelli et al. (2016) designed a novel MRI-based theranostic protocol for improving the anticancer efficacy of a Doxil-like liposomal formulation in which the intratumor drug release was promoted by the sequential application of low-intensity pulsed ultrasound. The tests on mice bearing a synergic breast cancer model showed a significant increase in the intratumor Doxil concentration resulting in the complete regression of the lesion. The application of focused ultrasound (capable of reversibly opening the blood–brain barrier (BBB) but not affecting normal tissues) introduced DOX-loaded cationic liposomes and showed pregnant inhibitory effectiveness on glioma cells in vitro and C6 glioma model in vivo and prolonged the animal survival period to 81.2 days (Lin et al., 2016).

PEGylated liposomal DOX suppressed tumor growth in a DOX-resistant breast cancer (MDA-MB-231) xenograft model. In hyperpermeable OSRC-2 tumors, the liposomes were distributed to deep tumor tissues where they delivered DOX to drug-resistant tumor cells, while extracellular matrix molecules rendered MDA-MB-231 tumors hypopermeable, which restrained the extent of the penetration and distribution of liposomes resulting in enhanced delivery of DOX to the vicinity of the tumor vasculature and excellent antiangiogenic effect (Kibria et al., 2016). Application of a novel PEGylated liposome system consisting of DOX and 5-FU optimally delivered in a liposome called DAFODIL resulted in >90% reduction in tumor growth of murine 4T1 mammary carcinoma in vivo and, at a convenient ratio of drugs, synergic effects of drugs were documented, the necessary doses being far below the maximum tolerable drug doses (Camacho et al., 2016). Zahmatkeshan et al. (2016) designed PEGylated anti-HER2 peptide liposomal formulations of DOX by an engineered breast tumor-targeting peptide ligand and found that with the increasing ligand density increased cytotoxicity and cell uptake in SKBR3 and TUBO cells which overexpress HER2 (but not in MDA-MB-231 cells with low HER2 expression profile) as well as higher anticancer activity in mice bearing TUBO breast cancer were observed. Jakoby et al. (2015) reported the use of primed antibody (anti-IGF1R)-anchor conjugates for specific postmodifications of the liposomal surface enabling selective binding to target cells which could facilitate the personalized approaches of commercially available liposomal drugs. The human atherosclerotic plaque-specific peptide-1 (AP1)-conjugated liposomal DOX preferentially killed interleukin-4 receptor-overexpressing CT26 cells, and its i.v. administration to mice efficiently inhibited tumor growth and decreased cardiotoxicity of DOX (Yang et al., 2015a). A low-molecular-weight heparin-coated DOX liposome developed for metastasis preventative therapy exhibited higher cytotoxicity than the unmodified ones as well as considerable tumor metastasis prevention by preventing tumor cell-platelet interactions and causing suppressed tumor growth (Chen et al., 2015a).

DOX and quinine hydrochloride (as a MDR reversing agent) coloaded liposomes exhibited high intracellular accumulation as well as potential of MDR reversal capability in MCF-7/ADR cells and at the ratio 0.5:1.0 (DOX:quinine hydrochloride) in liposomes, the apoptotic percentage increased to 62.2% (Shen and Qiu, 2016). The liposomes of dexamethasone palmitate and DOX coloaded liposomes modified with a sialic acid-octadecylamine conjugate exhibited better tumor growth suppression than single drug–loaded liposomes and induced "shedding" of cancerous tissue in mice (Sun et al., 2016a). DOX-loaded liposomes comodified with transferrin (Tf) and FA exhibited higher cytotoxic effects against HeLa and A2780-ADR ovarian carcinoma cells than untargeted or single-ligand targeted liposomes as well as higher tumor growth inhibition; a similar finding was obtained in a HeLa xenograft model in nude mice (Sriraman et al., 2016).

Combination therapy using Bip17 (QLPVM) penta-peptide functionalized liposomal TAM and DOX was reported as a possible rational strategy for the treatment of Luminal A breast cancer (Wang et al., 2016a).

Sardan et al. (2013) prepared cell-penetrating peptide amphiphile integrated liposomal systems for enhanced delivery of anticancer drugs (DOX HCl and PTX) to tumor cells by performing a noncovalent integration of a cell-penetrating arginine-rich peptide amphiphile into a liposomal formulation of negatively charged 1,2-dioleoyl-sn-glycero-3-[phospho-rac-(1-glycerol)] phospholipids in the presence of cholesterol, which, due to its amphipathic character, showed enhancement in liposomal uptake and enhanced efficacy of the drugs compared to the free drugs. The Tf and TAT (cell-penetrating peptide) comodified liposomes encapsulating both PTX and DOX and showing targeted delivery of both drugs for melanoma were characterized by enhanced targeting efficiency and increased therapeutic efficacy (Yuan et al., 2016; Chen et al., 2016b).

A neovessel-targeted PEGylated liposomal formulation of PTX prepared by integrating a RGD containing peptide coupled with a Lys-Gly-Gly (KGG)-palmitic acid conjugate to the lipid bilayers where the mean particle size of the liposomes was <100 nm and drug entrapment efficiency $>95\%$ was reported by Meng et al. (2011). The researchers observed greater cellular uptake in the integrin-targeted liposomal PTX-treated group compared with the Taxol-treated group and the treatment of mice bearing A549 tumors with these liposomes resulted in a lower tumor microvessel density than with Taxol-treatment.

By substituting a neutral distearoylglycerol anchor for the more commonly applied anionic DSPE anchor in preparing a PEGylated nanoliposomal formulation using triethylammonium sucrose octasulfate for encapsulation of VNB not only considerably improved efficacy in both a human colon carcinoma (HT-29) and a murine (C-26) colon carcinoma models was observed, but also PEGylated nanoliposomal VNB exhibited in vivo drug retention and it was highly active against human breast and lung tumor xenografts when VNB was stably encapsulated in liposomes using triethylammonium sucrose octasulfate with high EE (Drummond et al., 2009).

pH-sensitive liposomes prepared by surface decoration with 1,2-distearoyl-sn-glycero-3-phosphoethanolamine-*N*-[methoxy(polyethylene glycol)] (mPEG-DSPE) and stearoyl-poly(ethylene glycol)-poly(methacryloyl sulfadimethoxine) copolymer loaded with GEM were stable at pH 7.4 for several hours, while at pH 6.5 rapidly aggregated and at pH 6.5 displayed higher cytotoxicity than at pH 7.4 or compared to nonresponsive control liposomes at both incubation pH which indicates that this carrier can efficiently deliver the anticancer drug to the cytosolic compartment (Bersani et al., 2014).

Wen et al. (2013) prepared multifunctional liposomes loaded with quantum dots (QDs) and anticancer drugs CPT and irinotecan (IRI) for simultaneous bioimaging and drug delivery and compared physicochemical characteristics and theranostic efficiency of cationic, PEGylated, and deformable liposomes for their theranostic efficiency. The researchers found that QDs were situated within liposomal bilayers and the incorporation of PEG and Span 20 into the liposomes resulted in significantly increased fluidity of the bilayers. The best results were obtained with cationic liposomes which were found to exhibit superior activity in the cytotoxicity and cell migration assay and showed a significant fluorescence signal in melanoma cells after internalization, whereby, if intratumorally administered to a melanoma-bearing mouse, showed brightest fluorescence in tumors lasting up to 24 hours and intratumoral accumulation of CPT after loading with cationic liposomes increased to 50 nmol/g while that from the free control was only 35 nmol/g.

Resveratrol and 5-FU were successfully coencapsulated in a single PEGylated nanoliposome in which resveratrol was localized near the glycerol backbone of the liposomal membrane, while 5-FU was localized closer to the phosphate moiety which influenced the release kinetics of both drugs. The estimated growth inhibition expressed by IC_{50} values of the nanoformulation tested in vitro on a head and neck cancer cell line NT8e was found to be similar to that of free 5-FU and the coencapsulation of resveratrol and 5-FU in a liposomal nanocarrier improved the cytotoxicity in comparison with the free drug combination when tested in vitro (Mohan et al., 2014).

Vrignaud et al. (2012) entrapped erlotinib hydrochloride (ERLO) used for the treatment of pancreatic cancers by means of reverse micellar incorporation using a mix of sorbitan trioleate (Span 85) and Labrafac oil at a 1:1 (w/w) ratio and the efficacy of ERLO-loaded nanocarriers on BxPC-3 pancreatic adenocarcinoma cells in an in vitro experiment was comparable to free ERLO and led to a cell death rate of around 40%.

Tetrac-tagged PEGylated liposomes (TPLs), in which Tetrac, an antagonist that blocks the binding of thyroid hormone to integrin αV/β3, was covalently linked to the activated end of PEGylated lipid and preferentially distributed to tumor tissues after systemic administration in mice, while nontargeting PEGylated liposomes were distributed to tumors only at background levels. Encapsulation of edelfosine into TPLs resulted in significantly reduced survival of A375 tumor cells compared to cells treated with edelfosine in PEGylated liposomes and i.v. administration of the drug encapsulated TPLs significantly reduced the growth of tumors and prolonged the survival of A375-xenografted mice, providing 100% protection for up to 50 days and some protection until 66 days (0% survival endpoint) (Lee et al., 2012).

Liposomes with the main lipid component 1,2-distearoyl-sn-glycero-3-phosphatidylcholine (DSPC) loaded with aquated cisplatin (CDDP) were characterized with slow drug release (2% after 72 hours at 37°C), 500-fold reduced clearance, and an 100-fold increase in drug accumulation in the tumor compared to free CDDP and these liposomes with C18 acyl chain length were found to be more effective than liposomes with C16 or C14 hydrocarbon chain length. Moreover, 90% EMT6-AR1 multidrug-resistant mouse breast tumor growth inhibition was observed with liposomal formulation of DSPC containing aquated CDDP (Zhao et al., 2015a). The inhibitory effectiveness of CDDP-containing egg L-phosphatidylcholine liposomes with different amounts of cholesterol with the size ca. 100 nm against the growth of ovarian and breast human cancer cells characterized by different drug sensitivity was comparable or higher than that of the free drug (Ringhieri et al., 2016). PEGylated liposomal CDDP with 142 nm particle size and the zeta potential of −22 mV showed high retention capability and ca. 2-fold higher cytotoxicity than the pure drug (Poy et al., 2016). Liposomes modified by acetyl GA, a hepatic targeting ligand, with the size ranging from 120 to 180 nm and the zeta potential from −39.7 to −3.18 mV designed for CDDP delivery exhibited higher cytotoxicity against HepG2 cells, indicating their stronger affinity to these cancer cells, and accumulated in the liver in a greater extent than unmodified liposomes (Zhang et al., 2016b). By conjugation of anti-HER2-affibody to CDDP PEGylated liposomes, enhanced intracellular Pt concentration was obtained and at the application of these anti-HER2-affibody-targeted liposomes concentration necessary to kill SKBR 3 (HER2 +) cells was lower and therapeutic efficiency in HER2 + -TUBO models (a cloned cell line generated from a spontaneous mammary gland tumor from a BALB-*neu*T mouse highly expressing HER2 protein on the cell membrane) was greater (reflected in extended survival of animals) than at treatment with nontargeted liposomes (Alavizadeh et al., 2016). CDDP-incorporated liposomes conjugated with EGFR antibodies exhibited higher cytotoxicity in the cells expressing EGFR than unmodified liposomes in vitro and their use in combination with radiation retarded tumor growth in an A549 cell-derived xenograft tumor mouse model without exerting nephrotoxic effects (Jung et al., 2015).

CDDP-loaded thermosensitive liposomes prepared by reactions of dipalmitoyl phosphatidylcholine (DPPC) with metal ions such as Zn^{2+}, Cu^{2+}, Mn^{2+}, and Mg^{2+} characterized by high drug EE exhibited sustained and temperature-dependent drug release and showed higher cytotoxicity than CDDP solution (Liu et al., 2015b). CDDP-incorporated DPPC liposomes suppressed the cytotoxic effects caused by CDDP in the testicular tissue of adult Wistar rats and showed positive effects on the maintaining of the body weight of the animals (Favareto et al., 2015). The pH-responsive liposomes designed to release CDDP within the acidic tumor interstitium designed by Stras et al. (2016) were reported to improve the intratumoral distributions of the free drug resulting in enhanced efficacy against triple negative breast cancer solid tumors. Long-circulating and pH-sensitive liposomes containing CDDP exhibited good antitumor activity resulting in a decrease in tumor volume and development of less tumor necrotic areas in pancreatic cancer-bearing mice compared to the pure drug (Carlesso et al., 2016).

Zhang et al. (2012a) developed liposomes for loading both CDDP and QDs for both drug delivery and bioimaging. The prepared QD-liposomes (QDLs) composed of CdSe or CdSe/ZnS QDs had the size of about 100 nm and most QDs were located within the liposomal bilayers. The cytotoxicity of CDDP-loaded QDLs was higher than that of the free drug. Following i.v. administration of CdSe/ZnS QDLs to mice, CDDP accumulation was detected in the brain and skin, as verified by ex vivo imaging, and drug distribution and liposomal inclusion could reduce the reticuloendothelial system uptake of QDs and CDDP.

Miriplatin-loaded liposomes with the size of 99 nm and the zeta potential of −50 mV were found to induce cell apoptosis and their effects were similar to that of commercial CDDP and oxaliplatin injections (Liu et al., 2016a).

The i.v. application of oxaliplatin-loaded PEGylated liposomes was found to decrease the fraction of Kupffer cells without causing injury to hepatocytes, suggesting that this nanoformulation could be helpful for the treatment of metastatic hepatic cancer due to the direct killing of the cancer cells along with the reduction of tumor-supportive Kupffer cells (Ukawa et al., 2016).

The incorporation of rhodamine-B within ruthenium-loaded liposomes formed by the cationic 1,2-dioleyl-3-trimethylammoniumpropane chloride (DOTAP) showed fast cellular uptake in human carcinoma cells with strong fluorescence accumulation within the cells, indicating DOTAP as a very efficient nanocarrier for anticancer efficacy of ruthenium-based drugs stabilization in aqueous media and transport in cells and these formulations have proved to be over 20-fold more effective against MCF-7 and WiDr adenocarcinoma cells with respect to the nude organometallic ruthenium complex AziRu (Vitiello et al., 2015).

Cholesterol-based cationic liposomes evaluated as drug delivery vehicles for CUR exhibited two to 8-fold higher cytotoxicity against HeLa, A549, HepG2, K562 and 1301 cell lines than free CUR with ED_{50} values ranging from 2.5 to 10 μM, and an enhanced cellular uptake of CUR into tested cells was estimated (Apiratikul et al., 2013).

Liu et al. (2016b) reported the synthesis of the PEGylated betulinic acid (BA) liposomes with mean diameter of 142 nm, which accumulated effectively in tumor tissues, showed a better sustained drug release effect than BA liposomes in vitro and exhibited a better tumor inhibitory effect than free BA and BA liposomes in both in vitro and in vivo experiments.

Costa et al. (2013) developed a lipid-based nanocarrier for targeted delivery of nucleic acids to glioblastoma by covalently coupling of chlorotoxin (CTX; a peptide reported to bind selectively to glioma cells while showing no affinity for non-neoplastic cells) to liposomes encapsulating antisense oligonucleotides or siRNAs. The attachment of CTX onto the liposomal surface enhanced particle internalization into glioma cells, while no significant internalization was observed in noncancer cells and preliminary in vivo studies revealed that CTX enhanced particle internalization into established intracranial tumors, indicating that prepared NPs are suitable for targeted nucleic acid delivery to cancer cells.

The Lf nanoliposomes showing acceptable stability in simulated gastrointestinal juice at 37°C for 4 hours and decreasing the viability of cells in a dose-dependent manner were found to be fit for the oral administration of Lf and this could be a useful approach for Lf availability in tumor cells (Guan et al., 2012).

The homogenous magnetic NPs with the size about 10 nm encapsulated in PEGylated fluorescent liposomes giving magneto-liposomes (MLs) with a diameter approximately 130 nm after embedding with Tf MLs were applied in an in vitro BBB transmigration study in the presence or absence of external magnetic force. Comparing with magnetic force- or Tf receptor-mediated transportation alone, their synergy resulted in 50%–100% increased transmigration without affecting the BBB integrity and higher cellular uptake of ML particles was observed due to the synergic effect, indicating that multifunctional liposomal magnetic nanocarriers possess great potential in particle transmigration across the BBB and could be used in drug delivery to the brain (Ding et al., 2014a).

8.6 NANOCAPSULES

Lipid nanocapsules (LNCPs) based on a core–shell structure consisting of an oil-filled core with a surrounding polymer (natural or synthetic) layer are known to be promising vehicles for the delivery of hydrophobic drugs (Huynh et al., 2009; Kothamasu et al., 2012). Thus they were designed to deliver lipophilic drugs to cells with nontoxic excipients by mimicking lipoproteins (Lagarce and Passirani, 2016). In general, the oily core of LNCPs is formed of medium-chain triglycerides, while the surrounding membrane consists of a mixture of lecithin and a PEGylated surfactant. The nonionic surfactant of polyoxyethylene type situated on the LNCP surface enables the nanocapsules to exhibit a P-gp inhibitory effect harmonized with a stealth effect versus the complement system as well as mononuclear phagocyte system uptake. By grafting ligands to LNCP, active-targeting drug delivery could be obtained. The advantages of LNCPs include solvent free and soft-energy fabrication, high stability, significant drug encapsulation capacity, and sustained drug release at the site of action (Diaspro et al., 2002; Huynh et al., 2009; Kothamasu et al., 2012).

The LNCPs consisting of a medium-chain triglyceride liquid oil core surrounded by a β-cyclodextrin-poly(4-acryloylmorpholine) conjugate as a shell component loaded with 4-hydroxynonenal having particle size ca. 350 nm and a negative surface charge exhibited increased antiproliferative, proapoptotic, and distinct activity in several tumor cell lines from different tissues (Pizzimenti et al., 2015). Farace et al. (2016) investigated the immune cell impact of LNCPs having different coating (Pluronic, CS and PEG) and found that immunomodulation of innate immunity which can induce monocyte activations was induced by Pluronic NCPs, treatment with CS-NCPs exhibited immunomodulation in monocytes and T lymphocytes, while PEG-coated NCPs were completely inert. Sanchez-Moreno et al. (2015) investigated the effect of corona on therapeutic efficacy and macrophage uptake of LNCPs and found that using of nonionic poloxamer resulted in nanocapsules showing a significant reduction in the number of protein-binding sites, which can modulate NPs clearance by the immune system, however, results in lower uptake and, thus, lower therapeutic efficacy. A review paper summarizing the recent findings concerning nucleic acid delivery using LNCPs, which would be useful in treatment of glioma and melanoma and show multiple acquired resistances, which can be reversed by DNA transfection or siRNA interference, was published by Lagarce and Passirani (2016).

DTX cationic LNCPs with the particle size of 130−150 nm, the zeta potential of +72 mV, and >95% EE exhibited considerably higher drug release than DTX solution in vitro, showed superior antiangiogenic activity, and effectively inhibited tumor growth in melanoma cell line induced tumors in C57BL/6 mice as compared to the drug solution (Jain et al., 2016). DTX-loaded poly-L-asparagine NCPs with an oily reservoir enabling effective encapsulation of lipophilic drugs and having particle sizes ranging from 170 to 200 nm and zeta potential in the range from −20 to −40 mV were found to be able to interact with NCI-H460 human lung cancer cells and contributed to the enhanced cellular toxicity of the drug (Rivera-Rodriguez et al., 2013). DTX-loaded lipoproteinated NCPs reduced drug release in plasma and its distribution among organs and showed better anticancer activity than Taxotere administered intravenously in the metastatic lung cancer model in Severe Combined Immune Deficiency-beige mice, which could be attributed to the extravasation effect resulting in NCPs accumulation in lung tumors, where the absorption of NCPs was mediated by a lymphatic process and reinforced during their transit (Attili-Qadri et al., 2013). Encapsulation of DTX into NCPs, in which HA shell was formed around the oily core, with spherical particles of 200 nm and a negative zeta potential resulted in considerable cytotoxicity improvement compared to NCI-H460 cancer cells in vitro and these NCPs could also be transformed into a powder by freeze-drying (Oyarzun-Ampuero et al., 2013).

Subcutaneous administration of LNCPs loaded with lauroyl-modified GEM ($GemC_{12}$), which can efficiently target the monocytic (M-) subset of myeloid-derived suppressor cells (MDSCs), resulted in reduced percentage of spleen and tumor-infiltrating M-MDSCs in lymphoma and melanoma-bearing mice compared to the pure drug and a very low dose of this nanoformulation mitigated tumor-associated immunosuppression and increased the efficacy of adoptive T cell therapy, indicating that the monocyte-targeting properties of $GemC_{12}$ LNCPs could be utilized for immunomodulatory purposes (Wauthoz et al., 2015; Sasso et al., 2016). Injectable gel-like nanodelivery system containing $GemC_{12}$ LNCPs showed a sustained and prolonged in vitro release of the drug over 1 month. These LNCPs exhibited superior in vitro cytotoxic activity on U87 MG glioma cells compared to the parent hydrophilic drug and significantly reduced the size of a subcutaneous human glioblastoma multiforme tumor model (Bastiancich et al., 2016). Alginate gel-core LNCPs consisting of hydrogel islets in an oil-filled matrix and surrounded by a polymeric shell designed for codelivery of both hydrophilic and hydrophobic anticancer drugs, GEM and PTX, respectively, with the particle size of 172 nm and spherical morphology with a smooth surface secured sustained release of both drugs (Poudel et al., 2014).

Modified LNCPs coated with CS shell allocating PTX in the inner oily core, in which genetic material CpG was associated onto the surface, showing the mean size of 70 nm and the zeta potential of +25 mV induced superior apoptotic effect on GL261 glioma cells compared to blank systems and also the overall survival of mice treated with these multifunctional LNCPs showed considerable increase compared to Taxol or separated injection of PTX-loaded LNCP and CpG. Moreover, the reduction of tumor growth in the animals was observed, indicating that such multifunctional nanoformulation can potentiate both CpG immunopotency and PTX antitumor activity, causing improved drug delivery into the tumor microenvironment (Lollo et al., 2015). Groo et al. (2013) reported that PTX-loaded LNCPs with the diameter ranging from 25 to 110 nm represent a formulation suitable for oral delivery because they are stable in mucus, do not change mucus rheological properties, and at low concentrations improve PTX diffusion. Encapsulation of PTX into LNCPs improved drug oral bioavailability and after oral administration to rats, a 3-fold increase of the AUC compared to the control was estimated for both PTX-loaded LNCPs and PTX associates with verapamil, indicating that application of this nanoformulation enables to avoid the use of pharmacologically active P-gp inhibitors (Peltier et al., 2006). The iRGD-modified core–shell LNCPs loaded with PTX showed better antitumor activity than the unmodified ones due the homing effect of iRGD (amino acid sequence: CRGDKGPDC) to integrins, which transported the nanocapsules to the tumor tissues in an actively targeted way, which resulted in the considerably enhanced accumulation of these LNCPs in tumor tissue. Compared to Taxol, the PTX-loaded LNCPs also showed significantly longer circulation in blood as well as increased bioavailability (Jin et al., 2016).

A decrease in glioma cell viability after in vitro treatment with MTX-loaded LNCPs was observed due to cell cycle arrest resulting in apoptotic death of cells, and using an in vivo glioma model a decrease in the tumor size and an increase in apoptosis in the tumor microenvironment were confirmed. This treatment caused a decrease in the leukocyte number, but did not alter the toxicological tissue marker expression or metabolic parameters (Figueiro et al., 2015). For resistant breast cancer cells MDA-MB-231, LNCPs loaded with MTX diethyl ester were found to be more effective in the apoptosis induction compared to solution of the pure drug; however, S-phase cell cycle arrest was induced only by drug solution (Yurgel et al., 2014).

The ETP-loaded LNCPs of the average diameter ranging from 25 to 100 nm showed sustained drug release over a period of 1 week with $t_{50\%}$ 15.9 ± 2.8 hours. They were found to be internalized in C6, F98, and 9 L glioma cells in vitro, showed higher effectiveness in P-gp inhibiting activity than drug solution and size-dependent cytotoxicity, small NPs being the most active with estimated IC_{50} values ca. 9- (C6), 33- (F98), and 13-fold lower than those of pure ETP in individual cell types (Lamprecht and Benoit, 2006). ETP-loaded LNCPs with the size of 54 nm and the zeta potential of −5.8 ± 3.5 mV released 40% of the drug after 48 hours and showed enhanced activity against H209 cells with IC_{50} value 40-fold lower than that of the pure drug; however, their cytotoxicity against ETP-resistant H69AR cells did not exceed that of etoposide (Saliou et al., 2013).

Tretinoin-loaded spray-dried LNCPs were found to protect the drug against UV A radiation, indicating that the supramolecular structure of the lipid-core NCPs did not changed by the drying process (Marchiori et al., 2012).

Tretinoin-loaded lipid-core LNCPs induced more effective apoptosis than the free drug and caused cell cycle arrest at the G_1-phase in A549 cells, whereby they were found to be able to overcome the resistance of adenocarcinoma cell line A549 to treatment with tretinoin, indicating that such nanoformulation could be promising in lung cancer therapy (Schultze et al., 2014).

The in vitro cytotoxicity of platinum-based anticancer drugs such as CDDP could by significantly improved by using LNCPs, in which small aggregates of solid drug are covered by a lipid bilayer (Hamelers and De Kroon, 2007). LNCPs prepared by CDDP coating with a phospholipid bilayer (a 1:1 mixture of PC and negatively charged phosphatidylserine) released the drug in water with a $t_{1/2}$ of 6.5 hours at 37°C, but at 4°C they were stable for 8 days. On the other hand, the formulation lost over 80% of CDDP within minutes after resuspension in mouse serum, indicating its destabilizing effect on CDDP LNCPs, whereby using PEG-derivatized PE and cholesterol in the bilayer coat resulted in the extended lifetime of CDDP LNCPs in mouse serum up to ca. 1 hour (Velinova et al., 2004).

LNCPs loaded with organometallic anticancer drug ferrociphenol (Fc-diOH) exhibited cytostatic activity on an orthotopic glioma model showing statistically significant difference in survival time between the infusion of Fc-diOH-LNCPs (0.36 mg/rat) and blank LNCPs and strong antiproliferative effect in vitro as well as the reduction of tumor volume, which was not affected neither after protection of the phenol groups of Fc-diOH with an acetate group, indicating that acetate is easily cleaved by cell hydrolases (Allard et al., 2009). A novel ferrocenyl complex encapsulated into LNCPs with the particle size of 40 nm exhibited effective in vitro anticancer activity against malignant glioma 9 L cells ($IC_{50} = 0.1\ \mu M$) connected with an oxidative stress and a dose-dependent alteration of the cell cycle and considerable tumor growth inhibition in ectopic glioma-bearing rats accompanied with the reduction of proliferative cells number in tumors compared to the control group was observed after repeated i.v. injections of this nanoformulation (Laine et al., 2014).

Artesunate-loaded CS-coated LNCPs with spherical particle shape and the size of 161 nm and a positive charge of 13.2 ± 0.87 mV showing sustained drug release exhibited stronger anticancer activity than the free drug on breast cancer cell lines MCF-7 and MDA-MB-231, respectively (Tuan et al., 2015).

PEG-shedding LNCPs prepared by decorating short PEG chains with TAT peptide and then covered by pH-sensitive graft copolymers of long PEG chains, which were loaded with disulfiram, with particle sizes of 60–90 nm were found to exhibit higher intracellular uptake and antitumor activity at pH 6.5 than at pH 7.4, showed significantly improved pharmacokinetic profiles compared to the free drug and their delivery efficiency in tumor-bearing mice was by 496.5% higher than that of free fluorescent probe Cy5.5 and by 74.5% higher than that of LNCPs in tumors (Zhang et al., 2015b).

Acetyleugenol (AcE)-loaded lipid-core LNCPs consisting of PCL, capric/caprylic triglyceride, and sorbitan monostearate and stabilized by polysorbate 80 with the particle size of 210 nm and the zeta potential of −10.0 mV induced in vitro late apoptosis and necrosis of human SK-Mel-28 melanoma cells after treatment with $18-90 \times 10^9$ NPs/mL during 1 hour, and application of $3-18 \times 10^9$ NPs/mL during 48 hours resulted in reduced cell proliferation and delayed the cell cycle. Daily intraperitoneal or oral treatment (days 3–10 after tumor injection) with LNCPs or AcE-LNCPs (1×10^{12} LNCPs/day) reduced the volume of the tumor in contrast to AcE application at the same dose (50 mg/kg/day); however, intraperitoneal treatment caused toxicity (Drewes et al., 2016).

Quercetin-loaded folate-modified LNCPs having spherical shape, particle size of 36 nm, and the zeta potential of −4.76 mV showed higher in vitro antiproliferative activity against HeLa and MCF-7/MDR cells than the free drug and markedly inhibited tumor growth in H22 tumor-bearing mice compared to the control (Ding et al., 2014b).

Encapsulation of plitidepsin in NCPs, in which polyglutamic acid was electrostatically deposited onto the lipid core with the mean particle size of 200 nm and a negative zeta potential exhibited prolonged blood circulation and significantly reduced toxicity in mice and showed more than 3-fold higher maximum-tolerated dose compared to Cremophor EL plitidepsin solution (Gonzalo et al., 2013).

The tetrandrine–phospholipid complex loaded LNCPs with spherical particles of 40 nm released 25% of the drug in the first hour and subsequently 70% over 12 hours and their relative bioavailability represented 208% of drug bioavailability from tablets, suggesting a powerful increase in the oral absorption of tetrandrine (Zhao et al., 2013b).

8.6.1 SOLID LIPID NANOPARTICLES

SLNPs are submicron-sized lipid emulsions in which the liquid lipid (oil) has been substituted by a solid lipid. These submicron colloidal lipid carriers composed of physiological lipids are dispersed in water or in an aqueous surfactant solution and their drug-loading capacity is limited by the solubility of the drug in the lipid melt, the structure of the lipid matrix, and the polymeric state of the lipid matrix (Mukherjee et al., 2009). SLNPs containing similar lipid molecules in their lipid matrix (e.g., tripalmitin) are characterized with highly ordered crystal lattice showing only few imperfections which limits incorporation of drugs between fatty acid chains and reduces the loading capacity, which could be improved by the application of more complex lipids (Mukherjee et al., 2009; Schöler et al, 2002).

In a review paper dealing with drug targeting using SLNPs, Rostami et al. (2014) discussed various types of SLNPs (e.g., magnetic SLNPs, pH-sensitive SLNPs, cationic SLNPs, magnetic SLNPs, and antibody SLNPs) suitable for therapy. For example, magnetic SLNPs with the mean particle size ca. 180 nm fabricated using iron oxide cores embedded within a glyceryl trimyristate solid matrix decreased the cell viability of human HT-29 colon adenocarcinoma cells and demonstrated a promising decrease in cell viability after exposure to the alternating electromagnetic field (de Escalona et al., 2016). Geszke-Moritz and Moritz (2016) presented a review paper focused on the application of SLNPs as a carrier system enabling improvement of the therapeutic efficacy of drugs from various therapeutic groups, modulated drug release, and highlighted their specific features (e.g., possibility of surface modification, increased permeation through biological barriers, resistance to chemical degradation, and stimuli-responsiveness) that contribute to their ubiquitous application. Luo et al. (2015) reported that the CS coating of SLNPs contributed to higher stability under acidic condition by forming a thick layer around the lipid core and considerably improved the mucoadhesive property of SLNPs, suggesting that such nanoformulations would be suitable as potential oral drug delivery system for hydrophobic drugs. Sun et al. (2016b) reported that SLNPs loaded with gadolinium diethylenetriaminepentaacetic acid and octadecylamine fluorescein isothiocyanate exhibiting no significant cytotoxicity in human colon carcinoma cells (HT-29) and mouse colon carcinoma cells (CT26) could serve as a tumor-absorbable nanoscale contrast agent for early diagnosis of colorectal tumors using magnetic resonance colonography.

SLNPs containing hypoxia-responsive lipid and a peptide lipid conjugate were found to penetrate to the hypoxic regions of tumors and under low oxygen partial pressure the encapsulated drug was relapsed after reduction of the hypoxia-responsive lipid resulting in the lipid membrane destabilization, while under hypoxic conditions the peptide-decorated LNPs reduced the viability of pancreatic cancer cell spheroids (BxPC-3) to 35% (Kulkarni et al., 2016). Chai et al. (2016) investigated the molecular transport mechanisms of SLNPs crossing simulative intestinal epithelial cell monolayers (Caco-2 cell monolayers) and found that the endocytosis of the SLNPs was time-dependent, their delivery was energy-dependent, and their internalization was mediated by macropinocytosis pathway and clathrin- and caveolae (or lipid raft)-related routes. Moreover, the SLNPs transported to the basolateral side were intact, and the transport of NPs did not destroy the structure of tight junctions. Based on investigation performed with lung A549 cells and cervical HeLa cells, Shah et al. (2016) reported that cellular uptake of SA-based SLNPs was energy-dependent and the endocytosis of SLNPs depended primarily on clathrin-mediated mechanisms.

SLNPs coloaded with DOX and α-tocopherol succinate with the size range of 74–80 nm, negative zeta potential, and EE of 99% showed prominent cytotoxicity against drug-resistant cancer cells (MCF-7/ADR and NCI/ADR, respectively) and a greater penetration of DOX in monolayer and 3D spheroid cancer cell models (Oliveira et al., 2016a). They exhibited a controlled drug release and showed higher cytotoxicity in breast cancer cells than free DOX (Oliveira et al., 2016b). DOX lauroyl ester-loaded SLNPs inhibited proliferation of a pancreatic (CFPAC-1), a lung (A549), and two ovarian (A2780 and DOX-resistant A2780 res) cell lines to a greater extent than free DOX; their inhibitory effect was maintained at 10^{-5} mM concentration, and a 40% inhibitory effect on A2780 res cell line up to 10^{-4} mM concentration was estimated (Peira et al., 2016). The DOX-loaded galactosylated SLNPs with the particle size of 239 nm and the EE of 72.3% ± 0.9% exhibited higher cellular uptake, cytotoxicity and nuclear localization against adenocarcinomic human alveolar basal epithelial A549 cells than DOX-loaded SLNPs, indicating that galactosylated carriers are valuable for targeted delivery of drugs to ameliorate their therapeutic index (Jain et al., 2015). The Tf-conjugated DOX-loaded lipid-coated NPs were more efficiently endocytosed by A549 cells, exhibited stronger antiproliferative effect on A549 cells and more efficient inhibition of tumor spheroid growth than lipid-coated NPs or poly(lactic-co-glycolic acid) (PLGA) NPs, whereby they were found to be the most effective also in vivo in the tumor growth inhibition in A549 tumor-bearing mice (Guo et al., 2015). The Tf-modified DOX and enhanced green fluorescence protein plasmid (pEGFP) coencapsulated SLNPs with the particle size of 267 nm and 42 mV surface charge was found to be a multifunctional delivery system that targets lung cancer cells and can improve the efficacy of cancer therapy through the combination of gene therapy and chemotherapy (Han et al., 2014a). Treatment with squalenoyl-DOX nanoassemblies of 130 nm mean diameter resulted in 90% inhibition of M109 lung tumors, which did not respond to DOX treatment, and 95% reduction of MiaPaCa-2 pancreatic tumor xenografts in mice compared with tumors in the saline-treated mice. Treatment with this nanoformulation also considerably exceeded the tumor reduction achieved with free DOX (29%). Moreover, this nanoformulation exhibited 5-fold higher maximum-tolerated dose compared to free DOX and did not cause any myocardial lesions (Maksimenko et al., 2014). DOX-loaded SLNPs showing particle size 217 nm and EE 59.45% ± 1.75% exhibited significant enhancement in bioavailability and improved anticancer efficacy and safety profile in 7,12-dimethylbenzanthracene-induced breast cancer in Sprague Dawley rats compared to the free drug (Patro et al., 2013).

In DOX-loaded SLNPs, the drug maintained its cytotoxic activity toward glioma cell lines; moreover, increased permeation through the hCMEC/D3 cell monolayer was observed, indicating that such nanoformulation is suitable for glioblastoma treatment (Battaglia et al., 2014). DOX-loaded SLNPs also efficiently enhanced apoptotic cell death through the higher accumulation of drug in MCF-7/ADR cells (Kang et al., 2010). DOX-loaded SLNPs modified with CS oligosaccharide with the zeta potential of 30 mV significantly reduced the burst release, secured controlled drug release, and improved drug cellular uptake and cytotoxicity of DOX (Ying et al., 2011).

The pH-responsive cholesterol-PEG adduct-coated SLNPs carrying DOX capable to overcome MDR of breast cancer cells of the mean hydrodynamic diameter 100 nm displayed superior ability in inhibiting the proliferation of MCF-7/MDR cells and inhibited the growth of MCF-7/MDR xenografted tumor in vivo in BALB/c nude mice more effectively than the free drug and DOX was found to be accumulated predominantly in tumor via enhanced permeability and retention effect. The enhanced passive tumor accumulation was connected with the loose intercellular junctions of endothelial cells lining inside blood vessels at tumor site and the lack of lymphatic drainage (Chen et al., 2015b).

DOX-loaded SLNPs designed for topical delivery against skin cancer with the mean particle size of 92 nm and the EE of 86% showed superior cytotoxicity against murine melanoma (B16F10) cells in vitro and in vivo in melanoma induced BALB/c mice (Tupal et al., 2016). The beneficial contribution of iontophoresis application for topical cancer therapy was described by Huber et al. (2015). Iontophoretic treatment with DOX-loaded cationic SLNPs increased the penetration of DOX in the viable epidermis 50-fold and effectively inhibited tumor cell survival and tumor growth of squamous cell carcinoma induced in nude BALB/c mice which was accompanied by an increase in keratinization and consequent cell death.

Comparison of SLNPs loaded with DOX and with IDA showed that the in vitro IC_{50} value of DOX NPs was 9-fold lower than that of free DOX solution in resistant P388/ADR cell line and IC_{50} values of free IDA and IDA-loaded LNPs in P-gp-overexpressing P388/ADR and HCT-15 cells were comparable. In the in vivo P388/ADR leukemia mouse model, significantly greater median survival time was observed after treatment with DOX NPs compared to the free drug and controls, but the effectiveness of IDA LNPs and free IDA in P388 and P-gp-overexpressing HCT-15 mouse tumor models was comparable. Consequently, it could be concluded that DOX LNPs could overcome P-gp-mediated MDR both in vitro in P388/ADR leukemia cells and in vivo in the murine leukemia mouse model (Ma et al., 2009). Compared to the treatment with the drug solution, the IDA-loaded SLNPs, which were administered to rats either by the duodenal route or intravenously, showed lower concentration of IDA and its main metabolite idarubicinol in heart, lung, spleen, and kidneys, but IDA and idarubicinol were detected in brain indicating that SLNPs were able to pass the BBB. The IDA-loaded SLNPs acted as a prolonged release system for the drug and duodenal administration of SLNPs resulted in higher AUC compared to i.v. administration (Zara et al., 2002). Application of SLNPs coloaded with DOX and CUR confirmed the synergistic effect of both drugs on the apoptosis, proliferation, and angiogenesis of hepatocellular carcinoma, and it was suggested that CUR in this formulation might reverse the MDR (Zhao et al., 2015b).

DOX-entrapped SLNPs with grafted aprotinin and melanotransferrin antibody containing 40% (w/w) of 1,2-dipalmitoyl-sn-glycero-3-phosphorylglycerol in lipids secured the prevention of fast drug dissolution and did not induce a serious cytotoxicity to human brain-microvascular endothelial cells and human astrocyte compared to free DOX, showed good ability to infuse the BBB and inhibited glioblastoma multiforme (Kuo and Lee, 2016). It is important to note that in another paper, Kuo and Wang (2016) reported that melanotransferrin antibody and TAM conjugated on ETP-entrapped SLNPs showing antiproliferation against U87MG cells could enhance the delivery of ETP across the BBB to restrain brain tumor growth.

SLNPs conjugated with TAM and Lf used to carry carmustine across the BBB showed better antiproliferation activity against U87MG cells than formulations of carmustine-loaded SLNPs lacking modification with TAM and Lf (Kuo and Cheng, 2016). TAM-loaded SLNPs showed cytotoxic activity against MCF-7 and MDA-MB-231 cells with lower IC_{50} values than the free drug; moreover, it was observed that TAM-loaded SLNPs caused a reduction in the viability of both cancer cell lines, which lost their normal morphological characteristics, detached, aggregated, and subsequently developed apoptotic bodies (Abbasalipourkabir et al., 2016). TAM citrate-loaded SLNPs fabricated using glycerol monostearate or SA as a lipid matrix and Poloxamer 188 or Tween 80 as stabilizers were found to increase the oral bioavailability (up to 160.61%) of the drug in rats compared to the free drug and simultaneously preserved the in vitro antitumor activity (Hashem et al., 2014). Spherical TAM-loaded α-lipoic acid-stearylamine conjugate-based SLNPs with smooth surface and the mean particle size of 261.08 ± 2.13 nm, showing slow and pH-dependent drug release, exhibited a 1.5-fold increase in drug relative bioavailability compared to the suspension of the pure drug and decreased the hepatotoxicity of TAM in Sprague Dawley rats (Dhaundiyal et al., 2016).

Epirubicin-loaded SLNPs prepared as an inhalable formulation for the treatment of lung cancer proved a markedly higher drug concentration in plasma after inhalation than after treatment with the drug solution (Hu et al., 2010).

PTX-loaded and cross-linked solid lecithin-acrylamido-2-methylpropane sulfonate-based phospholipid NPs with a negative surface charge prepared by UV polymerization with the average size of 200 nm and the zeta potential of -70 mV exhibited more controlled release behavior with longer release time compared to the uncrosslinked ones and, based on the in vitro cytotoxicity test performed using MCF-7 human breast adenocarcinoma cell line, were considered to have a potential therapeutic effect for breast cancer treatments (Mutlu et al., 2016). Coating of PTX-loaded SLNPs with a PEG layer contributed to an increased intracellular concentration of the drug resulting in improved efficacy and it was supposed that such nanoformulation could considerably circumvent cancer cell resistance and increase patient survival and cure (Arranja et al., 2016). Spherical PTX-loaded mannosylated SLNPs with the mean particle size 254 nm and the zeta potential of 3.27 mV were found to deliver higher drug concentration in an alveolar cell site compared to nonfunctionalized SLNPs, indicating their potential use for lung cancer targeting (Sahu et al., 2015). PTX-loaded cationic SLNPs complexed with herceptin were found to exhibit higher in vitro toxicity against MDA-MB-453 cells compared to PTX-loaded anionic and cationic SLNPs (Buyukkoroglu et al., 2016). The Lf-modified SLNPs of PTX exhibited cytotoxicity on human bronchial epithelial cells BEAS-2B ex vivo and showed better anticancer activity than unmodified SLNPs and the pure drug as well as higher PTX concentrations accumulated in lungs in vivo (Pandey et al., 2015). The Tyr-3-octreotide-modified SLNPs loaded with PTX with the diameter <100 nm achieved a high and specific accumulation within C6 glioma cells, improved drug induced apoptosis and in both subcutaneous and orthotopic tumor models they were found to exhibit better antiglioma efficacy compared to the unmodified SLNPs and Taxol (Banerjee et al., 2016). The improved antiglioma efficiency could be connected with their tumor cell and neovasculature targeting efficiency and improved PTX accumulation at tumor site due to the modification of the

SLNPs with compound known as a ligand for somatostatin subtype 2 receptors. In preclinical human cancer xenograft mouse model studies the PTX-loaded LDL-mimetic SLNPs were found to exhibit pronounced in vivo stability, enhanced biocompatibility, and better antitumor activity than nanoparticular therapeutics in clinical use (Taxol and Genexol PM), whereby they yielded long-term complete responses (Kim et al., 2015).

The cytotoxicity of PEGylated trimyristin-based SLNPs loaded with DTX against various human and murine cancer cells in culture markedly exceeded that of Tween 80/ethanol solution with the solubilized drug and more effectively inhibited tumor growth in mouse model with preestablished tumors than the Tween formulation containing DTX, also exhibiting a higher drug concentration in tumor tissues and decreased its concentrations in the liver, spleen, heart, lung, and kidney (Naguib et al., 2014). DTX-loaded SLNPs were found to exhibit lower cytotoxicity than the drug commercial formulation Taxotere, induced the enhancement of apoptosis at 24 hours after treatment in vitro, notably suppressed tumor growth in a mice model with human xenograft breast cancer, and considerably lowered myelosuppression toxicity to bone marrow cells from mice (Yuan et al., 2014). DTX-loaded tristearin-based SLNPs surface-modified by Tween 80 or TPGS 1000 exhibited enhanced intestinal absorption, lymphatic uptake, and relative oral bioavailability of the drug in rats compared to Taxotere, the formulation with TPGS 1000 modification being more effective most likely due to better inhibition of drug efflux by TPGS 1000 along with intestinal lymphatic uptake (Cho et al., 2014). The spherical SLNPs containing DTX with the mean size of 180 nm applied at the concentration of 100 μM practically completely reduced the viability of colorectal (C-26) and malignant melanoma A375 cells after 48 and 72 hours, respectively, and showed higher inhibitory efficacy and survival in C-26-implanted BALB/c mice than Taxotere (Mosallaei et al., 2013). DTX-loaded SLNPs surface modified with adenosine exhibited sustained drug release and showed higher cytotoxicity than the free drug in MCF-7 (IC_{50}: 0.09 and 0.41 μg/mL, respectively) and DU-145 cells (IC_{50}: 0.22 and 0.44 μg/mL, respectively), indicating that adenosine would be a suitable ligand to target breast and prostate cancer (Swami et al., 2015). The Lf-conjugated SLNPs loaded with DTX exhibited higher apoptotic activity than the pure drug and the unmodified SLNPs and accumulated more in the brain than the marketed drug formulation, suggesting increased targeting potential for brain tumors (Singh et al., 2016). Photothermal sensitive SLNPs coloaded with DTX, oxidized single-walled carbon nanotubes and FA exhibited a higher antitumor efficacy on a human hepatic carcinoma cell line (SMMC-7721) than free DTX in vitro and more effectively reduced the relative tumor volume of sarcoma 180 (S180) tumor-bearing mice compared to control groups, and all these effects were much stronger at simultaneous application of 808 nm NIR laser (Zhu et al., 2014).

The GEM-loaded mannosylated SLNPs showed augmented drug delivery in lung adenocarcinoma cells A549, indicating improved therapeutic effectiveness (Soni et al., 2016). Fucose anchored MTX-loaded SLNPs showed a greater cellular uptake and better cytotoxicity at lower IC_{50} of MTX in breast cancer cells, increased apoptosis with altered lysosomal membrane permeability, and a better rate of degradation of lysosomal membrane in vitro, while in vivo they caused extensive reduction in tumor burden compared to the free drug due to enhanced MTX bioavailability with minimum secondary drug distribution in various organs (Garg et al., 2016). The i.v. administration of SLNPs loaded with anticancer drugs mitoxantrone (MTO), PTX, or MTX resulted in a considerable inhibition of tumor size in mice exceeding that caused by treatment with pure drugs (Zhuang et al., 2012). Lyophilized MTO-loaded SLNPs of 79 nm exhibited a markedly higher drug concentration in local lymph nodes and lower concentration in other tissues than the drug solution and the tests for lymph node metastases in mice confirmed a significant reduction of the average size of a lymph node after treatment with SLNPs (41.85 ± 27.42 mm^3) compared to the administration of the drug solution (119.32 ± 57.30 mm^3) or the placebo (186.83 ± 77.71 mm^3) (Lu et al., 2006).

After i.v. administration to male rats, spherical and uniform SLNPs loaded with CDDP of the mean particle size of ca. 121 nm, the zeta potential of -46.4 ± 10.3 mV and the EE of 82.3% showed targeted effect to the liver, while low drug concentration in the kidneys of rats was estimated (Tian et al., 2008). CDDP magnetic SLNPs prepared by film scattering ultrasonic technique in which magnetic NPs were combined with lipids during film formation and distributed in the middle of the lipid layer in SLNPs exhibited higher cytotoxicity against human cervical carcinoma SiHa cells than the pure drug, and at in vivo intratumoral administration, higher concentrations of CDDP in the tumor tissue were estimated than when treated with the free drug, particularly under application of external magnetic field (Zhao et al., 2015c). Metal ion (Zn^{2+}, Cu^{2+}, Mn^{2+}, and Mg^{2+}) introduction in SLNPs to modify the microstructure of the lipid layer resulted in an increase of the compact degree of the lipid molecules and, consequently, in an increased cisplatin EE and a decreased drug releasing rate due to electrostatic interactions between metal ions and phospholipid acyl and choline polar groups (Yang et al., 2016b).

SLNPs and PEGylated SLNPs loaded with 7-ethyl-10-hydroxycamptothecin with particle sizes 103 and 131 nm and zeta potential in the range from -5 to -15 mV showed markedly higher toxicity to C-26 and HT-116 cell lines after 48 hours of exposure than IRI, with the HT-116 cells being more sensitive (Mosallaei et al., 2016). CPT-loaded SLNPs prepared using lipid matrix based on trilaurin, Poloxamer 188, and PEGylated phospholipid as stabilizers exhibited significant tumor targeting and tumor growth inhibition after i.v. administration to mice with subcutaneous transplants of CT26 carcinoma cells and preinjection of blank SLNPs considerably reduced their accumulation in organs of the reticuloendothelial system (Jang et al., 2016).

The ETP-loaded SLNPs prepared using trimyristin showed increased AUC and $t_{1/2}$ and MRT for drug in plasma as well as an enhanced ratio of the drug that reached the highly perfused organs and improved the tumoricidal activity and the survival rate in animals also in vivo using a B16F10 mouse melanoma model (Athawale et al., 2014). Compared to the free drug, the ETP-loaded SLNPs with the particle size of 30–50 nm, zeta potential of −28.4 mV, and 36.91% drug loading were found to considerably enhance in vitro cytotoxicity against SGC7901 cells, induced higher apoptotic rates and cell cycle arrest effects as well as greater cellular uptake in SGC7901 cells, suggesting that this nanoformulation could be used to treat gastric carcinoma (Wang et al., 2014d).

Tf-conjugated SLNPs loaded with temozolomide with the mean particle size of 221 nm were found to be uptaken by the brain in a higher extent than the unconjugated ones (Jain et al., 2013).

Encapsulation of all-*trans*-retinoic acid (ATRA) in cholesteryl butyrate SLNPs resulted in better antileukemic efficacy and bioavailability compared with the pure drug (Silva et al., 2016) and ATRA-loaded SLNPs were also described as a promising dermal drug delivery formulation (Charoenputtakhun et al., 2014).

Spherical HA-functionalized LNPs loaded with vorinostat with the size ca. 100 nm and negative charge (about −9 mV) were characterized by high intracellular uptake in CD44 overexpressing cells, A549, and SCC-7 cells while in SCC-7 cells pretreated with HA or in MCF-7 cells with low overexpressed CD44 their uptake was reduced. The cytotoxicity of these SLNPs against A549 and SCC-7 cells exceeded that of the free drug and HA shell secured their longer blood circulation, higher plasma concentration, and bioavailability in rats (Tran et al., 2014a). Vorinostat-loaded SLNPs with the mean particle size of 100 nm were reported to be more efficient not only against sensitive MCF-7 and A549, but also against multidrug-resistant cancer cells MDA-MB-231 than the pure drug, and their i.v. administration to rats resulted in considerably prolonged drug plasma circulation time and decreased elimination rate constant (Tran et al., 2014b).

The tolerated dose of staurosporine encapsulated into SLNPs by BALB/c mice was increased 3-fold compared with the pure drug, and such nanoformulation could cause approximately complete growth inhibition of a MDR breast tumor in contrast to the pure drug, which showed moderate activity (Tang et al., 2016).

The TPGS-coated SLNPs loaded with *trans*-resveratrol exhibited a notably higher in vitro cytotoxicity against C6 glioma cells and 11.12-fold higher AUC than *trans*-resveratrol solution as well as 9.23-fold higher brain distribution compared to the pure drug with a plasma half-life being 9.37-fold higher than that of the drug solution suggesting that the nanoformulation could prolong systemic circulation and could be promising in the treatment of glioma (Vijayakumar et al., 2016).

SLNPs loaded with a novel lignan composition from *Cedrus deodara* (AP9-cd) showed cytotoxicity and apoptotic potential higher than that of pure AP9-cd by 30%–50%, exhibited enhanced antitumor activity in Ehrlich ascites tumor and induced apoptosis in human acute lymphoblastic leukemia MOLT-4 cells (Bhushan et al., 2016).

SLNPs loaded with asiatic acid (a pentacyclic triterpene found in *Centella asiatica* showing neuroprotective and anticancer activity against glioma) exhibited markedly higher cytotoxicity against U87MG cells than against human fetal glial normal cells (SVG P12) and displayed concentration-dependent apoptotic activity on glioma cells (Garanti et al., 2016).

8.7 NANOSTRUCTURED LIPID CARRIERS

NLCs, known as the second generation of SLNPs, are characterized by the lipid matrix having imperfect crystal or amorphous structure and are suitable for loading drugs in both the molecular form and in clustered aggregates at lattice imperfections. The number of imperfections in the core of a solid matrix increases with the incorporation of a liquid lipid to the solid matrix and the absence of a crystal structure results in enhanced drug loading and less pronounced drug expulsion. NLCs consisting of an unstructured solid lipid matrix representing a blend of solid- and liquid-phase lipids and an aqueous phase containing a surfactant(s) are generally solid at least at temperatures of 40°C (Martins et al., 2007; Zhang et al., 2008b; Beloqui et al., 2016). A comprehensive overview of the current state of the art of NLCs as controlled drug delivery systems for future clinics was presented by Beloqui et al. (2016). NLCs showed a higher cellular uptake and permeability across the barrier, but SLNPs could enter cells faster than NLCs. The internalization of lipid NPs occurs mainly through a clathrin-mediated endocytosis mechanism, although caveolae-mediated endocytosis is also involved in the uptake (Neves et al., 2016). The function of NLCs in cancer chemotherapy was analyzed by Selvamuthukumar and Velmurugan (2012).

NLCs, that is, biocompatible and/or biodegradable lipids-based NPs composed of solid and liquid lipids as a core matrix dispersed in surfactant solution, can improve the solubility of most of the hydrophobic cancer therapeutics and surface modification of these nanocarriers could result in overcoming drug resistance in cancer chemotherapy, site specific targeting for better efficacy, and reduced dose related toxicity (Rizwanullah et al., 2016). An NLC modified with a small peptide ligand Ala-Glu-Tyr-Leu-Arg (AEYLR) targeting EGFR showed obvious targeting to A549 xenograft tumor compared with the control group, indicating its potential to be a promising targeted delivery system for combination cancer chemotherapy to improve therapeutic efficacy and to minimize adverse effects (Han et al., 2014b). A multifunctional nanostructured NLC system for efficient delivery of an anticancer drug and siRNA directly into the lungs by inhalation designed by Taratula et al. (2013) left healthy lung tissues intact and also significantly decreased the exposure of healthy organs when compared with i.v. injection; moreover, it showed enhanced antitumor activity when compared with i.v. treatment.

DOX-loaded NLCs, the surface of which was modified with an alternating layer-by-layer assembly of polycation and polyanion polyelectrolytes and an additional coating with PEG, exhibited longer residence in the blood, lower accumulation in the liver, higher accumulation in tumors, and a significant tumor growth inhibition effect (Mussi et al., 2015). A Tf-modified DOX and pEGFP coencapsulated NLC with the particle size of 198 nm and +19 mV surface charge showed significantly greater gene transfection efficiency and enhanced antitumor activity in vivo evaluated in a mouse bearing A549 cells model compared to a corresponding SLNPs formulation. Hence, using Tf-modified NLC as a novel gene and a drug delivery system could represent a promising strategy for the treatment of lung cancer (Han et al., 2014c). NLC coloaded with DOX and docosahexaenoic acid with the size of ca. 80 nm, negative zeta potential, EE of 99%, and drug loading of 31 mg/g showed the same activity as the free drugs against MCF-7, but a stronger activity against MCF-7/ADR cells in which higher accumulation of encapsulated DOX was also estimated, suggesting a bypassing of P-gp bomb efflux (Mussi et al., 2014). DOX HCl-oleic acid conjugate loaded NLCs showing a pH-dependent drug release, which was fast at pH 3.8 and 5.7 and sustained at pH 7.4, exhibited cytotoxicity comparable to that of the pure drug and were favorably taken up by HCT-116 cells (Zhao et al., 2016). DOX and vincristine encapsulated into NLCs were found to be more effective in the treatment of Bcl cells through the targeted delivery of both drugs to lymph cancer animal model than the equivalent dose of free DOX and a single drug (DOX or VCR) formulation, whereby a significant reduction of the drug-associated systemic toxicity was also estimated (Dong et al., 2016).

Xu and Lee (2015) compared PTX-loaded SLNPs and NLCs from the context of drug delivery. Both nanoformulations showed cytotoxicity comparable to that of the commercial formulation and free PTX against human breast cancer cell line, MCF-7, however, higher anticancer activity against MDR cancer cell line, MCF-7/ADR, compared to the free PTX delivered in dimethyl sulfoxide, indicates that both SLNPs and NLCs would be effective carriers to avoid efflux pump expressed in MDR cancer cells and could be used as an effective delivery system for PTX. Cholesterol NLCs with various oleic acid content loaded with PTX were found to enter cancer cells selectively through the LDL receptor endocytic pathway. They were effective against HT-29 cells showing lower IC_{50} value after 72 hours exposure for NLC-born PTX (5.24 ± 0.96 ng/mL) than that of PTX solubilized in Cremophor EL (8.32 ± 1.35 ng/mL) (Emami et al., 2012). PTX encapsulated in stearyl-2-amino-2-deoxyglucose (2-DG)-modified NLC, in which 2-DG serving as a broad tumor targeting ligand was conjugated to the surface of NLC was found to actively and efficiently accumulate at the site of the tumor and exhibited good antitumor effectiveness and low toxicity in MCF-7 tumor-bearing mice (Chen et al., 2012). Shao et al. (2015) developed Tf-decorated NLCs as a multifunctional nanomedicine for codelivery of PTX and pEGFP showing low cytotoxicity, high gene transfection efficiency, and enhanced antitumor activity in vitro and in vivo and Tf decoration was found to enhance the active-targeting ability of the carriers to NCl-H460 cells. According to the researchers, this novel drug and gene delivery system offers a promising strategy for the treatment of lung cancer. An NLC modified with photo-responsive cell-penetrating peptides (pCPP) carrying PTX possessed suitable physiochemical properties such as small particle size, high drug EE, and good stability and their excellent penetration into HT-1080 multicellular tumor spheroids confirmed that the temporary mask of the photolabile-protective group in pCPP does not disturb the penetration ability of CPP in the tumor microenvironment with illumination. The triggered activation exhibited higher antitumor efficacy with the tumor spheroids than nonmodified NLC and Taxol (Yang et al., 2014). PTX-loaded NLC dually modified with photon-sensitive cell-penetrating peptides and peptide containing the asparagine-glycine-arginine (NGR) motif which selectively binds to CD13-positive tumors with the particle size ca. 100 nm and the drug EE >90% possessed the long-circulation characteristic with the $t_{1/2}$ of 6.1 ± 0.3 hours, while the tumor inhibition ratio significantly increased with the aid of NIR illumination at the tumor site and NGR (Yang et al., 2015b).

DTX-loaded NLCs were found to be stable in simulated gastrointestinal fluids, exhibited sustained drug release for 48 hours in vitro, and significantly improved drug absorption in the intestine and in vivo the estimated $AUC_{0\rightarrow 24}$ was ca. 4-fold higher than that of the drug solution; however, the blockage of the lymphatic transport pathway resulted in their markedly reduced oral bioavailability (Fang et al., 2015a). Cysteine-functionalized NLCs with the particle size <100 nm and the negative zeta potential of -13.72 ± 0.07 mV efficiently encapsulated DTX and exhibited significantly improved mucoadhesion properties compared with unmodified NLCs; they were found to be absorbed in the enterocytes via both endocytosis and passive transport and showed a 12-fold increase in the $AUC_{(0-t)}$ in vivo compared to the drug solution (Fang et al., 2015b). By encapsulation of a novel lipophilic oleate prodrug of DTX (DTX-OA) in NLC, in which the liquid lipid was used as a core matrix to enhance compatibility with DTX-OA, particles with the size about 100 nm and high drug-loading capacity were prepared and more easily accumulated in MDCK cells than DTX solution and showed 4- and 2-fold higher bioavailability than DTX solution and DTX-NLC, respectively. These findings suggest that such core-matched prodrug-NLC could be advantageous to facilitate the oral delivery of DTX (Sun et al., 2016c). The NLCs modified by an amphiphilic carboxymethylcellulose-graft histidine/methoxypolyethylene glycol (CMP) copolymer with particles of 89 nm and the negative zeta potential of -20.9 ± 1.11 mV, in which DTX was encapsulated, exhibited considerably reduced burst drug release at pH 7.4 (due to CMP coating) as well as longer retention times in blood and higher AUC in rats than drug-loaded NLCs without coating, which was reflected also in better antitumor effects in the tumor-bearing nude mice (Jiang et al., 2015). DTX-loaded NLCs revealed higher cytotoxicity against A549 cells by inducing more apoptosis and more G2/M arrest and more effectively inhibited tumor growth in Kunming mice bearing murine malignant melanoma (B16) compared to Duopafei (Liu et al., 2011). In the skin permeability test using vertical Franz-type diffusion cells, the DTX-nicotinamide complex NLCs transported drugs more easily than DTX-NLCs and an in vivo study showed that they were maintained most in the skin, suggesting that

DTX-nicotinamide complex NLCs can increase the permeation of DTX through the skin (Fan et al., 2013).

Spherical freeze-dried DTX-NLCs with 5% (w/w) DTX loading efficiency showed sustained drug release in vitro and their sustained release properties were estimated also after i.v. injection into rats. The AUC and MRT were increased ca. 5- and 3-fold compared with the drug solution and rats treated with these NLCs showed considerably higher DTX concentration in the lungs than rats treated with the DTX solution (Li et al., 2009). The investigation of codelivery of PTX and DOX by NLCs showed that PTX-DOX-NLCs exhibited the highest cytotoxic effect against NCL-H460 human non−small cell lung carcinoma (NSCLC) cells in vitro as compared to single drug delivery NLC and also in vivo the codelivery of PTX and DOX was characterized with high tumor-targeting capacity and strong antitumor activity on NSCLC animal models (Wang et al., 2016b). Also HA decorated NLCs were reported as a potential nanocarrier to codeliver baicalein and DOX for breast cancer therapy (Liu et al., 2016c). While DOX-loaded folated NLCs prepared using monostearin as a solid lipid and oleic acid as a liquid lipid matrix exhibited high cytotoxicity only in multidrug-resistant MCF-7/ADR cells, the PTX-loaded NLCs showed high cytotoxicities in both MCF-7 and MCF-7/ADR cells, exceeding that of the drug solution, and the reversal powers of the PTX-NLCs and DOX-NLCs were 34- and 6-fold, respectively. These drug-loaded nanoformulations exhibited also enhanced cytotoxicity against human ovarian cancer SKOV3 cells and their multidrug-resistant SKOV3-TR30 cells, whereby their reversal powers were 31-fold (PTX) and 2-fold (DOX), respectively (Zhang et al., 2008a).

CDDP-loaded NLCs modified by folate containing PEG-DSPE with the particle size of 143 nm and the surface charge of +25.7 mV were found to be efficient in selective drug delivery to cancer cells overexpressing FRs and these NLCs could perform targeted transfer of CDDP to the cervical cancer cells and enhance the antitumor capacity (Zhang et al., 2016c). Di et al. (2016) designed DOX and CDDP-loaded NLCs showing high cytotoxicity and synergistic effect of two drugs against MCF-7/ADR cells in vitro and confirmed their great antitumor activity using the murine bearing MCF-7/ADR cells model. HA-coated NLC loaded with both 5-FU-SA lipid conjugate and CDDP tested in vitro against human gastric cancer cell line BGC823 and in vivo using a mice bearing BGC823 cells model showed synergistic effect in combination therapy and displayed greater antitumor activity than the free drugs or uncoated NLC in vitro and in vivo (Qu et al., 2015b). Spherical oxaliplatin-loaded NLCs with the average diameter of ca. 111 nm, the zeta potential of −27.4 mV, and the EE of 77.4% ± 2.5% were found to exhibit sustained drug release in a buffer solution (pH 4.5) after a burst release in the initial phase in vitro (Zhou et al., 2010).

The IRI-loaded hyaluronated cationic NLCs for MDR modulation and CD44 targeting in colon adenocarcinoma was presented by Negi et al. (2015). The cationic core NLCs prepared by using Capmul MCM and Compritol as mix lipid phase had the particle size of 150 nm and 93.98% ± 2.5% EE and, following the decoration of the surface with HA, the particle size increased to 255 nm. The drug-loaded nanoformulation showed a slow and sustained release of the drug and strong uptake of NLCs by both CD44 overexpressing cell lines (HT-29 and Colo-320) in just 2 hours and the developed NPs exhibited enhanced cytotoxicity against non-MDR as well MDR cancer cells. Particles of an HA-coated IRI-loaded NLC (HA-NLC-Ir) prepared using Compritol ATO 888 and Capmul MCM C8 hybrid lipid mix of the size of 386 nm and the zeta potential of 19.7 ± 1.2 mV showed a slow and sustained drug release, and after 3 days of the treatment of CD44 expressing MDR cancer cells Colo-320, IC_{50} value was 4.56 ± 0.18 μM, while IC_{50} value estimated with uncoated NPs was 7.59 ± 0.12 μM. This NLC formulation was found to have a potential of inhibiting P-gp by affecting ATPase activity and MDR1 gene expression (Negi et al., 2014a). HA-NLC-Ir had significantly lower hemolytic potential as compared to uncoated NLC and showed reduced plasma protein interaction and low macrophage uptake. In vivo antitumor activity of HA-NLC-Ir studied using Ehrlich's Ascites Tumour allograft model was 5.8- and 2.6-fold higher as compared to the control and the free drug solution, respectively, and also reduced thrombocytopenia and neutropenia compared to the free drug were estimated (Negi et al., 2014b). HA-coated NLCs were found to be an effective carrier for delivering PTX to tumors overexpressing CD44 as well (Yang et al., 2013a).

Khajavinia et al. (2012) evaluated Tf-conjugated NLCs prepared using stearylamine (SA), dodecylamine or spermine loaded with ETP for targeting the drug to K562 acute myelogenous leukemia cells and estimated enhanced cellular uptake of Tf-modified NLCs compared to unconjugated ones, and superior cytotoxicity with IC_{50} value 15-fold lower than that of the free drug was shown by NPs with chemically coupled SA. ETP-loaded NLC prepared using monostearin, soybean oil, soya lecithin + PEG-40, and DSPE-PEG showed relative bioavailability 3.5-fold higher than that of the drug suspension and exhibited high cytotoxicity against human epithelial-like lung carcinoma cells, indicating that such nanoformulation could have promising perspective for oral delivery of ETP (Zhang et al., 2011). ETP-loaded NLCs with the particle size of 91 nm, the zeta potential of +23.1 mV, and the drug EE of 78% significantly enhanced in vitro cytotoxicity against human gastric cancer cells (SGC7901) and in vivo antitumor effect in mice bearing SGC7901 cells xenografts compared to the free ETP, indicating that such formulation could be used in the treatment of gastric carcinoma (Jiang et al., 2016). NLCs prepared using a mixture of a solid lipid (tripalmitin) with a liquid lipid (Captex 355 EP/NF or Miglyol 812) with spherical NPs of the diameter <300 nm loaded with sorafenib exhibited enhanced antitumor activity compared to that of the free drug and could be used in the treatment of hepatocellular carcinoma through parenteral administration (Bondi et al., 2015). A formulation of ETP-loaded FA decorated PEG-DSPE NLCs was found to be effective in selective drug delivery to CT26, SGC790 or NCI-H209 cells overexpressing FA receptors and showed enhanced antitumor activity in vivo using mice bearing CT26 cells xenografts (Zhang et al., 2016d).

10-Hydroxycamptothecin-loaded NLCs coated with xyloglucan (XG) prepared using soya oil and XG consisting of side chains with galactose residues, a terminal moiety that can be used to target the drug to hepatoma, showed superior cytotoxicity against the drug-resistant HepG2 cells and generated a higher therapeutic effect in vivo using a human tumor xenograft nude mouse model (implanted with HepG2 cells) indicating their great potential to overcome MDR of HepG2 cells (Liu et al., 2016d). QD-loaded NLCs (QDNLCs) containing lipophilic QDs, carboxyl-function QDs or PEG QDs developed by Hsu et al. (2013) with the size of 245 nm and CPT EE >99% showed higher cytotoxicity against melanoma cells than pure CPT. The highest cell internalization and in vivo fluorescence labeling were obtained with carboxylic acid-conjugated QDNLCs and CPT incorporation into QDNLCs resulted in 6.4-fold higher drug accumulation in melanomas, indicating to be efficient for both tumor imaging and drug delivery.

Decitabine (DCB)-loaded NLCs with particles showing spherical shape, smooth surface, size ca. 117 nm, zeta potential -31.8 ± 0.96 mV, and sustained drug release up to 24 hours ($80.23 \pm 4.67\%$) exhibited a 4-fold increase in the permeation of the drug compared with the plain drug solution in an ex vivo gut permeation study and were found to possess excellent cytotoxic activity against cancer cells indicating that this nanoformulation could be suitable for oral delivery of DCB (Neupane et al., 2014).

A polysorbate 80 coated cytarabine-NLC, in which the drug was entrapped inside the lipid as molecular dispersion, showed initial fast release followed by sustained release up to 72 hours and exhibited concentration-dependent cytotoxicity on leukemic EL-4 cells exceeding that of the drug solution, suggesting that this nanoformulation could be applied for effective treatment of meningeal leukemia (Sharma et al., 2011).

Spheroidal PEGylated NLCs for the i.v. delivery of 17-allylamino-17-demethoxygeldanamycin prepared using a mixture of glycerol monostearate and PEG2000-stearate as solid lipids and medium-chain triglyceride as a liquid lipid with the particle size of 189.4 nm, the zeta potential of -20.2 mV and 3.42% EE exhibited a biphasic release pattern with burst release followed by a sustained release and significantly higher in vitro antitumor efficacy as well as longer retention time in vivo than the drug solution (Wang et al., 2016c).

5-FU-loaded NLCs with the particle size of 208 nm and the zeta potential of -21.82 ± 0.40 mV showing spherical shape and dispersed in gel medium were found to improve the penetration of 5-FU through the *stratum corneum* with enormous retention and with minimal skin irritation compared to free 5-FU gel, which suggested their suitability for topical application (Rajinikanth and Chellian, 2016). The cytotoxic effect of NLCs loaded with 5-FU and retinoic acid-octadecylamine conjugate on colon cancer cells (HT-29) was nearly 2-fold of that of NLCs just loaded with 5-FU and more than 5-fold of that of free 5-FU, suggesting their potential in reducing 5-FU dose in colorectal cancer treatment (Varshosaz et al., 2012).

An in vitro study showed that MTX-loaded NLCs showed superior inhibitory effect on cell growth over the free drug especially in ovarian human cancer cells A2780 and a higher cytotoxic effect on human prostate cancer DU-145 cells at higher drug concentration, while MTX-free NLCs maintained cell viability (Abdelbary and Haider, 2013).

Nanostructured lipid-carrageenan hybrid carriers (NLCCs) designed for the oral delivery of water-soluble MTO hydrochloride exhibited approximately 3.5-fold higher apparent bioavailability compared to the drug solution and enhanced the antitumor efficacy against resistant MCF-7/MX cells, and they showed 9-fold higher relative cellular association than the drug solution in breast cancer resistance protein (BCRP) overexpressing MCF-7/MX cells, suggesting that the introduction of NLCCs resulted in reduced BCRP-mediated drug efflux (Ling et al., 2016).

Similar promising results were obtained also with MTO-loaded nanostructured lipid-dextran sulfate hybrid carriers designed to overcome tumor MDR of MTO hydrochloride (Zhang et al., 2012b).

Optimized DAC-loaded NLCs consisting of glyceryl palmitostearate and isopropyl myristate with the particle size of ca. 190 nm, the zeta potential of 43.5 ± 1.2 mV, the EE of 98%, and the drug loading of 14% showing an improved therapeutic profile of DAC were reported by Almoussalam and Zhu (2016).

Temozolomide- and DNA-loaded NLCs with the particle size of 179 nm, the surface charge of +23 mV, the gene-loading capacity of 91%, and the drug EE of 83% were found to transfer both the drug and the gene to the gliomatosis cerebri and enhance the antitumor capacity against U87MG cells and gene transfection efficacy suggesting that such nanoformulation could serve as a new promising strategy for the treatment of malignant gliomatosis cerebri (Chen et al., 2016c). The RGD-modified NLCs loaded temozolomide effectively inhibited U87MG cells in vitro and exhibited antitumor efficacy in vivo using mice bearing glioblastoma multiform model (Song et al., 2016).

The i.v. and oral administration of optimized imatinib-loaded NLCs with the particle size of 149 nm, the polydispersity index of 0.191 ± 0.017, the zeta potential of -23.0 ± 1.5 mV, the drug loading of $5.48\% \pm 0.01\%$, and the EE of $97.93\% \pm 0.03\%$ showing sustained release in vitro considerably increased in vivo bioavailability of the drug to rats; and in vitro cytotoxicity of this nanoformulation on NCI-H727 cells was also superior compared to the free drug (Gupta et al., 2015).

A freeze-dried TAM citrate-loaded NLCs (TAM-NLCs) formulated using long chain solid lipids and oils with the particle size of ca. 215 nm and drug EE >90% were reported to withstand various gastrointestinal tract media in the range from pH 1.2 to 7.4 and showed a sustained release pattern irrespective of the pH of the medium (Shete and Patravale, 2013). The TAM-NLCs showed improved activity against MCF-7 cells compared with the free drug; however, the activity was compromised against human estrogen receptor expressing breast cancer cells ZR-75-1 which could be attributed to its upregulation of MUC1 gene and the application of this nanoformulation resulted in an enhancement of the bioavailability ca. 3-fold and prolonged $t_{1/2}$ 7-fold. It was shown that the TAM-NLCs targeted intestinal lymphatic systems and the major uptake of the drug occurred via the lymphatic system (Shete et al., 2013).

Spherical NPs of vorinostat-loaded NLCs of 150 nm with the negative charge of −22 mV and the EE of 99% showed a sustained drug release profile and demonstrated higher cytotoxicity in A549, MCF-7, and SCC-7 cell lines and 4.4-fold enhancement of the AUC of vorinostat than the free drug (Tran et al., 2016).

ATRA-loaded NLC formulations exhibited a photoprotective property and showed higher cytotoxicity than the free drug and HL60 cells were more sensitive to ATRA than HepG2 cells (Chinsriwongkul et al., 2012).

Bleomycin sulfate loaded NLCs with the particles size of ca. 17 nm and the drug EE of 45.3% ± 3.4% showed not only enhanced cytotoxicity against HeLa cells with ca. 6-fold lower IC_{50} value but also ca. 4-fold higher drug concentration and a greater degree of apoptosis compared to the free drug as well as 3.4-fold higher bioavailability (Saini et al., 2014).

GA-modified CUR-loaded NLCs (CUR-GA-PEG-NLC) with the particle size ranging from 123.1 to 132.7 nm and the EE of 90.06–95.31% containing 10% (w/w) GA exhibited in vitro a significant cellular uptake and cytotoxicity against HepG2 cells compared with CUR-NLC and CUR-PEG-NLC, and it was shown that the overabundance of ligands at GA content >10% may block the combination of receptor and ligand due to steric hindrance (Chu et al., 2014).

Biocompatible and biodegradable quercetin-NLCs with the diameter of 32 nm excellently improved the aqueous solubility of quercetin (at least 1000-fold), dramatically increased cytotoxicity compared to free quercetin and increased quercetin uptake and induced apoptosis at 20 μM in MCF-7 and MDA-MB-231 breast cancer cells, indicating that such nanoformulation could be suitable for chemopreventive use in breast cancer (Sun et al., 2014).

Using a human colorectal adenocarcinoma (Caco-2) cell line, the IC_{50} concentration estimated for zerumbone (ZER)-loaded NLC at 24 hours was found to le lower (4.25 μM) than that of the free drug (23.75 μM), and the treated cells demonstrated a slightly slower progression of apoptosis, which could be due to the controlled release of ZER from the NLC matrix (Nathaniel et al., 2016).

8.8 LIPID–POLYMER HYBRID NANOPARTICLES

LPNPs are core–shell nanoparticle (NP) structures comprising polymer cores and lipid/lipid–PEG shells, which combine positive attributes of both liposomes and polymeric NPs, while excluding some of their shortages such as the structural integrity of the polymeric particles and the biomimetic properties of the liposome, and are characterized with good physical stability and biocompatibility (Zhang and Zhang, 2010; Hadinoto et al., 2013; Hallan et al., 2016). The hydrophobic polymeric core of LPNPs is suitable for encapsulating poorly water-soluble drugs, their hydrophilic polymeric shell showing antibiofouling properties contributes to the enhanced stability of LPNPs and systemic circulation half-life, while the lipid monolayer at the interface of the core and the shell acts as a molecular barrier supporting retention of drug inside the polymeric core, which is reflected in enhanced drug EE, increased drug-loading yield and controlled drug release (Zhang et al., 2008b). Factors affecting drug encapsulation and stability of LPNPs, such as the size of NPs, stability in salt solutions, drug lipid ionic interactions, and drug lipophilicity using PLGA and PC as the polymer and lipid models and fluoroquinolone antibiotics were investigated by Cheow and Hadinoto (2011). In a review paper Krishnamurthy et al. (2015) summarized formulation methods used to synthesize LPNPs, highlighted the strategies of their use to treat cancer, and pointed out challenges that must be overcome to realize their extensive use in the clinic. The different classes of nanohybrids characterized by a drug-loaded polymeric matrix core enclosed in a lipid shell and their main applications in drug delivery were described by Raemdonck et al. (2014) and Mandal et al. (2013).

Targeted ultrasmall LPNPs with sizes <25 nm loaded with DTX exhibited better efficacy in the treatment of mouse tumor model than the clinically available formulation of the drug (Dehaini et al., 2016). The LPNPs with coencapsulated DTX and CUR with the particle size of 169.6 nm and the zeta potential of 35.7 mV, respectively, exhibited synergistic effect of the two drugs in PC-3 human prostate carcinoma cells in vitro and their better antitumor efficiency was confirmed in vivo in mice bearing PC-3 tumor xenografts, while their application was not accompanied with obvious side effects (Yan et al., 2016). RGD-modified LPNPs consisting of PLGA NPs coated with lipids, which were loaded with DTX, were uptaken by C6 glioma cells to a greater extent, showed better antiproliferative activity and increased the median survival time of glioblastoma multiforme bearing rats more than PLGA NPs (Shi et al., 2015b). Also LPNPs modified by RGD peptide encapsulating CUR increased the uptake of CUR by HUVECs, effectively inhibited tumor growth in a subcutaneous B16 melanoma tumor model and increased the number of apoptotic cells, while reducing the number of microvessels and proliferation-positive cells (Zhao et al., 2014).

PTX-loaded LPNPs consisting of a PCL-PEG-PCL amphiphilic copolymer core, a lipid monolayer formed by DSPE-PEG2000 and FA as a targeting ligand administered intratumorally in EMT6 breast tumor model showed similar antitumor efficacy but lower toxicity than Taxol and more efficient inhibition of tumor growth than nontargeted PTX-loaded LPNPs (65.78% and 48.38%, respectively) (Zhang et al., 2015c). Stolzoff et al. (2015) designed pH-responsive LPNPs with a polymer core and PEGylated lipid-coating, which were able to expand in size from 100 to 1000 nm in a mildly acidic pH environment, because they combined the swelling behavior of the polymeric core with the improved colloidal stability and surface functionality of PEGylated liposomes. Moreover, the surface functionalization of these LPNPs allowed the incorporation of FA and FR-targeting and enabled the greater in vitro uptake and potency of encapsulated PTX compared to nontargeted LPNPs.

Also DOX-loaded PLGA-lipid NPs with FA targeting ligand showed a significantly higher uptake by folate-receptor-positive MCF-7 cells than PLGA-lipid NPs without folate and exhibited higher cytotoxicity in MCF-7 cells than nontargeted NPs (Zheng et al., 2015). Biodegradable core–shell copolymer-phospholipid NPs with a PLGA core and a DPPC shell coloaded with dihydroartemisinin and DOX were reported to be promising delivery systems to improve combination chemotherapy (Ma et al., 2015). Biodegradable folate-targeted LPNPs with a PLGA core, a soybean lecithin monolayer, a monomethoxy-PEG-*S-S*-hexadecyl reduction-sensitive shell, and a FA-targeted ligand loaded with DOX showed higher cell uptake as well as higher cytotoxicity against human epidermoid carcinoma in the oral cavity compared to nontargeted redox-sensitive and targeted redox-insensitive controls in vitro, while in the in vivo experiment they exhibited significant reduction of tumor growth and increased drug accumulation in the tumor (Wu et al., 2015). DOX-loaded LPNPs consisting of a PLGA polymeric core and FA-PEG-DSPE as a shell with the mean particle size of 185 nm, which released the drug in acidic environment faster than in the neutral one, showed higher cytotoxicity on KB, A549, MDA-MB-231, and MDA-MB-231/ADR cells than the free drug and naked PLGA NPs and were able to deliver more DOX to the nuclear of KB cells and MDA-MB-231/ADR cells and induced apoptosis in these cells, indicating that they could kill also drug-resistant tumor cells (Du et al., 2014). The combined treatment of laser irradiation with DOX and ICG (photothermal agent)-loaded PLGA lecithin PEG NPs designed for combined targeting cancer imaging and chemo-photothermal therapy resulted in synergistically induced apoptosis and death of both DOX-sensitive (MCF-7) and DOX-resistant (MCF-7/ADR) cells and the suppression of MCF-7 and MCF-7/ADR tumor growth in vivo, whereby a single dose of these LPNPs with laser irradiation was sufficient to prevent tumor recurrence (Zheng et al., 2013). Two-component reduction-sensitive LPNPs prepared using amphiphilic polymer 1,2-dilauroyl-sn-glycero-3-phosphorylethanolamine-*S-S*-methoxyPEG and PCL encapsulating DOX showing average particle size 100–120 nm and spherical shape exhibited in vitro higher antitumor activity against HeLa cells, HepG2 cells, and COS-7 cells and lower cardiac toxicity and kidney toxicity than the free drug. The cleavage of the disulfide bond in the polymer contributed to fast intracellular drug release from LPNPs and in vivo experiments with mice showed that 12 hours after injection, the free drug accumulated particularly in the heart and kidney, while LPNPs accumulated massively in the liver and spleen (Zhang et al., 2016e). The iRGD decorated LPNPs designed for codelivery of DOX and sorafenib exhibited synergistic cytotoxicity, proapoptotic ability, an enhanced internalization rate in human liver cancer HepG2 cells and extended circulation and bioavailability as well as considerably higher antitumor efficiency in hepatocellular carcinoma xenograft mouse models compared to free drugs (Zhang et al., 2016f). LPNPs encapsulating both DOX and 2′-deoxy-5-azacytidine, an epigenetic drug, which can cause reactive expression of tumor suppressor genes (TSG) in human cancers, considerably enhanced the sensitivity of cancer cells to DOX, resulting in cell growth rate inhibition and cell apoptosis induction and rescued the expression of silenced TSG in cancer cells (Su et al., 2013). Han et al. (2015) designed hybrid lipid-capped mesoporous silica NPs using a TPGS-containing lipid layer which were able to secure redox and pH-responsive release of DOX, released DOX within the cancerous cells and could avoid the premature leakage of drug before reaching the specific site of action. Moreover, they showed also improved uptake efficiency and intracellular accumulation in resistant MCF-7/ADR cells compared with the drug solution.

10-Hydroxycamptothecin-loaded LPNPs modified by cyclic RGD which could be used for targeted drug delivery to human breast cancer, were characterized with markedly higher ability of cell killing and endocytosis due to surface modification of the lipid layer (Yang et al., 2013b). In the presence of a remote radio frequency field, the CPT-loaded LPNPs containing magnetic beads for stimuli-responsive drug release showed a significant reduction of MT2 mouse breast cancer cell growth in vitro (Kong et al., 2013).

Biocompatible LPNPs prepared by coating cationic ε-polylysine copolymer NPs by a PEGylated lipid bilayer were tested for codelivering hypoxia-inducible factor 1α (HIF-1α) siRNA and GEM for pancreatic cancer treatment in subcutaneous and orthotopic tumor models. The nanoformulation was characterized with prolonged lifetime in bloodstream and improved drug release via the enhanced tumor vasculature effect in tumor tissues and effectively suppressed the HIF-1α expression both in vitro and in vivo, showing notable inhibition of tumor metastasis in an orthotopic tumor model (Zhao et al., 2015d).

Multiple layer-by-layer LPNPs loaded with three active antitumor constituents of FOLFIRINOX regimen (5-FU, IRI, oxaliplatin) showed higher antitumor efficacy with minimal side effects compared with FOLFIRINOX regimen alone, suggesting that they could be used for improved chemotherapeutic efficacy for the patients with pancreatic cancer (Li et al., 2015).

MTX-loaded spherical LPNPs with the size ranging from 150 to 300 nm released >80% of the drug at the end of seventh day by slow and sustained release and showed good in vitro cytotoxicity against breast cancer MCF-7 cells (Garg et al., 2015).

ERLO-loaded core–shell type LPNPs prepared using PCL as a biodegradable polymeric core and a phospholipid shell composed of hydrogenated soy PC and 1,2-distearoyl-sn-glycero-3-phosphoethanolamine-*N*-[methoxy(polyethylene glycol)-2000] with the average particle size of 170 nm showed enhanced uptake and efficacy in human lung adenocarcinoma cells (A549) in vitro compared to the free drug solution, indicating that they could be used in the therapy for the delivery of ERLO in non–small cell lung cancer (Mandal et al., 2016).

LPNPs modified by wheat germ agglutinin designed for the oral delivery of oridonin with the particle size of 327 nm and the zeta potential of -31.8 ± 1.04 mV were characterized by efficient binding to Caco-2 and HT-29-MTX cell lines and underwent receptor-mediated endocytosis, showed increased intestinal bioadhesion and cellular uptake and could be considered as a promising nanoformulation suitable to improve the oral delivery of poorly water-soluble drugs (Liu et al., 2016e).

LPNPs loaded with sirolimus or propolis known as antiproliferative drugs significantly reduced the proliferation of human aortic smooth muscle cells while maintaining high viability of human aortic endothelial primary cells (Bose et al., 2016).

LPNPs encapsulating CUR were found to modulate the vascular deposition of breast cancer cells, prevent metastasis, and limit the progression of the disease by modulating vascular inflammation and impairing circulating tumor cells arrest (Palange et al., 2014).

8.9 CONCLUSION

Based on statistics, tumors of various etiologies are one of the leading causes of morbidity and mortality worldwide, especially in high-income regions due to lifestyle (processed foods, stress, alcohol consumption, smoking, etc.) and polluted environment. Thus it can be stated that tumors represent an enormous disease burden in terms of human suffering and economic cost. Currently, drugs can be effective at the beginning of treatment, but cancer cells can develop resistance. In spite of the fact that the group of antineoplastic drugs is currently one of the fastest growing groups of medicaments including many different scaffolds, in general, most of antineoplastics show poor water solubility and, thus, low bioavailability. The constantly increasing number of cancer incidences requires not only the development of new effective anticancer drugs, but also innovative drug formulations contributing to increased drug solubility, prolonged duration of exposure, selective delivery of entrapped drug to the site of action, improved therapeutic index, and potential to overcome resistance to anticancer agents. Rapid development of various nanosystems has offered new opportunities for the treatment of malignancies. However, problems of development of antineoplastics, chemotherapy, and oncology in general are extensive, and, of course, it is not possible to capture all aspects of developments of various nano-anticancer agents in this brief survey. This chapter was aimed to inform about some innovative nano-based approaches to clinically or experimentally used anticancer drugs based on the application of lipid-based nanocarriers as a class of biocompatible and biodegradable vehicles, resulting in advantages, especially in the increased bioavailability and targeted biodistribution and, thus, better therapeutic efficiency of anticancer medicaments as well as decreased toxicity.

ACKNOWLEDGMENTS

This study was supported by the Ministry of Education, Youth and Sports of the Czech Republic (LO1305) and by SANOFI-AVENTIS Pharma Slovakia, s.r.o.

REFERENCES

Abbasalipourkabir, R., Salehzadeh, A., Abdullah, R., 2016. Tamoxifen-loaded solid lipid nanoparticles-induced apoptosis in breast cancer cell lines. J. Exp. Nanosci. 11, 161–174.

Abdelbary, G., Haider, M., 2013. In vitro characterization and growth inhibition effect of nanostructured lipid carriers for controlled delivery of methotrexate. Pharm. Dev. Technol. 18, 1159–1168.

Afzal, S.M., Shareef, M.Z., Dinesh, T., Kishan, V., 2016. Folate-PEG-decorated docetaxel lipid nanoemulsion for improved antitumor activity. Nanomedicine 11, 2171–2184.

Akbarzadeh, A., Rezaei-Sadabady, R., Davaran, S., Joo, S.W., Zarghami, N., Hanifehpour, Y., et al., 2013. Liposome: classification, preparation, and applications. Nanoscale Res. Lett. 8, 102.

Alanazi, F.K., Haq, N., Radwan, A.A., Alsarra, I.A., Shakeel, F., 2015. Formulation and evaluation of cholesterol-rich nanoemulsion (LDE) for drug delivery potential of cholesteryl-maleoyl-5-fluorouracil. Pharm. Dev. Technol. 20, 266–270.

Alavizadeh, S.H., Akhtari, J., Badiee, A., Golmohammadzadeh, S., Jaafari, M.R., 2016. Improved therapeutic activity of HER2 Affibody-targeted cisplatin liposomes in HER2-expressing breast tumor models. Exp. Opin. Drug. Deliv. 13, 325–336.

Alayoubi, A.Y., Anderson, J.F., Satyanarayanajois, S.D., Sylvester, P.W., Nazzal, S., 2013. Concurrent delivery of tocotrienols and simvastatin by lipid nanoemulsions potentiates their antitumor activity against human mammary adenocarcinoma cells. Eur. J. Pharm. Sci. 48, 385–392.

Alayoubi, A., Ayoub, N.M., Malaviya, A., Sylvester, P.W., Nazzal, S., 2014. Entrapment into nanoemulsions potentiates the anticancer activity of tocotrienols against the highly malignant (+SA) mouse mammary epithelial cells. J. Nanosci. Nanotechnol. 14, 4002–4005.

Al-Hazzani, A., Periyasamy, V., Subash-Babu, P., Alshatwi, A.A., 2012. Formulation of cashew nut shell liquid (CSNL) nanoemulsion, a potent inhibitor of human MCF-7 breast cancer cell proliferation. Med. Chem. Res. 21, 1384–1388.

Alkhatib, M.H., Alkreathy, H.M., Balamash, K.S., Abdu, F., 2016. Antitumor activity of doxorubicine-loaded nanoemulsion against Ehrlich ascites carcinoma-bearing mice. Trop. J. Pharm. Res. 15, 937–944.

Allard, E., Huynh, N.T., Vessieres, A., Pigeon, P., Jaouen, G., Benoit, J.P., et al., 2009. Dose effect activity of ferrocifen-loaded lipid nanocapsules on a 9L-glioma model. Int. J. Pharm. 379, 317–323.

Almoussalam, M., Zhu, H.J., 2016. Encapsulation of cancer therapeutic agent dacarbazine using nanostructured lipid carrier. J. Vis. Exp. 110, e53760.

Anton, N., Vandamme, T.F., 2011. Nano-emulsions and micro-emulsions: clarifications of the critical differences. Pharm. Res. 28, 978–985.

Apiratikul, N., Penglong, T., Suksen, K., Svasti, S., Chairoungdua, A., Yingyongnarongkul, B., 2013. In vitro delivery of curcumin with cholesterol-based cationic liposomes. Russ. J. Bioorg. Chem. 39, 444–450.

Arranja, A., Gouveia, L.F., Gener, P., Rafael, D.F., Pereira, C., Schwartz, S., et al., 2016. Self-assembly PEGylation assists SLN-paclitaxel delivery inducing cancer cell apoptosis upon internalization. Int. J. Pharm. 501, 180–189.

Athawale, R.B., Jain, D.S., Singh, K.K., Gude, R.P., 2014. Etoposide loaded solid lipid nanoparticles for curtailing B16F10 melanoma colonization in lung. Biomed. Pharmacother. 68, 231–240.

Attili-Qadri, S., Karra, N., Nemirovski, A., Schwob, O., Talmon, Y., Nassar, T., et al., 2013. Oral delivery system prolongs blood circulation of docetaxel nanocapsules via lymphatic absorption. Proc. Natl. Acad. Sci. U.S.A. 110, 17498–17503.

Bae, P.K., Jung, J., Lim, S.J., Kim, D., Kim, S.K., Chung, B.H., 2013. Bimodal perfluorocarbon nanoemulsions for nasopharyngeal carcinoma targeting. Mol. Imaging Biol. 15, 401–410.

Bahadori, F., Topcu, G., Eroglu, M.S., Onyuksel, H.A., 2014. New lipid-based nano formulation of vinorelbine. AAPS PharmSciTech 15, 1138–1148.

Banerjee, I., De, K., Mukherjee, D., Dey, G., Chattopadhyay, S., Mukherjee, M., et al., 2016. Paclitaxel-loaded solid lipid nanoparticles modified with Tyr-3-octreotide for enhanced anti-angiogenic and anti-glioma therapy. Acta Biomater. 38, 69–81.

Bastiancich, C., Vanvarenberg, K., Ucakar, B., Pitorre, M., Bastiat, G., Lagarce, F., et al., 2016. Lauroyl-gemcitabine-loaded lipid nanocapsule hydrogel for the treatment of glioblastoma. J. Controlled Release. 225, 283–293.

Battaglia, L., Gallarate, M., Peira, E., Chirio, D., Muntoni, E., Biasibetti, E., et al., 2014. Solid lipid nanoparticles for potential doxorubicin delivery in glioblastoma treatment: preliminary in vitro studies. J. Pharm. Sci. 103, 2157–2165.

Beloqui, A., Solinís, M.A., Rodríguez-Gascón, A., Almeida, A.J., Préat, V., 2016. Nanostructured lipid carriers: promising drug delivery systems for future clinics. Nanomedicine 12, 143–161.

Bersani, S., Vila-Caballer, M., Brazzale, C., Barattin, M., Salmaso, S., 2014. pH-sensitive stearoyl-PEG-poly(methacryloyl sulfadimethoxine) decorated liposomes for the delivery of gemcitabine to cancer cells. Eur. J. Pharm. Biopharm. 88, 670–682.

Bhat, M.A., Iqbal, M., Al-Dhfyan, A., Shakeel, F., 2015. Carvone Schiff base of isoniazid as a novel antitumor agent: nanoemulsion development and pharmacokinetic evaluation. J. Mol. Liq. 203, 111–119.

Bhushan, S., Kakkar, V., Pal, H.C., Mondhe, D.M., Kaur, I.P., 2016. The augmented anticancer potential of AP9-cd loaded solid lipid nanoparticles in human leukemia Molt-4 cells and experimental tumor. Chem. Biol. Interact. 244, 84–93.

Biswas, S., Deshpande, P.P., Perche, F., Dodwadkar, N.S., Sane, S.D., Torchilin, V.P., 2013. Octa-arginine-modified pegylated liposomal doxorubicin: an effective treatment strategy for non-small cell lung cancer. Cancer Lett. 335, 191–200.

Bondi, M.L., Botto, C., Amore, E., Emma, M.R., Augello, G., Craparo, E.F., et al., 2015. Lipid nanocarriers containing sorafenib inhibit colonies formation in human hepatocarcinoma cells. Int. J. Pharm. 493, 75–85.

Bose, R.J.C., Lee, S.H., Park, H., 2016. Lipid polymer hybrid nanospheres encapsulating antiproliferative agents for stent applications. J. Ind. Eng. Chem. 36, 284–292.

Bu, H.H., He, X.Y., Zhang, Z.W., Yin, Q., Yu, H.J., Li, Y.P., 2014. A TPGS-incorporating nanoemulsion of paclitaxel circumvents drug resistance in breast cancer. Int. J. Pharm. 471, 206–213.

Buyukkoroglu, G., Senel, B., Gezgin, S., Dinh, T., 2016. The simultaneous delivery of paclitaxel and Herceptin® using solid lipid nanoparticles: in vitro evaluation. J. Drug. Deliv. Sci. Technol. 35, 98–105.

Camacho, K.M., Menegatti, S., Vogus, D.R., Pusuluri, A., Fuchs, Z., Jarvis, M., et al., 2016. DAFODIL: a novel liposome-encapsulated synergistic combination of doxorubicin and 5FU for low dose chemotherapy. J. Controlled Release 229, 154–162.

Cao, X., Luo, J.W., Gong, T., Zhang, Z.R., Sun, X., Fu, Y., 2015. Coencapsulated doxorubicin and bromotetrandrine lipid nanoemulsions in reversing multidrug resistance in breast cancer in vitro and in vivo. Mol. Pharm. 12, 274–286.

Carlesso, F.N., Araujo, R.S., Fuscaldi, L.L., Miranda, S.E.M., Rubello, D., Teixeira, C.S., et al., 2016. Preliminary data of the antipancreatic tumor efficacy and toxicity of long-circulating and pH-sensitive liposomes containing cisplatin. Nucl. Med. Commun. 37, 727–734.

Chai, G.H., Xu, Y.K., Chen, S.Q., Cheng, B.L., Hu, F.Q., You, J.A., et al., 2016. Transport mechanisms of solid lipid nanoparticles across Caco-2 cell monolayers and their related cytotoxicology. ACS Appl. Mater. Interfaces 8, 5929–5940.

Chang, T.L., Zhan, H.L., Liang, D., Liang, J.F., 2015. Nanocrystal technology for drug formulation and delivery. Front. Chem. Sci. Eng. 9, 1–14.

Charoenputtakhun, P., Opanasopit, P., Rojanarata, T., Ngawhirunpat, T., 2014. All-*trans* retinoic acid-loaded lipid nanoparticles as a transdermal drug delivery carrier. Pharm. Dev. Technol. 19, 164–172.

Chen, J., Chen, H.Y., Cui, S.S., Xue, B., Tian, J.M., Achilefu, S., et al., 2012. Glucosamine derivative modified nanostructured lipid carriers for targeted tumor delivery. J. Mater. Chem. 22, 5770–5783.

Chen, Y., Peng, J.Q., Han, M.M., Omar, M., Hu, D.R., Ke, X., et al., 2015a. A low-molecular-weight heparin-coated doxorubicin-liposome for the prevention of melanoma metastasis. J. Drug Target 23, 335–346.

Chen, H.H., Huang, W.C., Chiang, W.H., Liu, T.I., Shen, M.Y., Hsu, Y.H., et al., 2015b. pH-Responsive therapeutic solid lipid nanoparticles for reducing P-glycoprotein-mediated drug efflux of multidrug resistant cancer cells. Int. J. Nanomed. 10, 5035–5048.

Chen, Q., Ding, H.W., Zhou, J.X., Zhao, X.F., Zhang, J.L., Yang, C.R., et al., 2016a. Novel glycyrrhetinic acid conjugated pH-sensitive liposomes for the delivery of doxorubicin and its antitumor activities. RSC Adv. 6, 17782–17791.

Chen, X.X., Yuan, M.Q., Zhang, Q.Y., Yang, Y.T., Gao, H.L., He, Q., 2016b. Synergistic combination of doxorubicin and paclitaxel delivered by blood brain barrier and glioma cells dual targeting liposomes for chemotherapy of brain glioma. Curr. Pharm. Biotechnol. 17, 636–650.

Chen, Z.H., Lai, X.L., Song, S.X., Zhu, X.G., Zhu, J.M., 2016c. Nanostructured lipid carriers based temozolomide and gene co-encapsulated nanomedicine for gliomatosis cerebri combination therapy. Drug. Deliv. 23, 1369–1373.

Cheow, W.S., Hadinoto, K., 2011. Factors affecting drug encapsulation and stability of lipid–polymer hybrid nanoparticles. Colloids Surf. B: Biointerfaces 85, 214–220.

Chinsriwongkul, A., Chareanputtakhun, P., Ngawhirunpat, T., Rojanarata, T., Sila-on, W., Ruktanonchai, U., et al., 2012. Nanostructured lipid carriers (NLC) for parenteral delivery of an anticancer drug. AAPS PharmSciTech 13, 150–158.

Cho, H.J., Park, J.W., Yoon, I.S., Kim, D.D., 2014. Surface-modified solid lipid nanoparticles for oral delivery of docetaxel: enhanced intestinal absorption and lymphatic uptake. Int. J. Nanomed. 9, 495–504.

Chu, Y., Li, D., Luo, Y.F., He, X.J., Jiang, M.Y., 2014. Preparation and in vitro evaluation of glycyrrhetinic acid-modified curcumin-loaded nanostructured lipid carriers. Molecules 19, 2445–2457.

Contente, T.C., Kretzer, I.F., Filippin-Monteiro, F.B., Maria, D.A., Maranhao, R.C., 2014. Association of daunorubicin to a lipid nanoemulsion that binds to low-density lipoprotein receptors enhances the antitumour action and decreases the toxicity of the drug in melanoma-bearing mice. J. Pharm. Pharmacol. 66, 1698–1709.

Costa, P.M., Cardoso, A.L., Mendonca, L.S., Serani, A., Custodia, C., Conceicao, M., et al., 2013. Tumor-targeted chlorotoxin-coupled nanoparticles for nucleic acid delivery to glioblastoma cells: a promising system for glioblastoma treatment. Mol. Ther. Nucleic Acids 2, e100.

Dagar, A., Kuzmis, A., Rubinstein, I., Sekosan, M., Onyuksel, H., 2012. VIP-targeted cytotoxic nanomedicine for breast cancer. Drug Deliv. Transl. Res. 2, 454–462.

de Escalona, M.M., Saez-Fernandez, E., Prados, J.C., Melguizo, C., Arias, J.L., 2016. Magnetic solid lipid nanoparticles in hyperthermia against colon cancer. Int. J. Pharm. 504, 11–19.

Dehaini, D., Fang, R.H., Luk, B.T., Pang, Z.Q., Hu, C.M.J., Kroll, A.V., et al., 2016. Ultrasmall lipid–polymer hybrid nanoparticles for tumor-penetrating drug delivery. Nanoscale 8, 14411–14419.

Dehghani, F., Farhadian, N., Golmohammadzadeh, S., Biriaee, A., Ebrahimi, M., Karimi, M., 2017. Preparation, characterization and in-vivo evaluation of microemulsions containing tamoxifen citrate anti-cancer drug. Eur. J. Pharm. Sci. 96, 479–489.

Desai, A., Vyas, T., Amiji, M., 2008. Cytotoxicity and apoptosis enhancement in brain tumor cells upon coadministration of paclitaxel and ceramide in nanoemulsion formulations. J. Pharm. Sci. 97, 2745–2756.

Deshpande, P.P., Biswas, S., Torchilin, V.P., 2013. Current trends in the use of liposomes for tumor targeting. Nanomedicine 8, 1509–1528.

De Smet, L., Colin, P., Ceelen, W., Bracke, M., Van Bocxlaer, J., Remon, J.P., et al., 2012. Development of a nanocrystalline Paclitaxel formulation for HIPEC treatment. Pharm. Res. 29, 2398–2406.

Dhaundiyal, A., Jena, S.K., Samal, S.K., Somvane, B., Chand, M., Sangamwarm, A.T., 2016. Alpha-lipoic acid-stearylamine conjugate-based solid lipid nanoparticles for tamoxifen delivery: formulation, optimization, in-vivo pharmacokinetic and hepatotoxicity study. J. Pharm. Pharmacol. 68, 1535–1550.

Di, H.F., Wu, H.Y., Gao, Y., Li, W.H., Zou, D.N., Dong, C.H., 2016. Doxorubicin- and cisplatin-loaded nanostructured lipid carriers for breast cancer combination chemotherapy. Drug. Dev. Ind. Pharm. 42, 2038–2043.

Diaspro, A., Krol, S., Cavalleri, O., Silvano, D., Gliozzi, A., 2002. Microscopical characterization of nanocapsules templated on ionic crystals and biological cells toward biomedical applications. IEEE Trans. Nanobiosci. 1, 110–115.

Dicheva, B.M., ten Hagen, T.L.M., Schipper, D., Seynhaeve, A.L.B., van Rhoon, G.C., Eggermont, A.M.M., et al., 2014. Targeted and heat-triggered doxorubicin delivery to tumors by dual targeted cationic thermosensitive liposomes. J. Controlled Release 195, 37–48.

Dicheva, B.M., Seynhaeve, A.L.B., Soulie, T., Eggermont, A.M.M., ten Hagen, T.L.M., Koning, G.A., 2016. Pharmacokinetics, tissue distribution and therapeutic effect of cationic thermosensitive liposomal doxorubicin upon mild hyperthermia. Pharm. Res. 33, 627–638.

Ding, H., Sagar, V., Agudelo, M., Pilakka-Kanthikeel, S., Atluri, V.S.R., Raymond, A., et al., 2014a. Enhanced blood–brain barrier transmigration using a novel transferrin embedded fluorescent magneto-liposome nanoformulation. Nanotechnology 25, 055101.

Ding, B.Y., Chen, P., Kong, Y., Zhai, Y.J., Pang, X., Dou, J., et al., 2014b. Preparation and evaluation of folate-modified lipid nanocapsules for quercetin delivery. J. Drug Target 22, 67–75.

Dong, X.Y., Wang, W., Qu, H., Han, D., Zheng, J.M., Sun, G.R., 2016. Targeted delivery of doxorubicin and vincristine to lymph cancer: evaluation of novel nanostructured lipid carriers in vitro and in vivo. Drug Deliv. 23, 1374–1378.

Drewes, C.C., Fiel, L.A., Bexiga, C.G., Asbahr, A.C.C., Uchiyama, M.K., Cogliati, B., et al., 2016. Novel therapeutic mechanisms determine the effectiveness of lipid-core nanocapsules on melanoma models. Int. J. Nanomed. 11, 1261–1279.

Drummond, D.C., Noble, C.O., Guo, Z.X., Hayes, M.E., Park, J.W., Ou, C.J., et al., 2009. Improved pharmacokinetics and efficacy of a highly stable nanoliposomal vinorelbine. J. Pharmacol. Exp. Ther. 328, 321–330.

Du, J.B., Song, Y.F., Ye, W.L., Cheng, Y., Cui, H., Liu, D.Z., et al., 2014. PEG-detachable lipid–polymer hybrid nanoparticle for delivery of chemotherapy drugs to cancer cells. Anti-Cancer Drug 25, 751–766.

Emami, J., Rezazadeh, M., Varshosaz, J., Tabbakhian, M., Aslani, A., 2012. Formulation of LDL targeted nanostructured lipid carriers loaded with paclitaxel: a detailed study of preparation, freeze drying condition, and in vitro cytotoxicity. J. Nanomater. 2012, 358782.

Fan, X.C., Chen, J.J., Shen, Q., 2013. Docetaxel-nicotinamide complex-loaded nanostructured lipid carriers for transdermal delivery. Int. J. Pharm. 458, 296–304.

Fang, J.Y., Hung, C.F., Hua, S.C., Hwang, T.L., 2009. Acoustically active per fluorocarbon nanoemulsions as drug delivery carriers for camptothecin: drug release and cytotoxicity against cancer cells. Ultrasonics 49, 39–46.

Fang, G.H., Tang, B., Chao, Y.H., Zhang, Y., Xu, H., Tang, X., 2015a. Improved oral bioavailability of docetaxel by nanostructured lipid carriers: in vitro characteristics, in vivo evaluation and intestinal transport studies. RSC Adv. 5, 96437–96447.

Fang, G.H., Tang, B., Chao, Y.H., Xu, H., Gou, J.X., Zhang, Y., et al., 2015b. Cysteine-functionalized nanostructured lipid carriers for oral delivery of docetaxel: a permeability and pharmacokinetic study. Mol. Pharm. 12, 2384–2395.

Farace, C., Sanchez-Moreno, P., Orecchioni, M., Manetti, R., Sgarrella, F., Asara, Y., et al., 2016. Immune cell impact of three differently coated lipid nanocapsules: pluronic, chitosan and polyethylene glycol. Sci. Rep. 6, 18423.

Favareto, A.P.A., Nogueira, L.C., Pontes, D.A., Mesquita, S.D.P., Cherici Camargo, I.C., 2015. Evaluation of testicular tissue of adult rats treated with cisplatin incorporated into the liposome. Microsc. Res. Tech. 78, 323–329.

Ferlay, J., Soerjomataram, I., Ervik, M., Dikshit, R., Eser, S., Mathers, C., et al., 2013. Cancer Incidence and Mortality Worldwide. GLOBOCAN, Lyon.

Figueiro, F., de Oliveira, C.R., Rockenbach, L., Mendes, F.B., Bergamin, L.S., Jandrey, E. H.F., et al., 2015. Pharmacological improvement and preclinical evaluation of methotrexate-loaded lipid-core nanocapsules in a glioblastoma model. J. Biomed. Nanotechnol. 11, 1808–1818.

Fofaria, N.M., Qhattal, H.S.S., Liu, X.L., Srivastava, S.K., 2016. Nanoemulsion formulations for anti-cancer agent piplartine—characterization, toxicological, pharmacokinetics and efficacy studies. Int. J. Pharm. 498, 12–22.

Ganta, S., Amiji, M., 2009. Coadministration of paclitaxel and curcumin in nanoemulsion formulations to overcome multidrug resistance in tumor cells. Mol. Pharm. 6, 928–939.

Ganta, S., Talekar, M., Singh, A., Coleman, T.P., Amiji, M.M., 2014. Nanoemulsions in translational research-opportunities and challenges in targeted cancer therapy. AAPS PharmSciTech 15, 694–708.

Ganta, S., Singh, A., Rawal, Y., Cacaccio, J., Patel, N.R., Kulkarni, P., et al., 2016. Formulation development of a novel targeted theranostic nanoemulsion of docetaxel to overcome multidrug resistance in ovarian cancer. Drug Deliv 23, 968–980.

Gao, L., Liu, G., Ma, J., Wang, X., Wang, F., Wang, H., et al., 2014. Paclitaxel nanosuspension coated with P-gp inhibitory surfactants: II. Ability to reverse the drug-resistance of H460 human lung cancer cells. Colloids Surf. B: Biointerfaces 117, 122–127.

Gaoe, H., Pang, Z.Q., Pan, S.Q., Cao, S.J., Yang, Z., Chen, C., et al., 2012. Anti-glioma effect and safety of docetaxel-loaded nanoemulsion. Arch. Pharm. Res. 35, 333–341.

Garanti, T., Stasik, A., Burrow, A.J., Alhnan, M.A., Wan, K.W., 2016. Anti-glioma activity and the mechanism of cellular uptake of asiatic acid-loaded solid lipid nanoparticles. Int. J. Pharm 500, 305–315.

Gardikis, K., Hatziantoniou, S., Bucos, M., Fessas, D., Signorelli, M., Felekis, T., et al., 2010. New drug delivery nanosystem combining liposomal and dendrimeric technology (liposomal locked-in dendrimers) for cancer therapy. J. Pharm. Sci. 99, 3561–3571.

Garg, N.K., Singh, B., Sharma, G., Kushwah, V., Tyagi, R.K., Jain, S., et al., 2015. Development and characterization of single step self-assembled lipid polymer hybrid nanoparticles for effective delivery of methotrexate. RSC Adv. 5, 62989–62999.

Garg, N., Singh, B., Jain, A., Nirbhavane, O., Sharma, R., Tyagi, R.K., et al., 2016. Fucose decorated solid-lipid nanocarriers mediate efficient delivery of methotrexate in breast cancer therapeutics. Colloids Surf. B: Biointerfaces 146, 114–126.

Geszke-Moritz, M., Moritz, M., 2016. Solid lipid nanoparticles as attractive drug vehicles: composition, properties and therapeutic strategies. Mater. Sci. Eng. C: Mater. Biol. Appl 68, 982–984.

Ghannam, M.M., El Gebaly, R., Fadel, M., 2016. Targeting doxorubicin encapsulated in stealth liposomes to solid tumors by non thermal diode laser. Lipids Health Dis 15, 68.

Gianella, A., Jarzyna, P.A., Mani, V., Ramachandran, S., Calcagno, C., Tang, J., et al., 2011. Multifunctional nanoemulsion platform for imaging guided therapy evaluated in experimental cancer. ACS Nano 5, 4422–4433.

Gill, K.K., Kaddoumi, A., Nazzal, S., 2015. PEG-lipid micelles as drug carriers: physiochemical attributes, formulation principles and biological implication. J. Drug Target. 23, 222–231.

Gonzalo, T., Lollo, G., Garcia-Fuentes, M., Torres, D., Correa, J., Riguera, R., et al., 2013. A new potential nano-oncological therapy based on polyamino acid nanocapsules. J. Controlled Release 169, 10–16.

Groo, A.C., Saulnier, P., Gimel, J.C., Gravier, J., Ailhas, C., Benoit, J.P., et al., 2013. Fate of paclitaxel lipid nanocapsules in intestinal mucus in view of their oral delivery. Int. J. Nanomed. 8, 4291–4302.

Guan, R.F., Ma, J.Q., Wu, Y.H., Lu, F., Xiao, C.G., Jiang, H., et al., 2012. Development and characterization of lactoferrin nanoliposome: cellular uptake and stability. Nanoscale Res. Lett. 7, 679.

Guo, Y.M., Liu, X., Sun, X., Zhang, Q., Gong, T., Zhang, Z.R., 2012. Mannosylated lipid nano-emulsions loaded with lycorine-oleic acid ionic complex for tumor cell-specific delivery. Theranostics 2, 1104–1114.

Guo, Y.J., Wang, L.J., Lv, P., Zhang, P., 2015. Transferrin-conjugated doxorubicin-loaded lipid-coated nanoparticles for the targeting and therapy of lung cancer. Oncol. Lett 9, 1065–1072.

Gupta, B., Poudel, B.K., Tran, T.H., Pradhan, R., Cho, H.J., Jeong, J.H., et al., 2015. Modulation of pharmacokinetic and cytotoxicity profile of imatinib base by employing optimized nanostructured lipid carriers. Pharm. Res. 32, 2912–2927.

Hadinoto, K., Sundaresan, A., Cheow, W.S., 2013. Lipid–polymer hybrid nanoparticles as a new generation therapeutic delivery platform: a review. Eur. J. Pharm. Biopharm 85, 427–443.

Hallan, S.S., Kaur, P., Kaur, V., Mishra, N., Vaidya, B., 2016. Lipid polymer hybrid as emerging tool in nanocarriers for oral drug delivery. Artif. Cells. Nanomed. Biotechnol. 44, 334–349.

Hamelers, I.H.L., De Kroon, A.I.P.M., 2007. Nanocapsules: a novel lipid formulation platform for platinum-based anti-cancer drugs. J. Liposome Res. 17, 183–189.

Han, Y., Zhang, P., Chen, Y.Y., Sun, J.P., Kong, F.S., 2014a. Co-delivery of plasmid DNA and doxorubicin by solid lipid nanoparticles for lung cancer therapy. Int. J. Mol. Med 34, 191–196.

Han, C.Y., Li, Y., Sun, M.S., Liu, C., Ma, X.X., Yang, X.G., et al., 2014b. Small peptide-modified nanostructured lipid carriers distribution and targeting to EGFR-overexpressing tumor in vivo. Artif. Cells Nanomed. Biotechnol. 42, 161–166.

Han, Y.Q., Zhang, Y., Li, D.N., Chen, Y.Y., Sun, J.P., Kong, F.S., 2014c. Transferrin-modified nanostructured lipid carriers as multifunctional nanomedicine for codelivery of DNA and doxorubicin. Int. J. Nanomed. 9, 4107–4115.

Han, N., Zhao, Q.F., Wan, L., Wang, Y., Gao, Y.K., Wang, P., et al., 2015. Hybrid lipid-capped mesoporous silica for stimuli-responsive drug release and overcoming multidrug resistance. ACS Appl. Mater. Interfaces 7, 3342–3351.

Hashem, F.M., Nasr, M., Khairy, A., 2014. In vitro cytotoxicity and bioavailability of solid lipid nanoparticles containing tamoxifen citrate. Pharm. Dev. Technol. 19, 824–832.

Hong, J.Y., Li, Y.H., Li, Y.J., Xiao, Y., Kuang, H.X., Wang, X.T., 2016. Annonaceous acetogenins nanosuspensions stabilized by PCL-PEG block polymer: significantly improved antitumor efficacy. Int. J. Nanomed. 11, 3239–3253.

Howell, M., Mallela, J., Wang, C., Ravi, S., Dixit, S., Garapati, U., et al., 2013. Manganese-loaded lipid-micellar theranostics for simultaneous drug and gene delivery to lungs. J. Controlled Release 167, 210–218.

Hsu, S.H., Wen, C.J., Al-Suwayeh, S.A., Huang, Y.J., Fang, J.Y., 2013. Formulation design and evaluation of quantum dot-loaded nanostructured lipid carriers for integrating bioimaging and anticancer therapy. Nanomedicine 8, 1253–1269.

Hu, L.D., Jia, Y.H., WenDing, 2010. Preparation and characterization of solid lipid nanoparticles loaded with epirubicin for pulmonary delivery. Pharmazie 65, 585–587.

Hua, S., Wu, S.Y., 2013. The use of lipid-based nanocarriers for targeted pain therapies. Front. Pharmacol. 4, 143.

Huang, R.F., Wei, Y.J., Inbaraj, B.S., Chen, B.H., 2015. Inhibition of colon cancer cell growth by nanoemulsion carrying gold nanoparticles and lycopene. Int. J. Nanomed. 10, 2823–2846.

Huber, L.A., Pereira, T.A., Ramos, D.N., Rezende, L.C.D., Emery, F.S., Sobral, L.M., et al., 2015. Topical skin cancer therapy using doxorubicin-loaded cationic lipid nanoparticles and iontophoresis. J. Biomed. Nanotechnol. 11, 1975–1988.

Huynh, N.T., Passirani, C., Saulnier, P., Benoit, J.P., 2009. Lipid nanocapsules: a new platform for nanomedicine. Int. J. Pharm. 379, 201–209.

Jacobs, C., Müller, R.H., 2002. Production and characterization of a budesonide nanosuspension for pulmonary administration. Pharm. Res 19, 189–194.

Jain, A., Singhai, P., Gurnany, E., Updhayay, S., Mody, N., 2013. Transferrin-tailored solid lipid nanoparticles as vectors for site-specific delivery of temozolomide to brain. J. Nanopart. Res. 15, 1518.

Jain, A., Kesharwani, P., Garg, N.K., Jain, A., Jain, S.A., Jain, A.K., et al., 2015. Galactose engineered solid lipid nanoparticles for targeted delivery of doxorubicin. Colloids Surf. B: Biointerfaces 134, 47–58.

Jain, A.S., Makhija, D.T., Goel, P.N., Shah, S.M., Nikam, Y., Gude, R.P., et al., 2016. Docetaxel in cationic lipid nanocapsules for enhanced in vivo activity. Pharm. Dev. Technol. 21, 76–85.

Jakoby, J., Beuschlein, F., Mentz, S., Hantel, C., Suess, R., 2015. Liposomal doxorubicin for active targeting: surface modification of the nanocarrier evaluated in vitro and in vivo—challenges and prospects. Oncotarget 6, 43698–43711.

Jampílek, J., Kráľová, K., 2017. Nano-antimicrobials: activity, benefits and weaknesses. In: Ficai, A., Grumezescu, A.M. (Eds.), Nanostructures in Therapeutic Medicine, Vol. 2—Nanostructures for Antimicrobial Therapy. Elsevier, London, pp. 23–54. (Chapter 2).

Jampílek, J., Kráľová, K., 2018. Application of nanobioformulations for controlled release and targeted biodistribution of drugs. In: Sharma, A.K., Keservani, R.K., Kesharwani, R.K. (Eds.), Nanobiomaterials: Applications in Drug Delivery. CRC Press, Warentown, pp. 131–208. (Part II, Chapter 5).

Jang, D.J., Moon, C., Oh, E., 2016. Improved tumor targeting and antitumor activity of camptothecin loaded solid lipid nanoparticles by preinjection of blank solid lipid nanoparticles. Biomed. Pharmacother. 80, 162–172.

Jiang, W.H., Wang, J., Yang, L., Jiang, X.W., Bai, Z.S., Wang, Z.R., et al., 2015. Nanostructured lipid carriers modified with PEGylated carboxymethylcellulose polymers for effective delivery of docetaxel. RSC Adv. 5, 90386–90395.

Jiang, H., Pei, L.N., Liu, N.N., Li, J.S., Li, Z.R., Zhang, S.M., 2016. Etoposide-loaded nanostructured lipid carriers for gastric cancer therapy. Drug Deliv. 23, 1379–1382.

Jin, Z., Lv, Y.Q., Cao, H., Yao, J., Zhou, J.P., He, W., et al., 2016. Core–shell nanocarriers with high paclitaxel loading for passive and active targeting. Sci. Rep. 6, 27559.

Jing, X.L., Deng, L., Gao, B.A., Xiao, L., Zhang, Y.Y., Ke, X.F., et al., 2014. A novel polyethylene glycol mediated lipid nanoemulsion as drug delivery carrier for paclitaxel. Nanomedicine 10, 371–380.

Joo, K.I., Xiao, L., Liu, S.L., Liu, Y.R., Lee, C.L., Conti, P.S., et al., 2013. Crosslinked multilamellar liposomes for controlled delivery of anticancer drugs. Biomaterials 34, 3098–3109.

Jordan, M., Nayel, A., Brownlow, B., Elbayoumi, T., 2012. Development and evaluation of tocopherol-rich argan oil-based nanoemulsions as vehicles possessing anticancer activity. J. Biomed. Nanotechnol. 8, 944–956.

Jung, J., Jeong, S.Y., Park, S.S., Shin, S.H., Ju, E.J., Choi, J., et al., 2015. A cisplatin-incorporated liposome that targets the epidermal growth factor receptor enhances radiotherapeutic efficacy without nephrotoxicity. Int. J. Oncol 46, 1268–1274.

Kakumanu, S., Tagne, J.B., Wilson, T.A., Nicolosi, R.J., 2011. A nanoemulsion formulation of dacarbazine reduces tumor size in a xenograft mouse epidermoid carcinoma model compared to dacarbazine suspension. Nanomedicine 7, 277–283.

Kamaly, N., Kalber, T., Thanou, M., Bell, J.D., Miller, A.D., 2009. Folate receptor targeted bimodal liposomes for tumor magnetic resonance imaging. Bioconjug. Chem 20, 648–655.

Kang, K.W., Chun, M.K., Kim, O., Subedi, R.K., Ahn, S.G., Yoon, J.H., et al., 2010. Doxorubicin-loaded solid lipid nanoparticles to overcome multidrug resistance in cancer therapy. Nanomedicine 6, 210–213.

Kauerová, T., Kos, J., Goněc, T., Jampílek, J., Kollár, P., 2016. Antiproliferative and pro-apoptotic effect of novel nitro-substituted hydroxynaphthanilides on human cancer cell lines. Int. J. Mol. Sci. 17, 1219.

Keykanlu, H.R.I., Zibaei, S., Ardjmand, M., Safekordi, A.A., 2016. Fluorocarbon nanostructures (PFOB-NEP) as camel milk lactoferrin and its anticancer effects on human breast cancer cell line MCF7. Bulg. Chem. Commun. 48, 323–331.

Khajavinia, A., Varshosaz, J., Dehkordi, A.J., 2012. Targeting etoposide to acute myelogenous leukaemia cells using nanostructured lipid carriers coated with transferrin. Nanotechnology 23, 405101.

Kibria, G., Hatakeyama, H., Sato, Y., Harashima, H., 2016. Anti-tumor effect via passive anti-angiogenesis of PEGylated liposomes encapsulating doxorubicin in drug resistant tumors. Int. J. Pharm. 509, 178–187.

Kim, J.H., Kim, Y., Bae, K.H., Park, T.G., Lee, J.H., Park, K., 2015. Tumor-targeted delivery of paclitaxel using low density lipoprotein-mimetic solid lipid nanoparticles. Mol. Pharm. 12, 1230–1241.

Kim, J.S., 2016. Liposomal drug delivery system. J. Pharm. Investig. 46, 387–392.

Kong, S.D., Sartor, M., Hu, C.M.J., Zhang, W.Z., Zhang, L.F., Jin, S.H., 2013. Magnetic field activated lipid–polymer hybrid nanoparticles for stimuli-responsive drug release. Acta Biomater. 9, 5447–5452.

Koning, G.A., Eggermont, A.M.M., Lindner, L.H., ten Hagen, T.L.M., 2010. Hyperthermia and thermosensitive liposomes for improved delivery of chemotherapeutic drugs to solid tumors. Pharm. Res. 27, 1750–1754.

Kopechek, J.A., Park, E., Mei, C.S., McDannold, N.J., Porter, T.M., 2013. Accumulation of phase-shift nanoemulsions to enhance MR-guided ultrasound-mediated tumor ablation in vivo. J. Healthc. Eng. 4, 109–126.

Kothamasu, P., Kanumur, H., Ravur, N., Maddu, C., Parasuramrajam, R., Thangavel, S., 2012. Nanocapsules: the weapons for novel drug delivery systems. BioImpacts 2, 71–81.

Král, V., Jampílek, J., Havlíček, J., Brusová, H., Pekárek, T., 2010. Dosage Forms of Tyrosine Kinase Inhibitors. WO/2010/081443 A2, July 22.

Král, V., Brusová, H., Jampílek, J., Havlíček, J., Pekárek, T., Tkadlecová, M., 2011. Imatinib Mesylate Polymorphs Generated by Crystallization in Aqueous Inorganic Salt Solutions. WO/2011/023146 A1, March 3.

Kretzer, I.F., Maria, D.A., Maranhao, R.C., 2012. Drug-targeting in combined cancer chemotherapy: tumor growth inhibition in mice by association of paclitaxel and etoposide with a cholesterol-rich nanoemulsion. Cell Oncol. 35, 451–460.

Kretzer, I.F., Maria, D.A., Guido, M.C., Contente, T.C., Maranhao, R.C., 2016. Simvastatin increases the antineoplastic actions of paclitaxel carried in lipid nanoemulsions in melanoma-bearing mice. Int. J. Nanomed. 11, 885–904.

Krishnamurthy, S., Vaiyapuri, R., Zhang, L.F., Chan, J.M., 2015. Lipid-coated polymeric nanoparticles for cancer drug delivery. Biomater. Sci. 3, 923–936.

Kulkarni, P., Haldar, M.K., Katti, P., Dawes, C., You, S., Choi, Y., et al., 2016. Hypoxia responsive, tumor penetrating lipid nanoparticles for delivery of chemotherapeutics to pancreatic cancer cell spheroids. Bioconjug. Chem. 27, 1830–1838.

Kumar, S., Sinha, V.R., 2016. Design, development and characterization of topical microemulsions of 5-fluorouracil for the treatment of non melanoma skin cancer and its precursor lesions. Anticancer Agents Med. Chem. 16, 259–268.

Kuo, Y.C., Cheng, S.J., 2016. Brain targeted delivery of carmustine using solid lipid nanoparticles modified with tamoxifen and lactoferrin for antitumor proliferation. Int. J. Pharm. 499, 10–19.

Kuo, Y.C., Lee, I.H., 2016. Delivery of doxorubicin to glioblastoma multiforme in vitro using solid lipid nanoparticles with surface aprotinin and melanotransferrin antibody for enhanced chemotherapy. J. Taiwan Inst. Chem. Eng. 61, 32–45.

Kuo, Y.C., Wang, I.H., 2016. Enhanced delivery of etoposide across the blood–brain barrier to restrain brain tumor growth using melanotransferrin antibody- and tamoxifen-conjugated solid lipid nanoparticles. J. Drug. Target. 24, 645–654.

Lagarce, F., Passirani, C., 2016. Nucleic-acid delivery using lipid nanocapsules. Curr. Pharm. Biotechnol 17, 723–727.

Laine, A.L., Clavreul, A., Rousseau, A., Tetaud, C., Vessieres, A., Garcion, E., et al., 2014. Inhibition of ectopic glioma tumor growth by a potent ferrocenyl drug loaded into stealth lipid nanocapsules. Nanomedicine 10, 1667–1677.

Lamprecht, A., Benoit, J.P., 2006. Etoposide nanocarriers suppress glioma cell growth by intracellular drug delivery and simultaneous P-glycoprotein inhibition. J. Controlled Release 112, 208–213.

Lanza, G.M., Winter, P.M., Caruthers, S.D., Hughes, M.S., Hu, G., Schmieder, A.H., et al., 2010. Theragnostics for tumor and plaque angiogenesis with perfluorocarbon nanoemulsions. Angiogenesis 13, 189–202.

Laskar, P., Samanta, S., Ghosh, S.K., Dey, J., 2014. In vitro evaluation of pH-sensitive cholesterol-containing stable polymeric micelles for delivery of camptothecin. J. Colloid Interface Sci. 430, 305–314.

Lee, S., Kim, J., Shim, G., Kim, S., Han, S.E., Kim, K., et al., 2012. Tetraiodothyroacetic acid-tagged liposomes for enhanced delivery of anticancer drug to tumor tissue via integrin receptor. J. Controlled Release 164, 213–220.

Lee, E.H., Kim, J.K., Lim, J.S., Lim, S.J., 2015. Enhancement of indocyanine green stability and cellular uptake by incorporating cationic lipid into indocyanine green-loaded nanoemulsions. Colloids Surf. B: Biointerfaces 136, 305–313.

Li, X.A., Wang, D.K., Zhang, J., Pan, W.S., 2009. Preparation and pharmacokinetics of docetaxel based on nanostructured lipid carriers. J. Pharm. Pharmacol 61, 1485–1492.

Li, X., Du, L., Wang, C.Y., Liu, Y., Mei, X.G., Jin, Y.G., 2011. Highly efficient and lowly toxic docetaxel nanoemulsions for intravenous injection to animals. Pharmazie 66, 479–483.

Li, D., Yang, K., Li, J.S., Ke, X.Y., Duan, Y., Du, R., et al., 2012. Antitumor efficacy of a novel CLA-PTX microemulsion against brain tumors: in vitro and in vivo findings. Int. J. Nanomed. 7, 6105–6114.

Li, W.J., Ng, K.Y., Heng, P.W.S., 2014. Development and evaluation of optimized sucrose ester stabilized oleanolic acid nanosuspensions prepared by wet ball milling with design of experiments. Biol. Pharm. Bull. 37, 926–937.

Li, F., Zhao, X., Wang, H., Zhao, R.F., Ji, T.J., Ren, H., et al., 2015. Multiple layer-by-layer lipid–polymer hybrid nanoparticles for improved FOLFIRINOX chemotherapy in pancreatic tumor models. Adv. Funct. Mater. 25, 788–798.

Li, F., Snow-Davis, C., Du, C., Bondarev, M.K., Saulsbury, M.D., Heyliger, S.O., 2016. Preparation and characterization of lipophilic doxorubicin pro-drug micelles. J. Vis. Exp. 114. Available from: https://doi.org/10.3791/54338.

Lin, Q., Mao, K.L., Tian, F.R., Yang, J.J., Chen, P.P., Xu, J., et al., 2016. Brain tumor-targeted delivery and therapy by focused ultrasound introduced doxorubicin-loaded cationic liposomes. Cancer Chemother. Pharmacol. 77, 269–280.

Lindner, L.H., Hossann, M., 2010. Factors affecting drug release from liposomes. Curr. Opin. Drug Disc 13, 111–123.

Ling, G.X., Zhang, T.H., Zhang, P., Sun, J., He, Z.G., 2016. Nanostructured lipid-carrageenan hybrid carriers (NLCCs) for controlled delivery of mitoxantrone hydrochloride to enhance anticancer activity bypassing the BCRP-mediated efflux. Drug Dev. Ind. Pharm. 42, 1351–1359.

Liu, D.H., Liu, Z.H., Wang, L.L., Zhang, C., Zhang, N., 2011. Nanostructured lipid carriers as novel carrier for parenteral delivery of docetaxel. Colloids Surf. B: Biointerfaces 85, 262–269.

Liu, Y., Xie, P.C., Zhang, D.R., Zhang, Q., 2012. A mini review of nanosuspensions development. J. Drug Target. 20, 209–223.

Liu, G.Y., Luo, Q.Q., Gao, H.Q., Chen, Y., Wei, X., Dai, H., et al., 2015a. Cell membrane-inspired polymeric micelles as carriers for drug delivery. Biomaterials 3, 490–499.

Liu, H., Zhang, Y.L., Han, Y.Z., Zhao, S., Wang, L., Zhang, Z.X., et al., 2015b. Characterization and cytotoxicity studies of DPPC:M^{2+} novel delivery system for cisplatin thermosensitivity liposome with improving loading efficiency. Colloid Surf. B Biointerfaces 131, 12–20.

Liu, S., Li, Y.F., Wang, X.W., Ma, J.A., Zhang, L.L., Xia, G.M., 2016a. Preparation, characterization, and antitumor activities of miriplatin-loaded liposomes. J. Pharm. Sci. 105, 78–87.

Liu, Y.P., Gao, D.W., Zhang, X.W., Liu, Z.W., Dai, K., Ji, B.S., et al., 2016b. Antitumor drug effect of betulinic acid mediated by polyethylene glycol modified liposomes. Mater. Sci. Eng. C: Mater. Biol. Appl 64, 124–132.

Liu, Q., Li, J., Pu, G.B., Zhang, F., Liu, H.Y., Zhang, Y.Q., 2016c. Co-delivery of baicalein and doxorubicin by hyaluronic acid decorated nanostructured lipid carriers for breast cancer therapy. Drug Deliv. 23, 1364–1368.

Liu, M., Chen, D.D., Mukerabigwi, J.F., Chen, S., Zhang, Y.N., Lei, S.J., et al., 2016d. Intracellular delivery of 10-hydroxycamptothecin with targeted nanostructured lipid carriers against multidrug resistance. J. Drug Target. 24, 433–440.

Liu, Y., Zhao, Y.Q., Liu, J.G., Zhang, M.Y., Yu, M.L., Feng, N.P., 2016e. Wheat germ agglutinin modification of lipid–polymer hybrid nanoparticles: enhanced cellular uptake and bioadhesion. RSC Adv. 6, 36125–36135.

Lollo, G., Vincent, M., Ullio-Gamboa, G., Lemaire, L., Franconi, F., Couez, D., et al., 2015. Development of multifunctional lipid nanocapsules for the co-delivery of paclitaxel and CpG-ODN in the treatment of glioblastoma. Int. J. Pharm. 495, 972–980.

Loureiro, A., Bernardes, G.J.L., Shimanovich, U., Sarria, M.P., Nogueira, E., Preto, A., et al., 2015. Folic acid-tagged protein nanoemulsions loaded with CORM-2 enhance the survival of mice bearing subcutaneous A20 lymphoma tumors. Nanomedicine 11, 1077–1083.

Lu, B., Xiong, S.B., Yang, H., Yin, X.D., Chao, R.B., 2006. Solid lipid nanoparticles of mitoxantrone for local injection against breast cancer and its lymph node metastases. Eur. J. Pharm. Sci. 28, 86–95.

Lu, L.Y., Liu, Y., Zhang, Z.F., Gou, X.J., Jiang, J.H., Zhang, J.Z., et al., 2015. Pomegranate seed oil exerts synergistic effects with *trans*-resveratrol in a self-nanoemulsifying drug delivery system. Biol. Pharm. Bull. 38, 1658–1662.

Luna, A.C.D., Saraiva, G.K.V., Ribeiro, O.M., Chierice, G.O., Neto, S.C., Cuccovia, I.M., et al., 2016. Potential antitumor activity of novel DODAC/PHO-S liposomes. Int. J. Nanomed. 11, 1577–1591.

Luo, Y.C., Teng, Z., Li, Y., Wang, Q., 2015. Solid lipid nanoparticles for oral drug delivery: chitosan coating improves stability, controlled delivery, mucoadhesion and cellular uptake. Carbohydr. Polym. 122, 221–229.

Lv, Q.Y., Shen, C.Y., Li, X.Y., Shen, B.D., Yu, C., Xu, P.H., et al., 2015. Mucoadhesive buccal films containing phospholipid-bile salts-mixed micelles as an effective carrier for cucurbitacin B delivery. Drug Deliv. 22, 351–358.

Ma, P., Dong, X.W., Swadley, C.L., Gupte, A., Leggas, M., Ledebur, H.C., et al., 2009. Development of idarubicin and doxorubicin solid lipid nanoparticles to overcome Pgp-mediated multiple drug resistance in leukemia. J. Biomed. Nanotechnol. 5, 151–161.

Ma, Y., Liu, D., Wang, D., Wang, Y.J., Fu, Q., Fallon, J.K., et al., 2014. Combinational delivery of hydrophobic and hydrophilic anticancer drugs in single nanoemulsions to treat MDR in cancer. Mol. Pharm. 11, 2623–2630.

Ma, W.C., Xu, A.R., Ying, J.Y., Li, B., Jin, Y., 2015. Biodegradable core–shell copolymer-phospholipid nanoparticles for combination chemotherapy: an in vitro study. J. Biomed. Nanotechnol. 11, 1193–1200.

Majeed, H., Antoniou, J., Fang, Z., 2014. Apoptotic effects of eugenol-loaded nanoemulsions in human colon and liver cancer cell lines. Asian Pac. J. Cancer Prev. 15, 9159–9164.

Maksimenko, A., Dosio, F., Mougin, J., Ferrero, A., Wack, S., Reddy, L.H., et al., 2014. A unique squalenoylated and nonpegylated doxorubicin nanomedicine with systemic long-circulating properties and anticancer activity. Proc. Natl. Acad. Sci. U.S.A. 111, E217–E226.

Mandal, B., Bhattacharjee, H., Mittal, N., Sah, H., Balabathula, P., Thoma, L.A., et al., 2013. Core–shell-type lipid–polymer hybrid nanoparticles as a drug delivery platform. Nanomedicine 9, 474–491.

Mandal, B., Mittal, N.K., Balabathula, P., Thoma, L.A., Wood, G.C., 2016. Development and in vitro evaluation of core–shell type lipid–polymer hybrid nanoparticles for the delivery of erlotinib in non-small cell lung cancer. Eur. J. Pharm. Sci. 81, 162–171.

Marchiori, M.C.L., Ourique, A.F., da Silva, C.D., Raffin, R.P., Pohlmann, A.R., Guterres, S.S., et al., 2012. Spray-dried powders containing tretinoin-loaded engineered lipid-core nanocapsules: development and photostability study. J. Nanosci. Nanotechnol. 12, 2059–2067.

Martins, S., Sarmento, B., Ferreira, D.C., Souto, E.B., 2007. Lipid-based colloidal carriers for peptide and protein delivery-liposomes versus lipid nanoparticles. Int. J. Nanomed. 2, 595–607.

Mei, L., Fu, L., Shi, K.R., Zhang, Q.Y., Liu, Y.Y., Tang, J., et al., 2014. Increased tumor targeted delivery using a multistage liposome system functionalized with RGD, TAT and cleavable PEG. Int. J. Pharm. 468, 26–38.

Meng, S.Y., Su, B., Li, W., Ding, Y.M., Tang, L., Zhou, W., et al., 2011. Integrin-targeted paclitaxel nanoliposomes for tumor therapy. Med. Oncol. 28, 1180–1187.

Merisko-Liversidge, E., Liversidge, G.G., Cooper, E.R., 2003. Nanosizing: a formulation approach for poorly-water-soluble compounds. Eur. J. Pharm. Sci 18, 113–120.

Mohan, P., Rapoport, N., 2010. Doxorubicin as a molecular nanotheranostic agent: effect of doxorubicin encapsulation in micelles or nanoemulsions on the ultrasound-mediated intracellular delivery and nuclear trafficking. Mol. Pharm. 7, 1959–1973.

Mohan, A., Narayanan, S., Sethuraman, S., Krishnan, U.M., 2014. Novel resveratrol and 5-fluorouracil coencapsulated in PEGylated nanoliposomes improve chemotherapeutic efficacy of combination against head and neck squamous cell carcinoma. BioMed Res. Int. 2014, 424239.

Mosallaei, N., Jaafari, M.R., Hanafi-Bojd, M.Y., Golmohammadzadeh, S., Malaekeh-Nikouei, B., 2013. Docetaxel-loaded solid lipid nanoparticles: preparation, characterization, in vitro, and in vivo evaluations. J. Pharm. Sci. 102, 1994–2004.

Mosallaei, N., Mahmoudi, A., Ghandehari, H., Yellepeddi, V.K., Jaafari, M.R., Malaekeh-Nikouei, B., 2016. Solid lipid nanoparticles containing 7-ethyl-10-hydroxycamptothecin (SN38): preparation, characterization, in vitro, and in vivo evaluations. Eur. J. Pharm. Biopharm. 104, 42–50.

Moura, J.A., Valduga, C.J., Tavares, E.R., Kretzer, I.F., Maria, D.A., Maranhao, R.C., 2011. Novel formulation of a methotrexate derivative with a lipid nanoemulsion. Int. J. Nanomed. 6, 2285–2295.

Mrozek-Wilczkiewicz, A., Kalinowski, D., Musioł, R., Finster, J., Szurko, A., Serafin, K., et al., 2010. Investigating anti-proliferative activity of styrylazanaphthalenes and azanaphthalenediones. Bioorg. Med. Chem. 18, 2664–2671.

Muehlmann, L.A., Rodrigues, M.C., Longo, J.P.F., Garcia, M.P., Py-Daniel, K.R., Veloso, A.B., et al., 2015. Aluminium-phthalocyanine chloride nanoemulsions for anticancer photodynamic therapy: development and in vitro activity against monolayers and spheroids of human mammary adenocarcinoma MCF-7 cells. J. Nanobiotechnol. 13, 36.

Mukherjee, S., Ray, S., Thakur, R.S., 2009. Solid lipid nanoparticles: a modern formulation approach in drug delivery system. Indian J. Pharm. Sci. 71, 349–358.

Musioł, R., Jampílek, J., Králová, K., Richardson, D.R., Kalinowski, D., Podeszwa, B., et al., 2007. Investigating biological activity spectrum for novel quinoline analogues. Bioorg. Med. Chem. 15, 1280–1288.

Mussi, S.V., Sawant, R., Perche, F., Oliveira, M.C., Azevedo, R.B., Ferreira, L.A.M., et al., 2014. Novel nanostructured lipid carrier co-loaded with doxorubicin and docosahexaenoic acid demonstrates enhanced in vitro activity and overcomes drug resistance in MCF-7/Adr cells. Pharm. Res. 31, 1882–1892.

Mussi, S.V., Parekh, G., Pattekari, P., Levchenko, T., Lvov, Y., Ferreira, L.A.M., et al., 2015. Improved pharmacokinetics and enhanced tumor growth inhibition using a nanostructured lipid carrier loaded with doxorubicin and modified with a layer-by-layer polyelectrolyte coating. Int. J. Pharm. 495, 186–193.

Mutlu, E.C., Bostan, M.S., Bahadori, F., Kocyigit, A., Oner, E.T., Eroglu, M.S., 2016. Lecithin-acrylamido-2-methylpropane sulfonate based crosslinked phospholipid nanoparticles as drug carrier. J. Appl. Polym. Sci. 133, 44105.

Muzammil, A.S., Naidu, V.G.M., Harishankar, N., Kishan, V., 2016. Albumin anchored docetaxel lipid nanoemulsion for improved targeting efficiency—preparation, characterization, cytotoxic, antitumor and in vivo imaging studies. Drug Deliv. 23, 1355–1363.

Naguib, Y.W., Rodriguez, B.L., Li, X.R., Hursting, S.D., Williams, R.O., Cui, Z.R., 2014. Solid lipid nanoparticle formulations of docetaxel prepared with high melting point triglycerides: in vitro and in vivo evaluation. Mol. Pharm. 11, 1239–1249.

Najlah, M., Kadam, A., Wan, K.W., Ahmed, W., Taylor, K.M.G., Elhissi, A.M.A., 2016. Novel paclitaxel formulations solubilized by parenteral nutrition nanoemulsions for application against glioma cell lines. Int. J. Pharm. 506, 102–109.

Natesan, S., Sugumaran, A., Ponnusamy, C., Jeevanesan, V., Girija, G., Palanichamy, R., 2014. Development and evaluation of magnetic microemulsion: tool for targeted delivery of camptothecin to BALB/c mice-bearing breast cancer. J. Drug Target. 22, 913–926.

Nathaniel, C., Elaine-Lee, Y.L., Yee, B.C., How, C.W., Yim, H.S., Rasadee, A., et al., 2016. Zerumbone-loaded nanostructured lipid carrier induces apoptosis in human colorectal adenocarcinoma (Caco-2) cell line. Nanosci. Nanotechnol. Lett. 8, 294–302.

Negi, L.M., Talegaonkar, S., Jaggi, M., Verma, A.K., Verma, R., Dobhal, S., et al., 2014a. Surface engineered nanostructured lipid carriers for targeting MDR tumor: Part I. Synthesis, characterization and in vitro investigation. Colloids Surf. B: Biointerfaces 123, 600–609.

Negi, L.M., Talegaonkar, S., Jaggi, M., Verma, A.K., Verma, R., Dobhal, S., et al., 2014b. Surface engineered nanostructured lipid carriers for targeting MDR tumor: Part II. In vivo biodistribution, pharmacodynamic and hematological toxicity studies. Colloids Surf. B: Biointerfaces 123, 610–615.

Negi, L.M., Jaggi, M., Joshi, V., Ronodip, K., Talegaonkar, S., 2015. Hyaluronic acid decorated lipid nanocarrier for MDR modulation and CD-44 targeting in colon adenocarcinoma. Int. J. Biol. Macromol. 72, 569–574.

Neupane, Y.R., Srivastava, M., Ahmad, N., Kumar, N., Bhatnagar, A., Kohli, K., 2014. Lipid based nanocarrier system for the potential oral delivery of decitabine: formulation design, characterization, ex vivo, and in vivo assessment. Int. J. Pharm. 477, 601–612.

Neves, A.R., Queiroz, J.F., Lima, S.A.C., Figueiredo, F., Fernandes, R., Reis, S., 2016. Cellular uptake and transcytosis of lipid-based nanoparticles across the intestinal barrier: relevance for oral drug delivery. J. Colloid Interface Sci. 463, 258–265.

Oktábec, Z., Jampílek, J., 2013. Brief history of chemotherapy. Chem. Listy 107, 150–158.

Oliveira, M.S., Aryasomayajula, B., Pattni, B., Mussi, S.V., Ferreira, L.A.M., Torchilin, V. P., 2016a. Solid lipid nanoparticles co-loaded with doxorubicin and α-tocopherol succinate are effective against drug-resistant cancer cells in monolayer and 3-D spheroid cancer cell models. Int. J. Pharm. 512, 292–300.

Oliveira, M.S., Mussi, S.V., Gomes, D.A., Yoshida, M.I., Frezard, F., Carregal, V.M., et al., 2016b. α-Tocopherol succinate improves encapsulation and anticancer activity of doxorubicin loaded in solid lipid nanoparticles. Colloids Surf. B: Biointerfaces 140, 246–253.

Opletalová, V., Kalinowski, D., Vejsová, M., Kuneš, J., Pour, M., Jampílek, J., et al., 2008. Identification and characterization of thiosemicarbazones with anti-fungal and antitumor effects: cellular iron-chelation mediating cytotoxic activity. Chem. Res. Toxicol. 21, 1878–1889.

Oyarzun-Ampuero, F.A., Rivera-Rodriguez, G.R., Alonso, M.J., Torres, D., 2013. Hyaluronan nanocapsules as a new vehicle for intracellular drug delivery. Eur. J. Pharm. Sci. 49, 483–490.

Ozpolat, B., Sood, A.K., Lopez-Berestein, G., 2014. Liposomal siRNA nanocarriers for cancer therapy. Adv. Drug Deliv. Rev. 66, 110–116.

Palange, A.L., Di Mascolo, D., Carallo, C., Gnasso, A., Decuzzi, P., 2014. Lipid–polymer nanoparticles encapsulating curcumin for modulating the vascular deposition of breast cancer cells. Nanomedicine 10, 991–1002.

Paliwal, S.R., Paliwal, R., Agrawal, G.P., Vyas, S.P., 2016. Hyaluronic acid modified pH-sensitive liposomes for targeted intracellular delivery of doxorubicin. J. Liposome Res. 26, 276–287.

Pandey, V., Gajbhiye, K.R., Soni, V., 2015. Lactoferrin-appended solid lipid nanoparticles of paclitaxel for effective management of bronchogenic carcinoma. Drug Deliv. 22, 199–205.

Patro, N.M., Devi, K., Pai, R.S., Suresh, S., 2013. Evaluation of bioavailability, efficacy, and safety profile of doxorubicin-loaded solid lipid nanoparticles. J. Nanopart. Res. 15, 2124.

Pawar, V.K., Gupta, S., Singh, Y., Meher, J.G., Sharma, K., Singh, P., et al., 2015. Pluronic F-127 stabilised docetaxel nanocrystals improve apoptosis by mitochondrial depolarization in breast cancer cells: pharmacokinetics and toxicity assessment. J. Biomed. Nanotechnol. 11, 1747–1763.

Peira, E., Chirio, D., Battaglia, L., Barge, A., Chegaev, K., Gigliotti, C.L., et al., 2016. Solid lipid nanoparticles carrying lipophilic derivatives of doxorubicin: preparation, characterization, and in vitro cytotoxicity studies. J. Microencapsul. 33, 381–390.

Peltier, S., Oger, J.M., Lagarce, F., Couet, W., Benoit, J.P., 2006. Enhanced oral paclitaxel bioavailability after administration of paclitaxel-loaded lipid nanocapsules. Pharm. Res. 23, 1243–1250.

Pentak, D., Kozik, V., Bąk, A., Dybał, P., Sochanik, A., Jampílek, J., 2016. Methotrexate and cytarabine—loaded nanocarriers for multidrug cancer therapy. Spectroscopic study. Molecules 21, 1689.

Periasamy, V.S., Athinarayanan, J., Alshatwi, A.A., 2016. Anticancer activity of an ultrasonic nanoemulsion formulation of *Nigella sativa* L. essential oil on human breast cancer cells. Ultrason. Sonochem. 31, 449–455.

Petersen, A.L., Hansen, A.E., Gabizon, A., Andresen, T.L., 2012. Liposome imaging agents in personalized medicine. Adv. Drug Deliv. Rev. 64, 1417–1435.

Pizzimenti, S., Daga, M., Ciamporcero, E., Toaldo, C., Pettazzoni, P., Osella-Abate, S., et al., 2015. Improved anti-tumoral therapeutic efficacy of 4-hydroxynonenal incorporated in novel lipid nanocapsules in 2D and 3D models. J. Biomed. Nanotechnol. 11, 2169–2185.

Podeszwa, B., Niedbala, H., Polański, J., Musioł, R., Tabak, D., Finster, J., et al., 2007. Investigating the antiproliferative activity of quinoline-5,8-dione analogues on tumour cell lines. Bioorg. Med. Chem. Lett. 17, 6138–6141.

Poudel, B.K., Pradhan, R., Gupa, B., Choi, J.Y., Yong, C.S., Kim, J.O., 2014. Preparation and characterization of alginate gel core–lipid nanocapsules for co-delivery of hydrophilic and hydrophobic anti-cancer drugs. J. Pharm. Investig. 44, 485–491.

Poy, D., Akbarzadeh, A., Shahmabadi, H.E., Ebrahimifar, M., Farhangi, A., Zarabi, M.F., et al., 2016. Preparation, characterization, and cytotoxic effects of liposomal nanoparticles containing cisplatin: an in vitro study. Chem. Biol. Drug. Des. 88, 568–573.

Qin, G.T., Li, Z., Xia, R., Li, F., O'Neill, B.E., Goodwin, J.T., et al., 2011. Partially polymerized liposomes: stable against leakage yet capable of instantaneous release for remote controlled drug delivery. Nanotechnology 22, 55605.

Qu, D., Ma, Y.H., Sun, W.J., Chen, Y., Zhou, J., Liu, C.Y., et al., 2015a. Microemulsion-based synergistic dual-drug codelivery system for enhanced apoptosis of tumor cells. Int. J. Nanomed. 10, 1173–1187.

Qu, C.Y., Zhou, M., Chen, Y.W., Chen, M.M., Shen, F., Xu, L.M., 2015b. Engineering of lipid prodrug-based, hyaluronic acid-decorated nanostructured lipid carriers platform for 5-fluorouracil and cisplatin combination gastric cancer therapy. Int. J. Nanomed. 10, 3911–3920.

Qu, D., Sun, W.J., Liu, M.J., Liu, Y.P., Zhou, J., Chen, Y., 2016. Bitargeted microemulsions based on coix seed ingredients for enhanced hepatic tumor delivery and synergistic therapy. Int. J. Pharm. 503, 90–101.

Raemdonck, K., Braeckmans, K., Demeester, J., De Smedt, S.C., 2014. Merging the best of both worlds: hybrid lipid-enveloped matrix nanocomposites in drug delivery. Chem. Soc. Rev. 43, 444–472.

Ragelle, H., Crauste-Manciet, S., Seguin, J., Brossard, D., Scherman, D., Arnaud, P., et al., 2012. Nanoemulsion formulation of fisetin improves bioavailability and antitumour activity in mice. Int. J. Pharm. 427, 452–459.

Rajinikanth, P.S., Chellian, J., 2016. Development and evaluation of nanostructured lipid carrier-based hydrogel for topical delivery of 5-fluorouracil. Int. J. Nanomed. 11, 5067–5077.

Rao, G.C., Kumar, M.S., Mathivavav, N., Rao, M.E., 2004. Nanosuspensions as the most promising approach in nanoparticulate drug delivery systems. Pharmazie 59, 5–9.

Rapoport, N., Nam, K.H., Gupta, R., Gao, Z.G., Mohan, P., Payne, A., et al., 2011. Ultrasound-mediated tumor imaging and nanotherapy using drug loaded, block copolymer stabilized perfluorocarbon nanoemulsions. J. Controlled Release 153, 4–15.

Rapoport, N., 2016. Drug-loaded perfluorocarbon nanodroplets for ultrasound-mediated drug delivery. In: Escoffre, J.M., Bouakaz, A. (Eds.), Therapeutic Ultrasound. Springer, Heidelberg, pp. 221–241.

Ravazzolo, E., Salmaso, S., Mastrotto, F., Bersani, S., Gallon, E., Caliceti, P., 2013. pH-responsive lipid core micelles for tumour targeting. Eur. J. Pharm. Biopharm. 83, 346–357.

Ren, S., Dai, Y., Li, C., Qiu, Z., Wang, X., Tian, F., et al., 2016. Pharmacokinetics and pharmacodynamics evaluation of a thermosensitive chitosan based hydrogel containing liposomal doxorubicin. Eur. J. Pharm. Sci. 92, 137–145.

Ringhieri, P., Pannunzio, A., Boccarelli, A., Morelli, G., Coluccia, M., Tesauro, D., 2016. Effect of cisplatin containing liposomes formulated by unsaturated chain-containing lipids on gynecological tumor cells. J. Liposome Res. 26, 307–312.

Rivera-Rodriguez, G.R., Alonso, M.J., Torres, D., 2013. Poly-L-asparagine nanocapsules as anticancer drug delivery vehicles. Eur. J. Pharm. Biopharm. 85, 481–487.

Rizwanullah, M., Ahmad, J., Amin, S., 2016. Nanostructured lipid carriers: a novel platform for chemotherapeutics. Curr. Drug Deliv. 13, 4–26.

Rizzitelli, S., Giustetto, P., Faletto, D., Castelli, D.D., Aime, S., Terreno, E., 2016. The release of doxorubicin from liposomes monitored by MRI and triggered by a combination of US stimuli led to a complete tumor regression in a breast cancer mouse model. J. Controlled Release 230, 57–63.

Rostami, E., Kashanian, S., Azandaryani, A.H., Faramarzi, H., Dolatabadi, J.E.N., Omidfar, K., 2014. Drug targeting using solid lipid nanoparticles. Chem. Phys. Lipids 181, 56–61.

Saadat, E., Amini, M., Khoshayand, M.R., Dinarvand, R., Dorkoosh, F.A., 2014. Synthesis and optimization of a novel polymeric micelle based on hyaluronic acid and phospholipids for delivery of paclitaxel, in vitro and in-vivo evaluation. Int. J. Pharm. 475, 163–173.

Sahu, P.K., Mishra, D.K., Jain, N., Rajoriya, V., Jain, A.K., 2015. Mannosylated solid lipid nanoparticles for lung-targeted delivery of paclitaxel. Drug Dev. Ind. Pharm. 41, 640–649.

Saini, J., Bansal, V., Chandra, A., Madan, J., Jain, U.K., Chandra, R., et al., 2014. Bleomycin sulphate loaded nanostructured lipid particles augment oral bioavailability, cytotoxicity and apoptosis in cervical cancer cells. Colloids Surf. B: Biointerfaces 118, 101–110.

Saliou, B., Thomas, O., Lautram, N., Clavreul, A., Hureaux, J., Urban, T., et al., 2013. Development and in vitro evaluation of a novel lipid nanocapsule formulation of etoposide. Eur. J. Pharm. Sci. 50, 172–180.

Sanchez-Moreno, P., Buzon, P., Boulaiz, H., Peula-Garcia, J.M., Ortega-Vinuesa, J.L., Luque, I., et al., 2015. Balancing the effect of corona on therapeutic efficacy and macrophage uptake of lipid nanocapsules. Biomaterials 61, 266–278.

Sandhu, P.S., Beg, S., Mehta, F., Singh, B., Trivedi, P., 2015. Novel dietary lipid-based self-nanoemulsifying drug delivery systems of paclitaxel with p-gp inhibitor: implications on cytotoxicity and biopharmaceutical performance. Exp. Opin. Drug. Deliv. 12, 1809–1822.

Sardan, M., Kilinc, M., Genc, R., Tekinay, A.B., Guler, M.O., 2013. Cell penetrating peptide amphiphile integrated liposomal systems for enhanced delivery of anticancer drugs to tumor cells. Faraday Discuss 166, 269–283.

Sasso, M.S., Lollo, G., Pitorre, M., Solito, S., Pinton, L., Valpione, S., et al., 2016. Low dose gemcitabine-loaded lipid nanocapsules target monocytic myeloid-derived suppressor cells and potentiate cancer immunotherapy. Biomaterials 96, 47–62.

Sawant, R.R., Torchilin, V.P., 2010. Multifunctionality of lipid-core micelles for drug delivery and tumour targeting. Mol. Membr. Biol. 27, 232–246.

Schöler, N., Hahn, H., Müller, R.H., Liesenfeld, O., 2002. Effect of lipid matrix and size of solid lipid nanoparticles (SLN) on the viability and cytokine production of macrophages. Int. J. Pharm. 231, 167–176.

Schultze, E., Ourique, A., Yurgel, V.C., Begnini, K.R., Thurow, H., de Leon, P.M.M., et al., 2014. Encapsulation in lipid-core nanocapsules overcomes lung cancer cell resistance to tretinoin. Eur. J. Pharm. Biopharm. 87, 55–63.

Selvamuthukumar, S., Velmurugan, R., 2012. Nanostructured lipid carriers: a potential drug carrier for cancer chemotherapy. Lipids Health Dis. 11, 159.

Shah, R.M., Rajasekaran, D., Ludford-Menting, M., Eldridge, D.S., Palombo, E.A., Harding, I.H., 2016. Transport of stearic acid-based solid lipid nanoparticles (SLNs) into human epithelial cells. Colloids Surf. B: Biointerfaces 140, 204–212.

Shakeel, F., Haq, N., Al-Dhfyan, A., Alanazi, F.K., Alsarra, I.A., 2014. Double w/o/w nanoemulsion of 5-fluorouracil for self-nanoemulsifying drug delivery system. J. Mol. Liq. 200, 183–190.

Shakeel, F., Alanazi, F.K., Raish, M., Haq, N., Radwan, A.A., Alsarra, I.A., 2015a. Pharmacokinetic and in vitro cytotoxic evaluation of cholesterol-rich nanoemulsion of cholesteryl-succinyl-5-fluorouracil. J. Mol. Liq. 211, 164–168.

Shakeel, F., Haq, N., Al-Dhfyan, A., Alanazi, F.K., Alsarra, I.A., 2015b. Chemoprevention of skin cancer using low HLB surfactant nanoemulsion of 5-fluorouracil: a preliminary study. Drug Deliv. 22, 573–580.

Shakeel, F., Iqbal, M., Ezzeldin, E., 2016. Bioavailability enhancement and pharmacokinetic profile of an anticancer drug ibrutinib by self-nanoemulsifying drug delivery system. J. Pharm. Pharmacol. 68, 772–780.

Shao, Z.Y., Shao, J.Y., Tan, B.X., Guan, S.H., Liu, Z.L., Zhao, Z.J., et al., 2015. Targeted lung cancer therapy: preparation and optimization of transferrin-decorated nanostructured lipid carriers as novel nanomedicine for co-delivery of anticancer drugs and DNA. Int. J. Nanomed. 10, 1223–1233.

Sharma, P., Dube, B., Sawant, K., 2011. Development and evaluation of nanostructured lipid carriers of cytarabine for treatment of meningeal leukemia. J. Nanosci. Nanotechnol. 11, 6676–6682.

Sharma, S., Verma, A., Teja, B.V., Shukla, P., Mishra, P.R., 2015. Development of stabilized paclitaxel nanocrystals: in-vitro and in-vivo efficacy studies. Eur. J. Pharm. Sci. 69, 51–60.

Shen, Q., Qiu, L., 2016. Reversal of P-glycoprotein-mediated multidrug resistance by doxorubicin and quinine co-loaded liposomes in tumor cells. J. Liposome Res. 27, 293–301.

Shen, R., Kim, J.J., Yao, M., Elbayoumi, T.A., 2016. Development and evaluation of vitamin E d-α-tocopheryl polyethylene glycol 1000 succinate-mixed polymeric phospholipid micelles of berberine as an anticancer nanopharmaceutical. Int. J. Nanomed. 11, 1687–1700.

Shete, H., Patravale, V., 2013. Long chain lipid based tamoxifen NLC. Part I: Preformulation studies, formulation development and physicochemical characterization. Int. J. Pharm. 454, 573–583.

Shete, H., Chatterjee, S., De, A., Patravale, V., 2013. Long chain lipid based tamoxifen NLC. Part II: Pharmacokinetic, biodistribution and in vitro anticancer efficacy studies. Int. J. Pharm. 454, 584–592.

Shi, Y.B., Li, J.C., Ren, Y., Wang, H.Q., Cong, Z.T., Wu, G.T., et al., 2015a. Pharmacokinetics and tissue distribution of emodin loaded nanoemulsion in rats. J. Drug Deliv. Sci. Technol. 30, 242–249.

Shi, K.R., Zhou, J., Zhang, Q.Y., Gao, H.L., Liu, Y.Y., Zong, T.L., et al., 2015b. Arginine-glycine-aspartic acid-modified lipid–polymer hybrid nanoparticles for docetaxel delivery in glioblastoma multiforme. J. Biomed. Nanotechnol. 11, 382–391.

Shim, G., Lee, S., Choi, J., Lee, S., Kim, C.W., Oh, Y.K., 2014. Liposomal co-delivery of omacetaxine mepesuccinate and doxorubicin for synergistic potentiation of antitumor activity. Pharm. Res. 31, 2178–2185.

Shinde, R.L., Jindal, A.B., Devarajan, P.V., 2011. Microemulsions and nanoemulsions for targeted drug delivery to the brain. Curr. Nanosci. 7, 119–133.

Silva, E.L., Lima, F.A., Carneiro, G., Ramos, J.P., Gomes, D.A., de Souza-Fagundes, E.M., et al., 2016. Improved in vitro antileukemic activity of all-*trans* retinoic acid loaded in cholesteryl butyrate solid lipid nanoparticles. J. Nanosci. Nanotechnol. 16, 1291–1300.

Singh, V., Bushettii, S.S., Raju, S.A., Ahmad, R., Singh, M., Bisht, A., 2011. Microemulsions as promising delivery systems: a review. Ind. J. Pharm. Edu. Res. 45, 392–401.

Singh, I., Swami, R., Pooja, D., Jeengar, M.K., Khan, W., Sistla, R., 2016. Lactoferrin bioconjugated solid lipid nanoparticles: a new drug delivery system for potential brain targeting. J. Drug Target. 24, 212–223.

Slomkowski, S., Alemán, J.V., Gilbert, R.G., Hess, M., Horie, K., Jones, R.G., et al., 2011. Terminology of polymers and polymerization processes in dispersed systems (IUPAC Recommendations 2011). Pure Appl. Chem. 83, 2229–2259.

Song, Z.M., Zhu, W.X., Liu, N., Yang, F.Y., Feng, R.L., 2014. Linolenic acid-modified PEG-PCL micelles for curcumin delivery. Int. J. Pharm. 471, 312–321.

Song, S.X., Mao, G.H., Du, J.S., Zhu, X.G., 2016. Novel RGD containing, temozolomide-loading nanostructured lipid carriers for glioblastoma multiforme chemotherapy. Drug Deliv. 23, 1404–1408.

Soni, N., Soni, N., Pandey, H., Maheshwari, R., Kesharwani, P., Tekade, R.K., 2016. Augmented delivery of gemcitabine in lung cancer cells exploring mannose anchored solid lipid nanoparticles. J. Colloid Interface Sci. 481, 107–116.

Souto, E.B., Nayak, A.P., Murthy, R.S.R., 2011. Lipid nanoemulsions for anti-cancer drug therapy. Pharmazie 66, 473–478.

Sriraman, S.K., Salzano, G., Sarisozen, C., Torchilin, V., 2016. Anti-cancer activity of doxorubicin-loaded liposomes co-modified with transferrin and folic acid. Eur. J. Pharm. Biopharm. 105, 40–49.

Steinhilber, D., Schubert-Zsilavecz, M., Roth, H.J., 2010. Medizinische Chemie: Targets, Arzneistoffe, Chemische Biologie. Deutscher Apotheker Verlag, Stutgart.

Stolzoff, M., Ekladious, I., Colby, A.H., Colson, Y.L., Porter, T.M., Grinstaff, M.W., 2015. Synthesis and characterization of hybrid polymer/lipid expansile nanoparticles: imparting surface functionality for targeting and stability. Biomacromolecules 16, 1958–1966.

Stras, S., Holleran, T., Howe, A., Sofou, S., 2016. Interstitial release of cisplatin from triggerable liposomes enhances efficacy against triple negative breast cancer solid tumor analogues. Mol. Pharm. 13, 3224–3233.

Su, X.W., Wang, Z.H., Li, L.L., Zheng, M.B., Zheng, C.F., Gong, P., et al., 2013. Lipid–polymer nanoparticles encapsulating doxorubicin and 2′-deoxy-5-azacytidine enhance the sensitivity of cancer cells to chemical therapeutics. Mol. Pharm. 10, 1901–1909.

Sun, M., Nie, S.F., Pan, X., Zhang, R.W., Fan, Z.Y., Wang, S., 2014. Quercetin-nanostructured lipid carriers: characteristics and anti-breast cancer activities in vitro. Colloids Surf. B: Biointerfaces 113, 15–24.

Sun, J., Song, Y.Z., Lu, M., Lin, X.Y., Liu, Y., Zhou, S.L., et al., 2016a. Evaluation of the antitumor effect of dexamethasone palmitate and doxorubicin co-loaded liposomes modified with a sialic acid-octadecylamine conjugate. Eur. J. Pharm. Sci. 93, 177–183.

Sun, J.H., Zhang, S.Z., Jiang, S.J., Bai, W.X., Liu, F., Yuan, H., et al., 2016b. Gadolinium-loaded solid lipid nanoparticles as a tumor-absorbable contrast agent for early diagnosis of colorectal tumors using magnetic resonance colonography. J. Biomed. Nanotechnol. 12, 1709–1723.

Sun, B.J., Luo, C., Li, L., Wang, M.L., Du, Y.G., Di, D.H., et al., 2016c. Core-matched encapsulation of an oleate prodrug into nanostructured lipid carriers with high drug loading capability to facilitate the oral delivery of docetaxel. Colloids Surf. B: Biointerfaces 143, 47–55.

Swami, R., Singh, I., Jeengar, M.K., Naidu, V.G.M., Khan, W., Sistla, R., 2015. Adenosine conjugated lipidic nanoparticles for enhanced tumor targeting. Int. J. Pharm. 486, 287–296.

Tagne, J.B., Kakurnanu, S., Nicolosi, R.J., 2008. Nanoemulsion preparations of the anticancer drug dacarbazine significantly increase its efficacy in a xenograft mouse melanoma model. Mol. Pharm. 5, 1055–1063.

Talekar, M., Ganta, S., Singh, A., Amiji, M., Kendall, J., Denny, W.A., et al., 2012. Phosphatidylinositol 3-kinase inhibitor (PIK75) containing surface functionalized nanoemulsion for enhanced drug delivery, cytotoxicity and pro-apoptotic activity in ovarian cancer cells. Pharm. Res. 29, 2874–2886.

Tang, X.J., Han, M., Yang, B., Shen, Y.Q., He, Z.G., Xu, D.H., et al., 2014. Nanocarrier improves the bioavailability, stability and antitumor activity of camptothecin. Int. J. Pharm. 477, 536–545.

Tang, W.L., Chen, W.C., Roy, A., Undzys, E., Li, S.D., 2016. A simple and improved active loading method to efficiently encapsulate staurosporine into lipid-based nanoparticles for enhanced therapy of multidrug resistant cancer. Pharm. Res. 33, 1104–1114.

Taratula, O., Kuzmov, A., Shah, M., Garbuzenko, O.B., Minko, T., 2013. Nanostructured lipid carriers as multifunctional nanomedicine platform for pulmonary co-delivery of anticancer drugs and siRNA. J. Controlled Release 171, 349–357.

Tian, J., Pang, X.J., Yu, K.F., Liu, L.J., Zhou, J., 2008. Preparation, characterization and in vivo distribution of solid lipid nanoparticles loaded with cisplatin. Pharmazie 63, 593–597.

Torchilin, V.P., 2007. Micellar nanocarriers: pharmaceutical perspectives. Pharm. Res. 24, 1–16.

Tran, T.H., Choi, J.Y., Ramasamy, T., Truong, D.H., Nguyen, C.N., Choi, H.G., et al., 2014a. Hyaluronic acid-coated solid lipid nanoparticles for targeted delivery of vorinostat to CD44 overexpressing cancer cells. Carbohydr. Polym. 114, 407–415.

Tran, T.H., Ramasamy, T., Truong, D.H., Shin, B.S., Choi, H.G., Yong, C.S., et al., 2014b. Development of vorinostat-loaded solid lipid nanoparticles to enhance pharmacokinetics and efficacy against multidrug-resistant cancer cells. Pharm. Res. 31, 1978–1988.

Tran, T.H., Chu, D.T., Truong, D.H., Tak, J.W., Jeong, J.H., Hoang, V.L., et al., 2016. Development of lipid nanoparticles for a histone deacetylases inhibitor as a promising anticancer therapeutic. Drug Deliv. 23, 1335–1343.

Tsai, Y.J., Chen, B.H., 2016. Preparation of catechin extracts and nanoemulsions from green tea leaf waste and their inhibition effect on prostate cancer cell PC-3. Int. J. Nanomed. 11, 1907–1926.

Tuan, T.H., Nguyen, T.D., Poudel, B.K., Nguyen, H.T., Kim, J.O., Yong, C.S., et al., 2015. Development and evaluation of artesunate-loaded chitosan-coated lipid nanocapsule as a potential drug delivery system against breast cancer. AAPS PharmSciTech 16, 1307–1316.

Tupal, A., Sabzichi, M., Ramezani, F., Kouhsoltani, M., Hamishehkar, H., 2016. Dermal delivery of doxorubicin-loaded solid lipid nanoparticles for the treatment of skin cancer. J. Microencapsul. 33, 372–380.

Ukawa, M., Fujiwara, Y., Ando, H., Shimizu, T., Ishida, T., 2016. Hepatic tumor metastases cause enhanced PEGylated liposome uptake by Kupffer cells. Biol. Pharm. Bull. 39, 215–220.

Urbinati, G., Marsaud, V., Renoir, J.M., 2012. Anticancer drugs in liposomal nanodevices: a target delivery for a targeted therapy. Curr. Top. Med. Chem. 12, 1693–1712.

Varshosaz, J., Hassanzadeh, F., Sadeghi, H., Andalib, S., 2012. Synthesis of octadecylamine-retinoic acid conjugate for enhanced cytotoxic effects of 5-FU using LDL targeted nanostructured lipid carriers. Eur. J. Med. Chem. 54, 429–438.

Varshosaz, J., Hassanzadeh, F., Sadeghi-Aliabadi, H., Firozian, F., 2014. Uptake of etoposide in CT-26 cells of colorectal cancer using folate targeted dextran stearate polymeric micelles. Biomed. Res. Int. 2014, 708593.

Varshosaz, J., Taymouri, S., Hassanzadeh, F., Javanmard, S.H., Rostami, M., 2015. Self-assembly micelles with lipid core of cholesterol for docetaxel delivery to B16F10 melanoma and HepG2 cells. J. Liposome Res. 25, 157–165.

Vecchione, R., Quagliariello, V., Calabria, D., Calcagno, V., De Luca, E., Iaffaioli, R.V., et al., 2016. Curcumin bioavailability from oil in water nano-emulsions: in vitro and in vivo study on the dimensional, compositional and interactional dependence. J. Controlled Release 233, 88–100.

Velinova, M.J., Staffhorst, R.W.H.M., Mulder, W.J.M., Dries, A.S., Jansen, B.A.J., de Kruijff, B., et al., 2004. Preparation and stability of lipid-coated nanocapsules of cisplatin: anionic phospholipid specificity. Biochim. Biophys. Acta: Biomembrane 1663, 135–142.

Verma, P., Meher, J.G., Asthana, S., Pawar, V.K., Chaurasia, M., Chourasia, M.K., 2016. Perspectives of nanoemulsion assisted oral delivery of docetaxel for improved chemotherapy of cancer. Drug Deliv. 23, 479–488.

Vijayakumar, M.R., Kumari, L., Patel, K.K., Vuddanda, P.R., Vajanthri, K.Y., Mahto, S. K., et al., 2016. Intravenous administration of *trans*-resveratrol-loaded TPGS-coated solid lipid nanoparticles for prolonged systemic circulation, passive brain targeting and improved in vitro cytotoxicity against C6 glioma cell lines. RSC Adv. 6, 50336–50348.

Vitiello, G., Luchini, A., D'Errico, G., Santamaria, R., Capuozzo, A., Irace, C., et al., 2015. Cationic liposomes as efficient nanocarriers for the drug delivery of an anticancer cholesterol-based ruthenium complex. J. Mater. Chem. B 3, 3011–3023.

Vrignaud, S., Hureaux, J., Wack, S., Benoit, J.P., Saulnier, P., 2012. Design, optimization and in vitro evaluation of reverse micelle-loaded lipid nanocarriers containing erlotinib hydrochloride. Int. J. Pharm. 436, 194–200.

Wan, K., Sun, L.L., Hu, X.Y., Yan, Z.J., Zhang, Y.H., Zhang, X., et al., 2016. Novel nanoemulsion based lipid nanosystems for favorable in vitro and in vivo characteristics of curcumin. Int. J. Pharm. 504, 80–88.

Wang, Y.G., Wang, R.Q., Lu, X.Y., Lu, W.L., Zhang, C.L., Liang, W., 2010. Pegylated phospholipids-based self-assembly with water-soluble drugs. Pharm. Res. 27, 361–370.

Wang, J., Fang, X.C., Liang, W., 2012a. Pegylated phospholipid micelles induce endoplasmic reticulum-dependent apoptosis of cancer cells but not normal cells. ACS Nano 6, 5018–5030.

Wang, Y.Z., Chen, L., Ding, Y.F., Yan, W.L., 2012b. Oxidized phospholipid based pH sensitive micelles for delivery of anthracyclines to resistant leukemia cells in vitro. Int. J. Pharm. 422, 409–417.

Wang, Z.H., Yu, Y., Dai, W.B., Lu, J.K., Cui, J.R., Wu, H.N., et al., 2012c. The use of a tumor metastasis targeting peptide to deliver doxorubicin-containing liposomes to highly metastatic cancer. Biomaterials 33, 8451–8460.

Wang, Y.C., Miao, X.Q., Sun, L., Song, J., Bi, C., Yang, X., et al., 2014a. Effects of nanosuspension formulations on transport, pharmacokinetics, in vivo targeting and efficacy for poorly water-soluble drugs. Curr. Pharm. Design 20, 454–473.

Wang, Y.Z., Fan, W., Dai, X., Katragadda, U., Mckinley, D., Teng, Q., et al., 2014b. Enhanced tumor delivery of gemcitabine via PEG-DSPE/TPGS mixed micelles. Mol. Pharm. 11, 1140–1150.

Wang, H.J., Shrestha, R., Zhang, Y., 2014c. Encapsulation of photosensitizers and upconversion nanocrystals in lipid micelles for photodynamic therapy. Part. Part. Syst. Charact. 31, 228–235.

Wang, J., Zhu, R.R., Sun, X.Y., Zhu, Y.J., Liu, H., Wang, S.L., 2014d. Intracellular uptake of etoposide-loaded solid lipid nanoparticles induces an enhancing inhibitory effect on gastric cancer through mitochondria-mediated apoptosis pathway. Int. J. Nanomed. 9, 3987–3998.

Wang, X.Y., Chen, X.H., Yang, X.C., Gao, W., He, B., Dai, W.B., et al., 2016a. A nanomedicine based combination therapy based on QLPVM peptide functionalized liposomal tamoxifen and doxorubicin against luminal A breast cancer. Nanomedicine 12, 387–397.

Wang, Y.W., Zhang, H.Y., Hao, J., Li, B., Li, M., Wang, X.W., 2016b. Lung cancer combination therapy: co-delivery of paclitaxel and doxorubicin by nanostructured lipid carriers for synergistic effect. Drug Deliv. 23, 1398–1403.

Wang, Z.Y., Wang, J.H., Yang, S.L., Hou, S.Y., 2016c. Construction and in vitro/in vivo evaluation of 17-allylamino-17-demethoxygeldanamycin (17AAG)-loaded PEGylated nanostructured lipid carriers. Drug Dev. Ind. Pharm. 42, 91–98.

Wauthoz, N., Bastiat, G., Moysan, E., Cieslak, A., Kondo, K., Zandecki, M., et al., 2015. Safe lipid nanocapsule-based gel technology to target lymph nodes and combat mediastinal metastases from an orthotopic non-small-cell lung cancer model in SCID-CB17 mice. Nanomedicine 11, 1237–1245.

Wen, C.J., Sung, C.T., Aljuffali, I.A., Huang, Y.J., Fang, J.Y., 2013. Nanocomposite liposomes containing quantum dots and anticancer drugs for bioimaging and therapeutic delivery: a comparison of cationic, PEGylated and deformable liposomes. Nanotechnology 24, 325101.

WHO, 2017. Cancer—Fact sheet (updated February 2017). <http://www.who.int/mediacentre/factsheets/fs297/en/>.

Willerding, L., Limmer, S., Hossann, M., Zengerle, A., Wachholz, K., ten Hagen, T.L.M., et al., 2016. Method of hyperthermia and tumor size influence effectiveness of doxorubicin release from thermosensitive liposomes in experimental tumors. J. Controlled Release 222, 47–55.

Wu, B., Yu, P., Cui, C., Wu, M., Zhang, Y., Liu, L., et al., 2015. Folate-containing reduction-sensitive lipid–polymer hybrid nanoparticles for targeted delivery of doxorubicin. Biomater. Sci. 3, 655–664.

Xiao, L., Xiong, X.Q., Sun, X.H., Zhu, Y.H., Yang, H., Chen, H.B., et al., 2011. Role of cellular uptake in the reversal of multidrug resistance by PEG-*b*-PLA polymeric micelles. Biomaterials 32, 5148–5157.

Xiao, Y.Y., Chen, X., Yang, L., Zhu, X.T., Zou, L., Meng, F.F., et al., 2013. Preparation and oral bioavailability study of curcuminoid-loaded microemulsion. J. Agric. Food. Chem. 61, 3654–3660.

Xing, H., Hwang, K., Lu, Y., 2016. Recent developments of liposomes as nanocarriers for theranostic applications. Theranostics 6, 1336–1352.

Xiong, P., Ding, A.G., Su, Z.Q., Shen, G.F., Chen, Y.Z., Zhang, S., 2015. The efficacy and hyperthermic release of doxorubicin from liposomal doxorubicin hydrochloride in rabbit VX2 tumours. Int. J. Hyperthermia 31, 900–908.

Xu, W., Lee, M.K., 2015. Development and evaluation of lipid nanoparticles for paclitaxel delivery: a comparison between solid lipid nanoparticles and nanostructured lipid carriers. J. Pharm. Investig. 45, 675–680.

Yadollahi, R., Vasilev, K., Spomenka Simovic, S., 2015. Nanosuspension technologies for delivery of poorly soluble drugs. J. Nanomater 2015, 13 pp., ID 216375.

Yan, J.K., Wang, Y.Z., Zhang, X.F., Liu, S.D., Tian, C., Wang, H.W., 2016. Targeted nanomedicine for prostate cancer therapy: docetaxel and curcumin co-encapsulated lipid–polymer hybrid nanoparticles for the enhanced anti-tumor activity in vitro and in vivo. Drug Deliv. 23, 1757–1762.

Yang, X.Y., Li, Y.X., Li, M., Zhang, L., Feng, L.X., Zhang, N., 2013a. Hyaluronic acid-coated nanostructured lipid carriers for targeting paclitaxel to cancer. Cancer Lett. 334, 338–345.

Yang, Z., Luo, X.G., Zhang, X.F., Liu, J., Jiang, Q., 2013b. Targeted delivery of 10-hydroxycamptothecin to human breast cancers by cyclic RGD-modified lipid–polymer hybrid nanoparticles. Biomed. Mater. 8, 025012.

Yang, Y., Yang, Y.F., Xie, X.Y., Cai, X.S., Mei, X.G., 2014. Preparation and characterization of photo-responsive cell-penetrating peptide-mediated nanostructured lipid carrier. J. Drug Target. 22, 891–900.

Yang, C.Y., Liu, H.W., Tsai, Y.C., Tseng, J.Y., Liang, S.C., Chen, C.Y., et al., 2015a. Interleukin-4 receptor-targeted liposomal doxorubicin as a model for enhancing cellular uptake and antitumor efficacy in murine colorectal cancer. Cancer Biol. Ther. 16, 1641–1650.

Yang, Y.F., Xie, X.Y., Yang, Y., Zhang, H., Mei, X.G., 2015b. Photo-responsive and NGR-mediated multifunctional nanostructured lipid carrier for tumor-specific therapy. J. Pharm. Sci. 104, 1328–1339.

Yang, S.M., Zhang, B., Gong, X.W., Wang, T.Q., Liu, Y.J., Zhang, N., 2016a. In vivo biodistribution, biocompatibility, and efficacy of sorafenib-loaded lipid-based nanosuspensions evaluated experimentally in cancer. Int. J. Nanomed. 11, 2329–2343.

Yang, C.Q., Lv, J., Lv, T., Pan, Y.H., Han, Y.Z., Zhao, S., et al., 2016b. Metal ion-assisted drug-loading model for novel delivery system of cisplatin solid lipid nanoparticles with improving loading efficiency and sustained release. J. Microencapsul. 33, 292–298.

Yin, T.J., Dong, L.H., Cui, B., Wang, L., Yin, L.F., Zhou, J.P., et al., 2015. A toxic organic solvent-free technology for the preparation of PEGylated paclitaxel nanosuspension based on human serum albumin for effective cancer therapy. Int. J. Nanomed. 10, 7397–7412.

Yin, T.J., Cai, H., Liu, J.Y., Cui, B., Wang, L., Yin, L.F., et al., 2016. Biological evaluation of PEG modified nanosuspensions based on human serum albumin for tumor targeted delivery of paclitaxel. Eur. J. Pharm. Sci. 83, 79–87.

Ying, X.Y., Cui, D., Yu, L.A., Du, Y.Z., 2011. Solid lipid nanoparticles modified with chitosan oligosaccharides for the controlled release of doxorubicin. Carbohydr. Polym. 84, 1357–1364.

Yu, W.Y., Zhang, N., 2009. Surface modification of nanocarriers for cancer therapy. Curr. Nanosci. 5, 123–134.

Yuan, Q., Han, J., Cong, W.S., Ge, Y., Ma, D.D., Dai, Z.X., et al., 2014. Docetaxel-loaded solid lipid nanoparticles suppress breast cancer cells growth with reduced myelosuppression toxicity. Int. J. Nanomed. 9, 4829–4846.

Yuan, M.Q., Qiu, Y., Zhang, L., Gao, H.L., He, Q., 2016. Targeted delivery of transferrin and TAT co-modified liposomes encapsulating both paclitaxel and doxorubicin for melanoma. Drug Deliv. 23, 1171–1183.

Yurgel, V.C., Oliveira, C.P., Begnini, K.R., Schultze, E., Thurow, H.S., Leon, P.M.M., et al., 2014. Methotrexate diethyl ester-loaded lipid-core nanocapsules in aqueous solution increased antineoplastic effects in resistant breast cancer cell line. Int. J. Nanomed. 9, 1583–1591.

Yutani, R., Komori, Y., Takeuchi, A., Teraoka, R., Kitagawa, S., 2016. Prominent efficiency in skin delivery of resveratrol by novel sucrose oleate microemulsion. J. Pharm. Pharmacol. 68, 46–55.

Zahmatkeshan, M., Gheybi, F., Rezayat, S.M., Jaafari, M.R., 2016. Improved drug delivery and therapeutic efficacy of PEGylated liposomal doxorubicin by targeting anti-HER2 peptide in murine breast tumor model. Eur. J. Pharm. Sci. 86, 125–135.

Zara, G.P., Bargoni, A., Cavalli, R., Fundaro, A., Vighetto, D., Gasco, M.R., 2002. Pharmacokinetics and tissue distribution of idarubicin-loaded solid lipid nanoparticles after duodenal administration to rats. J. Pharm. Sci. 91, 1324–1333.

Zhang, X.G., Miao, J., Dai, Y.Q., Du, Y.Z., Yuan, H., Hu, F.Q., 2008a. Reversal activity of nanostructured lipid carriers loading cytotoxic drug in multi-drug resistant cancer cells. Int. J. Pharm. 361, 239–244.

Zhang, L.F., Chan, J.M., Gu, F.X., Rhee, J.W., Wang, A.Z., Radovic-Moreno, A.F., et al., 2008b. Self-assembled lipid–polymer hybrid nanoparticles: a robust drug delivery platform. ACS Nano 2, 1696–1702.

Zhang, L., Zhang, L.F., 2010. Lipid–polymer hybrid nanoparticles: synthesis, characterization and applications. Nano Life 01, 163.

Zhang, T.C., Chen, J.N., Zhang, Y., Shen, Q., Pan, W.S., 2011. Characterization and evaluation of nanostructured lipid carrier as a vehicle for oral delivery of etoposide. Eur. J. Pharm. Sci. 43, 174–179.

Zhang, L.W., Wen, C.J., Al-Suwayeh, S.A., Yen, T.C., Fang, J.Y., 2012a. Cisplatin and quantum dots encapsulated in liposomes as multifunctional nanocarriers for theranostic use in brain and skin. J. Nanopart. Res. 14, 882.

Zhang, P., Ling, G.X., Pan, X.L., Sun, J., Zhang, T.H., Pu, X.H., et al., 2012b. Novel nanostructured lipid-dextran sulfate hybrid carriers overcome tumor multidrug resistance of mitoxantrone hydrochloride. Nanomedicine 8, 185–193.

Zhang, P., Kopechek, J.A., Porter, T.M., 2013. The impact of vaporized nanoemulsions on ultrasound-mediated ablation. J. Ther. Ultrasound 1, 2.

Zhang, J., Li, M., Liu, Z., Wang, L.L., Liu, Y.J., Zhang, N., 2014. Preclinical studies of *N*-3-*O*-toluyl-fluorouracil-loaded lipid-based nanosuspensions in H-22-bearing mice. Int. J. Nanomed. 9, 2741–2751.

Zhang, H., Hu, H.X., Zhang, H.R., Dai, W.B., Wang, X.L., Wang, X.Q., et al., 2015a. Effects of PEGylated paclitaxel nanocrystals on breast cancer and its lung metastasis. Nanoscale 7, 10790–10800.

Zhang, L., Tian, B., Li, Y., Lei, T., Meng, J., Yang, L., et al., 2015b. A copper-mediated disulfiram-loaded pH-triggered PEG-shedding TAT peptide-modified lipid nanocapsules for use in tumor therapy. ACS Appl. Mater. Interfaces 7, 25147–25161.

Zhang, L.H., Zhu, D.W., Dong, X., Sun, H.F., Song, C.X., Wang, C., et al., 2015c. Folate-modified lipid–polymer hybrid nanoparticles for targeted paclitaxel delivery. Int. J. Nanomed. 10, 2101–2114.

Zhang, T.P., Dong, D., Lu, D.Y., Wang, S., Wu, B.J., 2016a. Cremophor EL-based nanoemulsion enhances transcellular permeation of emodin through glucuronidation reduction in UGT1A1-overexpressing MDCKII cells. Int. J. Pharm. 501, 190–198.

Zhang, L.J., Kuang, Y., Liu, J., Liu, Z.L., Huang, S.W., Zhuo, R.X., 2016b. Long circulating anionic liposomes for hepatic targeted delivery of cisplatin. RSC Adv. 6, 76905–76914.

Zhang, G., Liu, F., Jia, E., Jiam, L., Zhang, Y., 2016c. Folate-modified, cisplatin-loaded lipid carriers for cervical cancer chemotherapy. Drug Deliv. 23, 1393–1397.

Zhang, S.M., Lu, C.X., Zhang, X.L., Li, J.S., Jiang, H., 2016d. Targeted delivery of etoposide to cancer cells by folate-modified nanostructured lipid drug delivery system. Drug Deliv. 23, 1838–1845.

Zhang, L.J., Wu, B., Zhou, W., Wang, C.X., Wang, Q., Yu, H., et al., 2016e. Two-component reduction-sensitive lipid–polymer hybrid nanoparticles for triggered drug release and enhanced in vitro and in vivo anti-tumor efficacy. Biomater. Sci. 5, 98–110.

Zhang, J.M., Hu, J., Chan, H.F., Skibba, M., Liang, G., Chen, M.W., 2016f. iRGD decorated lipid–polymer hybrid nanoparticles for targeted co-delivery of doxorubicin and sorafenib to enhance anti-hepatocellular carcinoma efficacy. Nanomedicine 12, 1303–1311.

Zhao, H., Lu, H., Gong, T., Zhang, Z.R., 2013a. Nanoemulsion loaded with lycobetaine-oleic acid ionic complex: physicochemical characteristics, in vitro, in vivo evaluation, and antitumor activity. Int. J. Nanomed. 8, 1959–1973.

Zhao, Y.Q., Wang, L.P., Ma, C., Zhao, K., Liu, Y., Feng, N.P., 2013b. Preparation and characterization of tetrandrine-phospholipid complex loaded lipid nanocapsules as potential oral carriers. Int. J. Nanomed. 8, 4169–4181.

Zhao, Y.B., Lin, D.Y., Wu, F.B., Guo, L., He, G., Ouyang, L., et al., 2014. Discovery and in vivo evaluation of novel RGD-modified lipid–polymer hybrid nanoparticles for targeted drug delivery. Int. J. Mol. Sci. 15, 17565–17576.

Zhao, Y.C., May, J.P., Chen, I.W., Undzys, E., Li, S.D., 2015a. A study of liposomal formulations to improve the delivery of aquated cisplatin to a multidrug resistant tumor. Pharm. Res. 32, 3261–3268.

Zhao, X.J., Chen, Q., Li, Y.S., Tang, H.B., Liu, W., Yang, X.L., 2015b. Doxorubicin and curcumin co-delivery by lipid nanoparticles for enhanced treatment of diethylnitrosamine-induced hepatocellular carcinoma in mice. Eur. J. Pharm. Biopharm. 93, 27–36.

Zhao, S., Zhang, Y.L., Han, Y.Z., Wang, J., Yang, J., 2015c. Preparation and characterization of cisplatin magnetic solid lipid nanoparticles (MSLNs): effects of loading procedures of Fe_3O_4 nanoparticles. Pharm. Res. 32, 482–491.

Zhao, X., Li, F., Li, Y.Y., Wang, H., Ren, H., Chen, J., et al., 2015d. Co-delivery of HIF1α siRNA and gemcitabine via biocompatible lipid–polymer hybrid nanoparticles for effective treatment of pancreatic cancer. Biomaterials 46, 13–25.

Zhao, S.N., Van Minh, L., Li, N., Gararnus, V.M., Handge, U.A., Liu, J.W., et al., 2016. Doxorubicin hydrochloride-oleic acid conjugate loaded nanostructured lipid carriers for tumor specific drug release. Colloids Surf. B: Biointerfaces 145, 95–103.

Zheng, W.W., Zhao, L., Wei, Y.M., Ye, Y., Xiao, S.H., 2010. Preparation and the in vitro evaluation of nanoemulsion system for the transdermal delivery of granisetron hydrochloride. Chem. Pharm. Bull. 1 58, 1015–1019.

Zheng, M.B., Yue, C.X., Ma, Y.F., Gong, P., Zhao, P.F., Zheng, C.F., et al., 2013. Single-step assembly of DOX/ICG loaded lipid–polymer nanoparticles for highly effective chemo-photothermal combination therapy. ACS Nano 7, 2056–2067.

Zheng, M.B., Gong, P., Zheng, C.F., Zhao, P.F., Luo, Z.Y., Ma, Y.F., et al., 2015. Lipid–polymer nanoparticles for folate-receptor targeting delivery of doxorubicin. J. Nanosci. Nanotechnol. 15, 4792–4798.

Zheng, N., Gao, Y., Ji, H., Wu, L., Qi, X., Liu, X., et al., 2016. Vitamin E derivative-based multifunctional nanoemulsions for overcoming multidrug resistance in cancer. J. Drug Target. 24, 663–669.

Zhou, H., Qiu, L.P., Yan, X.X., Li, L., Li, X., Wang, L., et al., 2010. Preparation and characterization of oxaliplatin-loaded nanostructured lipid carriers. Yaoxue Xuebao 45, 1177–1182.

Zhu, L., Kate, P., Torchilin, V.P., 2012. Matrix metalloprotease 2-responsive multifunctional liposomal nanocarrier for enhanced tumor targeting. ACS Nano 6, 3491–3498.

Zhu, X., Huang, S., Huang, H., Zhang, Y., Xie, Y., Hou, L., et al., 2014. In vitro and in vivo anti-cancer effects of targeting and photothermal sensitive solid lipid nanoparticles. J. Drug Target 22, 822–828.

Zhuang, Y.G., Xu, B., Huang, F., Wu, J.J., Chen, S., 2012. Solid lipid nanoparticles of anticancer drugs against MCF-7 cell line and a murine breast cancer model. Pharmazie 67, 925–929.

Zou, H., Li, L., Carcedo, I.G., Xu, Z.P., Monteiro, M., Gu, W.Y., 2016. Synergistic inhibition of colon cancer cell growth with nanoemulsion-loaded paclitaxel and PI3K/mTOR dual inhibitor BEZ235 through apoptosis. Int. J. Nanomed. 11, 1947–1958.

CHAPTER

Solid lipid nanoparticles for skin and drug delivery: methods of preparation and characterization techniques and applications

9

Amaldoss M J Newton[1,2] and Sukhjinder Kaur[3]

[1]*School of Materials Science and Engineering, UNSW Sydney, NSW, Australia* [2]*Lowy Cancer Research Center, UNSW Sydney, NSW, Australia* [3]*Rayat-Bahra Institute of Pharmacy, Rayat-Bahra University, Hoshiarpur, Punjab, India*

9.1 INTRODUCTION

Nanomaterial demonstrates new characteristics which are different from the original base material from which it is derived. Materials exhibit novel characteristics when in the nano-range. Over 4000 years ago, Egyptians, Greeks, and Romans used nanotechnology in hair dye preparations and perfumery with different odors from natural origins (Eglash, 2011). This suggests that nanotechnology has been used in cosmetic preparations for at least 2000 years.

The concept of nanotechnology was developed 40 years ago and extends its web almost in all fields. Nanotechnology is the hot research platform in cosmetics, healthcare, communication, bioinformatics, food processing, biomedical, drug delivery, and nutraceuticals (Duncan, 2011).

The field of cosmetics had developed gradually. Currently, the cosmetics industry is at its peak in the market share and cosmeceutical products are very popular in which the skin-care market is the largest growing sector (Fulekar, 2010). Skin-care products are designed to help protect the skin and prevent it from aging.

The word "cosmeceutical" defines a product that fits in the niche between a drug and cosmetics (Friends of the Earth Report, 2006). Currently, cosmeceutical formulations are expanded from skin to body care. Cosmeceutical preparations related to photo-aging, hyperpigmentation, wrinkles, and hair damage are in frontline and have widespread usage (Müller et al., 2002). The cosmeceutical industry has demonstrated high growth prospects for the coming years (Brandt et al., 2011) The annual growth of the cosmeceutical industry is expected to be 7.7% by

Nanoarchitectonics in Biomedicine. DOI: https://doi.org/10.1016/B978-0-12-816200-2.00015-3

2017 and the global cosmeceutical market will reach $31.84 billion (RNCOS E-Services Pvt. Ltd, 2016). Nanotechnology-based projects are changing the manufacturing trend of the companies from cosmeceuticals' to nanocosmeceuticals'. The reports forecasted $42.4 billion of share by 2018 (GBI Research, 2018).

9.1.1 NANOTECHNOLOGY IN COSMECEUTICALS

Nanotechnology applied to cosmeceuticals results in so-called nanocosmeceuticals and was first aimed at making the incense of perfumes last longer and for "intelligent" sunscreen agents to protect the skin. Nanocosmeceuticals are also used in the preparation of deep-acting antiaging creams to reverse the effects of age as well as moisturizing preparations to prevent the dehydration of skin for a longer period of time (Singh et al., 2013). The formulation of nanocosmeceuticals may be a nanoemulsion which is a transparent or near-transparent liquid with unique tactile and texture properties used for various skin care applications (Padamwar and Pokharkar, 2006), nanocapsules which contains antioxidants, vitamins, anti bacterial and essential oils for skin-care products, and nanopigments which increase the efficiency of sunscreen products, (liposomes which are a vesicular-structured envelope with light- or oxygen-sensitive cosmetic material). These and other categories such as niosomes, nanocrystals, Solid lipid nanoparticles (SLNs), carbon nanotubes, fullerenes, dendrimers, may contribute much in nanocosmeceuticals' in future (Mu and Sprando, 2010). This chapter focusses mainly on the role of SLNs on the skin. The chapter also elaborately discusses the various preparation methods of SLNs, advantages, disadvantages, mechanisms of nanocosmeceutical penetration through skin, characterization, evaluation, recent patents, and future prospects of SLNs as potential nanotechnological products (Mukharjee et al., 2009)

9.1.1.1 Advantages

1. Nanotechnology in cosmetics is used to make the fragrances of perfumes and deodorants last longer.
2. Nanotechnology-based creams such as sunscreen and antiaging creams work more efficiently than conventional creams. Nanotechnology in cosmetics accommodates a multicomponent system into a single preparation (Mehnart and Mader, 2011)
3. Nanotechnology in hair-care preparations used to prevent hair from turning gray and hair loss can be achieved by incorporating vitamins and antioxidants in the preparations.
4. UV protection is made more efficient by combining organic sunscreens such as 2-hydroxy 4-methoxy benzophenones in the nanocosmetic creams.
5. Herbal preparations prepared by nanotechnology are more suitable for hyper-allergic skin conditions and are less irritant and more easily adjustable to the skin (Rupenagunta et al., 2011; Fahr and Liu, 2007).

9.1.1.2 Disadvantages

1. The smaller particles demonstrates the greater reactivity and chemically produced a greater number of reactive oxygen species (ROS).
2. The higher the production of ROS leads to oxidative stress, inflammation, and damage to proteins and membranes of DNA. This causes serious health-related issues (Schwarz et al., 1994).
3. Nanomaterial can be toxic to human tissues and cell cultures through oxidative stress which leads to cell death.
4. Photo-activated titanium dioxide nanoparticles are toxic to skin fibroblasts, nucleic acid, and human colon carcinoma cells.
5. The inhaled nanoparticles and ultrafine nanoparticles of quartz, minerals, dust, coal, silicates, and asbestos lead to inflammation, fibrosis, cytotoxicity, and malignancies.
6. The safety of nanotechnology is unclear because of the ability of these particles to be absorbed into skin cells and through the eyes and nose which may lead to serious complications.
7. There is the possibility of nano-based skin formulations that act on the skin and penetrate it to enter the blood circulation due to their size. The mechanisms and range of particles that act only on the skin and not enter the blood circulation are yet to be studied (Sharma et al., 2011).
8. Although the market potential is tremendous, the safety profile of nanoformulations cannot be overlooked since this technology is in an early stage. Research and development should focus on the disadvantages and safety of nanoparticles.

9.2 SKIN PENETRATION OF NANOPARTICLES

Targeting drug molecules through organs such as skin is a challenging concept in pharmaceutical formulations.

9.2.1 ANATOMY OF SKIN

Skin is the body's largest organ and has a surface area of 1.8 m^2. Skin is an attractive route for drug delivery, but the structure of skin limits the passage of molecules through it because of its property as the body's protective layer. Skin is divided into three major portions, the epidermis, dermis, and hypodermis. The epidermis is also known as nonviable epidermis called stratum corneum (SC). The dermis is also called viable dermis which has sweat glands with complex structures, hair follicles, and blood vessels. The SC, also called the horny layer, is the main skin barrier and is made up of keratinized cells (Tak et al., 2015). The intercellular spaces of the SC are filled with a bilayer which contains nonpolar lipids consisting of ceramides (47%), free fatty acids (9%), esters of free fatty acids, cholesterol (27%), and some amount of sulfates.

9.2.2 MECHANISM OF TOPICAL ABSORPTION OF CONVENTIONAL COSMECEUTICALS

There are some passive routes by which the molecules can cross the SC those are called intracellular, transcellular and appendageal routes. Intracellular routes involve the drug transport through corneocytes and intervening lipids which are more suitable for lipophilic compounds. The intercellular route is the major pathway for most of the drugs in which the molecules can pass through the aqueous regions in the lipid bilayer (Pierfrancesco, 2010). The transport of molecules through hair follicles and sebaceous glands is known as the appendageal routes. It also demonstrates the iontophoretic shunt through which charged molecules and the particulate system can penetrate.

A variety of cosmeceutical products can be applied to the skin for various applications. The transport of nanoparticles through the skin is influenced by the nature and physicochemical properties of the nanoparticles, vehicles used, and the condition of the skin (Lohani et al., 2014). The dermal absorption of nanoparticles does not occur immediately, but may happen under certain conditions. Cosmetics preparations are exclusively used for healthy skin, but rarely it may be applied to nonhealthy skin. Under such conditions, the barrier properties of the skin may be disturbed. The research results revealed that nanoparticles are passed through hair follicles and skin pores and deposited in minimum amount of space behind the SC (Ribier and Biatry, 2000)

9.3 NANOPARTICLES USED IN COSMECEUTICALS

The very low drug absorption through skin can be improved by systems such as lipid nanoparticles, liposomes, and microemulsions. SLNs have been demonstrated to be very promising carriers for skin absorption (Schueller and Romanowski, 2006) and offer superior stability compared to liposomes (Manjunath et al., 2005; Müller et al., 2002). Before going into the details of SNPs, we discuss some of the significant, commonly employed nanoparticles in skin drug delivery, including liposomes, nanoemulsions, nanocapsules, nanocrystals, nanosilver, nanogold, dendrimers, cubosomes, hydrogels, and buckyballs.

9.3.1 LIPOSOMES

These are the well-known formulations which are concentric bilayer vesicles in which aqueous spaces are enclosed by a lipid bilayer composed of synthetic and natural phospholipids from the GRAS (generally regarded as safe) category. The liposomes may fuse with the bilayer present in the cell membrane and deliver the encapsulated content to the skin or cells. This makes them more useful drug delivery systems in the skin due to their ease of preparation, increased and rapid absorption by the skin, and continuous release of the active ingredient for a prolonged period of time. The next generation of liposomes are known as transferosomes (Cevc, 1996; Thong et al., 2007), niosomes (Uchegbua and Vyas, 1998; Balakrishnan et al., 2009), and ethosomes (Touitou et al., 2000) which clearly established that liposomes are considered highly versatile nanovesicular systems, and some of the main research studies in this area include:

Vyas et al. (2013) developed a topical liposomal gel formulation of caffeine for anticellulite activity by using the thin film-hydration technique, a well suited approach for the preparation of stable liposomes. In this research, liposomes are prepared for dispensing caffeine in the carbopol gel. The results demonstrated that liposomal gel containing caffeine is one of the most successful formulations for the reduction of cellulite deposits in the human body.

Sampathi et al. (2014) developed a liposome entrapped hydrogel complex system of itraconazole for enhanced transdermal permeation by using thin-film hydration technique. It was found that liposomes in hydrogel complex systems, improved the skin permeability of the drug compared to control with higher flux and permeability coefficient. These results indicate that liposome in hydrogel system can function as a probable drug delivery system to enhance transdermal permeation of water-insoluble itraconazole for treating skin infections.

Bharadwaj et al. (2011) developed an acyclovir (ACV)-loaded liposome-based gel. In this study, liposomes were prepared by using the natural solid lipid phase. The prepared formulations were reported to cause less skin irritancy, greater skin tolerance, and controlled drug release of ACV than available commercial products. The results of the characterization, evaluation, and safety of prepared liposomes confirmed that natural lipids are suitable and compatible novel excipients in the preparation of liposomes.

Clares et al. (2014) prepared liposomes for retinyl palmitate (RP), an effect on skin permeation. The aim of this study was to develop biocompatible lipid-based nanocarriers for RP to improve its skin delivery, photostability, and biocompatibility, and to avoid undesirable topical side effects

9.3.2 NANOCAPSULE

These are submicron particles that consist of an aqueous or oily phase of polymer enclosed by a polymeric capsule. With respect to conventional emulsions, it was observed that penetration of UV filter octyl methoxycinnamate was decreased by the use of nanocapsules in pig skin (Hwang and Kim, 2008).

Paolino et al. (2012) developed nanocapsules of gemcitabine (GEM) for the treatment of human cancers. The biological activity of NanoGEM was compared and tested with the free drug in vitro and in vivo. NanoGEM was made of polylactic acid and allowed the entrapment of a great amount of GEM. The incorporation of nanocapsules enhanced the half-life of the drug in plasma and also allowed larger accumulation of GEM inside the tumorous cells.

Kizuka et al. (2012) developed a Fullerene-based carbon Nano-capsules, nanotubes and nano-whiskers by applying the liquid–liquid interfacial precipitation method by heating at 1173K, the Nano-whiskers were changed to Fe_3C-encapsulated carbon nanocapsules and carbon nanotubes.

Esmaeili et al. (2014) developed polyamide nanocapsules of *Aloe vera* L. delivery and studied in vivo by an emulsion diffusion technique. Nanocapsules containing. *Aloe vera* L. were administered to rats and the effects were compared with a control. The results demonstrated that in the *Aloe vera* L. group, the effect was higher.

El-Gogary et al. (2014) developed polyethylene glycol-conjugated polymeric nanocapsules for site-specific delivery of Quercetin to folate-expressing cancer cells in vitro and in vivo. The passive tumor accumulation was confirmed upon intravenous administration in HeLa or IGROV-1 tumor-bearing mice.

Bouchemal et al. (2004) used polycondensation combined with a spontaneous emulsification technique for the development of polyurethane and poly(ether urethane) which were found to be a successful method for the preparation of nanocapsules.

9.3.3 SOLID LIPID NANOPARTICLES

SLNs are droplets of lipids that are made up of oil and are solid at body temperature. They are stabilized by surfactants and can protect active ingredients against degradation by encapsulating them and providing the controlled release of the drug administered topically over a prolonged period of time. They have also been used to improve drug penetration through the SC (Müller et al., 2002). In cosmetics, SLNs provides UV-resistant properties which were further enhanced when molecular sunscreen was added and tested. When 3,4,5-trimethoxybenzoylchitin (a good UV absorber) was added to SLNs, it resulted in enhanced UV blocking (Wissing and Müller, 2003). In vivo studies demonstrated that a formulation containing SLNs demonstrated better skin hydration than a placebo (Song and Liu, 2005). SLNs are the colloidal carriers with submicron size ranging from 50 to 1000 nm and consist of physiological lipid, which are dispersed in water or an aqueous solution of surfactants (Ekambaram et al., 2012). They can be manufactured from natural or synthetic ingredients and provide drug delivery systems with the least toxicity (Sinha et al., 2010). SLNs enhance the efficient penetration of drugs into anatomical barriers, have the desired sustained-release profile, and allow stability of the drug. SLN's are potential carriers for intravenous applications (Kaur et al., 2008) as well as to improve delivery of pharmaceuticals, nutraceuticals, and other materials. Some common lipids and surfactants used in the manufacturing of SLNs along with their hydrophilic lipophilic balance (HLB) value (Mahajan et al., 2015) are listed in Table 9.1.

Jain et al. (2011) developed ACV-loaded SLNs and studied the potential of dermal delivery. The study proved that SLNs are beneficial for improved dermal delivery of ACV through the skin for herpes simplex infection.

Shah et al. (2007) used the simple emulsification solvent diffusion technique for the preparation of SLNs of Tretinoin (TRE). This study was conducted to evaluate the potential of SLN-based gels in improving the dermal delivery of drugs. The skin irritation studies of this research work revealed that SLN is less irritating than the marketed TRE cream. The study report revealed that SLNs have the equivalent permeation profile through rat skin as that of the marketed TRE cream.

Jenning et al. (2000) prepared SLNs of vitamin A for topical use. The preparation demonstrated burst-release and sustained-release profiles due to the SLNs. The release of vitamin A was controlled for up to 12–18 hours.

Chen et al. (2012) developed tripterine-loaded nanostructured lipid carriers (NLCs) and evaluated in vitro skin permeation and in vivo pharmacodynamics parameters. The study concluded that the surface charge of NLCs has great impact on the parameters, such as skin permeation and pharmacodynamics of the

Table 9.1 Lipids Used in the Preparation of Solid Lipid Nanoparticles

Lipids	HLB Value	Surfactants	HLB Value
Acylglycerols	3.7	Phospholipids	14
Glycerol Monooleate	3.4	Soya lecithin	8
Sorbitan monooleate	4.3	Egg lecithin	4.7
Sorbitan Monostearate	4.7	Poloxamer 182	7
Triacylglycerol		Poloxamer 188	29
Tricaprin	3.4	Poloxamer 407	22
Trimyristin	3	Poloxamer 908	30
Ceresin	8	Sorbitan tristearate	2.1
Dimethicone	5	Phosphatidylcholine	7.5
Cetyl Ester Waxes NF	10	Bile salt	
Capric triglycerides	11	Sodium Cholate	18
Cetyl Palmitate	10	Sodium glycocholate	16
Cyclodextrin	8	Sodium taurocholate	15.6
Tocopherols	6	Sodium taurodeoxycholate	16
Fatty acids		Alcohols	
Stearic acid	15.0	Isopropyl myristate, Isopropyl palmitate	11.5
Palmitic acid	15.6		

loaded drug tripterine. The study confirmed that tripterine-loaded NLCs are efficient in enhancing the percutaneous penetration and antimelanoma efficacy.

Sanad et al. (2010) formulated novel oxybenzone-loaded NLCs. NLCs of oxybenzone were prepared via the solvent diffusion method. The study revealed that oxybenzone demonstrated in vitro sun enhanced protection factor and erythemal UVA protection factor into six-fold and eight-fold, respectively, with the further advantage of reducing the side effects of free oxybenzone contents with less irritation potential.

9.3.4 NANOCRYSTALS

Nanocrystals comprise of an aggregate of several hundred to tens of thousands of atoms which are further combined to form a "cluster." Average sizes of the aggregates lie between 10 and 400 nm. The physical and chemical properties of nanocrystals are somewhere between that of bulk solids and molecules this helps the passage through skin effectively (Petersen, 2008).

> Ai et al. (2014) developed a group of nanocrystals of N-doped TiO2 nanocrystals with exposed {001} facets. The results demonstrated that the nanocrystals confirmed the higher photocatalytic activity than P25. The higher photocatalytic activity also favored by the synergistic effects of various factors, such as better crystallinity, efficient light response characteristics, and high reactivity of {001} facets.

Kumar et al. (2013) formulated cellulose nanocrystals (CNCs) and poly(vinyl alcohol) (PVA) based on sugarcane bagasse to fabricate biodegradable porous foam bionanocomposites. The objective of this preparation is for the thermally-induced phase separation (freeze-drying) technique that influences the formulation of the porous composite which has higher potential in biomedical applications.

Taheri and Mohhamadi (2014) formulated CNCs of hydroquinone for topical delivery. The hydroquinone-CNCs followed the sustained release delivery and last longer in action.

Amna et al. (2014) formulated virgin olive oil-polyurethane micro/nanofibers decorated with copper oxide nanocrystals for biomedical applications. The study on the interaction of cells with the scaffold by culturing NIH 3T3 fibroblasts for the assessment of viability, proliferation pattern, and growth potential suggested that the prepared nanofibers had proved their positive potential in wound dressing and would be beneficial at every stage during the wound-healing process. Thus, these nanofibers provide diverse biomedical applications, such as tissue regeneration, stem repair, and wound healing.

Mauludin and Müller (2013) studied the physicochemical properties of hesperidin nanocrystals. Hesperidin nanocrystals were generated by high-pressure homogenization. Hesperidin nanocrystals influences the saturation solubility and dissolution rates compared to the basic raw material since solubility and dissolution rate are the rate limiting factors for the basic material.

9.3.5 NANOGOLD AND NANOSILVER

Silver nanoparticles are coming up as a potential concept in cosmetics as well as in skin-care formulations. The antibacterial properties of nanosilver extends its potential to a wide range of applications. Many companies and manufacturers are producing various antiperspirants and deodorants with nanosilver claiming up to 24-hours protection against bacterial sweating. Nanogold is also used as a promising material against bacterial infections and is used widely for oral hygiene, particularly in forming eminent ingredients for various medical toothpastes (Petersen, 2008).

Zhou et al. (2012) synthesized nanosilver/gelatin/carboxymethyl chitosan hydrogel by reduction using radiations and crosslinking at ambient temperature. The comprehensive study suggested that the prepared formulations had potential antibacterial activity in wound dressing.

Kumar et al. (2012) formulated silver nanoparticles using the leaf extracts of partheniumhysterophorus. This method is called green synthesis which is claimed to be the fastest synthesis process of nanoparticles compared to chemical methods, and could be used in areas like cosmetics, foods, medicals, and biotechnology. Biological methods are also eco-friendly and easy to perform.

Verma et al. (2014) synthesized and characterized gold nanoparticles used as diagnostic tools in different therapies, and, in particular, for peste des petit ruminants PPR; (ovine rinderpest). In this research, gold nanoparticles were synthesized by the reduction of Au + 3 (HAuCl) to AuO. The uniform gold nanoparticles were synthesized by reduction of citrate. The study was carried out to analyze the effects of salt concentration, trisodium citrate concentration, and mixing rate on size distribution at optimum pH and temperature. The gold nanoparticles prepared by this method were used for the preparation of gold-labeled lateral flow strip for PPR diagnosis.

Taufikurohmah et al. (2013) studied a pre-clinic test of Nano-gold in Musmusculus fibroblast proliferation and collagen biosynthesis. The test was performed in vivo using mice (Mus musculus) to assess proliferation of cells and enhancing the collagen quantity in the skin. The test result of antiaging activity using mice in vivo demonstrated that the quantity of collagen and fibroblasts increased in the treatment of mice through nanogold infiltration. Overall the result concluded that the nanogold acts as an antiaging tool by promoting fibroblasts proliferation and increasing collagen biosynthesis.

9.3.6 CUBOSOMES

Cubosomes comprise of submicron, discrete, nanostructured particles that are in the bicontinuous cubic liquid crystalline phase (Spicer et al., 2003). Cubosomes can exist at any level of dilution and provides a large surface area and low viscosity. When the water and a microstructure are mixed at a certain ratio with liquid crystalline particles of certain surfactants, it forms the self-assembly of cubosomes. They can carry both hydrophilic and hydrophobic molecules and possess high heat-stability (Kesselman et al., 2007). They became the desired choice in skin applications as well as in other forms of drug delivery because of their controlled release through functionalization, and by combining capacity with low-cost raw materials.

Esposito et al. (2005) developed cubosome dispersions for drug delivery to the skin by using drug indomethacin. Dispersions were produced by emulsification and homogenization of monooleine (MO) and poloxamer in water. The MO dispersions exhibited a lower flux with respect to the analogous formulation that contained the free drug in the aqueous phase and to the control formulation based on carbomer gel. Reflectance spectroscopy demonstrates that the incorporation of indomethacin into MO dispersions produces a prolonged release.

Morsi et al. (2014) developed silver sulfadiazine-based cubosome hydrogels for the topical treatment of burns. For the optimal cubogel formulae, histopathological evaluation was conducted in vivo on rats to predict the effectiveness of the fresh cubogels with respect to a commercial cream (Dermazin). The in vivo histopathological study results proved that prepared cubogels were effective in the treatment of deep, second-degree burns which may allow improved compliance of patients while the healing results were good with fewer side effects to that of commercial products in the market.

Yang et al. (2012) developed amphotericin B-loaded (Am-B) cubosomes through the sol-emulsion technology for enhancing the oral bio availability. The acute nephrotoxicity, bioavailability and tissue distribution study of AmB-loaded cubosomes were orally evaluated in rats. SAXS and cryo-TEM exhibited AmB-loaded cubosomes as a bicontinuous cubic liquid crystalline phase that has Pn_3m geometry. The results of in vitro release, encapsulation efficiency, and stability studies in gastrointestinal fluid demonstrated that AmB was introduced and encapsulated successfully in cubosomes. AmB-loaded cubosomal formulation orally administrated in rats did not show nephrotoxicity and its relative bioavailability was almost 285% compared to that of Fungizone. The AmB-loaded cubosomal formulation provided an effective approach for improving the oral bioavailability of AmB.

Bei et al. (2009) prepared dacarbazine-loaded cubosomes and studied the influence of formulation variable. The purpose of this study was to investigate the combined influence of three-level, three-factor variables on the formulation of dacarbazine (a water-soluble drug) loaded cubosomes. The Box–Behnken design was used to obtain a second-order polynomial equation with interaction terms to predict response values.

9.3.7 NIOSOMES

Niosomes are vesicles of nonionic surfactants which are made up of nonionic surfactants (Anisha et al., 2010). The morphology consists of an aqueous cavity in the center surrounded by layers of nonionic surfactant in the lamellar phase

(Sankhyan and Pawar, 2012). It is a nanostructure in the size range of 100 nm–2 nm. In comparison to liposomes, they have better penetration, high-entrapment efficiency, improved chemical stability, and low cost. They are widely used as they increase the residence time of the active ingredients in the SC as well as epidermis and lower systemic absorption (Bei et al., 2010). Niosomes also provide site-specific target delivery of drugs (Anisha et al., 2010).

Desai et al. (2011) developed an antifungal topical niosomal gel formulation. The research revealed that the release from the niosomal gel was highly prolonged when corelated to conventional gel and displayed a two-fold increase in the drug deposition in the skin compared to conventional gel. The stability studies depicted vesicles had greater stability at 4°C followed by 25°C. The investigation concluded that the developed niosomal gel formulation of nystatin showed great potential in treatments by providing a prolonged release profile for fungal infection.

Sundaresan et al. (2012) evaluated aceclofenac-niosomes developed by several techniques. The in vitro release studies indicated that all the formulation processes decreased release for 24 hours, but the thin-film hydration method and reverse-phase evaporation method was found to be most satisfactory in relation to niosomes particle size, drug entrapment efficiency, in vitro drug release, and the release mechanism following zero order kinetics.

Hazel et al. (2012) developed and evaluated norfloxacin-loaded maltodextrin-based proniosomes. The kinetic analysis showed that the drug release followed super case II transport. Proniosomes showed stability for 90 days when compared with reconstituted niosomes by storing the formulation at refrigerator condition.

Madhavi et al. (2013) formulated and evaluated metformin-based niosomes. The study was an attempt to prepare a niosomal drug delivery system and its in vitro performance was evaluated. The different types of surfactants were used to prepare the formulations. This study of metformin noisome provides the most promising bioavailability of metformin.

Kumar et al. (2013) formulated econazoleniosomes via the thin-film hydration technique by changing the cholesterol to surfactant ratios in the following manner (1:1, 1:2, 1:3, 1:4). Each formulation was evaluated for drug release.

9.3.8 MICROSPONGES

Microsponges are tiny, sponge-like spherical particles composed of a myriad of interconnecting voids within the noncollapsable structure with a large porous surface (Bhimavarapu et al., 2013). Microsponges are polymeric microspheres that are mostly used for topical delivery (Orlu et al., 2006). They have large void spaces to entrap active ingredients such as emollients, fragrances, sunscreens, essential oils, and antifungal, and antiinflammatory drugs, etc. (Embil and Nacht, 1996).

Abdelmalak and El-Menshawe (2012) developed topical fluconazole (FLZ) microsponge-loaded hydrogel using a quasi-emulsion solvent diffusion method. A selected FLZ microsponge (FLZ and EC in 1:1 ratio and prepared using 0.75% PVA and methylene chloride) was introduced in a carbopol gel formulation and evaluated for the in vitro release profile. The final developed microsponges were observed to be spherical and porous. There was no interaction between drug and polymer molecules. The drug release via cellulose dialysis membrane showed fickion release.

9.3.9 NANOEMULSIONS

Nanoemulsion consists of a dispersion system that includes nanoscale droplets of one liquid within another. The structure can be manipulated or changed based on the method of preparation as they are metastable systems and are prepared by using GRAS products and good with respect to safety purposes. As they are very small in size, they possess high stability and better suitability to carry active ingredients (Fred et al., 2006; Sonneville-Aubrun et al., 2004).

Recent Patent Review on Nanoparticles-Based Skin Preparation

Title	Publication No.	Publication Date	Applicant
Thermal treatment of acne with nanoparticles	US 8834933B2	September 16, 2014	Sienna Labs. Inc
Novel nanoparticles formulation for skin delivery	US20120195957A1	August 2, 2012	Mandip Singh Sachdeva, Ram Patolla
Nanoparticles dispersion for detection of skin health conditions and diagnostic kit thereof	W02014108758A1	July 17, 2014	ITC Limited
Polymeric nanoparticles for photosensitizers	WO2013/020204	February 14, 2013	Quest Pharmatech Inc.
Mitadichol liquid and gel nanoparticles formulation	US20140275285A1	September 18, 2014	NanoRx, NC
Virucidal materials	W02007093808A2	August 23, 2007	Queen Mary and Westfield College
Compositions containing lipid micro or nanoparticles for the enhancement of the dermal action of solid particles	W02010051918A2	May 14, 2010	PharmasolGmBH
Virion derived protein nanoparticles for delivering diagnostic or therapeutic agents for treatment of skin-related diseases	W02013009717A1	January 17, 2013	Los PinosElisabet De

(Continued)

Continued

Title	Publication No.	Publication Date	Applicant
Compositions of empty nanoparticles and their use for treating dermatological conditions	EP2667860A1	December 4, 2013	AnteriosInc
Protein nanocarriers for topical delivery	US20120195947A1	August 2, 2012	Omatanu P. Perumel
Triggered cargo release from nanoparticles stabilized liposomes	EP2545533A1	January 16, 2013	The Reagent of the University of California
SLN of Roxithromycin for hair loss and acne	W02014077712A1	May 22, 2014	GdanskiUniversytet

Patents on Nanocosmetics

Title	Publication No.	Publication Date	Applicant
Nanoparticles for cosmetic applications	W0201210638A2	August 2, 2012	Yissum Research Development Company of the Hebrew University of Jerusalem Ltd
Hair removal with nanoparticles	US8821941B2	September 2, 2014	Sienna Labs. Inc
Nanoparticles based on poly(lactic glycolic) acid for cosmetics application	US20130266625A1	October 10, 2013	Yissum Research Development Company of the Hebrew University of Jerusalem Ltd
Screening cosmetic composition one nano-pigment of metallic oxide and one fat-soluble screening polymer	US5939053A	August 17, 1999	L'oreal
Cosmetic pigment composition containing gold or silver nanoparticles	EP1909745A1	April 16, 2008	Korea research institute of bioscience and biotechnology
Use of Nanoscale deodrants	EP1239823B1	June 16, 2004	Cognis Deutschland GmbH&CoKG
Metal oxide nanocomposites for UV Protection	US20130022655A1	January 24, 2013	BASF SE

(*Continued*)

Continued

Title	Publication No.	Publication Date	Applicant
Nanodiamond UV protectant Formulation	US20090220556A1	September 3, 2009	International technology center
Nanocrystals for use in topical cosmetic formulation and thereof method of production	EP2099420A1	September 16, 2009	Abbott GmbH & Co. KG
Gel technology suitable for use in cosmetic compositions	US20100266649A1	October 21, 2010	Avon Products, Inc

9.4 METHODS OF PREPARATION

9.4.1 HIGH-PRESSURE HOMOGENIZATION

This is a reliable and highly applicable technique which is used for the production of SLNs (Almeida and Eliana, 2007). This procedure involves high-pressure homogenization which entails pushing the liquid or content using high pressure through a narrow gap in the range of few microns (pressure 100–2000 bar) (Mehlen et al., 2013). The fluid is accelerated through that small orifice at a velocity of 1000 km/h. During this process, shear stress and forces through the cavity break the particles down to the submicron range. Generally, during the preparation 5%–10% of lipid content is used, but it can be extended up to 40% (Muller et al., 2011). High-pressure homogenization is mostly adopted in various research works which can be further classified into Hot Homogenization and Cold Homogenization (Mehnert and Mader, 2010).

9.4.1.1 Hot homogenization

As the name implies, hot homogenization involves the application of heat. The temperature range should be above the melting point of lipid present in the formulation. The hot homogenization process involves two phases (Mishra et al., 2010):

1. Drug in melted lipid in a homogeneous state by stirring
2. Hot aqueous surfactant mixture.

Both the phases should be maintained at a temperature in the range of 60°C–70°C. The primary is prepared by mixing the lipid phase into the aqueous phase by an overhead stirrer. The obtained primary emulsion is further subjected to high shear homogenization (Ekambaram et al., 2012). The resultant oil in water emulsion is cooled to room temperature that leads to the formulation of solids such as crystals and SLNs. Small particle size is achieved via a high processing temperature by which the viscosity of lipid phase is lowered and this favors the formation of nanosized range particles (Manjunath and Venkateswarlu, 2012). Care should be taken in maintaining the temperature as high temperatures lead to the degradation of the drug and the whole system. The whole system should not be exposed to more than 3–5 temperature or pressure cycles. Increased homogenization or temperature, or both, leads to the formation of particles of increased size due to the increased kinetic energy of molecules present in the hot system. The pressure cycle usually ranges between 500 and 1500 rpm. For nanoparticle preparation, stirring was maintained for at least 5–10 minutes with the rpm ranging from 2000 to 20,000 (Hu et al., 2012). The increase in the rpm or the time of stirring may not significantly affect the particle size, but it may cause minor variations in the poly-dispersibility index (PDI).

9.4.1.2 Cold homogenization

Cold homogenization is an alternative to hot homogenization to overcome the disadvantages related to hot homogenization (Mishra et al., 2010). As the latter is not suitable for thermolabile drugs and it may cause temperature-linked degradation issues, drug loss into the aqueous phase, and partitioning of the drug. Also, a critical issue involved in hot homogenization is that in this process the drug-containing lipid melt is cooled rapidly by using dry ice or liquid nitrogen (Manjunath and Venkateswarlu, 2012). This favors the homogenised distribution of the drug in the lipid matrix. The drug incorporated in lipid matrixes were subjected to pulverization which created the particles of the micron or nano-range and this pulverized dispersion is mixed with the cold water containing the surfactant solution. The primary emulsion or free suspension is homogenized under chilled conditions or at room temperature. In this process, the temperature is controlled as compared to hot homogenization technique. The cold homogenization produces particles of a larger size range and broader size distribution. Hot temperatures are not completely avoided in the cold homogenization technique since the lipid requires a melting process as a beginning step; however, the overall thermal exposure is less in this method compared to the hot homogenization method.

Advantages

Works well at laboratory scale. Cost effective.

Disadvantages:

Energy consuming.
Biomolecules may get damaged when performed at laboratory scale (Hu et al., 2012).
Unproven scalability.

9.4.2 ULTRASONICATION AND HIGH SHEAR HOMOGENIZATION

SLNs can be prepared by either ultrasonication or the high shear homogenization technique, sometimes both methods are required to obtain particles in the nano-sized range. Both are laboratory scale techniques (Trotta et al., 2003) and can be applicable for smaller volumes of preparations such as 500 mL to 1 L. For the larger volumes or bulk preparations laboratory-scale sonicators or stirrers are not sufficient to achieve uniform particle size. Along with this approach, temperature applications may be required to achieve the stable nanoparticles (Table 9.2).

9.4.2.1 Advantages

Shear stress is reduced.
Simple and easy to prepare.
May give better results in less viscous formulations (Hu et al., 2012).

9.4.2.2 Disadvantages

Potential physical stability issues such as particle growth upon storage
Metal contamination.
Difficulty in scale up.

9.4.3 SOLVENT EVAPORATION

This method involves dissolving the lipophilic material in a water-immiscible organic solvent such as cyclohexane or ether, etc. (Yuan et al., 2008). The solvent phase is then emulsified in the aqueous phase by the addition of smaller quantities. The solvents are subjected to evaluation which leads to the precipitation of the lipid in the aqueous phase. The resultant final lipid phase containing nanoparticles may be further subjected to high-pressure homogenization to achieve nanoparticles of the lower size range (Wissing et al., 2010). The organic solvent can be removed by using rota-evaporator under reduced pressure (40–60 bar). There may be a requirement of emulsifiers such as polysorbates and phosphatidylcholine and also cosurfactants such as poloxamer and sodium mono-octyl phosphate. The emulsifiers and coemulsifiers maintains the whole potential of the system and supports the stability of emulsions by favoring the zeta potential in favor of single charge (Kuo and Chen, 2009).

9.4.3.1 Advantages

Commercially successful method.
Suitable for scale-up.
Continuous process.

9.4.3.2 Disadvantages

High energy is applied.
PDI may be high.
Biomolecules may get damaged during the process.

Table 9.2 Summary of different homogenization techniques used in SLNs synthesis (Battaglia et al., 2014).

Technique	Solvent Used	Instrumentation Required	Working Temperature	Optimum Conditions
High-pressure homogenization	Various lipids and surfactants	High-pressure homogenizer	5°C–10°C upon lipid mp	Cavitation forces
High shear homogenization	Various lipids and surfactants	High shear homogenizer	5°C–10°C upon lipid mp	–
Hot homogenization	Lipids, oils, waxes, alcohols, and surfactants	High shear homogenizer	30°C or above	–
Cold homogenization	Lipids and surfactants	High shear homogenizer	Low temp	–
Ultrasound homogenization	Lipids, oils, and waxes	Ultrasound apparatus	5°C–10°C upon lipid mp	Ultrasound treatment
Melt dispersion	Waxes and oils	High shear homogenizer	5°C–10°C upon lipid mp	–
Solvent evaporation from emulsions	Chlorinated solvents	High shear/pressure homogenizer	25°C	–
Solvent diffusion from emulsions	Partially water-miscible solvents	High shear/pressure homogenizer	40°C–50°C	–
PGSS	–	GAMA apparatus	5°C–10°C upon lipid mp	Pressure >74 bar
SFEE	Chlorinated solvents	High-pressure homogenizer/ SFEE apparatus	>31°C	Pressure >74 bar
Cryogenic micronisation	–		−80°C	–
Membrane contractor method	–	Membrane contractor	5°C–10°C upon lipid mp	–
Spray-drying	Ethanol	Spray-drier	70°C	–
Electrospray	Aliphatic Alcohols	Electrospray apparatus	25°C	High voltage
Spray-congealing	–	Spray-congealer	5°C–10°C upon lipid mp	–

PGSS, *Particle gas-saturated solution;* SFEE, *supercritical fluid extraction of emulsion.*

9.4.4 SOLVENT EMULSIFICATION-DIFFUSION METHOD

SLNs can also be prepared via the solvent emulsification-diffusion technique. In this method, the size of the particles in the preparation is greatly influenced by lipid content or concentration in the organic phase, type of emulsifier, and the HLB value (Kuo and Chen, 2009). This method is advantageous because high temperatures are avoided. In this method the lipid base is dissolved in a suitable solvent such as toluene, chloroform, or dichloromethane. The solvent phase containing lipid is added dropwise into the aqueous phase containing the cosurfactant under homogenization by suitable homogenizers. The resultant homogenized mixture is subjected to rota-evaporation under reduced pressure. The nanoparticles are formed in dispersions of the precipitations (Wissing et al., 2010).

9.4.5 SUPERCRITICAL FLUID TECHNIQUE

The supercritical fluid technique is used for various applications such as extraction, chromatography, and material processing in the preparation of SLNs. Currently, two common ways for particle formulation in supercritical fluid technique are the supercritical antisolvent (SAS) process and rapid expansion of supercritical solution (RESS) (Freitas and Muller, 2012).

In the SAS process, the solute is dissolved in a suitable organic solvent and then the solution is sprayed into an excess of anti-solvent. The solubility of the solute in the antisolvent is limited. Only the organic solvent is miscible with the antisolvent. During this process high level of mass transfer occurs which leads to the formation of solid particles. Another process is RESS which involves two steps, the first step being the extraction process where the supercritical fluid is saturated by dissolving the substrate. The second step entails depressurization being applied to the solution by passing it through a nozzle. This decreases the solvent power and the respective temperature of the liquid leads to the precipitation of the solute (Lv et al., 2009). This method is called supercritical because the fluid's pressure and temperature exceed the respective critical value. In that condition, the capacity of the fluid is increased to dissolve the high amount of solute added to it. This technology involves several processes for the production of SLNs other than RESS such as the particle gas-saturated solution (PGSS), aerosol solvent extraction solvent, and supercritical fluid extraction of emulsion (SFEE).

9.4.5.1 Advantages

This method minimises the solvent usage.

SLNs can be prepared by the rapid expansion of supercritical fluid carbon dioxide solution method by (RESS).

99.99% of carbon dioxide solvent is used for this method.

Particles are collected as a dry powder instead of a suspension.

Pressure and temperature stress is less.

9.4.5.2 Disadvantages

Cost ineffective.

9.4.6 MICROEMULSION METHOD

SLNs can be prepared by diluting the microemulsion. Microemulsion is a two-phase system that entails an inner and outer phase. Microemulsions can be prepared by stirring the mixture at 65°C–70°C. The substances such as low melting point, fatty acids, emulsifiers (polysorbates), and coemulsifiers (butanol and water) are required for the synthesis of nanoemulsion. The microemulsion in the hot-stage is dispersed into cold water under stirring conditions. The temperature of the cold water is constantly maintained between 2°C and 3°C. The typical volume of the hot microemulsion and the cold water should be in the ratio of low microemulsion content dispersed in 10–50 folds of the aqueous phase with respect to microemulsion volume. The dilution process is determined by the composition of the microemulsion. The first step involves hot microemulsion dispersed in cold water and can be done via mechanical stirring. The microemulsion itself is already in the lowest size range. With in this method nanoparticles can be produced by solvents which dissolve rapidly in the aqueous phase such as hydrophilic solvents, while more lipophilic solvents produce larger-sized nanoparticles (Cortesi et al., 2002).

9.4.6.1 Advantages

Low mechanical stress is applied.
Theoretically stable, but may vary according to the formula.

9.4.6.2 Disadvantages

Extremely sensitive to change.
Particles achieved are in low concentration

9.4.7 MELT DISPERSION TECHNIQUE

In this method, the solid lipid is melted (70°C–80°C) with an organic solvent under magnetic stirring conditions. In another phase, the drug-containing surfactant solution of the aqueous phase is heated at 70°C–80°C. Both the lipid phase melted in the organic solvent and aqueous phase should be maintained at the same temperature. Stirring should be continued for a few hours at moderately high rpm of 1500–2500. The organic phase containing lipid melt is then introduced dropwise into the aqueous phase. The stirring is continued at the same level and then the whole system is cooled to room temperature which produces the SLNs. This method is better than ultrasonication, but solvent emulsification is still a better method than MDT (Govender et al., 2015).

9.4.8 PRECIPITATION METHOD

In this method, the lipid is dissolved in a suitable organic solvent such as chloroform, or acetone, etc. Then this solution is subjected to sonication for the complete solubilization of the liquid for at least 5–10 minutes with the help of aluminum foil to prevent the evaporation of the organic solvent. The prepared organic lipid mixture is to be injected into an external aqueous phase containing the cosurfactant such as Pluronic F-68 (0.5–1 %w/v) and Tween 80 (0.01–0.1% w/v) with the help of a syringe in a dropwise manner (Lv et al., 2009). The whole system is kept under high-speed mechanical stirring of 500–1500 rpm to get the desired nanodispersion After the complete addition of organic lipid phase into the aqueous phase the nanodispersion is transferred to magnetic stirring conditions (500 rpm) for at least 4 hours in order to evaporate the organic solvents and, finally, the same dispersion is subjected to rota-evaporation under reduced pressure at 60°C for the removal of traces of organic solvents present in it.

9.5 CHARACTERIZATION AND EVALUATION OF NANOFORMULATIONS

As soon as the nanoparticles are prepared, they are evaluated for their characteristic size, shape, entrapment of the drug inside the SLNs, and the release profile of the drug from SLNs.

9.5.1 PRELIMINARY PARTICLE SIZE DETERMINATION BY PHOTON CORRELATION SPECTROSCOPY (E.G., ZETA SIZER)

Photon Correlation Spectroscopy (PCS) is the most widely used principle in the study of polymeric nanoparticles. PCS is also known by several names such as Quasi-Elastic Light Scanning, Dynamic Light Scattering (DLS), and Intensity Flucham Spectroscopy in the past, but, currently, DLS is the most-used term. The Z-Average (d.nm) value and PDI is determined by zeta sizer (e.g., Malvern Nano S90). This technique measures the diffusion of particles moving under Brownian motion and converts this to size and a size distribution using the Stokes-Einstein relationship. The analysis is performed at a scattering angle of 90 degrees by using DLS, which is widely used to determine the size of Brownian nanoparticles in colloidal suspensions in the nano and submicron ranges. When a monochromatic light (laser) falls onto a solution of spherical particles in Brownian motion causes a Doppler shift and changing the wavelength of the incoming light. This change is related to the size of the particle. It is possible to extract the size distribution and give a description of the particles in motion in the medium by measuring the diffusion coefficient of the particle and using the autocorrelation function. The colloid particles are in random motion called Brownian Motion due to the collision of molecules in the liquid. The scattered light from the diffusivity particles fluctuate with time and carry in the details of the diffusion coefficient of the particles.

DLS requires particles of approximately less than a micron in suspension form. The sample of nanoparticles containing suspension or the nanoparticles dispersed in the suitable organic or aqueous phase is placed in the optical path of the laser. The laser light then falls on the particles in Brownian motion and get scattered. This scattered light is captured by the photomultiplier tube detector and the rate of diffusion is determined which reveals the velocity of the moving particles and average hydrodynamic particle size called the Z average. Z average can be calculated by intensity weight basis using Stokes-Einstein equation. The principle of the measurement is based on the basic concept of measuring smaller particles more rapidly than the larger particles. Usually, Z average is reported using PDI which is the indicator of the broadness of particle size distribution. The sample preparation is an important step in the DLS analysis and sometimes dilution of the sample is required to limit the effect of particle–particle interaction. Care should be taken during analysis so as not to destabilize the dispersed system.

The particle size of SLNs is determined by DLS at an angle of 90 degrees and at a temperature of 25°C using a sample appropriately diluted with distilled water. SLNs are operated with a shaft pump and stirrer to disperse the nanoparticles continuously within the measurement zone (Meyer and Heinzelmann, 1992). Nanoparticles are characterized by distribution width (D10, D50, and D90), mean particle size, span, and uniformity. The experiment has to be reported in triplicate.

9.5.1.1 Size and surface charge analysis

The laser light scattering instruments (e.g., DelsaNano Common; Beckman Coulter) are used for particle size analyses based on volume distribution at a fixed angle of 90 degrees. Zeta potential is also measured by using the same instrument under the same conditions. Previously, the samples are diluted appropriately and then kept in a low-conductivity Zeta cell in order to meet the instrumental conditions and, then, the readings are taken. The surface charge is indirectly measured by zeta potential. The measurement of the zeta potential allows for predictions about the storage stability of the colloidal dispersion (Ekambaram et al., 2012).

9.5.1.2 Drug-polymer compatibility study

Possible drug-polymer interaction is studied by using differential scanning calorimetry (DSC), Fourier transform infrared (FTIR) spectroscopy, and X-ray diffraction (XRD) (Rabinarayan and Padilama, 2010).

9.5.1.3 Fourier transform infrared spectroscopy study

FTIR spectroscopy is used to predict molecular interactions occurring between a drug and excipients. FTIR spectra show the compatibility of the drug and excipient in the form of physical mixing. FTIR spectra at room temperature are recorded with the help of FTIR from 4000 to 400 as scanning range between wave number (cm^{-1}) and % transmittance. KBr discs (2 mg sample in 200 mg KBr) are used to prepare the samples with the hydrostatic press at a force of 5 cm^{-2} and the resolution of 4 cm^{-2}. Measurements are carried out using the attenuated total reflectance technique. Experiments need to be repeated to confirm the reproducibility. The major sharp and significant peaks (functional groups) of the drug and the drug with excipients are identified in the results which helped to determine the chemical compatibility between the different chemicals and drugs used in the preparation of SLNs. The FTIR on the developed SLNs is also recommended.

9.5.1.4 Differential scanning calorimetry

DSC is a thermoanalytical technique in which the difference in the amount of heat required to increase the temperature of a sample and reference is measured as a function of temperature. DSC studies are used to evaluate the melting and crystalline behaviors of polymeric membranes. DSC thermograms analysis is carried out on nano samples which are put into aluminum pans (Muller, 1998). Standard procedures are used to calibrate temperature as well as energy scales. The calorimeter is calibrated for temperature and heat flow accuracy using a temperature range of 25°C–500°C at a heating range of 10°C/min in a N_2 atmosphere (Drake et al., 1989). DSC studies helps to under stand the thermal behaviours of the nano particles and the influence of temperature in particle size and stability of the SLNs.

9.5.1.5 X-ray diffraction study

X-ray powder diffraction (XRD) is a rapid analytical technique primarily used for phase identification of a crystalline material and can provide information on unit cell dimensions. The analyzed material is finely ground, homogenized, and the average bulk composition is determined. Usually a wide angle XRD is performed by using Bruker AXS D8Advance a synchrotron light beam with a wavelength of 1.5406 Å at 40 kV and 40 mA. The scattering intensity is recorded by Si(Li)PSD detector. These studies are used to investigate the crystallinity of the sample. Samples are mounted on a sample holder and XRD patterns are recorded in the range 2 theta between 3° and 50° at the speed of 5°/min. The XRD patterns of the drug, final, and dummy formulation need to be studied (Muller and Runge, 1998). Polymorphic transition happened in the SLNs can be determined by the XRD. The information about the crystallisation behaviour of the lipids and their influence on the stability of the SLNs can be derived from the XRD studies. XRD also gives the information about the polymorphic changes happens to the drugs present in the SLNs.

9.5.1.6 Transmission electron microscopy

Transmission electron microscopy (TEM) is a technique in which a beam of electrons is transmitted through an ultra-thin specimen, interacting with the specimen as it passes through it. TEM is used to characterize surface morphology and particle size. SLN is placed on the carbon-coated copper grid. The sample is previously diluted with HPLC grade water and subjected to sonication for 2 minutes. The sample is examined under a TEM, operated at 200 kV.

9.5.1.7 Atomic force microscopy

Atomic microscopy (AFM) is one of the most-used scanning probe microscopy methods. If the nanoparticles are spherical in shape and deposited on an ideally flat substrate, the size of such nanoparticles can be determined easily by AFM. This is done by simply measuring the nanoparticles image height. However, if the particles are deposited on a rough surface, particle size analysis is not straightforward and many questions will be raised about the implementation of AFM image software (Vasilev et al., 2013). There are many factors such as agglomeration, and self-ordering on the substrate that also influence the AFM analysis. The AFM nanoparticle samples can be prepared by skin-off coating through a simple homemade apparatus. The nanoparticles are dispersed in ethylene glycol and water respectively and dried after deposition. After the sample preparation analysis and data monitoring can be performed by following steps:

1. Modeling of the rough surface.
2. Simulation of particle deposition on the surface.
3. Creation of AFM images by tip-sample convulsion.
4. Statistical analysis using AFM images and data processing software.

The advantages of AFM include accuracy within the 0.1 nm scale. AFM can also characterize properties other than topographic such as magnetic and mechanical properties. AFM can image organic and inorganic molecules. AFM can assist the study of the interaction between nanoparticles and biomolecules.

AFM provides better results than other light scattering techniques such as DLS. AFM can characterize shape as well as size, which is not possible via DLS (Hoo et al., 2008).

In conclusion, AFM gives a strong contrast report on any material (whereas electron microscopy gives the poor contrast) which is a highly suitable technique to characterize a wide variety of nanoparticles including SLNs.

9.5.2 DETERMINATION OF DRUG LOADING IN SOLID LIPID NANOPARTICLES

9.5.2.1 Drug entrapment efficiency

10 mL of the sample is subjected to an entrapment efficiency test to check the percentage of the incorporated drug. The percentage entrapment efficiency is determined spectrophotometrically at fixed nm using UV–visible Spectrophotometer. Previously, 100 mL of nanoparticles are prepared and subjected to centrifugation at 5000 rpm (having 4626 rcf) for 45 min. The amount of

free drug was detected in the supernatant and the amount of loaded drug was determined as a result (Jenning et al., 2000; Müller et al., 2002).

Percentage entrapment efficiency can be calculated by using the equation:

$$EE\% = \frac{\text{Total amount of drug-amount of unbound drug}}{\text{Total amount of drug}} \times 100$$

9.5.2.2 In vitro method of drug release from solid lipid nanoparticles

It is challenging to develop a suitable in vitro drug release testing method for nanoparticles (Junemann and Dressman, 2012; Wallace et al., 2012). Due to the difficulty in isolating nanoparticles from the releasing medium, lack of clarity in release mechanisms of the SLNs. The complexity and variability of a nanoparticulate system such as controlled release SLN nanosuspensions, cell targeting nanoparticles, and polymeric nanoparticles, are some factors causing difficulty in designing the ideal model for studying in vitro drug release pattern, however, certain laboratory methods include:

In vitro drug release studies by membrane diffusion technique

The membrane diffusion method is also called dialysis. It is the most widely used method in the in vitro release of nanoparticulate systems. In this method, the nanoparticles are separated from the release medium through the dialysis membrane, they are permeable to medium but not to SLNs. Dialysis method is reported by many researchers to study the release profile of liposomes (Bhardwaj and Burgess, 2010; Xu et al., 2012), emulsion (Chidambaram and Burgess, 1999), polymeric nanoparticles (Muthu and Singh, 2009; Avgoustakis et al., 2002), and SLNs.

In vitro drug release studies can be performed by using the dialysis membrane diffusion technique. The dialysis membrane of MWCO 12,000–14,000 is used as this retains nanoemulsion, but allows the free drug to release in the media. An accurately weighed amount equivalent to 1 mg sample is transferred to a 2.5 cm diameter glass tube having a length 10 cm and is open at both the ends, its lower end covered with the membrane pre-soaked in distilled water. The cylinder is fitted to the shaft and immersed in the beaker of the dissolution apparatus USP (e.g., Pharma Test, Hamburg, Germany) containing 50 mL of media as receiver fluid. This volume provided complete sink conditions for the sample. The entire system is kept at 37°C ± 0.5°C with continuous stirring at 50 rpm. At different time intervals, samples are withdrawn and replaced by fresh buffer.

Dialysis method

There are three types of dialysis methods, i.e., dialysis sac method, reverse dialysis sac method, and side-by-side dialysis method.

In the dialysis sac method, the SLNs are kept inside the dialysis sac and then dipped into a reservoir container containing suitable media such as buffers, distilled water, SNF, STF, or other media mimicking body conditions. The buffers are chosen according to the application of the polymers and their intended use (Bangham, 1963). The dialysis sac dipped in the reservoir container is subjected to stirring by using a shaking water bath. The dialysis sac method can also be done by USP dissolution apparatus.

The reverse dialysis sac method is employed for nanomaterials containing emulsions and colloidal systems. Recently this method was modified into glass basket dialysis method. In this method, a glass basket is sealed with a dialysis membrane at the bottom which replaces the regular USP apparatus. The parameter such as buffer media, agitation, and sink conditions can be maintained as recommended in membrane diffusion technique, however, there is a possibility of disturbing sink conditions in this procedure (Mezei and Gulasekharam, 1980)

There are other methods for studying drug release profiles of nanoparticles such as centrifugation, ultracentrifugation influenced sample, and separation methods which have the disadvantages of separating nanoparticles from a medium which leads to erroneous results. Another method is the continuous flow method which was originally designed for modified release oral dosage. However, in the particle size of less than 100 nm range, it is very difficult to test it under USP IV apparatus due to the clogging of nanoparticles in the filters which leads to slow-flow rates and resulting erroneous data, therefore, this method should be avoided in drug release studies of SLN.

9.5.2.3 In vivo method of drug release studies from solid lipid nanoparticles

Successfully prepared SLNs can be studied for in vivo release studies of the drug in the skin.

Skin penetration studies

Skin penetration study can be conducted in animals such as Wistar rats, albino rats, guinea pigs, rabbits, and any other suitable animal models recommended and approved by the ethics committee (Marzulli et al., 1969). The prepared SLNs are incorporated in suitable vehicles such as lotions and creams that can be used further for skin penetration studies. The abdomen of the animal full thickness of the skin is to be shaved carefully by using an electric razor. Care must be taken to prevent damage to the SC. An accurate amount of SLNs containing creams are applied evenly on the skin surface. Time intervals of 30 minutes, 3 h, 6 h, and 24 h of exposure to the SLNs are recorded from the animal models (Vogel and Turtletaub, 1991). After the predetermined time period, the animals were sacrificed by an excess inhalation of ether and the skin was carefully cut into pieces ($1.5\ cm^2 \times 1.5\ cm^2$). The skin sample are then rinsed with distilled water and preserved until the analysis is performed (Vogel, 1992).

The controls were prepared by using an aqueous solution of Rhodamine B which is used to determine the direction of flow and transport which is used widely in biotechnological applications due to its fluorescence characteristics. Rhodamine B can be easily detected by fluorescence microscopy, fluorescence correlation spectroscopy, and ELISA. The skin has to be cut into vertical slices with 15–30 μm thickness by a freezing microtome, and stored at 4°C to be analyzed within 24 hours. These skin slices can be investigated by both normal and fluorescence microscope. The images of normal light and fluorescence microscopy of the same skin area and the distribution of the dye in the skin can be studied qualitatively.

Tape stripping technique

This is the simplest method for reducing the barrier imposed by SC. In this, an adhesive tape is used to remove the layer of corneocytes (Menon et al., 1998). The adhesive tapes are cut, weighed and applied to the preselected and cleaned skin area, preferably for an hour (Denda et al., 1996). Usually, at least about 30 stripes are needed to strip off most of the horny skin layer. The multiple stripes remove the substantial skin barriers which are evidenced by the 20–25-fold increase in transepidermal water loss (Van der Walk and Maibach, 1990; Pierard-Franchimont, and Pierard, 1987). The nano-formulation to be tested is applied to the skin for the particular duration as required and then the tapes were removed and weighed which contains SC cells (Wagner et al., 2003). The amount of SC cells removed are studied through the difference of weight. The quantification of the drug present in the tapes can be studied by suitable analytical methods. The tape stripping is useful in the determination of kinetics and penetration potential of different permeants. The core of permeants can also be studied from SC via noninvasive methods (Inoue and Aramaki, 2006) is useful in evaluating the various nanopreparations applied on the skin.

9.5.2.4 Drug release kinetics and mechanism of drug release

To characterize the mechanism of drug release the cumulative amount of drug release from the nanoemulsion formulation at different time intervals are: fitted to cumulative drug release versus time (zero-order kinetic model); log cumulative of percentage drug remaining versus time (first-order kinetic model); and cumulative percentage drug release versus square root of time (Higuchi model). The zero-order rate describes the system where the drug release rate is independent of its concentration and the first-order describes the release of the system where the release rate is concentration-dependent. The Higuchi model describes the release of the drug from nanoemulsion as a square root of the time-dependent process based on Fickian diffusion (Müller-Goymann, 2004).

The drug release data of SLNs generally fits with Korsmeyer–Peppas' power equation:

$$\mathrm{Mt} = \mathrm{M1}\frac{1}{4}\,Ktn$$

where Mt = M1 is the fraction of drug released at time t, in which K is constant, and n represents the release exponent indicative of mechanism of drug release.

When $n = 0.5$ there is Fickian diffusion, $0.5 > n < 1.0$ indicates nonFickian diffusion, and $n = 1.0$ Case II diffusion (Fahr and Liu, 2007).

9.5.3 STABILITY STUDIES

The physical and chemical stability of SLNs are regulated by two visibly different processes, i.e., the SLN suspension's ability to retain homogeneity (suspension stability) and preventing the carrier from recrystallization (carrier stability) (Müller and Dingler, 1998).

The successful preparation of SLNs can be confirmed by suitable stability study methods. The SLNs for gene delivery have to be studied in storage conditions such as −20°C, 5°C, and 24°C (Martins et al., 2007). The biotechnology of the products is usually stable at the lower temperatures such as −20°C. As per the general guidelines, SLNs have to be studied for their stability under different temperatures and humidity conditions. The prepared SLNs can be packed in a suitable glass container and closed tightly. The containers have to be flushed with nitrogen before depositing the SLNs into it (Muller et al., 2005). The filled containers are then stored at a different temperatures such as 4°C ± 3°C for freezing conditions, 25°C ± 2°C/60% ± 5% RH for room temperature and 40°C ± 2°C/75% ± 5% RH as per ICH guidelines. After the appropriate storage period of 1−3 months under the recommended storage conditions, the containers are taken out and the samples are tested on various parameters such as size, zeta potential, % drug entrapment, and drug release profile to determine the possible agglomeration and drug-loading profile. The data must be compared with the initial data so that the product can be declared stable or unstable. Long-term stability studies for 12-24 months are highly recommended for SLNs to get detailed information of the stability profile of the SLN.

9.6 TOXICOLOGICAL CONSIDERATIONS OF SOLID LIPID NANOPARTICLES

Appropriate toxicological studies of SLNs can be designed based on their intended use, exposure levels, the degree of concerns, and ingredients present in the formulations as well as their toxicity profiles. The ingredients used for the preparation of nanoparticles and their physicochemical properties, impurity, and purity also influence the toxic profiles of SLNs (Takeda et al., 2009). Interactions such as drug−excipient, excipient−excipient and ingredient packaging also influences toxicity profile (W. H. De Jong). The existed toxicity studies applied to the pharmaceutical formulations cannot be used directly for SLNs as nanoformulations demonstrates the different physico -chemical and therapeutic properties which is different from the pharmaceutical formulations. The existed method should be appropriately modified with respect to the nano-materials. The existed study method available for soluble drugs cannot be applicable for insoluble nanomaterials so that it has to be modified appropriately for testing insoluble or partially soluble SLNs (Dunford et al., 1997; Arora et al., 2012). The toxicology studies of normal nanoparticles and agglomerated nanoparticles have to be studied separately because they have different chemical and physiological properties. SLNs have a tendency to bind with various substances including proteins which leads to altered biological properties, so must be addressed properly (De Jong and Borm, 2008). The existed metric system of mg/L or mg/kg cannot be applicable for nanoparticles which should rather be studied in terms of particle number and surface area. However, lipid nanoparticles are well-tolerated in the living system since most of the lipids used in the SLNs are sourced from the physiological compounds which can under goes the regular metabolic pathways of the body. To evaluate the toxicological profile of SLNs, nanotoxicology studies are required for cytotoxicity and genotoxicity analyses (Poon and Burd, 2004).

9.7 APPLICATIONS OF SOLID LIPIDS NANOPARTICLES

Currently, SLNs are widely used and with a fast-increasing pace and have a wide range of applications in various fields like medicine, therapeutic treatments, biotechnology, cosmetics, bioinformatics, agriculture, microbiology, and others. Some of the main uses of SLNs in medical treatments with a special focus on skincare and cosmetics are listed below:

9.7.1 SLNs FOR SKIN CARE

Skin is considered a less permeable organ which offers lower risk during the delivery of drugs through it and is considered as the fundamental route of entry for nanoparticles, both intentional and unintentional contact. The transport of molecules through the skin can be enhanced by altering or damaging the SC by using a substance such as solvents, surfactants, and enhancers. The various applications of SLNs are discussed below:

9.7.1.1 Photoprotection of skin

Photoprotection is a core area of commercial skin products. There are several photoprotective products from leading companies available in the market. Photprotection products also have a commendable market share. There are many products in the pipeline for protection against UV radiation as sunscreen agents. UV radiation and sun exposure is linked to skin cancer, epithelial skin damage, and melanoma (Bangham, 1963). It has been proven that sunscreen agents are playing a potential role in preventing these skin issues. Sunscreens, creams, and products are also an integral part of antiaging products. There are three basic photoprotecting methods in use, i.e., antioxidants, repair mechanical stimulators, and physical photon blockers (Mezei and Gulasekharam, 1980). A large proportion of the literature reports that nanoparticles can penetrate human hair upper layer follicles and the upper layers of SC. However, they can not penetrate the barrier of the intact skin and viable epidermis. SLNs for skin protection are performing better than the conventional photoprotection products.

9.7.1.2 Solid lipid nanoparticles as barrier creams

The barrier function of the SC may not be sufficient to protect the skin from allergens or other pathological conditions such as dermatitis and allergies due to chemotherapy. In such situations supporting the barrier function of the skin is required. SLNs were found to be more beneficial than regular creams as the SLNs are more capable of protecting the skin from dehydration and prevent related risks such as eczema (Kaur and Agrawal, 2007). SLN-containing creams high lipid contents are potentially more effective than regular moisturizers.

9.7.1.3 Solid lipid nanoparticles in sebaceous glands targeting

The sebaceous gland opens its ducts at the hair follicle canal. It is possible to target the molecules through this canal to treat diseases such as acne and rosacea to achieve the beneficial effects of follicular penetration of topically applied creams loaded with an active therapeutic moiety.

9.7.1.4 Solid lipid nanoparticles in the treatment of hair diseases

Particulate drug delivery systems in the form of nanosized particles, especially SLNs, play a superior role in drug delivery for the treatment of alopecia and androgenetic than aqueous or alcoholic solutions of the same drug moiety . SLNs penetrate better the barriers and enter the openings of hair follicles and act as a depot for the drug while releasing the drug slowly into the hair follicle (Lasic, 1998).

9.7.1.5 Solid lipid nanoparticles in hair care

RBC Life science launched a nanomedical "Citrus Mint Shampoo" with a conditioner using nanocluster technology. The products suggest that the nanoclusters provide hair with a healthier shine for a longer period of time without damaging the hair (Müller-Goymann, 2004).

9.7.1.6 Solid lipid nanoparticles in cosmetics

SLNs are the next generation drug delivery systems after liposomes in skin-care and cosmetic preparations due to their less-systemic side effects and high-penetration potentials. The SC of the skin is a potential barrier for xenobiotics to enter viable skin to treat skin related diseases. The SLNs as lipid based drug delivery system improves the skin penetration since the epidermal lipids are found high in penetration barrier. The particulate system containing the lipid carrier is attached to the skin surface and exchanged between the outermost layers of the SC. The hydration of the SC may be increased by reduced corneocytes packing and expands the inner corneocytes gaps. This facilitates the drug penetration into the deeper skin. Nanoparticulate systems are 15-fold more occlusive than microparticulate systems. The SLNs have the particles smaller than 400 nm and containing more than 35% of lipids in over all content demonstrates the higher skin penetration and therapeutic efficacy.

9.7.1.7 Solid lipid nanoparticles in topical use

SLNs have been widely applied in the preparation of sunscreens and as an active carrier for molecular sunscreens and UV blockers. UV filters such as titanium dioxide and zinc oxide are superior in nanoform and constitute a transparent layer rather than a white layer over the skin for sun screening as it is more effective and convenient for usage. The in vivo study of skin hydration was increased by 31% after the administration of SLNs (4% for four weeks) compared to conventional creams. SLNs and NLCs have proven to control the release of occlusive topicals. A SLNs containing a vitamin A and glyceryl behenate demonstrated the better localisation of Vitamin a in the the upper layers of the skin as compared to the conventional formulation.

A common application of NLCs and SLNs for topical administration is their ability to accommodate drugs such as trampoline, imidazole antifungals, isotretinoin, ketoconazole, flurbiprofen, and glucocorticoids. The penetration of podophyllotoxin-SLN along the skin surface to the SC is useful in epidermal targeting. Vitamin A-loaded nanoparticles can be prepared by using glyceryl behenate as this method is useful in the sustained drug release dosage to improve skin penetration of the drug. The isotretinoin-loaded nanoparticles formulated for topical delivery by the hot homogenization technique increase the accumulative uptake of isotretinoin in the skin. Production of flurbiprofen-loaded SLN gels possess potential benefits in the delivery of drugs to the site of action through the skin and producing higher tissue concentrations. Polyacrylamide, glycerol, and water are used to prepare this type of SLN gel.

9.7.1.8 Solid lipid nanoparticles in transdermal delivery

The transdermal drug delivery of hydrophilic or lipophilic drugs are highly limited by strong lipophilic SC which blocks the permeation of hydrophilic molecules passing through it. Likewise, it also retains the lipophilic drugs within the layer and not allowing them to pass further. Nanoparticles have been explored as a strong drug carrier for transdermal drug delivery system (TDDS). It has been reported in much of the literature that the substance in nanoparticles can penetrate the lipophilic SC better due to the result of follicular targeting. The nanoparticle drug delivery of calcium carbonate SLN, NLC and polymeric nanoparticles are suitable to be incorporated in TDDS. Recently, polymeric nanoparticles combining transdermal drug delivery with the support of electroporation was developed for insulin delivery through the skin.

9.7.2 APPLICATION IN CANCER THERAPY

Over the past few decades, nanomedicines have impacted the chemotherapeutic approach significantly. Many chemotherapeutic agents have been converted into nanoparticles by encapsulating the chemotherapeutic drug in the nanoparticles. Chemotherapeutic agents encapsulated or loaded in nanoparticles proved to have tremendous potential in treating cancer. The targeting of the chemotherapeutic agent can be done successfully by using the ligands or by using a magnetic material such as Fe_2O_3 or Fe_3O_4, also called maghemite and magnetite. The ligand guides the nanoparticles containing chemotherapeutic agents to be delivered to cancer sites and binds to the cancer cells and acts better (Lu et al. 2009). Magnetic nanoparticles are already being used in the treatment of cancer by means of hyperthermia to kill particular cancer cells. A newer strategy that entails incorporating the chemotherapeutic agent in the magnetic nanoparticles for the treatment of cancer gives a double impact on hyperthermia and chemotherapy on cancer cells, and were found to be beneficial in the treatment of various types of cancers. Combining hyperthermia and chemotherapy is described as chemothermia which is under intensive research for treating various type of cancers such as pancreatic, colon, breast, and bladder cancer. The outcome of this research demonstrated the significant improvement and efficiency of anticancer drugs. The serious issues related to chemotherapy, such as side effects, can be reduced by the nanoparticle approach (Lu et al., 2009; Paliwal et al., 2009). Many types of nanoparticles can be used for the treatment of cancers including SLNs. SLNs are used as a potential nanocarrier for the encapsulation of chemotherapeutic agents and also improves the stability and efficacy of the encapsulated anticancer drug. Moreover, SLNs offer a high-drug loading capacity similar to polymeric nanoparticles. The improved pharmacokinetic properties and less in vivo toxicity are further advantages of SLNs which makes them a versatile nanodrug carrier for anticancer drugs. Usually, there are several challenges encountered with an anticancer drug such as determination of normal tissue toxicity, decreased specificity and stability, and the high incidence of resistance-related issues can be significantly overcome by delivering these agents through SLNs. The chance of rapid removal of nanoparticles by macrophages of RES (Reticuloendothelial System) is one of the obstacles in targeting the drugs to the particular tissues such as bone marrow and solid tumors.

9.7.2.1 Solid lipid nanopartcles in a solid tumor

SLNs are potential drug carriers for the treatment of various types of solid tumors. The drug tamoxifen incorporated in the SLN and administered through IV for breast cancer demonstrated prolonged drug release, higher permeability, and good retention. Drugs such as methotrexate can be loaded successfully in the SLNs for the treatment of different types of cancers. Camptothecin can be successfully loaded in the SLNs (as like methotrexate) for treating cancers. The SLNs can be used for the delivery of doxorubicin for the treatment of breast cancer (Paliwal et al., 2009).

Andrographolide is a potential cancer therapeutic agent isolated from Andrographis Paniculata Andrographolide and inhibited the in vitro proliferation of different tumor cell lines related to the various type of cancers (Rajugopal et al., 2003). An SLN of andrographolide was encapsulated into the SLN and for improving the aqueous solubility and lipophilicity. The antitumor activity studies on animals demonstrated tremendous potential against tumor cells.

9.7.2.2 Solid lipid nanoparticles in breast cancer

Mitoxantrone-loaded SLN was found to be beneficial in the treatment of breast cancer. Mitoxantrone can be given as infusions to reduce the toxic effects and improves the safety and bioavailability of the drug in the form of SLNs. Doxorubicin is known to be effective in the treatment of breast cancer when delivered through SLNs. Carrier SLNs of anti-cancer drugs against MCF-7 cells and murine breast cancer model was reported as a promising anticancer treatment for breast cancer (Lua et al., 2006). Drugs such as mitoxantrone, methotrexate, and paclitaxel demonstrated having a toxic effect on human breast cancer cell lines when prepared as SLNs.

9.7.2.3 Solid lipid nanoparticles in prostate cancer

Docetaxel nanoparticles demonstrated their potential against prostate cancer when prepared with prostate specific membrane antigen (PSMA). PSMA is a clinically validated tumor antigen expressed in the prostate cancer style (Hrkach et al., 2012).

9.7.2.4 Solid lipid nanoparticles in colon cancer

5-Fluorouracil is a drug widely used in SLN drug delivery systems by using different ingredients such as triglycerides and soy lecithin (ELdelbyassin et al., 2010) which are potential agents against colon cancer cells. The study found that SLNs proved high uptake capacity by the colon cancer cells. Oxaliplatin SLNs for colon cancer are found to be a potential carrier for colorectal tumors (Tumnala et al., 2014).

9.7.2.5 Solid lipid nanoparticles in pancreatic cancer

Pancreatic cancer is known for the poor survival rate and the over all survival period of the patient is ranging from few months to five years. Many nanoparticles have been prepared to combat pancreatic cancer by using drugs such as curcumin, sulforaphane, 5-FU, and GEM and demonstrated the promising results in pre[HYPHEN]clinical studies as compared to the conventional chemotherapy (Lu et al., 2009).

9.7.3 SOLID LIPID NANOPARTCLES IN GENE THERAPY

Drug delivery research is dominated by gene therapy which is a new paradigm and plays a potential role in the treatment of many diseases including genetic disorders, cystic fibrosis, hemophilia, and many somatic diseases such as tumor neurodegenerative diseases and viral infections such as AIDS (2006).

The nanovehicles are successfully employed as a carrier for delivering cytotoxic drugs to the particular tissues and are more efficient and have fewer side effects. Adequate nanoemulsions are used with ultrasound to deliver the genes into the particular in vitro or in vivo study-based cells. The nanoscale nature of nanoparticles noninvasively enter the largest tissues, such as tumor tissue, and deliver the drugs into the cells. Moreover, SLNs are more permeable through the cell membrane which can also be influenced by extreme fields such as ultrasound or magnetic field. SLNs can be utilized successfully as a vector for the delivery of peptide material/biomolecules such as DNA, plasmid DNA, and nucleic acids. The quality of exchange can be enhanced by the incorporation of the polar HIV-1 peptide (TAT2) done by SLN. SLNs are potential tools for gene therapy for in vivo protein expression by intravenous administration. The SLNs are proven to have potential to deliver DNA to induce the expression of a foreign protein (del Pozo et al., 2010). SLNs are shown to be beneficial in the nonviral vector for gene therapy. Retinal pigment epithelial cells are usually employed for diseases related to the retina. SLNs are successfully used for retinal gene therapy (del Pozo et al., 2008)

9.7.4 SOLID LIPID NANOPARTICLES IN TUBERCULOSIS

Antituberculosis drugs (ATD) can be successfully loaded into SLNs such as rifampin, isoniazid, and pyrazinamide. Many studies reported that SLNs have great potential in the treatment of tuberculosis by reducing the frequency of dosage and increasing the efficiency of ATD (Gohla, 2000; Lang et al.,1999).

9.7.5 SOLID LIPID NANOPARTCLES IN PARASITIC DISEASES

Parasitic diseases such as malaria, leishmaniasis, and trypanosomiasis can be treated successfully using SLNs. Antiparasitic drugs are the only choice to overcome this type of infection. These infections do not elicit a proper immune response so it is difficult to develop vaccines for them. SLNs and NLCs, due to their narrow size and special structures, are emerging as a drug delivery system for these infectious diseases. Nanocarriers such as liposomes, polymeric nanoparticles, and SLNs are improving the efficacy and tolerability of antiparasitic drugs (Gohla, 2000).

9.7.6 SOLID LIPID NANOPARTCLES IN BRAIN-TARGETING

The blood–brain barrier (BBB) is a dynamic, impermeable, highly selective barrier which limits the passage into the brain of various molecules such as pathogenic organisms, harmful substances, and xenobiotics. BBB allows the permeation of nutrients for proper function. It has already been reported that lipid-soluble molecules with M.W of 450 daltons can pass through the BBB by passive diffusion. More than 99% of drugs cannot cross the BBB and enter the CNS instead SLNs are promising drug delivery carriers to the brain. If SLNs demonstrates the size below 50 nm and formulated with lipid carrier having the melting point slightly higher than the body temperature maintains the solid state after administration in side the body. Thereby, the SLNs retains the physical integrity and delivers the drug to the brain. Due to their lipophilicity and small size, SLNs are less cytotoxic and easily avoid P-GP afflux of brain endothelial cells when coated with polysorbate-80. Polysorbate-80 improves the brain-specific delivery by its high affinity to adsorb plasma-apo-lipoproteins on the surface and recognizing LDL receptors, followed by the uptake of the endothelial cell lining of brain tissues.

9.7.7 OTHER APPLICATIONS

SLNs are useful in various other applications such as ocular drug delivery, pulmonary drug delivery, and vaccine drug delivery to the lungs. SLNs can be used as transfection agents for nervous applications. SLNs have also been found to be beneficial in treating rheumatoid arthritis, skin hyperpigmentation, and atopic dermatitis which can be reviewed from various literature.

9.8 CONCLUSION

SLNs are versatile drug carriers which can be declared as the next-generation novel carriers. SLNs are advantageously easy to prepare, safe, biocompatible, and capable of crossing the BBB. SLNs are highly recommended for skin-related formulations and nanocosmeceuticals due to their efficient permeability through the skin. Various skin preparations are in the pipeline based on SLNs strategy which may appear in the market in the near future. The preparation methods of SLNs, characterization, and release studies are already well established, but there is a need for improved data related to the safety, toxicity, and tolerability of nanoparticles. Although they are biocompatible, in future more authenticated data on these areas are likely to be published which may take SLNs to next level of application and commerce.

REFERENCES

Almeida, A.J., Eliana, S., 2007. Adv. Drug Deliv. Rev. 5 (9), 479–490.

Anisha, S., Kumar, S.P., Kumar, G.V., Garima, G., 2010. Approaches used for penetration enhancement in transdermal drug delivery system. Int. J. Pharm. Sci. 2 (3), 708–716.

Arora, S., Rajwade, J.M., Paknikar, K.M., 2012. Nanotoxicology and in vitro studies: the need of the hour. Toxicol. Appl. Pharmacol. 258 (2), 151–165.

Avgoustakis, K., Beletsi, A., Panagi, Z., Klepetsanis, P., Karydas, A.G., Ithakissios, D.S., 2002. PLGA-mPEG nanoparticles of cisplatin: in vitro nanoparticle degradation, in vitro drug release and in vivo drug residence in blood properties. J. Control Release 79 (1–3), 123–135 [PubMed].

Balakrishnan, P., Shanmugam, S., Lee, W.S., Lee, W.M., 2009. Formulation and *in vitro* assessment of minoxidil niosomes for enhanced skin delivery. Int. J. Pharm. 377, 1–8 [PubMed].

Bangham, A.D., 1963. Physical structure and behavior of lipids and lipid enzymes. Adv. Lipid Res. 64, 65–104.

Battaglia, L., Gallarate, M., Panciani, P.P., Ugazio, E., Sapino, S., Peira, E., et al., 2014. Techniques for the preparation Of solid lipid nano and microparticles, Application of Nanotechnology in Drug Delivery, Ali Demir Sezer, IntechOpen, http://dx.doi.org/10.5772/58405.

Bei, D., Meng, J., Youan, B.-B.C., 2010. Engineering nanomedicines for improved melanoma therapy: progress and promises. Nanomedicine 5 (9), 1385–1399.

Bhardwaj, U., Burgess, D.J., 2010. Physicochemical properties of extruded and non-extruded liposomes containing the hydrophobic drug dexamethasone. Int. J. Pharm. 388 (1–2), 181–189 [PubMed].

Bhimavarapu, R., RamaDevi, R., Nissankararao, S., Devarapalli, C., Paparaju, S., 2013. Microsponges: as a novel imperative for drug delivery system. Res. J. Pharm. Technol. 6, 842–848.

Brandt, F.S., Cazzaniga, A., Hann, M., 2011. Cosmeceuticals: current trends and market analysis. Semin. Cutan. Med. Surg. 30 (3), 141–143.

Cevc, G., 1996. Transfersomes, liposomes and other lipid suspensions on the skin: permeation enhancement, vesicle penetration, and transdermal drug delivery. Crit. Rev. Ther. Drug Carrier Syst. 13, 257–388 [PubMed].

Chidambaram, N., Burgess, D.J., 1999. A novel in vitro release method for submicron sized dispersed systems. AAPS Pharm. Sci. 1 (3), E11 [PubMed] compounds. Biomaterials 2002; 23: 2283-94.

Cortesi R., Esposito E., Luca G., Nastruzzi C., 2002. Production of Lipospheres as Carriers For Bioactive.

De Jong, W.H., Borm, P.J.A., 2008. Drug delivery and nanoparticles: applications and hazards. Int. J. Nanomed. 3 (2), 133–149.

Denda, M., Wood, L.C., Emami, S., Calhoun, C., Brown, B.E., Elias, P.M., et al., 1996. The epidermal hyperplasia associated with repeated barrier disruption by acetone treatment or tape stripping cannot be attributed to increased water loss. Arch. Dermatol. Res. 288, 230–238 [PubMed].

Drake, B., Prater, C.B., Weisenhorn, A.L., Gould, S.A.C., Albrecht, T.R., Quate, C.F., et al., 1989. Imagingcrystals polymers and process in water with the AFM. Science 243, 1586–1589.

Duncan, T.V., July 2011. Applications of Nanotechnology in Food Packaging and Food Safety: Barrier Materials, Antimicrobials and Sensors.

Dunford, R., Salinaro, A., Cai, L., et al., 1997. Chemical oxidation and DNA damage catalysed by inorganic sunscreen ingredients. FEBS Lett. 418, 87–90.

Eglash, R., 2011. Nanotechnology and traditional knowledge systems. In: Maclurcan, D., Radywyl, N. (Eds.), Nanotechnology and Global Sustainability. CRC Press, pp. 45–68.

Ekambaram, P.A., Sathali, A.H., Priyanka, K., 2012. Solid lipid nanoparticles: a review. Sci. Rev. Chem. Commun. 2 (1), 80–102.

Embil, K., Nacht, S., 1996. The microsponge delivery system (MDS): a topical delivery system with reduced irritancy incorporating multiple triggering mechanisms for the release of actives. J. Microencapsul. 13, 575–588.

Fahr, A., Liu, X., 2007. Drug delivery strategies for poorly water soluble drugs. Expert. Opin. Drug Deliv. 4 (4), 403–416.

Fred, Z., Esther, B., Daniel, S., Christina, L., Franz, S. Preparation and Properties of Coenzyme Q10 Nanoemulsions, Cosmetic Science Technology. (Last accessed on 2006). Available from: <http://www.mibellebiochemistry.com/pdfs/Preparation_and_Properties_of_Coenzyme_Q10_Nanemulsions_Cosm_Sci_Technol_2006.pdf>.

Freitas, C., Muller, R.H., 2012. Spray-drying of solid lipid nanoparticles (SLN-TM). Eur. J. Pharm. BioPharm 46, 145–151.

Friends of the Earth Report. Nanomaterials, Sunscreens and Cosmetics: Small Ingredients Big Risks. (Last accessed on 2006). Available from: <http://www.nano.foe.org.au. http://www.foe.org>.

Fulekar, M.H., 2010. Nanotechnology: Importance and Application. IK International Publishing House, New Delhi, India.

GBI Research. Cosmeceuticals Market to 2018—Technological Advances and Consumer Awareness Boost Commercial Potential for Innovative and Premium-Priced Products. <http://www.researchandmarkets.com/reports/2393091/cosmeceuticals_market_to_2018_technological>.

Gohla, S., 2000. Eur. J. Pharm. Biopharm. 50, 161–177.

Govender, T., Choonara, Y.E., Kumar, P., du Toit, L.C., Modi, G., Naidoo, D., et al., 2015. Polymers 7, 1145–1176. Available from: https://doi.org/10.3390/polym7061145. A Novel Melt-Dispersion Technique for Simplistic Preparationof Chlorpromazine-Loaded Polycaprolactone Nanocapsules.

Hoo, C.M., Starostin, N., West, P., et al., 2008. J Nanopart. Res. 10 (Suppl 1), 89. Available from: https://doi.org/10.1007/s11051-008-9435-7.

Hu, F.Q., Yuan, H., Zhang, H.H., Fang, M., et al., 2012. Preparation of solid lipid nanoparticles with clobetasol propionate by a novel solvent diffusion method in aqueous system and physicochemical characterization. Int. J. Pharma. 239, 121–128.

Hwang, S.L., Kim, J.C., 2008. *In vivo* hair growth promotion effects of cosmetic preparations containing hinokitiol-loaded poly(epsilon-caprolacton) nanocapsules. J. Microencapsul. 25, 351–356 [PubMed].

Inoue, J., Aramaki, Y., 2006. Toll-like receptor 9 induced by tape-stripping triggers on effective immune response with CpG-oligodeoxynucleotides. Vaccine 25, 1007–1013 [PubMed].

Jenning, V., Gysler, A., Schafer-Korting, M., Gohla, S., 2000. Vitamin A loaded solid lipid nanoparticles for topical use: occlusive properties and drug targeting to the upper skin. Eur. J. Pharm. Biopharm. 49, 211–218.

Junemann, D., Dressman, J., 2012. Analytical methods for dissolution testing of nanosized drugs. J. Pharm. Pharmcol. 64 (7), 931–943 [PubMed].

Kaur, I.P., Agrawal, R., 2007. Nanotechnology: a new paradigm in cosmeceuticals. Recent Patents Drug Deli. Formul. 1 (2), 171–182.

Kaur, I.P., Bhandari, R., Bhandari, S.K., 2008. J. Controlled Release 127, 97–109.

Kesselman, E., Efrat, R., Garti, N., Danino, D. Formation of Cubosomes as Vehicles of Biologically Active Substances. (Last accessed on 2007). Available from <http://www.materials.technion.ac.il/ism/Docs/2007/LifeAbstracts/Poster/E_Kesselman.pdf>.

Kuo, Y.C., Chen, H.H., 2009. Entrapment and Release of saquinavir using novel cationic solid lipid nanoparticles. Int. J. Pharm. 365, 206–213.

Lang, S., Lu, L.F., Cai, Y., Zhu, J.B., Liang, B.W., Yang, C.Z., 1999. J. Controlled Release 59, 299–307.

Lasic, D.D., 1998. Novel applications of liposomes. Trends Biotechnol. 16 (7), 307–321.

Lohani, A., Verma, A., Joshi, H., Yadav, N., Karki, N., May 2014. Nanotechnology-Based Cosmeceuticals, ISRN Dermatology, vol. 2014, Article ID 843687, 14 pages, 2014. Available from https://doi.org/10.1155/2014/843687.

Lu, Q.Z., Yu, A., Xi, Y., Li, H., Song, Z., Cui, J., et al., 2009. Int. J. Pharm. 372, 191–198.

Lua, B., Xionga, S.-B., Yanga, H., Yina, X.-D., Chaoa, R.-B., 2006. Eur. J. Pharmaceutical. Sci. 28 (1–2), 86–95.

Lv, Q., Yu, A., Li, H., Song, Z., et al., 2009. Development and evaluation of penciclovir-loaded solid lipid nanoparticles for topical delivery. Int. J. Pharm. 372, 191–198.

Mahajan, P.S., Mahajan, K.B., Darekar, A.B., September 2015. A Review on Solid Lipid Nanoparticle (SLN): An Advanced Treatment Modality.

Manjunath, K., Venkateswarlu, V., 2012. Preparation, Characterization and In Vitro Rel. Kinetics of Clozapine Solid Lipid Nanoparticles. J. Controlled Release 95, 627–638.

Martins, S., Sarmento, B., Ferreira, D.C., Souto, E.B., 2007. Stability Studies and Characterization of SLN Formulations.

Marzulli, F.N., Brown, D.W.C., Maibach, H., 1969. Techniques for studying skin penetration. Toxicol. Appl. Pharmacal. (Suppl. 3), 76.

Mehlen, A.Z., Schwarz, C., Mehnart, W., et al., 2013. Eur. J. Pharm. Biopharm. 45, 149–155.

Mehnart, W., Mader, K., 2011. Problem associated with SLN. Adv. Drug Delivery Rev. 1165–1195.

Mehnert, W., Mader, K., 2010. Solid lipid nanoparticles-Production, characterization and applications. Adv. Drug Del. Rev. 47, 165–196.

Menon, G.K., Feingold, K.R., Elias, P.M., 1998. Lamellar body secretory response to barrier dis-ruption. J. Invest. Dermatol. 98, 279–289 [PubMed].

Meyer, E., Heinzelmann, H., 1992. Scanning force microscopy. In: Wiesendanger, R., Guntherodt, H.J. (Eds.), Scanning Tunneling Microscopy II, Surface Science. Springer Verlag, New York, pp. 99–149.

Mezei, M., Gulasekharam, V., 1980. Liposomes – a selective drug delivery system for the topical route of administration. I. Lotion dosage form. Life Sci. 26 (18), 1473–1477.

Mishra, B., Patel, B.B., Tiwar, S., 2010. Colloidal Nano carrier. Nano Med. 6 (1), 9–24.

Mu, L., Sprando, R.L., 2010. Application of nanotechnology in cosmetics. Pharm. Res. 27 (8), 1746–1749.

Mukharjee, S., Ray, S., Thakur, R.S., 2009. Ind. J. Pharm. 349–358.

Müller, R.H., Dingler, A., 1998. The next generation after the liposomes: solid lipid nanoparticles (SLN, Lipopearls) as dermal carrier in cosmetics. Eurocosmetics 7/8, 19–26.

Muller, R.H., Runge, S.A., 1998. Solid lipid nanoparticles (SLN) for controlled drug delivery. In: Benita, S. (Ed.), Submicron Emulsions in Drug Targeting and Delivery. Harwood Academic Publishers, Amsterdam, pp. 219–234.

Müller, R.H., Radtke, M., Wissing, S.A., 2002. Solid lipid nanoparticles (SLN) and nanostructured lipid carriers (NLC) in cosmetic and dermatological preparations. Adv. Drug Deliv. Rev. 54 (Suppl 1), S131–S155.

Muller, R.H., et al., 2005. Production of biofunctionalized solid lipid nanoparticles for site-specific drug delivery. Biol. Pharm. Nanomater. 291–307. Challa Kumar (1).

Muller, R.H., Mader, K., Gohla, S., et al., 2011. Eur. J. Pharm. Biopharm. 50 (1), 161–177.

Müller-Goymann, C.C., 2004. Physicochemical characterization of colloidal drug delivery systems such as reverse micelles, vesicles, liquid crystals and nanoparticles for topical administration. Eur. J. Pharm. Biopharm. 58 (2), 343–356.

Muthu, M.S., Singh, S., 2009. Poly(D,L-lactide) nanosuspensions of risperidone for parenteral delivery: formulation and in-vitro evaluation. Curr. Drug Deliv. 6 (1), 62–68 [PubMed].

Orlu, M., Cevher, E., Araman, A., 2006. Design and evaluation of colon specific drug delivery system containing flubiprofen microsponges. Int. J. Pharm. 318, 103–117.

Padamwar, M.N., Pokharkar, V.B., 2006. Development of vitamin loaded topical liposomal formulation using factorial design approach: drug deposition and stability. Int. J. Pharm. 320 (1–2), 37–44.

Paliwal, R., Rai, S., Vaidya, B., Khatri, K., Goyal, A.K., Mishra, N., et al., 2009. Nanomed. Nanotechnol. Biol. Med. 5 (2), 184–191.

Petersen, R., 2008. US Patent 60/866233 Nanocrystals for Use in Topical Cosmetic Formulations and Method of Production Thereof. Abbott GmbH and Co.

Pierard-Franchimont, C., Pierard, G.E., 1987. Assessment of aging and actinic damages by cyanoacrylate skin surface strippings. Am. J. Dermatopathol. 9, 500–509 [PubMed].

Pierfrancesco, M., 2010. Use and potential of nanotechnology in cosmetic dermatology. Clin. Cosmet. Investig. Dermatol. 3, 5–13.

Poon, V.K.M., Burd, A., 2004. In vitro cytotoxity of silver: implication for clinical wound care. Burns 30 (2), 140–147.

Rabinarayan, P., Padilama, S., 2010. Production of solid lipid nanoparticles-drug loading and release mechanism. J. Chem. Pharm. Res. 2 (1), 211–227.

Ribier, A., Biatry B., May 2000. Cosmetic or Dermatologic Oil/Water Dispersion Stabilized With Cubic Gel Particles and Method of Preparation, European Patent 0711540B1.

RNCOS E-Services Pvt. Ltd., 2016. Global Cosmeceuticals Market Outlook. <http://www.giiresearch.com/report/rnc263147-global-cosmeceuticals-marketoutlook.html>.

Rupenagunta, A., Somasundaram, I., Ravichandiram, V., Kausalya, J., Senthilnathan, B., 2011. Solid lipid nanopar-ticles—a versatile carrier system. J. Pharm. Res. 4 (7), 2069–2075.

Sankhyan, A., Pawar, P., 2012. Recent trends in noisome as vesicular drug delivery system. J. Appl. Pharm. Sci. 2, 20–32.

Schueller, R., Romanowski, P. Emerging Technologies and the Future of Cosmetic Science, Cosmetics and Toiletries. (Last accessed on 2006); Available from: <http://www.specialchem4cosmetics.com/services/articles.aspx?id = 957>.

Schwarz, C., Mehnert, W., Lucks, J.S., Muller, R.H., 1994. Solid lipid nanoparticles (SLN) for controlled drug deli-very I. Production, characterization and sterilization. J. Controlled Release 30 (1), 83–96.

Sharma, V.K., Diwan, A., Sardana, S., Dhall, V., 2011. Int. J. Res. Pharm. 2 (3), 450–461.

Singh, R., Tiwari, S., Tawaniya, J., 2013. Review on nanotechnology with several aspects. Int. J. Res. Comput. Eng. Electron. 2 (3), 1–8.

Sinha, V.R., Srivastava, S., Goel, H., Jindal, V., 2010. Solid lipid nanoparticles (SLN'S) – trends and implications in drug targeting. Int. J. Adv. Pharm. 1, 212–238.

Song, C., Liu, S., 2005. A new healthy sunscreen system for human: solid lipid nanoparticles as carrier for 3,4,5-trimethoxybenzoylchitin and the improvement by adding vitamin E. Int. J. Biol. Macromol. 36, 116–119 [PubMed].

Sonneville-Aubrun, O., Simonnet, J.T., L'Alloret, F., 2004. Nanoemulsions: a new vehicle for skincare products. Adv. Colloid Interface Sci. 108–109, 145–149 [PubMed].

Spicer, P.T., Lynch, M.L., Visscher, M., Hoath, S., 2003. Bicontinuous cubic liquid crystalline phase and cubosome personal care delivery systems. In: Rosen, M. (Ed.), Personal Care Delivery Systems and Formulations. Noyes Publishing, Berkshire, UK.

Tak, Y.K., Pal, S., Naoghare, P.K., Rangasamy, S., Song, J.M., 2015. Shape-Dependent Skin Penetration of Silver Nanoparticles: Does It Really Matter? www.nature.com/scientificreports. Sci. Rep.

Takeda, K., Suzuki, K.-I., Ishihara, A., et al., 2009. Nanoparticles transferred from pregnant mice to their off-spring can damage the genital and cranial nerve systems. J. Health Sci. 55 (1), 95–102.

Thong, H.Y., Zhai, H., Maibach, H.I., 2007. Percutaneous penetration enhancers: an overview. Skin Pharmacol. Physiol. 20, 272–282 [PubMed].

Touitou, E., Dayan, N., Bergelson, L., Godin, B., Eliaz, M., 2000. Ethosomes-novel vesicular carriers for enhanced delivery: characterization and skin penetration properties. J. Controlled Release 65, 403–418 [PubMed].

Trotta, M., Debernardi, F., Caputo, R.O., et al., 2003. Preparation of solid lipid nanoparticles by a solvent emulsification–diffusion technique. Int. J. Pharm. 257, 153–160.

Uchegbua, I.F., Vyas, S.P., 1998. Non-ionic surfactant based vesicles (niosomes) in drug delivery. Int. J. Pharm. 172, 33–70.

Van der Walk, P.G., Maibach, H.I., 1990. A functional study of the skin barrier to evaporative water loss by means of repeated cellophane-tape stripping. Clin. Exp. Dermatol. 15, 180–182.

Vasilev, C., Brindley, A.A., Olsen, J.D., Saer, R.G., Beatty, J.T., Hunter, C.N., 2013. Nano-mechanical mapping of the interactions between surface-bound RC-LH1-PufX core complexes and cytochrome c 2 attached to an AFM probe. Photosynth Res. 120 (1-2), 169–180.

Vogel, J.S., 1992. Radiocarbon 34, 344–350.

Vogel, J.S., Turtletaub, K.W., 1991. Trends Anal. Chem. 11, 142–149.

Wagner, H., Kostka, K.H., Lehr, C.M., Schaefer, U.F., 2003. pH profiles in human skin: influence of two in vitro test systems for drug delivery testing. Eur. J. Pharm. Biopharm. 55, 57–65 [PubMed].

Wallace, S.J., Li, J., Nation, R.L., Boyd, B.J., 2012. Drug release from nanomedicines: selection of appropriate encapsulation and release methodology. Drug Deliv. Transl. Res. 2 (4), 284–292 [PubMed].

Wissing, S.A., Müller, R.H., 2003. Cosmetic applications for solid lipid nanoparticles (SLN). Int. J. Pharm. 254, 65–68 [PubMed].

Wissing, S.A., Kayser, O., Muller, R.H., et al., 2010. Solid lipid nanoparticles for parenteral drug delivery. Adv. Drug Deliv. Rev. 56, 1257–1272.

Xu, X., Khan, M.A., Burgess, D.J., 2012. A two-stage reverse dialysis in vitro dissolution testing method for passive targeted liposomes. Int. J. Pharm. 426 (1–2), 211–218 [PubMed].

Yuan, H., Huang, L., Du, Y., Ying, X., You, J., Hu, F., et al., 2008. Solid lipid nanoparticles prepared by solvent diffusion method in a nanoreactor system. Colloids Surf. B: Biointerface 61, 132–137.

FURTHER READING

Cosmeceuticals: Products and Global Markets, <http://www.bccresearch.com/market-research/advanced-materials/cosmeceuticals-global-markets-avm099a.html>.

Law 360, 2011. Nano-Cosmetics: Beyond Skin Deep. (Last accessed on 2011). Available from: <http://www.shb.com/newsevents/2011/NanoCosmeticsBeyondSkinDeep.pdf>.

Mehnart, W., Mader, K., 2001. Adv. Drug Deliv. Rev. 47, 165–196.

Yu Kyung Tak, Sukdeb Pal, Pravin K. Naoghare, Sabarinathan Rangasamy & Joon Myong Song, Shape-Dependent Skin Penetration of Silver Nanoparticles: Does It Really Matter? Scientific Reports volume 5, Article number: 16908 (2015)

Nanoemulsion based on Phosphoric Acid Fatty Acid Esters and Its Uses in the Cosmetics, Dermatological, Pharmaceutical, and/or Ophthalmological Fields. L'Oréal, US Patent 6274150.

Nanotechnology in Cosmetics Analysed. (Last accessed on 2010). Available from: <http://www.personalcaremagazine.com/Story.aspx?Story = 7419>.

SCCP (Safety Committee on Consumer Products). Opinion on Safety of Nanomaterials in Cosmetic Products. (Last accessed on 2007). Available from: <http://www.ec.europa.eu/health/ph_risk/committees/04_sccp/docs/sccp_o_123.pdf>.

U.S. Food and Drug Administration. Cosmetics Q&A: FDA's Authority. <http://www.fda.gov/Cosmetics/ResourcesForYou/Consumers/CosmeticsQA/ucm135709.htm>.

U.S. Food and Drug Administration. Is it a Cosmetic, A Drug, or Both? (Or is it Soap?), <http://www.fda.gov/cosmetics/guidancecomplianceregulatoryinformation/ucm074201.htm>.

CHAPTER

A new approach for drug targeting to the central nervous system: lipid nanoparticles

10

Berrin Küçüktürkmen and Asuman Bozkır

Department of Pharmaceutical Technology, Ankara University Faculty of Pharmacy, Ankara, Turkey

10.1 INTRODUCTION

When compared to the diseases of the peripheral organs, diagnosis of central nervous system (CNS) diseases is rather difficult, and any CNS injury resulting in neuronal death can not be regenerative. Most of the existing drugs are not effective in the treatment of CNS diseases because of their failure in reaching the target area in the CNS. The reasons for the failure of drugs in reaching their target areas in the CNS include the first passage metabolism of the drugs in the systemic circulation, uptake by the other tissues in the body, and the blood–brain barrier (BBB) that physically and chemically inhibit the passage of the drugs to brain parenchyma (Stockwell et al., 2014). Methods that can potentially increase the passage of drugs in effective concentrations are attracting great attention and the most recent studies carried out aim at targeting drugs to the brain (Gastaldi et al., 2014).

Nanoparticles comprise particles in nano dimensions with distinct properties and functions. Since the minimal diameter of blood capillaries is within the 6–9 μm, drug carrier systems in nano dimensions are capable of reaching organs through the blood circulation. The average size of the cells in human body is 10–20 μm, therefore, uptake of drugs via drug carriers in nano dimensions is also possible. However, nanoparticles are foreign particles for the body defense system and they can easily be opsonized from the circulation before they reach their targets and removed from the circulation through phagocytosis. Based on this, stability of nanoparticles in blood can be increased by modifying their size, charge and surface characteristics, and their uptake by the reticuloendothelial system (RES) in the liver and spleen and glomerular excretion by the kidney can be prevented. Nanoparticles in sizes around 100 nm with neutral surfaces generally stay in the blood circulation for longer periods (Craparo et al., 2011). Together with this, the functionalization of nanoparticle surfaces with targeting ligands can provide

Nanoarchitectonics in Biomedicine. DOI: https://doi.org/10.1016/B978-0-12-816200-2.00014-1

passage through the BBB (Li et al., 2016). The number of products based on microparticles or nanoparticles is limited because of the cytotoxicity of existing polymeric particles and the lack of suitable large-scale production methods. Lipid nanoparticles developed as an alternative to polymeric nanoparticles constitute a new industrial area considered by the pharmaceutical industry because of the biocompatibility of the lipids included in their structures and the possibility of large-scale production. Solid lipid nanoparticles (SLNs), the first-produced lipid nanoparticles, are the dispersions in nano dimension of lipids such as triglycerides, fatty acids, or paraffin which are solid at body temperature and stabilized with surfactants. Excipients used in the preparation of SLNs are included in the GRAS (generally recognized as safe) class as approved by the US Food and Drug Administration (FDA) (Kumar and Randhawa, 2013). With the purpose of coping with the disadvantages of SLNs including their low capacity of drug-load or the observation of drug expulsion during storage, a new generation of lipid nanoparticles called nanostructured lipid carriers (NLCs) have been developed. NLCs include liquid lipids in their structure together with solid lipids. Lipid-drug conjugates (LDCs) have been developed to cope with the low drug-load capacities for hydrophilic drugs in SLNs and NLCs. Following the preparation of the insoluble conjugate of active ingredient with lipid, the LDC nanoparticles are created using a suitable method. Lipid nanocapsules (LNCs), which are another type of lipid nanoparticles, comprise a liquid or semiliquid core and an external layer of solid lipid. In contrast with SLNs and NLCs, LNCs have a core-shell structure. Improvement of the pharmacokinetic profiles of drugs and providing for the passage of drugs to the brain in high concentrations are hoped for with these new lipid nanoparticle formulations (Kaur et al., 2008; Gastaldi et al., 2014). Surfaces of lipid nanoparticles are modified with targeting agents with the purpose of increasing their passage to the brain. Polyethylene glycol (PEG), polysorbate 80, transferrin (Tf), folate, cell penetrating peptides, monoclonal antibodies, apolipoprotein E, and positively charged lipids and surfactants are frequently used to functionalize the lipid nanoparticle surface in order to target drugs to the brain.

The importance of the administration route of drugs is also great in ensuring the passage of drugs to the brain in effective concentrations. Drugs are administered directly into the brain using intracerebroventricular and intraparenchymal administration methods. In this way, the BBB will not cause any hindrance and high drug concentrations can be achieved. However, both of these methods are invasive and drug diffusion from the injection site is not easy. Intraperitoneal, intravenous or subcutaneous administration methods are less invasive; however, most of the drugs will not pass the BBB following parenteral administration. Intranasal drug administration has been investigated frequently in recent years since it is not invasive and it has the capacity of bypassing the BBB through olfactory and trigeminal pathways (Devkar et al., 2014; Gabal et al., 2014; Gartziandia et al., 2015; Sharma et al., 2016).

The successful findings achieved in preclinical studies targeting the treatment of CNS diseases are highly important since they ensure the shift to clinical studies. Different strategies to deliver drugs to the brain have been developed.

In this chapter, we discuss the lipid nanoparticles targeted to the brain for the treatment of CNS diseases as well as summarizing some recent studies.

10.2 CHALLENGES IN DRUG DELIVERY TO THE CENTRAL NERVOUS SYSTEM

There are two regions in the brain that ensure the passage of the blood components to the CNS; BBB and the blood–cerebrospinal fluid barrier (BCSFB). BCSFB is created by the epithelial cells of the choroid plexus that produce the cerebrospinal fluid. Although BCSFB is another barrier against the passage of drugs to the CNS, it is about 5000 times smaller than the BBB. Therefore, BBB is the key barrier against the passage of drugs to the CNS (Stockwell et al., 2014). BBB is a semipermeable membrane that protects the brain from harmful substances and toxins that circulate within the blood while allowing the passage of nutrients (Gentry et al., 1999).

10.2.1 STRUCTURE OF THE BRAIN–BLOOD BARRIER

Existence of a selective barrier between blood and brain was first discovered by Paul-Ehrlich in 1880. In 1900, Lewandowsky reported that there was a mechanical membrane between blood and brain that diminished the effectiveness in the CNS of some drugs administered intravenously. Goldman used tripane blue to stain the brain and spinal cord tissues in 1909 and 1913 and showed that there was no such barrier between the cerebrospinal fluid and the brain. The studies carried out with ferritine and horseradish peroxidase using electron microscope showed that BBB was localized in the tight junctions between the adjacent brain endothelial cells (Zlokovic, 2008).

Cerebral microvessels comprise of four types of cells: endothelial cells, pericytes, astrocytes, and cells providing for the neuronal stimulation of endothelial cells. BBB inhibits the passage to the brain both physically (tight junctions) and through metabolic barriers (enzymes). The tight junctions located between the endothelial cells that make the endothelial BBB constitute an uninterrupted cellular membrane preventing the passage of molecules between blood and intracellular fluid in the brain. Tight junctions are made from Claudin (which is called "the triple membrane protein"), occludin, and junction adhesion molecules together with many other peripheral proteins that contribute to the junction complex (Zlokovic, 2008). The tight junctions located between the endothelial cells that make the BBB create a high electrical resistance and prevent the penetration of potentially therapeutical agents. While this electrical resistance is between 3 and 33 Ω/cm^2 in tissues, it reaches 1500–2000 Ω/cm^2 in the BBB. There are no gaps between these endothelial cells that make up the BBB, they have no pinocytic activity, they have more mitochondria compared to the other endothelial cells in the body, and their metabolic activity is high (Oldendorf et al., 1977; Visser et al., 2005; Brasnjevic et al., 2009). Enzymatic barriers found in the cerebral endothelium have active phase 1 (cytochrome P450, flavine-linked oxygenases, monoamine oxydases, reductases and hydrolases) and phase 2 (conjugating enzymes, sulfo-transferases, GSH-transferases, UDP-glucuronozyl transferases) enzymes. These enzymatic barriers in the cerebral endothelium are capable of metabolizing many substances and drugs that have passed through the BBB. There are many enzymes that are specific for brain capillaries. Enzymes including γ-glutamyl transpeptidase (γ-GTP), alkalene phosphatase, and aromatic acid decarboxylase are found in high concentrations in the cerebral microvessels, while they are found either in very low concentrations or not at all in nonneural capillaries. Functioning of γ-GTP and alkalen phosphatase in the luminal side of the endothelium and functioning of Na^+-K^+-ATPase and sodium-dependent neutral aminoacid carrier system in the abluminal side of the endothelium and provides the membranous side of the endothelium with polar characteristics (Begley and Brightman, 2003).

BBB prevents the passage of compounds from blood to the extracellular environment of the brain. While 98% of active ingredients with small molecules are unable to pass the BBB, almost the entire amount of the active ingredients with large molecules are unable to pass through the BBB (Jaypraka et al., 2009). Any active ingredient must be soluble in lipids, nonionized in the physiologic pH, have small molecular weight, and binding to serum proteins must be low in order to pass the BBB.

10.2.2 MECHANISMS OF PASSAGE THROUGH THE BRAIN–BLOOD BARRIER

A schematic representation of the mechanisms of passage through the BBB is given in the Fig. 10.1.

10.2.2.1 Paracellular and transcellular diffusion

Paracellular diffusion is a nonsaturable and noncompetitive process that occurs between cells. Tight junctions found between the brain endothelial cells greatly inhibit paracellular diffusion. Small molecules that are soluble only in water can pass through the BBB in small amounts. Transcellular diffusion occurs through the cells, and likewise, is nonsaturable and noncompetitive. Substances must be highly lipophilic and must have low molecular weight to pass through the BBB with transcellular diffusion (Gastaldi et al., 2014). Immunocytes can pass through the CNS endothelium with paracellular and transcellular mechanisms (Begley and Brightman, 2003).

10.2.2.2 Passive diffusion

Passive diffusion is a process based on the movement of molecules from high concentration to low concentration without using energy. Passive diffusion occurs

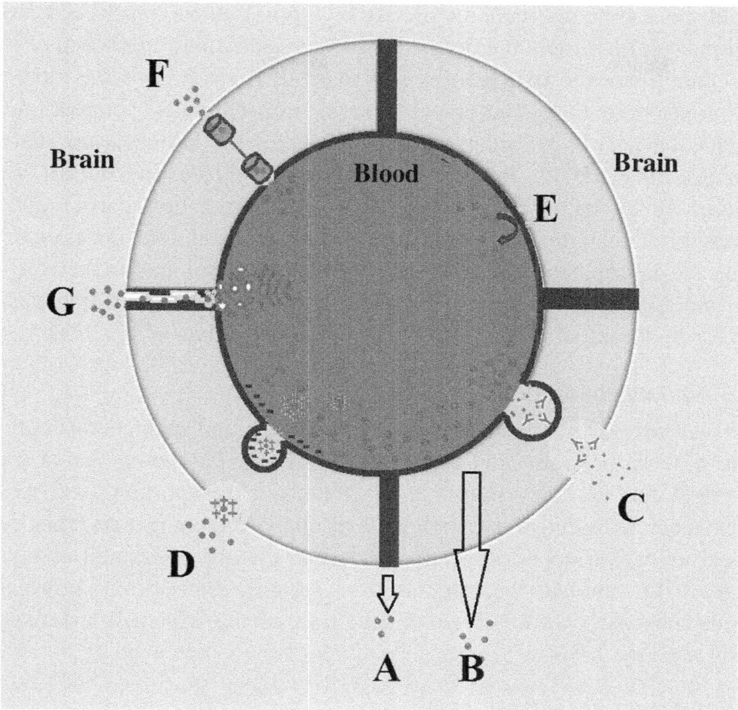

FIGURE 10.1

Mechanisms of transport of molecules into the brain through the blood–brain barrier: (A) paracellular diffusion; (B) transcellular diffusion; (C) receptor-mediated endocytosis; (D) adsorptive endocytosis; (E) carrier-mediated efflux; (F) carrier-mediated transport; and (G) focused ultrasound-mediated BBB disruption. *BBB*, Brain–blood barrier.

via the transcellular route because of the tight junctions between the endothelial cells in the brain (Goyal et al., 2014). Some lipophilic substances (such as alcohol, narcotics, and anticonvulsants) can passively diffuse through the endothelial cells and enter the CNS (Begley and Brightman, 2003).

10.2.2.3 Carrier-mediated efflux

The P-glycoprotein (P-gp), which is an adenosine triphosphate-binding cassette (ABC) transporter, multiple drug resistance protein (MRP), multi-specific organic anion transporter (MOAT), and breast cancer-resistant protein (BCRP) constitute the efflux mechanisms responsible for the back transport of most of the compounds outside the CNS and development of multiple drug resistance. ABC transporters come in the first place among the efflux pump transporters and they are capable of hydrolyzing ATP that provides energy for the movement of substrates against the concentration gradient. P-gp is the first drug efflux transporter detected in the endothelial cells of the BBB. P-gp is an active efflux transporter for both lipids and hydrophobic drugs. It has been stated that many active ingredients that cannot pass the BBB can be substrates of P-gp. P-gp is the most important ABC transporter that is an apical-to-basal trans-epithelial permeation barrier and inhibits the access of numerous active ingredients to the brain (Pardridge, 1997; Tahara et al., 2011; Punitha and Srivastava, 2013). In contrast with the P-gp, MRP1 functions as a basal-to-apical drug-permeation barrier and is an efflux transporter for amphifilic molecules. The organic anion transporters found in the surface epithelium of the choroid plexus play an important role in the uptake of amphifilic and hydrophilic anions, while organic cation transporters do the same for hydrophilic organic cations (Punitha and Srivastava, 2013).

10.2.2.4 Carrier-mediated transport

The BBB has several transport systems for nutrients and endogenous substances (including glucose or aminoacid transporter systems). Passive or active transport is provided in the BBB for nutrients or aminoacid transporting systems. Some polar substances including monocarboxylates, hexoses, aminoacids, nucleosides, glutathione, and small peptides are transported in this way (Gastaldi et al., 2014). It is targeted to simulate the structures of systems transporting nutrients and endogenous substances for the drugs or drug transporting systems to increase their passage through the BBB.

10.2.2.5 Adsorptive endocytosis

Adsorptive-mediated endocytosis systems are based on the electrostatic interaction between the positively charged group on the particle and the negative charge on the plasma membrane. Anionic regions are located in the luminal sides of the capillaries depending on the sialic acid residues of glycoproteins on the plasma membrane (Beduneau et al., 2007). Targeting of the active ingredients to the brain is possible by using adsorptive-mediated endocytosis systems. Efforts are being exerted to ensure the passage of the cationic lipid nanoparticles prepared using cationic lipids or surfactants. Because most biological membranes are negatively charged, adsorptive-mediated targeting has poor selectivity, especially in systemic administration.

10.2.2.6 Receptor-mediated endocytosis

Receptor-mediated endocytosis occurs with the binding of ligands to the specific receptors located in the luminal membrane of the BBB. There are many endogenous transport systems in the BBB. For example, glucose, which is the major energy source of the brain is transported by glucose transporter-1 receptor, while aminoacids are transported by the L-type aminoacid transporter-1 receptor. Thiamin, biotin, folic acid, vitamin B12, Tf, and neuropeptides are transported by their respective specific receptors, while insulin-like growth factors IGF-I and IGF-II and leptin are transported by the endogenous receptors located in the brain vascular endothelium constituting the BBB (Punitha and Srivastava, 2013).

The receptor-mediated endocytosis systems used to administer active ingredients to the brain are based on both endogenous and chimeric ligands. Insulin, Tf, and folic acid are endogenous ligands with high affinity for brain and tumor cells. Their biocompatibility and nonimmunologic nature are advantageous. Insulin has some disadvantages such as its short half-life and causing hypoglycemia at high concentrations. Tf is a monomeric glycoprotein that can carry one or two iron atoms. It is highly expressed in brain capillary endothelium and the surfaces of some rapidly reproducing cells like tumor cells of the brain (Beduneau et al., 2007). Folic acid receptors are commonly expressed in brain and choroid plexus cells. Folic acid receptors are expressed particularly in most cancer cells originating from the epithelium and play an important role in the survival and reproduction of cells (Punitha and Srivastava, 2013). Therefore, targeting drugs to folic acid receptors especially in oncologic diseases is promising.

10.2.3 STRATEGIES TO INCREASE THE PASSAGE THROUGH THE BRAIN–BLOOD BARRIER

Strategies aimed at increasing the passage through the BBB are shown in Table 10.1.

10.2.3.1 Direct injection into the central nervous system

In this technique, the drug is directly injected into the brain. It requires craniotomy for intracerebroventricular infusion and intracerebral drug administration. In case of intracerebroventricular drug delivery, high concentrations of the drug can be transported in the peripheral blood stream, but drug distribution to the targeted tissues is limited (Soni et al., 2016).

10.2.3.2 Disruption of the brain–blood barrier

In order to increase the permeability of the BBB, techniques have been developed to temporarily disrupt the tight junctions, allowing drugs to leak into the CNS. One technique uses an osmotic shock to shrink the endothelial cells and disrupt the tight junctions. With this purpose, a hyperosmotic solution of mannitol or arabinose is injected to the carotid artery and rapid opening of the BBB is ensured

Table 10.1 Strategies Used for Improving Brain–Blood Barrier (BBB) Penetration

Invasive Approach	Pharmacological Approach	Physiological Approach
• Direct injection into the CNS • Disruption of the BBB	• Improve lipophilicity • Use of prodrugs • Use of drug carriers • Inhibition of proteolytic enzymes or efflux pumps	• Receptor-mediated transcytosis/chimeric peptide technology • Use of CNS transport systems

by shrinking as a result of dehydration. In this way, substances that normally cannot pass through the BBB can diffuse freely within a period of a few hours (Jaypraka et al., 2009). The osmotic opening of the BBB has been used successfully to administer chemotherapeutic drugs in patients with cerebral lymphomas and CNS tumors. Disadvantages of the use of this technique include the problems encountered in relation with the permeability increase and excessive adverse effects (including seizures, transient exacerbation of existing neurological diseases and transient neurological problems, brain edema, etc.) (Brasnjevic et al., 2009).

Focused ultrasound-mediated drug administration is another technique that disrupts the BBB temporarily and ensures administration of drugs into the CNS. Focused ultrasound-mediated drug administration uses microbubbles activated by ultrasound waves to provide mechanical power in the target regions on the BBB. Microbubbles are 1–5 μm sized gas-filled vesicles stabilized by phospholipids, proteins, or polymers and facilitate reversible disruption of the BBB. Microbubbles and the drug are injected in a certain volume to the systemic circulation. Ultrasound waves are focused at the target area on the BBB and disrupts it. Ultrasound waves will cause oscillation of the microbubbles and create mechanical power on the target area on the BBB, which allows penetration of the drug to the target tissue. Microbubbles will commonly diffuse only to the site on the barrier disrupted by ultrasound. Although focused ultrasound-mediated drug administration ensures administration of unmodified drugs to the brain in high concentrations, it can also result in the penetration of toxic substances such as bacteria or antibodies to the brain (Stockwell et al., 2014; Upadhyay, 2014; Dasgupta et al., 2016).

10.2.3.3 Increasing the lipophilicity

Lipophilic properties of substances and their permeability through the BBB are directly proportional. Hydrophobic analogs of small-sized hydrophilic substances penetrate through the BBB more easily (Brasnjevic et al., 2009). It has been stated that increasing the lipophilic properties for molecules smaller than 400 Da will increase permeability through the BBB (Gentry et al., 1999). Lipophilic properties can be increased by reducing the hydrogen-binding capacity or by adding lipophilic groups. Methylation can decrease the hydrogen-binding potential of the active ingredients and, therefore, can increase diffusion through the membrane. Halogenation of the active ingredients can increase lipophilic properties and, consequently, passage through the BBB. Unexpected results have also been observed by increasing the lipophilic properties. Binding to plasma proteins and distribution volume are increased by increasing the lipophilic properties and pharmacokinetic parameters are affected (Patel, 2007). The rate of blockage with the enzyme cytochrome P450 is increased. In conclusion, both increasing the lipophilic properties and optimization of pharmacokinetic parameters need to be optimized.

10.2.3.4 Prodrug administration

Prodrugs are inactive compounds created by chemical modification of biologically active compounds. Water-soluble parts of the active ingredient that are unable to penetrate the BBB are transformed into prodrugs soluble in lipid (Brasnjevic et al., 2009). Prodrugs generally transform into the active form in single step hydrolysis of the modified groups. There are significant limitations of the prodrug strategy. Unwanted increases in the uptake by other tissues are observed with the increase of lipophilic properties. Low selectivity and high transformation into reactive metabolites often narrows the therapeutical indices of active ingredients prepared in the form of prodrugs. In the study carried out by Liu et al. (2012), NLCs have been developed for the administration of diacetyl apomorphine and diisobutyryl apomorphine in the form of the prodrug apomorphine diester. It was observed that the prodrugs underwent biotransformation in plasma and brain extracts, and that diacetyl apomorphine degraded faster than diisobutyryl apomorphine. Reduced active ingredient release has been ensured through the combination of prodrug and NLC strategies.

10.2.3.5 Use of drug carrier systems

Colloidal drug carriers such as micelles, liposomes, nanoparticles, and emulsions are being used to increase the concentrations of drugs passing through the BBB to the brain. By the specific administration of active ingredients to cells or tissues with the use of these colloidal systems, it is also possible to control the release rate in target regions. Drug administration frequency and adverse effects can be reduced by using drug carrier systems and physicochemical properties of the active ingredient can be masked to increase the passage through the biologic membrane and bioavailability. Furthermore, the active ingredient entrapped within the drug carrier system can be protected against enzymatic degradation. One unit of the drug transporter system can entrap multiple active molecules (Kabanov and Batrakova, 2004; Jatariu et al., 2010) to ensure administration of high amounts of active ingredient per receptor using one targeting group. Such drug transporter systems injected systemically must be biodegradable, nonimmunogenic, and stable in blood (Beduneau et al., 2007; Celia et al., 2011).

10.2.3.6 Inhibition of proteolytic enzymes and/or efflux pumps

Uptake of drugs by the brain can also be increased by decreasing the excretion rates of drugs from the plasma and brain. This can be ensured via the inhibition of proteolytic enzymes and/or activity of the efflux pumps. Either inhibitors specific for efflux pumps are developed to avoid the activity of the efflux pumps, or analogs of the drugs that do not react with the efflux pumps are developed to increase the entry of substrates into the CNS (Pardridge, 1997). With the purpose of reducing the interaction with the efflux pumps and increasing the penetration of the drugs to the brain, conjugation with PEG, encapsulation within drug carrier systems, or use of pluronic block components have been investigated and contributions to the inhibition of drug efflux systems have been reported (Kabanov and Batrakova, 2004).

PEG is a nontoxic, nonimmunogenic, and nonantigenic polymer that highly dissolves in water and is approved by the FDA. Molecules with different pharmacokinetic and pharmacodynamic properties and with the same biologic activity are created through PEGylation. In this process, one PEG polymer binds to the amino-, thiol-, hydroxy-, or amide-group of the target molecule. The created molecule will have greater molecular weight and removal by the kidneys and RES will be delayed and, consequently, their half-life in the body will be longer. Due to the elongation of the half-life of the molecule, the administration frequency of the drug can be reduced and patient compliance can be provided. Modification of the nanoparticle surfaces with PEG will inhibit interaction with blood components and decrease the adsorption of plasma proteins. Protein immunogenicity will decrease, stability against metabolic enzymes will increase, and degradation will decrease. PEGylation will create a protective shield effect and this, in turn, will reduce both degradation of the molecule with trypsin, chymotrypsin, and nucleases and decreases immunogenicity of the molecule by inhibiting its antigenicity. PEG conjugation will lengthen the periods of stay of the drugs in the body since it inhibits the transport by the immune system (Reddy, 2000). It is possible to adsorb the PEG chains physically on the nanoparticle surfaces or it can bind with covalent bonds. However, the PEG layer can dissolve in the blood circulation following the adsorption of the PEG group on the surface and opsonins can bind to the gaps thus created. PEGylated copolymers or lipids can be used as starting materials to obtain fully covalent-bounded nanocarriers containing PEG (Craparo et al., 2011). The thermal and pH stability of the branched-chain PEG is higher and PEG molecules with a branched chain are more resistant against proteolytic degradation compared to the linear chains (Harris et al., 2001). Coating nanoparticles with PEG particle degradation during storage will be reduced due to the steric obstacle created by the polymer chain and the stability of particles will increase (Mandal et al., 2013). PEGylation has an important place in drug administration because of these highly advantageous features.

10.2.3.7 Chimeric peptide technology

Administration of drugs to the brain using chimeric peptide technology is realized by the method of receptor-mediated endocytosis. The chimeric peptide will bind to the luminal plasma membrane of the endothelial cells through the interaction of the carrier system with the cellular surface receptor. As a result of this binding, the complex will be taken in with endocytosis. The chimeric peptide will pass to the extracellular fluid of the brain. The bonds between the drug and the carrier system will be broken by the enzymes and the active ingredient will be set free to interact with its specific receptor on the brain cell. In addition to the drug–receptor interaction, structure–activity relationship, binding to the target receptor, and in vivo behavior must also be studied. Binding of the active ingredient to the chimeric peptide can be made easier with avidin-biotin technology. The active ingredient is monobiotinized and formulated within a vial. On the other hand, the streptavidin monoclonal antibody is conjugated with the stable sulfur-ether bond. Contents of the two vials are mixed to form the chimeric peptide.

The biotinized drug will rapidly be conjugated because of the high affinity of avidin for biotin (Pardridge, 2005).

10.2.3.8 Use of the transport systems found in the central nervous system

There are numerous transport systems in the CNS such as the hexose transport systems, peptide transport systems, aminoacid transport systems, and nucleoside transport systems. Making use of these transporters to increase the drug passage to the brain is included in the potential strategies. Physicochemical properties of the active ingredients can be modified to resemble nutrients (aminoacids, hexose, or nucleosides) with the purpose of increasing the passage of active ingredients to the brain. Endogenous substances can also be used as transporters to increase the passage of active ingredients through the BBB.

10.3 LIPID NANOPARTICLES

10.3.1 INTRODUCTION TO LIPID NANOPARTICLES

Lipid nanoparticles have many features that make them superior to polymeric nanoparticles and other traditional drug delivery systems. Basic advantages of lipid nanoparticles include the biocompatibility of lipids used in the preparation of lipid nanoparticles, their low cytotoxicity, possibility of large-scale production, and the absence of organic solvents (Battaglia and Gallarate, 2012). The high-pressure homogenization (HPH) method, one of the lipid nanoparticle preparation methods, is an affordable method and is repeatable. This method allows the large-scale production of lipid nanoparticle formulations. Stability of lipid nanoparticles are good compared to the other systems (Blasi et al., 2007; Kaur et al., 2008). In addition to allowing controlled release, lack of toxicity is one of the advantages provided by the lipid nanoparticles in parenteral administrations. With this purpose, glycerides comprising fatty acids and emulgators accepted for parenteral applications (lecithin, polysorbate 80, poloxamer 188) have been used in the preparation of lipid nanoparticles (Müller et al., 2000; Wissing et al., 2004).

10.3.2 TYPES AND APPLICATIONS OF LIPID NANOPARTICLES

Many different types of lipid nanoparticles have been developed such as SLNs, NLCs, LDCs, and LNCs (Fig. 10.2). The formulation that was discovered first and known best are SLNs. SLNs were developed in the early 1990s as a transporter system as an alternative for emulsions, liposomes, and polymeric nanoparticles. SLNs include lipids that are solid at room temperature and body temperature. Pure triglycerides (e.g., tristearin, tripalmitin, trimyristin, tricaprin, trilaurin), diglycerides (e.g., glyceryl behenate), monoglycerides (e.g., glycerol monostearate), fatty acids (e.g., stearic acid), long-chain alcohols (e.g., cetyl alcohol),

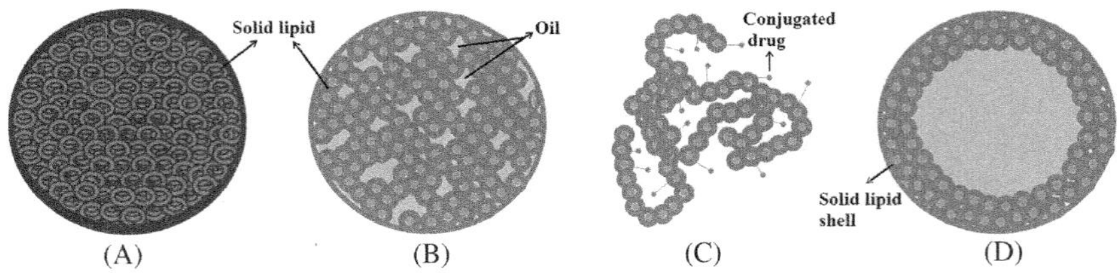

FIGURE 10.2

Different types of lipid nanoparticles. (A) SLN; (B) NLC; (C) LDC; and (D) LNC. *SLN*, Solid lipid nanoparticle; *NLC*, nanostructured lipid carrier; *LDC*, Lipid-drug conjugate; *LNC*, Lipid nanocapsule.

paraffins (e.g., wax, cetyl palmitate), and steroids (e.g., cholesterol) have been used as lipids (Kumar and Randhawa, 2013; Fonseca-Santos et al., 2015). Commonly used surfactants include poloxamers, bile salts, cetyltrimethylammonium bromide, soya lecithin, phosphatidylcholine, polyethyleneglycols, polyvinylalcohols, Tween 80, Tween 60, brij 30, brij 98, and brij 76. SLNs can also be prepared without using organic solvents. Excipients used in the preparation of SLNs are included in the GRAS status and are approved by the FDA (Kumar and Randhawa, 2013). They can be administered through injection or via intraocular or transdermal routes or through inhalation (Stockwell et al., 2014).

Low drug-loading capacity and observation of drug expulsion because of crystallization of the lipid matrix or lipid polymorphism are the potential disadvantages of SLNs. A new generation of nanoparticles called NLCs have been developed with the purpose of coping with these limitations. NLCs are produced from mixtures of solid and liquid lipids; however, they are still solid in body temperature. Compared to SLNs, NLCs have greater drug-loading capacity and display stability for longer periods under storage conditions (Craparo et al., 2011). Different types of NLCs are obtained based on the composition of the lipid mixture and production method: imperfect, amorphous, and multiple types. In the imperfect type, lipid crystallization is reduced by small amounts of liquid lipid. In the amorphous type, the lipid matrix is solid, but it is in the amorphous form. In the multiple type, the solid lipid matrix contains tiny compartments. This can be achieved by mixing solid lipid with high amounts of liquid lipid. Such changes in the lipid matrix aim at increasing drug encapsulation and reducing drug expulsion during storage (Battaglia and Gallarate, 2012).

LDCs have been developed to cope with the low drug-loading capacities of SLNs and NLCs. In the production of LDCs, the insoluble drug-lipid conjugate will be prepared with salt formation (e.g., with a fatty acid) or covalent bonds (e.g., with esters or ethers) in the first step. The LDC thus obtained is mixed with an aqueous surfactant solution, and the nanoparticle formulation is prepared using a suitable method (Gastaldi et al., 2014). Olbrich et al. (2004) have prepared LDC nanoparticles for the administration to the brain of diminazene diaceturate, which is a hydrophilic antitrypanosomal drug, by combining the drug with stearic acid and oleic acid. In the same study, it was found that cytotoxicity of LDC nanoparticles containing stearic acid and polysorbate 80 was high in comparison to SLNs and parenteral fatty emulsions produced from triglycerides or waxes.

LNCs, which are another type of lipid nanoparticles, comprise an internal core that is liquid or semiliquid at room temperature and an external solid lipid layer. In contrast with SLNs and NLCs, LNCs have a core-shell structure (Battaglia and Gallarate, 2012).

10.3.3 THE METHODS CURRENTLY IN USE FOR LIPID NANOPARTICLE PRODUCTION

SLNs are spherical particles in nanometer size prepared by dispersing solid/liquid lipid particles in water or aqueous surfactant solutions. The active ingredient can be dissolved or dispersed within the solid lipid matrix. SLNs or NLCs can be prepared with different methods using lipids, surfactants, or water/solvents. Methods used in the preparation of lipid nanoparticles can be listed as the HPH, high shear homogenization and ultrasonication technique, the solvent evaporation method, the solvent emulsification-diffusion method, the microemulsion method, the solvent injection method, the coaservation technique, the supercritical fluid (SCF) method, the membrane-based method, and the phase-inversion temperature (PIT) (Kumar and Randhawa, 2013; Fonseca-Santos et al., 2015). The method used to prepare lipid nanoparticles plays an important role in determining the structure and physical properties of the particles formed. There are many kinetic factors effective on the formation of particles including nucleation and precipitation mechanisms, rate of solvent extraction, effects of surfactants, and distribution of the drug within the nanoparticle. Therefore, structures of particles prepared from a drug solved in melted lipid and particles prepared from a solution containing drugs and lipids together will not be the same, even if the compositions are the same (Chattopadhyay et al., 2007).

The most important exigency for the introduction of nanoparticulate drug carriers in the drug market is the realization of their production in large scales. The main method used for the production of SLNs, NLCs, and LDCs with this purpose is HPH. HPH can be performed at high temperatures and below room temperature, which are called the hot and cold homogenization techniques, respectively (Fig. 10.3). In the hot homogenization, lipid will be heated to a temperature greater than its melting temperature, and the drug will be dissolved or dispersed within the melted lipid. The aqueous phase containing the emulsifier will also be heated to the same temperature separately. Lipid phase and aqueous phase will be mixed with a high-speed mixer to form the preemulsion. This emulsion will be passed through the HPH device a few times and cooled down (Müller et al., 2000; Blasi et al., 2007; Kamboj et al., 2010). The hot homogenization technique was used when preparing muco-adhesive NLCs loaded with ondansetron hydrochloride for intranasal

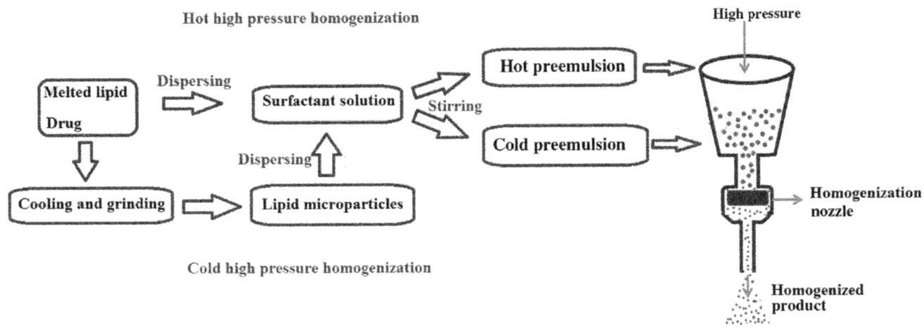

FIGURE 10.3

Production of lipid nanoparticles using the high-pressure homogenization technique.

drug administration to the brain by Devkar et al. (2014). While production of lipid nanoparticles with HPH is suitable for industrial production, the high process temperature and cavitation forces are critical process parameters that should be taken into consideration for powerful production. Since the period of exposure to high temperatures is very short in the hot homogenization technique, it can also be suitable for some active ingredients that are sensitive to high temperatures. The cold homogenization technique has been developed for active ingredients that are very sensitive for high temperatures. In cold homogenization, the melted lipid containing the active ingredient will be cooled down and lipid microparticles (50–100 μm in size) will be grounded. These lipid microparticles will be dispersed in a cold surfactant solution. The suspension formed will be homogenized at room temperature or under room temperature. The abrasion force created will degrade the lipid microparticles to SLNs (Müller et al., 2000; Kamboj et al., 2010). When producing LDC nanoparticles for diminazene, which is a hydrophilic antitrypanosomal drug, Olbrich et al. (2004) have used the HPH method. With this purpose, binary fatty salts of diminazene were prepared as the LDC bulk material in the first step. Drug and the fatty acid were dissolved within ethanol and the solvent was vaporized under low pressure at 40°C. This LDC bulk material was dispersed in 1% polysorbate 80 and 2.25% glycerol solution when preparing the LDC nanoparticles and was homogenized with HPH at 1500-bar pressure.

HPH devices push the liquid through a narrow hole with a diameter of several microns under high pressure (100–2000 bar). The liquid moves with a very high-speed within a very short period (over 1000 km/h). The very high shear stress and abrasive forces will degrade the particles to submicron sizes (Kamboj et al., 2010). A major drawback of the HPH method is the use of high temperatures and high pressures. Alternative lipid nanoparticle production methods have been developed for drugs that are not resistant against high temperature and high pressures.

In the high shear homogenization and ultrasonication method, lipids are heated to temperatures over their melting points; the aqueous phase-containing surfactant at the same temperature is added to the lipid phase and mixed at high speed. The mixture is homogenized in the sonicator by keeping the temperature constant.

Lipid nanoparticles are obtained when the mixture is cooled down to room temperature. SLNs and NLCs loaded with clotrimazole were prepared with the high shear homogenization and ultrasonication technique in a study by Das et al. (2012). In another study, Martins et al. (2013) prepared the SLN formulations loaded with camptothecin using the high shear homogenization and ultrasonication technique.

Alternative methods based on emulsions can be prepared using an organic solvent (volatile or partially mixed with water) that dissolve lipids. After forming O/W or W/O/W emulsions, nanoparticles are formed by evaporating the solvent (solvent evaporation method) or water diffusion (solvent diffusion method) (Battaglia and Gallarate 2012). Jose et al. (2014) have prepared resveratrol-encapsulated SLNs with high speed homogenization and ultrasonic treatment with the emulsion solvent evaporation method.

In the microemulsion method, lipids are heated over their melting points and microemulsion is formed by mixing it with the aqueous phase-containing surfactant and cosurfactant. Then, the microemulsion formed is diluted with cold water to precipitate the SLNs (Kuo and Chung, 2011).

In the solvent injection method, lipid will be dissolved in an organic solvent that mixes with water (like ethanol). The lipid phase and the aqueous phase will be heated to a temperature over the melting point of the lipid. The lipid phase will be injected to the aqueous phase through a syringe. Lipids will precipitate in the form of nanoparticles upon contact with water (Hansraj et al., 2015). In situ gelling NLCs with encapsulated rivastigmine have been prepared via the ethanol injection method and intranasal administration of NLCs to the brain has been studied (Wavikar and Vavia, 2015). In another study, LNCs loaded with curcumin for the treatment of glioblastoma have been prepared with the solvent injection method using acetone as the organic solvent (Zanotto-Filho et al., 2013).

The coacervation technique is a solvent-free technique developed to prepare SLNs based on fatty acids. The fatty acid alkaline salt micellar solution is prepared in the presence of a suitable polymeric stabilizer. When pH of the solution is lowered to acidic values, fatty acids will precipitate depending on the proton exchange between the fatty acid sodium salt and acidic solution. This process is called "coacervation." In this technique, lipophilic drugs can be dissolved in a micellar solution and encapsulated within nanoparticles before the acidification of the solution (Chirio et al., 2014; Battaglia et al., 2014). Chirio et al. (2014) have prepared behenic acid SLNs loaded with paclitaxel using the coacervation technique. Battaglia et al. (2014) prepared SLNs with the fatty acid coacervation technique to increase doxorubicin permeation in an artificial BBB model. Doxorubicin was encapsulated in SLNs using negatively charged surfactants as counter-ions using the hydrophobic ion pairing method. Permeation through the artificial BBB model increased when the drug was encapsulated within the SLN. Likewise, it has been stated that quantities of hydrophilic drugs loaded within SLNs can be increased, increasing the lipophilic properties with hydrophobic ion pairing.

Nanoparticle production with SCF technology is gradually attracting more attention since it is an effective and environment-friendly technique. SCF is formed at critical pressure and temperature. Densities of liquid and gaseous phases are similar over this critical point and they do not separate and form a homogeneous environment. The most commonly used SCF is carbon dioxide with a low critical point of 31°C and 74 bar. Carbon dioxide is preferred because it is not flammable, is nontoxic, mixable with many organic solvents, entails low cost, and allows for easy recycling following the process. It is more rapidly diffused than conventional solvents as it is a small and linear molecule. SCF technology is advantageous because of its capability for producing nanoparticles without organic solvents and because it is a one-step process (Battaglia and Gallarate, 2012; Santo et al., 2013). Acevedo-Morantes et al. (2013) have prepared SLNs loaded with camptothecin using SCF technology. For this purpose, Compritol 888 ATO and camptothecin were dissolved in dichloromethane, mixed for 30 minutes, and then sonicated for 1 hour. The mixture was then dissolved in liquefied CO_2 in the saturation channel at 35°C and under a maximum pressure of 100 bar. High temperature and pressure conditions were provided and an ultrasonic probe was placed in the saturation channel for increased mass transfer. Then the organic solvent in CO_2 was expanded using an atomizer. SLNs were obtained through the re-precipitation of the mixture. The SCF extraction of emulsions method, however, is the method for the production of lipid nanosuspensions by SCF extraction of the organic solvent from O/W emulsions. In this method, production of nanoparticles is realized by combining the emulsion systems with SCF extraction. The very short period required for the precipitation of particles ensures that the process is more controllable and provides uniform distribution of particle size. Interaction between the SCF and lipid during extraction can cause reduction of the glass transmission temperature or suppression of the melting point and can result in solid particles with different structures and physical properties (Chattopadhyay et al., 2007).

Another method developed for the production of SLNs is the use of membrane contactors. Lipid is heated over its melting point and passed through the membrane under pressure and droplets are formed. The aqueous phase will circulate within the membrane module and will collect the droplets formed outside the pores. Dispersion is cooled down to room temperature to obtain SLNs (Charcosset et al., 2005).

The PIT method is frequently preferred for the production of LNCs. The PIT concept uses the specific properties of some polyethoxylated surfactants to modify their affinity for water and lipid as a function of temperature. The use of such a surfactant in the PIT nanoemulsion preparation method leads to the conversion of the emulsion type from oil-in-water (o/w) to water-in-oil (w/o) when the temperature is raised above PIT. Three main components are used when preparing LNCs using the PIT method: oil phase, water phase, and nonionic surfactant. All the components are mixed using a magnetic stirrer and heated up from room temperature to a temperature over the PIT to obtain W/O emulsion. O/W emulsion is formed when the emulsion is cooled down and the temperature falls below the PIT. With the purpose of reducing the droplet size of the emulsion obtained, several temperature cycles are performed within the phase-inversion zone. Then, the system is cooled down rapidly to precipitate LNCs or is diluted with cold water (Battaglia and Gallarate, 2012). Resnier et al. (2013) have loaded EGFR siRNA to LNCs using cationic liposomes using the PIT method. Lipoflex was used in place of water at the final stage to obtain LNCs loaded with siRNA. Bastiancich et al. (2016) produced LNC hydrogel containing anticancer prodrug lauroyl-gemcitabine with phase-inversion process to treat glioblastoma multiforme. Similarly, the LNC formulation of NFL-TBS.40-63 peptide targeting neural stem-cells was prepared by the phase-inversion process by Carradori et al. (2016).

10.3.4 PHYSICOCHEMICAL CHARACTERIZATION OF LIPID NANOPARTICLES

Proper characterization of lipid nanoparticles is highly important for determining product quality and stability. There are many important characterization parameters including particle size, distribution of particle size, zeta potential, encapsulation efficiency, in vitro release, morphology, and sterilization.

10.3.4.1 Particle size, particle size distribution, and zeta potential

Particle size, particle size distribution, and zeta potential are important parameters that must be studied to create convenient nanoparticle systems. Photon correlation spectroscopy is the most frequently used method to determine particle size since it provides quick results. Photon correlation spectroscopy measures the fluctuations in the density of the scattered light related to the movements of particles (Yoon et al., 2013). It is possible to calculate the mean particle size and polydispersity index using computer programs. In this method, microbial contamination, crystallization of the particle content, and secondary particle agglomerates can cause erroneous results through the formation of differing particle sizes. Confirmation of results with other suitable methods such as electron microscopy is considered beneficial to avoid such erroneous results (Kaur et al., 2008). Since lipid nanoparticles constitute a polydispersed system, management of the polydispersity index is essential for the evaluation of particle size distribution. The nanoparticle dispersion is generally accepted as homogenous when the polydispersity index value is less than 0.3 (Yoon et al., 2013). Zeta potential is the measure of the electrostatic or push-and-pull forces between particles. Zeta potential analysis is performed to determine the surface charge on nanoparticles and gives an idea about the stability of the particles in the solution. It has been stated that colloidal systems do not display aggregation and can remain stable to be dispersed depending on the pushing forces on particles when the zeta potential value is at a minimum of ± 30 mV (Celia et al., 2011; Xie et al., 2011).

Kaur et al. (2008) have reported that SLNs smaller than 200 nm remained in the blood circulation for longer periods and, therefore, the contact period with the BBB is lengthened. Likewise, Patel et al. (2013) stated that uptake of SLNs with sizes ranging between 120 and 200 nm by RES cells was not easy. In another study, it was reported that colloidal systems with mean particle sizes of less than 100 nm and have particle size distribution within a narrow range is optimal for administration to the brain. Colloidal systems did not display aggregation when the zeta potential value is minimum ± 30 mV and formation of stable nano-suspension was ensured (Celia et al., 2011).

10.3.4.2 Drug encapsulation

High encapsulation efficiency indicates that a large portion of the drug added to the formulation is effectively loaded into the nanoparticle. Determining the quantity of the drug loaded into the nanoparticles is important to provide therapeutical efficiency and is necessary to determine the in vitro release profile of the drug from the system. The drug amount encapsulated within the unit weight of nanoparticles can be determined by separating the unloaded drug in the aqueous environment. This process can be carried out using ultracentrifuge, filtration with centrifuge, or gel permeation chromatography. Various filter centrifuge tubes of millipore and sartorius are used for filtration with centrifuge. Zanotto-Filho et al. (2013) have used the ultrafiltration/centrifugation technique to determine the encapsulation efficiency of curcumin in NLCs and the encapsulation efficiency of curcumin was found to be 100%. The quantity of the drug loaded into the nanoparticle system is calculated by determining the amount of the drug in the filtrate and subtracting it from total amount of the drug or, alternatively, the nanoparticle sediment is dissolved within a suitable solvent and the amount of the drug is analyzed. Solubility of the drug in the melted lipid, miscibility of the drug solution with melted lipid, chemical and physical structure of the solid lipid matrix, and polymorphic status of the lipid material are factors that determine the capacity of the drug to be loaded into the lipid (Kaur et al., 2008).

The encapsulation of drugs in SLNs has been studied in detail. Drugs can be dispersed in the lipid matrix homogenously or can concentrate in the shell or core of the nanoparticles. In the homogenous matrix model, the drug is dispersed into the matrix in molecular form or amorphous clusters. In the drug-enriched shell-type, the drug is located on the external shell and surrounds the lipid core. This is often seen in the hot HPH method during the cooling process when phase separation is present between the drug and lipid. A drug-enriched core-type, however, forms when the drug starts precipitating before the lipid. This often happens in the hot HPH method when the concentration of the drug in the melted lipid is close to saturation concentration. A significant amount of the drug is on the surface of the particles and this causes burst release. In fact, the major portion of the drug is not actually loaded to the lipid matrix as it is absorbed on the surface. Therefore, nanoparticles must be washed with a suitable solution to remove the drug adsorbed on the external surface after precipitating the nanoparticles in order to determine the drug encapsulation accurately (Battaglia and Gallarate, 2012).

10.3.4.3 In vitro release

In vitro release of the drug from the nanoparticle system is important in terms of ensuring the ideal release profile. Modification of the in vitro release profile of a drug optimizes the blood concentration and bioavailability. In addition to ensuring patient compliance, it is also possible to achieve an adequate pharmacokinetic profile with this method. Extension of the drug release period in drug administration to the brain ensures a sufficient period of time that will allow diffusion of the drug to specific regions (Celia et al., 2011). The two important production factors influencing the in vitro drug release profile from SLNs are the surfactant concentration and process temperature. Drugs pass to the aqueous phase from the lipid phase during the nanoparticle production using the hot HPH method. The higher the process temperature and surfactant concentration, the higher the solubility of the drug will be in the aqueous phase. Solubility of the drug in the aqueous phase will decrease with the decreasing temperature of the aqueous phase during the cooling process and re-partition of the drug in the lipid phase will be observed. When the recrystallization temperature of the lipid is reached, the solid lipid core will start to form together with the drug within the lipid phase. Since the lipid phase has solidified, no more drugs can enter. Therefore, the drug will concentrate on the surface of particles. While the drug found on the external shell and particle surface will rapidly be released, the drug encapsulated in the particle core will be released within a longer period (Battaglia and Gallarate, 2012).

Dialysis tubing and Franz diffusion cell methods are the most commonly used in vitro release methods (Kumar et al., 2007; Chan et al., 2009; Nair et al., 2011; Ghadiri et al., 2012; Belletti et al., 2012; Küçüktürkmen et al., 2016). Providing the sink condition during the experiment is essential. In the dialysis tubing method, SLN dispersion is placed within the prewashed dialysis bag and the bag is closed hermetically. Then, the dialysis bag is subjected to dialysis against a suitable dissolution environment. Samples are collected at certain time points from the dissolution environment and drug amount will be analyzed with a suitable analytic method. In the Franz diffusion cell method, however, SLN dispersion will be placed in the donor chamber of the Franz diffusion cell separated with a cellophane membrane. Dispersion will be dialyzed at a certain temperature against a suitable dissolution environment. Samples are collected at certain time points from the dissolution environment and the drug amount will be analyzed using a suitable analytic method (Kaur et al., 2008).

10.3.4.4 Morphological properties

Morphological properties of nanoparticles can be determined with scanning electron microscopy (SEM), transmission electron microscopy (TEM), and atomic force microscopy (AFM). Cryogenic field emission SEM is generally preferred in the analysis of lipid nanoparticles with SEM. While SEM makes use of electron transmission through the sample surface, TEM uses electron transmission through the sample. With the purpose of having better images of the shell-core structure, samples are often stained with uranyl acetate, osmium tetroxide, or phosphotungstic acid in TEM. While two-dimensional images are obtained with the electron microscopy, AFM provides a three-dimensional surface profile (Yoon et al., 2013; Mandal et al., 2013).

10.3.4.5 Crystallinity

Differential scanning calorimetry (DSC) and X-ray diffractometry (XRD) are the most frequently used techniques to determine the crystal status and polymorphic behavior of lipid nanoparticle components. DSC is commonly used in thermal analysis to image the endothermic processes (melting, solid–solid phase transmission) and exothermic processes (crystallization and oxidative decomposition). DSC analysis has been found to be rather useful for preliminary formulation studies because it allows to see the potential drug–excipient and excipient–excipient interactions. At the same time, with DSC analysis it is possible to determine the form of the drug that has been loaded into the nanoparticles (Kushwaha et al., 2013; Christoper et al., 2014). Compounds with specific crystal structures can be recognized with XRD. Lengths of the long and short intervals of the lipid cage can be determined based on the design of the XRD profiles (Yoon et al., 2013).

10.3.4.6 Sterilization

Sterilization of nanoparticles prepared for parenteral administration is important. European Pharmacopeia allows five types of sterilization methods including steam sterilization, sterilization with dry heat, sterilization with ionizing radiation, and sterilization with gases and filtration. The normal procedure for SLNs includes the preparation of SLNs under aseptic conditions and application of a suitable sterilization method such as filtration, autoclaving, or γ-irradiation. In the filtration method, the pore size of the filter must be 0.2 μm or smaller. Therefore, it is not possible to apply to nanoparticles with sizes larger than 0.2 μm. Steam sterilization is carried out in the autoclave at a minimum temperature of 121°C for a period of 15 minutes. Autoclaving can affect the physical stability of nanoparticles and can cause increases in particle size depending on aggregation. In spite of this, autoclaving is recommended for SLN suspensions with low lipid concentrations because no significant changes are observed in this method (Blasi et al., 2007). Sterilization with ionizing radiation is carried out in the form of gamma irradiation of the product from a suitable source (such as ^{60}Co) or exposing it to ionizing radiation in the form of electron irradiation from an electron beam. In the past three decades, most of the publications on sterilization with radiation discuss the determination of physicochemical properties of therapeutical substances or aqueous solutions exposed to gamma radiation or electron beams. Researchers are using new instrumental methods to study the effects of sterilization and are obtaining exact and repeatable results. Methods used recently include X-ray diffraction, thermal methods, and resonance methods. Considering the available results, it has been reported that most of the drugs in solid phase do not impair with ionizing radiation, and the amount of the drug either decreases with a very small amount or does not change at all. Together with this, biological studies on drugs exposed to sterilization with radiation are inadequate and this limits the applicability of this method. The main issue in the evaluation of sterilization with radiation is the study of by-products forming during the process. Safe and successful use of sterilization with radiation requires more detailed studies with biological methods showing, particularly, the pharmacologic activity and toxic effects of radiologic decomposition products (Marciniec and Dettlaff, 2008). It has been reported that sterilization with γ-irradiation can result in the production of radicals that affect the chemical modification of SLNs (Blasi et al., 2007).

10.3.5 FORMULATION DEVELOPMENT FOR DRUG TARGETING TO THE CENTRAL NERVOUS SYSTEM

With the purpose of ensuring the passage of drugs through the BBB, many strategies have been developed, many drug administration systems have been studied, and many drugs have been encapsulated in lipid nanoparticles in order to direct them to the brain (Table 10.2).

It has been stated that nanoparticles with particle sizes smaller than 200 nm can remain in the blood for a longer period and can come into contact with the BBB more, while the drug concentration taken in to the brain can increase. Furthermore, rendering the nanoparticle surface "sterically stable" by coating with hydrophilic polymers and surfactants can inhibit their recognition by RES and reduce the uptake to the liver and can ensure the stay of nanoparticles in the circulation for longer periods. PEG is the most commonly used polymer for this purpose. Coating nanoparticles with PEG, which is hydrophilic in structure, has reduced renal ultrafiltration and has increased the concentration transmitted to the brain (Kaur et al., 2008). Antitumoral activity of ferrociphenol-loaded LNCs and DSPE-mPEG2000 coated ferrociphenol-loaded LNCs has been studied with a convection-enhanced delivery technique or intracarotid injections on the orthotopic gliosarcoma model. In the treatment given with the convection-enhanced delivery technique, LNCs loaded with ferrociphenol, extended the survival period compared to the untreated group. In contrast, LNCs coated with DSPE-mPEG2000 and loaded with ferrociphenol had no effects on the mean survival period. This failure has been attributed to the inhibition of the interaction between the drug-loaded particles and tumor cell surface by the long-chain PEG found on the nanoparticle surface. Later, LNCs coated with hyperosmolar DSPE-mPEG2000 and loaded with ferrociphenol were administered through intracarotid injection and a significant increase in the mean survival rate was obtained (Huynh et al., 2012). This study shows the importance of the administration route on the efficiency of developed formulations. In another study, NLCs either with a cationic charge or with DSPE–PEG were developed for the administration of

Table 10.2 Properties of Different Lipid Nanoparticles Used for Central Nervous System Targeting

Type	Synthesis Method	Lipid Composition	Drug	Targeting Ligand	Particle Size (nm)	Reference
SLN	Microemulsion	Compritol	Quercetin	Polysorbate 80	159	Dhawan et al. (2011)
SLN	Ultrasonication	Tristearin, soy lecithin, stearic acid	Methotrexate	BSA	94.5	Agarwal et al. (2011b)
SLN	Microemulsion	Dynasan 114, palmitic acid	Saquinavir	83-14MAb, Poloxamer 407 and Polysorbate 80	120–450	Kuo and Ko (2013)
SLN	HPH	Cetyl palmitate, Dynasan 114, Witepsol E85	Camptothecin	Polysorbate 80, Polysorbate 60	91.3–158.2	Martins et al. (2013)
SLN	Emulsification and solvent evaporation	Glyceryl monostearate, soy lecithin	Docetaxel and ketoconazole	Stearic acid, folic acid	80.8	Venishetty et al. (2013a)
SLN	Emulsification and solvent evaporation	Monostearin, soy lecithin	Docetaxel	Beta-hydroxybutyric acid	76.4	Venishetty et al. (2013b)
SLN	Coacervation	Sodium behenate	Doxorubicin	–	278	Battaglia et al. (2014)
SLN	High shear homogenization and ultrasonication	Compritol 888 ATO	Resveratrol	–	248.3	Jose et al. (2014)
SLN	Coacervation	Behenic acid	Paclitaxel	Stearylamine, glycol chitosan	299.2–984.9	Chirio et al. (2014)
SLN	Microemulsion	Dynasan 114, cacao butter, stearic acid	Etoposide	5-HT moduline	110–450	Kuo and Hong (2014)
SLN	Solvent injection	Tripalmitin	Sumatriptan succinate	–	236.2	Hansraj et al. (2015)
SLN	Microemulsion	Cardiolipin, Compritol 888 ATO, stearic acid	Etoposide	83-14MAb and AEGFR	170	Kuo and Lee (2016)
SLN	Emulsification and solvent diffusion	Behenic acid, tripalmitin, cacao butter	Carmustine	Tamoxifen and Lf	105–185	Kuo and Cheng (2016)
SLN	Emulsification and solvent diffusion	1,2-Dipalmitoyl-*sn*-glycero-3-phosphocholine, cacao butter, hexadecanoic acid	Doxorubicin	Aprotinin, melanotransferrin	140.9–191.7	Kuo and Lee (2016)

SLN	Emulsification and solvent evaporation	Glyceryl monostearate, stearic acid, soy lecithin	Docetaxel	Lf	121	Singh et al. (2016)
SLN	High shear homogenization and ultrasonication	Cetyl palmitate	Resveratrol	Apolipoprotein E	167.8–217.1	Neves et al. (2016)
SLN	Emulsification and solvent evaporation	Stearic acid, soy lecithin	Vincristine, temozolomide	–	179.1	Wu et al. (2016)
NLC	HPH	Precirol ATO 5, Transcutol HP	Didanosine	Solutol HS 15, Tween 80 and Lutrol F68	148	Kasongo et al. (2011)
NLC	Ultrasonication	Precirol ATO 5, Miglyol	Curcumin	–	162.4	Puglia et al. (2012)
NLC	Sonication	Tripalmitin, gelucire, hydrogenated soybean phosphatidylcholine	Baicalein	–	102.5	Tsai et al. (2012)
NLC	Ultrasonication	Cetyl palmitate, sesame oil, Myverol	Diacetyl apomorphine, diisobutyryl apomorphine	DSPE–PEG	250.1	Liu et al. (2012)
NLC	Ultrasonication	Glyceryl monostearate, Capryol PGMC	Duloxetine	–	–	Alam et al. (2013)
NLC	HPH	Glycerol monostearate, Capryol90	Ondansetron hydrochloride	–	92.3	Devkar et al. (2014)
NLC	Hot high shear homogenization	Compritol 888 ATO, Labrafac Lipophile WL1349, Stearylamine	Ropinirole hydrochloride	–	160–175	Gabal et al. (2014)
NLC	Ultrasonication	Precirol ATO 5, Dynasan 114, Miglyol	Human insulin-like growth factor-I	Chitosan	114	Gartziandia et al. (2015)
NLC	Ethanol injection	Glyceryl monosterate, Capmul MCM C8, Lecithin	Rivastigmine	–	123.2	Wavikar and Vavia (2015)
NLC	Solvent evaporation	Cholesterol oleate, glycerol trioleate	Curcumin	Polysorbate 80	90.5	Meng et al. (2016)

(Continued)

Table 10.2 Properties of Different Lipid Nanoparticles Used for Central Nervous System Targeting *Continued*

Type	Synthesis Method	Lipid Composition	Drug	Targeting Ligand	Particle Size (nm)	Reference
NLC	Emulsification and solvent evaporation	Cholesterol, triolein	Paclitaxel	Tf	205.4	Emami et al. (2016)
NLC	Emulsification and ultrasonication	Cetyl palmitate, octyldodecanol	Embelin	—	152	Sharma et al. (2016)
NLC	HPH	Tripalmitin, Oleic acid	Curcumin	—	214	Chen et al. (2016)
NLC	Emulsification and solvent diffusion	Compritol 888 ATO, Cremophor ELP, soybean phosphatidylcholine	Vincristine, temozolomide	—	117.4	Wu et al. (2016)
LNC	PIT	Solutol_ HS15, Lipoid_ S75-3, Labrafac WL1349	Ferrociphenol	NFL-TBS.40–63, OX26-MAb	45.7–63.7	Laine et al. (2012)
LNC	PIT	Solutol HS15, Lipoid, Labrafac	Ferrociphenol	MIAMI cells	76	Roger et al. (2012)
LNC	PIT	Solutol HS15, Lipoid, Labrafac	Ferrociphenol	DSPE-mPEG2000	49	Huynh et al. (2012)
LNC	Solvent injection	Grape seed oil	Curcumine	—	196	Zanotto-Filho et al. (2013)
LNC	PIT	Labrafac WL 1349, Lipoid S75-3, Kolliphor HS 15	EGFR siRNA	—	<150	Resnier et al. (2013)
LNC	PIT	Labrafac, Kolliphor	Lauroyl-gemcitabine	—	69	Bastiancich et al. (2016)
LNC	PIT	Solutol HS15, Lipoid, Labrafac	—	NFL-TBS.40–63 peptide	66	Carradori et al. (2016)

apomorphine diester prodrugs, diacetyl apomorphine, and diisobutyryl apomorphine to the brain. It was seen in bioimaging studies carried out on mice that DSPE–PEG-NLCs mostly accumulated in the brain. Widespread bio-distribution was also obtained with cationic NLCs. After bioimaging studies, it has been reported that delivery of prodrugs to the brain and release of active drugs there is possible with loading into the lipid nanoparticles (Liu et al., 2012).

Among the surfactants used to coat the surfaces of nanoparticles, only particles coated with polysorbate had pharmacologic effects on the CNS. Coating with poloxamers (184, 188, 388, or 407), poloxamine 908, cremophors (EZ or RH40) or polyoxyethylene (23)-laurylether was not found to be effective. Polysorbates are being studied widely because of their high potential in the administration of SLNs to the brain (Kaur et al., 2008). Polysorbate 80 ensures the adsorption of apolipoprotein on the nanoparticle surface and makes it resemble a natural lipoprotein and ensures its uptake in the brain with receptor-mediated endocytosis. It has also been found that polysorbate 80 increases the drug administration to the brain through the inhibition of P-gp that is expressed in endothelial cells of brain capillaries (Meng et al., 2016). Passage of drugs through the BBB has been shown in many studies when they are loaded on polysorbate 80 coated nanoparticles. SLNs coated with polysorbate 80 have been prepared by Kakkar et al. (2013) to allow oral administration of curcumin, which is an antioxidant molecule, to the brain. Brain concentration of curcumin was found to be 30 times higher with curcumin loaded SLNs than that of free curcumin. As regards the transmission of SLNs loaded with curcumin, it has been reported that what inhibited the first passage effect of the liver was small particle size and nonionic surfactants (such as polysorbate 80 and lecithin) that increased the permeation through the intestines. It has been reported that coating SLNs with polysorbate 80 increased the endocytosis of SLNs in the brain by increasing the absorption of ApoE on the nanoparticle surface. In another study, the delivery of polysorbate 80 coated SLNs to the brain was investigated by Goppert and Muller (2005). They found that Apo Cy and Apo CII inhibited the binding of lipoproteins containing Apo E to LDL receptors, and, therefore, high Apo E/ Apo CII adsorbed on the particle surface for targeting to the brain. It has been seen that low ApoCII adsorption reduces the recognition of SLNs stabilized with Polysorbate 80 by macrophages. Martins et al. (2013) have prepared cetyl palmitate-based SLNs loaded with camptothecin to increase the uptake of camptothecin to the brain and were coated with polysorbate 60 or 80. In comparison with camptothecin in the form of suspension, it was seen that intravenous administration of SLNs loaded with camptothecin passed the BBB and were well-targeted to the brain. Dhawan et al. (2011) prepared quercetin-loaded SLN formulations with the purpose of increasing transmission through the BBB for Alzheimer's disease treatment. Polysorbate 80 was selected as a surfactant when preparing SLNs. The reasons for selecting Polysorbate 80 has been listed as its good acceptability for intravenous applications, its high effectiveness in the loading of active drugs, and its capability of targeting the SLNs to the brain because of its hydrophilic structure. SLNs containing polysorbate 80 in the formulation and without additional coating with polysorbate 80 have performed better. The reasons for this have been explained with the aggregation of particles because of the additional coating with polysorbate 80, increase of the particle size, and inhibition of the entry of particles into the CNS. Meng et al. (2016) have produced NLCs coated with polysorbate 80 and loaded with curcumin and studied the passage through the BBB. Cellular uptake of NLCs coated with polysorbate 80 by the brain endothelial cells and macrophages and in vivo transmission to the brain in rats have been studied. Significant increase in the cellular uptake of NLCs coated with polysorbate 80 by the brain endothelial cells following incubation with plasma have confirmed the adsorption to the NLC surface of apolipoprotein which mediates the receptor-mediated endocytosis of nanoparticles to the brain. It has been shown that NLCs coated with polysorbate 80 localize in the coronal areas in the brain.

Even if SLNs are capable of passing through the BBB, their concentrations reaching the BBB can be low if they are not targeted. Surfaces of nanoparticles that remain in the circulation for long periods of time can be modified with ligands specifically binding to the surface epitopes or receptors in the target areas. Binding suitable ligands to nanoparticle surfaces will ensure the increase of selectivity of nanoparticles for the target areas. Modification of the nanoparticle surface with specific ligands will increase concentration of nanoparticles in the BBB (Kaur et al., 2008). Many ligands including Tf, lactoferrin (Lf), folic acid, 83-14 monoclonal antibody (8314MAb), beta-hydroxybutyric acid, and apolipoprotein E have been studied in relation with the targeting of the SLNs to the brain.

Tf receptors are commonly used receptors for brain targeting. Tf receptors provide for the transmission of iron-saturated Tf from blood to brain through receptor-mediated endocytosis. Emami et al. (2016) developed NLCs conjugated with Tf for the administration of paclitaxel to the brain. The optimized paclitaxel-NLCs were conjugated with Tf as the targeting ligand and the in vivo cytotoxicity of the formulation on U-87 brain cancer cells were studied. NLCs conjugated with Tf displayed high cytotoxic activity compared to nontargeted NLCs and a free drug. Lf also belongs to the Tf family and ensures the transport of Lf through the BBB by way of Lf receptors found in the BBB. SLNs conjugated with tamoxifen and Lf have been prepared with the purpose of reducing proliferation in glioblastoma multiforme. The anticancer drug carmustine was loaded to the prepared SLNs and transmission through the BBB was studied. SLNs loaded with carmustine and modified with Tamoxifen and Lf have increased the permeability of the drug through the BBB and inhibited the proliferation of glioblastoma multiforme cells (Kuo and Cheng, 2016). In another study, SLNs conjugated with aprotinin and melanotransferrin were prepared to facilitate the passage of doxorubicin through the BBB for the chemotherapy of glioblastoma multiforme. The prepared SLNs increased the passage of doxorubicin through the BBB and inhibited the development of glioblastoma multiforme. Singh et al. (2016) prepared SLNs loaded with docetaxel and conjugated with Lf with the emulsification and solvent evaporation method. Surfaces of SLNs were modified with Lf to increase the uptake in the brain. Cytotoxicity studies showed that the effects of SLNs conjugated with Lf increased. Ligand receptor saturation studies and the significant increase in the cellular uptake have shown the importance of Lf conjugation. Together with this, it has been characterized with apoptosis studies that docetaxel encapsulated in SLNs fully preserved its activity and action mechanism. Pharmacokinetic and tissue distribution studies have shown that SLNs conjugated with Lf increased the docetaxel concentration in the brain compared to SLNs not conjugated with Lf and the marketed formulation.

Surface of the LNC formulation containing ferrociphenol prepared by Laine et al. (2012) was modified with cell-internalizing peptide NFL-TBS.40−63 interacting with tubulin-binding sites or murine monoclonal antibodies (OX26-MAb) targeting the Tf receptors. The surface modification carried out with OX26 or the peptide did not change the in vitro antiproliferative effect of ferrociphenol LNCs on 9 L cells and IC50 values were found within the same range. Likewise, the in vivo intracerebral administration of LNCs without surface modification did not elongate the mean survival period of rats compared to untreated rats. Following the intracarotid administration to rats, ferrociphenol LNCs with surface modification carried out with peptide improved the survival period, while no such result was obtained with OX26-MAb modification. Carradori et al. (2016) prepared LNC formulations using NFL-TBS.40−63 peptides with selective affinity for neural stem cells with the purpose of targeting these cells. NFL-TBS.40−63 is cell penetrating with 24 synthetic aminoacids and have been adsorbed on the LNC with electrostatic interaction. Targeting efficiency of NFL-TBS.40−63 has been evaluated on the neural stem cells of the subventricular zone (brain) and central canal (spinal cord). The targeted LNC formulation was selectively internalized by the neural stem cells; however, it did not interact with the neural stem cells of the spinal cord. In this study, in vivo results were consistent with the in vitro results and it was concluded that targeted NLCs are capable of transmitting bioactive molecules to the neural stem cells of the brain selectively.

In a study by Venishetty et al. (2013a), a docetaxel and ketoconazole combination was loaded to SLNs and the SLN surface was modified with stearic acid and folic acid conjugation. In comparison with the unmodified SLNs, the surface modification and encapsulation of docetaxel together with ketoconazole resulted in an increase in the uptake of docetaxel to the brain. Presence of folic acid on the SLN surface ensured the uptake of nanoparticles by the folate receptors like normal folic acid and increased the uptake of docetaxel to the brain. BBB integrity studies showed that SLNs caused no changes on the integrity of BBB.

Kuo and Ko (2013) prepared SLNs modified with 8314MAb with the purpose of increasing the brain-targeted administration saquinavir. 8314MAb is a peptidomimetic monoclonal insulin-like antibody with high affinity for brain capillaries and strongly binds to the subunit of human insulin receptors. It was seen that modified SLNs were biocompatible with the human brain-microvascular endothelial cells (HBMECs). 8314MAb and polysorbate 80 on the SLN surface stimulated endocytosis by the HBMECs and increased the permeation of nanoparticles through the BBB. In another study, Kuo and Lee (2016) synthesized SLNs conjugated with surface 8314MAb and anti-epithelial growth factor receptor (AEGFR) and loaded with etoposide. Passage of the synthesized SLNs through the BBB and their effects on the inhibition of glioblastoma multiforme were studied. SLNs were targeted to insulin receptors on HBMEC by conjugation with 8314MAb for passage through the BBB. Together with this, SLNs were conjugated with AEGFR to target them to EGFR on glioblastoma multiforme cells for an antiproliferative effect on glioblastoma multiforme cells. 8314MAb on SLNs displayed high affinity for insulin receptors on HBMECs and significantly increased the permeation of etoposide through the BBB. Binding SLNs with AEGFR inhibited the proliferation of U87MG cells.

Resnier et al. (2013) loaded EGFR siRNA to LNCs using cationic liposomes. An effective electrostatic interaction was ensured between lipids and siRNA thanks to this nanosystem created with the union of DOTAP/DOPE siRNA cationic lipoplexes and LNCs, and the high transfection capacity of cationic lipids and stealth characteristic of LNCs were made use of. LNCs loaded with siRNA have been tested on glioma cells with the purpose of improving the effective transmission of siRNA to the cytoplasm and effects of siRNA on EGFR was studied on U87MG human glioma cells. It was concluded, based on the analyzed protein levels, that siRNA entered the cytoplasm together with LNCs and a high-level inhibition was observed.

Venishetty et al. (2013b) prepared SLNs conjugated with beta-hydroxybutyric acid and loaded with docetaxel. Beta-hydroxybutyric acid is a new ligand expressed in brain endothelial cells and is transported by monocarboxylic acid transporter. Surface modification of SLNs loaded with docetaxel with beta-hydroxybutyric acid increased the uptake of docetaxel to the brain.

Resveratrol-loaded SLNs has been functionalized with apolipoprotein E that is transported with LDL receptors that are overexpressed in the BBB. Resveratrol-loaded SLNs functionalized with apolipoprotein E significantly increased permeability through Human Cerebral Microvascular Endothelial Cell Line hCMEC/D3 (Neves et al., 2016).

Another important strategy used to improve transmission through the BBB is the use of positively charged nanoparticles. While the lipid core ensures the loading of hydrophobic drugs, the cationic surface facilitates endocytosis. Together with this, the positive charge can modify the drug encapsulation and release. The positive charge on the surface of the lipid nanoparticles facilitates the interaction with the cell membrane and increases the cellular permeability and transfection capability. Cationic Bovine Serum Albumin (BSA) is a ligand with high affinity for the brain. Agarwal et al. (2011a) prepared doxorubicin-loaded SLNs conjugated with cationic BSA to study the transmission through the BBB. Uptake of BSA-conjugated SLNs to the brain capillary endothelial cells was higher compared to doxorubicin-loaded SLNs. In another study, cationic BSA was used as the ligand to increase the passage of SLNs loaded with methotrexate through the BBB. In vitro transport studies on brain capillary endothelial cells have shown that cationic BSA-conjugated SLNs underwent significant transcytosis. Cationic BSA-conjugated SLNs were selectively internalized by brain capillaries and cultured human neuroglial tumor cells better compared to SLNs not conjugated SLNs and methotrexate'a alone. In conclusion, it has been reported that SLNs loaded with methotrexate and conjugated with cationic BSA is promising for brain cancer chemotherapy because of their potential for passing through the BBB (Agarwal et al., 2011b). In the study by Gabal et al. (2014), the effects of the surface charge of the nanoparticle on the intranasal administration of ropinirole hydrochloride, which is a hydrophilic drug, to the brain were investigated. With this purpose, anionic and cationic NLCs were prepared and were administered intranasally through an in situ gel and the behavior in the brain and plasma were studied. Anionic NLCs caused mild and reversible inflammation in the nasal epithelium, while cationic NLCs destroyed the enclosing nasal mucosal epithelium. Both anionic and cationic NLCs increased the bioavailability compared to the intranasal solution. The maximum concentration of cationic NLCs in the brain was found significantly higher than that of the anionic NLCs. It has been observed that effectiveness of anionic NLCs on the drug targets in the brain when administered in the form of in situ gel was 1.2-fold greater than that of cationic NLCs.

Roger et al. (2012) presented a new option of treatment of brain tumors combining the cellular transporters and nanoparticles. With this purpose "Marrow-isolated adult multilineage inducible" (MIAMI) cells that feature aggregation in brain tumors and is the subpopulation of mesenchymal stromal cells that were used. Use of the cytotoxic effect obtained by adding LNCs containing an organometallic complex (ferrociphenol) to MIAMI cells in the treatment of brain tumors have been investigated. MIAMI cells have internalized Ferrociphenol-LNCs and this internalization has not resulted in the death of MIAMI cells. On the other hand, MIAMI cells loaded with ferrociphenol-LNCs have shown an in vitro cytotoxic effect on U87MG glioma cells. This cytotoxic effect was proven with the in vivo intratumoral injection of MIAMI cells loaded with ferrociphenol-LNC to the heterotopic U87MG glioma model.

10.4 CONCLUSIONS

The BBB prevents entry of most drugs to the brain and the basic problem in the treatment of CNS diseases is the inability of drugs in reaching the brain in amounts required for therapeutical effect. Nanomaterials have important advantages including the administration of drugs targeted to cells or tissues and providing for long-term effects in target tissues and the ability of passing through the physiological barriers. Numerous studies have been carried out in recent years to develop lipid nanoparticle formulations for many drugs that are unable to pass through the BBB and to increase the therapeutical efficacy. Lipid nanoparticles greatly attract the attention of the pharmaceutical industry because of the biocompatibility of the lipids included in their structure, possibility of production without organic solvents, and allowing large-scale production with the HPH method. With the purpose of treating the CNS diseases, surfaces of lipid nanoparticle formulations can be modified with ligands, PEGylation, or other vectors to ensure passage of the existing drugs through the BBB. Furthermore, combination of lipid nanoparticles with cellular vectors is a new promising research area in the treatment of brain diseases. It is hoped that this method will improve the pharmacokinetic profiles of drugs and achieve higher drug concentrations in the brain through the new drug formulations prepared in the form of lipid nanoparticles for the treatment of CNS diseases.

REFERENCES

Acevedo-Morantes, C.Y., Acevedo-Morantes, M.T., Suleiman-Rosado, D., Ramírez-Vick, J.E., 2013. Evaluation of the cytotoxic effect of camptothecin solid lipid nanoparticles on MCF7 cells. Drug Deliv. 20 (8), 338–348.

Agarwal, A., Agrawal, H., Tiwari, S., Jain, S., Agrawal, G.P., 2011a. Cationic ligand appended nanoconstructs: a prospective strategy for brain targeting. Int. J. Pharm. 421 (1), 189–201.

Agarwal, A., Majumder, S., Agrawal, H., Majumdar, S., Agrawal, G.P., 2011b. Cationized albumin conjugated solid lipid nanoparticles as vectors for brain delivery of an anti-cancer drug. Curr. Nanosci. 7, 71–80.

Alam, M.I., Baboota, S., Ahuja, A., Ali, M., Ali, J., Sahni, J.K., 2013. Intranasal infusion of nanostructured lipid carriers (NLC) containing CNS acting drug and estimation in brain and blood. Drug Deliv. 20 (6), 247–251.

Bastiancich, C., Vanvarenberg, K., Ucakar, B., Pitorre, M., Bastiat, G., Lagarce, F., et al., 2016. Lauroyl-gemcitabine-loaded lipid nanocapsule hydrogel for the treatment of glioblastoma. J. Control. Release 225, 283–293.

Battaglia, L., Gallarate, M., 2012. Lipid nanoparticles: state of the art, new preparation methods and challenges in drug delivery. Expert Opin. Drug Deliv. 9 (5), 497–508.

Battaglia, L., Gallarate, M., Peira, E., Chirio, D., Muntoni, E., Biasibetti, E., et al., 2014. Solid lipid nanoparticles for potential doxorubicin delivery in glioblastoma treatment: preliminary in vitro studies. J. Pharm. Sci. 103 (7), 2157–2165.

Beduneau, A., Saulnier, P., Benoit, J.P., 2007. Active targetting of brain tumors using nanocarriers. Biomaterials 28, 4947–4967.

Begley, D.J., Brightman, M.W., 2003. Structural and functional aspects of the blood brain barrier. Prog. Drug Res. 61, 39–78.

Belletti, D., Tosi, G., Forni, F., Gamberini, M.C., Baraldi, C., Vandelli, M.A., et al., 2012. Chemico-physical investigation of tenofovir loaded polymeric nanoparticles. Int. J. Pharm. 436, 753–763.

Blasi, P., Giovagnoli, S., Schoubben, A., Ricci, M., Rossi, C., 2007. Solid lipid nanoparticles for targeted brain drug delivery. Adv. Drug Deliv. Rev. 59, 454–477.

Brasnjevic, I., Steinbusch, H.W.M., Schmitz, C., Martıinez, P.M., 2009. Delivery of peptide and protein drugs over the blood–brain barrier. Prog. Neurobiol. 87, 212–251.

Carradori, D., Saulnier, P., Préat, V., des Rieux, A., Eyer, J., 2016. NFL-lipid nanocapsules for brain neural stem cell targeting in vitro and in vivo. J. Control. Release 238, 253–262.

Celia, C., Cosco, D., Paolino, D., Fresta, M., 2011. Nanoparticulate devices for brain drug delivery. Med. Res. Rev. 31 (5), 716–756.

Chan, J.M., Zhang, L., Yuet, K.P., Liao, G., Rhee, J.W., Langer, R., et al., 2009. PLGA-lecithin-PEG core-shell nanoparticles for controlled drug delivery. Biomaterials 30, 1627–1630.

Charcosset, C., El-Harati, A., Fessi, H., 2005. Preparation of solid lipid nanoparticles using a membrane contactor. J. Control. Release 108 (1), 112–120.

Chattopadhyay, P., Shekunov, B.Y., Yim, D., Cipolla, D., Boyd, B., Farr, S., 2007. Production of solid lipid nanoparticle suspensions using supercritical fluid extraction of emulsions (SFEE) for pulmonary delivery using the AERx system. Adv. Drug Deliv. Rev. 59 (6), 444–453.

Chen, Y., Pan, L., Jiang, M., Li, D., Jin, L., 2016. Nanostructured lipid carriers enhance the bioavailability and brain cancer inhibitory efficacy of curcumin both in vitro and in vivo. Drug Deliv. 23 (4), 1383–1392.

Chirio, D., Gallarate, M., Peira, E., Battaglia, L., Muntoni, E., Riganti, C., et al., 2014. Positive-charged solid lipid nanoparticles as paclitaxel drug delivery system in glioblastoma treatment. Eur. J. Pharm. Biopharm. 88 (3), 746–758.

Christoper, G.V.P., Raghavan, C.V., Siddharth, K., Kumar, M.S.S., Prasad, R.H., 2014. Formulation and optimization of coated PLGA – Zidovudine nanoparticles using factorial design and in vitro in vivo evaluations to determine brain targeting efficiency. Saudi Pharm. J. 22 (2), 133–140.

Craparo, E.F., Bondı, M.L., Pitarresi, G., Cavallaro, G., 2011. Nanoparticulate systems for drug delivery and targeting to the central nervous system. CNS Neurosci. Therap. 17 (6), 670–677.

Das, S., Ng, W.K., Tan, R.B., 2012. Are nanostructured lipid carriers (NLCs) better than solid lipid nanoparticles (SLNs): development, characterizations and comparative evaluations of clotrimazole-loaded SLNs and NLCs? Eur. J. Pharm. Sci. 47 (1), 139–151.

Dasgupta, A., Liu, M., Ojha, T., Storm, G., Kiessling, F., Lammers, T., 2016. Ultrasound-mediated drug delivery to the brain: principles, progress and prospects. Drug Discov. Today Technol. 20, 41–48.

Devkar, T.B., Tekade, A.R., Khandelwal, K.R., 2014. Surface engineered nanostructured lipid carriers for efficient nose to brain delivery of ondansetron HCl using *Delonix regia* gum as a natural mucoadhesive polymer. Colloids Surf. B. Biointerfaces 122, 143–150.

Dhawan, S., Kapil, R., Singh, B., 2011. Formulation development and systematic optimization of solid lipid nanoparticles of quercetin for improved brain delivery. J. Pharm. Pharmacol. 63 (3), 342–351.

Emami, J., Rezazadeh, M., Sadeghi, H., Khadivar, K., 2016. Development and optimization of transferrin-conjugated nanostructured lipid carriers for brain delivery of paclitaxel using Box–Behnken design. Pharm. Dev. Technol. 30, 1–13.

Fonseca-Santos, B., Gremião, M.P., Chorilli, M., 2015. Nanotechnology-based drug delivery systems for the treatment of Alzheimer's disease. Int. J. Nanomed. 10, 4981–5003.

Gabal, Y.M., Kamel, A.O., Sammour, O.A., Elshafeey, A.H., 2014. Effect of surface charge on the brain delivery of nanostructured lipid carriers in situ gels via the nasal route. Int. J. Pharm. 473 (1–2), 442–457.

Gartziandia, O., Herran, E., Pedraz, J.L., Carro, E., Igartua, M., Hernandez, R.M., 2015. Chitosan coated nanostructured lipid carriers for brain delivery of proteins by intranasal administration. Colloids Surf. B Biointerfaces 134, 304–313.

Gastaldi, L., Battaglia, L., Peira, E., Chirio, D., Muntoni, E., Solazzi, I., et al., 2014. Solid lipid nanoparticles as vehicles of drugs to the brain: current state of the art. Eur. J. Pharm. Biopharm. 87 (3), 433–444.

Gentry, C.L., Egleton, R.D., Gillespie, T., Abbruscato, T.J., Bechowski, H.B., Hruby, V.J., et al., 1999. The effect of halogenation on blood–brain barrier permeability of a novel peptide drug. Peptides 20, 1229–1238.

Ghadiri, M., Fatemi, S., Vatanara, A., Doroud, D., Najafabadi, A.R., Darabi, M., et al., 2012. Loading hydrophilic drug in solid lipid media as nanoparticles: statistical modeling of entrapment efficiency and particle size. Int. J. Pharm. 424 (1–2), 128–137.

Goppert, T.M., Muller, R.H., 2005. Polysorbate-stabilized solid lipid nanoparticles as colloidal carriers for intravenous targeting of drugs to the brain: comparison of plasma protein adsorption patterns. J. Drug Target. 13, 179–187.

Goyal, K., Koul, V., Singh, Y., Anand, A., 2014. Targeted drug delivery to central nervous system (CNS) for the treatment of neurodegenerative disorders: trends and advances. Cent. Nerv. Syst. Agents Med. Chem. 14 (1), 43–59.

Hansraj, G.P., Singh, S.K., Kumar, P., 2015. Sumatriptan succinate loaded chitosan solid lipid nanoparticles for enhanced anti-migraine potential. Int. J. Biol. Macromol. 81, 467–476.

Harris, J.M., Martin, N.E., Modi, M., 2001. Pegylation: a novel process for modifying pharmacokinetics. Clin. Pharmacokinet. 40, 539–551.

Huynh, N.T., Passirani, C., Allard-Vannier, E., Lemaire, L., Roux, J., Garcion, E., et al., 2012. Administration-dependent efficacy of ferrociphenol lipid nanocapsules for the treatment of intracranial 9L rat gliosarcoma. Int. J. Pharm. 423 (1), 55–62.

Jatariu, A.N., Popa, M., Peptu, C.A., 2010. Different particulate systems-bypass the biological barriers. J. Drug Target. 18 (4), 243–253.

Jaypraka, M., Ramnar, J., Shah, S., 2009. An overview on brain targeting drug delivery system. Pharm. Rev. 7 (1).

Jose, S., Anju, S.S., Cinu, T.A., Aleykutty, N.A., Thomas, S., Souto, E.B., 2014. In vivo pharmacokinetics and biodistribution of resveratrol-loaded solid lipid nanoparticles for brain delivery. Int. J. Pharm. 474 (1–2), 6–13.

Kabanov, A.V., Batrakova, E.V., 2004. New technologies for drug delivery across the blood brain barrier. Curr. Pharm. Des. 10, 1355–1363.

Kakkar, V., Mishra, A.K., Chuttani, K., Kaur, I.P., 2013. Proof of concept studies to confirm the delivery of curcumin loaded solid lipid nanoparticles (C-SLNs) to brain. Int. J. Pharm. 448, 354–359.

Kamboj, S., Bala, S., Nair, A.B., 2010. Solid lipid nanoparticles: an effective lipid based technology for poorly water soluble drugs. Int. J. Pharm. Sci. Rev. Res. 5 (2), 78–90.

Kasongo, K.W., Jansch, M., Müller, R.H., Walker, R.B., 2011. Evaluation of the in vitro differential protein adsorption patterns of didanosine-loaded nanostructured lipid carriers (NLCs) for potential targeting to the brain. J. Liposome Res. 21 (3), 245–254.

Kaur, I.P., Bhandari, R., Bhandari, S., Kakkar, V., 2008. Potential of solid lipid nanoparticles in brain targeting. J. Control. Release 127 (2), 97–109.

Kumar, S., Randhawa, J.K., 2013. High melting lipid based approach for drug delivery: solid lipid nanoparticles. Mater. Sci. Eng. C Mater. Biol. Appl. 33, 1842–1852.

Kumar, V.V., Chandrasekar, D., Ramakrishna, S., Kishan, V., Rao, Y.M., Diwan, P.V., 2007. Development and evaluation of nitrendipine loaded solid lipid nanoparticles: influence of wax and glyceride lipids on plasma pharmacokinetics. Int. J. Pharm. 335 (1–2), 167–175.

Kuo, Y.C., Cheng, S.J., 2016. Brain targeted delivery of carmustine using solid lipid nanoparticles modified with tamoxifen and lactoferrin for antitumor proliferation. Int. J. Pharm. 499 (1–2), 10–19.

Kuo, Y.C., Chung, J.F., 2011. Physicochemical properties of nevirapine-loaded solid lipid nanoparticles and nanostructured lipid carriers. Colloids Surf. B: Biointerfaces 83 (2), 299–306.

Kuo, Y.C., Hong, T.Y., 2014. Delivering etoposide to the brain using catanionic solid lipid nanoparticles with surface 5-HT-moduline. Int. J. Pharm. 465, 132–142.

Kuo, Y.C., Ko, H.F., 2013. Targeting delivery of saquinavir to the brain using 83-14 monoclonal antibody-grafted solid lipid nanoparticles. Biomaterials 34 (20), 4818–4830.

Kuo, Y.C., Lee, C.H., 2016. Dual targeting of solid lipid nanoparticles grafted with 83-14 MAb and anti-EGF receptor for malignant brain tumor therapy. Life Sci. 146, 222–231.

Kushwaha, A.K., Vuddanda, P.R., Karunanidhi, P., Singh, S.K., Singh, S., 2013. Development and evaluation of solid lipid nanoparticles of raloxifene hydrochloride for enhanced bioavailability. Biomed. Res. Int. 2013, 584549.

Küçüktürkmen, B., Devrim, B., Saka, O.M., Yilmaz, Ş., Arsoy, T., Bozkir, A., 2016. Co-delivery of pemetrexed and miR-21 antisense oligonucleotide by lipid-polymer hybrid nanoparticles and effects on glioblastoma cells. Drug Dev. Ind. Pharm. 29, 1–10.

Laine, A.L., Huynh, N.T., Clavreul, A., Balzeau, J., Béjaud, J., Vessieres, A., et al., 2012. Brain tumour targeting strategies via coated ferrociphenol lipid nanocapsules. Eur. J. Pharm. Biopharm. 81 (3), 690–693.

Li, X., Tsibouklis, J., Weng, T., Zhang, B., Yin, G., Feng, G., et al., 2016. Nano carriers for drug transport across the blood–brain barrier. J. Drug Target. 19, 1–12.

Liu, K.S., Wen, C.J., Yen, T.C., Sung, K.C., Ku, M.C., Wang, J.J., et al., 2012. Combined strategies of apomorphine diester prodrugs and nanostructured lipid carriers for efficient brain targeting. Nanotechnology 23 (9), 095103.

Mandal, B., Bhattacharjee, H., Mittal, N., Sah, H., Balabathula, P., Thoma, L.A., et al., 2013. Core-shell-type lipid-polymer hybrid nanoparticles as a drug delivery platform. Nanomedicine 9, 474–491.

Marciniec, B., Dettlaff, K., 2008. Radiation sterilization of drugs. In: Trends in Radiation Sterilization of Healthcare Products. International Atomic Energy Agency, Vienna, pp. 187–223.

Martins, S.M., Sarmento, B., Nunes, C., Lúcio, M., Reis, S., Ferreira, D.C., 2013. Brain targeting effect of camptothecin-loaded solid lipid nanoparticles in rat after intravenous administration. Eur. J. Pharm. Biopharm. 85 (3 Pt A), 488–502.

Meng, F., Asghar, S., Xu, Y., Wang, J., Jin, X., Wang, Z., et al., 2016. Design and evaluation of lipoprotein resembling curcumin-encapsulated protein-free nanostructured lipid carrier for brain targeting. Int. J. Pharm. 506 (1–2), 46–56.

Müller, R.H., Mäder, K., Gohla, S., 2000. Solid lipid nanoparticles (SLN) for controlled drug delivery - a review of the state of the art. Eur. J. Pharm. Biopharm. 50 (1), 161–177.

Nair, R., Vishnu Priya, K., Arun Kumar, K.S., Badivaddin, T.M.D., Sevukarajan, M., 2011. Formulation and evaluation of solid lipid nanoparticles of water soluble drug: isoniazid. J. Pharm. Sci. Res. 3 (5), 1256–1264.

Neves, A.R., Queiroz, J.F., Reis, S., 2016. Brain-targeted delivery of resveratrol using solid lipid nanoparticles functionalized with apolipoprotein E. J. Nanobiotechnology 14, 27.

Olbrich, C., Gessner, A., Schröder, W., Kayser, O., Müller, R.H., 2004. Lipid-drug conjugate nanoparticles of the hydrophilic drug diminazene-cytotoxicity testing and mouse serum adsorption. J. Control. Release 96 (3), 425–435.

Oldendorf, W.H., Cornford, M.E., Brown, W.J., 1977. The large apparent work capability of the blood–brain barrier: a study of the mitochondrial content of capillary endothelial cells in brain and other tissues of the rat. Ann. Neurol. 1, 409–417.

Pardridge, W.M., 1997. Drug delivery to the brain. J. Cerebr. Blood Flow Metab. 17, 713–731.

Pardridge, W.M., 2005. Drug and gene targetting to the brain via blood–brain barrier receptor-mediated transport systems. Int. Congr. Ser. 1277, 49–62.

Patel, T.B., 2007. Recent trends in brain targeted drug delivery systems: an overview. Pharm. Rev. 5 (2).

Patel, M., Souto, E.B., Singh, K.K., 2013. Advances in brain drug targeting and delivery: limitations and challenges of solid lipid nanoparticles. Expert Opin. Drug Deliv. 10 (7), 889–905.

Puglia, C., Frasca, G., Musumeci, T., Rizza, L., Puglisi, G., Bonina, F., et al., 2012. Curcumin loaded NLC induces histone hypoacetylation in the CNS after intraperitoneal administration in mice. Eur. J. Pharm. Biopharm. 81 (2), 288–293.

Punitha, A.D., Srivastava, A.K., 2013. CNS drug targeting: have we travelled in right path? J. Drug Target. 21 (9), 787–800.

Reddy, K., 2000. Controlled release, pegylation, liposomal formulation: new mechanisms in the delivery of injectable drugs. Ann. Pharmacother. 34, 915–923.

Resnier, P., David, S., Lautram, N., Delcroix, G.J., Clavreul, A., Benoit, J.P., et al., 2013. EGFR siRNA lipid nanocapsules efficiently transfect glioma cells in vitro. Int. J. Pharm. 454 (2), 748–755.

Roger, M., Clavreul, A., Huynh, N.T., Passirani, C., Schiller, P., Vessières, A., et al., 2012. Ferrociphenol lipid nanocapsule delivery by mesenchymal stromal cells in brain tumor therapy. Int. J. Pharm. 423 (1), 63–68.

Santo, I.E., Pedro, A.S., Fialho, R., Cabral-Albuquerque, E., 2013. Characteristics of lipid micro- and nanoparticles based on supercritical formation for potential pharmaceutical application. Nanoscale Res. Lett. 8 (1), 386.

Sharma, N., Bhandari, S., Deshmukh, R., Yadav, A.K., Mishra, N., 2016. Development and characterization of embelin-loaded nanolipid carriers for brain targeting. Artif. Cells Nanomed. Biotechnol. 25, 1–5.

Singh, I., Swami, R., Pooja, D., Jeengar, M.K., Khan, W., Sistla, R., 2016. Lactoferrin bioconjugated solid lipid nanoparticles: a new drug delivery system for potential brain targeting. J. Drug Target. 24 (3), 212–223.

Soni, S., Ruhela, R.K., Medhi, B., 2016. Nanomedicine in central nervous system (CNS) disorders: a present and future prospective. Adv. Pharm. Bull. 6 (3), 319–335.

Stockwell, J., Abdi, N., Lu, X., Maheshwari, O., Taghibiglou, C., 2014. Novel central nervous system drug delivery systems. Chem. Biol. Drug Des. 83 (5), 507–520.

Tahara, K., Kato, Y., Yamamoto, H., Kreuter, J., Kawashima, Y., 2011. Intracellular drug delivery using polysorbate 80 modified poly(D,L-lactide-coglycolide) nanospheres to glioblastoma cells. J. Microencapsul. 28 (1), 29–36.

Tsai, M.J., Wu, P.C., Huang, Y.B., Chang, J.S., Lin, C.L., Tsai, Y.H., et al., 2012. Baicalein loaded in tocol nanostructured lipid carriers (tocol NLCs) for enhanced stability and brain targeting. Int. J. Pharm. 423, 461–470.

Upadhyay, R.K., 2014. Drug delivery systems, CNS protection, and the blood brain barrier. Biomed. Res. Int. 2014, 869269.

Venishetty, V.K., Komuravelli, R., Kuncha, M., Sistla, R., Diwan, P.V., 2013a. Increased brain uptake of docetaxel and ketoconazole loaded folate-grafted solid lipid nanoparticles. Nanomedicine 9 (1), 111–121.

Venishetty, V.K., Samala, R., Komuravelli, R., Kuncha, M., Sistla, R., Diwan, P.V., 2013b. β-Hydroxybutyric acid grafted solid lipid nanoparticles: a novel strategy to improve drug delivery to brain. Nanomedicine 9 (3), 388–397.

Visser, C.C., Stevanovic, S., Voorwinden, L.H., Bloois, L.V., Gaillard, P.J., Danhof, M., et al., 2005. Targeting liposomes with protein drugs to the blood–brain barrier in vitro. Eur. J. Pharm. Sci. 25, 299–305.

Wavikar, P.R., Vavia, P.R., 2015. Rivastigmine-loaded in situ gelling nanostructured lipid carriers for nose to brain delivery. J. Liposome Res. 25 (2), 141–149.

Wissing, S.A., Kayser, O., Müller, R.H., 2004. Solid lipid nanoparticles for parenteral drug delivery. Adv. Drug Deliv. Rev. 56, 1257–1272.

Wu, M., Fan, Y., Lv, S., Xiao, B., Ye, M., Zhu, X., 2016. Vincristine and temozolamide combined chemotherapy for the treatment of glioma: a comparison of solid lipid nanoparticles and nanostructured lipid carriers for dual drugs delivery. Drug Deliv. 23 (8), 2720–2725.

Xie, S., Zhu, L., Dong, Z., Wang, X., Wang, Y., Li, X., et al., 2011. Preparation, characterization and pharmacokinetics of enrofloxacin-loaded solid lipid nanoparticles: influences of fatty acids. Colloids Surf. B Biointerfaces 83 (2), 382–387.

Yoon, G., Park, J.W., Yoon, I., 2013. Solid lipid nanoparticles (SLNs) and nanostructured lipid carriers (NLCs): recent advances in drug delivery. J. Pharm. Invest. 43, 353–362.

Zanotto-Filho, A., Coradini, K., Braganhol, E., Schröder, R., de Oliveira, C.M., Simões-Pires, A., et al., 2013. Curcumin-loaded lipid-core nanocapsules as a strategy to improve pharmacological efficacy of curcumin in glioma treatment. Eur. J. Pharm. Biopharm. 83 (2), 156–167.

Zlokovic, B.V., 2008. The blood–brain barrier in health and chronic neurodegenerative disorders. Neuron 57, 178–201.

FURTHER READING

Küçüktürkmen, B., Bozkır, A., 2018. Development and characterization of cationic solid lipid nanoparticles for co-delivery of pemetrexed and miR-21 antisense oligonucleotide to glioblastoma cells. Drug Dev. Ind. Pharm. 44 (2), 306–315.

Küçüktürkmen, B., Devrim, B., Saka, O.M., Yilmaz, Ş., Arsoy, T., Bozkir, A., 2017. Co-delivery of pemetrexed and miR-21 antisense oligonucleotide by lipid-polymer hybrid nanoparticles and effects on glioblastoma cells. Drug Dev. Ind. Pharm. 43 (1), 12–21.

Yung-Chih, Kuo, I-Hsuan, Lee, 2016. Delivery of doxorubicin to glioblastoma multiforme in vitro using solid lipid nanoparticles with surface aprotinin and melanotransferrin antibody for enhanced chemotherapy. J. Taiwan Inst. Chem. Eng. 61, 32–45.

CHAPTER

Nanoparticles for improvement in oral bioavailability

11

Naseem Ahmad Charoo[1], Ziyaur Rahman[2] and Mansoor A. Khan[2]

[1]Zeino Pharma LLC, Khalifa Industrial Zone, Abu Dhabi, United Arab Emirates [2]Irma Lerma Rangel College of Pharmacy, Texas A&M Health Science Center, College Station, TX, United States

11.1 INTRODUCTION

The oral route of drug administration is noninvasive, cost-effective, and convenient for chronic therapy. Flexibility in dosing and the ease of administration renders it preferable to patients and physicians alike. The bioavailability of drugs administered through the oral route depends primarily on its solubility in the gastrointestinal (GI) fluids and permeability across the GI tract (GIT) barrier. Drug solubility, in turn, is determined by the physicochemical attributes of the drug's substance which may also influence permeability across the GI barrier. Additionally, oral bioavailability is limited by efflux transport and gut/hepatic metabolism.

A substantial number of newly developed drug candidates fall in the "poorly soluble" category. Attaining therapeutically relevant systemic concentration of these candidates is highly challenging. Performing toxicological studies, which generally require higher doses than therapeutic doses, is an even more arduous task. Although various formulation approaches have been suggested to overcome the poor solubility of drug substances, their relevance is seriously limited by the toxicity of excipients employed in their preparation, which may potentially confound the drug's toxicity. Nanoparticle technology, on the other hand, has received wide acceptance as it involves the use of a minimum number and quantity of excipients to enhance a drug's solubility. The improvement in solubility is derived from the large surface to volume ratio resulting from nanosizing. The increased surface area alters physical, chemical, and biological properties which, in turn, improves drug solubility and permeability and, thus, modifies pharmacokinetics and pharmacodynamics response. The oral bioavailability benefits from the increased dissolution rate, according to the Noyes–Whitney equation, as well as from the increased saturation solubility (Moghimi et al., 2012). This chapter will focus on the solubility, dissolution, biodistribution, cellular uptake, and oral bioavailability aspects of nanoparticles.

Nanoarchitectonics in Biomedicine. DOI: https://doi.org/10.1016/B978-0-12-816200-2.00006-2

11.2 CELLULAR UPTAKE AND BIODISTRIBUTION—PHYSICOCHEMICAL PROPERTIES OF NANOPARTICLES

11.2.1 CELLULAR UPTAKE MECHANISM

Transcellular and paracellular routes are the primary routes of absorption for orally administered drugs. However, the paracellular route is not of much significance in the case of nanoparticles as it is unlikely that particles greater than 1 nm in size would breach tight protein junctions. However, junctional proteins can be modified reversibly to open membrane barriers for improving oral drug delivery. The agents used for this purpose can interact either directly with tight junction proteins and surface receptors or indirectly enhance the delivery by chelating calcium, thereby disrupting the junctions. To develop orally efficient nanoparticles it is important to discern their interaction with cellular systems for understanding the mechanism underlying the cellular uptake of nanoparticles (Komarova and Malik, 2010).

Nanoparticle-based drug delivery is developed with the objective of precisely delivering the drug at the specific target organs, tissues, and cells. To accomplish this, the translocation of nanoparticles across the plasma membrane barrier assumes high importance. The plasma membrane is vital for the normal functioning of cells. It also takes part in critical functions such as cellular adhesion, communication, and division. Endocytosis is pivotal in regulating these activities (Chauhan and Jain, 2013). Endocytosis is a cellular uptake mechanism which involves either pinocytosis or phagocytosis. The process entails the formation of new vesicles from the cell membrane accompanied by the incorporation of proteins, lipids, and extracellular fluids. Pinocytosis and phagocytosis involve the uptake of fluids and large solid particles, respectively. The type of endocytosis process employed is determined by the size of the endocytic vesicle, the nature of the material, and the mechanism of vesicle formation. The heterogeneity in endocytic pathways provides an opportunity in internalization of packets of material to different intracellular locations.

Phagocytosis is performed by monocytes/macrophages, neutrophils, and dendritic cells which are often referred to as professional phagocytes. Generally, the cells/particles are first opsonized, that is, enveloped by proteins (opsonins) such as immunoglobulins, which causes them to be recognized by the phagocytes prior to phagocytosis (Aderem and Underhill, 1999). Subsequently, the interaction of receptors on the cell surface and the ligands (opsonins) trigger cellular uptake of macromolecules. Some of the important receptors involved in phagocytosis are the Fc receptor family for IgG (FcγRI, FcγRIIA, and FcγRIIA), the complement receptors (CR1, CR3, and CR4), and $\alpha5\beta1$ integrin (Underhill and Goodridge, 2012). Similarly, opsonins of significance include antibodies, acetylcholine, fibronectin, and laminin. In addition to opsonization, the interaction between cells and nanoparticle surfaces can also be triggered by attractive forces like van der

Waals, electrostatic, ionic, and hydrophobic/hydrophilic interaction (Owens and Peppas, 2006). Cells or particles of up to tens of micrometers can be incorporated through phagocytosis.

The endocytic uptake of particles is highly influenced by the particle size, shape, composition, surface coating, charge, cellular environment, interaction with other cells, age, and type of cells (Chithrani et al., 2006; Rejman et al., 2004; Sun et al., 2005; Nativo et al., 2008). The optimum particle size required for the efficient endocytic process is determined by the cell type. For example, alveolar macrophages ingest particles in the 3–6 μm range, whereas, peripheral blood mononuclear cells are reported to engulf particles in the 0.3–1.1 μm range (Hirota et al., 2007; Pratten and Lloyd, 1986). The contact angle between the particle and membrane surface has been shown to have a significant impact on phagocytosis. Various shapes of nanoparticles would generate different angles and consequently affect the endocytic uptake process differently (Sharma et al., 2010). For instance, when an elliptical disk-shaped particle attaches to the membrane perpendicularly along its long axis, the contact angle is small and the particle is engulfed after it is covered by the membrane symmetrically (Champion and Mitragotri, 2006). Similarly, prolate ellipsoids having major axis 0.35–2 μm and minor axis 0.2–2 μm exhibited highest attachment rate, but the slowest internalization rate in comparison to spheroidal particles with a radius of 0.26–1.8 μm and oblate ellipsoidal nanoparticles having major axis 0.35–2.5 μm and minor axis 0.2–2 μm (Toy et al., 2014). On the other hand, when the short axis is perpendicular to the membrane, the contact angle and the number of contact points with the membrane increase, resulting in the failure of engulfment. Thus a high rate of attachment does not necessarily translate into a high internalization rate. However, within 2–3 μm long axis size range, maximum attachment to cells take place in spite of the large contact angle, leading to successful engulfment (Doshi and Mitragotri, 2010). Due to the symmetrical shape of spherical particles, their internalization is independent from the contact angle.

In contrast to phagocytosis, pinocytosis is carried out by all types of cells and, depending on the type of cells, occurs through four different mechanisms (Yameen et al., 2014):

- micropinocytosis,
- clathrin-mediated endocytosis,
- caveolae-mediated endocytosis, and
- clathrin-independent and caveolae-independent endocytosis.

Refer to Table 11.1 for an overview of the differences between these four procedures.

Prior to their interaction with the cell membrane upon entering into a biological environment, nanoparticles are exposed to body fluids which contain a variety of biomolecules such as proteins, sugars, and lipids. Subsequently, nanoparticles are coated with biomolecules to form a structure known as the protein corona

Table 11.1 Pinocytosis Uptake Pathways

Macropinocytosis	Clathrin-Mediated Endocytosis	Caveolae-Mediated Endocytosis	Clathrin- and Caveolae-Independent Endocytos
Formation			
• Activation of macropinocytosis is independent of the direct receptor action or ligand molecule • The actin driven process leads to the formation of protrusions at the outer cell membrane. Protrusions fuse with the cell membrane and form macropinosomes and internalize large volumes of extracellular fluid • The size of macropinosomes is >1 μm	• Occurs by specific receptor–ligand interaction or nonspecific endocytosis • In specific clarithrin mediated endocytosis coated pits are formed by transmembrane receptors and objects coated with cytosolic coat proteins (clarithrin). Assisted by GTPase dynamin protein, coated pits retract to form vesicles. The dynamin protein encircles the neck of newly formed invagination Nonspecific endocytosis involves hydrophobic interactions with cell membrane • Vesicles are 100–120 nm in size	• The process is highly regulated and involves complex signaling pathways • The caveolae form flask shaped invaginations (static structures) in the interior plasma membrane, comprising of glycophosphingolipids and high amount of cholesterol. Caveolin-1 is a dimeric protein and is responsible for giving shape and structure to the vesicle. The fission of caveolae from the membrane is mediated by GTPase dynamin • The size of vesicles is approximately 50–100 nm	• Works independent of receptors or material st • Does not require the cc proteins for vesicle formation and internalization; however actin and actin-related proteins are essential elements for vesicle formation
Fate			
• The macropinosomes are sensitive to cytoplasmic pH. They may be acidified, shrink or may fuse with lysosomal compartment or even recycle their content to the surface • In macrophages, macropinosomes may endure a fate similar to that of endosomes	• Post release, auxilin and HSC70 dependent proteins mediate disassembling of the clathrin pit • The uncoated vesicles are led to either early endosomes or recycled to the plasma membrane surface • The vesicles may also be transferred to more mature endosomes and later to lysosomes and multivesicular bodies	• The coated proteins known as cavins together with caveolins regulate the caveolae formation and their disposition • Endothelial cell caveolae participate in transendothelial transport, which is being explored for nanoparticle delivery to subendothelial tissues • Initially the endocytosed material is confined in caveosomes. The neutral pH of caveosomes prevents the exposure to hydrolytic environment of lysosomes. The transport of caveosomes to the Golgi Apparatus and Endoplasmic Reticulum can be explored for the targeted delivery of therapeutic agents	• Once inside, cargos co be delivered to the early endosomes, and subsequently might be transferred to late endosomes and lysoso • Cargo can also find wa into the trans-Golgi net or may be recycled bac the plasma membrane
Examples			
• Apoptotic cell fragments, viruses and bacteria are ingested by this route	• Uptake of low-density lipoprotein and transfer occurs via this route • Chitosan and PLGA also use clathrin dependent endocytosis	• Used for internalizing folic acid, albumin and cholesterol • Caveolae participate in different cellular functions such as cholesterol homeostasis, protein endocytosis, and signal transduction • Caveolae are present in abundance in smooth muscles, adipocytes, endothelial cells and fibroblasts	• Polyplexes of self-bran and trisaccharide-substituted chitosan oligomer nanoparticles been explored for the delivery of; • DNA • Cowpea mosaic virus • Vaccine development • In vivo vascular imaging • Tissue-targeted deliver

HSC70, *Heat shock cognate 70;* PLGA, *poly(lactide-co-glycolide).*

(Treuel and Nienhaus, 2012). Cell surface receptors, thus, do not interact with bare particles directly. This coating can have a remarkable impact on the nanoparticle uptake process. In pinocytosis, an invagination forms in the cell membrane which finally disengages to form a vesicle incorporating internalized materials. The vesicles possess receptor proteins for recognizing specific chemical groups present on the molecules to be internalized. Hence, if protein adsorbed on nanoparticle activates cell surface receptors, they will enhance the uptake process. In contrast, proteins exhibiting weak affinity toward cell surface receptors would reduce the uptake process. This property has been explored for enhancing drug delivery by coating nanoparticle with ligands that bind to specific receptors and, hence, enhance their internalization. For instance, conjugation of lectins to polymeric nanoparticles augments their clathrin-mediated uptake across the intestinal mucosa (Zhang et al., 2006). Similarly, cobalamin-coated nanoparticles containing insulin administered orally were reported to produce a pharmacological response greater than 80% of the response achieved by the subcutaneous route due to binding of cobalamine to the intrinsic factor in GIT (Pharmaceuticals, Inc., 2010).

On the other hand, exocytosis involves the fusion of inner vesicles with the plasma membrane and the transport of molecules to an extracellular space (Doherty and McMahon, 2009). Both endocytosis and exocytosis are dynamic and well-regulated (Steinman et al., 1983). Since nanoparticles are cleared mainly by the liver and spleen, their exocytosis from cells in general, and liver and spleen in particular, is of paramount importance for the biosafety of these particles. Smaller nanoparticles with a slower wrapping time and rod-shaped nanoparticles exhibit more rapid exocytosis than spherical particles. The impact of various factors on exocytosis has been covered aptly by Oh et al. (Oh and Park, 2014).

11.2.2 BIODISTRIBUTION

Water-insoluble polymers forming stable nanoparticles are likely to exhibit better absorption characteristics than less stable nanoparticles formed from polyelectrolyte complexes owing to their rapid dissociation even though they dissociate incompletely. The absorption of the polymer is determined by its physicochemical properties such as molecular weight, hydrophobicity, and conformation. The physicochemical properties of nanoparticles including size, surface charge, and surface chemistry are not only important for absorption, but also influence their pharmacokinetics and biodistribution which, in turn, determine the therapeutic effect and toxicity potential of the nanoparticles (Plapied et al., 2011).

The uptake of nanoparticles occurs through M (microfold) cells or absorptive enterocytes. Uptake by M cells leads to their transcytosis near the immune cells, and likely, delivery to the gut associated lymphoid tissue and lymphoid cells (Brayden et al., 2005). On the other hand, uptake by absorptive enterocytes guides them into the blood circulation. M cells and absorptive enterocytes show enhanced import potential for hydrophobic and hydrophilic nanoparticles, respectively (LeFevre et al., 1985; des Rieux et al., 2005). The benefits of nanoparticles in disease treatment can be realized only if they stay in the blood stream long enough to reach the target therapeutic site of action. However, opsonization of nanoparticulate drug carriers paves the way for their removal by the mononuclear phagocytic system (MPS), which is also referred to as the reticuloendothelial system (RES). The bare nanoparticles (not PEGylated) are removed more rapidly by MPS organs (liver and spleen) owing to the lack of stealth properties. The stealth nanoparticles, on the contrary, demonstrate favorable biodistribution profiles (Panagi et al., 2001). Therefore PEGylation is commonly employed to mitigate the MPS clearance. Polyethylene glycol (PEG) forms a hydrated phase on the surface of nanoparticles and prevents their uptake by macrophage cells. Nevertheless, it is worthwhile to note that the amount of PEG that can be used in the formulation is limited by its toxicity (Zhang et al., 2014). For PEGylated nanoparticles, the rate of clearance and final biodistribution depends on several factors besides particle size. Moghimi et al. reported faster clearance of particles (both PEGylated or non-PEGylated) of hydrodynamic radii of more than 200 nm in comparison to particles with radii less than 200 nm (Moghimi et al., 2001). Likewise, PEGylated nanoparticles of hydrodynamic radii of less than 150 nm exhibited high uptake in the bone marrow of rabbits, whereas particles of 250 nm size were sequestered mainly in the spleen and liver (Porter et al., 1992). This suggests a filtration mechanism in the removal of larger particles by the spleen and liver whereby small particles sift through to the bone marrow (Moghimi et al., 1993). The other hypothesis to explain the differences in the uptake and biodistribution of stealth particles is the specificity of opsonins to only a certain type of phagocyte. For example, Moghimi and Patel hypothesized that the high accumulation of cholesterol-rich liposomes in the spleen was due to the presence of opsonins specific to splenic phagocytes (Moghimi and Patel, 1988). On the contrary, kupffer cell specific opsonins might possess large binding sites for binding large particles that are subsequently sequestered in the liver. The other factors affecting the final biodistribution and clearance rate of nanoparticles include thickness, charge, surface density, functional group, and conformation characteristics of PEG. Currently, there is limited understanding of the influence of these factors on biodistribution and clearance owing to the high complexity involved in the selection of animal models (intra- and interspecies variation), and the variability associated with the raw materials used to manufacture the nanoparticles (Li and Huang, 2008; Porter et al., 1992). In spite of these limitations, two generalizations can be made from most of the biodistribution studies. Firstly, larger molecular weight PEG polymer nanoparticles exhibit longer blood circulation half-lives in vivo (Gref et al., 1995) and, secondly, the uncoated nanoparticles accumulate mainly in the liver and spleen, while coated (PEGylated) nanoparticles distribute primarily in the spleen. (Owens and Peppas, 2006)

The aspect ratio (AR) of nanoparticles also affects partitioning during biodistribution and plays an important role in the organ-specific uptake of nanoparticles. Shukla et al. (2015) prepared two sets of tobacco mosaic virus (TMV) based on nanoparticle formulations with a constant diameter and equal surface, but varying AR, to compare their biodistribution and tumor targeting properties. Stealth TMV formulations were prepared by PEGylation to subdue immune regulation, whereas targeted formulations were prepared by displaying the integrin-targeting cyclic peptide ligand [RGD(Arg-Gly-Asp)] by means of an intervening PEG spacer. PEGylated nanoparticles exhibited longer residence time in circulation and accumulated passively in tumors by the enhanced permeability and retention (EPR) effect. Lowest AR (AR:3.50) particles, owing to their capability to diffuse through the endothelium and into the tumor, accumulated in tumor cells in higher numbers than higher AR particles. In contrast, RGD-labeled particles with the lowest AR were cleared rapidly by macrophages and were less efficient in tumor targeting. Mid-sized AR particles ($130 \times 18\ nm^2$, AR: 7) exhibited the highest tumor accumulation as they could evade the immune system (Shukla et al., 2015). The optimum size of nanoparticles intended for targeting tumors is around 20 nm (Wittrup et al., 2012). Spherical particles exhibit preferential uptake by macrophages in comparison to long nanochains possessing a high AR. The lower liver uptake of long nanochains correlates with their prolonged residence in the blood, resulting in higher extravasation into tumors. For instance, gold nanorods of $10 \times 45\ nm^2$ size having AR 4.5 exhibited less uptake in the liver and spleen in comparison to gold nanospheres (Arnida et al., 2011).

Gold nanoparticles demonstrate excellent reproducibility and flexibility which has been increasingly utilized in drug delivery, especially for delivering chemotherapeutic agents. Their therapeutic effect and systemic toxicity can be optimized by manipulating their surface–particle chemistry. For instance, the surface chemistry of gemcitabine gold nanoparticle conjugate was modified with PEG, resulting in an increase in bioavailability without an increase in the accumulation in clearance organs like liver (Kudgus et al., 2014). The accumulation of gold nanoparticles in the liver increased from 50% to 99% when their size was increased from 1.4 to 200 nm. The accumulation in skeleton and other organs (kidney, heart, brain, uterus, skin, and soft tissue) showed little size dependency in the size range 18–200 nm. On the other hand, with a decrease in size (high volumetric specific surface area) in the range of 1.4 to 5 nm, gold nanoparticles exhibited a linear increase and linear decrease, respectively, in its accumulation in other organs and liver. The differently charged gold nanoparticles also exhibited dissimilar accumulations in various organs (Hirn et al., 2011). For example, negatively charged gold nanoparticles of 2.8 nm size accumulated more significantly in liver than positively charged particles. Conversely, in the spleen, positively charged nanoparticles showed higher accumulation than negatively charged particles (Hirn et al., 2011).

11.3 SOLUBILITY, DRUG RELEASE, AND STABILITY IN THE GASTROINTESTINAL TRACT

11.3.1 SOLUBILITY AND EFFECT ON DISSOLUTION RATE

For a drug to get absorbed, it must be soluble in the biological fluids and permeable across the physiological barriers. It is estimated that about 40% of the new drug candidates discovered by combinatorial screening programs endure poor solubility. The conventional methods employed to enhance the solubility of drug candidates encompass salt formation, particle size reduction, prodrug approach, cosolvency, and crystal habit alteration (Rahman et al., 2012a,b). However, these approaches are complicated, unpredictable, and do not always yield the desired results. Consequently, development is either halted or the molecule is developed with suboptimal properties such as poor bioavailability. The nanoparticle formulation represents a promising approach for overcoming poor solubility in such instances. Besides improving drug solubility, nanoparticles can also enhance drug stability, performance, and efficacy appreciably (Kesisoglou et al., 2007).

According to the Nernst–Brunner/Noyes–Whitney equation, dissolution rate (d_c/d_t) is directly proportional to the effective surface area (A) of the drug particles in contact with GI fluids and saturation solubility of the drug in solution in the diffusion layer (C_s).

$$\frac{d_c}{d_t} = \frac{D \cdot A}{h}\left(C_s - \frac{X_d}{V}\right) \tag{11.1}$$

where D represents the diffusion coefficient of the drug in the GI media and h is the thickness of the diffusion layer around each drug particle. X_d and V are the amount dissolved and volume of fluid available for dissolution, respectively (Amidon et al., 1995; Dressman et al., 1998). The solubility of the drug substance is inherently determined by its physical characteristics, such as polymorphism and salt/base form. On the other hand, the surface area of a drug is more closely related to its bulk properties which can be manipulated to enhance the dissolution rate of poorly soluble drug substances. Founded on these principles, nanosizing of particles has been extensively explored to improve oral bioavailability of biopharmaceutics classification system (BCS) class II and IV drugs owing to the enormous surface available for dissolution. The increase in dissolution rate of nanoparticles is more pronounced when compared to microparticles (Merisko-Liversidge and Liversidge, 2008) (Table 11.2). For example, aprepitant surface area increased 41.5-fold when its particle size was reduced from 5 μm to 120 nm (Kesisoglou et al., 2007). Similarly, freeze-dried finasteride nanoparticles developed from nanosuspension formulations exhibited a 2.5-fold increase in drug dissolution compared to the pure drug (Ahmed, 2016). The increase in dissolution rate resulted in considerable improvement in bioavailability with $AUC_{0-\infty}$ (area under the plasma concentration vs time curve) values with 1976.18, 2343.27, and 2932.68 ng/mL min recorded, respectively, for the pure drug suspension,

Drug Substance	Nanocarrier	Aqueous Solubility (μg/mL)		Dissolution		Particle Size (nm)	Poly-dispersity Index	Zeta Potential (mV)	References
		Pure drug substance	Nano-particle	Medium	% Drug Release				
Acetylpuerarin	PLGA nanoparticles	–	–	Phosphate buffered saline (pH 7.4) containing 1% polysorbate 80 (w/v)	38.0% in 4 h[a]	145.00	0.15	− 14.81	Sun et al. (2016)
Ergosterol		–	–	pH 6.8 medium	54.6% in 120 h[a]	156.90	0.08	− 19.27	Zhang et al. (2016)
Ezetimibe	Solvent evaporated solid dispersion	1.70	189.80	pH 6.8 medium	55.0% in 30 min versus 25.0% for free drug	6000.00	< 0.29	–	Rashid et al. (2015)
Lovastatin	Nanostructured lipid carriers	0.40	–	Phosphate buffer solution (pH 7.4)	20.0%–30.0% in 5 h[a] versus > 90.0% in 5 h for suspension formulation	23.50	0.17	− 31.20	Zhou and Zhou (2015)
Prednisolone	Tablets containing chitosan nanoparticles	350.00	970.00	Phosphate buffer solution (pH 6.8)	> 80.0% in 30 min	130.00	0.31	–	Chen et al. (2015)
Stiripentol	Nanoemulsions	49.20	–	Phosphate buffer solution (pH 6.8)	11.6% in 24 h[a]	142.50	0.19	− 33.70	Lu et al. (2015)
Torcetrapib	Solid lipid nanoparticles	< 0.04	–	Phosphate buffer solution (pH 6.8)	> 90.0% release in 30 min versus <20.0% release in 24 h by powder formulation	155.00	0.30	− 28.60	Liu et al. (2015b)
Fenretinide	Hydrophilic nanoparticles	0.005	–	Simulated intestinal fluid (pH 6.8)	40.0% in 15 min	200.00	0.22	–	Ledet et al. (2015)
Celecoxib	Solid dispersion nanoparticles with D-α-tocopheryl polyethylene glycol 1000 succinate	3.90	6522.30	pH 6.8 medium	> 60.0% in 2 h[a]	< 300.00	–	–	Ha et al. (2014)
Nifedipine	Self-nanoemulsifying drug delivery system	5.80	–	Simulated gastric fluid (pH 1.2)	> 80.0% in 60 min versus < 20.0% in 1 h by powder	132.90	0.18	–	Weerapol et al. (2014)
Sirolimus	Solid self-microemulsifying drug delivery system	2.60	–	Simulated gastric fluid (pH 1.2)	> 50.0% in 2 h[a]	200.00	–	–	Cho et al. (2015)
Pranlukast hemihydrates	Self-microemulsifying drug delivery system	0.01	5.70	Simulated intestinal fluid (pH 6.8)	> 95.0% in 5 min	< 10.00	–	–	Baek et al. (2013)
Telmisartan	Nanocrystals	< 10.00	> 150.00	Phosphate buffer solution (pH 6.8)	99.0% in 60 min versus < 5.0% for powder	82.60	0.47	7.94	Bajaj et al. (2012)

PLGA, *Poly(lactide-co-glycolide).*
[a]*Indicating sustained release.*

microparticles, and nanoparticles (Ahmed, 2016). Similar improvements in oral bioavailability have been shown for other drugs by nanonization (Quinn et al., 2012; Wu et al., 2004).

Nanonization may also increase the saturation solubility of the drug according to the Ostwald–Freundlich equation (Eq. 11.2), which would provide further impetus to drug bioavailability (Kesisoglou et al., 2007). With the increase in saturation solubility, dissolution pressure increases owing to the strong surface curvature of the nanoparticles (according to Prandtl Eq. 11.3), resulting in a higher concentration gradient between epithelial cells and the mesenteric circulation. The reduction in particle size together with the increasing curvature of nanoparticles also reduces the thickness of the diffusion layer, which further augments the concentration gradient.

$$\frac{S(d)}{S_0} = \exp\frac{\gamma V_m}{RTd} \tag{11.2}$$

where $S(d)$ represents the solubility of crystals with diameter (d) at temperature (T), molar volume (V_m), surface tension (γ) and gas constant (R). S_0 is the solubility of the bulk material. Keeping all the terms constant, the solubility would increase with the decrease in particle size. Nevertheless, for the ratio $S(d)/S_0$ to exceed 1, the exponential term must be much smaller than 1, which is possible only when the particle size is in the nano range.

$$h_H A = k\left(\frac{\sqrt{L}}{\sqrt{V}}\right) \tag{11.3}$$

where L is length of the surface in the direction of flow, k is a constant, V represents relative velocity of the flowing liquid against a flat surface, and h_H denotes the hydrodynamic boundary layer thickness.

Muller and Peters observed a greater increase in solubility than predicted by the Ostwald–Freundlich equation. They propounded the creation of high-energy surfaces on the drug crystals during nanomilling and exposure of inner lipophilic surfaces of the crystal to the aqueous dispersion medium, resulting in a change in surface tension and, consequently, the saturated solubility (Muller and Peters, 1998). The presence of surfactants in nanoparticles would further aid this process. Some reports even suggest a change in the effective intestinal permeability (P_{eff}) apparently to decrease the thickness of the unstirred water layer (Sugano, 2010). The aqueous diffusion layer or "unstirred" water layer is present adjacent to the intestinal membrane and comprises of water and mucus. The slow diffusion across the unstirred water layer may limit transcellular transport of lipophilic drug candidates. The improvement in permeability through the unstirred water layer adds to the total P_{eff} of a compound which translates into an improvement in oral bioavailability (Sugano, 2009). The different $S(d)$ values shown by the polymorphic forms are also potentially caused by the difference in interfacial energy and packing density (V_m).

Considering these facts, nanoparticle formulations are considered efficient drug delivery carriers and, additionally, boast a high-drug loading capacity (Xia et al., 2014). In the case of nanocrystals, up to 100% drug loading capacity is reported which is far superior than polymeric nanoparticles, liposomes, and lipid-based nanoparticles (Hu et al., 2008). The nanocrystal formulations of sirolimus (Rapamune tablets), aprepitant (Emend), fenofibrate (TriCor tablets), megestrol acetate (Megace ES), and many other drug candidates are commercially available. Rapamune tablets exhibited about 21%–27% increase in bioavailability in comparison to Rapamune solution (Gülsün et al., 2009; Junghanns and Müller, 2008). Telmisartan, a BCS class II drug with an aqueous solubility of 9.9 μg/mL only, was formulated as nanocrystals by the evaporative antisolvent method. The saturation solubility increased 3.74- and 5.04-fold in fasted and fed simulated intestinal fluid media, respectively. The nanosuspension coated on pellets exhibited a release of 85% in water, 95% in 0.1 N HCl, and 75% in phosphate buffer within 30 minutes (Bajaj et al., 2012).

The presence of a drug in its amorphous state in nanoparticles is a valuable strategy to improve drug solubility further. However, the major challenge in this approach is to prevent the polymorphic conversion during the drug's shelf life. Polymers have been purposely used to provide steric stability (Cho et al., 2015; Fule and Amin, 2014). For instance, polyvinyl acetate stabilized furosemide (BCS class II drug) nanosuspensions prepared by the precipitation method using dimethyl sulfoxide and water as solvent and antisolvent, respectively, released more than 81% drug in 4 hours in phosphate buffer of pH 6.5, containing 0.5% sodium lauryl sulfate. The saturation solubility of furosemide increased significantly in simulated gastric fluid of pH 1.2 from 12 to 438.38 μg/mL in nanosuspension possibly due to an increase in the surface area and existence of the drug in the amorphous state (Sahu and Das, 2014). Similarly, indomethacin nanoparticles prepared by a controlled pH, coprecipitation method yielded amorphous drug nanoparticles in polyvinyl pyrrolidine matrix. The nanoparticles released 45% of the drug within 30 minutes, whereas micronized drug–polyvinyl pyrrolidone physical mixture released only 10% of the drug during the same period. The hydrophilic nature of these polymers might also contribute in enhancing the drug's dissolution (Mokarram et al., 2010). Similarly, the improvement in aqueous solubility, dissolution, and bioavailability of amorphous ezetimibe formulated in solid self-nanoemulsifying drug delivery system and the solvent evaporated solid dispersion formulations has also been reported (Rashid et al., 2015; Nazzal et al., 2002).

Suitable stabilizers are often required to inhibit Ostwald aggregation and, hence, mitigate the adverse influence on drug solubility. The cationic charge polymers by virtue of their adsorption on the surface of particles are very effective in preventing particle aggregation. For instance, low cationic density nanocrystals of atorvastatin prepared using chitosan exhibited better stability characteristics at elevated temperature than nanocrystals prepared with labrasol only. The solubility increased by 60-fold in comparison to pure atorvastatin, which was ascribed to the presence of surfactant as well as chitosan, a cationic charge polymer. The nanocrystals not only enhanced the drug dissolution, but also sustained it for a longer duration (Kurakula et al., 2015). The GI environment is a stumbling block to oral drug delivery owing to the hostile acidic environment of the stomach, degradative enzymes, and mucus barrier secreted by the epithelium. Additionally, as a result of rapid clearance rates, foreign particles are removed efficiently, thereby limiting the residence time of orally administered conventional nanoparticles. In this context, slow-releasing nanoparticle formulations allow the drug to bypass the harsh gastric environment. Consequently, the relative bioavailability of acid labile drugs would increase in comparison to conventional formulations. For instance, nanoemulsion formulation of stiripentol, an acid labile drug, prepared from poly (ethylene glycol) monooleate/medium-chain triglycerides released only 5.23% of the drug in 0.1 M HCl. Further, only 11.56% of the drug was released in a pH 6.8 buffer within 24 hours, implying sustained drug release. As a result, the relative bioavailability of stiripentol nanoemulsion improved significantly up to 206.2% relative to the conventional suspension formulation (Lu et al., 2015).

Nanoemulsions are an attractive choice for delivery of lipophilic drug candidates. Nanoemulsions consist of oil and aqueous phases, stabilized by an emulsifying agent, surfactant, or a cosurfactant. To obtain oil droplets in the nanometer scale range, high energy is required for their preparation. The dissolution rate of the drug in aqueous media is determined by its intrinsic solubility, an interfacial barrier, transfer rate constant, and the droplet size. The droplet size of the formulations has been propounded as a significant factor in increasing a drug's bioavailability. On the other hand, the surface charge might not significantly affect absolute bioavailability. For instance, Devalapally et al. reported higher bioavailability of nanoemulsion formulations, irrespective of the presence of cosurfactants stearylamine and deoxycholic acid, in comparison to other formulations prepared in various vehicles. The nanoemulsion droplets varied in size from 238 to 344 nm, with the surface charges of −24.8, −26.5, and +27.8 mV for nanoemulsion, nanoemulsion-deoxycholic acid, and nanoemulsion-stearylamine formulations, respectively (Devalapally et al., 2013). Similarly, nanogel–drug conjugates comprising of small hydrophilic particles with a hydrophobic cholesterol core have been employed for the encapsulation of poorly soluble hydrophobic drugs. This approach prevents exposure to toxic solvents or solubilizers used currently to solubilize poorly soluble drugs. This strategy has also been applied to develop dual-drug nanoformulations of a polymeric nanogel conjugate with floxuridine and paclitaxel, or 17-AAG, for combination therapy. Both dual nanodrug combinations effectively inhibited tumor growth after oral administration in comparison to floxuridine nanogel. Further, the combinations were very stable, releasing only 0.4%–0.6% of the drug in gastric conditions after 6 hours incubation. The second drug release was evidently faster than conjugated floxuridine for an extended period of time, suggesting the importance of the second drug release kinetics on the anticancer effect of dual nanodrugs after accumulation in tumors (Senanayake et al., 2014).

11.3.2 STABILITY IN GASTROINTESTINAL TRACT

Ostwald's ripening is a thermodynamically driven spontaneous process triggered by nanosizing. The nanoparticles tend to agglomerate to form bigger sized particles to attain thermodynamic stability. Larger particles have lower surface to volume ratio and represent the lower energy state in comparison to the high-energy state of nanoparticles. Therefore nanoparticles tend to attach to the surface of larger particles, and this tendency increases at elevated temperature conditions resulting in a pronounced aggregation of particles. For instance, the Eudragit RLPO-based nanosuspensions of glimepiride exhibited better stability characteristics in refrigerated conditions in comparison to nanosuspensions stored at high temperatures and humidity conditions of 40 ± 2°C/75 ± 5% RH (Yadav et al., 2012).

Nanoemulsions improve the polydispersity index of formulations suggesting improvement in physical stability of formulations. The polydispersity index is used as an indicator of the broadness of a particle's size distribution. Higher values (>0.7) indicate broad particle size distribution. It is defined as the ratio of weight average molecular weight (M_w) to number average molecular weight (M_n).

$$\text{Polydispersity index} = \frac{M_w}{M_n} \tag{11.4}$$

Narrow polydispersity index values indicate homogenous dispersions and processes that result in the formation of homogenous dispersion and may prevent Ostwald ripening of drug substances on storage, especially in the case of amorphous drug substances which show a high tendency for agglomeration (Sahu and Das, 2014; Cho et al., 2015).

Zeta potential is also a frequently used parameter to predict colloidal stability. Nanoparticles having Zeta potential values of more than +25 mV or less than −25 mV are generally stable, whereas, dispersion formulations with a low zeta potential value will eventually aggregate owing to Van Der Waal interparticle attractions. Nanoparticles with Zeta potential values between ±5 and ±15 mV show limited flocculation while particles with values between 0 and ±5 mV exhibit maximum flocculation (Schwarz et al., 1994). Yet, lower Zeta potential values should not be inferred to cause complete instability. The lack of necessary repulsion required to keep the system in a deflocculated state for a long time can be compensated by using steric stabilizers such as surfactants. For instance, Tween 80 was reported to impart steric stability to solid lipid nanoparticle and nanostructured lipid carrier formulations (Rahman et al., 2010a; Thatipamula et al., 2011).

Particle aggregation is generally more prominent in aqueous nanosuspension formulations than in a solidified state like freeze-dried nanoparticles. Moreover, hydrolysis further compounds the instability issues seen in liquid formulations. The steric stabilizer molecules are effective in counterbalancing the high surface energy of the particles in the liquid state, but their effect decreases in the solid state. Nevertheless, cryoprotectant agents (such as mannitol) have demonstrated excellent crystal growth inhibition properties during freeze drying. Polymers also prevent crystal growth in freeze-dried products due to their ability to form a protective coat on the nanoparticles (Gao et al., 2010; Wang et al., 2013).

Nanoemulsion formulations also improve the biochemical stability of acid labile drugs and consequently improves their bioavailability. For example, the encapsulation of stiripentol, an acid labile drug, into the inner oil phase of nanoemulsion formulation shielded it from an acid attack, which was confirmed by the large amount of drug (98%) retained in the formulation (Lu et al., 2015). In comparison, more than 45% drug was degraded after four hours of exposure to acidic conditions. Similarly, varying pH conditions and monovalent and bivalent electrolytes may influence the stability of nanosuspension formulations considerably. The electrolytes disrupt the electrostatic barrier layer surrounding the nanoparticles, and according to the Schulze–Hardy rule, bivalent ions are more disruptive than monovalent ions. In this context, polymer barriers again show potential for stabilizing nanoparticles against electrolyte-mediated destabilization. For instance, the polymeric/surfactant interfacial barrier was found to be resilient toward destabilization of telmisartan nanoparticles in the presence of electrolytes NaCl and $CaCl_2$ (Bajaj et al., 2012).

11.4 SURFACE MODIFICATIONS OF NANOPARTICLES

11.4.1 CRITERIA FOR SURFACE MODIFICATION

The nanoparticles owe their unique properties to their distinctive surface characteristics. These characteristics are not only responsible for their medical application, but also have a bearing on the potential toxicity of nanoparticles. Bare nanoparticles are swiftly coated with proteins (opsonins) upon contact with body fluids, evoking a different biological response and eventually leading to their elimination by the MPS (Garnet and Kallinteri, 2006). To be effective, nanoparticles must have a long circulation half-life to earn more time to cross barriers like the blood–brain barrier (BBB). This is achieved through the modulation of the surface properties of nanoparticles by attaching different coupling agents, making them amenable to binding different molecules. Consequently, the opsonization process subsides, leading to the prolongation of the circulation half-life of nanoparticles. Hydrophilic polymer coatings of chitosan, PEG, and polycaprolactone have been explored for retarding opsonization and, thus, the mononuclear phagocyte uptake. The stealth vehicles thus formed display a prolonged circulation time. Surface modification with polymers additionally improves aqueous stability and decreases particle aggregation (Cho et al., 2008). PEG has been extensively used due to its solubility in both aqueous and organic solvents. PEG-coated nanoparticles have shown a reduced tendency for phagocytosis across all macrophage phenotypes. Even derivatives of PEG such as PEG-silanes which are employed to modify nanozeolites' surface through a process called silanization are purposefully effective. PEG-coated nanoparticles have also demonstrated enhanced transport within the brain interstitium. Besides improvement in pharmacokinetics response, stealth coatings with targeting moieties remarkably improve BBB transport (Backmann et al., 2010).

Nanocarriers enter the cells primarily via an endocytic pathway, which also happens to be the major limiting factor for delivering drugs to intracellular targets. The endocytic pathway involves the formation of endosomes that undergo a maturation phase from early endosome to late endosomes with a shift in pH from 6 to 5, respectively. During the late phase, endosomes merge with lysosomes, rendering the encapsulated drug accessible to digestive enzymes such as proteases, glycosidase, nucleases, and lipases present in lysozymes.

The uptake of nanoparticles by intestinal epithelial cells is significantly impacted by the surface charge of nanoparticles. Positively charged particles show more affinity toward the negatively charged cell membrane, ensuing fusion of nanoparticles with the endosomal membrane culminates in the release of cargo into the cytoplasm. Hence, modifications of nanoparticle surface have been explored extensively for improving absorption and target drug delivery (des Rieux et al., 2006). For instance, nanoparticles modified with didodecyldimethylammonium bromide, a cationic surfactant, imparted the positive charge to the nanoparticle surface which resulted in increased retention time at the cell surface. The particle uptake was enhanced and, consequently, drug bioavailability increased. (Mei et al., 2009). Surface charge manipulation has been investigated for target drug delivery to cancerous cells. Cancerous cells are negatively charged owing to the overexpression of negatively charged phospholipids on their cell membranes relative to normal cells and, consequently, they are likely to interact efficiently with positively charged nanoparticles. For example, positively charged zinc oxide nanoparticles exhibited better selective cytotoxic potential against cancer cells in comparison to negatively charged polyacrylic acid coated zinc oxide nanoparticles. However, cationic nanoparticles exhibited a faster aggregation rate and dissolution, which might reduce their bioavailability and limit their potential as anticancer agents. In contrast, negatively charged nanoparticles were less selective against cancer cells, but showed desirable properties, particularly lower dissolution and better dispersion properties (Wingett et al., 2016).

An encounter with digestive enzymes may also be avoided by the proton sponge effect. Based on osmotic action, the proton sponge phenomenon helps in lysis of the endosomes via pH buffering effect, resulting in the liberation of the cargo in the cytoplasm. In this respect, secondary and tertiary amines with pKa values near to endosomal pH are frequently used as surface modulators owing to their ability to accept H^+ ions at the acidic pH of the endosomes. The internal acidic pH of endosomes is regulated by specific endosomal H^+ transmembrane pumps and the same carrier pathway is utilized by chlorine ions. The amines prevent acidification of vesicles and, therefore, block the action of the pump through negative molecular feedback inhibition mechanism. Higher molecular weight amines have demonstrated a superior proton sponge effect (Parodi et al., 2015).

The acidic pH of endosomes is also employed to propel the interaction of peptides with the lipid bilayer of endosomal vesicles. Some naturally occurring proteins have fusogenic or membrane pore forming properties. The acidic pH of endosomes prompts alteration in the conformation of these proteins, resulting in interactions between the peptide and the lipid bilayer of endosomes. At acidic pH, the amino acid residues may accept H^+ ions causing peptides to obtain an amphipathic α-helical conformation. Consequently, strong interaction of peptides with phospholipid membranes may result in the formation of pores in membranes or may even cause membrane fusion or lysis. Listeriolysin O, a toxin produced by the bacterium *Listeria monocytogenes*, has been demonstrated to induce pore formation in cholesterol-containing lipid bilayers at acidic pH bringing about the release of liposomal nanocarriers. Some cationic lipids have been shown to facilitate the release of nanocarriers from endosomes by interacting with negatively charged lipids of the cytosolic layer of the endosomes. However, their fusogenic effect is moderate and helper lipids are required to promote the effect (Kullberg et al., 2012).

Cell membrane coatings have also been utilized to counter opsonization. These coatings provide unique advantages of preserving the natural structure of cell membranes. Red blood cells (RBCs) coated nanoparticles have been significantly explored as they are readily available. The RBC membrane coating averts the nonspecific adsorption of proteins. Additionally, the self-signaling property of the membrane protein "CD47" prevents clearance of nanoparticles by phagocytes, thereby, prolonging the blood circulation time of the nanoparticles. Similarly, membranes of leukocytes and platelets have been investigated as nanoparticle coating agents for targeting injured tissues (Zhou et al., 2016).

Cobalt-based nanoparticles, especially cobalt oxide nanoparticles, have also shown promise in biomedical applications (Wang et al., 2005) due to their unique size- and shape-dependent properties.

11.4.2 BIOCOMPATIBILITY AND BIOMEDICAL APPLICATIONS

Some nanoparticle designs, such as carbon-based nanoparticles, have been consistently associated with cytotoxic effects in several in vitro and in vivo assays (Zhao et al., 2008). As an illustration, carbon nanotubes are known to induce mesothelioma, akin to the toxicity associated with a naturally occurring carcinogenic mineral, asbestos (Donaldson et al., 2010). Certain reports indicate the predominant role of particle shape in toxicity inducement, suggesting a critical role of particle morphology in cell toxicity (Johnston et al., 2010). Similarly, some studies clearly delineate the effect of the size of nanoparticles on biocompatibility. For example, gold nanoparticles having a diameter of 1.4 nm were linked to cell toxicity, whereas 15 nm particles were devoid of cell toxicity (Pan et al., 2007). In this context, the surface modification of nanoparticles with suitable additives would increase selectivity and efficiency and, consequently, minimize the potential toxicity to cells. For instance, surface functionalization of inorganic particles with 2,3-dimercaptosuccinic acid (DMSA) demonstrated improved biocompatibility (Auffan et al., 2006). Silva et al. also reported improved biocompatibility of gold and maghemite nanoparticles functionalized with DMSA without any notable toxic effects in human mesenchymal stem cells (Sun et al., 2016). In another study, the harmful effects of iron oxide were also reported to reduce upon coating with a polymer coating in in vitro and in vivo studies (Gupta and Wells, 2004).

Gold nanoparticles possess some distinctive characteristics such as small size, low toxicity, biocompatibility, and its surface being amenable to modification, thus, favoring application of gold nanoparticles in the biomedical and biological field such as drug delivery for cancer chemotherapy, sensing, imaging and radiotherapy (Podsiadlo et al., 2008). The uptake and toxicity of gold nanoparticles are largely dependent on its surface properties. The active surface of gold particles adapts to the inclusion of inorganic and organic molecule stabilizers in its formulation (Podsiadlo et al., 2008). The ligand or surfactants used to engineer gold nanoparticles are often added in high quantities. Any unutilized quantity may remain in free form after synthesis. The access quantity of these additives must be removed to prevent an impact on the stability, purity, and reactivity of the resulting nanoparticles (Sapsford et al., 2011). For example, unbound cetyltrimethylammonium bromide (CTAB) molecules left after preparation of CTAB-capped gold nanorods were reportedly cytotoxic to H-29 cells. In comparison, gold nanorods without CTAB did not display similar cytotoxicity against HT-29 cells (Alkilany et al., 2009). CTAB is customarily employed to prepare gold nanorods of a precise length and AR. However, toxicity limits its potential use in biomedicine applications. CTAB is cytotoxic above 1–10 μM concentration, and additionally, particles stabilized with CTAB exhibit low colloidal stability in aqueous salt solution, incompatibility with other solvents and instability during shelf life (Ray, 2009). The ensuing crystallization of CTAB leads to the loss of morphology (Iqbal and Tae, 2006). Alternative approaches based on either replacing CTAB or functionalizing CTAB-coated gold nanoparticles with additives such as polymers, peptides, surfactants, and lipids have been investigated to improve stability and biocompatibility (Tebbe et al., 2015). To this end, oleic acid-based surfactants or end-thiolated CTAB have been frequently used (Vigderman et al., 2011). Polymers like PEG may form a hydrophilic protective coat over nanoparticles which can drive off opsonin protein absorption, contributing to prolonged circulation half-life. PEGylated nanoparticles have also demonstrated low cytotoxicity on several cell lines including HeLa, EMT-6, and CT26 cells (Niidome et al., 2006; Liu et al., 2010). Jing et al. constructed a multifunctional nanoagent formulation by modifying hollow prussian blue nanoparticles using hyaluronic acid grafting PEG. The nanoparticles were subsequently loaded with 10-hydroxycamptothecin for tumor-targeted thermochemotherapy. The nanoagent displayed excellent colloidal stability, long blood circulation half-life, and the potential for selective targeting to cancer cells. The good photothermal efficiency and light-triggered drug release manifested a significantly high inhibitory effect on cancer cells. Furthermore, nanoagents exhibited excellent biocompatibility characteristics vis a vis no noted alteration in body weight, histopathology, and blood biochemical parameters (Jing et al., 2016). Similarly, molecularly imprinted polymeric nanoparticles were reportedly modified with polymerizable PEG. The modification led to decrease in surface energy and thereupon improvement in dispersibility, storage, and colloidal stability in comparison to unmodified particles (Moczko et al., 2013). Proteins, due to their well-defined structure and presence in various chemical functional groups, demonstrate a high binding affinity for metal surfaces which makes them suitable multifunctional coating vehicles. Their high molecular weight further contributes in imparting electrosteric stabilization and pH-responsiveness characteristics (Tebbe et al., 2015).

Impurities emanating from the synthesis of carbon nanotubes have been associated with toxic effects. This requires improvements in synthetic methods and processes. Methods including surface ligand exchange, chelation of metal components, high-temperature thermal treatments, overcoating, field flow fraction, and electrophoresis have been explored (Alkilany et al., 2009; Hull et al., 2009). Impurities severely influence the function and other properties of nanomaterials, suggesting changes in the structure of the surface ligands which may further aggravate the toxicity. Harper et al. described the preparation of highly purified terminally modified glutathione-coated gold nanoparticles with excellent biocompatibility characteristics for imaging and metal chelation applications. The authors propounded the removal of impurities together with the ultracentrifugation step before and after the coupling reaction with the amino acid sequences as an effective approach for enhancing biocompatibility (Harper et al., 2014).

Silicon dioxide, an inorganic-based nanomaterial, has several attractive properties making it suitable for various biomedical applications. Apparently it has produced dose-dependent cytotoxicity in a variety of in vitro systems. In this regard, unmodified silicon dioxide nanoparticles have shown higher genotoxicity, lung, and immunomodulatory effects (Klein et al., 2012; Landsiedel et al., 2010). Therefore the application of silica nanoparticles depends largely on the ability of surface modifications to limit toxicity. For instance, amino and phosphate surface modifications reportedly reduced the proinflammatory and immunomodulatory allergic effect of silica nanoparticles, whereas, PEG was reportedly ineffective in mitigating the toxicity of silica particles (Marzaioli et al., 2014). Mesoporous silica displays favorable chemical properties, thermal stability, and biocompatibility. It is endowed with a unique mesoporous structure which is responsible for efficient drug loading and controlled release. The pore size and porosity are highly flexible and can be varied according to the size requirements of the drug substance. The flexible pores also prevent premature drug release and, hence, the degradation of drugs prior to reaching their target. Further, their active surface presents an opportunity to attach different ligands and therapeutic molecules (Kwon et al., 2013).

Zinc oxide nanoparticles have shown considerable promise for multiple applications in biomedical engineering. They are biocompatible, economical, and easy to manufacture. Additionally, its antibacterial properties support their use as a prophylactic agent (Taccola et al., 2011). Akin to the zinc oxide, silver nanoparticles also express antibacterial properties and are frequently used in consumer products. Capping agents are often added to prevent aggregation of nanoparticles, however, they may also influence the physicochemical properties. In this context, potent capping agents like polyvinyl pyrrolidone strongly bind to the silver core and provide steric stabilization (Tolaymat et al., 2010).

Magnetic iron oxide nanoparticles are also increasingly used in biomedical applications in cell labeling, magnetic hyperthermia, magnetic resonance imaging, and drug delivery because of their chemical stability, low toxicity, and strong magnetic sensitivity (Huang et al., 2012). However, the hydrophobic residue such as oleic acid on their surfaces stemming from their synthetic route decreases the solubility and limits their potential as nanocarriers. Both synthetic polymers and naturally occurring proteins have been investigated as potential agents for surface modification. In illustration of this fact, Huang et al. coated iron oxide particles with casein protein which was obtained from milk by exchange encapsulation method. The process utilized ionic oligosaccharide for ligand exchange and subsequent encapsulation of nanoparticles with casein. The particles displayed excellent water solubility and colloidal stability in the pH range of 5.0–8.0. The particles also exhibited acceptable biocompatibility characteristics (Huang et al., 2013).

Similarly, polysaccharides such as dextran and heparin have been reported to produce stealth coated nanocarriers to counter opsonization. Their easy availability, biocompatibility, and well-established modification schemes have led to their widespread use in controlled drug delivery. Further, by interacting specifically with various receptors on target cells, they show better cellular uptake in comparison to bare nanoparticles (Passirani et al., 1998).

The surface-modified phosphonomethyl iminodiacetic acid coated cobalt oxide nanoparticles with an average size of less than 100 nm demonstrated enhanced uptake by cancer cells, resulting in apoptosis of cancer cell (Chattopadhyay et al., 2012). Similarly, positively charged didodecyldimethylammonium bromide coated poly(lactide-*co*-glycolide)-D-α-tocopheryl PEG 1000 succinate nanoparticles showed higher levels of cellular uptake than that of didodecyldimethylammonium bromide modified poly(lactide-*co*-glycolide) (PLGA) nanoparticles and unmodified PLGA-D-α-tocopheryl PEG 1000 succinate nanoparticles. The positive charge of nanoparticles ensured prolonged retention time at the cell surface, resulting in improvement in bioavailability. Further, D-α-tocopheryl PEG 1000 succinate, a water-soluble derivative of natural vitamin E, reportedly improved drug permeability through the cell membrane by inhibiting P-glycoprotein-mediated multidrug resistance in tumor cells (Chen et al., 2011).

11.5 IMPROVEMENT IN ORAL BIOAVAILABILITY AND PERFORMANCE

11.5.1 ABSORPTION, DISTRIBUTION, AND ELIMINATION

A drug is confronted with many profound challenges when administered through the oral route, chiefly among these is to surmount the GIT barrier. The hostile environment of GIT with wide-ranging pH variation (from 1 to 8) can induce drug degradation via oxidation, deamidation, or hydrolysis, thereby causing significant loss of drug activity. Additionally, GIT enzymes like lipases, proteases, and nucleases may also degrade the drugs prior to absorption. Even if these barriers are overcome, the drug must traverse the epithelial barrier for absorption to take place (Sood and Panchagnula, 2001). The barrier comprises of the mucus layer, epithelial cell monolayer, and lamina propria which consists of blood and lymph vessels that are responsible for transporting the drug to other parts of the body after absorption.

Epithelial cells are coated with a mucus layer, which is a complex hydrogel material. Secreted by goblet cells, mucus comprises of salts, proteins, carbohydrates, lipids, and antibodies. Many compounds, bacteria, and cellular waste are mechanically entrapped in the mucus and may even link with it through hydrogen bonding, van der Waals forces, and hydrophobic or electrostatic interactions (Ensign et al., 2012). The thickness of the mucus layer varies across the GIT, depending on diet and may loosely or firmly cling to the epithelium. In humans, mucus is substantially thick, about 100 μm. With a turnover of approximately 4–6 hours, particles must traverse the entire mucin layer within this time period to reach the absorptive surface (Atuma et al., 2001). Particle size exerts considerable influence on mobility through the mucin layer. Smaller particles tend to move quickly through the mucus layer and avoid elimination by mucilliary clearance, whereas particles of size 1 μm or greater are virtually immobile in the mucus layer. Mucus production is considerably less in the small intestine, which allows greater interaction with the GIT immune cells, resulting in the uptake of particles as large as 1 μm (Fig. 11.1). The basic functions of mucus include lubrication of food particles to facilitate transport along the GIT, to protect epithelial surfaces by trapping pathogens and foreign particulates, and rapidly clearing them subsequently (Cone, 2009).

Only when a drug crosses the mucus barrier does it hit the target absorptive epithelial cells (enterocytes) which are compactly held in monolayers. These cells also represent the major transport hurdle for a drug to cross from the intestinal lumen to the lamina propria (Banan, 2005). The P-gp efflux pump and first-pass metabolism further reduce drug bioavailability. Moreover, the immune system (which is closely associated with the epithelium) offers additional resistance to the drug absorption (DeMario and Ratain, 1998). Lamina propria is closely guarded by immune cells like T cells, macrophages, and dendritic cells.

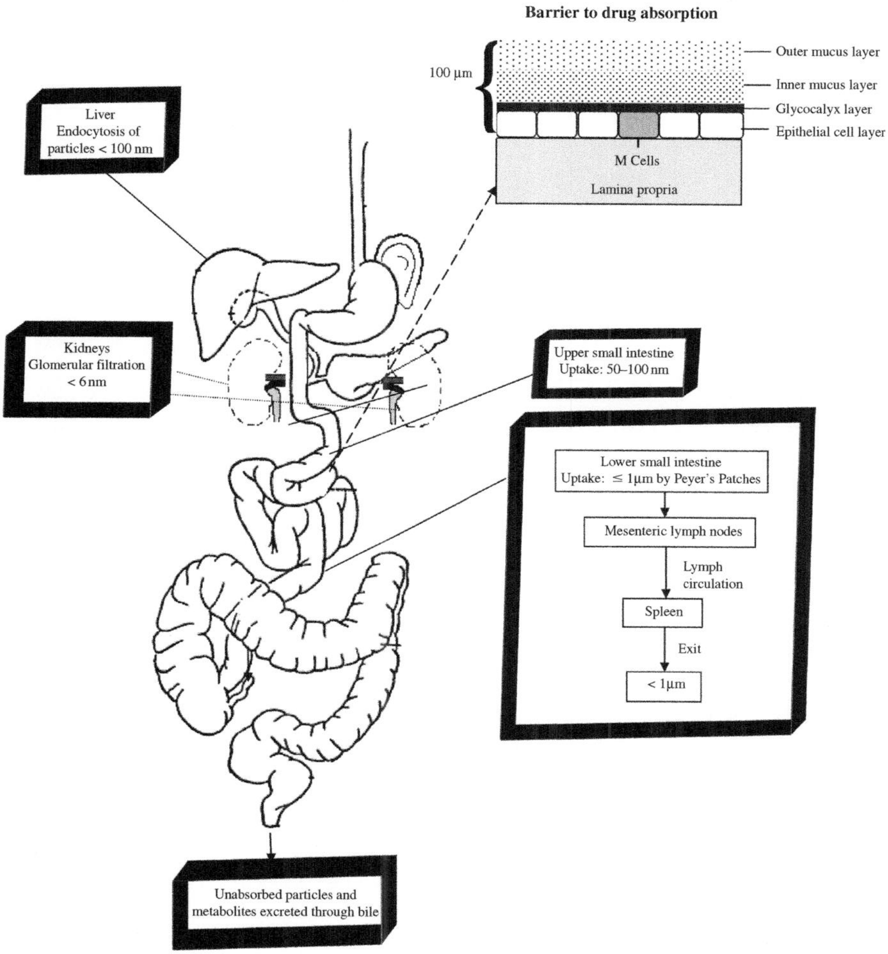

FIGURE 11.1

Anatomy of drug absorption from nanoparticles.

Nanoparticles must also overcome the MPS while releasing their drug load (Macdonald and Monteleone, 2005).

In light of the above, the nanotechnology approach has gained much importance on account of the fact that nanoencapsulation presents an opportunity to enhance the oral bioavailability resulting from improvements in the solubility, permeability, and stability characteristics of the drug (Table 11.3).

To evade GIT barrier challenges, polymeric nanoparticles have shown the potential to deliver drug candidates, especially biologic substances, through the oral route. Polymeric nanoparticles offer several benefits to oral drug administration, as shown in Fig. 11.2. They can further shield the drug from recognition by

Table 11.3 Improvement in Pharmacokinetic Parameters of Nanoparticles

No.	Drug Substance	Carrier	Species	$AUC_{0-\infty}$ (μg h/mL)		C_{max} (μg/mL)		T_{max} (h)		Elimination Half-Life ($t_{1/2}$) (h)		References
				Control Formulation	Nanoparticle Formulation	Control Formulation	Nanoparticle Formulation	Control Formulation	Nanoparticle Formulation	Control Formulation	Nanoparticle Formulation	
1	Acetylpuerarin	PLGA nanoparticles	Wistar rats	2.24[a]	6.18	0.73[a]	1.30	1.80[a]	1.10	1.41[a]	3.07	Sun et al. (2016)
2	Ergosterol	PLGA nanoparticles	Rats	6.35[b,c]	31.12[b,d]	0.57[b]	0.93	8.00[b]	13.20	4.07[b]	20.61	Zhang et al. (2016)
3	BMK-20113 (hepatitis C virus NS5A inhibitor)	Liquid crystal nanoparticles	Sprague Dawley rats	762.00[e,f]	1070.00[e]	149.00[g,f]	306.00[g]	2.40[f]	1.30	2.23[f]	2.06	Lee et al. (2016)
4	Ezetimibe	Solvent evaporated solid dispersion	Male Sprague Dawley rats	3.14[b]	5.58	0.30[b]	0.98	1.40[b]	0.80	16.96[b]	20.07	Rashid et al. (2015)
5	Lovastatin	Nanostructured lipid carriers	Wistar rats	0.93[a]	3.72	0.11[a]	0.22	–	–	1.82[a]	7.21	Zhou and Zhou (2015)
6	Megestrol acetate	SAS process	Sprague Dawley rats	2.03[b]	8.12	0.23[b]	1.23	2.00[b]	1.00	–	–	Ha et al. (2015)
7	Docetaxel	Solid lipid nanoparticles surface modified by D-alpha-tocopheryl poly(ethylene glycol 1000) succinate	Male Sprague Dawley rats	0.06[h]	0.22	0.06[h]	0.13	0.25[h]	0.75	–	–	Cho et al. (2014)
8	Genistein	Polymer-based nanoparticles	Sprague Dawley rats	10.8[i,a,d]	26.10[i,d]	2.17[a]	5.32	2.80[a]	2.50	4.24[a]	4.60	Tang et al. (2011)
9	Celecoxib	Polymer-based nanoparticles	Healthy Human volunteers	10.08[j]	13.71	0.80[j]	2.49	3.00[j]	0.80	–	–	Morgen et al. (2012)
10	Miconazole	Solid lipid nanoparticles	Albino male rabbits	0.12[j]	0.29	0.009[j]	0.01	1.50[j]	4.10	–	–	Aljaeid and Hosny (2016)
11	Carvedilol	Nanosuspensions	Male beagle dogs	4,657.86[k]	9,948.70	1,062.61[k]	2,224.58	2.80[k]	0.90	–	–	Liu et al. (2015a)
12	Zidovudine	Lactoferrin nanoparticles	Male rats	63.48[l]	251.57	37.67[l]	49.20	1.00[l]	2.00	1.759[l]	3.07	Kumar et al. (2015)

13	Fenofibrate	Self-microemulsifying drug delivery systemcomprising of Myritol 318 an Tween 80	Healthy Human volunteers	87.03[m]	106.84	3.27[m]	3.74	2.75[m]	1.80	26.43[m]	32.04	Lin et al. (2011)
14		Electrosprayed nanospherule	Male Sprague Dawley rats	17.70[n]	41.00	2.40[n]	5.10	1.60[n]	1.80	6.10[n]	5.30	Yousaf et al. (2016)
15	Itraconazole	Solid dispersion	Beagle dogs	3.90[o]	4.66	0.31[o]	0.42	1.83[o]	2.20	–	–	Yin et al. (2015)
16	Torcetrapib	Solid lipid dispersions	Rats	0.68[p,b]	7.90[i]	0.08[b]	2.02	6.00[b]	1.00	–	–	Liu et al. (2015b)
17	Valsartan	Composite nanoparticles	Male Sprague-Dawley rats	16.0[q]	73.80[q]	1.90	13.60	1.90	0.50	–	–	Kim and Baek (2014)
18	Sirolimus	Solid dispersion	Rats	0.06[r,d]	0.46[d]	0.01[r]	0.10	3.80[r]	1.20			Cho et al. (2015)
19	Posaconazole	Amorphous solid dispersion	Wistar rats	4.71[s,c]	10.16[c]	0.13[s]	0.32	5.00[s]	1.00	25.58[s]	25.12	Fule and Amin (2014)

AUC, *area under the plasma concentration versus time curve;* PLGA, *Poly(lactide-co-glycolide);* SAS, *supercritical antisolvent.*
[a] *Suspension formulation.*
[b] *Drug powder.*
[c] $AUC_{0-72\ h}$.
[d] $AUC_{0\rightarrow 12\ h}$.
[e] *h nmol/L.*
[f] *Host–guest complex.*
[g] *nmol/L.*
[h] *Taxotere formulation.*
[i] *μg/mL μh.*
[j] *Commercial capsule formulation.*
[k] *Commercial tablets.*
[l] *Soluble formulations.*
[m] *Tricor tablets.*
[n] *Powder formulations.*
[o] *Sporanox capsules.*
[p] *μm h.*
[q] $AUC_{0-24\ h}$.
[r] *Physical mixture.*
[s] *Commercial capsule (Noxafil).*

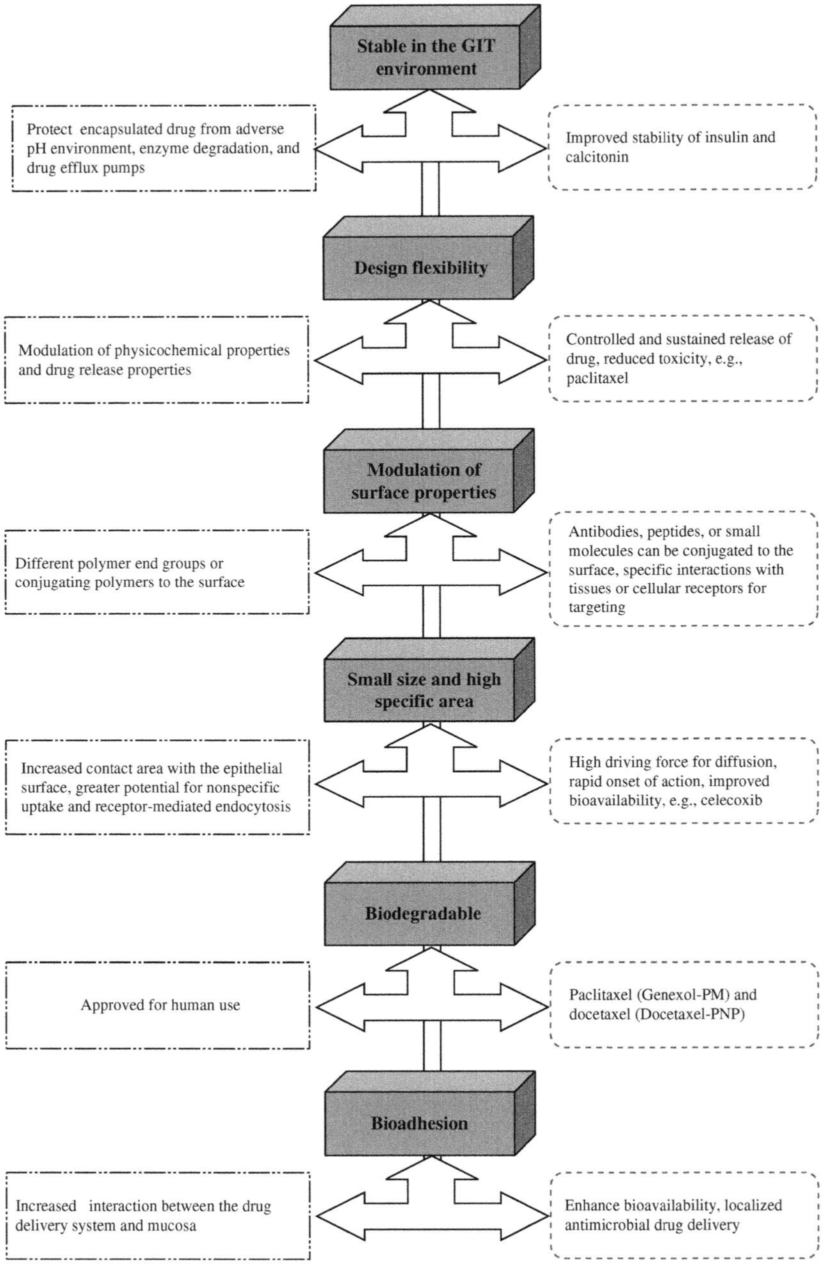

FIGURE 11.2

Desired attributes of polymeric nanoparticles.

P-gp and intra-enterocytic CYP450 (Boddupalli et al., 2010; Takeuchi et al., 2005). Various alternative approaches have also been explored to surmount the GIT barrier. Among these procedures, one of the most frequently used strategies is to employ mucoadhesives. Mucoadhesives adhere to the mucus layer and increase the drug concentration near epithelial cells in addition to causing a reversible change in compact epithelial junctions. However, compromising the tight junction integrity seriously restricts the utility of this approach owing to safety concerns. Furthermore, transport through tight junctions is limited to small molecules only (Yeh et al., 2011; Sonaje et al., 2011).

The transcytosis pathway employed by M cells to transport antigens across the epithelium can be utilized to overcome barrier properties of GIT. Targeting transcytosis pathways of M cells would facilitate movement across the barrier and allow access to the systemic circulation (Macdonald and Monteleone, 2005; Jang et al., 2004). Nanoparticles may adhere to the luminal surface of M cells and deliver the drug directly by transcytosis. Adhesion is facilitated by a thin glycocalyx layer, a glycoprotein-polysaccharide layer surrounding the M cells (Jang et al., 2004). This assumes importance in invasive therapies such as chemotherapy. The adverse effects associated with these therapies (such as mucositis, ulceration of the GIT, and diarrhea) could be significantly reduced. However, the Peyer's Patches represent only a modest fraction of the intestinal surface area and, therefore, have limited absorption capacity. Additionally, the immune cells associated with M cells might uptake nanoparticles before reaching the systemic circulation (Kammona and Kiparissides, 2012).

Preferential GI uptake of particles of a certain size has also been investigated in pursuit of achieving target therapeutic concentration. Particles in the size range 50–100 nm are preferably taken up by enterocytes in the upper small intestine, whereas larger particles reaching 1 μm are taken up by Peyer's Patches in the lower small intestine. Following uptake by Peyer's Patches, larger particles near diameter the 1 μm or less are translocated to the mesenteric lymph nodes and eventually gain access to the spleen via lymphatic circulation. Particle size is also indispensable to their exit from the spleen. While particles of 1 μm size are sequestered in the spleen, particles less than 1 μm are able to egress the spleen with relative ease (des Rieux et al., 2007; Desai et al., 1996).

After endocytosis, the transport of nanoparticles present in the lamina propria depends on their hydrophobicity. Usually, water-soluble substances are transported to distant tissues directly via the blood circulation system, whereas, lipophilic compounds drain through the lymphatic pathway before entering the blood circulation (Navarro et al., 2014). As an illustration of the influence of the transport pathway on pharmacokinetics, the anticancer activity of orally administered docetaxel improved significantly compared to an intravenously administered solution. The higher drug blood levels of docetaxel achieved through the oral route were ascribed to its absorption via the lymphatic route (Cho et al., 2014). The lymphatic uptake limits the hepatic input, thereby evading the first-pass effect and protecting the drug from CYP3A4 degradation in the gut and systemic circulation (Attili-Qadri et al., 2013.). A very subtle and important generalization has been recognized in the case of polystyrene nanoparticles. Smaller particles are preferentially taken rather than the larger sized particles and nonionized particles are taken up more than carboxylated ones and 407 poloxamer-coated more than 188 poloxamer-coated particles across the GIT (Hillery and Florence, 1996). Typically, particles smaller than 70 nm accumulate in the liver, whereas, larger particles (200 nm) accumulate in the spleen (Liu et al., 1992). The particles have been shown to distribute in small and large intestinal walls, irrespective of surface charge which has been related to their uptake by the lymphoid tissue associated with these organs (Baek et al., 2012; Cho et al., 2013).

Polymeric nanoparticles may delay T_{max} (time to achieve maximum drug concentration in the systemic circulation) of encapsulated drugs and prolong circulation half-life, resulting in enhanced bioavailability (Table 11.3). The possible explanation for this observation may be due to the delayed drug release or direct uptake from the GIT. Improved bioavailability is also assisted by reduced drug clearance. Furthermore, the ability of these nanocarriers to target specific cells reduces adverse effects. Small and long circulating nanoparticles are distributed considerably to tissues with discontinuous endothelium, such as the liver. The process appears to be competing in nature, depending primarily on nanoparticle properties, the vascular permeability, and blood supply to the tissues. While the uptake by hepatocytes is invaluable in the treatment of many infections and metabolic disorders that are associated with the liver, the therapeutic delivery to extrahepatic tissues becomes challenging as rapid distribution to the liver must be avoided (Liu et al., 2014). PLGA is increasingly distributed in the brain and has been reported as an excellent vehicle for molecules employed in the treatment of high-grade glioma. Though the mechanism is not yet entirely understood, some reports indicate nanoparticles' mediated improvement in permeability and inhibition of P-gp-mediated efflux (Zhang et al., 2016; Joshi et al., 2014)

The excretion process of nanoparticles entails the elimination of material waste or metabolites. The uptake of nanoparticles by the MPS implies that they bear a high potential to induce liver and spleen toxicity. In this context, surface-modified nanoparticles advantageously facilitate excretion through the hepatobiliary excretion pathway. The size of fenestration of the endothelium of the liver is approximately 100 nm, suggesting favorable endocytosis of particles less than 100 nm (Seo et al., 2015). Three hydrodynamic diameter thresholds, specifically the glomerular wall filtration (8 nm), rapid extravascular equilibration (5 nm), and EPR (20 nm), assume high relevance in pharmacokinetics. Particles with a hydrodynamic diameter of less than 6 nm undergo glomerular filtration with relative ease. In contrast, the glomerular wall is impervious to particles larger than 8 nm. Consequently, larger particles must decompose or degrade prior to their clearance through urine. Further, the negative charge of the glomerular wall would ensure selective filtration of positively charged particles, whereas, negatively charged particles may exhibit a prolonged circulation time. Additionally, negative particles may accumulate in the kidney and cause marked toxicity. However, some reports seem to contradict the influence of surface charge, and even suggest a lack of any impact on the excretion kinetics due to surface charge (Choi and Choy, 2014).

With regard to renal clearance of nanoparticles, it must be noted that owing to its predictable behavior, renal clearance is still considered favorable over RES clearance. Paradoxically, the effect of particle size on toxicity seems unpredictable. Particles with 1.4 and 13 nm diameter are associated with acute toxicity, whereas other particle size ranges do not seem to cause any fundamental toxicity in short-term studies. The unabsorbed particles and metabolites excreted through bile are finally excreted via feces. Both fecal and biliary excretion routes, thus, play a significant part in the elimination of nanoparticles, irrespective of route of administration, particle size, surface charge, or gender (Perrault et al., 2009).

Some reports suggest an effect of particle shape on residence time in the cytoplasm. Nanoparticles with sharp corners and edges exhibited a long residence time in cytoplasm regardless of their surface compositions, charges, materials, and sizes. On the other hand, spherical nanoparticles after endocytosis remained in endosomes, evolved with endosome maturation, and were easily excreted via exocytosis (Chu et al., 2014).

11.5.2 ORAL BIOAVAILABILITY

Oral bioavailability (F) can be expressed mathematically by:

$$F = F_a \times F_g \times F_h \tag{11.5}$$

where F_a, F_g, and F_h represent the fraction of drug absorbed, fraction escaping metabolism in the GIT and fraction escaping first-pass hepatic metabolism, respectively. (Kwan, 1997). Absorption is one of the critical parameters that determines the bioavailability of orally administered drugs. After oral administration, the rate and extent of drug absorption from nanocarriers largely depend on their stability, dissolution of the drug in the GIT, and permeability. The next section will cover some commonly employed approaches based on nanotechnology for enhancing oral bioavailability of poorly soluble drugs (Table 11.3).

Self-microemulsifying drug delivery systems (SMEDDS) are designed to form an o/w microemulsion on dilution with an aqueous phase or GIT fluids upon oral administration, forming droplets in the nanosize range with high surface to volume ratio. The presence of the drug in dissolved form enhances drug absorption. Additionally, the surfactant present in SMEDDS might alter the permeability of cell membrane and hence increase drug permeability. Moreover, surfactants are absorbed at the interface, due to their amphiphilic nature, which may reduce interfacial surface tension and enhance drug penetration into the epithelial cells. The polar group of surfactants interact with the polar head groups present in the lipid bilayers, thereby modifying hydrogen bonding and ionic forces. Likewise, nonpolar chains (by virtue of their association with the lipophilic tails of the lipid bilayers) disrupt the lipid-packing arrangement. Also, the presence of lipid in a SMEDDS formulation may act as the precursor for the formation of lipoproteins and chylomicrons which, in turn, may promote the drug absorption via the lymphatic pathway, thereby enhancing oral drug bioavailability by reducing hepatic first-pass metabolism (Hauss, 2007; Shah et al., 2007). SMEDDS have also attracted attention due to its reproducible drug release, which is less dependent on the physiological state of GIT. The evidence of their success lies in the commercial availability of cyclosporine A, ritonavir, and saquinavir formulations (Wu et al., 2015). Improvement in oral absorption and bioavailability of oridonin, simvastatin, and silymarin is also reported (Yang et al., 2013).

Polymeric nanoparticles improve the oral bioavailability of many drugs, mainly due to protection of the drug from the hostile environment of GIT and interaction with mucosal layers. The drug release from polymeric nanoparticles depends on the physicochemical properties of drug substances, type of polymer and its physicochemical properties, the particle size and morphology. Among the polymer-based drug deliveries, the PLGA-based nanoparticulate drug delivery systems have generated lots of interest owing to their excellent biocompatibility and predictable biodegradability characteristics. Their release rate can be modified by altering the ratio of lactide and glycolide in PLGA polymer (Rahman et al., 2010b). Al-Quadeib et al. (2015) reported the notable impact of the percentage of PEG in the PLGA–PEG diblock copolymer and the surfactant/stabilizer on the in vitro drug release, entrapment efficiency, and mean particle size. The intestinal uptake and bioavailability of gemcitabine HCl was enhanced when it was administered in the form of a polymeric nanoparticle. In the case of cyclosporine, not only in vivo release was sustained, but small-sized PLGA nanoparticles were able to reduce nephrotoxicity compared with microemulsion formulation.

Lipid-based nanoparticles combine the advantages of both nanoparticles and lipid-based formulations. They are prepared preferably by a low energy phase inversion temperature procedure. In addition to low cost, the process time is short and use of organic solvents can be avoided. Excipient-related biotoxicity can notably be reduced by using biodegradable and biocompatible lipids and surfactants. Moreover, lipid nanocapsules not only improve drug solubility and bioavailability, but also enhance drug stabilization. Furthermore, by PEGylation of the particle surface, interaction with serum components can be significantly prevented, which may improve the pharmacokinetics profile and extend residence time. For instance, the improvement in bioavailability of solid lipid formulations of piribedil, vinpocetine, and cyclosporine A are reported (Swathi et al., 2010). Accumulation of drugs in tumor tissues may also increase many-fold resulting in the enhancement of therapeutic effects. Consequently, lipid-based nanoscale drug delivery systems have demonstrated superior oral bioavailability enhancing capability for BCS class IV drug candidates over polymeric nanoparticles. Besides these factors, the improvement in oral bioavailability is also facilitated by the formation of ultrafine dispersion within nanometer range which endows them with a huge surface area and increased adhesion to the gut wall. Further, lipid nanocarriers are transformed into a series of secondary derivatives by the actions of lipase, endogenous bile salts, and phospholipids present in GIT. These secondary derivatives are extremely useful in solubilizing the poorly soluble drugs. Additionally, lipid-based formulations may utilize intestinal lymphatic drug transport system and extend the later absorption phase of drugs by avoiding first-pass metabolism (Wang et al., 2014).

The unique characteristics of lipid-based liquid crystalline nanoparticles (LCNPs) to encapsulate drugs with a different polarity inside the discrete/bicontinuous aqueous channel and lipid domains is advantageously employed to improve the bioavailability of both hydrophilic and hydrophobic drugs. High-energy fragmentation techniques, such as ultrasonication, microfluidization, or homogenization are commonly used techniques to disperse the nonlamellar liquid crystalline matrix (three-dimensional nanostructured crystalline discrete structures, such as cubic or hexagonal phase) into the water phase, whereby LCNPs self-assemble to form polar amphiphilic lipids. Besides this, LCNPs are thermodynamically stable and possess a huge surface area. Over and above, they are biocompatible, digestible, and bioadhesive. The excipients employed in their manufacture are nontoxic in nature. Yet, they offer the same level of protection against harsh GI conditions as provided by polymeric nanoparticles. The toxicity associated with a drug can be mitigated further by their ability to sustain drug release. Notwithstanding, the massive energy input required in their manufacturing increases the cost of production. As an example, the oral bioavailability of paclitaxel-loaded LCNPs demonstrated an increase of 2.1-folds in comparison to a commercial formulation, Taxol (Zeng et al., 2012).

The amorphous solid dispersion (ASD) is among the proven procedures for enhancing the solubility of poorly soluble drug candidates (Cho et al., 2015; Rahman et al., 2014). They are prepared primarily by solvent evaporation and/or a melting process. Supercritical fluid technology based on solvent evaporation is commonly employed to prepare ASD nanoparticles with improved flow properties in a single-step production process. The process is performed at low temperature to avoid drug degradation that may occur at the high temperatures. The requirement for organic solvents is also low. Improvement in solubility, dissolution, and oral bioavailability of poorly water-soluble itraconazole, dutasteride, megestrol acetate, and valsartan have been reported (Kim and Baek, 2014; Ha et al., 2014, 2015). Evidently, the electrospray technique is a very promising approach in preparing nanospherule formulations with improved aqueous solubility and bioavailability due to excellent wetting properties, particle size reduction, and the amorphous state of drug substance. Fenofibrate nanospherule formulations prepared with polyvinylpyrrolidone and Labrafil M 2125 as carriers exhibited a 2.5-fold increase in bioavailability in comparison to a free drug (Yousaf et al., 2016). The antisolvent precipitation method has also been demonstrated to generate ASD. The advantages associated with this method include high-drug loading, simple equipment and process, and, additionally, does not subject the drug to thermal and mechanical stresses associated with other methods. Importantly, Liu et al. have demonstrated that the drug release can be modulated from the solid lipid matrix by controlling the precipitation temperature during the formation of the drug-lipid particles. The increase of 6- to 11-fold in oral bioavailability was recorded for torcetrapib when compared to oral drug powder formulations (Liu et al., 2015b). It is crucial to recognize the inherent thermodynamic instability of an amorphous form. To maintain the drug in an amorphous state, it is mandatory to use stabilizing agents like polymers to stabilize the drug during its shelf life as well as to maintain the supersaturated concentration of the drug during dissolution. Rightly selected polymers can stabilize the drug either by forming secondary interactions with the drug or by reducing the molecular mobility due to their antiplasticization effect (Verma and Rudraraju, 2015).

11.6 SUMMARY

Oral absorption and the bioavailability of a drug is dictated by its fundamental physicochemical properties such as solubility, stability, and permeability in the GIT. Nanoparticles are increasingly recognized as efficient drug delivery vehicles for poorly soluble drug molecules as well as for the target delivery to organs and tissues. Nanoparticles allow modulation of drug release by varying formulation composition and alteration of physicochemical properties to suit the particular therapy requirement. Bare nanoparticles are readily removed by the MPS. The majority of the work pertaining to the surface modification of nanoparticles, therefore, focuses on avoiding their removal by the MPS as well as endowing them with target delivery attributes. The prolonged circulation half-life earns them more time to cross barriers like the BBB. Coating nanoparticles with receptor specific ligands also contributes to reducing the drug-related adverse effects. One of the approaches to promote selective uptake is to charge the nanoparticle surface. Recent studies demonstrate their potential use for overcoming multidrug resistance in cancer treatment. Nanoparticles may also shield the drug from recognition by P-gp and intra-enterocytic CYP450 and contribute in enhanced bioavailability. Considering these benefits, the number of commercially available nanotechnology-based products has increased significantly, with many more products still being in the development phase.

REFERENCES

Pharmaceuticals, Inc., 2010. Access Pharmaceuticals (ACCP)'s cobalamin oral insulin achieves greater than 80% oral bioavailability in preclinical models. Access Pharmaceuticals, Inc., +1-214-905-5100; or Investor Relations, Donald C. Weinberger or Alisa Steinberg of Wolfe Axelrod Weinberger Assoc. LLC, +1-212-370-4500. <https://www.marketscreener.com/ACCESS-PH-183181/news/ACCESS-PH-Access-Pharmaceutical-s-Cobalamin-Oral-Insulin-Achieves-Greater-Than-80-Oral-Bioavailab-13308880/>.

Aderem, A., Underhill, D.M., 1999. Mechanisms of phagocytosis in macrophages. Annu. Rev. Immunol. 17, 593–623.

Ahmed, T.A., 2016. Preparation of finasteride capsules-loaded drug nanoparticles: formulation, optimization, in vitro, and pharmacokinetic evaluation. Int. J. Nanomedicine 11, 515–527.

Aljaeid, B.M., Hosny, K.M., 2016. Miconazole-loaded solid lipid nanoparticles: formulation and evaluation of a novel formula with high bioavailability and antifungal activity. Int. J. Nanomed. 11, 441–447.

Alkilany, A.M., Nagaria, P.K., Hexel, C.R., Shaw, T.J., Murphy, C.J., Wyatt, M.D., 2009. Cellular uptake and cytotoxicity of gold nanorods: molecular origin of cytotoxicity and surface effects. Small. 5, 701–708.

Al-Quadeib, B.T., Radwan, M.A., Siller, L., Horrocks, B., Wright, M.C., 2015. Stealth amphotericin B nanoparticles for oral drug delivery: in vitro optimization. Saudi Pharm. J. 23 (3), 290–302.

Amidon, G.L., Lennernas, H., Shah, V.P., Crison, J.R., 1995. A theoretical basis for a biopharmaceutic drug classification: the correlation of in vitro drug product dissolution and in vivo bioavailability. Pharm. Res. 12, 413–420.

Arnida, Janat-Amsbury, M.M., Ray, A., Peterson, C.M., Ghandehari, H., 2011. Geometry and surface characteristics of gold nanoparticles influence their biodistribution and uptake by macrophages. Eur. J. Pharm. Biopharm. 77 (3), 417–423.

Attili-Qadri, S., Karra, N., Nemirovski, A., Schwob, O., Talmon, Y., Nassar, T., et al., 2013. Oral delivery system prolongs blood circulation of docetaxel nanocapsules via lymphatic absorption. Proc. Natl. Acad. Sci. U.S.A. 110 (43), 17498–17503.

Atuma, C., Strugala, V., Allen, A., Holm, L., 2001. The adherent gastrointestinal mucus gel layer: thickness and physical state in vivo. Am. J. Physiol-Gastrointest. Liver Physiol. 280, G922–G929.

Auffan, M., Decome, L., Rose, J., Orsiere, T., De Meo, M., Briois, V., et al., 2006. In vitro interactions between DMSA-coated maghemite nanoparticles and human fibroblasts: a physicochemical and cyto-genotoxical study. Environ. Sci. Technol. 40, 4367–4373.

Backmann, N., Kappeler, N., Braun, T., Huber, F., Lang, H.P., Gerber, C., et al., 2010. Sensing surface PEGylation with microcantilevers. Belstein. J. Nanotechnol. 1, 3–13.

Baek, M., Chung, H.E., Yu, J., Lee, J.A., Kim, T.H., Oh, J.M., et al., 2012. Pharmacokinetics, tissue distribution, and excretion of zinc oxide nanoparticles. Int. J. Nanomed. 7, 3081–3097.

Baek, M.K., Lee, J.H., Cho, Y.H., Kim, H.H., Lee, G.W., 2013. Self-microemulsifying drug-delivery system for improved oral bioavailability of pranlukast hemihydrate: preparation and evaluation. Int. J. Nanomed. 8, 167–176.

Bajaj, A., Rao, M.R.P., Pardeshi, A., Sali, D., 2012. Nanocrystallization by evaporative antisolvent technique for solubility and bioavailability enhancement of telmisartan. AAPS PharmSciTech. 13 (4), 1331–1340.

Banan, A., 2005. Theta isoform of protein kinase C alters barrier function in intestinal epithelium through modulation of distinct claudin isotypes: a novel mechanism for regulation of permeability. J. Pharmacol. Exp. Ther. 313, 962–982.

Boddupalli, B.M., Mohammed, Z.N.K., Nath, R.A., Banji, D., 2010. Mucoadhesive drug delivery system: an overview. J. Adv. Pharm. Technol. Res. 1, 381–387.

Brayden, D.J., Jepson, M.A., Baird, A.W., 2005. Keynote review: intestinal Peyer's patch M cells and oral vaccine targeting. Drug Discov. Today 10, 1145–1157.

Champion, J.A., Mitragotri, S., 2006. Role of target geometry in phagocytosis. Proc. Natl. Acad. Sci. U.S.A. 103, 4930–4934.

Chattopadhyay, S., Chakraborty, S.P., Laha, D., Baral, R., Pramanik, P., Roy, S., 2012. Surface-modified cobalt oxide nanoparticles: new opportunities for anti-cancer drug development. Cancer Nanotechnol. 3 (1–6), 13–23.

Chauhan, V.P., Jain, R.K., 2013. Strategies for advancing cancer nanomedicine. Nat. Mater. 12, 958–962.

Chen, H., Zheng, Y.I., Tian, G., Tian, Y., Zeng, X., Liu, G., et al., 2011. Oral delivery of DMAB-modified docetaxel-loaded PLGA-TPGS nanoparticles for cancer chemotherapy. Nanoscale Res. Lett. 6 (1), 4.

Chen, Y.D., Liang, Z.Y., Cen, Y.Y., Zhang, H., Han, M.G., Tian, Y.Q., et al., 2015. Development of oral dispersible tablets containing prednisolone nanoparticles for the management of pediatric asthma. Drug Des. Dev. Ther. 9, 5815–5825.

Chithrani, B.D., Ghazani, A.A., Chan, W.C.W., 2006. Determining the size and shape dependence of gold nanoparticle uptake into mammalian cells. Nano Lett. 6, 662–668.

Cho, H.J., Park, J.W., Yoon, I.S., Kim, D.D., 2014. Surface-modified solid lipid nanoparticles for oral delivery of docetaxel: enhanced intestinal absorption and lymphatic uptake. Int. J. Nanomed. 9, 495–504.

Cho, K., Wang, X., Nie, S., Chen, Z.G., Shin, D.M., 2008. Therapeutic nanoparticles for drug delivery in cancer. Clin. Cancer Res. 14, 1310.

Cho, W.S., Kang, B.C., Lee, J.K., Jeong, J., Che, J.H., Seok, S.H., 2013. Comparative absorption, distribution, and excretion of titanium dioxide and zinc oxide nanoparticles after repeated oral administration. Part Fibre Toxicol. 10, 9.

Cho, Y., Ha, E.S., Baek, I.W., Kim, M.S., Cho, C.W., Hwang, S.J., 2015. Enhanced supersaturation and oral absorption of sirolimus using an amorphous solid dispersion based on Eudragit® E. Molecules 20 (6), 9496–9509.

Choi, S.J., Choy, J.H., 2014. Biokinetics of zinc oxide nanoparticles: toxicokinetics, biological fates, and protein interaction. Int. J. Nanomed. 9 (Suppl. 2), 261–269.

Chu, Z., Zhang, S., Zhang, B., Zhang, C., Fang, C.Y., Rehor, I., et al., 2014. Unambiguous observation of shape effects on cellular fate of nanoparticles. Sci. Rep. 4, 4495.

Cone, R.A., 2009. Barrier properties of mucus. Adv. Drug Deliv. Rev. 61, 75–85.

DeMario, M.D., Ratain, M.J., 1998. Oral chemotherapy: rationale and future directions. J. Clin. Oncol. 16, 2557–2567.

des Rieux, A., Ragnarsson, E.G., Gullberg, E., Préat, V., Schneider, Y.J., Artursson, P., 2005. Transport of nanoparticles across an in vitro model of the human intestinal follicle associated epithelium. Eur. J. Pharm. Sci. 25 (4–5), 455–465.

des Rieux, A., Fievez, V., Garinot, M., Schneider, Y.J., Préat, V., 2006. Nanoparticles as potential oral delivery systems of proteins and vaccines: a mechanistic approach. J. Controlled Release 116 (1), 1–27.

des Rieux, A., Fievez, V., Théate, I., Mast, J., Préat, V., Schneider, Y.J., 2007. An improved in vitro model of human intestinal follicle-associated epithelium to study nanoparticle transport by M cells. Eur. J. Pharm. Sci. 30, 380–391.

Desai, M.P., Labhasetwar, V., Amidon, G.L., Levy, R.J., 1996. Gastrointestinal uptake of biodegradable microparticles: effect of particle size. Pharm. Res. 13, 1838–1845.

Devalapally, H., Silchenko, S., Zhou, F., McDade, J., Goloverda, G., Owen, A., et al., 2013. Evaluation of a nanoemulsion formulation strategy for oral bioavailability enhancement of danazol in rats and dogs. J. Pharm. Sci. 102 (10), 3808–3815.

Doherty, G.J., McMahon, H.T., 2009. Mechanisms of endocytosis. Annu. Rev. Biochem. 78, 857–902.
Donaldson, K., Murphy, F.A., Duffin, R., Poland, C.A., 2010. Asbestos, carbon nanotubes and the pleural mesothelium: a review of the hypothesis regarding the role of long fibre retention in the parietal pleura, inflammation and mesothelioma. Part. Fibre. Toxicol. 7, 5.
Doshi, N., Mitragotri, S., 2010. Macrophages recognize size and shape of their targets. PLoS One 5, e10051.
Dressman, J.B., Amidon, G.L., Reppas, C., Shah, V.P., 1998. Dissolution testing as a prognostic tool for oral drug adsorption: immediate release dosage forms. Pharm. Res. 15, 11–22.
Ensign, L.M., Cone, R., Hanes, J., 2012. Oral drug delivery with polymeric nanoparticles: the gastrointestinal mucus barriers. Adv. Drug Deliv. Rev. 64, 557–570.
Fule, R., Amin, P., 2014. Hot melt extruded amorphous solid dispersion of posaconazole with improved bioavailability: investigating drug-polymer miscibility with advanced characterisation. Biomed. Res. Int. 2014, 146781.
Gao, Y., Qian, S., Zhang, J., 2010. Physicochemical and pharmacokinetic characterization of a spray-dried cefpodoxime proxetil nanosuspension. Chem. Pharm. Bull. 58 (7), 912–917.
Garnet, M.C., Kallinteri, P., 2006. Nanomedicines and nanotoxicology: some physiological principles. Occup. Med. 56, 307–311.
Gref, R., Domb, A., Quellec, P., Blunk, T., Muller, R.H., Verbavatz, J.M., et al., 1995. The controlled intravenous delivery of drugs using PEG-coated sterically stabilized nanospheres. Adv. Drug Deliv. Rev. 16, 215–233.
Gülsün, T., Gürsoy, R.N., Öner, L., 2009. Nanocrystal technology for oral delivery of poorly water-soluble drugs. FABAD J. Pharm. Sci. 34, 55–65.
Gupta, A.K., Wells, S., 2004. Surface-modified superparamagnetic nanoparticles for drug delivery: preparation, characterization, and cytotoxicity studies. IEEE Trans. Nanobiosci. 3 (1), 66–73.
Ha, E.S., Choo, G.H., Baek, I.H., Kim, M.S., 2014. Formulation, characterization, and in vivo evaluation of celecoxib-PVP solid dispersion nanoparticles using supercritical antisolvent process. Molecules. 19 (12), 20325–20339.
Ha, E.S., Kim, J.S., Baek, I.H., Yoo, J.W., Jung, Y., Moon, Y.R., et al., 2015. Development of megestrol acetate solid dispersion nanoparticles for enhanced oral delivery by using a supercritical antisolvent process. Drug Des. Dev. Ther. 9, 4269–4277.
Harper, B., Sinche, F., Wu, R.H., Gowrishankar, M., Marquart, G., Mackiewicz, M., et al., 2014. The impact of surface ligands and synthesis method on the toxicity of glutathione-coated gold nanoparticles. Nanomaterials 4 (2), 355–371.
Hauss, D.J., 2007. Oral lipid-based formulations. Adv. Drug Del. Rev. 59, 667–676.
Hillery, A.M., Florence, A.T., 1996. The effect of adsorbed poloxamer 188 and 407 surfactants on the intestinal uptake of 60-nm polystyrene particles after oral administration in the rat. Int. J. Pharm. 132, 123–130.
Hirn, S., Semmler-Behnke, M., Schleh, C., Wenk, A., Lipka, J., Schäffler, M., et al., 2011. Particle size-dependent and surface charge-dependent biodistribution of gold nanoparticles after intravenous administration. Eur. J. Pharm. Biopharm. 77 (3), 407–416.
Hirota, K., Hasegawa, T., Hinata, H., Ito, F., Inagawa, H., Kochi, C., et al., 2007. Optimum conditions for efficient phagocytosis of rifampicin loaded PLGA microspheres by alveolar macrophages. J. Controlled Release 119, 69–76.

Hu, L., Huang, B., Zuo, M., Guo, R., Wei, H., 2008. Preparation of the phycoerythrin subunit liposome in a photodynamic experiment on liver cancer cells. Acta Pharmacol. Sin. 29, 1539–1546.

Huang, J., Zhong, X., Wang, L., Yang, L., Mao, H., 2012. Improving the magnetic resonance imaging contrast and detection methods with engineered magnetic nanoparticles. Theranostics. 2, 86–102.

Huang, J., Wang, L., Lin, R., Wang, A.Y., Yang, L., Kuang, M., et al., 2013. Casein-coated iron oxide nanoparticles for high mri contrast enhancement and efficient cell targeting. ACS. Appl. Mater. Interfaces 5 (11), 4632–4639.

Hull, M.S., Kennedy, A.J., Steevens, J.A., Bednar, A.J., Weiss Jr., C.A., et al., 2009. Release of metal impurities from carbon nanomaterials influences aquatic toxicity. Environ. Sci. Technol. 43, 4169–4174.

Iqbal, M., Tae, G., 2006. Unstable reshaping of gold nanorods prepared by a wet chemical method in the presence of silver nitrate. J. Nanosci. Nanotechnol. 6, 3355–3359.

Jang, M.H., Kweon, M.N., Iwatani, K., Yamamoto, M., Terahara, K., Sasakawa, C., et al., 2004. Intestinal villous M cells: an antigen entry site in the mucosal epithelium. Proc. Natl. Acad. Sci. U.S.A. 101, 6110–6115.

Jing, L., Shao, S., Wang, Y., Yang, Y., Yue, X., Dai, Z., 2016. Hyaluronic acid modified hollow prussian blue nanoparticles loading 10-hydroxycamptothecin for targeting thermochemotherapy of cancer. Theranostics 6 (1), 40–53.

Johnston, H.J., Hutchison, G.R., Christensen, F.M., 2010. A critical review of the biological mechanisms underlying the in vivo and in vitro toxicity of carbon nanotubes: the contribution of physicochemical characteristics. Nanotoxicology 4 (2), 207–246.

Joshi, G., Kumar, A., Sawant, K., 2014. Enhanced bioavailability and intestinal uptake of gemcitabine HCl loaded PLGA nanoparticles after oraldelivery. Eur. J. Pharm. Sci. 60, 80–89.

Junghanns, J.A.H., Müller, R.H., 2008. Nanocrystal technology, drug delivery and clinical applications. Int. J. Nanomed. 3 (3), 295–310.

Kammona, O., Kiparissides, C., 2012. Recent advances in nanocarrier-based mucosal delivery of biomolecules. J. Controlled Release 161, 781–794.

Kesisoglou, F., Panmai, S., Wu, Y., 2007. Nanosizing—oral formulation development and biopharmaceutical evaluation. Adv. Drug Deliv. Rev. 59 (7), 631–644.

Kim, M.S., Baek, I., 2014. Fabrication and evaluation of valsartan–polymer–surfactant composite nanoparticles by using the supercritical antisolvent process. Int. J. Nanomed. 9, 5167–5176.

Klein, C.L., Wiench, K., Wiemann, M., Ma-Hock, L., van Ravenzwaay, B., Landsiedel, R., 2012. Hazard identification of inhaled nanomaterials: making use of short-term inhalation studies. Arch. Toxicol. 86, 1137–1151.

Komarova, Y., Malik, A.B., 2010. Regulation of endothelial permeability via paracellular and transcellular transport pathways. Annu. Rev. Physiol. 72, 463–493.

Kudgus, R.A., Walden, C.A., McGovern, R.M., Reid, J.M., Robertson, J.D., Mukherjee, P., 2014. Tuning pharmacokinetics and biodistribution of a targeted drug delivery system through incorporation of a passive targeting component. Sci. Rep. 4, 5669.

Kullberg, M., Mann, K., Anchordoquy, T.J., 2012. Targeting Her-2 + breast cancer cells with bleomycin immunoliposomes linked to LLO. Mol. Pharm. 9 (7), 2000–2008.

Kumar, P., Lakshmi, Y.S., Bhaskar, C., Golla, K., Kondapi, A.K., 2015. Improved safety, bioavailability and pharmacokinetics of zidovudine through lactoferrin nanoparticles during oral administration in rats. PLoS One 10 (10), e0140399.

Kurakula, M., El-Helw, A.M., Sobahi, T.R., Abdelaal, M.Y., 2015. Chitosan based atorvastatin nanocrystals: effect of cationic charge on particle size, formulation stability, and in-vivo efficacy. Int. J. Nanomed. 10, 321−334.

Kwan, K.C., 1997. Oral bioavailability and first-pass effects. Drug Metab. Dispos. 25 (12), 1329−1336.

Kwon, S., Singh, R.K., Perez, R.A., Abou Neel, E.A., Kim, J.W., Chrzanowski, W., 2013. Silica-based mesoporous nanoparticles for controlled drug delivery. J. Tissue Eng. 4, 2041731413503357.

Landsiedel, R., Ma-Hock, L., Kroll, A., 2010. Testing metal-oxide nanomaterials for human safety. Adv. Mater. 22, 2601−2627.

Lee, D., Park, J.S., Bae, I.H., Lee, Y., Kim, B.M., 2016. Liquid crystal nanoparticle formulation as an oral drug delivery system for liver-specific distribution. Int. J. Nanomed. 11, 853−871.

Ledet, G.A., Graves, R.A., Glotser, E.Y., Mandal, T.K., Bostanian, L.A., 2015. Preparation and in vitro evaluation of hydrophilic fenretinide nanoparticles. Int. J. Pharm. 479 (2), 329−337.

LeFevre, M.E., Warren, J.B., Joel, D.D., 1985. Particles and macrophages in murine Peyer's patches. Exp. Cell Biol. 53 (3), 121−129.

Li, S.D., Huang, L., 2008. Pharmacokinetics and biodistribution of nanoparticles. Mol. Pharm. 5, 496−504.

Lin, Y.M., Wu, J.Y., Chen, Y.C., Su, Y.D., Ke, W.T., Ho, H.O., et al., 2011. In situ formation of nanocrystals from a self-microemulsifying drug delivery system to enhance oral bioavailability of fenofibrate. Int. J Nanomed. 6, 2445−2457.

Liu, C.J., Wang, C.H., Chen, S., 2010. Enhancement of cell radiation sensitivity by pegylated gold nanoparticles. Phys. Med. Biol. 55 (4), 931−945.

Liu, D., Mori, A., Huang, L., 1992. Role of liposome size and RES blockade in controlling biodistribution and tumor uptake of GM1-containing liposomes. Biochim. Biophys. Acta. 1104 (1), 95−101.

Liu, D., Pan, H., He, F., Wang, X., Li, J., Yang, X., et al., 2015a. Effect of particle size on oral absorption of carvedilol nanosuspensions: in vitro and in vivo evaluation. Int. J. Nanomed. 10, 6425−6434.

Liu, Y., Hu, Y., Huang, L., 2014. Influence of polyethylene glycol density and surface lipid on pharmacokinetics and biodistribution of lipid-calcium-phosphate nanoparticles. Biomaterials 35 (9), 3027−3034.

Liu, Y., Salituro, G.M., Lee, K., Bak, A., Leung, D.H., 2015b. Modulating drug release and enhancing the oral bioavailability of torcetrapib with solid lipid dispersion formulations. AAPS PharmSciTech. 16 (5), 1091−1100.

Lu, R., Liu, S., Wang, Q., Li, X., 2015. Nanoemulsions as novel oral carriers of stiripentol: insights into the protective effect and absorption enhancement. Int. J. Nanomed. 10, 4937−4946.

Macdonald, T., Monteleone, G., 2005. Immunity, inflammation, and allergy in the gut. Science. 307, 1920−1925.

Marzaioli, V., Aguilar-Pimentel, J.A., Weichenmeier, I., Luxenhofer, G., Wiemann, M., Landsiedel, R., et al., 2014. Surface modifications of silica nanoparticles are crucial for their inert versus proinflammatory and immunomodulatory properties. Int. J. Nanomed. 9, 2815−2832.

Mei, L., Sun, H., Song, C., 2009. Local delivery of modified paclitaxel-loaded poly (ε-caprolactone)/pluronic F68 nanoparticles for long-term inhibition of hyperplasia. J. Pharm. Sci. 98 (6), 2040−2050.

Merisko-Liversidge, E.M., Liversidge, G.G., 2008. Drug nanoparticles: formulating poorly water-soluble compounds. Toxicol. Pathol. 36, 43–48.

Moczko, E., Guerreiro, A., Piletska, E., Piletsky, S., 2013. PEG-stabilized core-shell surface-imprinted nanoparticles. Langmuir. 29 (31), 9891–9896.

Moghimi, S.M., Hedeman, H., Muir, I.S., Illum, L., Davis, S.S., 1993. An investigation of the filtration capacity and the fate of large filtered sterically-stabilized microspheres in rat spleen. Biochin. Biophys. Acta 1157, 233–240.

Moghimi, S.M., Patel, H.M., 1988. Tissue specific opsonins for phagocytic cells and their different affinity for cholesterol-rich liposomes. FEBS Lett. 233, 143–147.

Moghimi, S.M., Hunter, A.C., Murray, J.C., 2001. Long-circulating and target-specific nanoparticles: theory to practice. Pharmacol. Rev. 53, 283–318.

Moghimi, S.M., Hunter, A.C., Andresen, T.L., 2012. Factors controlling nanoparticle pharmacokinetics: an integrated analysis and perspective. Annu. Rev. Pharmacol. Toxicol. 52, 481–503.

Mokarram, R., Zadeh, K., Keshavarz, M., Ahmadi, A., Mohtat, B., 2010. Preparation and in-vitro evaluation of indomethacin nanoparticles. Daru 18 (3), 185–192.

Morgen, M., Bloom, C., Beyerinck, R., Bello, A., Song, W., Wilkinson, K., et al., 2012. Polymeric nanoparticles for increased oral bioavailability and rapid absorption using celecoxib as a model of a low-solubility, high-permeability drug. Pharm. Res. 29 (2), 427–440.

Muller, R.H., Peters, K., 1998. Nanosuspensions for the formulation of poorly soluble drugs I. Preparation by a size-reduction technique. Int. J. Pharm. 160, 229–237.

Nativo, P., Prior, I.A., Brust, M., 2008. Uptake and intracellular fate of surface-modified gold nanoparticles. ACS Nano 2, 1639–1644.

Navarro, S.M., Darensbourg, C., Cross, L., Stout, R., Coulon, D., Astete, C.E., et al., 2014. Biodistribution of PLGA and PLGA/chitosan nanoparticles after repeat-dose oral delivery in F344 rats for 7 days. Ther. Deliv. 5 (11), 1191–1201.

Nazzal, S., Smalyukh, I.I., Lavrentovich, O.D., Khan, M.A., 2002. Preparation and invitro characterization of a eutectic-based semisolid self-nanoemulsified drug delivery system (snedds) of ubiquinone: mechanism and progress of emulsion formation. Int. J. Pharm. 235 (1–2), 247–265.

Niidome, T., Yamagata, M., Okamoto, Y., 2006. PEG-modified gold nanorods with a stealth character for in vivo applications. J. Controlled Release 114 (3), 343–347.

Oh, N., Park, J.H., 2014. Endocytosis and exocytosis of nanoparticles in mammalian cells. Int. J. Nanomed. 9 (Suppl. 1), 51–63.

Owens, D.E., Peppas, N.A., 2006. Opsonization, biodistribution, and pharmacokinetics of polymeric nanoparticles. Int. J. Pharm. 307, 93–102.

Pan, Y., Neuss, S., Leifert, A., 2007. Size-dependent cytotoxicity of gold nanoparticles. Small. 3 (11), 1941–1949.

Panagi, Z., Beletsi, A., Evangelatos, G., Livaniou, E., Ithakissios, D.S., Avgoustakis, K., 2001. Effect of dose on the biodistribution and pharmacokinetics of PLGA and PLGA-mPEG nanoparticles. Int. J. Pharm. 221, 143–152.

Parodi, A., Corbo, C., Cevenini, A., Molinaro, R., Palomba, R., Pandolfi, L., et al., 2015. Enabling cytoplasmic delivery and organelle targeting by surface modification of nanocarriers. Nanomedicine (Lond). 10 (12), 1923–1940.

Passirani, C., Barratt, G., Devissaguet, J.P., Labarre, D., 1998. Interactions of nanoparticles bearing heparin or dextran covalently bound to poly(methyl methacrylate) with the complement system. Life Sci. 62 (8), 775–785.

Perrault, S.D., Walkey, C., Jennings, T., Fischer, H.C., Chan, W.C., 2009. Mediating tumor targeting efficiency of nanoparticles through design. Nano Lett. 9 (5), 1909–1915.

Plapied, L., Duhem, N., des Rieux, A., Préat, V., 2011. Fate of polymeric nanocarriers for oral drug delivery. Curr. Opin. Colloid Interface Sci. 16, 228–237.

Podsiadlo, P., Sinani, V.A., Bahng, J.H., Kam, N.W.S., Lee, J., Kotov, N.A., 2008. Gold nanoparticles enhance the anti-leukemia action of a 6-mercaptopurine chemotherapeutic agent. Langmuir. 24 (2), 568–574.

Porter, C.J.H., Moghimi, S.M., Illum, L., Davis, S.S., 1992. The polyoxyethylene polyoxypropylene block copolymer poloxamer-407 selectively redirects intravenously injected microspheres to sinusoidal endothelial-cells of rabbit bone-marrow. FEBS Lett. 305, 62–66.

Pratten, M.K., Lloyd, J.B., 1986. Pinocytosis and phagocytosis: the effect of size of a particulate substrate on its mode of capture by rat peritoneal macrophages cultured in vitro. Biochim. Biophys. Acta 881, 307–313.

Quinn, K., Gullapalli, R.P., Merisko-Liversidge, E., Goldbach, E., Wong, A., Liversidge, G. G., et al., 2012. A formulation strategy for gamma secretase inhibitor ELND006, a BCS class II compound: development of a nanosuspension formulation with improved oral bioavailability and reduced food effects in dogs. J Pharm. Sci. 101 (4), 1462–1474.

Rahman, Z., Zidan, A., Khan, M., 2010a. Non-destructive methods of characterization of risperidone solid lipid nanoparticles. Eur. J. Pharm. Biopharm. 76 (1), 127–137.

Rahman, Z., Zidan, A., Khan, M., 2010b. Understanding the quality of protein loaded PLGA nanoparticles variability by Plackett–Burman design. Int. J. Pharm. 389 (1–2), 186–194.

Rahman, Z., Samy, R., Sayeed, V.A., Khan, M.A., 2012a. Physicochemical and mechanical properties of carbamazepine cocrystals with saccharin. Pharm. Dev. Technol. 17 (4), 457–465.

Rahman, Z., Zidan, A.S., Samy, R., Sayeed, V.A., Khan, M.A., 2012b. Improvement of physicochemical properties of an antiepileptic drug by salt engineering. AAPS PharmSciTech. 3 (3), 793–801.

Rahman, Z., Siddiqui, A., Bykadi, S., Khan, M.A., 2014. Determination of tacrolimus crystalline fraction in the commercial immediate release amorphous soliddispersion products by a standardized X-ray powder diffraction method with chemometrics. Int. J. Pharm. 475 (1–2), 462–470.

Ray, P.C., Yu, H., Fu, P.P., 2009. Toxicity and Environmental Risks of Nanomaterials: Challenges and Future Needs. J. Environ. Sci. Health, Part C: Environ. Carcinog. Ecotoxicol. Rev. 27, 1–35.

Rashid, R., Kim, D.W., Yousaf, A.M., Mustapha, O., Din, F., Park, J.H., et al., 2015. Comparative study on solid self-nanoemulsifying drug delivery and solid dispersion system for enhanced solubility and bioavailability of ezetimibe. Int. J. Nanomed. 10, 6147–6159.

Rejman, J., Oberle, V., Zuhorn, I., Hoekstra, D., 2004. Sizedependent internalization of particles via the pathways of clathrin-and caveolae-mediated endocytosis. Biochem. J. 377, 159–169.

Sahu, B.P., Das, M.K., 2014. Nanoprecipitation with sonication for enhancement of oral bioavailability of furosemide. Acta Pol. Pharm. Drug Res. 71 (1), 129–137.

Sapsford, K.E., Tyner, K.M., Dair, B.J., Deschamps, J.R., Medintz, I.L., 2011. Analyzing nanomaterial bioconjugates: a review of current and emerging purification and characterization techniques. Anal. Chem. 83, 4453–4488.

Schwarz, C., Mehnert, W., Lucks, J.S., Müller, R.H., 1994. Solid lipid nanoparticles (SLN) for controlled drug delivery. I. Production, characterization and sterilization. J. Controlled Release 30 (1), 83–96.

Senanayake, T.H., Lu, Y., Bohling, A., Raja, S., Band, H., Vinogradov, S.V., 2014. Encapsulation of poorly soluble drugs in polymer-drug conjugates: effect of dual-drug nanoformulations on cancer therapy. Pharm. Res. 31 (6), 1605–1615.

Seo, H.J., Nam, S.H., Im, H.J., Park, J.Y., Lee, J.Y., Yoo, B., et al., 2015. Rapid hepatobiliary excretion of micelle-encapsulated/radiolabeled upconverting nanoparticles as an integrated form. Sci. Rep. 5, 15685.

Shah, R.B., Zidan, A.S., Funck, T., Tawakkul, M.A., Nguyenpho, A., Khan, M.A., 2007. Quality by design: characterization of self-nano-emulsified drug delivery systems (SNEDDs) using ultrasonicresonator technology. Int. J. Pharm. 341 (1–2), 189–194.

Sharma, G., Valenta, D.T., Altman, Y., Harvey, S., Xie, H., Mitragotri, S., et al., 2010. Polymer particle shape independently influences binding and internalization by macrophages. J. Controlled Release 147 (3), 408–412.

Shukla, S., Eber, F.J., Nagarajan, A.S., DiFranco, N.A., Schmidt, N., Wen, A.M., et al., 2015. The impact of aspect ratio on the biodistribution and tumor homing of rigid soft-matter nanorods. Adv. Healthc. Mater. 4 (6), 874–882.

Sonaje, K., Lin, K.J., Tseng, M.T., Wey, S.P., Su, F.Y., Chuang, E.Y., et al., 2011. Effects of chitosan-nanoparticle-mediated tight junction opening on the oral absorption of endotoxins. Biomaterials 32, 8712–8721.

Sood, A., Panchagnula, R., 2001. Peroral route: an opportunity for protein and peptide drug delivery. Chem. Rev-Columb. 101, 3275–3304.

Steinman, R.M., Mellman, I.S., Muller, W.A., Cohn, Z.A., 1983. Endocytosis and the recycling of plasma membrane. J. Cell Biol. 96, 1–27.

Sugano, K., 2009. Theoretical investigation of passive intestinal membrane permeability using Monte Carlo method to generate drug like molecule population. Int. J. Pharm. 373, 55–61.

Sugano, K., 2010. Possible reduction of effective thickness of intestinal unstirred water layer by particle drifting effect. Int. J. Pharm. 387, 103–109.

Sun, D., Xue, A., Zhang, B., Xue, X., Zhang, J., Liu, W., 2016. Enhanced oral bioavailability of acetylpuerarin by poly(lactide-*co*-glycolide) nanoparticles optimized using uniform design combined with response surface methodology. Drug Des. Dev. Ther. 10, 2029–2039.

Sun, X.K., Rossin, R., Turner, J.L., Becker, M.L., Joralemon, M.J., Welch, M.J., et al., 2005. An assessment of the effects of shell cross-linked nanoparticle size, core composition, and surface PEGylation on in vivo biodistribution. Biomacromolecules 6, 2541–2554.

Swathi, G., Prasanthi, N., Manikiran, S., Ramarao, N., 2010. Solid lipid nanoparticles: colloidal carrier systems for drug delivery. Int. J. Pharm. Sci. Res. 1 (12), 1–16.

Taccola, L., Raffa, V., Riggio, C., Vittorio, O., Iorio, M.C., 2011. Zinc oxide nanoparticles as selective killers of proliferating cells. Int. J. Nanomed. 6, 1129–1140.

Takeuchi, H., Thongborisute, J., Matsui, Y., Sugihara, H., Yamamoto, H., Kawashima, Y., 2005. Novel mucoadhesion tests for polymers and polymer-coated particles to design optimal mucoadhesive drug delivery systems. Adv. Drug Deliv. Rev. 57, 1583–1594.

Tang, J., Xu, N., Ji, H., Liu, H., Wang, Z., Wu, L., 2011. Eudragit nanoparticles containing genistein: formulation, development, and bioavailability assessment. Int. J. Nanomed. 6, 2429–2435.

Tebbe, M., Kuttner, C., Männel, M., Fery, A., Chanana, M., 2015. Colloidally stable and surfactant-free protein-coated gold nanorods in biological media. ACS. Appl. Mater. Interf. 7 (10), 5984–5991.

Thatipamula, R., Palem, C., Gannu, R., Mudragada, S., Yamsani, M., 2011. Formulation and in vitro characterization of domperidone loaded solid lipid nanoparticles and nanostructured lipid carriers. Daru 19 (1), 23–32.

Tolaymat, T.M., El Badawy, A.M., Genaidy, A., Scheckel, K.G., Luxton, T.P., Suidan, M., 2010. An evidence-based environmental perspective of manufactured silver nanoparticle in syntheses and applications: a systematic review and critical appraisal of peer-reviewed scientific papers. Sci. Total Environ. 408 (5), 999–1006.

Toy, R., Peiris, P.M., Ghaghada, K.B., Karathanasis, E., 2014. Shaping cancer nanomedicine: the effect of particle shape on the in vivo journey of nanoparticles. Nanomedicine (Lond). 9 (1), 121–134.

Treuel, L., Nienhaus, G.U., 2012. Toward a molecular understanding of nanoparticle–protein interactions. Biophys. Rev. 4, 137–147.

Underhill, D.M., Goodridge, H.S., 2012. Information processing during phagocytosis. Nat. Rev. Immunol. 12, 492–502.

Verma, S., Rudraraju, V.S., 2015. Wetting kinetics: an alternative approach towards understanding the enhanced dissolution rate for amorphous solid dispersion of a poorly soluble drug. AAPS PharmSciTech. 16 (5), 1079–1090.

Vigderman, L., Manna, P., Zubarev, E.R., 2011. Quantitative replacement of cetyl trimethylammonium bromide by cationic thiol ligands on the surface of gold nanorods and their extremely large uptake by cancer cells. Angew. Chem., Int. Ed. 51, 636–641.

Wang, K., Xu, J.J., Chen, H.Y., 2005. A novel glucose biosensor based on the nanoscaled cobalt phthalocyanine-glucose oxidase biocomposite. Biosens. Bioelectron. 20, 1388–1396.

Wang, K., Qi, J., Weng, T., Tian, Z., Lu, Y., Hu, K., et al., 2014. Enhancement of oral bioavailability of cyclosporine A: comparison of various nanoscale drug-delivery systems. Int. J. Nanomed. 9, 4991–4999.

Wang, Y., Zheng, Y., Zhang, L., Wang, Q., Zhang, D., 2013. Stability of nanosuspensions in drug delivery. J. Controlled Release 172 (3), 1126–1141.

Weerapol, Y., Limmatvapirat, S., Nunthanid, J., Sriamornsak, P., 2014. Self-nanoemulsifying drug delivery system of nifedipine: impact of hydrophilic-lipophilic balance and molecular structure of mixed surfactants. AAPS PharmSciTech. 15 (2), 456–464.

Wingett, D., Louka, P., Anders, C.B., Zhang, J., Punnoose, A., 2016. A role of ZnO nanoparticle electrostatic properties in cancer cell cytotoxicity. Nanotechnol. Sci. Appl. 9, 29–45.

Wittrup, K.D., Thurber, G.M., Schmidt, M.M., Rhoden, J.J., 2012. Practical theoretic guidance for the design of tumor-targeting agents. Methods Enzymol. 503, 255–268.

Wu, L., Qiao, Y., Wang, L., Guo, J., Wang, G., He, W., et al., 2015. A self-microemulsifying drug delivery system (SMEDDS) for a novel medicative compound against depression: a preparation and bioavailability study in rats. AAPS PharmSciTech. 16 (5), 1051–1058.

Wu, Y., Loper, A., Landis, E., Hettrick, L., Novak, L., Lynn, K., et al., 2004. The role of biopharmaceutics in the development of a clinical nanoparticle formulation of MK-0869: a beagle dog model predicts improved bioavailability and diminished food effect on absorption in human. Int. J. Pharm. 285 (1–2), 135–146.

Xia, D., Gan, Y., Cui, F., 2014. Application of precipitation methods for the production of water-insoluble drug nanocrystals: production techniques and stability of nanocrystals. Curr. Pharm. Des. 20, 408–435.

Yadav, S.K., Mishra, S., Mishra, B., 2012. Eudragit-based nanosuspension of poorly water-soluble drug: formulation and in vitro–in vivo evaluation. AAPS. PharmSciTech. 13 (4), 1031–1044.

Yameen, B., Choi, W., Vilos, C., Swami, A., Shi, J., Farokhzad, O.C., 2014. Insight into nanoparticle cellular uptake and intracellular targeting. J. Controlled Release 190, 485–499.

Yang, R., Huang, X., Dou, J., Zhai, G., Su, L., 2013. Self-microemulsifying drug delivery system for improved oral bioavailability of oleanolic acid: design and evaluation. Int. J. Nanomed. 8, 2917–2926.

Yeh, T.H., Hsu, L.W., Tseng, M.T., Lee, P.L., Sonjae, K., Ho, Y.C., et al., 2011. Mechanism and consequence of chitosan-mediated reversible epithelial tight junction opening. Biomaterials 32, 6164–6173.

Yin, X., Daintree, L.S., Ding, S., Ledger, D.M., Wang, B.W., Zhao, W., et al., 2015. Itraconazole solid dispersion prepared by a supercritical fluid technique: preparation, in vitro characterization, and bioavailability in beagle dogs. Drug Des. Dev. Ther. 9, 2801–2810.

Yousaf, A.M., Mustapha, O., Kim, D.W., Kim, D.S., Kim, K.S., Jin, S.G., et al., 2016. Novel electrosprayed nanospherules for enhanced aqueous solubility and oral bioavailability of poorly water-soluble fenofibrate. Int. J. Nanomed. 11, 213–221.

Zeng, N., Gao, X., Hu, Q., Song, Q., Xia, H., Liu, Z., et al., 2012. Lipid-based liquid crystalline nanoparticles as oral drug delivery vehicles for poorly water-soluble drugs: cellular interaction and in vivo absorption. Int. J. Nanomed. 7, 3703–3718.

Zhang, H., Firempong, C.K., Wang, Y., Xu, W., Wang, M., Cao, X., et al., 2016. Ergosterol-loaded poly(lactide-*co*-glycolide) nanoparticles with enhanced in vitro antitumor activity and oral bioavailability. Acta Pharmacol. Sin. 37 (6), 834–844.

Zhang, N., Ping, Q., Huang, G., Xu, W., Cheng, Y., Han, X., 2006. Lectin-modified solid lipid nanoparticles as carriers for oral administration of insulin. Int. J. Pharm. 327 (1–2), 153–159.

Zhang, X., Wang, H., Ma, Z., Wu, B., 2014. Effects of pharmaceutical PEGylation on drug metabolism and its clinical concerns. Expert Opin. Drug Metab. Toxicol. 10 (12), 1691–1702.

Zhao, Y., Xing, G., Chai, Z., 2008. Nanotoxicology: are carbon nanotubes safe? Nat. Nanotechnol. 3 (4), 191–192.

Zhou, H., Fan, Z., Lemons, P.K., Cheng, H., 2016. A facile approach to functionalize cell membrane-coated nanoparticles. Theranostics 6 (7), 1012–1022.

Zhou, J., Zhou, D., 2015. Improvement of oral bioavailability of lovastatin by using nanostructured lipid carriers. Drug Des. Dev. Ther. 9, 5269–5275.

FURTHER READING

Florence, A.T., Hussain, N., 2001. Transcytosis of nanoparticle and dendrimer delivery systems: evolving vistas. Adv. Drug Deliv. Rev. 50 (Suppl. 1), S69–S89.

Italia, J.L., Bhatt, D.K., Bhardwaj, V., Tikoo, K., Kumar, M.N., 2007. PLGA nanoparticles for oral delivery of cyclosporine: nephrotoxicity and pharmacokinetic studies in comparison to Sandimmune Neoral. J. Controlled Release 119 (2), 197–206.

CHAPTER

Targeted delivery systems for treatment of pancreatic cancer

12

Melek Karaca and Yıldız Özsoy

Department of Pharmaceutical Technology, Faculty of Pharmacy, Istanbul University, Istanbul, Turkey

12.1 INTRODUCTION

Pancreatic ductal adenocarcinoma (PDAC) is an aggressive solid organ tumor and is the seventh leading cause of cancer-related deaths with a low five-year survival rate of 6%. Smoking, age, family history, and diseases such as gestational diabetes, pancreatitis, and diabetes mellitus are the risk factors that often lead to the development of PDAC (Michaud, 2004).

Different approaches are used for the treatment of PDAC patients. Primary tumor disposal by robot-assisted, laparoscopic, or open respectable surgery is the most efficient approach because all of affected tissue can be removed, whereas other approaches can only reign over a certain percentage of cells. However, the overall survival after diagnosis and subsequent respectable surgery is five years in 5%–20% of patients due to the local invasions and further metastasis (Abrams et al., 2009; Hubenak et al., 2014). Radiotherapy, as a noninvasive treatment, has emerged as an alternative to treat PDAC. Known that it is a simple application of irradiation with high-energy radiation such as X-rays, γ rays, electrons and protons, which results in DNA damage and further cell killing. However, the possibility of impairing healthy tissues creates a dilemma for this treatment. Despite the obstacles faced with radiation therapy, it provides fast and easy improvement (Hubenak et al., 2014; Neuman et al., 2008).

To overcome the limited efficacy of current treatment approaches, interests have been focused on drug therapy. Since the 1990s, gemcitabine has been the upper-most choice as a result of a clinical study outcome where states increased overall survival by gemcitabine compared to treatment with 5-flurouracil (5-FU). In this clinical study on 123 patients conducted by Burris et al. (1997), gemcitabine-treated patients' median survival was found to be 5.65 months whereas 5-FU treatment resulted in 4.41 months . Until the 2000s, many combination therapy approaches were assessed to attain an acceptable cure effect. However, all trials except one failed to show a statistically increased survival rate over gemcitabine therapy alone.

Nanoarchitectonics in Biomedicine. DOI: https://doi.org/10.1016/B978-0-12-816200-2.00007-4

Considering the multiple factors that manage PDAC development, combination therapy strategies have attracted more attention. FOLFIRINOX is an example of a combination therapy option evaluated in the PRODIGE4/ACCORD11 clinical study. It shows multiple effects as it consists of four different active compounds (5-FU, leucovorin, irinotecan, and oxaliplatin). The PRODIGE4/ACCORD11 clinical study demonstrated significant survival advantage over gemcitabine treatment (Ko, 2015).

In 2013, a new treatment regimen was introduced. Gemcitabine plus nanoparticle albumin bound (nab) paclitaxel (PTX) provides the delivery of PTX without using toxic and immunogenic solvents (Ko, 2015; Von Hoff et al., 2011).

Currently, 5-FU and radiotherapy combination treatment, gemcitabine treatment, human albumin nanoparticles bound to PTX, nab- PTX (Abraxane), and FOLFIRINOX composed of fluorouracil, irinotecan, oxaliplatin, and folinic acid are the standard chemotherapeutic regiments for metastatic PDAC treatment (Neuman et al., 2008). The development of first-line chemotherapy options for the treatment of metastatic pancreatic cancer is listed in Table 12.1.

Despite several advancements, treatment success still remains a challenge. Chemotherapy requires multiple stages during treatment and the efficacy is limited due to the narrow therapeutic index of agents. Proliferative normal tissues, such as bone marrow and gastrointestinal mucosa, lymphoid tissue, fetus, and hair follicles are susceptible to the cytotoxic effect of chemotherapeutic agents that leads to severe side effects due to the nonspecificity of anticancer agents. Taking

Table 12.1 List of First-Line Chemotherapy for Metastatic Pancreatic Cancer Treatment

Agents	Comparison	Median Overall Survival	Results	Reference
5-FU	5-FU combined with radiotherapy versus only radiotherapy	10.4 versus 6.3 months	More side effects with combination treatment	Moertel et al. (1969)
Gemcitabine	Gemcitabine versus 5-FU	5.65 versus 4.41 months	Higher mitigation of disease-related symptoms when compared to 5-FU	Burris et al. (1997)
FOLFIRINOX	FOLFIRINOX versus Gemcitabine	11.1 versus 6.8 months	More side effects compared to single gemcitabine treatment	Vaccaro et al. (2011)
Nab-paclitaxel	Nab-paclitaxel combined with gemcitabine versus only gemcitabine	8.5 versus 6.7 months	Higher clinical response compared to single treatment with gemcitabine	Von Hoff et al. (2013)

5-FU, *Fluorouracil.*

these concerns into account, optimization of selectivity is an important solution to achieve a successful treatment paradigm. Strategies to deliver chemotherapy agents to the tumor site have been widely investigated to increase their anticancer effect and to reduce their associated toxicity.

Nanotechnology has arisen as an alternative approach to treat cancer by providing improved drug delivery. Nanotechnology has several advantages such as the higher accumulation of drugs in the tumoral area, improved solubility, reduced toxicity, and the control of release kinetic of anticancer agents which protects them from degradation and, therefore, allows an increased half-life.

Delivery systems such as polymeric nanoparticles, micelles, and liposomes were introduced to improve drug concentration at the related site by means of their ability to passively target pancreatic cancer cells while remaining undisturbed by the reticuloendothelial system (RES). Targeting cancer cells actively via nano-sized delivery systems which have the ability to bind to receptors or antigens on pancreatic cancer cells is another approach to obtain improved therapeutic effects as well as to improve the drug concentration at tumor site.

In this context, drug delivery technologies for passive targeting and active targeting with potential targets are reviewed with examples from recent research studies. The pharmaceutical nanocarriers investigated in these studies are illustrated in Fig. 12.1.

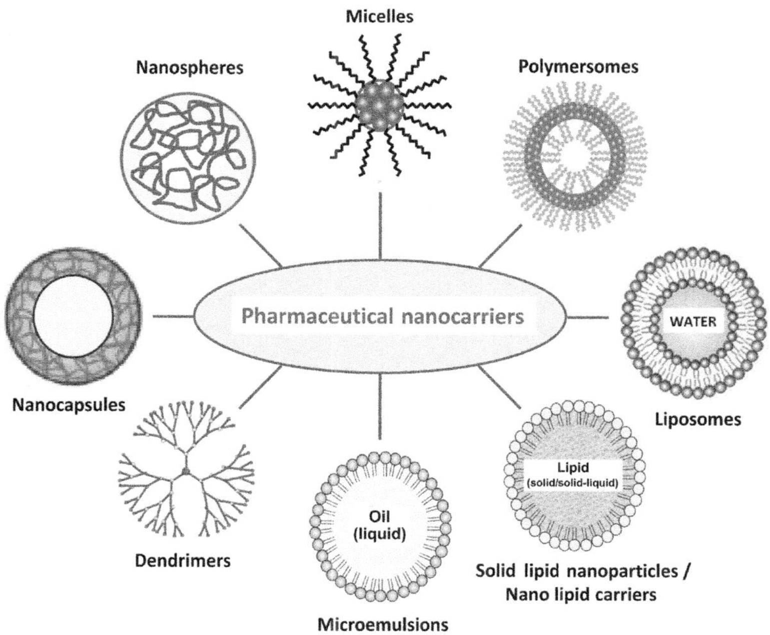

FIGURE 12.1

The illustrations of pharmaceutical nanocarriers developed in research studies for pancreas cancer treatments.

12.2 PASSIVE TARGETING AND DRUG DELIVERY SYSTEMS

Passive targeting can be defined as the increased accumulation of nanocarriers within the tumoral milieu due to the altered tumoral microenvironment structure and the characteristics of nanocarriers such as shape, size, and surface charge (Bazak et al., 2014; Au et al., 2016). The Enhanced Permeation and Retention (EPR) effect relies on the formation of new blood vessels as a result of growth factor production due to the presence of insufficient oxygen (Bates et al., 2002). During the angiogenesis process, a manner of discrete epithelium structure and basal membrane deficiency occurs and this fact enables the formation of fenestrations in the capillaries with a size in the range of 200–2000 nm (Hobbs et al., 1998). Due to the defection in lymphatic activity, normal extracellular flow could not be maintained and allows for the accumulation of macromolecules as well as nanoparticles in the tumoral area. Presence of necrotic domains, relevance of delivery systems to vascular wall in terms of vessel architecture, presence of fenestrations, the flow rate of blood, and the concentration present in blood mainly govern extravasation. Besides tumoral biological factors, factors related to colloidal systems such as their circulation time, size, charge, shape, and hydrophobicity (like surface characteristics) can affect extravasation. It has been shown that an increment of phagocytic cells can lead to higher accumulation of colloids in the tumor area (Zamboni et al., 2011).

However, although the EPR effect provides great opportunity from bloodstream to target site, nanocarriers confront some issues affecting on the treatment. The macrophages present in blood are the “soldiers” of our defensive system (i.e., RES) and carry out the cleaning of undefined molecules and the drug delivery systems identified as foreign structures in the body. The second challenge for these molecules is the filtration phenomenon in the spleen and kidneys. The third challenge is to succeed at arriving at the target site by diffusing through cell membrane and endocytosis (Estanqueiro et al., 2015). However, the thick structure formed by the extracellular matrix (ECM) strongly disrupts the delivery of nanoparticle systems. The ECM of the tumor environment consists of collagen, fibronectin, hyaluronic acid (HA) and proteoglycans, proteins, and cellular debris that govern the fate of the colloids motion.

Various drug delivery technologies have been studied in order to overcome these challenges.

Polymeric nanoparticles are submicron-sized delivery systems formed by biodegradable or nonbiodegradable polymers and have long been used for diagnosis and cure purposes (Youns et al., 2011; Dimou et al., 2012). Zhang et al. reported a study in which hedgehog inhibitor cyclopamine was applied to Capan-2 cells tumor xenograft-bearing mouse models to observe the increased accumulation of PTX nanoparticles formed with biodegradable poly(ethylene glycol)-poly(lactic acid) (PEG-PLA) polymer having 115 nm particle size. The results showed promising warrant for the importance of ECM regulation in treatment that changing tumor biology can help to improve the EPR effect that enables the accumulation of macromolecules and nanomaterials within the tumor (Zhang et al., 2016).

Albumin-based passive targeting is already known as the commercialized product Abraxana which has been used in cancer treatment for so long. Between 100 and 200 nm chemotherapy drug-albumin-nanoparticles were studied in clinical trials ABI-008, ABI-009, ABI-0010, and ABI-0011. A clinical study on docetaxel-albumin nanoparticles demonstrated excellent efficacy in monkeys and rats (D'Cruz et al., 2010).

Uptake and transcytosis of PTX formulations—nab-PTX and Cremophor EL (CrEL) and ethanol-PTX formulations—were compared. Nab-PTX showed superior uptake and transport by human vascular endothelial cells compared to conventional formulation. Studies on human tumor xenografts (MIA PaCa-2 tumor-bearing mice) to assess the distribution pattern confirmed that nab-PTX formulation has stable distribution as well as efficacy and a safety profile. It was reported that mitotic arrest was obtained significantly more with the nab-PTX formulation. Nab-PTX provided drug localization in consistency with the dose applied, whereas CrEL-PTX was found to be unstable and unpredictive (Chen et al., 2015).

PTX loaded mPEG-PLGA-PLL-cRGD nanoparticles was dispersed in F-127, F-68, HPMC, methyl cellulose and sodium alginate (SA) mixture thermosensitive gel to maintain sustained release and excellent antitumor effect and fewer toxic effect in PTX resistant Aspc-1 tumor-bearing nude mice. The nanoparticle size was defined around 133 nm. The gel system remained for more than 50 days in tumoral area (Shen et al., 2015).

Micelles are self-assembled polymeric core-shell structures allowing encapsulation of lipophilic drugs into its hydrophobic core. Generally, micelle polymers having a PEG group as a hydrophilic block and polyester or poly(amino acids) are the most reported. Poly(ethylene oxide)-*co*-poly(L-lactide) (PEG-PLLA), poly (ethylene oxide)-*co*-poly(D,L-lactide) (PEG-PDLA), or poly(ethylene oxide)-*co*-poly(caprolactone) (PEG-PCL) are pH sensitive and biodegradable polymers. The micellar structures formed by these polymers tend to release the loaded drug in an acidic environment of endosomes and lysosomes following their cellular uptake through endocytosis (Gupta et al., 2015).

So far, only a few micelle formulations have reached to clinical trials. Doxorubicin (DOX) conjugation to PEG-poly(aspartic acid) and following encapsulation of free DOX into PEG-Poly(aspartic acid)-DOX micelles was investigated in clinical trial for pancreatic cancer treatment (Matsumura et al., 2004). PTX was loaded into poly(ethylene oxide)-*co*-poly(D,L-lactide) (PEO2000-*co*-PDLA1750) micelles commercial name as Genexol-PM has demonstrated to have a favorable effect in the treatment of breast, lung, and pancreatic cancer in clinical studies (Kim et al., 2007; Saif et al., 2010). Perfluorocarbon (PFC) compound used in the formation of micelles was suggested by Gupta and his group (Gupta et al., 2015). PFC is considered to serve a hydrophobic wall, thus, increasing the stability of micellar carriers. It was reported that this effect could be gained by introducing oil or hydrophobic drugs as well (Rapoport, 1999; Shin et al., 2012). It has also been reported that a low amount of PFC is solubilized and increases the stability of micelles. However, when the PFC concentration increases more than its solubility, micellar structures turn into nanodroplets. The therapeutic efficacy and systemic toxicity of PTX loaded PEG-PDLA micelles and PEG-PDLA nanoemulsions compared in vivo resulted in both effective therapeutic activities, but with lower systemic toxicity with nanoemulsion formulation (Gupta et al., 2015).

Encapsulation of cyclopamine into oligo (ε-caprolactone) grafted–triethoxysilane tethered poly[oligo (ethylene glycol) monomethyl ether methacrylate]$_{31}$-*b*-poly(2-hydroxyethyl methacrylate)$_{26}$ [P(OEGMA)$_{31}$-*b*-P(HEMA-CL$_5$-Si)]$_{26}$ block copolymer micelles by ionizing Cs-137 radiation showing the damage to stellate cells was demonstrated with stroma production. In vitro studies showed that system potentiated radio response in MIA PaCa-2 and HPSC cells, but not in Panc-1 cells (Zhao et al., 2015). In order to provide a prolonged release of gemcitabine the micellar formulation approach has been used.

Gemcitabine (dFdC), as a first-line treatment for pancreatic cancer, is a hydrophilic pyrimidine analogue and shows its anticancer effect by blocking nucleic acid synthesis. Its hydrophilic nature obstructs the diffusion process through the cell membrane. Moreover, following administration, it can easily be metabolized into its metabolite dFdU by cytidine deaminase. dFdU is known to be an inactive molecule and is transmitted out of the cell (Duangjai et al., 2014; Immordino et al., 2004). Therefore, in order to overcome these obstacles both conjugation and encapsulation strategies have been studied for the effective delivery of gemcitabine. The conjugation of PEG to the drug in order to improve the efficacy, increase its half-life, and for its elevated uptake has been discovered. It has been studied that the SQdFdC form of gemcitabine showed improved pharmacokinetic profiles and elevated accumulation in resistant cancer cells (Bildstein et al., 2010). However, the presence of RES led to the removal of these delivery ways through phagocytic cells; therefore, the high concentration application caused to increased toxicity which required to investigate alternative ways for the delivery. Chitkara et al. (2013) have conjugated gemcitabine to methoxypoly-(ethylene)-block-poly(2-methyl-2-carboxyl-propylene carbonate) (mPEG-*b*-PCC) copolymer, which self-assembled into micelles and resulted in its improved plasma stability, enhanced internalization, and apoptosis.

The study was further continued with a micellar mixture formulation to deliver gemcitabine with a Hedgehog inhibitor GDC-0449. For this purpose, methoxy PEG-block-poly(2-methyl-2-carboxyl-propylenecarbonate) graft-dodecanol (mPEG-*b*-PCC-*g*-DC) amphiphilic copolymer was used to conjugate gemcitabine and encapsulating a Hh inhibitor, vismodegib (GDC-0449). Vismodegib (GDC-0449) encapsulation into micellar core and gemcitabine conjugated amphiphilic polymer micelle is depicted in Fig. 12.2. Almost 80% of the encapsulated GDC-0449 and 19% conjugated GEM were released in vitro at pH 5.5 over 48 hours in a sustained manner. The system showed increased treatment efficacy attributed to

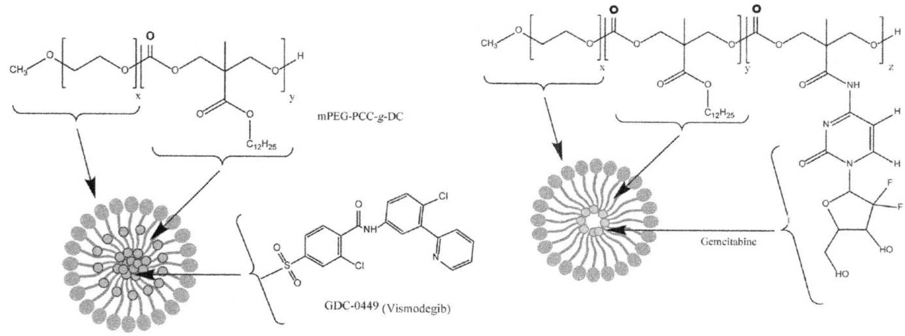

FIGURE 12.2

Schematic illustration of vismodegib(GDC-0449)-loaded micelles (left) and gemcitabine conjugated micelles (right).

Table 12.2 Liposome Formulations Reached in Clinical Assessments

Name	Drug	Diameter (nm)	Reference
Lipoplatin	Cisplatin	110	Boulikas (2009)
Lipoxal	Oxaliplatin	100	Stathopoulos et al. (2006)
EndoTAG-1	Paclitaxel	180–200	Awada et al. (2014)
Caelyx/Doxil	Doxurubisin	100	O'Brien (2004)
ONCO-TCS	Vincristine	120	Vincristine Liposomal ssss-INEX, 2004

the micellar delivery that provided sustained release and protection of both drugs from the physiological environment (Karaca et al., 2016).

Liposomes are self-assembled lipid-based small vesicles composed of two layers providing encapsulation of both hydrophilic and hydrophobic drugs. Their size changes with respect to the preparation strategy as well as their composition, starting from around 50 nm to more than 1 um (Yang et al., 2011). Although, they have the ability to carry hydrophobic drugs, due to the destabilization effect and small area limitation, liposomes are considered as hydrophilic drug carriers. These carriers demonstrate additional advantages in terms of biocompatibility as they are generally considered to be safe. However, it has been noted that intravenous administration of liposomal formulations may cause hypersensitivity reactions (Pérez-Herrero and Fernández-Medarde, 2015).

Clinical applications of liposomes are well-recognized since 1995 with the approval of the first liposomal chemotherapeutic drug, Doxil (Doxorubisin HCl) by the US Food and Drug Administration (FDA) (Northfelt et al., 1998). Advancements in liposomal drug delivery framework have given a chance to target particular agents for the treatment of pancreatic cancer. Clinical assessments have been undertaken with a few liposomal formulations recently. Liposome formulations which have reached clinical assessment are listed in Table 12.2.

Platinum-based cancer treatment has turned into the standard first-line option for some solid tumors after the clinical achievement of cisplatin (Wheate et al., 2010). Cisplatin and oxaliplatin are broadly used for treatment notwithstanding various hindrances. A randomized phase III trial, platin derivatives plus gemcitabine was compared to gemcitabine alone and did not demonstrate any improvement in terms of survival compared to single gemcitabine treatment for pancreatic cancer treatment (Colucci et al., 2010).

Special tumor uptake has been achieved by liposomal formulation of cisplatin. Lipoplatin proposes a noteworthy capability that can kill both tumor cells and endothelial cells of tumor vasculature (Stathopoulos, 2010). Lipoxal is another platin-based liposomal formulation carrying oxaliplatin. It has demonstrated to have a superior effect compared to free oxaliplatin in terms of reduced adverse reactions and myelotoxicity. EndoTAG-1 (lipid-PTX complex) is another liposomal formulation which targets negatively charged endothelial cells through its positively charged nature (Eichhorn et al., 2010). A phase II study has been conducted to evaluate its superiority over gemcitabine alone in 200 patients with locally advanced and metastatic pancreatic cancer. Combination treatment demonstrated 4.1−4.6 months survival compared to 2.7 months median survival maintained with single gemcitabine treatment (Hofheinz et al., 2004).

The activity of plant alkaloid curcumin derivative 3,5-bis (2-fluorobenzylidene)-4-piperidone (EF24) has great anticancer effects, as proved previously (Mosley et al., 2007). In the literature it is reported that the preclinical evaluation of PEGylated liposomal EF24 formulation formed by 1-Palmitoyl-2-oleoyl-sn-glycero-3-phosphocholine (POPC), cholesterol and 1,2-distearoyl-sn-glycero-3-phosphoethanolamine-*N*-[PEG-2000] (PEG-DSPE) have a molar ratio of 65:30:5 mol%. The formulation inhibited the NF-kappaB activation and suppressed pancreatic cancer tumor growth in MIAPaCa cell-bearing mice after treatment intravenously. The formulation was found to be promising due to its acceptable safety profile (Bisht et al., 2016).

Fang et al. conjugated vitamin E succinate and N4-amino group of pyrimidine ring of gemcitabine and formed an amide bond resulting in the formation of VES-Gemcitabine prodrug. Furthermore, they prepared a nanocapsule formulation and studied cellular uptake, cytotoxicity, and intracellular drug release in BxPC-3 pancreatic cancer cells (Fang et al., 2015). Xu et al. used D-alpha-tocopheryl PEG succinate (TPGS) and prepared coassembled nanoformulation of TPGS/VES-dFdC with a molar ratio ranging between 0.5/1 and 1.5/1 by the codissolution-coprecipitation method to increase the stability of VES-dFdC prodrug. In vivo efficacy studies on BxPC-3 tumor-bearing nude mice showed that 0.2 mmol/kg of TPGS/VES-dFdC (1/1) nanoformulation demonstrated 4.7-fold higher tumor growth inhibition compared to free dFdC (Xu et al., 2015).

FOLFIRINOX showed superior therapeutic results (31.6%) compared to gemcitabine alone (9.4%). However, the toxicity problem of irinote can causes neutropenia, necrosis, steatosis, vomiting, and diarrhea in the patient that calls for new delivery ways to overcome irinotecan toxicity. Therefore, Liu et al. established lipid bilayer coated custom-designed mesoporous silica nanoparticles in which the lipid bilayer was used to load irinotecan through proton gradient (via the weak basic properties of irinotecan ($pK_a = 8.1$) that enables higher dose encapsulation. KPC-derived pancreatic cancer tumor-bearing mice were used to evaluate the in vivo treatment efficacy. An equivalent liposome formulation was involved for comparison and the system developed by Liu et al. was found to have more impact in terms of activating tumor killing and suppressing metastasis. Moreover, the physical and other biological characteristics such as stability, toxicity, drug loading, and release profile capacity were superior over equivalent liposome formulations (Liu et al., 2016).

12.3 ACTIVE TARGETING AND DRUG DELIVERY SYSTEMS

Active targeting, also known as ligand-mediated targeting, is a phenomenon of nanoparticle modification with targeting ligands that allow for the delivery of macromolecular drugs by increasing the alliance to tumor cells. Some specific molecules or receptors are overexpressed in particular diseases. The ligands to be conjugated to nanoparticles are decided based on those particular molecules or receptors (Kamaly et al., 2012; Bertrand et al., 2014). The ligand, its density and size of the particulate system are the main factors that affect the efficacy of active targeting. Proteins, peptides, nucleic acids, sugars, and vitamins are the ligands and proteins, sugars, or lipids are the target molecules (Bertrand et al., 2014). The success of active targeting can be defined by the biodistribution of the system and specificity to diseased tissue. Two types of ligand conjugations are available: (1) direct conjugation of ligand to polymer and then preparation of formulation; and (2) ligand conjugation after preparing the formulation. Native proteins are not convenient for direct conjugation with polymers, whereas postconjugation is suitable with all types of ligands such as antibodies, proteins, peptides, aptamers, and small molecules, etc.

The coupling strategies are the general approaches for the conjugation. Usually, exploiting the activators such as *N*-hydroxysuccinimide or 1-Ethyl-3-(3-dimethylaminopropyl) form the bond between the nucleophilic group of ligands and NPs are the general manner of conjugation. The thiol group of existing polymers is convenient for maleimide group coupling.

In addition, the physicochemical properties such as size, charge, shape, and hydrophobicity of NPs have a significant effect on the distribution and blood profile.

12.3.1 POTENTIAL TARGET MECHANISMS AND DELIVERY STRATEGIES

Some of the targets for specific delivery to pancreatic cancer cells are listed in Table 12.3 and the delivery systems used to target these molecules are summarized next.

Table 12.3 Potential Targets and Targeting Agents for Effective Drug Delivery for Pancreas Cancer

Target Mechanism	Definition	Reference
CD44	CD44 is a cell surface receptor has an important role in angiogenesis, cell proliferation, differentiation, and cell motility. Targeting with hyaluronic acid is a promising approach due to not only its biologic compatibility but also its selectivity to CD44 receptor excessed existence in cancer cells	Choi et al. (2012)
Vitamin H	Vitamin H is needed for extreme proliferation of cancer cells, thus, vitamin H receptors are over expressed in tumors	Taheri et al. (2011a,b)
Folic acid	Folic acid is a pivotal component in cells as it is necessary for critical vital missions such as DNA and RNA synthesis, amino acid metabolism and methylation reactions. Folic acid receptors are found to be overexpressed in some tumors such as ovarian, kidney, lung, brain, endometrial, colorectal, pancreatic, gastric, prostate, testicular, bladder, head and neck, breast cancer, and nonsmall lung cancer	Assaraf et al. (2014)
Vintafolide	Vintafolide is conjugate form of folic acid and DAVLBH known to be a *vinca* alkaloid showing its effect by microtubule polymerization corruption and apoptosis stimulation. A disulfide linker binds DAVLBH and folic acid and provides drug release to cancer cells. The targeting is achieved by the presence of folic acid enabling receptor-mediated endocytosis through the FRs present on the surface of cancer cells	Pribble and Edelman (2012)
IMGN853	IMGN853 is another FRα targeting drug under investigation in Phase 1 study on the treatment of endometrial cancer and epithelial ovarian cancer. System is composed of a cytotoxic drug maytansinoid and targeting agent folic acid, showing its effect by suppressing microtubule polymerization	Moore et al. (2015)
Transferrin	Transferrin is an 80 kDa plasma glycoprotein has a critical role in cellular transports. Transferrin receptors located on the cell surface can bind to iron and internalized by receptor mediated endocytosis	Qian (2002)
EGFR	EGFR as trans membrane glycoprotein is a member of the tyrosine kinase family of growth factors receptors. Several studies have demonstrated that EGFR is overexpressed in pancreatic cancer leads to more advanced disease, poor survival, and the presence of metastases	Oliverira-Cunha et al. (2011)

(*Continued*)

Table 12.3 Potential Targets and Targeting Agents for Effective Drug Delivery for Pancreas Cancer *Continued*

Target Mechanism	Definition	Reference
EphA2 receptor	EphA2 receptor is a member of the tyrosine kinase receptors family found to be overexpressed in cancer cells. Its mission is to form interaction between cancer cells, stroma, and vascular cells	Tandon et al. (2011)
Acridine orange	Acridine orange has an intrinsic affinity toward DNA due to intercalation	Anajafi et al. (2016)
Antigens	Specific antibody attachment to delivery vehicles can be a way to target particular antigens exists on the cell surface	Friedman et al. (2013)
Integrin and neuropilin receptors	Integrins are located on the cell surface composing of α and aβ-subunits. Their duty is to attach the cells to the extracellular matrix, maintain cell adhesion, and signal transduction. $\alpha\nu\beta3$ integrins are found to be overexpressed in various types of tumors	Kulkarni et al. (2016)
MMPs	MMPs are promising since they are found to be overexpressed in pancreatic cancer tumors and its expression is related to migration, invasion, and metastasis	Grünwald et al. (2016)

DAVLBH, *Desacetylvinlastine hydrazide;* FR, *folate receptor;* EGFR, *epidermal growth factor receptor; MMPs, matrix metalloproteinases.*

CD44 targeting with HA-SMA (hylaluronic acid-styrene-maleic acid) self-assembling nanomicelles for the delivery of encapsulated anticancer agent 3,4-difluorobenzylidene curcumin (CDF) has been studied on pancreatic cancer stem cell and, resulted in better internalization by receptor mediated endocytosis (Kesharwani et al., 2015b). *Nanostructured lipid carriers* (NLCs) are colloidal drug carriers prepared by solid and liquid lipids and surfactants, providing higher drug loading compared to other lipid carriers. In a study conducted by Lu et al., stearic acid was conjugated to gemcitabine and formed lipid prodrug. Further, HA-amino acid-baicalein (HA-AA-BCL) were synthesized and combination delivery of GEM and BCL was maintained by GEM-SA- and HA-AA-BCL loaded NLCs prepared by a solvent evaporation method with sizes between 100 and 200 nm. Intravenous injection of NLCs to AsPC1 cell-bearing C57BL/6 mice models resulted in significant tumor inhibition compared to GEM-BCL NLCs without HA conjugation which the proves importance of active targeting during treatment (Lu et al., 2017).

In another study conducted by Kesharwani et al., dendrimer drug delivery systems were used. *Dendrimers* have a three-dimensional globular shape with sizes ranging from 1 to 100 nm characterized with branches. Poly(amidoamine) (PAMAM) dendrimer is known to be most investigated for gene and drug delivery. CDF is formulated as CD44-targeted PAMAM dendrimer formulation and evaluated in pancreatic cancer treatment. The system has a particle size of approximately 9.3 nm and a negatively charged surface. The targeted dendrimer formulation of CDF inhibited the proliferation of cancer stem like cells. Compared to nontargeted PAMAM-CDF formulation, targeted formulation showed 1.71-fold increase in IC50 value (Kesharwani et al., 2015a).

Vitamin H targeting has been investigated in a study in which human serum *albumin nanoparticles* with targeting moiety Vitamin H were used to deliver methotrexate for the treatment of breast cancer (Taheri et al., 2011a,b).

Folate receptors (FR) have four well-known isoforms (α, β, χ, and δ). FRα-expressing cells demonstrate the optimal response to folic acid targeted therapy. Currently, there are three medicinal agents targeting folic acid receptors and these are being investigated under clinical studies. Farletuzumab, also known as MORAb-003, is a monoclonal antibody and showed excellent behavior in the treatment of ovarian tumor xenografts (Teng et al., 2012). A phase II study was conducted with farletuzumab and carboplatin/taxane for the treatment of platinum-sensitive ovarium, fallopian tube, and primary peritoneal cancers. Results proved that this combination improved the therapy response. However, subsequent phase III trials have failed due to unmet findings. A placebo-controlled phase II trial is under exploration on FRα overexpressing-lung cancer patients (Thomas et al., 2013). Taking into account folic acid's (FA) high affinity to bind FR, folate-chitosan-gemcitabine (FA-Chi-Gem) core-shell nanoparticles with a size ranging in 200–300 nm were studied on pancreatic cancer cells. The study showed that active targeting was provided via FA-FR binding after internalization of nanoparticles via endocytosis chitosan was dissolved and gemcitabine as successfully released to cytoplasm (Xu et al., 2013).

A targeted poly(lactic-*co*-glycolic acid) (PLGA) *nanoparticular system* carrying bortezomib, a proteasome inhibitor, has been investigated on pancreatic cancer cells. The study proved that the presence of transferrin ligand significantly increased the delivery of PLGA nanoparticles to pancreatic cancer cells. Not only was the limitation of bortezomid due to its toxicity overcome by the polymeric drug delivery system, but the targeting aim was also achieved (Frasco et al., 2015). In a recent study, the neural activity role on the progression of pancreatic cancer was investigated. For this purpose, ferritin nanoparticles were prepared to deliver carbachol-muscarinic agonist and atropine-muscarinic antagonist neural drugs for the activation and blocking of neural recess in pancreatic cancer. The nanoparticles target the pancreatic tumor via passive targeting of the EPR effect of tumors and active targeting via the transferrin receptor (Lei et al., 2016).

Epidermal growth factor receptor (EGFR) targeting by type B gelatin PEG *nanoparticles* for the delivery of gemcitabine have been developed by Singh et al. and tested in vitro on a Panc-1 pancreatic cancer cell line. Succinimidyl (3-[2-pyridyldithio]-propionate (SPDP) was conjugated to gemcitabine and further conjugated with thiolated type B gelatin to form Gem-Gel conjugate. Gem-SPDP showed the lowest IC50 value ($8.39 \pm 1.79\ \mu M$) due to interaction between the succinimidyl group and cell surface proteins. Although it showed the lowest IC50 value, its ability to bind any protein in vivo makes it not a good option for treatment. Therefore, EGFR-targeting nanoparticles were considered to be reasonable with an IC50 value of 17.08 ± 2.32 μM). Orthotropic PANC-1 bearing SCID beige mice were used to conduct in vivo experiments in which EGFR-targeted gelatine nanoparticles were found to be presentable to deliver gemcitabine that resulted in significant inhibition of tumor growth compared to a gemcitabine solution (Singh et al., 2016).

Quinn et al. reported a study where they conjugated agent YNH (of AA sequence YSAYPDSVP (Nle) (Hsr) S, where Nle and Hsr represent L-norleucine and L-homo-serine, respectively) to form YNH-L2-GEM and agent 123B9 to form 123B9-L2-GEM for pancreatic cancer treatment by targeting EphA2 receptors. In vivo studies on pancreatic cancer xenograft using a MIA PaCa-2 cell line showed that this targeting approach significantly increased the survival time in mice compared to the single gemcitabine treatment group (Quinn et al., 2016).

Targeting via acridine orange has been recently studied beacause of its high affinity to DNA leads to nuclear localizing. For this purpose, *polymersomes* formed from amphiphilic polymers consisting of a hydrophilic inner part and a bilayer membrane that can release encapsulated drug through different stimuli mechanisms such as light, magnetic, redox, pH, or temperature were used (Anajafi et al., 2016). A fluorescent polymersome formulation encapsulating both gemcitabine and doxorubusin has been reported by Anajafi et al. Reduction sensitive copolymer PEG (1900)-*S*-*S*-PLA (6100) and another amphiphilic copolymer with azide group at the hydrophilic terminus N_3-PEG (1900)-PLA (6100) were used to form polymersomes. Disulfide group was planned to impart redox-sensitive property. The in vitro cell culture experiments showed the higher cell killing activity of drug delivery systems in PANC-1 monolayer and three-dimensional spheroid cultures (Anajafi et al., 2016).

Targeting particular antigens existing on the cell surface by attaching specific antibodies as targeting moieties has been thoroughly investigated over the past decade. One recent study conducted by Spadavecchia et al. investigated intravenous DOX delivery for treatment of pancreatic cancer via DOX-loaded and anti-Kv11.1 polyclonal antibody [pAb] targeting dicarboxylic acid-terminated PEG, *(PEG)-gold nanoparticles* (AuNPs). Anti-Kv11.1-pAb is specific to human ether-*a*-go-go related gene 1 hERG1 as it is aberrantly expressed in pancreatic cancer. Despite the toxicity problem of hERG-1 blockers, this platform offered higher safety. The system provided prolonged release up to seven days and the EC50 value of nanocarriers were found to be reduced 30-fold compared to free DOX (Spadavecchia et al., 2016). It is also important to note that the development of gold nanoparticles is considered to be promising step in cancer treatment. Aurimune (CYT-6091) consisting of PEGylated (polyethylene gycole) gold nanoparticles was used in cancer treatment through the delivery of tumor necrosis factor-alpha (Cytimmune.com, 2015; Libutti et al., 2010).

Peptide-based targeting of overexpressed integrin and neuropilin receptors by *lipid nanoparticles* (LNs) has been studied by Kulkarni et al. recently (Kulkarni et al. 2016). LNs emerged as a result of research to develop alternate approaches for nanoparticles based on lipid components other than phospholipids in order to control drug release and delivery of therapeutics, which may not efficiently load into liposomes. Compared with other drug-delivery systems, LNs have been developed more recently and are potentially attractive, marketable choices due to their natural components and easily scaled-up synthesis processes.

Hexynoic acid conjugated iRGD peptide was synthesized and conjugated to saturated lipid DSPE-PEG-N3 further incorporated into LNs for the assessment in pancreatic cancer treatment. Gemcitabine was encapsulated into the iRGD peptide functionalized, hypoxia-responsive LNs. It was reported that 65% of the encapsulated drug was released under hypoxic conditions in 2 hours. Monolayer and three-dimensional spheroid cultures of BxPC-3 pancreatic cancer cells were used to evaluate the effectiveness of LNs. Viability of cells in monolayer cultures decreased in the presence of LNs encapsulating gemcitabine under both normoxic and hypoxic conditions as iRGD peptide on the surface allowed the LNs to penetrate deeper and deliver the anticancer drug to the hypoxic cores (Kulkarni et al., 2016).

cRGD conjugated gemcitabine loaded *human serum albumin nanoparticles* (cRGD-Gem-HSA-NPs) have been developed and evaluated the effect in vitro on BxPC-3 pancreatic cancer cell lines. Targeted NPs was compared to free drug and it was confirmed that ligand targeted NPs demonstrated the greatest inhibitory effect that establish the therapeutic potential of formulation (Yu et al., 2016).

Targeting matrix metalloproteinase (MMPs) is promising since they are found to be overexpressed in pancreatic cancer tumors and its expression related to migration, invasion, and metastasis. The overexpression of this specific protein makes it a favorable target for successful delivery. Thus, a MMP-9-cleavable linker mediating the specific removal of a PEG shield from a PLGA-*b*-PEG-based polymeric nanocarrier leading to specific uptake of small *polymeric nanoparticles* with their cargo into cells has been developed (Grünwald et al., 2016).

It should be noted that active targeting is more effective than passive targeting. Firstly, active targeting allows delivery systems to be accumulated in the tumor environment faster. Secondly, the drug amount that arrives to the tumor is much more than that delivered via passively targeted systems. Thirdly, active targeting has the possibility to overcome multidrug resistance (Torchilin, 2010).

12.4 GENE THERAPY BY miRNAs AND SMALL INTERFERING RNAS WITH POLYMERIC AND LIPIDIC DELIVERY SYSTEMS

Somatic or inherited genetic mutations were well-recognized in pancreatic cancer development and open the door for gene therapy. miRNAs are composed of

20–25 nucleotides, known as single striated noncoding small RNAs. They are considered to be gene regulators as their mission is cleavage and/or translational repression of targeted mRNA (Kim, 2005). It has been well-proven that miRNAs play a role in cell proliferation, apoptosis, differentiation, and organ development (Brennecke et al., 2003). However, the mechanism of regulation changes of miRNAs is still not clear and may be attributed to transcription factor aberrance, mutations, or deletions of gene or epigenetic alterations (Chitkara et al., 2015). miRNAs have an important role in pancreatic cancer phenomena as they can act as both tumor suppressor and oncogene. Therefore, the treatment can be conducted with two different approaches: (1) the substitution therapy; and (2) one that relies on the inhibition of particular miRNA which has a key role in the development or aggressiveness of disease. It has been proven in several studies that overexpression of miR-17-p, miR-21, miR-10b, miR-196a, miR-21, miR-205, miR-181b, invasion (miR-21), metastasis (miR- 10b) miR-218, miR-21, and EMT (miR-200c, let-7) is related to pancreatic cancer development (Chitkara et al., 2015). miR-155 is found to be overexpressed in pancreatic cancer tissue specimens which is also relevant with metastasis through lymph nodes as well as invasion and migration (Li and Sarkar, 2016). MiR-221 has been investigated and its increased levels are found to be associated with the proliferation of tumors (Sarkar et al., 2013). miR-145, miR-146a, and miR-34 are the other miRNAs which have an important effect on the development and aggressiveness of pancreatic cancer. Therefore, targeting these particular miRNAs with additional exposure to anticancer drugs could be a promising approach for treatment. miR-150, tumor-suppressive miRNA, has demonstrated satisfied encapsulation efficacy, release and successful intracellular fate by administration in a nanoparticular system that resulted in suppressed proliferation, clonogenicity, and invasion of pancreatic cancer (Arora et al., 2014).

Tumor-suppressor miRNAs normally remain active in cells and have a duty to prevent the progression of oncogenic actions. miR200 a/b/c has an ability to inhibit stem cell factors such as Sox2, SIP 1, and ZEB-1 (Cheng et al., 2012). The Let-7 miRNA family is another miRNA that demonstrates suppressor action in several cancers such as breast, lung, prostate, pancreatic, ovarian, and melanoma. Besides the promising role of tumor suppressors in the development of cancers, their oncogenes have a contradictory function. PDAC is regulated by the activity of miR-21, miR-155, miR-196a, miR-210, and miR-221 as their overexpression is categorized with PDAC. On the other hand, the expression of some critical miRNAs is important. As an example, MiR-26a is identified as a tumor suppressor in pancreatic cancer cells and its expression can reduce cell proliferation, colony formation, and cell migration in BxPc-3 pancreatic cancer cells (Batchu et al., 2015).

Genetic studies have proven the K-ras gene activation is involved in more than 85% pancreatic cancer cases while p16 and TP53 genes are inhibited in 95% of pancreatic cancers and the SMAD4 gene was found to be absent in 55% of cases. Compelling evidence proved the link between miRNA expression and

PDAC previously. miR-21 as the most-cited miRNA was found to be related to desperate response in cancers and its overexpression has been defined in pancreatic cancer tumors, lesions, and cell lines. Among 80 of pancreatic cancer tissue samples, 63 of them (79%) was found to express miR-21, whereas 12 of benign pancreases and 12 of chronic pancreatitis showed modest expression. This study further proved that higher miR-21 expression is related to cancer deaths (Dillhoff et al., 2008). Cell line studies to investigate the role of miR-21 has proven that miR-21 has an significant impact on the proliferation, invasion, and chemoresistance to gemcitabine and the development of PanIN in K-ras mutant mouse models, whereas decreased miR-21 levels were found to have a role on effective anticancer treatment in vitro (du Rieu et al., 2010; Hwang et al., 2010). Treatment efficacy comparison of standard gemcitabine therapy and miR-21 therapy was assessed by Sicard et al. LV (GFP) and LV (a/miR-21) vectors were injected into pancreatic tumors and a comparison group was intraperitoneally injected with 125 mg/kg high dosage of gemcitabine twice a week. miR-21 inhibition strongly suppressed pancreatic tumor growth compared to LV (GFP)-transduced tumors. The next step was applying both gemcitabine and miR-21 to exploit the angiogenesis effect of miR-21 depletion. Due to angiogenesis as a result of targeting miR-21, it was suggested that gemcitabine delivery could be maximized (Sicard et al., 2013).

Small interfering RNA (siRNA)-based therapy has several superiorities such as providing high safety, potency, target specificity, and rapid development. Since siRNA shows its action without interacting with DNA, the safety profile is found to be acceptable as there would not be any mutation and teratogenicity case because of treatment (Zuckerman and Davis, 2015; Xu and Wang, 2015). Silenseed Ltd. has worked on a system named LOcal Drug EluteR (LODER) that is composed of a biodegradable polymeric matrix PLGA with a molecular weight >50 kDa which can be loaded with a drug and release siRNA. siG12D-LODER system releases siRNA to the target gene (G12D-mutated KRAS) for a long time. It is applied by implanting it into the pancreatic tumor and has already passed a phase 1/2a clinical trial (Golan et al., 2015). It has been evaluated that the safety and toxicity profile of siG212D-LODER on 150 rats after subcutaneous implantation resulting in a high safety profile as it does not demonstrate any local or systemic adverse effects (Ramot et al., 2016).

The biological pathway of applied therapeutic miRNA is tough as they have to transit the blood environment, tumor tissue, and tumor cell milieu and final cytosol arrival. It is a challenge to penetrate into cells due to their negative charge and high molecular weight which results in problematic in vivo stability, lower cellular uptake, unexpected targets, and ad hoc distribution. It has been reported in several studies that siRNAs are very unstable as they are directly disposed within 5 minutes following their intravenous administration (Takei et al., 2004; van de Water et al., 2006). The reticular endothelial system (RES) and immunologic defense system are the restrictive mechanisms that cause degradation and elimination of miRNAs. miRNA complexes can reach the cell surface by initially crossing the vascular endothelium wall and further the ECM. Following arrival to the ECM, miRNA complexes should reach the cytoplasm; however, due to the negatively charged cytoplasm membrane, oligonucleotides face a problem to cross it. The cellular uptake of miRNAs is conducted mainly by macropinocytosis and clathrin-mediated endocytosis. However, the reality of being degraded inside lysosome brings the requirement of endosomal escape in order to be effective inside the cells.

Facilitation of the delivery of miRNAs via complex formation by *lipid-based carriers* is a promising way as lipid carriers have long been investigated as a delivery vehicle owing to their high penetration capability into cell membranes (Ozpolat et al., 2014; Piao et al., 2012). Cationic lipids provide electrostatic interaction between their positively charged hydrophilic group and negatively charged miRNA. Moreover, the existence of the hydoxyl group in the cationic group of lipids helps transfection.

Dimethyl dioctadecyl ammonium bromide is a cationic lipid-containing quaternary ammonium head group has been investigated in a study (Piao et al., 2012) in which it was conjugated with Chol and D-alpha tocopheryl PEG 1000 succinate (TPGS) at a molar of 60:35:5 for the delivery of synthetic pre-miR-107 to treat head and neck squamous cell carcinoma. Lipid-based carrier greatly suppressed the oncogenic protein kinase Cε, cyclin-dependent kinase and hypoxia-inducible factor 1-b (HIF-1b). In addition to suppression, the tumor progression has been regressed as 45.2% compared to the control group (Piao et al., 2012). Another cationic lipid-containing quaternary ammonium headgroup is *N*-(1,2,3-dioleoyloxy)propyl-*N*,*N*,*N*-trimethyl-ammonium chloride (DOTMA)-based cationic lipoplexes which were studied for the delivery of miR-29b in the treatment of nonsmall cell lung cancer (NSCLC) and the control group was defined as transfection agent siPORT NeoFX. Lipoplexes composed of DOTMA and Chol and TPGS (49.5:49.5:1) miR-29blipoplexes demonstrated 60% inhibition of tumor growth against the control group (Wu et al., 2013).

Investigations on neutral *lipid emulsions* to deliver miRNAs have long been conducted since the presence of a positive charge in cationic lipids restricts its pertinence. A study on prostate cancer has been reported in which neutral lipid emulsion was used to deliver miR-34a systemically resulting in suppressed metastasis and increased overall survival of animal models (Liu et al., 2011). In another study, miR-34a has been delivered to NOD-SCID mouse models for the treatment of lymphoma, resulting in 76% suppression of tumor growth with no adverse effects which makes the lipid emulsions considered to be safe (Craig et al., 2012).

Polyethyleneimine-based cationic polymer is the most popular delivery intermediate of plasmid DNA and siRNA (Baumann and Winkler, 2014). The delivery of miR-145 by polyurethane-polyethylene imine copolymer (PU-PEI) has been studied for the treatment of lung adenocarcinoma generated by cancer stem cells (Chiou et al., 2012). In another study, miR-145 delivery by PU-PEI has been investigated in which glioblastoma tumor genesis was regressed (Yang et al., 2012). *Silica nanoparticles* was investigated for miRNA delivery and greatly helped to generate apoptosis in resistant glioma cells (Bertucci et al., 2015).

Another way to deliver siRNA and miRNA is by conjugating lipids or receptor-binding molecules to nucleic acid. Conjugation of asialoglyco protein receptor ligand N-acetylgalactosamine (GalNAc) has been studied for the delivery of anti-miR-122 (Baumann and Winkler, 2014).

MiR-205 is a tumor suppressor miRNA and was found to be downregulated in pancreatic cancer cells and leads to gemcitabine resistance. It has been worked on *micellar* gemcitabine conjugated cationic copolymers forming complex with miR-205, exhibiting great efficacy in pancreatic cancer ectopic tumor model. It was reported that gemcitabine release maintained for more than 10 days and transfection efficiency was found to be more than 90% (Mittal et al., 2014). miR-let7b, another tumor-suppressor miRNA, is known to be downregulated in pancreatic cancer. It can target K-RAS, MUC4, NCOA3, HMGA2, and TGFβR1-like tumor-endorsing genes (Patel et al., 2014; Kumar et al., 2015). A study Kumar et al. studied the delivery of both miR-let7b and hedgehog inhibitor vismodegib loaded methoxy PEG-block-poly(2-methyl-2-carboxyl-propylene carbonate-graft-dodecanol-graft tetraethylpentamine) (mPEG-b-PCC-g-DC-g-TEPA). The evaluation on subcutaneous pancreatic tumor-bearing mice resulted in effective inhibition of tumor growth (Kumar et al., 2015).

Absence of miR-145 is characterized with pancreatic cancer. The normalization of miR-145 expression is associated with repairment of various oncogenes such as MUC13 upregulated in pancreatic cancer. *Magnetic nanoparticle*-based systems have emerged as a delivery tool for genes and drugs. Iron oxide nanoparticles have been developed for the delivery of miR-145. The study proved that miR-145 delivery was successfully achieved by miR-145-MNPF and decreased MUC13 expression in pancreatic cancer cells. This delivery system provided the sustained release of miR-145 as well as having a prolonged effect on target protein MUC13 (Setua et al., 2016).

miR-150-loaded PLGA/Polyethylenimine *nanoparticles* were developed for the intracellular delivery of miR-150 to inhibit its target gene MUC4. The treatment showed a great inhibition effect on pancreatic tumor cell proliferation and clonogenicity. miR-150 is known to be a tumor suppressor miRNA in pancreatic cancer (Srivastava et al., 2014). The negative charge, hydrophilic character, degradability by nuclease, and ineffective uptake of miRNA by cells a requirement for suitable delivery approaches (Arora et al., 2014).

CC9 peptide (CRGDKGPDC) was attached to β-cyclodextrin-polyethylenimine (PEI-CD) or physically mixed with cationic PEI-CD vectors to deliver miR-34a. CC9 is targeted to tumors through the RGD motif and, furthermore, CRGDK occurs due to enzymatic cleavage by protease. CRGDK binds to neuropilin-1 that activates tumoral penetration. Hu and co-workers showed that treatment with this nanoparticle system increased the miR-34a level in PANC-1 pancreatic cancer cell line and miR-34a suppressed the expression of target E2F3, Bcl-2, c-myc, and cyclin D1 genes which have impact on apoptosis, proliferation, and migration (Hu et al., 2013).

12.5 CONCLUSION

Conventional treatment options are no longer attractive due to formulation and toxicity problems as well as the difficulties in arriving at the tumoral area. Thus, over the past few decades new strategies have been under investigation by researchers. Novel approaches such as using drug delivery technologies, molecular active targeting of the tumors by targeting ligands, or genetic battle with the altered miRNAs to modulate their expression profiles stand out for the treatment pancreatic cancer. However, PDAC is very complex tumor type due to the presence of massive stromal consisting of various cell types, matrix structures, and signaling pathways that lead to debates whether this structure promotes or suppresses the carcinogenesis. Therefore, understanding the pancreatic cancer biology is pivotal. Still more studies are necessary to confirm the improved efficacy and safety of these novel approaches.

REFERENCES

Abrams, R.A., Lowy, A.M., O'Reilly, E.M., Wolff, R.A., Picozzi, V.J., Pisters, P.W., 2009. Combined modality treatment of resectable and borderline resectable pancreas cancer: expert consensus statement. Ann. Surg. Oncol. 16, 1751–1756.

Anajafi, T., Scott, M.D., You, S., Yang, X., Choi, Y., Qian, S.Y., et al., 2016. Acridine orange conjugated polymersomes for simultaneous nuclear delivery of gemcitabine and doxorubicin to pancreatic cancer cells. Bioconjug. Chem. 27, 762–771.

Arora, S., Swaminathan, S.K., Kirtane, A., Srivastava, S.K., Bhardwaj, A., Singh, S., et al., 2014. Synthesis, characterization, and evaluation of poly(D,L-lactide-*co*-glycolide)-based nanoformulation of miRNA-150: potential implications for pancreatic cancer therapy. Int. J. Nanomed. 9, 2933–2942.

Assaraf, Y.G., Leamon, C.P., Reddy, J.A., 2014. The folate receptor as a rational therapeutic target for personalized cancer treatment. Drug Resist. Updates 17, 89–95.

Au, J., Yeung, B., Wientjes, M., Lu, Z., Wientjes, M., 2016. Delivery of cancer therapeutics to extracellular and intracellular targets: determinants, barriers, challenges and opportunities. Adv. Drug Deliv. Rev. 97, 280–301.

Awada, A., Bondarenko, I., Bonneterre, J., Nowara, E., Ferrero, J., Bakshi, A., et al., 2014. A randomized controlled phase II trial of a novel composition of paclitaxel embedded into neutral and cationic lipids targeting tumour endothelial cells in advanced triple-negative breast cancer (TNBC). Ann. Oncol. 25, 824–831.

Batchu, R., Gruzdyn, O., Qazi, A., Kaur, J., Mahmud, E., Weaver, D., et al., 2015. Enhanced phosphorylation of p53 by microRNA-26a leading to growth inhibition of pancreatic cancer. Surgery 158, 981–987.

Bates, D.O., Hillman, N.J., Williams, B., Neal, C.R., Pocock, T.M., 2002. Regulation of microvascular permeability by vascular endothelial growth factors. J. Anat. 200, 581–597.

Baumann, V., Winkler, J., 2014. miRNA-based therapies: strategies and delivery platforms for oligonucleotide and non-oligonucleotide agents. Future Med. Chem. 6, 1967–1984.

Bazak, R., Houri, M., El Achy, S., Hussein, W., Refaat, T., 2014. Passive targeting of nanoparticles to cancer: a comprehensive review of the literature. Mol. Clin. Oncol. 2, 904–908.

Bertrand, N., Wu, J., Xu, X., Kamaly, N., Farokhzad, O.C., 2014. Cancer Nanotechnology: the impact of passive and active targeting in the era of modern cancer biology. Adv. Drug Deliv. Rev. 66, 2–25.

Bertucci, A., Prasetyanto, E., Septiadi, D., Manicardi, A., Brognara, E., Gambari, R., et al., 2015. Combined delivery of temozolomide and anti-miR221 PNA using mesoporous silica nanoparticles induces apoptosis in resistant glioma cells. Small 11, 5687–5695.

Bildstein, L., Dubernet, C., Marsaud, V., Chacun, H., Nicolas, V., Gueutin, C., et al., 2010. Transmembrane diffusion of gemcitabine by a nanoparticulate squalenoyl prodrug: an original drug delivery pathway. J. Controlled Release 147, 163–170.

Bisht, S., Schlesinger, M., Rupp, A., Schubert, R., Nolting, J., Wenzel, J., et al., 2016. A liposomal formulation of the synthetic curcumin analog EF24 (Lipo-EF24) inhibits pancreatic cancer progression: towards future combination therapies. J. Nanobiotechnol. 14, 57.

Boulikas, T., 2009. Clinical overview on Lipoplatin: a successful liposomal formulation of cisplatin. Expert Opin. Invest. Drugs 18, 1197–1218.

Brennecke, J., Hipfner, D., Stark, A., Russell, R., Cohen, S., 2003. Bantam encodes a developmentally regulated microrna that controls cell proliferation and regulates the proapoptotic gene hid in drosophila. Cell 113, 25–36.

Burris, H.A., Moore, M.J., Andersen, J., Green, M.R., Rothenberg, M.L., Modiano, M.R., et al., 1997. Improvements in survival and clinical benefit with gemcitabine as first-line therapy for patients with advanced pancreas cancer: a randomized trial. J. Clin. Oncol. 15, 2403–2413.

Chen, N., Brachmann, C., Liu, X., Pierce, D.W., Dey, J., Kerwin, W.S., et al., 2015. Albumin-bound nanoparticle (nab) paclitaxel exhibits enhanced paclitaxel tissue distribution and tumour penetration. Cancer Chemother. Pharmacol. 76, 699–712.

Cheng, H., Shi, S., Cai, X., Long, J., Xu, J., Liu, C., et al., 2012. microRNA signature for human pancreatic cancer invasion and metastasis. Exp. Ther. Med. 4, 181–187.

Chiou, G., Cherng, J., Hsu, H., Wang, M., Tsai, C., Lu, K., et al., 2012. Cationic polyurethanes-short branch PEI-mediated delivery of Mir145 inhibited epithelial–mesenchymal transdifferentiation and cancer stem-like properties and in lung adenocarcinoma. J. Controlled Release 159, 240–250.

Chitkara, D., Mittal, A., Behrman, S., Kumar, N., Mahato, R., 2013. Self-assembling, amphiphilic polymer–gemcitabine conjugate shows enhanced antitumour efficacy against human pancreatic adenocarcinoma. Bioconjug. Chem. 24, 1161–1173.

Chitkara, D., Mittal, A., Mahato, R., 2015. miRNAs in pancreatic cancer: therapeutic potential, delivery challenges and strategies. Adv. Drug Deliv. Rev. 81, 34–52.

Choi, K.Y., Saravanakumar, G., Park, J.H., Park, K., 2012. Hyaluronic acid-based nanocarriers for intracellular targeting: interfacial interactions with proteins in cancer. Colloids Surf., B: Biointerfaces. 99, 82–94.

Colucci, G., Labianca, R., Di Costanzo, F., Gebbia, V., Cartenì, G., Massidda, B., et al., 2010. Randomized phase III trial of gemcitabine plus cisplatin compared with single-agent gemcitabine as first line treatment of patients with advanced pancreatic cancer: the GIP-1 study. J. Clin. Oncol. 28, 1645–1651.

Craig, V., Tzankov, A., Flori, M., Schmid, C., Bader, A., Müller, A., 2012. Systemic microRNA-34a delivery induces apoptosis and abrogates growth of diffuse large B-cell lymphoma in vivo. Leukemia 26, 2421–2424.

Cytimmune.com, 2015. Welcome to CytImmune.com. CytImmune.com [online]. Available from: <http://www.cytimmune.com/#pipeline> (accessed August 2015).

D'Cruz, O., Piacente, M., Trieu, V., Tao, C., Desai, N., 2010. Cardiovascular and CNS safety profile of ABI-013, a novel nanoparticle albumin-bound (nab) analog of docetaxel. In: AACR 101st Annual Meeting, Washington, DC., Abstract 2617.

Dillhoff, M., Liu, J., Frankel, W., Croce, C., Bloomston, M., 2008. MicroRNA-21 is overexpressed in pancreatic cancer and a potential predictor of survival. J. Gastrointest. Surg. 12, 2171–2176.

Dimou, A., Syrigos, K., Saif, M., 2012. Overcoming the stromal barrier: technologies to optimize drug delivery in pancreatic cancer. Ther. Adv. Med. Oncol. 4, 271–279.

du Rieu, M., Torrisani, J., Selves, J., Al Saati, T., Souque, A., Dufresne, M., et al., 2010. MicroRNA-21 is induced early in pancreatic ductal adenocarcinoma precursor lesions. Clin. Chem. 56, 603–612.

Duangjai, A., Luo, K., Zhou, Y., Yang, J., Kopecek, J., 2014. Combination cytotoxicity of backbone degradable HPMA copolymer gemcitabine and platinum conjugates toward human ovarian carcinoma cells. Eur. J. Pharm. Biopharm. 87, 187–196.

Eichhorn, M.E., Ischenko, I., Luedemenn, S., Strieth, S., Papyan, A., Werner, A., et al., 2010. Vascular targeting by EndoTAG-1 enhances therapeutic efficacy of conventional chemotherapy in lung and pancreatic cancer. Int. J. Cancer 126, 1235–1245.

Estanqueiro, M., Amaral, M., Conceição, J., Sousa Lobo, J., 2015. Nanotechnological carriers for cancer chemotherapy: the state of the art. Colloids Surf., B: Biointerfaces 126, 631–648.

Fang, Y., Du, F., Xu, Y., Meng, H., Huang, J., Zhang, X., et al., 2015. Enhanced cellular uptake and intracellular drug controlled release of VESylated gemcitabine prodrug nanocapsules. Colloids Surf., B: Biointerfaces 128, 357–362.

Frasco, M.F., Almeida, G.M., Santos-Silva, F., Pereira, M.C., Coelho, M.A., 2015. Transferrin surface-modified PLGA nanoparticles-mediated delivery of a proteasome inhibitor to human pancreatic cancer cells. J. Biomed. Mater. Res. A 103, 1476–1484.

Friedman, A., Claypool, S., Liu, R., 2013. The smart targeting of nanoparticles. Curr. Pharm. Des. 19, 6315–6329.

Golan, T., Khvalevsky, E., Hubert, A., Gabai, R., Hen, N., Segal, A., et al., 2015. RNAi therapy targeting KRAS in combination with chemotherapy for locally advanced pancreatic cancer patients. Oncotarget 6, 24560–24570.

Grünwald, B., Vandooren, J., Locatelli, E., Fiten, P., Opdenakker, G., Proost, P., et al., 2016. Matrix metalloproteinase-9 (MMP-9) as an activator of nanosystems for targeted drug delivery in pancreatic cancer. J. Controlled Release 239, 39–48.

Gupta, R., Shea, J., Scafe, C., Shurlygina, A., Rapoport, N., 2015. Polymeric micelles and nanoemulsions as drug carriers: therapeutic efficacy, toxicity, and drug resistance. J. Controlled Release 212, 70–77.

Hobbs, S.K., Monsky, W.L., Yuan, F., Roberts, W.G., Griffith, L., Torchilin, V.P., et al., 1998. Regulation of transport pathways in tumour vessels: role of tumour type and microenvironment. Proc. Natl. Acad. Sci. U.S.A. 95, 4607–4612.

Hofheinz, R.D., Willer, A., Weisser, A., Gnad, U., Saussele, S., Kreil, S., et al., 2004. Pegylated liposomal doxorubicin in combination with mitomycin C, infusional 5-fluorouracil and sodium folinic acid. A phase-I-study in patients with upper gastrointestinal cancer. Br. J. Cancer 90, 1893–1897.

Hu, Q., Jiang, Q., Jin, X., Shen, J., Wang, K., Li, Y., et al., 2013. Cationic microRNA-delivering nanovectors with bifunctional peptides for efficient treatment of PANC-1 xenograft model. Biomaterials 34, 2265–2276.

Hubenak, J.R., Zhang, Q., Branch, C.D., Kronowitz, S.J., 2014. Mechanisms of injury to normal tissue after radiotherapy: a review. Plast. Reconstr. Surg. 133, 49e−56e.

Hwang, J., Voortman, J., Giovannetti, E., Steinberg, S., Leon, L., Kim, Y., et al., 2010. Identification of microRNA-21 as a biomarker for chemoresistance and clinical outcome following adjuvant therapy in resectable pancreatic cancer. PLoS One 5, e10630.

Immordino, M.L., Brusa, P., Rocco, F., Arpicco, S., Ceruti, M., Cattel, L., 2004. Preparation, characterization, cytotoxicity and pharmacokinetics of liposomes containing lipophilic gemcitabine prodrugs. J. Controlled Release 100, 331−346.

Kamaly, N., Xiao, Z., Valencia, P.M., Radovic-Moreno, A.F., Farokhzad, O.C., 2012. Targeted polymeric therapeutic nanoparticles: design, development and clinical translation. Chem. Soc. Rev. 41, 2971−3010.

Karaca, M., Dutta, R., Ozsoy, Y., Mahato, R., 2016. Micelle mixtures for coadministration of gemcitabine and GDC-0449 to treat pancreatic cancer. Mol. Pharm. 13, 1822−1832.

Kesharwani, P., Xie, L., Banerjee, S., Mao, G., Padhye, S., Sarkar, F.H., et al., 2015a. Hyaluronic acid-conjugated polyamidoamine dendrimers for targeted delivery of 3,4-difluorobenzylidene curcumin to CD44 overexpressing pancreatic cancer cells. Colloids Surf., B: Biointerfaces 136, 413−423.

Kesharwani, P., Banerjee, S., Padhye, S., Sarkar, F., Iyer, A., 2015b. Hyaluronic acid engineered nanomicelles loaded with 3,4-difluorobenzylidene curcumin for targeted killing of CD44 + stem-like pancreatic cancer cells. Biomacromolecules 16, 3042−3053.

Kim, V., 2005. MicroRNA biogenesis: coordinated cropping and dicing. Nat. Rev. Mol. Cell Biol. 6, 376−385.

Kim, J.G., Sohn, S.K., Song, H.S., Kwon, K.Y., Do, Y.R., Lee, K.H., et al., 2007. Multicenter phase II study of weekly paclitaxel plus cisplatin combination chemotherapy in patients with advanced gastric cancer. Cancer Chemother. Pharmacol. 60, 863−869.

Ko, A.H., 2015. Progress in the treatment of metastatic pancreatic cancer and the search for next opportunities. J. Clin. Oncol. 33, 1779−1786.

Kulkarni, P., Haldar, M., Katti, P., Dawes, C., You, S., Choi, Y., et al., 2016. Hypoxia responsive, tumour penetrating lipid nanoparticles for delivery of chemotherapeutics to pancreatic cancer cell spheroids. Bioconjug. Chem. 27, 1830−1838.

Kumar, V., Mondal, G., Slavik, P., Rachagani, S., Batra, S., Mahato, R., 2015. Codelivery of small molecule hedgehog inhibitor and miRNA for treating pancreatic cancer. Mol. Pharm. 12, 1289−1298.

Lei, Y., Hamada, Y., Li, J., Cong, L., Wang, N., Li, Y., et al., 2016. Targeted tumour delivery and controlled release of neuronal drugs with ferritin nanoparticles to regulate pancreatic cancer progression. J. Controlled Release 232, 131−142.

Li, Y., Sarkar, F., 2016. MicroRNA targeted therapeutic approach for pancreatic cancer. Int. J. Biol. Sci. 12, 326−337.

Libutti, S., Paciotti, G., Byrnes, A., Alexander, H., Gannon, W., Walker, M., et al., 2010. Phase I and pharmacokinetic studies of CYT-6091, a novel PEGylated colloidal gold-rhTNF nanomedicine. Clin. Cancer Res. 16, 6139−6149.

Liu, C., Kelnar, K., Liu, B., Chen, X., Calhoun-Davis, T., Li, H., et al., 2011. The microRNA miR-34a inhibits prostate cancer stem cells and metastasis by directly repressing CD44. Nat. Med. 17, 211−215.

Liu, X., Situ, A., Kang, Y., Villabroza, K.R., Liao, Y., Chang, C.H., et al., 2016. Irinotecan delivery by lipid-coated mesoporous silica nanoparticles shows improved efficacy and safety over liposomes for pancreatic cancer. ACS Nano 10, 2702−2715.

Lu, Z., Su, J., Li, Z., Zhan, Y., Ye, D., 2017. Hyaluronic acid-coated, prodrug-based nanostructured lipid carriers for enhanced pancreatic cancer therapy. Drug Dev. Ind. Pharm. 43, 160–170.

Matsumura, Y., Hamaguchi, T., Ura, T., Muro, K., Yamada, Y., Shimada, Y., et al., 2004. Phase I clinical trial and pharmacokinetic evaluation of NK911, a micelle-encapsulated doxorubicin. Br. J. Cancer 91, 1775–1781.

Michaud, D.S., 2004. Epidemiology of pancreatic cancer. Minerva Chir. 59, 99–111.

Mittal, A., Chitkara, D., Behrman, S., Mahato, R., 2014. Efficacy of gemcitabine conjugated and miRNA-205 complexed micelles for treatment of advanced pancreatic cancer. Biomaterials 35, 7077–7087.

Moertel, C.G., Childs, D.S., Reitemeier, R.J., Colby, M.Y., Holbrook, M.A., 1969. Combined 5-fluorouracil and supervoltage radiation therapy of locally unresectable gastrointestinal cancer. Lancet 2, 865–867.

Moore, K., Martin, N., Seward, L.P., Bauer, S.M., O'Malley, T.M., Perez, D.M., et al., 2015. Preliminary single agent activity of IMGN853, a folate receptor alpha (FRα)-targeting antibody-drug conjugate (ADC), in platinum-resistant epithelial ovarian cancer (EOC) Patients (pts): Phase I Trial. J. Clin. Oncol. 33 (suppl; abstr 5518).

Mosley, C.A., Liotta, D.C., Snyder, J.P., 2007. Highly active anticancer curcumin analogues. Adv. Exp. Med. Biol. 595, 77–103.

Neuman, D., Ostrowski, A.D., Mikhailovsky, A.A., Absalonson, R.O., Strouse, G.F., Ford, P.C., 2008. Quantum dot fluorescence quenching pathways with Cr(III) complexes. photosensitized NO production from trans-Cr(cyclam)(ONO)2 + . J. Am. Chem. Soc. 130, 168–175.

Northfelt, D.W., Dezube, B.J., Thommes, J.A., Miller, B.J., Fischl, M.A., Friedman-Kien, A., et al., 1998. Pegylated-liposomal doxorubicin versus doxorubicin, bleomycin, and vincristine in the treatment of AIDS-related Kaposi's sarcoma: results of a randomised phase III clinical trial. J. Clin. Oncol. 16, 2445–2451.

O'Brien, M., 2004. Reduced cardiotoxicity and comparable efficacy in a phase III trial of pegylated liposomal doxorubicin HCl (CAELYXTM/Doxil") versus conventional doxorubicin for first-line treatment of metastatic breast cancer. Ann. Oncol. 15, 440–449.

Oliveira-Cunha, M., Newman, W., Siriwardena, A., 2011. Epidermal growth factor receptor in pancreatic cancer. Cancers 3, 1513–1526.

Ozpolat, B., Sood, A., Lopez-Berestein, G., 2014. Liposomal siRNA nanocarriers for cancer therapy. Adv. Drug Deliv. Rev. 66, 110–116.

Patel, K., Kollory, A., Takashima, A., Sarkar, S., Faller, D., Ghosh, S., 2014. MicroRNA let-7 downregulates STAT3 phosphorylation in pancreatic cancer cells by increasing SOCS3 expression. Cancer Lett. 347, 54–64.

Pérez-Herrero, E., Fernández-Medarde, A., 2015. Advanced targeted therapies in cancer: drug nanocarriers, the future of chemotherapy. Eur. J. Pharm. Biopharm. 93, 52–79.

Piao, L., Zhang, M., Datta, J., Xie, X., Su, T., Li, H., et al., 2012. Lipid-based nanoparticle delivery of Pre-miR-107 inhibits the tumourigenicity of head and neck squamous cell carcinoma. Mol. Ther. 20, 1261–1269.

Pribble, P., Edelman, M.J., 2012. EC145: a novel targeted agent for adenocarcinoma of the lung. Expert Opin. Invest. Drugs. 21, 755–761.

Qian, Z., 2002. Targeted drug delivery via the transferrin receptor-mediated endocytosis pathway. Pharmacol. Rev. 54, 561–587.

Quinn, B., Wang, S., Barile, E., Das, S., Emdad, L., Sarkar, D., et al., 2016. Therapy of pancreatic cancer via an EphA2 receptor-targeted delivery of gemcitabine. Oncotarget 7, 17103–17110.

Ramot, Y., Rotkopf, S., Gabai, R., Zorde Khvalevsky, E., Muravnik, S., Marzoli, G., et al., 2016. Preclinical safety evaluation in rats of a polymeric matrix containing an siRNA drug used as a local and prolonged delivery system for pancreatic cancer therapy. Toxicol. Pathol. 44, 856–865.

Rapoport, N., 1999. Stabilization and activation of pluronic micelles for tumor-targeted drug delivery. Colloids Surf., B: Biointerfaces. 16, 93–111.

Saif, M.W., Podoltsev, N.A., Rubin, M.S., Figueroa, J.A., Lee, M.Y., Kwon, J., et al., 2010. Phase II clinical trial of paclitaxel loaded polymeric micelle in patients with advanced pancreatic cancer. Cancer Invest. 28, 186–194.

Sarkar, S., Dubaybo, H., Ali, S., Goncalves, P., Kollepara, S.L., Sethi, S., et al., 2013. Down-regulation of miR-221 inhibits proliferation of pancreatic cancer cells through up-regulation of PTEN, p27kip1, p57kip2, and PUMA. Am. J. Cancer Res. 3, 465–477.

Setua, S., Khan, S., Yallapu, M., Behrman, S., Sikander, M., Khan, S., et al., 2016. Restitution of tumour suppressor microrna-145 using magnetic nanoformulation for pancreatic cancer therapy. J. Gastrointest. Surg. 21, 94–105.

Shen, M., Xu, Y.Y., Sun, Y., Han, B.S., Duan, Y.R., 2015. Preparation of a thermosensitive gel composed of a mPEG-PLGA-PLL-cRGD nanodrug delivery system for pancreatic tumour therapy. CS Appl. Mater. Interfaces 7, 20530–20537.

Shin, H.C., Cho, H., Lai, T.C., Kozak, K.R., Kolesar, J.M., Kwon, G.S., 2012. Pharmacokinetic study of 3-in-1 poly(ethylene glycol)-block-poly(D,L-lactic acid) micelles carrying paclitaxel, 17-allylamino-17-demethoxygeldanamycin, and rapamycin. J. Controlled Release 163, 93–99.

Sicard, F., Gayral, M., Lulka, H., Buscail, L., Cordelier, P., 2013. Targeting miR-21 for the therapy of pancreatic cancer. Mol. Ther. 21, 986–994.

Singh, A., Xu, J., Mattheolabakis, G., Amiji, M., 2016. EGFR-targeted gelatin nanoparticles for systemic administration of gemcitabine in an orthotopic pancreatic cancer model. Nanomedicine 12, 589–600.

Spadavecchia, J., Movia, D., Moore, C., Manus Maguire, C., Moustaoui, H., Casale, S., et al., 2016. Targeted polyethylene glycol gold nanoparticles for the treatment of pancreatic cancer: from synthesis to proof-of-concept in vitro studies. Int. J. Nanomed. 11, 791–822.

Srivastava, S., Arora, S., Singh, S., Bhardwaj, A., Averett, C., Singh, A., 2014. MicroRNAs in pancreatic malignancy: progress and promises. Cancer Lett. 347, 167–174.

Stathopoulos, G.P., 2010. Liposomal cisplatin: a new cisplatin formulation. Anticancer Drugs. 21, 732–736.

Stathopoulos, G.P., Boulikas, T., Kourvetaris, A., Stathopoulos, J., 2006. Liposomal oxaliplatin in the treatment of advanced cancer: a phase I study. Anticancer Res. 26, 1489–1493.

Taheri, A., Dinarvand, R., Atyabi, F., Nouri, F., Ahadi, F., Ghahremani, M.H., et al., 2011a. Targeted delivery of methotrexate to tumour cells using biotin functionalized methotrexate-human serum albumin conjugated nanoparticles. J. Biomed. Nanotechnol. 7, 743–753.

Taheri, A., Dinarvand, R., Nouri, F.S., Khorramizadeh, M.R., Borougeni, A.T., Mansoori, P., et al., 2011b. Use of biotin targeted methotrexate-human serum albumin conjugated nanoparticles to enhance methotrexate antitumour efficacy. Int. J. Nanomed. 6, 1863–1874.

Takei, Y., Kadomatsu, K., Yuzawa, Y., Matsuo, S., Muramatsu, T., 2004. A small interfering RNA targeting vascular endothelial growth factor as cancer therapeutics. Cancer Res. 64, 3365–3370.

Tandon, M., Vemula, S.V., Mittal, S.K., 2011. Emerging strategies for EphA2 receptor targeting for cancer therapeutics. Expert Opin. Ther. Targets 15, 31–51.

Teng, L., Xie, J., Teng, L., Lee, R.J., 2012. Clinical translation of folate receptor-targeted therapeutics. Expert Opin. Drug Deliv. 9, 901–908.

Thomas, A., Maltzman, J., Hassan, R., 2013. Farletuzumab in lung cancer. Lung Cancer 80, 15–18.

Torchilin, V., 2010. Passive and active drug targeting: drug delivery to tumours as an example. Handb. Exp. Pharmacol. 197, 3–53.

Vaccaro, V., Sperduti, I., Milella, M., 2011. FOLFIRINOX versus gemcitabine for metastatic pancreatic cancer. N. Engl. J. Med. 365, 768–769.

van de Water, F.M., Boerman, O.C., Wouterse, A.C., Peters, J.G., Russel, F.G., Masereeuw, R., 2006. Intravenously administered short interfering RNA accumulates in the kidney and selectively suppresses gene function in renal proximal tubules. Drug Metab. Dispos. 34, 1393–1397.

Von Hoff, D.D., Ramanathan, R.K., Borad, M.J., Laheru, D.A., Smith, L.S., Wood, T.E., et al., 2011. Gemcitabine plus nab-paclitaxel is an active regimen in patients with advanced pancreatic cancer: a phase I/II trial. J. Clin. Oncol. 29, 4548–4554.

Von Hoff, D.D., Ervin, T., Arena, F.P., Chiorean, E.G., Infante, J., Moore, M., et al., 2013. Increased survival in pancreatic cancer with nab-paclitaxel plus gemcitabine. N. Engl. J. Med. 369, 1691–1703.

Wheate, N.J., Walker, S., Craig, G.E., Oun, R., 2010. The status of platinum anticancer drugs in the clinic and in clinical trials. Dalton Trans. 39, 8113–8127.

Wu, Y., Crawford, M., Mao, Y., Lee, R., Davis, I., Elton, T., et al., 2013. Therapeutic delivery of microRNA-29b by cationic lipoplexes for lung cancer. Mol. Ther. Nucleic Acids. 2, e84.

Xu, C., Wang, J., 2015. Delivery systems for siRNA drug development in cancer therapy. Asian J. Pharm. Sci. 10, 1–12.

Xu, S., Xu, Q., Zhou, J., Wang, J., Zhang, N., Zhang, L., 2013. Preparation and characterization of folatee-chitosan-gemcitabine core–shell nanoparticles for potential tumour-targeted drug delivery. J. Nanosci. Nanotechnol. 13, 129–138.

Xu, Y., Meng, H., Du, F., Lu, W., Liu, S., Huang, J., et al., 2015. Preparation of intravenous injection nanoformulation of VESylated gemcitabine by co-assembly with TPGS and its anti-tumour activity in pancreatic tumour-bearing mice. Int. J. Pharm. 495, 792–797.

Yang, F., Jin, C., Jiang, Y., Li, J., Di, Y., Ni, Q., et al., 2011. Liposome based delivery systems in pancreatic cancer treatment: from bench to bedside. Cancer Treat. Rev. 37, 633–642.

Yang, Y., Chien, Y., Chiou, G., Cherng, J., Wang, M., Lo, W., et al., 2012. Inhibition of cancer stem cell-like properties and reduced chemoradioresistance of glioblastoma using microRNA145 with cationic polyurethane-short branch PEI. Biomaterials 33, 1462–1476.

Youns, M., Hoheisel, D.J., Efferth, T., 2011. Therapeutic and diagnostic applications of nanoparticles. Curr. Drug Targets 12, 357–365.

Yu, X., Song, Y., Di, Y., He, H., Fu, D., Jin, C., 2016. Enhanced tumour targeting of cRGD peptide-conjugated albumin nanoparticles in the BxPC-3 cell line. Sci. Rep. 6, 31539.

Zamboni, W.C., Eiseman, J.L., Strychor, S., Rice, P.M., Joseph, E., Zamboni, B.A., et al., 2011. Tumour disposition of pegylated liposomal CKD-602 and the reticuloendothelial system in preclinical tumour models. J. Liposome Res. 21, 70–80.

Zhang, B., Jiang, T., Shen, S., She, X., Tuo, Y., Hu, Y., et al., 2016. Cyclopamine disrupts tumour extracellular matrix and improves the distribution and efficacy of nanotherapeutics in pancreatic cancer. Biomaterials 103, 12–21.

Zhao, J., Wu, C., Abbruzzese, J., Hwang, R.F., Li, C., 2015. Cyclopamine-loaded core-crosslinked polymeric micelles enhance radiation response in pancreatic cancer and pancreatic stellate cells. Mol. Pharm. 12, 2093–2100.

Zuckerman, J., Davis, M., 2015. Clinical experiences with systemically administered siRNA-based therapeutics in cancer. Nat. Rev. Drug Discov. 14, 843–856.

CHAPTER

13 Targeted drug delivery and gene therapy through natural biodegradable nanostructures in pharmaceuticals

V.N. Pandey[1], Neha Tiwari[1], V.S. Pandey[1], Anita Rao[1] and Ishwar Das[2]

[1]Experimental Botany and Nutraceutical Laboratory, Department of Botany, DDU Gorakhpur University, Gorakhpur, India [2]Department of Chemistry, DDU Gorakhpur University, Gorakhpur, India

13.1 INTRODUCTION

Over the past three decades, nanoparticles (NPs) have been increasingly used for a variety of applications, preferably for delivery of therapeutic and diagnostic medicines (Lohcharoenkal et al., 2014). Drug delivery may be defined as the process of administration of pharmaceutical compounds in order to achieve a therapeutic effect in humans or animals (Tiwari et al., 2012). The nanostructured materials with sizes in the 1−100 nm range are capable of possessing a high degree of interaction with the cell membrane and proteins. The natural nanostructures include most of the biologically active molecules and agents (Mohanraj and Chem, 2006). There are several well-recognized drugs with high therapeutic efficacy, but provide insufficient therapeutic indices due to insufficient delivery. By increasing the specificity due to targeting of drugs to a particular tissue, intracellular compartment or cell, controlling over release kinetics and protecting the therapeutic agents the therapeutic index of traditional and novel drugs could be improved (Swain et al., 2005). Currently, NPs have promising applications in delivering a range of molecules to desired sites in the body. The physical penetration of lipid bilayer membrane through these NPs opens the possibility of organelle-specific targeting (Zhang et al., 2015). Natural biomolecules like proteins and polysaccharides are suitable and better alternatives to synthetic polymers due to their safety, biocompatibility, and biodegradability (Salatin et al., 2015; Elsabahy and Wooley, 2012). The incorporation of drugs into polymeric material facilitates the predefined release of drugs in a reproducible manner for prolonged duration. The drug delivery through biodegradable polymers are gaining much interest due to their stability against forces generated during aerosolization,

Nanoarchitectonics in Biomedicine. DOI: https://doi.org/10.1016/B978-0-12-816200-2.00012-8

predetermined release of drugs along with its degradation within an acceptable period of time. The control of particle size, surface properties, and release of pharmacologically active agents in order to fulfill the site-specific action at the therapeutically optimal rate and dose regimen are the major aims of designing NPs (Salauti and Anangari, 2014). One of the advantages of the use of NPs in drug delivery systems is that their particle morphology, size, and surface charge can easily be controlled. By using modified cell degradation properties, the properties of particles can be altered. The physical properties viz., zeta potential, size, and drug release profile of NPs can be turned by selecting different polymer lengths and organic solvents during synthesis. The surface modification of polymeric NPs can be done either by drug moieties or target ligand (Zhang et al., 2010). Due to better drug loading capacity, biocompatibility and possible less opsonization by Reticuloendothelial System through aqueous steric barrier; naturally derived hydrophilic polymers are considered to be more efficient (Liu et al., 2008). Interestingly, they have the abilities of gelation and emulsification as well as water binding capacity (Elzoghby et al., 2011, 2012).

13.2 METHODS FOR PREPARATION OF NANOPARTICLES

There are two main methods used for the preparation of NPs: two-step procedure and one-step procedure. In the two-step procedure, the first step involves the generation of nanodroplets of definite sizes and organic compounds (polymer, lipid, monomer) that are solubilized and in the second step, the precipitation, gelation or polymerization of materials is carried out. The one-step procedure does not require the emulsification of NPs, as can be seen in Fig. 13.1 and Table 13.1.

13.3 FACTORS INFLUENCING TARGETED DRUG DELIVERY THROUGH NANOPARTICLES

13.3.1 PARTICLE SIZE AND SHAPE

The drug loading, drug release, stability of NPs, biological fate, toxicity, and in vivo distribution as well as the targeting ability of the NPs is influenced by the particle size and size distribution (Singh and Lillard, 2009). The majority of the cells uptake submicron NPs (Zauner et al., 2001). Due to larger surface to volume ratio, the smaller-sized NPs have a faster drug release rate and in the case of larger NPs, more of the drug is entrapped per particle, hence, providing slower release. Thus, the drug release rate can be controlled by adjusting the size of the NPs. Greater membrane wrapping time is required for rod shaped and elongated NPs. So, spherical-shaped NPs of similar size are taken up and internalized in the cells more easily than rod-shaped NPs of similar sizes. There is a strong impact

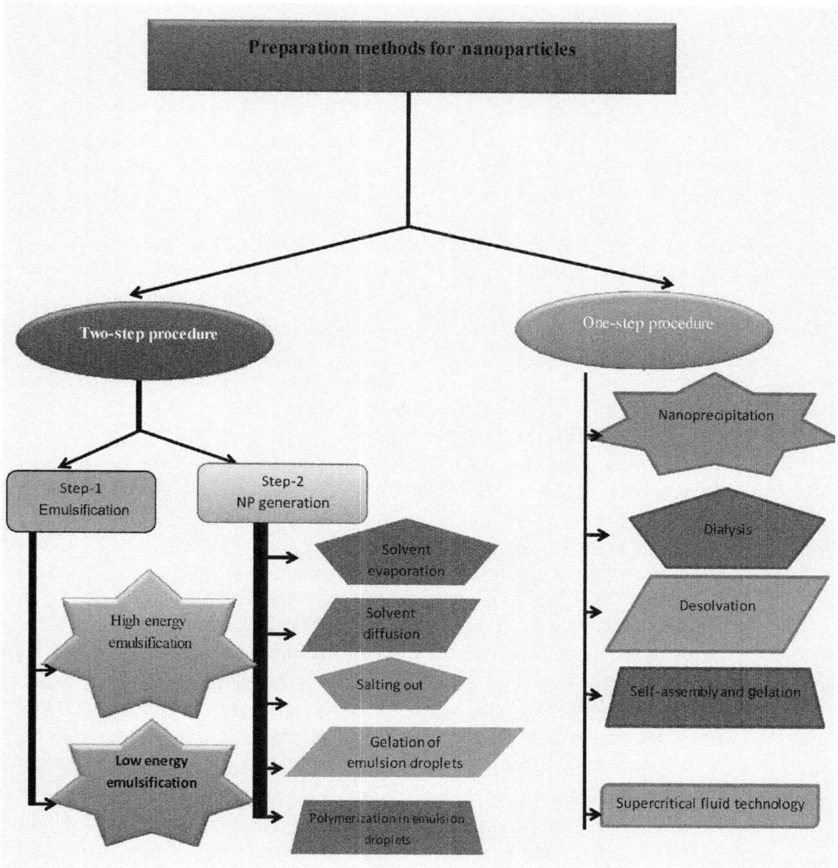

FIGURE 13.1

Various methods employed for nanoparticle preparation.

of NP shape and size in binding, activation of membrane receptors, and protein expression (Jiang et al., 2008). NPs of 50 nm are more efficient in uptake (Chitrani et al., 2006; Chitrani and Chan, 2007).

13.3.2 SURFACE PROPERTIES AND SURFACE MODIFICATION

The role of surface charge in the internalization of the gold NPs was investigated by Cho et al. The gold core of the NPs was selectively dissolved when I_2/KI etchant is used (Cho et al., 2009). The negatively charged groups present on the cell surface such as sialic acid can bind easily with the cationic particles. This binding facilitates the easy penetration and cellular internalization (Alberts et al., 2002). In addition to this, the cell-fusogenic or cell-penetrating biological ligands play an important role in assisting the nanomaterial to reach the cytosol or nucleus of the cells.

Table 13.1 Some Methods Used for Nanoparticle Preparation

S. No.	Preparation Method	Description	Reference
1.	Solvent evaporation	This method is most commonly used in nanoparticle preparation. This process requires an aqueous phase for the emulsification of the polymer solution followed by evaporation of the polymer solvent resulting in polymer precipitation as nanospheres. After ultracentrifugation and washing with distilled water the nanoparticles are collected	Song et al. (1997)
2.	Solvent diffusion	This method is a modified form of the solvent evaporation method and is also called the solvent displacement method. The emulsion preparation consists of two immiscible phases, the continuous phase consisting of oil saturated with water, and an oil-dispersed one composed of water saturated with the polymer. The solvent saturated with water and water saturated with the solvent at the bottom and the top of the resulting system can be collected	Moinard-Checot et al. (2006, 2008)
3.	Salting out	In this method, the polymer is emulsified with a solvent-like acetone that is miscible with water. The strong salting-out effect is provoked which modifies the solubility of water with the solvent when the high content of salt or sucrose is dissolved in the aqueous phase and emulsified with the help of the artifice used	Allouche (2013)
4.	Nanoprecipitation	In this method, the polar, water-miscible solvents (acetone, ethanol, methanol, acetonitrile) are used to dissolve polymers and drugs. This solution is then added drop-by-drop into the aqueous solution with the surfactant. resulting in the formation of nanoparticles. The solvent is then removed from the solution	Govender et al. (1999)
5.	Desolvation	This method involves the slow addition of desolvation factors like salts, alcohols, or other solvents for the precipitation of the polymers and is generally employed for the production of protein nanoparticles, often followed by a crosslinking step to stabilize the nanoparticles	Marty et al. (1978), Kommareddy and Amiji (2007)
6.	Dialysis	In this method, the polymer is initially dissolved in an organic solvent and then placed inside a dialysis tube with the proper molecular weight cut-off. Finally, the displacement of the solvent inside the membrane occurs followed by progressive aggregation of polymers due to solubility loss and formation of the homogenous suspension of nanoparticles	Nagavarma et al. (2012)
7.	Self-assembly and gelation	This involves polyelectrolytes complexation, "Lock and Key" nanogels and ion gelation methods. In polyelectrolyte complexation, the association of oppositely charged macromolecules occurs when mixed in specific conditions, resulting in the formation of organic nanoparticles. In the case of the "Lock and Key" concept, the nanospheres are formed due to the neutral association of macromolecules of certain types. The ion gelation method is used with biopolymers (especially charged polysaccharides) in an aqueous medium	Allouche (2013)
8.	Supercritical fluid technology	This is considered to be environmentally benign and is a safer method for nanoparticle preparation. It includes two principles: rapid expansion of supercritical fluid and rapid expansion of supercritical solution into a liquid solution. In first method, a solution is prepared by dissolving the polymer in a supercritical fluid followed by rapid expansion of the solution across an orifice or a capillary nozzle into ambient air. The second method involves the expansion of the supercritical solution into a liquid solvent instead of ambient air	Nagavarma et al. (2012)

The surface nonmodified NPs are rapidly opsonized and cleared by Mononuclear Phagocyte System when entering the blood stream. The surface modification of NPs plays an important role in successful drug targeting by minimizing opsonization and prolonging the in vivo circulation of NPs. The surface modification of NPs is achieved by using hydrophilic polymers/surfactant NP coating or by formulating the NPs with different types of biodegradable copolymers with hydrophilic characteristics like polyethylene glycol (PEG), poloxamine, and polyethylene oxide, etc. Choi et al. (2011) observed that the tumor-targeting ability of hyaluronic acid (HA) NPs is improved by PEGylation and the accumulation of PEGylated NPs in tumor tissue is 1.6-fold higher than bare HA NPs. Similarly, the cellular uptake of Coumarin-6- and docetaxel-encapsulated poly(D, L-lactide-*co*-glycolide) NPs was also enhanced when the surface of the NPs were modified with the surfactants such as Polysorbate-80 (Tween80), Poloxamer 188 (F68), and Poloxamer 407 (F127) (Kulkarni and Feng, 2011).

13.3.3 DRUG LOADING AND DRUG RELEASE

The negatively charged lipid bilayer of the cell membrane has higher affinity for the positively charged NPs or, in other words, the positively charged NPs have higher efficiency for cell membranes with negatively charged lipid bilayers. Therefore, they can easily get adsorbed to the surface. More anticancer and small molecules are loaded by amphiphilic polymeric NPs containing a higher degree of substitution in hydrophobic groups. The drug release rate through the NPs depends on the solubility of the drug, drug diffusion through NPs matrix, degradation or erosion of the NP matrix, desorption of the surface-bound or adsorbed drug, and the combined effect of the diffusion and erosion processes. The drug release rate mainly depends on the diffusion (Niwa et al., 1993). In case of uniformly distributed/dissolved drug i.e., in a matrix device, if the matrix degradation is slower than drug diffusion, then the drug release occurs by the process of diffusion. The nanocarrier system plays an important role in increasing drug solubility and, thereby, enhances the release rate. A much faster dissolution rate is achieved by Simvastatin-loaded SHMC (highly ordered mesoporous carbon) samples synthesized by the nanocasting technique with shorter t_{max} and higher C_{max} (Zhang et al., 2013a).

13.4 SOME NATURAL NANOSTRUCTURES IN DRUG DELIVERY

13.4.1 PROTEINS AS A NANOCARRIER IN DRUG DELIVERY

Proteins are the class of natural molecules that have unique functionalities and potential applications in both biological as well as material fields (Jahanshahi and Babaei, 2008). Protein-derived nanomaterials are nonantigenic, amphiphilic, metabolizable, inexpensive, biocompatible, biodegradable, nontoxic, and can be easily amenable for surface modification and covalent attachment of drugs and ligands. These protein-based NPs have profound applications as potential macromolecular carriers for pulmonary delivery of protein therapeutics. Recent advancements in transmembrane drug delivery of proteins and peptide drugs by Trans-acting activator of Transcription (TAT)-mediated transduction have opened new horizons for the treatment of cancer with high target specificity and reduced side effects and toxicity. These groups of proteins possess cell membrane penetrating properties with enhanced ability to cross plasma membrane in a receptor-independent fashion. Moreover, it eliminates the difficulties of gene therapy relying on viral (adenoviral) and nonviral vectors related to the lack of target specificity and toxicity problems (Wadia and Dowdy, 2005). Liu et al. reported the novel biocompatible, biodegradable protein-polymer conjugate preparation via maleimide-sulfhudryl coupling reaction by covalently linking the tailor-made hydrophobic maleimide-functionalized poly(ε-caprolactone) to hydrophilic bovine serum albumin (BSA). These negatively charged, self-assembled vesicles further enhance the cellular uptake via linking targeting ligand cetuximab (Liu et al., 2014).

13.4.2 ALBUMINS

Albumins are the family of globular proteins that are soluble in water and diluted salt solutions. These are commonly found in milk, egg white, and blood serum. It regulates the colloidal osmotic pressure of blood having an isoelectric point of 4.9 (Albert, 2010). It also works as a carrier for the molecules of low water solubility such as bile salts, apoproteins, and lipid-soluble hormones along with some drugs like phenytoin, clofibrate, warfarin, and phenobutazone. The small-sized (50–300 nm) albumin NPs prepared by controlled desolvation, coacervation, or emulsion formation methods have better patient acceptance, compliance, and controlled release properties than liposomes (Arshady, 1989). Albumin-based nanocarriers are one of the most promising strategies for the delivery of Doxorubicin (DOX), an antitumor drug. An increase in the maximum tolerated dose of DOX-loaded NPs (NP-DOX) was observed from 10 to over 30 mg/kg when compared with free DOX (Yuan et al., 2013). Elblbesy demonstrated the hemocompatibility of bovine albumin NPs (BANs). With the increase in BAN concentration, the prothrombin time and erythrocyte sedimentation rate decreased. The study concluded that the BANs have biomedical applications and could be used safely without any abnormal effect (Elblbesy, 2016). A significant inhibition and regression of the tumor (breast cancer cell line SK-BR-3) volume was noticed when noscapine loaded on glutaraldehye crosslinked human serum albumin (HSA) NPs and delivered to tumor cells without any detectable toxicity in cell or tissue (Sebak et al., 2010). The HSA NPs are safe and feasible in functionalization. Watcharin et al. investigated the application of gadolinium diethylene triaminepenta acetic acid (Gd-DTPA) conjugated HSA NPs as a T1 MRI of hepatocellular carcinoma.

These NPs are stable in vivo and readily taken up by Huh-7 hepatocellular carcinoma cells and less than 5% of the Gd(III) was released from the particles after 24 hours incubation in blood serum (Watcharin et al., 2014). The amino-oxyacetic acid-aclacinomycin (ACM) conjugate with self-assembled HSA NPs constructed using tris(2-carboxyethyl) phosphine showed drug loading efficacy of 7.4% having fourfold higher tumor targeting capability on S180-tumor bearing mice in comparison to free ACM (Gong et al., 2016) (Table 13.2).

Ovalbumin is a glycoprotein found in egg white consisting 60%–65% of the total egg protein having a molecular weight of 45,000 Da. Chemically, it is a polypeptide chain with two phosphate groups per mole and a side chain of mannose and glucosamine residues, as shown in Fig. 13.2. It is widely used in drug delivery for controlled drug release due to its temperature and pH-sensitive properties (Oakenfull et al., 1997). Hariyadi et al. prepared unlyophilized and lyophilized ovalbumin-loaded alginate microspheres using the aerosolization technique with $BaCl_2$ and $CaCl_2$ as the crosslinking agent. The entrapment efficiency of lyophilized ovalbumin-loaded alginate microspheres using 10% lactose lyoprotectant and maltodextrin were 46% and 54% respectively while unlyophilized ovalbumin-loaded alginate microspheres using crosslinkers $BaCl_2$ and $CaCl_2$ produced high entrapment efficiency of 80% and loading of above 65%. These ovalbumin-loaded alginate microspheres possess high stability and are able to deliver antigen to the target sites (Hariyadi et al., 2015).

13.4.3 GELATIN

Gelatin is a protein present in collagen obtained from animal raw materials (i.e., skin, bones, and connective tissues) as shown in Fig. 13.3B. It is colorless, translucent, flavorless, odorless, temperature-sensitive protein that is soluble in hot water, acetic acid, and glycerol. It forms a gel within the pH range of 4–8 at a minimum concentration of 0.5%. Studies have shown that gelatin can be used as a vehicle for drug and peptide delivery due to its biocompatibility (Silvio et al., 1994). The gelatin NPs are employed as size-controllable porogen and used as a surface coating agent. It forms composite scaffolds when mixed with synthetic or natural biopolymers (Su and Wang, 2015). A solvent-free emulsion polymerization method was used to prepare spherical nanogels using gelatin *N,N*-ethylenebisacrylamide and sodium methacrylate. Sunflower oil with varying amounts of lecithin is used for the continuous phase. The toxicity of these nanogels was tested and the results revealed that these NPs are biocompatible, highly stable, and nontoxic to human bone marrow mesenchymal stromal cells with enzymatic stability in the gastric tract. These are the most versatile natural biopolymers widely used in the drug delivery system (Curcio et al., 2013). Gelatin is more fruitful over synthetic polymers due to the presence of amino acid sequences such as Asg-Gly-Asp that improve its final biological performance and also help in cell recognition (Wang et al., 2012). The antitumor and antimetastatic effects of synthesized gelatin-DOX and PEGylated gelatin-DOX (PGD) NPs of 135 and

Table 13.2 Some Natural Nanoparticle (NP) Systems and their Application in Drug Delivery

NP System	Drug Loaded	Morphology	Preparation Method	Techniques Used	Aims	Applications	Reference
Gelatin NP	Terbutaline sulfate	<500 nm, spherical	Double Desolvation Method	PCS, Zeta-Potential, TEM	In vitro and in vivo release study of synthesized NPs to make fairly stable NPs with high entrapment efficiency	Pulmonary drug delivery	Bharti et al. (2014)
Alginate-gelatin nanohybrid	Doxorubicin	400–600 nm	Single O/W emulsification method	SEM, FTIR, potential metastasize analyzer	To develop novel NP hybrid and to study its drug release profile	Tissue engineering and regenerative medicine	Lee et al. (2014)
Human serum albumin	Noscapine	150–300 nm	pH-coacervation method	SEM, particle size analyzer, Zeta potential	To enhance drug bioavailability	Breast cancer treatment	Sebak et al. (2010)
Zein NP	5-FU	800–900 nm, spherical	Liquid-liquid dispersion method	SEM, TEM, FTIR, TGA	To improve biocompatibility and cytotoxicity of drug-loaded NPs	In tumor targeting, bioimaging, drug delivery	Aswathy et al. (2012)
Zein-pectin NP	Quercetin	56 ± 178 nm, spherical	Ultrasonication method	UV–vis absorption Spectra, FTIR	To develop biodegradable, nontoxic NPs	As a carrier in drug targeting	Dhanya et al. (2012)
Zein-pectin NPs	DHC	256 ± 30 nm, spherical	pH-controlled nanoprecipitation method	Particle size analyzer, Zeta potential, AFM, DSC	To investigate influence of formulation parameter on NPs preparation	NPs used for drug encapsulation	Podaralla and Perumal (2012)
Chitosan-pectin NP	Curcumin	± 40 nm, spherical	PBS buffer method	TEM, SEM, XRD, FTIR	To develop NPs for curcumin delivery and to study its in vitro and in vivo release kinetics	In delivery of curcumin (a potential drug to cure many diseases)	Andriani et al., (2015)
Eudragit S100 coated Citrus-pectin NP	5-FU	218.12 ± 10.25 nm, spherical	Solvent evaporation method	PCS, zeta potential, TEM, SEM	To develop NPs for delivery of 5-FU for treatment of colorectal cancer	Colon targeting	Subudhi et al. (2015)
Pectin-Bora rice nanopheres	Glipizide	Spherical, rough and porous	Ionotropic gelation method	FTIR, XRD, UV–vis spectroscopy, DSC	To develop NP for controlled release of drug	Colon targeted drug delivery	Ramteke and Nath (2014)

PCS, *Photon correlation spectroscopy;* O/W, *oil in water;* FTIR, *Fourier transform infrared spectroscopy;* 5-FU, *5-fluorouracil;* DHC, *6,7-dihydrocoumarin;* PBS, *phosphate-*

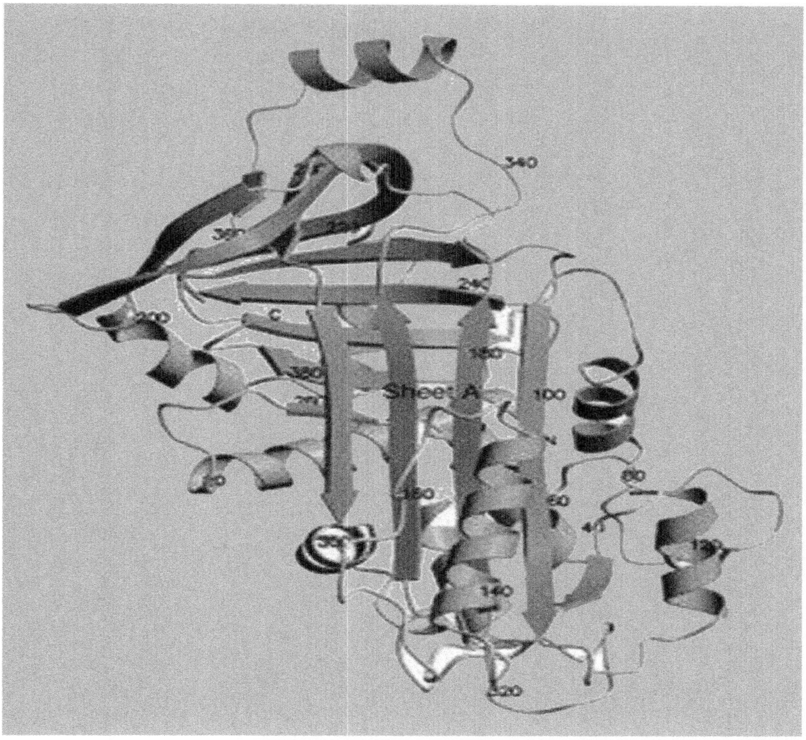

FIGURE 13.2

Crystal structure of ovalbumin (Roth, 2012).

250 nm size was compared by Lee et al. DOX is a drug used in cancer chemotherapy. The results showed that PGD significantly inhibited tumor growth and is a potential cancer therapy (Lee et al., 2006). Gelatin NPs of mean diameter 200–300 nm with an entrapment efficiency of 87.4 were prepared using two-step desolvation method used for the delivery of BSA. The release of protein drugs and, consequently, the prolonged bioavailability of the drugs can be sustained using these protein-containing NPs (Azimi et al., 2014). The intranasal administration of osteopontin through gelatin microspheres showed neuroprotective effects in the postischemic brain of rats. The gelatin nanocarriers provide a new platform for the intranasal treatment of ischemic stroke and other neurologic disorders (Joachim et al., 2014). Khan et al. demonstrated the drug delivery of cytarabine through genipin-modified gelatin nanocarriers. The influence of various factors such as pH and simulated physiological fluids as a release medium on the release profile of cytarabine was also studied. These NPs have effective biocompatible properties with desired drug release profiles and its chemical structure influences its releasing capacity (Khan et al., 2016). The chemical modification of gelatin-based hydrogels for improving drug release rate using gluteraldeyde

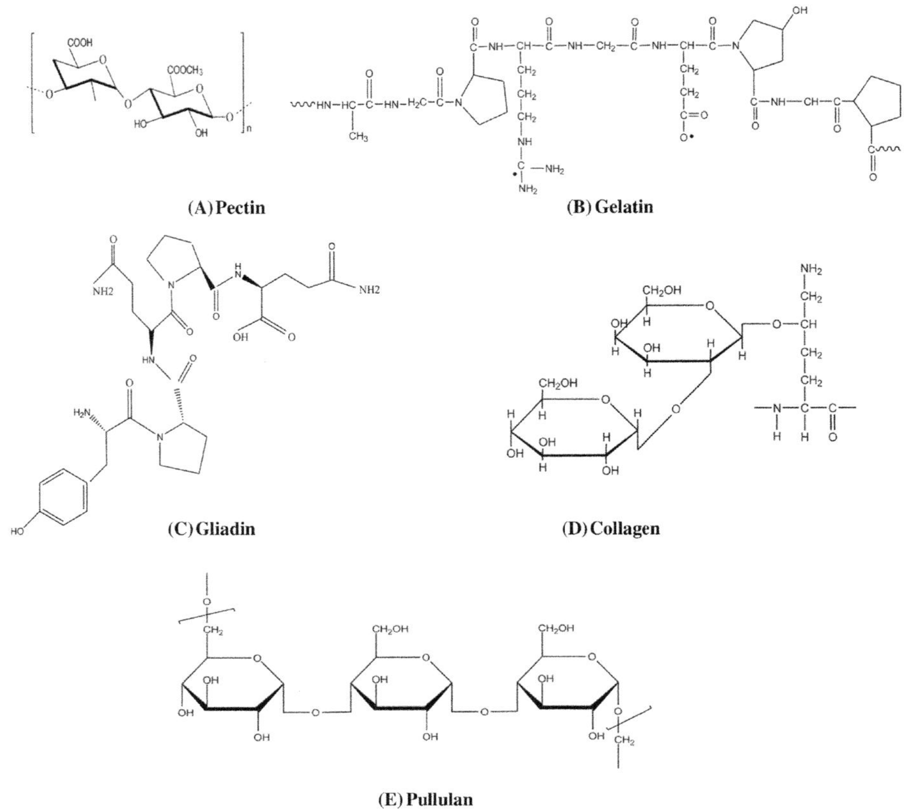

FIGURE 13.3

Chemical structure of different nanostructures involved in drug delivery: (A) Pectin, (B) gelatin, (C) gliadin, (D) collagen, (E) pullulan, (F) chitosan, (G) alginate, (H) starch, (I) xanthan gum, and (J) hyaluronic acid.

(Ciobanu et al., 2014) and genipin (Wu et al., 2013) have also been reported. The genipin-crosslinked gelatin NPs of 150 nm were prepared by the emulsion cross-linking method to load ciplastin. The crosslinking of the gelatin with genipin was confirmed by Fourier Transform Infrared Spectroscopy (FTIR) with greater drug releasing ability at acidic and alkaline pH (Khan et al., 2015). 3-Carboxyphenylboronic acid modified, phenylboronic acid decorated gelatin NPs were prepared for tumor-targeted drug delivery in SH-SY5Y cells, H22 cells and HepG2 cells were used to examine in vitro cellular uptake and cytotoxicity of DOX hydrochloride loaded NPs and the NPs showed improved tumor accumulation and antitumor effect (Wang et al., 2016). Suarasan et al. (2016) examined the anticancerous activity against MCF-7 breast cancer cells of DOX loaded gelatin-coated gold NPs at different pH values (7.4, 5.3, and 4.6) and temperature (22°C–45°C). The electrostatic bond between alginate and gelatin is responsible

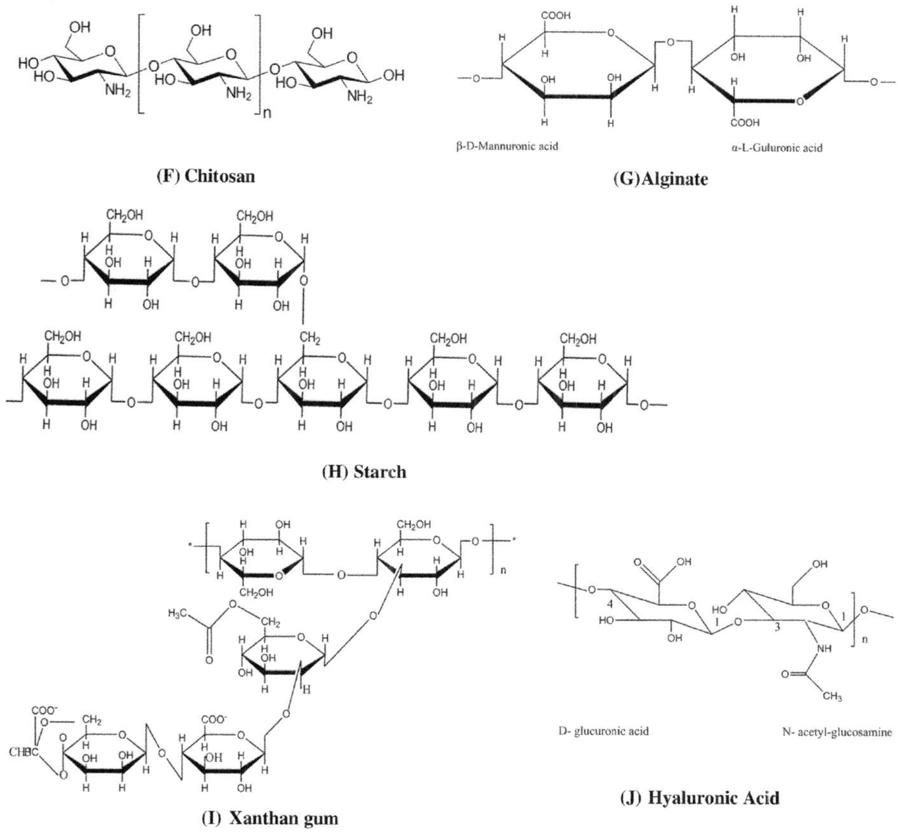

FIGURE 13.3

(Continued)

for the controlled release of the drug from alginate-gelatin nanohybrids, therefore, showing potential to deliver drug and therapeutic agents (Lee et al., 2014).

13.4.4 GLIADIN

Gliadin, as shown in Fig. 13.3C, is glutenin insoluble but water soluble proteins present in cereals and are suitable for preparation of mucoadhesive NPs. The carbazole-loaded gliadin NPs showed bioadhesive and absorptive properties in the upper gastrointestinal region in both nonhardened and crosslinked forms (Arangoa et al., 2001). The Gliadin-based NPs hold versatility and have properties suitable for the delivery of bioactive molecules for pharmaceutical applications. Gliadin contains neutral and lipophilic amino acid residues. The neutral amino acid present in this protein can easily interact with mucosa through hydrogen bonding while lipophilc residues promote hydrophobic interactions with biological tissues. Gulfam et al. electrosprayed gliadin and gliadin–gelatin composite NPs for the delivery and controlled release of cyclophosphamide, an anticancerous drug. 7% of gliadin-based NPs have an average size of 218.66 ± 5.1 nm and could achieve over 72.02% ± 5.6% drug loading. The cyclophosphamide-loaded 7% gliadin added to the breast cancer cell culture for 24 hours and examined through Western blotting analysis, causes apoptosis (Gulfam et al., 2012). Lectin-conjugated gliadin-bearing NPs bearing acetohydroxamic acid prepared by the desolvation method is used to locate and anchor a drug delivery system on the carbohydrate receptors of *Helicobacter pylori*. The mannose-specific Con A-Gliadin NPs (GNP) and alpha (L)-fucose specific UEA-GNP showed a strong agglutination pattern and the formulations inhibited the binding of *H. pylori* strains with human stomach cells (Umamaheshwari and Jain, 2003). The spherical-shaped gliadin NPs in the size range of 250–500 nm with a positive zeta potential of 22.8 mV containing clarithromycin (an antimicrobial drug) was prepared using pluronic F-68 as a stabilizing agent showed maximum percentage entrapment and yield of about 73% and 88%, respectively, with a sustained release behavior of 37 ± 1° both in phosphate saline (pH 7.4) and simulated gastric fluid (pH 1.2). The residence time of clarithromycin in the stomach was increased when it was administered in the form of mucoadhesive NPs with higher antibacterial activity (Ramteke et al., 2006). Kajal and Mishra (2011) prepared tetanus toxoid and ovalbumin-loaded gliadin NPs and found that approximately 50% w/w of antigen were size stable over 3 weeks of testing.

13.4.5 ZEIN PROTEIN

Small drug molecules can be delivered through oral routes using zein microspheres (Reddy and Yang, 2011). The NPs prepared from yellow zein using casein as stabilizers was reported by Patel et al. (2010), while lecithin and pluronic F68 stabilizers for the preparation of yellow zein NPs by a pH-controlled nanoprecipitation method was also studied (Satheesh and Perumal, 2010). The treatment of parasitic diseases in dogs through zein-based microspheres for oral administration of ivermectin was investigated by Gong et al. (2011). Satheesh and Perumel also examined the influence of pH of aqueous phase, buffer type, ionic strength, and zein concentration and surfactant on particle size, zeta potential, and polydispersity index of DHC-loaded zein-NPs. pH close to isoelectric point results in the formation of smaller NPs. The DHC-loaded zein-NPs constructed using citrate buffer (pH 7.4) is better in inhibiting particle aggregation during lyophilization. The particle size of NPs increases with the increase in zein concentration with an encapsulation efficiency of 78% ± 7% (Satheesh and Perumel, 2012). The adsorption of specific salt ions like Na^+ and Cl^- on DHC-loaded zein-NP surface also alters its zeta potential (Fairhurst and Lee, 2011).

13.4.6 SOY PROTEIN

Soy protein is obtained from soybeans and contains at least 60%–70% of the total soybean protein (Shewry et al., 1995). It possesses sorption, gelation, adhesion, ligand-binding, and film formation properties (Kinsella, 1979). Vaz et al. investigated a soy protein drug delivery matrix system by melt-processing techniques with an encapsulated drug Theophylline (TH). The drug release through these matrix systems was controlled with constant diffusion coefficients and stationary boundary conditions at isotonic saline solution (ISS) buffered at pH 5.0. Polymer dissolution and swelling effected drug release and was found to be faster compared to ISS-200 mM acetate buffer at pH 5.0 (Vaz et al., 2003). A melt extrusion technique was used to prepare soy protein/MMT-Montmorillonite (MMT) nanocomposites having different percentages of MMT and the characterization of the NPs was done by using XRD, SEM, TEM, and TGA techniques. The MMT content was directly proportional to the degree of exfoliation. The controlled and sustained drug release through these nanocomposites was confirmed by blending it with ofloxacin and the release was noticed to be faster at pH 3.4 (Nayak et al., 2011). Genipin-crosslinked soy–whey–based NPs were prepared using the desolvation method and the amine—amine crosslinking reaction to load Atorvastatin (ATR). The encapsulation of the ATR into the NPs was confirmed by X-ray diffractometry spectra whereas the spherical shape of the NPs was revealed by TEM. The ATR-loaded soy protein isolate (SPI)/whey protein concentrate (WPC) showed a significant reduction in cholesterol level in comparison to SPI and WPC individually with high levels of cellular uptake and cytocompatibility (Kanoujia et al., 2016). Deng et al. worked on the bioavailability of β-carotene using SPI. The hydrophobic interaction is mainly responsible for the complexation of the β-carotene on the surface of the SPI NPs and this complexation resulted in interparticle aggregation in a concentration-dependent manner. The result suggested that the SPI exhibits good potential as a carrier to enhance the water dispersibility, stability, and bioaccessibility of β-carotenes (Deng et al., 2016). Jin et al. investigated riboflavin delivery through self-assembled modified soy protein and dextran nanogels. The modified soy protein-dextran nanogels synthesis were characterized and confirmed by FTIR, X-ray photoelectron microscopy and ζ-potential studies. The NPs have core shell structure, 32–40 nm size and spherical in shape, confirmed by TEM. At 250 μg/mL riboflavin concentration, the encapsulation efficiency of the nanogels was found to be up to 65.9% (Jin et al., 2016).

13.4.7 LEGUMIN

Legumin is a storage protein with a complex globular structure consisting of six pairs of subunits with each subunit made up of disulfide-linked acid and basic polypeptide with molecular weights of 40 and 80 kDa, respectively. It belongs to the group of 11S globulins commonly found in pea seeds (*Pisum sativum*) having molecular weight of 300–400 kDa. A pH-coacervation method with glutaraldehyde as crosslinkers was used to prepare 250 nm legumin NPs. Rapid degradation of NPs under acidic pH conditions (pH 5.5, 37°C) was observed while quite stable legumin NPs were obtained at pH close to neutrality. These NPs can be used as nanocarriers to deliver drugs in cases of transdermal drug delivery (Gueguen et al., 1995). The synthesis of glutaraldehyde crosslinked legumin NPs by the pH-coacervation method was studied by Mirshahi et al. These NPs were found to be quite stable in phosphate-buffered saline (PBS). The half-life (t_{50}) of the NPs was increased with an increase in glutaraldehyde concentration (Mirshahi et al., 1996). Mirshahi et al. also carried out studies on adaptive immune response of legumin NPs in rats. A strong expression of antibodies against the legumin protein was observed when sera from rats immunized with legumin. On the other hand, when the serum of rats inoculated with glutaraldehyde crosslinked legumin NPs did not contain antibodies. This result suggests that the reduction on the antigenic epitopes of legumin protein is induced by glutaraldehyde which is used during the crosslinking step (Mirshahi et al., 2002).

13.4.8 SILK

Recently, there have been a number of studies on silk-based NPs due to its capacity of maintaining the functional drugs, tunable degradation, and biocompatibility. Small molecules, biomolecules, and anticancer drugs can be successfully delivered through silk fibroin NPs (Mottaghitalab et al., 2015). Wonqpinyochit et al. (2015) investigated drug delivery through PEGylated silk NPs and its application in treatment of human breast cancer. Xiao et al. directly prepared the silk NPs by the modified dissolution process. The prepared NPs showed improved pH–dependent drug release behavior, homogenous morphology, and amorphous structure as well as improved dispersion in a PBS solution and water with higher cytotoxicity against cancerous cells in comparison to traditionally prepared silk NPs (Xiao et al., 2016). Methotrexate loaded silk fibroin-albumin blended NPs prepared by using the desolvation method showed higher drug-loading efficacy of 15%–24% and entrapment efficacy of 83%–87%. These nanocarriers have potential applications for delivery of drugs and other hydrophobic therapeutic agents to specific sites (Subia and Kundu, 2013). The folate-conjugated silk fibroin NPs are effective in checking the effect of proinflammatory responses by cytokines such as TNF-α, IL-1β, and nitric oxide by stimulating macrophages. DOX-loaded, folate-conjugated, silk fibroin NPs are used in treatment of cancer cells (Subia et al., 2014).

13.4.9 COLLAGEN

Collagen, as shown in Fig. 13.3D, is a major structural protein and constituent of skin, ligaments, tendons, and cornea, etc. Collagen is considered to be a favorable matrix for onsite drug delivery due to its biocompatibility, high safety profile, and immunological properties. It has a wide range of applications in the treatment and prophylaxis of bone and soft tissue infections, ophthalmic, periodontal problems, wound healing, and for the delivery of antibiotics such as gentamicin (Ruszczak and Friess, 2003). The collagen molecules are bound together by intermolecular and intramolecular covalent crosslinking. The intermolecular crosslinking provides molecular stability to the collagen (Brown and Timple, 1995; Todhunter et al., 1994). The physical properties of a collagen hydrogel (i.e., mesh size) can easily be controlled. This property of collagen is used in preparing 340 nm collagen-based NPs using lipid vesicle cages which help in controlling particle dimension as well as in protecting its environment, providing site-specific delivery (Papi et al., 2011). Villegas et al. constructed hybrid collagenase nanocapsules for enhanced nanocarrier penetration in cancerous cells. These novel nanodevices trigger their controlled and sustained release under mild acidic conditions and its coating also protects the housed enzyme from proteolytic attack. These collagen-based NPs are another candidate for drug delivery with high penetration efficacy in human osteosarcoma cells (Villegas et al., 2015).

13.4.10 PULLULAN

Pullulan-based NPs are widely used in drug delivery (Fawzia et al., 2012) and coating applications (San Juan et al., 2009). It is a biodegradable, water soluble polysaccharide consisting of maltotriose repeating units containing α-(1→4) and α-(1→6) glycosidic linkages in a 2:1 ratio, as shown in Fig. 13.3E. Shen et al. investigated the efficiency of the epirubicin-loaded cholesterol-modified pullulan self-aggregated NPs (EPI-CHSPNs). The characterization of the synthesized NPs was done by TEM, DLS, and zeta potential. The EPI-CHSPNs exhibited a longer half-life time $t(1/2)$, higher plasma drug concentration, and larger area under-the-curve (Shen et al., 2014). More recently, Dionsio et al. modified pullulan for obtaining charged derivatives for the production of pullulan-based nanocarriers by polyelectrolyte complexation. The FTIR analysis confirmed the presence of the charged groups. The (1)H- and (13)C NMR (COSY, HSQC-DEPT, HSQC-TOCSY, and HMBC) experiments were used for the assessment of the reaction position (Dionsio et al., 2016). The 200–450 nm spherical-shaped pullulan acetate NPs (PANs) with low zeta potential in distilled water as well as in 10% FEBS prepared by solvent diffusion method to load epirubicin and determination of the acetylation degree was done by proton nuclear magnetic resonance (^{1}H NMR) i.e., found to be 2.6 (Tang et al., 2010). The Hydrophobic-Cholesterol-modified amino pullulan (CHAP) conjugates were prepared by grafting ethylene-diamine and hydrophobic groups. The degree of substitution of the cholesterol in the three types of NPs—i.e., CHAP, cholesterol-modified pullulan (CHP), and cholesterol-modified carboxyethyl pullulan (CHCP)—was 2.92%, 3.11%, and 3.46%, respectively. The encapsulation efficiency and drug-loading capacity of CHAP-NPs was 70.8% and 12.3%, respectively. The CHAP-NPs have promising applications in tumor treatment and are found to be cytotoxic to HeLa cells when loaded with mitoxantrone (Tao et al., 2016).

13.4.11 CHITOSAN

Negatively charged and hydrophobic molecules like DNA do not cross the biological membrane and are quickly degraded by nucleases (Vasir and Labhasetwar, 2006). However, a sufficient amount of DNA would be required for effective delivery because the naked plasmid DNA possesses low transfection efficiency due to dilution and degradation in the intestinal tract (Mao et al., 2010). The mucoadhesive properties of chitosan (CS) (Fig. 13.3F) lead to its significant application in the field of oral delivery. The presence of mucus over the intestinal epithelium raises the residence time of mucoadhesive particles on the cell surface. Paclitaxel (Px)-loaded *N*-acetylhistidine-glycol (NAcHis-GC) CS conjugates for the intracytoplasmic delivery and were effective in inducing growth arrest of cancerous cells. The self-assembled NPs of 150–250 nm diameter were found to be hydrophobic at neutral pH. At acidic pH, the protonation of imidazole group of NAcHis results in the breakdown of the hydrophilic/hydrophobic balance and, finally, the NPs are dissembled. Therefore, pH plays an important role in the self-assembling of NPs (Cho et al., 2006). The 181–363 nm curcumin loaded bovine–serum–albumin–CS NPs suppresses the nuclear transcription factor NF-KB, thereby, blocking the proliferation of human colorectal adeno-carcinoma (DLD-1) cancer cells. The characterization of the prepared nanoformulations was done by surface, morphology, particle size, FTIR, UV–vis spectrophotometer, polydispersity index, confocal microscopy, and in vitro cytotoxicity studies (Rajan et al., 2016).

13.4.12 ALGINATE

Alginate, as shown in Fig. 13.3G, is a natural anionic polymer obtained from brown seaweed and is widely used in biomedical applications and drug delivery due to its low toxicity, biocompatibility, and relatively low cost. The therapeutic efficacy and biocompatibility of NPs in nanoparticulate drug delivery systems could also be improved by surface functionalization. The construction of pH responsive mesoporous silica NPs was achieved by polyelectrolyte multilayered coating of alginate/chitosan around it. Since, the release of an anticancerous drug DOX was pH-dependent. The in vitro evaluation of this drug on HeLa cells showed the sustained intracellular release of DOX and prolonged accumulation in the nucleus (Feng et al., 2014).

13.4.13 DEXTRAN

Dextran is neutral, biodegradable, naturally branched polysaccharide made up of glucose molecules with α 1, 6-glycosidic linkages, as shown in Fig. 13.3G.

DOX-loaded PEG-modified dextran (Zhang et al., 2013b) and pH-responsive dextran-*g*-cholesterol micelle exhibited an effective and controlled drug release capability (Yao et al., 2014). The bioreducible carboxymethyl dextran (CMD) derivatives chemically modified with litocholic acid (LCA) through disulfide linkages was developed by Thambi et al., and self-assembled 163 to 242 nm dextran NPs were formed due to the hydrophobic nature of LCA that allows the conjugates (CMD-SS-LCAs) to form NPs. When the DOX is loaded on these NPs, the NPs exhibit high toxicity to SCC7 cancer cells; therefore, they can be effectively used to deliver drug for cancer therapy (Thambi et al., 2014). Wasiak et al. reported that dextran NPs are a promising nanodrug carrier. The synthesis of 100 nm dextran NPs was achieved by using 70 kDa dextran with a 5% degree of oxidation of polysaccharide chain and 50% substitution with dodecylamine. The DOX can be loaded on these dextran NPs that allow the release of the drug at lower pH (Wasiak et al., 2016).

13.4.14 STARCH

Biopolymer-based nanomaterials (specially polysaccharides) are the key components in food industrial formulations and have extensive applications as gelling, emulsifying, and/or stabilizing agents (Bouyer et al., 2012). Starch is a polymeric carbohydrate produced by many plants, rice, potato, wheat, and corn, etc. It is composed of glucose molecules and consists of two polysaccharides-amylose and amylopectin, as shown in Fig. 13.3H. Amylose (15–20 wt%) is smaller than amylopectin (80%–85%) having molecular weight range of 10^5–10^6 g/mol and 10^6–10^7 g/mol, respectively. It is soluble in cold water due to having a large particle size and can be dispersed in water by gelatinization at high temperatures. Due to its polymeric structure, biocompatibility and eco-friendliness it is widely used in the field of drug delivery (Ellis et al., 1998). The 160 ± 5 nm monodispersed, spherical hydroxyethyl starch (HES) NPs prepared using the two-step crosslinking precipitation route have better in vitro efficacy in its nanoencapsulated form than in its bare form. The nanoencapsulated form of HES is advantageous for parenteral delivery at high concentration (0.05–1 mg/ml) in vitro and in vivo as proved by a panel of hemo/cytocompatibility assay (Narayanan et al., 2015). The starch functionalized grapheme nanosheets (starch-GNS) prepared via the reduction of exfoliated grapheme oxides by soluble starch is of particular interest in the field of biomedicine. The hydroxycamptothecin loaded starch-GNS showed high loading capacity and is highly toxic to SW-620 cancer cells while, before loading, the starch GNS was nontoxic to SW-620 cells even at 200 μg/mL concentration (Liu et al., 2015). The curcumin-loaded starch-maleate (CurSM) NPs with a particle size range of 30–110 nm exhibited a high rate of water solubility of 6.0×10^{-2} mg/g; about 300 times higher than free curcumin (Pang et al., 2015).

13.4.15 PECTIN

Pectin, as shown in Fig. 13.3A, is an excellent biodegradable and biocompatible polymer of natural origin and forms the structural component of the plant cell wall. It is widely used to deliver drugs through oral and vaginal routes. Pectins can easily pass through tissue interstice and are easily absorbed by particular cells and removed by phagocytes due to its ultra-small size. It is used for controlled delivery of various drugs and for protein targeting. Pectin and pectin oligosaccharides result in the induction of the apoptosis in human adenocarcinoma cells. The colon-targeted drug delivery of TH was done through pectin when grafted with poly (*N*-isopropylacrylamide). Pectin-4-aminothiophinole (ATP) conjugated microparticles for colon-specific drug delivery were reported by Perera et al. (Perera et al., 2010). The disintegration behavior, drug loading, drug-release behavior, particle size, and its impact on Caco-2 cells was also studied. Ciprofloxacin-loaded HSA-pectin NPs prepared by the pH-coacervation method were reported by Kumar and Jain. They studied the suppression of agglomeration in HSA NPs (Kumar and Jain, 2007). An ionotropic method was used to prepare thiolated pectin-based NPs using timolol maleate as a model drug and magnesium chloride as ionic crosslinker. The thiolated pectin-based NPs have potential applications in ocular drug delivery (Sharma et al., 2012).

13.4.16 XANTHAN GUM

Xanthan gum is a natural polysaccharide, shown in Fig. 13.3I, and is produced by the bacterium *Xanthomonas campestris* (García-ochoa et al., 2000). Gohel et al. (2002) worked on the development of a modified release drug delivery system of diltiazem HCl with the use of a blend containing xanthan gum and hydroxypropylmethyl cellulose. Marianecci et al. (2011) worked on the preparation of Xanthan Gum (XG) hydrogels blended with biopolymer locust bean gum (LBG) and these polymeric system LBG/XGs were found to be convenient for the delivery of ibuprofen and caffeine when taken as model drugs in the experiment. Reddy et al. investigated the efficacy of insulin-loaded CS-xanthan gum NPs for oral delivery. An ionotropic gelation method used for preparation of CS-xanthan NPs and sodium-tripolyphosphate (TPP) was used as the crosslinking agent. The in vivo studies revealed the maximum reduction of blood glucose level to be 52% (Reddy et al., 2012).

13.4.17 HYALURONIC ACID

HA is widely distributed throughout the epithelial, connective, and neural tissues. It is an anionic nonsulfated glucosaminoglycan and is composed of repeating polymeric disaccharides of D-gluconic acid and *N*-acetyl-D-glucosamine which is linked together by a glucoronidic β (1→3) bond (Fig. 13.3J) (Papakonstantinou et al., 2012). The production of HA primarily occurs in mesenchymal cells, but also by other cell types (Papakonstantinou et al.,1995; Lee and Spicer, 2000). HA-CS NPs are constructed by the ion gelification technique for the delivery of two proangiogenic growth factors, i.e., vascular endothelial growth factor (VEGF) and platelet-derived growth factor (PDGF-BB). The entrapment efficiency of HA/CS NPs for both growth factors VEGF and PDGF is 94% and 54%, respectively. HA/CS NPs showed sustained release of PDGF-BB for 1 week while VEGF was completely released within 24 hours (Parajò et al., 2010). Chlorine6-loaded HA NPs (Ce6-HANPs), when injected intravenously into tumor-bearing mice, showed that the drug loaded NPs efficiently reached the tumor tissue via the passive targeting mechanism through the receptor-mediated endocytosis specifically enter the tumor cells. The endocytosis of the NPs is based on the interaction between HA NPs and CD44 (the HA receptor on the tumor cell surface). This Ce6-HANPs could be successfully employed as a carrier for in vivo photodynamic tumor imaging and therapy (Yoon et al., 2012). Han et al. investigated the therapeutic effect of DOX-loaded bioreducible shell-crosslinked HA NPs (DOX-HA-ss-NPs) for cancer therapy. The drug release by DOX-HA-ss-NPs is greatly retarded under physiological pH 7.4 and the presence of glutathione markedly enhanced the drug release rate (Han et al., 2015). Wan et al. (2016) studied the therapeutic effect of cargo-loaded DOX and verapamil (VER) HA functionalized uniform mesoporous carbon spheres (UMCS) on an in vivo HCT-116 tumor in BALB/c nude mice. HA-UMCS NPs open a new door for the targeted co-delivery of drugs to tumor overexpressing CD44 receptors.

13.5 SOME PROMISING DRUG DELIVERIES

13.5.1 BRAIN-TARGETED DRUG DELIVERY

Brain-targeted drug delivery systems have shown increasingly attractive options, particularly in the treatment of central nervous system(CNS)–related diseases as there are a variety of physiological, biochemical, and metabolic barriers in the delivery of drug molecules to the brain that together form the blood–brain barrier (BBB), blood–tumor barrier, and blood–cerebrospinal-fluid barrier. Drug delivery directly to the brain interstitium through this method provides new hope to overcome these barriers. It is also essential to make drugs with high lipid solubility, positive charge, and low molecular size for achieving success in transversing the BBB (Banerjee et al., 1996). The nonimmunogenic drug delivery systems facilitate the uptake of the drug by interacting with the receptors present on the surface of the BBB. The modification of the surface of the drug is done with the help of hydrophilic polymers such as PEG, which is helpful in increasing the circulation time of the delivery system. *p*-Glycoprotein is an ATP-dependent efflux transporter that plays an important role in checking the drug entry into the brain. By masking or inhibiting binding to the cellular efflux system (i.e., *p*-glycoprotein) passage through the BBB can be achieved. The entrapment of a drug like Px in the NPs significantly improves the drug uptake in the brain and toxicity towards the *p*-glycoprotein (a multidrug resistant protein) which is highly expressed in drug-resistant tumor cells (Koziara et al., 2004). Targeted nanocarriers [viz., liposome, polymeric NPs, and solid lipid NPs (SLNs)] are also used to transport and deliver protein drugs (Solaro et al., 2010). Lipid-based nanocarriers are a new approach in the drug-delivery system for the delivery and transport of therapeutic agents to the CNS. Lipid-based nanocarriers include liposomes, lipid drug conjugates, SLN, micro- or nanoemulsions, and nanostructured lipid carriers. The liposome conjugated voltage-dependent anionic channel (VDAC) and pro-apoptotic Bak are the two mitochondrial outer membrane proteins which activate the intrinsic apoptotic pathways to HCT116p53$^{+/+}$ and HCT116p53$^{-/-}$, helpful in cancer therapy (Liguori et al., 2008). A significant hypoglycemic effect, low blood glucose level, and higher pharmacological activity were obtained when insulin encapsulated in liposome nanocarriers are were in the body.

13.5.2 OCULAR DRUG DELIVERY

Drug delivery through the ocular route is an interesting, but challenging, field because only 1%–2% of the drug is available in the eye for therapeutic action while rest of the drug is drained out through the nasolachrymal drainage system and other ocular physiological barriers. In spite of the low bioavailability and pulsed release of the drug, it is expected that direct application of the drug to the eye will give maximum responses with reduced frequency of administration (Fawzia et al., 2012). The aim of the ophthalmic drug delivery system focuses on the maximizing drug delivery absorption rather than minimizing systemic absorption (Järvinen et al., 1995). The use of CS-based NPs for ocular drug delivery was also reported (Paolicelli et al., 2009). Selvaraj et al. prepared CS microspheres by ion gelation of the CS with TPP anions for ocular drug delivery of acyclovir which is concerned with the effective management of various ocular viral diseases. The size of the microspheres is between 2 and 12 μm having encapsulation and loading capacity of 52%–78% and 12%–26%, respectively. The result suggested that the acyclovir-loaded CS microspheres are an efficient vehicle for the management of ocular viral infections (Selvaraj et al., 2012). The topical ophthalmic formulations containing brimonidine-loaded bioadhesive cationic CS or anionic alginate NPs for the sustained release of brimonidine was prepared by Ibrahim et al. The spherical-shaped nanoformulations had desirable sizes of 115.67 ± 3.58 and 157.67 ± 5.53 nm; encapsulation efficiency of $73.34\% \pm 2.05\%$ and $70.40\% \pm 2.77\%$; drug loading $11.81\% \pm 0.67\%$ and $13.14\% \pm 0.90\%$; zeta potential $+35.27 \pm 3.39$ and -37.8 ± 3.77 mV, and yield $87.91\% \pm 5.92\%$ and $76.53\% \pm 3.32\%$, respectively (Ibrahim et al., 2015). An antiglaucoma drug, acetazolamide-loaded polymeric film was also prepared for sustained glaucoma drug delivery and have excellent potential to be developed into drug-loaded contact lenses with fewer drug-induced side effects (Dhesingh and Prakash, 2016). Levofloxacin encapsulated poly (lactic-*co*-glycolic acid) NPs were developed by

Gupta et al. The 190–195 nm spherical-shaped NPs have antimicrobial efficacy against *Staphylococcus aureus* due to having good spread and good retention over the precorneal area. Stability studies revealed a degradation constant less than 5×10^{-4} with a shelf life of 2 years (Gupta et al., 2011). The study was also conducted on polysaccharide-based biodegradable nanocarriers for the treatment of ocular disorders (Formica et al., 2015; Maria et al., 2015). In the case of gelatin nanocarriers, by changing their electrical and physical properties, the ocular delivery system can be optimized (Young et al., 2005). The crosslinking in the gelatin chains provides stability towards thermal degradation. Pilocarpine HCl and Hydrocortisone-encapsulated gelatin NPs for topical ophthalmic applications were studied by Vandervoort and Ludwig. These NPs possess prolonged residence time at the ocular surface; therefore, they could be effective polymeric nanocarriers for ophthalmic drug delivery (Vandervoort and Ludwig, 2004).

13.5.3 NASAL DRUG DELIVERY

Nose to brain drug delivery has attracted a great deal of interest due to the fast onset of action as the route possesses higher vascularization and permeability. It also protects the drugs from enzymatic and chemical degradation. This route is noninvasive and enhances patient compliance as the drug enters the CNS by passing the BBB during transportation (Djupesland et al., 2014; Talegonkar and Mishra, 2004). Bhavna et al. reported the preparation and characterization of CS NPs for nasal delivery of a cholinesterase inhibitor (Bhavna et al., 2007). Baltzley et al. performed an in vivo study on intranasal delivery of the drug olanzapine (OZ) through CS NPs. Results revealed enhanced systemic absorption with $51 \pm 11.2\%$ absolute bioavailability by OZ-loaded CS NPs compared to $28 \pm 6.7\%$ after administration of a pure OZ solution (Baltzley et al., 2014). Ruby and Pandey prepared memantine hydrochloride-loaded CS NPs by ion gelation of CS NPs with a TPP anion at different concentrations of CS to TPP (i.e., 2:1, 3:1, 4:1, and 5:1) with varying stirring time (2, 4, 6, and 8 hours), and stirring speed (500, 1000, 1500, 2000 rpm). The most suitable speed for CS NP formulation to achieve effective NP sizes of 129 nm is 100 rpm, the suitable concentration of CS and TPP is 2:1, and time is 4 hours. These NPs have potential application in nasal drug delivery of memantine hydrochloride (Ruby and Pandey, 2015).

13.5.4 ORAL DRUG DELIVERY

Drug delivery through the oral route still holds much attention as it is easier, safer, patient-friendly, and a large population immunization could be easily achievable (Galindo-Rodriguez et al., 2005). The mucoadhesive properties of CS lead to its significant application in the field of oral drug delivery. The presence of mucus over the intestinal epithelium raises the residence time of mucoadhesive particles on the cell surface. Jain et al. worked on the synthesis of spherical-shaped gelatin coated hybrid lipid NPs for the oral drug delivery of Amphotericin

B (AmB). The gelatin-coated hybrid lipid NPs improved the oral bioavailability of AmB with a particle size of 50.61 ± 2.20 nm and polydispersity index (PDI) of 0.274 ± 0.008. The orientation of lecithin (located in the core) and gelatin (exterior coat) was confirmed by CLSM and FRET. The oral bioavailability of the drug was increased up to 4.69-fold when Am B was incorporated in the gelatin-coated hybrid lipid NPs in comparison to free drugs (Jain et al., 2012).

13.5.5 DERMAL AND TRANSDERMAL DRUG DELIVERY

Dermal and transdermal drug delivery are attractive controlled-release devices to deliver conventional hydrophobic small molecule drugs, hydrophilic drugs, and macromolecules, etc., having several benefits such as providing a steady plasma profile for sustained release of drugs and reducing dose schedule and side effects (Teo et al., 2006). It also enhances patient compliance, eliminates the first-pass metabolism effect of drugs, and lowers the amount of drug administration; hence, it is safer in hepato-compromised patients (Paudel et al., 2010). The transdermal drug delivery system also ensures continued drug delivery and continuous input of drugs with short biological half-lives as the drug is kept in contact with the skin surface by an adhesive layer (Park et al., 2008). The other advantage of transdermal drug delivery is the avoidance of the risks of painful hypodermic injections that generate medical waste and risks of transmission of dangerous communicable diseases by needle reuse, especially in developing countries (Miller and Pisani, 1999). Additionally, the permeability of lipid-based NPs is improved when dissolved in the lipids of stratum corneum (Fang et al., 2001). During the past few decades, certain studies have shown the importance of the transfollicular pathway, particularly for preferential penetration of particulate substances. The delivery of particles through this mode is considered to be safer from stratum corneum turns and also persists for several days before they are cleared (Badri et al., 2014).

13.5.6 PULMONARY DRUG DELIVERY

Drug delivery through the pulmonary route permits rapid drug delivery and improved bioavailability containing large surface area for solute transport (Patton and Byron, 2007; Laube, 2005). The pulmonary drug delivery system is considered as a noninvasive, nonsystemic delivery platform for both local and systemic delivery of drugs and it directly targets the disorders related to lungs. The pulmonary route is a good option for drug delivery due to high permeability and large absorptive surface area of the lungs having extremely thin absorptive and mucosal membrane (approximately 70–140 nm^2 absorptive surface area in adult humans) with a good blood supply. Most of the therapeutics and macromolecules are absorbed through alveolar epithelium of the distal lungs (Patil and Sarasija, 2012). Bharti et al. performed an in vivo study using a rat model for indicating the localization of albumin NPs in the lungs of rats for pulmonary drug delivery. Albumin NPs with a 81.6–148.4 nm size range have entrapment efficiency of

61.74%–73.45%. The results indicated that albumin NPs of sizes less than 500 nm were localized effectively in the pulmonary route to treat various lung-related diseases such as chronic obstructive pulmonary diseases and asthma, etc. (Bharti et al., 2015). The inhalable alginate NPs as an antitubercular drug carrier against experimental tuberculosis were constructed by Zahoor et al. A relative higher bioavailability was found in the case of the pulmonary route in comparison to the oral delivery of the drug (Zahoor et al., 2005).

13.6 DRUG DELIVERY APPLICATIONS

Nanomaterials have wide range of applications in medicine including drug delivery, in vivo and in vitro diagnostics, development of improved biocompatible materials and in nutraceuticals (De Jong and Borm, 2008). NPs have potential applications in gene delivery for the treatment of cancer, pulmonary, genetic, and other infectious diseases. The clinical applications of NPs in the diagnosis and treatment of renal diseases was discussed by Brede and Labhasetwar (2013). Natural polymers and their derivatives are promising nanocarriers for oral, transdermal, mucosal, colonic, and differently targeted therapeutic agents, gene/vaccine, and anticancer drug delivery. Mamo et al. (2010) reviewed the effective treatment and prevention of HIV/AIDS by various nanoscale drug delivery systems. The major impact of the NP drug delivery system is on the treatment of cancers and other chronic diseases.

13.6.1 CANCER THERAPY THROUGH NANOSTRUCTURES

NPs loaded with drugs are passively used to target cancer tissues due to the differences in the characteristics of the blood vessels in the cancer tissue and rich blood flow. The tumor growth arises from cancer stem cells (CSCs), which is responsible for eliciting resistance against various conventional drugs. The mutated stem cells give rise to CSCs (Kumar and Kutty, 2012; Ma et al., 2014). The CSCs show a high level of expression of aldehyde dehydrogenase (AlDH) associated with chemoresistance and the development of tumerogicity (Li et al., 2015). The higher expression of AlDHs and increased tumeroginicity is involved in the development of various types of cancers like melanoma (Yue et al., 2015), non-small cell lung cancer (Shao et al., 2014), and cervical cancer (Liu and Zheng, 2013). CD44 is a transmembrane glycoprotein which is involved in various important functions such as cell division, cell migration, cell adhesion, and cell signaling (Basakran, 2015). Apart from this, the CD44 markers also play an important role in chemoresistance, tumor initiation, and metastasis (Yan et al., 2015). The NPs prepared from polymer-encapsulated curcumin are used in the case of brain tumor stem cells. A decrease in growth and clonogenicity was also observed after treatment with NPs. Along with enhanced cell cycle arrest, apoptosis and bioavailability of curcumin was also observed. The $CD133^+$ stem cell population is gradually reduced (>50% reduction). The conjugated anti-CD133 antibody and SN-38-loaded NPs suppresses tumor growth and retards its recurrence by efficiently binding to HCT116 cells which overexpresses CD133 glycoprotein. The targeted NPs CD133Ab-NPs-SN cells gave better results than nontargeted NPs NPs-SN-38 (Ning et al., 2016). Low-dose decitabine (DAC) encapsulated NPs are helpful in the treatment of breast cancer cells. The combined effect of NP-DOX and NPs loaded with low-dose DAC (NPDAC) reduces the amount of CSCs and, in turn, high activity of AlDH is also reduced. A downregulation in expression of DNMT1 and DNMT3b in a MB-MDA-231 xenograft murine model with enhanced caspase-9 expression was also noted (Li et al., 2015). Thus, the NPs mask the effect of CSCs by targeting various specific markers like AlDH, CD44, CD90, and CD133 as well as interfering with Notch, Hedgehog, and TNF-β signaling pathways which are involved in the function and maintenance of CSCs (Lu et al., 2016). The coumarin-conjugated NPs are used in the treatment of tumor cells by increasing penetration capacity through matrix gel. 3D cell colonies were successfully developed from 95-D, U87 and HCT116 cell lines after seven day culture in the collagen matrix. This form of drug transport is a new step toward the development of novel drug delivery systems for cancer therapy (Le et al., 2016).

13.6.2 GENE THERAPY THROUGH NANOSTRUCTURES

Gene therapy is a medical intervention that uses genes to treat or prevent diseases. Gene therapy is not only applicable for genetic deficiencies, but also for other chronic and complicated diseases like AIDS, cancer, diabetes, coronary, and artery diseases, etc. (Hendriks et al., 2004) The role of polymeric NPs (PNPs) as gene expression vectors was investigated by Vasir and Labhasetwar (2006). Genetic materials and drugs are easily incorporated into polymeric NPs due to their unique nanostructures. Additionally, the modification of NPs with different functional groups allows efficient penetration and specific targeting. The PNPs synthesized with nontoxic, biodegradable, and biocompatible materials are efficient candidates for gene delivery. They have high stability with controlled release of bioactive ingredients. The polymers with a positive charge (i.e., cationic polymers) are more efficient nonviral vectors because the positively charged groups present in the backbone of polymers interact with the negative charge of the anionic genetic materials (Fishbein et al., 2010). Similarly, the role of albumin–CS core–shell NPs in gene delivery was investigated by Karimi et al. (2013).

13.7 SUMMARY

This chapter discusses the developments in drug delivery through natural nanostructures in medicine. The preparation methods of NPs and factors effecting drug delivery through them is also provided.

ACKNOWLEDGMENTS

Authors are thankful to the Head, Department of Botany, DDU Gorakhpur University, Gorakhpur (India) for providing necessary facilities and DST, New Delhi for providing Inspire Fellowship to Neha Tiwari. Professor Ishwar Das is thankful to UGC, New Delhi for the award of UGC Emeritus Fellowship.

REFERENCES

Albert, F., 2010. Albumin usage in clinical medicine: tradition or therapeutic? Tranfusion Med. Rev. 24 (1), 53–63. Available from: https://doi.org/10.1016/j.tmrv.2009.09.005.

Alberts, A.J.B., Lewis, J., Raff, M., Roberts, K., Walter, P., 2002. Molecular biology of the cell. Garland Science. Taylor and Francis Group, New York.

Allouche, J., 2013. Synthesis of organic and bioorganic nanoparticle: An overview of the preparation methods. In: Brayneretal, R. (Ed.), Nanomaterials: A Danger or a Promise. Springer-Verlag, London 2013. Available from: https://doi.org/10.1007/578-1-4471-4213-2.

Andriani, Y., Grasianto, Siswanta, Mudasir, 2015. Glutaraldehyde–cross linked chitosan pectin nanoparticles as a potential carrier for curcumin delivery and its in vitro release study. Int. J. Drug Deliv. 7 (3), 167–173.

Arangoa, M.A., Campanero, M.A., Renedo, M.J., Ponchel, G., Irache, J.M., 2001. Gliadin nanoparticles as carrier for the oral administration of lipophilic drugs: relationships between bioadhesion and pharmacokinetics. Pharm. Rev 18 (11), 1521–1527.

Arshady, R., 1989. Preparation of microspheres and microcapsules by interfacial polycondensation techniques. J. Microcapsul 6 (1), 13–28.

Aswathy, R.G., Sivakumar, B., Brahatheeswaran, D., Fukuda, T., Yoshida, Y., Maekawa, T., et al., 2012. Biocompatible fluorescent zein nanoparticles for simultaneous bioimaging and drug delivery application. Adv. Nat. Sci.: Nanosci. Nanotechnol 3 (2), 025006. Available from: https://doi.org/10.1088/2043-6262/3/2/025006.

Azimi, B., Nourpanah, P., Rabiee, M., Arbab, S., 2014. Producing gelatin nanoparticles as delivery system for bovine serum albumin. Iran. Biomed. J 18 (1), 34–40. Available from: https://doi.org/10.6091/ibj.1242.2013.

Badri, W., Eddabra, R., Fessi, H., Abdelhamid, E., 2014. Biodegradable polymer based nanoparticles: Dermal and Transdermal drug delivery. J. Colloid. Sci. Biotechnol 3, 141–149.

Baltzley, S., Mohammad, A., Malkawi, A.H., Al-Ghananeem, A.M., 2014. Intranasal drug delivery of olanzapine-loaded chitosan nanoparticles. AAPS PharmSciTech 15 (6), 1598–1602.

Banerjee, G., Nandi, G., Mahato, S.B., Pakrashi, A., Basu, M.K., 1996. Drug delivery system: targeting of pentamidines to specific sites using sugar liposomes. J. Antimicrob. Chemother. 38, 145–150.

Basakran, N.S., 2015. CD44 as a potential diagnostic tumor marker. Saudi Med. J. 36 (3), 273–279.

Bharti, N., Harikumar, S.L., Shishu, Buddiraja, A., 2014. Formulation and evaluation of gelatin nanoparticles for pulmonary drug delivery. World J. Pharm. Pharmac. Sci. 3 (12), 733–744.

Bharti, N., Harikumar, S.L., Shishu, Buddiraja, A., 2015. Development and characterization of albumin nanoparticles for pulmonary drug delivery. World J. Pharm Sci. 3 (1), 86–92.

Bhavna, Sharma, V., Ali, M., Baboota, S., Ali, J., 2007. Preparation and characterization of chitosan nanoparticles for nose to brain delivery of a cholinesterase inhibitor. Indian J. Pharm. 69 (5), 712–713. Available from: https://doi.org/10.4103/0250-474X.38488.

Bouyer, E., Mekhloufi, G., Rosilio, V., Grossiord, J.-L., Agnely, F., 2012. Proteins, polysaccharides and their complexes used as stabilizers for emulsions: alternatives to synthetic surfactants in the pharmaceutical field. Int. J. Pharm. 436 (1-2), 359–378.

Brede, C., Labhasetwar, V., 2013. Applications of nanoparticles in the detection and treatment of kidney diseases. Adv. Chronic Kidney Dis. 20 (6). Available from: https://doi.org/10.1053/j.ackd.2013.07.006. PMCID:PMC3824259.

Brown, J.C., Timple, R., 1995. The collagen superfamily. Int. Arch. Allergy Immunol. 107, 484–490.

Chitrani, B.D., Chan, W.C., 2007. Elucidating the mechanism of cellular uptake and removal of protein-coated gold nanoparticles of different sizes and shapes. Nano Lett. 7 (6), 1542–1550.

Chitrani, B.D., Ghajani, A.A., Chan, W.C., 2006. Determining the size and shape dependence of gold nanoparticle uptake into mammalian cells. Nano Lett. 6 (4), 662–668. Available from: https://doi.org/10.1021/nl052396o.

Cho, E.C., Xie, J.W., Wurm, P.A., Xia, Y., 2009. Understanding the role of surface charges in cellular adsorption versus internalization by selectively removing gold nanoparticles on the cell surface with a I2/KI etchant. Nano Letter 9 (3), 1080–1084.

Cho, V.W., Park, J.S., Han, T.H., Lee, K.Y., Han, S.S., Hwang, J.J., et al., 2006. J. Controlled Release 115, 37.

Choi, K.Y., Min, K.H., Yoon, H.Y., Kim, K., Park, J.H., Kwon, I.C., et al., 2011. PEGylation of hyaluronic acid nanoparticles improves tumor targetability in vivo. Biomaterials 32 (7), 1880–1889. Available from: https://doi.org/10.1016/j.biomaterials.2010.11.010.

Ciobanu, B.C., Cadinoiu, A.N., Popa, M., Desbriéres, Peptu C.A., 2014. Modulated release from liposome entrapped in chitosan/gelatin hydrogels. Mater. Sci. Eng. C 43, 383–391.

Curcio, M., Altimari, I., Spizzirri, U.G., Cirillo, G., Vittorio, O., Puoci, F., et al., 2013. Biodegradable gelatin-based nanospheres as pH responsive drug delivery systems. J. Nanopart. 15–1581. Available from: https://doi.org/10.1007/s11051-013-1581-x.

De Jong, W.H., Borm, P.J.A., 2008. Drug delivery and nanoparticles: applications and hazards. Int. J. Nanomed 3 (2), 133–149. PMCID:PMC2527668.

Deng, X.-X., Zhang, N., Tang, C.-H., 2016. Soy protein isolate as a nanocarrier for enhanced water dispersibility, stability and bioaccessibility of β-carotene. J. Sci. Food Agric. Available from: https://doi.org/10.1002/jsfa.8033.

Dhanya, A.T., Haridas, K.R., Divia, N., Sudheesh, S.J., 2012. Development of zein-pectin nanoparticles as a drug carrier. Int. J. Drug Deliv. 4 (2), 147–152.

Dhesingh, R.S., Prakash, M., 2016. Nanoparticle modified drug loaded biodegradable polymeric contact lenses for sustained ocular drug delivery. Curr. Drug Deliv 13. Available from: https://doi.org/10.2174/15672018136661018153547.

Dionsio, M., Braz, L., Corvo, M., Lourençο, J.P., Grenha, A., Rosa da Costa, A.M., 2016. Charged pullulan derivatives for the development of nanocarriers by polyelectrolyte complexation. Int. J. Biol. Macromol. 86, 129–138. Available from: https://doi.org/10.1016/j.ijbiomac.2016.01.054.

Djupesland, P.G., Messina, J.C., Mahmud, R.A., 2014. The nasal approach to delivering treatment for brain diseases: an anatomic, physiologic and delivery technology overview. Ther. Deliv. 5, 709–733.

Elblbesy, M.A., 2016. Hemocompatibility of albumin nanoparticles as a drug delivery system—an in vitro study. J. Biomater. Nanobiotechnol 7 (2). Available from: https://doi.org/10.4236/jbnb.2016.72008. Article ID: 65368.

Ellis, R.P., Cochrane, M.P., Dale, M.F.B., Duffus, C.M., Lynn, A., Morrison, I.M., et al., 1998. Starch production and industrial use. J. Sci. Food Agric. 77, 289–311.

Elsabahy, M., Wooley, K.L., 2012. Design of polymeric nanoparticles for biomedical delivery applications. Chem. Soc. Rev. 41 (7), 2545–2561. Available from: https://doi.org/10.1039/c2cs15327k.

Elzoghby, A.O., Abo El-Fotoh, W.S., Elgindy, N.A., 2011. Casein-based formulations as promising controlled release drug delivery systems. J. Controlled Release 153 (3), 206–216.

Elzoghby, A.O., Samy, W.M., Elgindy, N.A., 2012. Albumin-based nanoparticles as potential controlled release drug delivery systems. J. Controlled Release 157, 168–182.

Fairhurst, D., Lee, R.W., 2011. The zeta potential and its use in pharmaceutical applications—Part-1: Charged interfaces in polar and non-polar media and the concept of zeta potential. Drug Dev. Deliv. 11, 60–64.

Fang, J., Hong, C., Chiu, W., Wang, Y., 2001. Effect of liposomes and niosomes on skin permeation of enoxacin. J. Pharm. 219 (1-2), 61–72.

Fawzia, H., Mona, E.-M., Shaheer, M., 2012. Ocular drug delivery and the importance of microemulsion as a potential delivery system. Int. J. Pharm. Chem. Sci 1 (2), 723–737.

Feng, W., Nie, W., He, C., Zhou, Z., Chen, L., Qiu, K., et al., 2014. Effect of pH-responsive alginate/chitosan coating on delivery efficiency, cellular uptake and biodistribution of mesoporous silica nanoparticles based nanocarriers. ACS Appl. Mater. Interfaces 6 (11), 8447–8460.

Fishbein, I., Chorny, M., Levy, R.J., 2010. Site-specific gene therapy for cardiovascular disease. Curr. Opin. Drug Discov. Dev. 13 (2), 203–213.

Formica, M.L., Calles, J.A., Palma, S.D., 2015. Polysaccharide-based nanocarriers for ocular drug delivery. Curr. Pharm. Des 21 (33), 4851–4868.

Galindo-Rodriguez, S.A., Allemann, E., Fessi, H., Doelker, E., 2005. Polymeric nanoparticles for oral delivery of drugs and vaccines; a critical evaluation of in vivo studies. Crit. Rev. Ther. Drug Carrier Syst 22, 419–464.

García-ochoa, F., Santos, V.E., Casas, J.A., Gómez, E., 2000. Xanthan gum: production, recovery and properties. Biotechnol. Adv. 18, 549–579.

Gohel, M.C., Amin, A.F., Patel, K.V., Panchal, M.K., 2002. Studies in release behavior of diltiazem HCl from matrix tablets containing (hydroxypropyl)methyl cellulose and xanthan gum. Boll. Chim. Farm. 141 (1), 21–28.

Gong, G., Liu, W., Wang, S., 2016. Self-assembled albumin nanoparticles as a nanocarrier for aclacinomycin A. Nanotechnology 27 (46). Available from: http://iopscience.iop.org/article/10.1088/0957-4484/27/46/465602.

Gong, S.J., Sun, S.X., Sun, Q.S., Wang, J.Y., Liu, X.M., Liu, G.Y., 2011. Tablets based on compressed zein microspheres for sustained oral administration: design, pharmacokinetics and clinical study. J. Biomater. Appl. 26, 195–208.

Govender, T., Stolnik, S., Garnett, M.C., Illum, L., Davis, S.S., 1999. PLGA nanoparticles prepared by nanoprecipitation: drug loading and release studies of a water soluble drug. J. Controlled Release 57 (2), 171–185.

Gueguen, J., Orecchioni, A.-M., Irache, J.M., Bergougnoux, L., Ezpeleta, I., 1995. Optimization and in vitro stability of legumin nanoparticles obtained by a coacervation method. Int. J. Pharm. 126, 103–109.

Gulfam, M., Kim, J.E., Lee, J.M., Ku, B., Chung, B.H., Chung, B.G., 2012. Anticancer drug-loaded gliadin nanoparticles induce apoptosis in breast cancer cells. Langmuir 28 (21), 8216–8223. Available from: https://doi.org/10.1021/la300691n.

Gupta, H., Aqil, M., Khar, R.K., Ali, A., Bhatnagar, A., Mittal, G., 2011. Biodegradable levofloxacin nanoparticles for sustained ocular drug delivery. J. Drug Target. 19 (6), 409–417. Available from: https://doi.org/10.3109/1061186X.2010.504268.

Han, H.S., Thambi, T., Choi, K.Y., Son, S., Ko, H., Lee, M.C., et al., 2015. Bioreducible shell-cross-linked hyaluronic acid nanoparticles for tumor-targeted drug delivery. Biomacromolecules 16 (2), 447–456. Available from: https://doi.org/10.1021/bm5017755.

Hariyadi, D.M., Purwanti, T., Kusumawati, I., Nirmala, R.N., Maindra, H.M.C., 2015. Physical characterization and in vivo study of ovalbumin encapsulated in alginate microspheres. Int. J. Drug Deliv. Technol. 5 (2), 48–53.

Hendriks, W.T., Ruitenberg, M.J., Blits, B., Boer, G.J., Vernaagen, J., 2004. Viral vector-mediated gene transfer of neurotrophins to promote regeneration of the injured spinal cord. Progr. Brain Res 146, 451–476.

Ibrahim, M.M., Abd-Elgawad, A.-E.H., Soliman, O.A.-E., Jablonski, M.M., 2015. Natural bioadhesive bioderagable nanoparticle-based topical ophthalmic formulations for management of glaucoma. Transl. Vis. Sci. Technol. 4 (3), 1–13. Article 12.

Jahanshahi, M., Babaei, Z., 2008. Protein nanoparticle: a unique system as drug delivery vehicles. African J. Biotechnol. 7 (25), 4926–4934. Available from: https://doi.org/10.4314/ajb.v7i25.59701.

Jain, S., Valvi, P.U., Swarnakar, N.K., Thanki, K., 2012. Gelatin coated hybrid lipid nanoparticles for oral delivery of amphotericin B. Mol. Pharm. 9 (9), 2542–2553. Available from: https://doi.org/10.1021/mp300320d.

Jiang, W., Kim, B.Y., Rutka, J.T., Chan, W.C., 2008. Nanoparticle-mediated cellular response is size dependent. Nat. Nanotechnol. 3 (3), 145–150. Available from: https://doi.org/10.1038/nnano.2008.30.

Jin, B., Zhou, X., Li, X., Lin, W., Chen, G., Qiu, R., 2016. Self-assembled modified soy protein/dextran nanogel induced by the ultrasonication as a delivery vehicle for riboflavin. Molecules 21, 282. Available from: https://doi.org/10.3390/molecules. 21030282.

Joachim, E., Kim, I.-D., Jin, Y., Kim, K., Lee, J.-K., Choi, H., 2014. Gelatin nanoparticles enhance the neuroprotective effects of intranasally administered osteopontin in rat. Drug Deliv. Transl. Res 4 (5), 395–399. Available from: https://doi.org/10.1007/s13346-014-0208-9.

Järvinen, K., Järvinen, T., Urtti, A., 1995. Ocular absorption following topical delivery. Adv. Drug Deliv. Rev. 16 (1), 3–19.

Kajal, H., Mishra, A., 2011. Preparation of tetanus toxoid and ovalbumin loaded gliadin nanoparticles for oral immunization. J. Biomed. Nanotechnol. 7 (1), 211–212. Article ID 21485875.

Kanoujia, J., Singh, M., Singh, P., Parashar, P., Tripathi, C.B., Arya, M., et al., 2016. RSC Adv. 6, 93275–93287. Available from: https://doi.org/10.1039/C6RA16830B.

Karimi, M., Avci, P., Mobasseri, R., Hamblin, M., Naderi-Manesh, H., 2013. The novel albumin-chitosan core-shell nanoparticles for gene delivery: preparation, optimization and cell uptake investigation. J. Nanopart. Res. 15 (5), 1–14.

Khan, H., Shukla, R.N., Bajpai, A.K., 2015. Investigation of in vitro release dynamics of ciplastin from genipin crosslinked gelatin nanocarriers. Drug Deliv. Lett 5 (2), 100–108. Available from: https://doi.org/10.2174/2210303105666150 6.

Khan, H., Shukla, R.N., Bajpai, A.K., 2016. Genipin-modified gelatin nanocarriers as swelling controlled drug delivery system for in vitro release of cytarabine. Mater. Sci. Eng. C: Mater. Biol. Appl. 61, 457–465. Available from: https://doi.org/10.1016/j.msec.2015.12.085.

Kinsella, J.E., 1979. Functional properties of soy proteins. J. Am. Oil Chem. Soc. 56 (3), 242–258.

Kommareddy, S., Amiji, M.M., 2007. Protein nanospheres for gene delivery: preparation and in vitro transfection studies with gelatin nanoparticles. Gene Transfer. Cold Spring Harbor Laboratory Press, New York, pp. 527–540.

Koziara, J.M., Lockman, P.R., Allen, D.D., et al., 2004. Paclitaxel nanoparticles for the potential treatment of brain tumors. J. Controlled Release 99, 259–269.

Kulkarni, S.A., Feng, S.-S., 2011. Effects of surface modification on delivery efficiency of biodegradable nanoparticles across the blood-brain barrier. Nanomedicine (London) 6 (2), 377–394. Available from: https://doi.org/10.2217/nnm.10.131.

Kumar, D.H., Kutty, M.K., 2012. Review of stem cell deregulation and breast cancer: an emerging hypothesis. Indian J. Pathol. Microbiol. 55, 147–153. Available from: https://doi.org/10.4103/0377-4929.97842.

Kumar, P.V., Jain, N.K., 2007. Suppression of agglomeration of ciprofloxacin-loaded human serum albumin nanoparticles. AAPS PharmSciTech 8 (1), E1–E6.

Laube, B.L., 2005. The expending role of aerosols in systemic drug delivery, gene delivery and vaccination. Respir. Care 50, 116–176.

Le, V.M., Lang, M.D., Shi, W.B., Liu, J.W., 2016. A collagen-based multicellular tumor spheroid model for evaluation of the efficacy of nanoparticle drug delivery. Artif. Cells Nanomed. Biotechnol. 44 (2), 540–544. Available from: https://doi.org/10.3109/21691401.2014.968820.

Lee, E.M., Singh, D., Singh, D., Choi, S.M., Zo, S.M., Park, S.J., et al., 2014. Novel alginate-gelatin hybrid nanoparticle for drug delivery and tissue engineering applications. J. Nanomater. 2014, 1–7. Available from: https://doi.org/10.1155/2014/124236. Article ID 124236.

Lee, G.V., Park, K., Nam, J.H., Kim, S.V., Byun, Y., 2006. Anti-tumor and anti-metastatic effects of gelatin-doxorubicin and PEGylated gelatin-doxorubicin nanoparticles in SCC7 bearing mice. J. Drug Target. 14 (10), 707–716. Available from: https://doi.org/10.1080/10611860600935701.

Lee, J.Y., Spicer, A.P., 2000. Hyaluronan: a multifunctional, megadalton, stealth molecule. Curr. Opin. Cell Biol. 12, 581–586. Available from: https://doi.org/10.1016/S0955-0674(00)00135-6. PMID:10978893.

Li, S.Y., Sun, R., Wang, H.X., Shen, S., Liu, Y., Du, X.J., et al., 2015. Combinational therapy with epigenetic-targeted and chemotherapeutic drugs delivered by nanoparticles to enhance the chemotherapy response and overcome the resistance by breast cancer stem cells. J. Controlled Release 10 (205), 7–14. Available from: https://doi.org/10.1016/j.jconrel.2014.11.011.

Liguori, L., Marques, B., Villegas-Mendez, A., Rothe, R., Lenormand, J.-L., 2008. Liposomes-mediated drug delivery of pro-apoptotic therapeutic membrane proteins. J. Controlled Release 126, 217–227.

Liu, K., Wang, Y., Li, H., Duan, Y., 2015. A facile one-pot synthesis of starch functionalized graphene as nano-carrier for pH sensitive and starch-mediated drug delivery. Colloids Surf., B: Biointerfaces 128, 86–93.

Liu, S.Y., Zheng, P.S., 2013. High aldehyde dehydrogenase activity identifies cancer stem cells in human cervical cancer. Oncotarget 4, 2462–2475. Available from: https://doi.org/10.18632/oncotarget.1578.

Liu, Z., Jiao, Y., Wang, Y., Zhou, C., Zhang, Z., 2008. Polysaccharides-based nanoparticles as drug delivery systems. Adv. Drug Deliv. Rev 60 (15), 1650–1662.

Liu, Z., Dong, C., Wang, X., Wang, H., Li, W., Tan, J., et al., 2014. Self-assembled biodegradable protein-polymer vesicle as a tumor-targeted nanocarrier. ACS Appl. Mater. Interfaces 6 (4), 2393–2400. Available from: https://doi.org/10.1021/am404734c.

Lohcharoenkal, W., Wang, L., Chen, Y.C., Rojanasakul, Y., 2014. Protein nanoparticles as drug delivery carriers for cancer therapy. BioMed Res. Int. 2014, Article ID 180549, 12 pages.

Lu, B., Huang, X., Mo, J., Zhao, W., 2016. Drug delivery using nanoparticles for cancer stem-like cell targeting. Front. Pharmacol., doi: 10.3389/fpar.2016.00084.

Ma, R., Bonnefond, S., Morshed, S.A., Latif, R., Davies, T.F., 2014. Stemness is derived from thyroid cancer cells. Front. Endocrinol. 5, 114. Available from: https://doi.org/10.3389/fendo.2014.00114.

Mamo, T., Moseman, E.A., Kolishetti, N., Salvador-Morales, C., Shi, J., Kuritzkes, D.R., et al., 2010. Emerging nanotechnology approaches for HIV/AIDS treatment and prevention. Nanomedicine (London) 5 (2), 269–285.

Mao, S., Sun, W., Kissel, T., 2010. Chitosan-based formulations for delivery of DNA and siRNA. Adv. Drug Deliv. Rev 62, 12–27.

Maria, L.F., Javier, A.C., Santiago, D.P., 2015. Polysaccharide-based nanocarriers for ocular drug delivery. Curr. Pharm. Des. 21 (33), 4851–4868.

Marianecci, C., Carafa, M., Di Marzio, L., Rinaldi, F., Di Meo, C., Matricardi, P., et al., 2011. A new vesicle loaded hydrogel system suitable for topical applications: preparation and characterization. J. Pharm. Pharm. Sci. 14 (3), 336–346.

Marty, J.J., Oppenheim, R.C., Speiser, P., 1978. Nanoparticles—a new colloidal drug delivery system. Pharm. Acta Helv 53 (1), 17–23.

Miller, M.A., Pisani, E., 1999. The cost of unsafe injections. Bull. World Health Organ. 77, 808–811.

Mirshahi, T., Irache, J.M., Gueguen, J., Orecchioni, A.M., 1996. Development of drug delivery systems from vegetal proteins: legumin nanoparticles. Drug Dev. Ind. Pharm. 22 (8), 841–846. Available from: https://doi.org/10.3109/03639049609065914.

Mirshahi, T., Irache, J.M., Nicholas, C., Mirshahi, M., Faure, J.P., Gueguen, J., et al., 2002. Adaptive immune responses of legumin nanoparticles. J. Drug Target 10 (8), 625–631.

Mohanraj, V.J., Chem, Y., 2006. Nanoparticles—a review. Trop. J. Pharm. Res 5 (1), 561–573.

Moinard-Checot, D., Chevalier, Y., Briançon, S., Fessi, H., Guinebretiere, S., 2006. Nanoparticles for drug delivery: review of the formulation and process difficulties illustrated by the emulsion-diffusion process. J. Nanosci. Nanotechnol. 6, 2664–2681.

Moinard-Checot, D., Chevalier, Y., Briançon, S., Beney, L., Fessi, H., 2008. Mechanism of nanocapsules formation by the emulsion-diffusion process. J. Colloid Interface Sci. 317 (2), 458–468.

Mottaghitalab, F., Farokhi, M., Shokrgozar, M.A., Atyabi, F., Hosseinkhani, H., 2015. Silk fibroin nanoparticle as a novel drug delivery system. J. Controlled Release 206, 161–176.

Nagavarma, B.V.N.M., Yadav, H.K.S., Ayaz, A., Vasudha, L.S., Shivakumar, H.G., 2012. Different techniques for preparation of polymeric nanoparticles—a review. Asian J. Pharm. Clin. Res. 5 (3), 16–23.

Narayanan, D., Nair, S., Menon, D., 2015. A systematic evaluation of hydroxyethyl starch as a potential nanocarrier for parenteral drug delivery. Int. J. Biol. Macromol. 74, 575–584. Available from: https://doi.org/10.1016/j.ijbiomac.2014.12.012.

Nayak, P., Sahoo, S.K., Behera, A., Nanda, P.K., Nayak, P.L., Guru, B.C., 2011. Synthesis and characterization of soy protein isolate/MMT nanocomposite film for the control release of the drug ofloxacin. World J. Nanosci. Eng. 1, 27–36. Available from: https://doi.org/10.4236/wjnse.2011.12005.

Ning, S.-T., Lee, S.-Y., Wei, M.-F., Peng, C.-L., Lin, S.Y.-F., Tsai, M.-H., et al., 2016. Targeting colorectal cancer stem-like cells with anti-CD133 antibody-conjugated SN-38 nanoparticles. ACS Appl. Mater. Interface 8 (28), 17793–17804. Available from: https://doi.org/10.1021/acsami.6b04403.

Niwa, T., Takeuchi, H., Hino, T., Kunou, N., Kawashima, Y., 1993. Preparation of biodegradable nanospheres of water-soluble and insoluble drugs with D,L-lactide/glycolide copolymer by a novel spontaneous emulsification solvent diffusion method and the drug release behavior. J. Controlled Release 25 (1-2), 89–98.

Oakenfull, D., Pearce, J., Burley, R.W., 1997. In: Damodaran, S., Paraf, A. (Eds.), Food Proteins and Their Applications. Marcel Dekker, New York, pp. 111–142.

Pang, S.C., Chin, S.F., Nadirah, A., Tay, S.H., Yazld, S.N.A.M., 2015. Fabrication of polysaccharide-based nanoparticles as drug delivery nanocarriers. ECS Trans. 66 (37), 15–32.

Paolicelli, P., de la Fuente, M., Sánchez, A., Seizo, B., Alonso, M.J., 2009. Chitosan nanoparticles for drug delivery to the eye. Expert Opin. Drug Deliv. 6 (3), 239–253. Available from: https://doi.org/10.1517/17425240902762818.

Papakonstantinou, E., Karakiulakis, G., Roth, M., Block, L.H., 1995. Platelet-derived growth factor stimulates the secretion of hyaluronic acid by proliferating human vascular smooth muscle cells. Proc. Natl. Acad. Sci. U.S.A. 92, 9881–9885. Available from: https://doi.org/10.1073/pnas.92.21.9881. PMID:7568237.

Papakonstantinou, E., Roth, M., Karakiulakis, G., 2012. Hyaluronic acid—a key molecule in skin aging. Dermato-Endocrinology 4 (3), 253–258.

Papi, M., Palmieri, V., Maulucci, G., Arcovito, G., Greco, E., Quintiliani, G., et al., 2011. Controlled self assembly of collagen nanoparticles. J. Nanopart. Res. 13, 6141–6147.

Parajò, Y., D'Angelo, I., Welle, A., Garcia-Fuentes, M., Alonso, M.J., 2010. Hyaluronic acid/Chitosan nanoparticles as delivery vehicle for VEGF and PDGF-BB. Drug Deliv 17 (8), 596–604. Available from: https://doi.org/10.3109/10717544.2010.509357.

Park, J.H., Lee, S., Kim, J.H., Park, K., Kim, K., Kwon, I.C., 2008. Polymeric nanomedicine for cancer therapy. Progr. Polym. Sci. 33 (1), 113–137.

Patel, A.R., Elisabeth, C.M., Bouwens, M., Velikov, K.P., 2010. Sodium caseinate stabilized zein colloidal particles. J. Agric. Food Chem. 58, 12497–12503.

Patil, J.S., Sarasija, S., 2012. Pulmonary drug delivery strategies: a concise, systematic review. Lung India 29 (1), 44–49. Available from: https://doi.org/10.4103/0970-2113.92361.

Patton, J.S., Byron, P.R., 2007. Inhaling medicines: delivering drugs to the body through lungs. Nat. Rev. Drug Discov. 6, 67–74.

Paudel, K.S., Milewski, M., Swadley, C.L., Brogden, N.K., Ghosh, P., Stinchcomb, A.L., 2010. Challenges and opportunities in dermal/transdermal delivery. Ther. Deliv. 1 (1), 109–131.

Perera, G., Barthelmes, J., Bernkop-Schnurch, A., 2010. A novel pectin-4-aminothiophenole conjugate microparticles for colon-specific drug delivery. J. Controlled Release 145, 240–246.

Podaralla, S., Perumal, O., 2012. Influence of formulation factors on the preparation of zein nanoparticles. AAPS PharmSciTech 13 (3), 919–927. Available from: https://doi.org/10.1208/s12249-012-9816-1.

Rajan, S.S., Pandian, A., Palaniappan, T., 2016. Curcumin loaded in bovine serum albumin-chitosan derived nanoparticles for targeted drug delivery. Bull. Mat. Sci 39 (3), 811–817.

Ramteke, K.H., Nath, L., 2014. Formulation, evaluation and optimization of pectin-bora rice beads for colon targeted drug delivery system. Adv. Pharm. Bull 4 (2), 167–177. Available from: https://doi.org/10.5681/apb.2014.025.

Ramteke, S., Maheshwari, R.B.U., Jain, N.K., 2006. Clarithromycin based oral sustained release nanoparticulate drug delivery system. Ind. J. Pharm. Sci. 68 (4), 479–484.

Reddy, M.M., Shanmugam, V., Kaza, R., 2012. Design and characterization of insulin nanoparticles for oral delivery. Int. J. Innov. Pharm. 3 (3), 238–243.

Reddy, N., Yang, Y., 2011. Potential of plant proteins for medical applications. Trends Biotechnol 29 (10), 490–498. Available from: https://doi.org/10.1016/tibtech.2011.05.003.

Roth, K., 2012. Boiled Eggs: Soft and Hard-Part 1. Chemie in unserer Zeit/Wiley-VCH, Wiley-VCH Verlag GmbH and Co. KGaA, Weinheim. doi: 10.1002/chemv.201200018.

Ruby, J.J., Pandey, V.P., 2015. Chitosan nanoparticles as a nasal drug delivery for memantine hydrochloride. Int. J. Pharm. Pharm. Sci 7 (1), 34–37.

Ruszczak, Z., Friess, W., 2003. Collagen as a carrier for on-site delivery of antibacterial drugs. Adv. Drug Deliv. Rev 55 (12), 1679–1698.

Salatin, S., Jelvehgari, M., Maleki-Dizaj, S., Adibkia, K., 2015. A sight on protein-based nanoparticles as drug/gene delivery systems. Ther. Deliv. 6 (8), 1017–1029. Available from: https://doi.org/10.4155/tde.15.28.

Salauti, M., Anangari, A., 2014. Nanoparticle based drug delivery systems for treatment of infectious diseases. Appl. Nanotechnol. Drug Deliv. 155–192. Available from: https://doi.org/10.5772/58423.

San Juan, A., Bala, M., Hlawaty, H., Portes, P., Vranckx, R., Feldman, L.J., et al., 2009. Development of a functionalized polymer for stent coating in the arterial delivery of small interfering RNA. Biomacromolecules 10 (11), 3074–3080.

Satheesh, P., Perumal, O., 2010. Preparation of zein nanoparticles by pH controlled nanoprecipitation method. J. Biomed. Biotechnol. 6, 312–317.

Satheesh, P., Perumel, O., 2012. Influence of formulation factors on the preparation of zein nanoparticles. AAPS PharmSciTech 13 (3), 919–927.

Sebak, S., Mirzaei, M., Malhotra, M., Kulamarva, A., Prakash, S., 2010. Human serum albumin nanoparticles as an efficient noscapine drug delivery system for potential use in breast cancer: preparation and in vitro analysis. Int. J. Nanomed. 5, 525–532. Available from: https://doi.org/10.2147/IJN.S10443.

Selvaraj, S., Karthikeyan, J., Saravanakumar, N., 2012. Chitosan loaded microspheres as an ocular delivery system for acyclovir. Int. J. Pharm. Pharm. Sci. 4 (1), 125–132.

Shao, C., Sullivan, J.P., Girard, L., Augustyn, A., Yenerall, P., Rodriguez-Canales, J., et al., 2014. Essential role of aldehyde dehydrogenase 1A3 for maintenance of non-small cell lung cancer stem cells is associated with the STAT3 pathway. Clin. Cancer Res. 20, 4154–4166. Available from: https://doi.org/10.1158/1078-0432.CCR-13-3292.

Sharma, R., Ahuja, M., Kaur, H., 2012. Thiolated pectin nanoparticles: preparation, characterization and ex vivo corneal permeation study. Carbohydr. Polym. 87, 1606–1610.

Shen, S., Li, H., Yang, W., 2014. The preliminary evaluation on cholesterol-modified pullulan as a drug nanocarrier. Drug Deliv. 21 (7), 501–508. Available from: https://doi.org/10.3109/10717544.2014.895068.

Shewry, P.R., Napier, J.A., Tatham, A.S., 1995. Seed storage proteins: structures and biosynthesis. Plant Cell 7, 945–956.

Silvio, L.D., Courteney-Harris, R.G., Downes, S., 1994. The use of gelatin as a vehicle for drug and peptide delivery. J. Mater. Sci.: Mater. Med 5 (11), 819–823.

Singh, R., Lillard Jr, J.W., 2009. Nanoparticle-based targeted drug delivery. Exp. Mol. Pathol. 86 (3), 215–223. Available from: https://doi.org/10.1016/j.yexmp.2008.12.004.

Solaro, R., Chiellini, F., Battisti, A., 2010. Targeted delivery of protein drugs by nanocarriers. Materials 3, 1928–1980. Available from: https://doi.org/10.3390/ma3031928.

Song, C.X., Labhasetwar, V., Murphy, H., Qu, X., Humphrey, W.R., Shebuski, R.J., et al., 1997. Formulation and characterization of biodegradable nanoparticles for intravascular local drug delivery. J. Controlled Release 43 (2), 197–212. Available from: https://doi.org/10.1016/S0168-3659(96)01484-8.

Su, K., Wang, C., 2015. Recent advances in the use of gelatin in biomedical research. Biotechnol. Lett. 37 (11), 2139–2145.

Suarasan, S., Focsan, S., Potara, M., Florea, A., Maniu, D., 2016. Doxorubicin-incorporated nanotherapeutic delivery system based on gelatin-coated gold nanoparticles: formulation, drug release and multimodal imaging of cellular internalization. ACS Appl. Mater. Interfaces 8 (35), 22900–22913. Available from: https://doi.org/10.1021/acsami.6b07583.

Subia, B., Kundu, S.C., 2013. Drug loading and release on tumor cells using silkfibroin-albumin nanoparticles as carrier. Nanotechnology 24, 035103. Available from: http://iopscience.iop.org/0957-4484/24/3/035103.

Subia, B., Chandra, S., Talukdar, S., Kundu, S.C., 2014. Folate conjugated silk fibroin nanocarriers for targeted drug delivery. Integr. Biol. (Camb.) 6 (2), 203–214.

Subudhi, M.B., Jain, A., Hurkat, P., Shilpi, S., Gulbake, A., Jain, S.K., 2015. Eudragit S100 coated citrus-pectin nanoparticles for colon targeting of 5-fluorouracil. Materials 8, 832–849. Available from: https://doi.org/10.3390/ma8030832.

Swain, S.N., Rao, K.K., Nayak, P.L., 2005. Biodegradable polymers: IV. Spectral, thermal and mechanical properties of cross-linked soy protein concentrate. Polym. Int. 54 (5), 739–743. Available from: https://doi.org/10.1002/pi.1735.

Talegonkar, S., Mishra, P.R., 2004. Intranasal delivery: an approach to bypass blood-brain barrier. Indian J. Pharmacol. 36, 140–147.

Tang, H.-B., Li, L., Chen, H., Zhou, Z.-M., Chen, H.-L., Li, X.-M., et al., 2010. Stability and in vivo evaluation of pullulan acetate as a drug nanocarrier. Drug Deliv 17 (7), 552–558.

Tao, X., Xie, Y., Zhang, Q., Qiu, X., Yuan, L., Wen, Y., et al., 2016. Cholesterol-modified amino-pullulan nanoparticles as a drug delivery carrier: comparative study of cholesterol-modified carboxyethyl pullulan and pullulan nanoparticles. Nanomaterials 6, 165.

Teo, L.A., Shearwood, C., Chye, Ng. K., Lu, J., Moochhala, S., 2006. Transdermal microneedles for drug delivery applications. Mater. Sci. Eng.: B. 132, 151–154.

Thambi, T., You, D.G., Han, H.S., Deepagan, V.G., Jeon, S.M., Suh, Y.D., et al., 2014. Bioreducible carboxymethyl dextran nanoparticle for tumor targeted drug delivery. Adv. Healthcare Mater. 3 (11), 1829–1838. Available from: https://doi.org/10.1002/adhm.201300691.Epub2014Apr17.

Tiwari, G., Tiwari, R., Srivastawa, B., Bhati, L., Pandey, S., Pandey, P., et al., 2012. Drug delivery systems: an updated review. Int. J. Pharm. Investig. 2 (1), 2–11. Available from: https://doi.org/10.4103/2230-973X.96920.

Todhunter, R.J., Wooton, J.A.M., Lust, G., Minor, R.R., 1994. Structure of equine type I and type II collagens. Am. J. Vet. Res. 55 (3), 425–431.

Umamaheshwari, R.B., Jain, N.K., 2003. Receptor mediated targeting of lectin conjugated gliadin nanoparticles in the treatment of *Helicobacter pylori*. J. Drug Target. 11 (7), 415–423.

Vandervoort, J., Ludwig, A., 2004. Preparation and evaluation of drug-loaded gelatin nanoparticles for topical ophthalmic use. Eur. J. Pharm. Biopharm. 57, 251.

Vasir, J.K., Labhasetwar, V., 2006. Polymeric nanoparticles for gene delivery. Expert Opin. Drug Deliv. 3 (3), 325–344. Available from: https://doi.org/10.1517/17425247.3.3.325.

Vaz, C.M., Doeveren, P.F., Reis, R.L., Cunha, A.M., 2003. Soy matrix drug delivery systems obtained by melt-processing techniques. Biomacromolecules 4 (6), 1520–1529. Available from: https://doi.org/10.1021/bm034050i.

Villegas, M.R., Baeza, A., Vallet-Regi, M., 2015. Hybrid collagenase nanocapsules for enhanced nanocarrier penetration in tumoral tissues. ACS Appl. Mater. Interfaces 7 (43), 24075–24081. Available from: https://doi.org/10.1021/acsami.5b07116.

Wadia, J.S., Dowdy, S.F., 2005. Transmembrane delivery of protein and peptide drugs by TAT-mediated transduction in the treatment of cancer. Adv. Drug Deliv. Rev 57 (4), 579–596.

Wan, L., Jiao, J., Cui, Y., Guo, J., Han, N., Di, D., et al., 2016. Hyaluronic acid modified mesoporous carbon nanoparticles for targeted drug delivery to CD44-overexpressing cancer cells. Nanotechnology 27 (13), 135102. Available from: https://doi.org/10.1088/0957-4484/27/13/135102.

Wang, H., Boerman, O.C., Sariibrahimoglu, K., Li, Y., Jansen, J.A., Leeuwenburgh, S.C. G., 2012. Comparison of micro- vs nanostructured colloidal gelatin gels for sustained delivery of osteogenic proteins: bone morphogenetic protein-2 and alkaline phosphatase. Biomaterials 33 (33), 8695–8703.

Wang, X., Wei, B., Cheng, X., Wang, J., Tang, R., 2016. Phenylboronic acid–decorated gelatin nanoparticles for enhanced tumor targeting and penetration. Nanotechnology 27 (38). Available from: https://doi.org/10.1088/0957-4484/27/38/385101.

Wasiak, I., kulikowska, A., Janczewka, M., Michalak, M., Cymerman, I.A., Nagalski, A., et al., 2016. Dextran nanoparticle synthesis and properties. PLoS One 11 (1), e0146237. Available from: https://doi.org/10.1371/journal.pone.0146237.

Watcharin, W., Schmithals, C., Pleli, T., köberle, V., Korkusuz, H., Huebner, F., et al., 2014. Biodegradable human serum albumin nanoparticles as contrast agents for the detection of hepatocellular carcinoma by magnetic resonance imaging. Eur. J. Pharm. Biopharm. 87 (1), 132–141. Available from: https://doi.org/10.1016/j.ejpb.2013.12.010.acs.biomac.5b01003.

Wonqpinyochit, T., Uhlmann, P., Urquhart, A.J., Seib, F.B., 2015. PEGylated silk nanoparticles for anticancer drug delivery. Biomacromolecules 16 (11), 3712–3722. Available from: https://doi.org/10.1021/.

Wu, T., Zhang, Q., Ren, W., Yi, X., Zhou, Z., Peng, X., et al., 2013. Controlled release of gentamicin from gelatin/genipin reinforced beta-tricalciumphosphate scaffold for the treatment of osteomyelitis. J. Mater. Chem. B 1 (26), 3304–3313.

Xiao, L., Lu, G., Lu, Q., Kaplan, D.L., 2016. Direct formation of silk nanoparticles for drug delivery. ACS Biomater. Sci. Eng. 2 (11), 2050–2057.

Yan, Y., Zuo, X., Wei, D., 2015. Concise review: emerging role of CD44 in cancer stem cells: a promising biomarker and therapeutic target. Stem Cells Trans. Med 4, 1033–1043. Available from: https://doi.org/10.5966/sctm.2015-0048.

Yao, X., Chen, L., Chen, X., He, C., Zheng, H., Chen, X., 2014. Intercellular pH-responsive histidine modified dextran-*G*-cholesterol micelle for anticancer drug delivery. Colloids Surf., B: Biointerfaces 121, 36–43.

Yoon, H.Y., Koo, H., Choi, K.Y., Lee, S.J., Kim, K., Kwon, I.C., et al., 2012. Tumor-targeting hyaluronic acid nanoparticles for photodynamic imaging and therapy. Biomaterials 33, 3980–3989. Available from: https://doi.org/10.1016/j.biomaterials.2012.02.016.

Young, S., Wong, M., Tabata, Y., Mikos, A.G., 2005. Gelatin as a delivery vehicle for the controlled release of bioactive molecules. J. Controlled Release 109, 256.

Yuan, A., Wu, J., Song, C., Tanq, X., Qiao, Q., Zhao, L., et al., 2013. A novel self-assembly albumin nanocarrier for reducing doxorubicin-mediated cardiotoxicity. J. Pharm. Sci. 102 (5), 1626–1635. Available from: https://doi.org/10.1002/jps.23455. Epub 2013 Feb 19.

Yue, L., Huang, Z.M., Fong, S., Leong, S., Jakowatz, J.G., Charruyer-Reinwald, A., et al., 2015. Targeting ALDH1 to decrease tumerogenicity, growth and metastasis of human melanoma. Melanoma Res. 25 (2), 138–148. Available from: https://doi.org/10.1097/CMR.0000000000000144.

Zahoor, A., Sharma, S., Khuller, G.K., 2005. Inhaleable alginate nanoparticles as antitubercular drug carrier against experimental tuberculosis. Int. J. Antimicrob. Agents 26, 298–303.

Zauner, W., Farrow, N.A., Haines, A.M., 2001. In vitro uptake of polystyrene microspheres: effect of particle size, cell line and cell density. J. Controlled Release 71 (1), 39–51.

Zhang, H., Ji, Q., Huang, C., Zhang, S., Yuan, B., Yang, K., et al., 2015. Cooperative transmembrane penetration of nanoparticles. Sci. Rep. 5, 10525. Available from: https://doi.org/10.1038/srep10525.

Zhang, L., Pornpattanangkul, D., Hu, C.-M.J., Huang, C.M., 2010. Development of nanoparticles for antimicrobial drug delivery. Curr. Med. Chem 17 (6), 585–594.

Zhang, Y., Wang, H., Gao, C., Li, X., Li, L., 2013a. Highly ordered mesoporous carbon nanomatrix as a new approach to improve the oral absorption of the lipophilc drug. Simvastatin. Eur. J. Pharm. Sci. 49 (5), 864–872.

Zhang, Z., Chen, X., Chen, L., Yu, S., Cao, Y., He, C., et al., 2013b. Intracellular pH-sensitive PEG-block-acetylated dextrans as efficient drug delivery platforms. ACS Appl. Mater. Interfaces 5, 10760–10766.

FURTHER READING

Assaf, S.M., Abul-Haija, Y.M., Fares, M.M., 2011. Versatile pectin grafted poly(*N*-isopropylacrylamide): modulated targeted drug release. J. Macromol. Sci., A 48, 493–502.

Fujioka-Kobayashi, M., Ota, M.S., Shimoda, A., Nakahama, K.I., Akiyoshi, K., Miyamoto, Y., et al., 2012. Cholesteryl group- and cryloyl group bearing pullulan nanogels to deliver BMP2 and FGF18 for bone tissue engineering. Biomaterials 33 (30), 7613–7620.

Ge, Y., Wang, J., Shi, Z., Yin, J., 2012. Gelatin-assisted fabrication of water-dispensable graphene and its inorganic analogues. J. Mater. Chem. 22, 17619–17624.

Girardin, F., 2006. Membrane transporter proteins: a challenge for CNS drug development. Dialog. Clin. Neurosci 8, 311–321.

Hamman, J., 2010. Chitosan based polyelectrolyte complexes as potential carrier material in drug delivery systems. Mar. Drugs 8 (4), 1305–1322.

CHAPTER

Site-specific drug delivery, targeting, and gene therapy 14

Tripti Shukla[1], Neeraj Upmanyu[1], Sharad P. Pandey[2] and M.S. Sudheesh[3]

[1]School of Pharmacy and Research, Peoples University, Bhopal, India [2]Department of Pharmacy, SGSITS, Indore, India [3]VNS Institute of Pharmacy, Bhopal, India

14.1 INTRODUCTION

Drug delivery systems and their development as targeted or site-specific delivery systems have fascinated the formulation scientists as a tool to overcome the drawbacks related to drugs and increase safety and efficacy with better patient compliance. William Brokedon started the journey of the modern drug delivery system by preparing compressed tablets more than 150 years ago, and medicine has progressed far since that initial step. Continuous need of better patient-compliant formulation, safety and efficacy-related issues, bioavailability, dosage regimen, toxicity of the drug, and high dosing frequency, etc., are some major issues which draw the attention of formulation scientists to develop a system which can overcome such issues.

It is a well-known fact that every drug needs to bind a specific receptor or needs to cross the cell barrier for better and effective action. At the same time, the correct amount of the drug at the correct site is also needed to avoid toxicity and achieve better bioavailability with increased therapeutic response. Generally, it has been observed that the drug loaded to any conventional system and administered by any route has several limitations like pH-dependent degradation, high protein binding, and unusual distribution through the entire body limiting to the action of drugs and increases its undesired effect on the various organs. Research also shows that a number of organic molecules are no longer used due to their toxicity and there are several potent molecules which have been removed from the market due to unwanted effects on the major organs. Several coxib derivatives like rofecoxib, melecoxib, and valdecoxib, etc., are good examples of drugs removed from the market due to their hepato-toxic and nephron-toxic behavior even though they are potent pain relievers.

Over past two decades, pharmaceutical and medicine systems have experienced tremendous change and advancement, and can be said to be in a phase of transition from conventional to novel drug delivery systems. Cost effectiveness,

Nanoarchitectonics in Biomedicine. DOI: https://doi.org/10.1016/B978-0-12-816200-2.00013-X

bioavailability issues, and toxicological effects related to certain drugs have drawn the attention of scientists to develop a site-specific delivery system to target a few specific tissues, especially in case of cytotoxic drugs. But, it has also been observed that either current systems are providing the symptomatic relief in majority of cases or it is leading to the toxic effects. Genetic diseases are largely untreatable and only symptomatic relief can be provided. The treatment of diseases at root level, especially those associated with genetics as the result of mutation or deletion of genes which lead to the impairment of normal metabolic pathways, have directed the current medicine system toward a new era, that is, gene therapy. The present techniques for gene therapy, that is, use of a gene gun and viral vectors, have certain limitations which can be minimized by using a polymeric system to deliver the gene (Vyas and Khar, 2004).

Over the past 50 years, pharmaceutical research scientists have done significant work regarding the development of a drug delivery system which has a targeted effect at the effector organ for the treatment of some specific diseases such as carcinomas, sarcomas, benign tumors, liver problems, etc. Targeting the particular organ, tissue, or cell decreases the toxicity due to a specific molecule and, at the same time, also increases the effectiveness and safety of the drug. In principle, targeting involves the modification of the pharmacokinetic behavior and cellular distribution of a drug by altering either the chemical properties of the drug or via the use of various carrier systems like liposome, niosomes, solid lipid nanoparticles, and transferosomes, etc. Selective targeting of a drug to a specific site allows the maximum concentration of the drug to the target site, therefore, increasing the efficacy. The majority of research related to targeting drugs is associated with cancer in the quest to decrease the toxicity of anticancer molecules. The blood brain barrier (BBB) is the major hindrance in the reach of drugs for brain tumors, which have been overcome by the use of specifically designed carrier molecules which have the ability to cross the BBB. Recent research has also revealed the advent of new chemical entities (NCEs) which are more potent, but have limitations related to their distribution and their toxic effect on healthy tissues (Molema and Meijer, 2001).

By incorporating or attaching NCEs with specific carriers, these problems could be overcome to a great extent through the considerable alteration in the drug's pharmacokinetic properties. Similarly, therapeutic macromolecules, gene transcription/translation modulating agents such as antisense oligonucleotides and genes may also be effectively used in the field of drug development. For these treatment modalities to become a major success, the delivery and/or targeting of these compounds will be an essential component (Molema and Meijer, 2001).

The basic aim of this chapter is to discuss the various concepts and systems of targeting and site-specific delivery, which has come into existence in the past few decades with a special focus on gene therapy. We have addressed the different types of carriers and their physiochemical and biological characteristics which are needed to develop site-specific drug delivery or for targeting. Furthermore, some formulations intended to lead the targeting or site-specific delivery which are either in the phase of clinical investigation or have been approved for clinical use are also addressed. A special concern has been given to the various carriers and their mechanisms, including viral and nonviral vectors with a special mention of polymeric systems used to deliver the gene to a specific site.

14.2 SITE-SPECIFIC, TARGETED DRUG DELIVERY SYSTEM

Targeting of any tissue or organ is based on the risk–benefit ratio associated with the drug and the condition that the targeted organ is distinguishable from the other organs with proper biochemical structure and reach. Over the past 20 years, several approaches have been used by various scientists to target specific organs, including:

1. Passive targeting
   a. Enhanced permeability retention (EPR)
   b. Localized delivery
2. Active targeting
   a. First-order targeting
   b. Second-order targeting
   c. Third-order targeting
3. Inverse targeting
4. Ligand-mediated targeting
5. Dual targeting
6. Double targeting
7. Combination targeting
8. Physical targeting

14.2.1 PASSIVE TARGETING

Passive targeting may be considered as the natural response of the body toward the molecule which has to be target. In this condition the inherent system of our body directs the movement and accumulation of the drug to a particular tissue or organ. Generally, this system is based on the deliverance of therapeutics by the macrophage system or monocyte-phagocytic system of other cells depending on the size and character of the therapeutic agent. Through this, the drug carriers get depleted over time, releasing the drug either into circulation or the target organ (Vyas and Khar, 2004).

Engulfment of biomolecules and colloidal carriers by the macrophage system has provided a tremendous opportunity for formulation scientists to achieve the targeting of such organs which utilize macrophages, that is, organs of the reticuloendothelial system, and can effectively be used to treat liver-, lung-, or brain-related diseases. At the same time, it could also be helpful to treat diseases related to the storage of lysosome enzymes (Vyas and Khar, 2004).

14.2.1.1 Enhanced permeability retention

Enhanced permeability and retention effect in the abnormal tumor tissue in comparison to normal tissue due to hypervascularization could be exploited for the targeting of macromolecules to the tumor (Fig. 14.1B). From the literature, it is evident that aberrant vascular architecture, stimulating extravasation within tumor tissues due to excessive production of vascular permeability factors, and the lack of lymphatic drainage creates a major opportunity for the targeting of drugs to such tissues (Greish, 2010; Azzopardi et al., 2013).

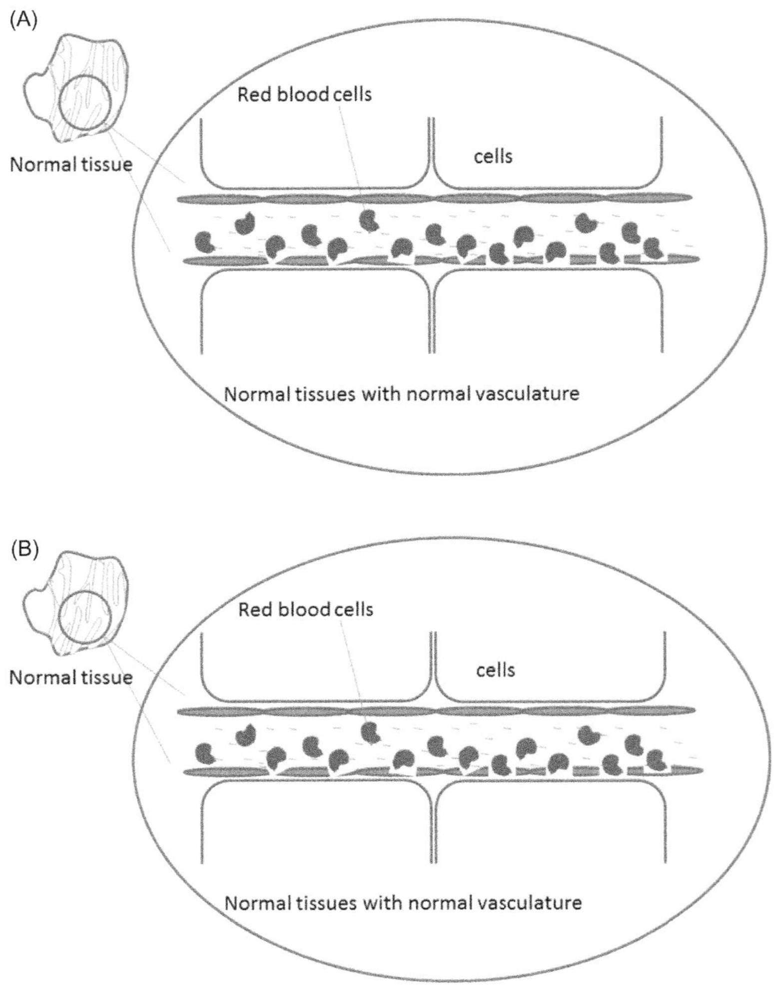

FIGURE 14.1

(A) Normal tissue with normal vasculature. (B) Enhanced permeability retention in tumor with abnormal vasculature.

A major limitation inherent to most conventional anticancer chemotherapeutic agents is their lack of tumor selectivity. One way to achieve selective drug targeting to solid tumors is to exploit the abnormalities of the tumor's vasculature, namely hypervascularization, aberrant vascular architecture, extensive production of vascular permeability factors stimulating extravasation within tumor tissues, and lack of lymphatic drainage. Maeda et al. have extensively studied tumor vascular abnormalities in terms of active and selective delivery of anticancer drugs to tumor tissues, notably defining the EPR effect of macromolecular drugs in solid tumors. Due to their large molecular size, nanosized macromolecular anticancer drugs administered intravenously (i.v.) escape renal clearance. Often they cannot penetrate the tight endothelial junctions of normal blood vessels, but they can extravasate in tumor vasculature and become trapped in the tumor vicinity. Over time, the tumor concentration will build up reaching several folds higher than that of the plasma due to the lack of efficient lymphatic drainage in the solid tumor; an ideal application for EPR-based selective anticancer drug delivery (Fang et al., 2011).

Vascular permeability enhancement, high vascular density, and rapid angiogenesis are some common factors that are responsible for the enhanced EPR effect (Maeda et al., 2009).

One thing which differentiates tumor cells from normal infected cells is the dysfunctioning of the lymphatic tissues, and EPR will not be protective due to rapid lymphatic clearance. Instead, some researchers have demonstrated early and significant macromolecular accumulation at infection sites (Evans et al., 1973; Laverman et al., 2001).

The smooth muscle layer in normal blood vessels responds toward vascular mediators (bradykinin, acetylcholine, NO, and calcium) through the receptors present at the surface leading to the maintenance of a constant blood-flow volume. Suzuki et al. have reported that a significant rise in blood pressure for a specific period of time by using angiotensin II (AT II) into tumor-bearing rats caused 2- to 6-fold selective increase in tumor blood flow, while the blood flow in normal organs and tissues remained constant regardless of the induced blood pressure. A further study has shown the significant rise of chromium labeled albumin retention in mice models having hypertension due to Angiotensin II infusion (Suzuki et al., 1981).

14.2.1.2 Localized or physical delivery

As the name indicates, this type of delivery is intended to deliver the drug at a specific, localized position. The best example of such a formulation may be considered as the enteric-coated tablets intended to deliver the drug into the intestine. Localized delivery of the desired molecule or formulation by means of injections or catheters to a specific position may be considered a more promising targeted delivery system. The major application of such a delivery system is to provide the localized effect of the therapeutic agent to inhibit restenosis (Labhasetwar et al., 1998; Song et al., 1998).

Anti-infective agents or drugs used to treat alopecia can also be administered directly to the hair follicles using a transdermal delivery system. Administration of nanoparticulate matters by using an aerosol may provide better treatment for asthma-like conditions and management aid in the case of rejection during lung transplant. Use of aerosolized cyclosporine by inhalation is the best example of such delivery for treating lung rejection conditions with greater efficacy. The intravitreal administration of drug-loaded nanomaterial is one of the most patient-compliant and efficacious ways to treat eye diseases especially, for example, glaucoma (Mainardes et al., 2005). The intravitreal administration of biodegradable polymer-based nanoparticulate materials have shown a better accumulation in retinal epithelium becoming a better aid for treating retinitis (Song et al., 1998).

14.2.2 ACTIVE TARGETING

Active targeting is a noninvasive way used to transport the drug to the target cells, tissue, or organ by using various influx transport systems, such as the carrier-mediated transport system, receptor-mediated endocytosis (Fig. 14.2A),

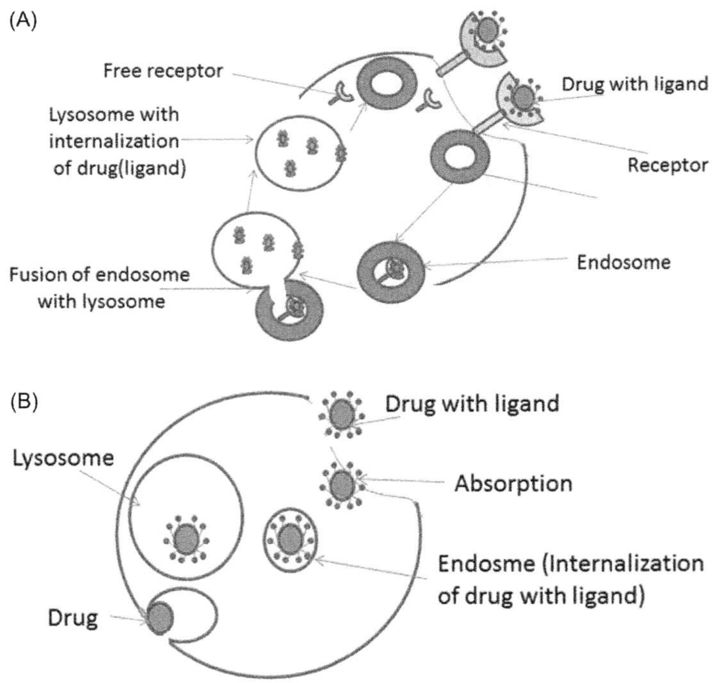

FIGURE 14.2

(A) Diagrammatic representation of receptor-mediated endocytosis. (B) Diagrammatic representation of absorption-mediated endocytosis.

and absorptive-mediated endocytosis (Fig. 14.2B). Delivery of higher therapeutic concentrations of the drug at the site of action is the first and foremost characteristic of active targeting, which can be achieved by target-based specific delivery of the drug. For the carrier-mediated transport system, pairing of drug carriers like liposome or other particulate nan-carriers with a ligand-engineered homing device leads to the specific targeting of the selected cells. Targeting ligands can be broadly classified as proteins (mainly antibodies and their fragments, such as TAT), nucleic acids (aptamers), or other receptor ligands (peptides, vitamins, and carbohydrates).

There are several approaches for the active targeting of drugs and may be further classified on the basis of three levels of targeting.

First-order targeting (organ level)
Second-order targeting (cellular level)
Third-order targeting (intracellular)

14.2.2.1 First-order targeting

Targeting of a drug moiety attached with carrier molecules to the capillary bed of the desired site, organ, or tissue (such as plural cavity, lymphatic, lungs, joint, eyes, etc.) is referred to as first-order targeting. Some researchers also categorize it as passive targeting (Vyas and Khar, 2008).

It has been noticed that liposomes get oozed out from the cell junction due to increased vascular permeability in tumorous cells rather than normal cells which have tight endothelial junctions. Oozing of liposomes, its movement through circulation of its entrapment is clearly governed by size. Larger liposomes get removed by mechanical filtration through the lung and smaller vesicles up to 150 nm are removed by the macrophages or its activity toward the MPS working as a natural targeting system for the liver, spleen, etc. Even smaller liposomes could effectively be used to target a tumor system due to its ability to ooze out through the vascular junction in diseased cells (Vyas and Khar, 2008).

14.2.2.2 Second-order targeting

When a delivery system is designed in such a way that it releases a drug to a particular cell within an organ or tissue is referred to as second-order or cellular targeting. It could easily be done by attaching an antibody to the delivery system which specifically recognizes and gets attached to a specific antigen on a cell surface. Monoclonal antibodies attached to the surface of polylactic-*co*-glycolic acid (PLGA) is one of the better examples which can be used to treat breast cancer (Vyas and Khar, 2008; Kocbek et al., 2007).

The approach of cellular targeting could also be explained by the selective delivery of dexamethasone coupled with mannosylated albumin (Dexa 5-Man10-HSA) to the Kupffer cells leading to a reduction of intrahepatic reactive oxygen species in vivo and tumor necrosis factor-α production in vitro and decreased depletion of glycogen in vivo, indicating effective pharmacologic targeting (Melgert et al., 2001).

14.2.2.3 Third-order targeting

Formulations which are designed to deliver the drug specifically to the intracellular system are termed third-order or subcellular targeting. Subcellular targeting is mostly achieved by modifying the delivery in such a way that it could easily penetrate the target cells or easily facilitated to move through the cellular membrane, and may be a ligand-mediated system or receptor-mediated system.

There are several biosynthetic and endocytic trafficking systems such as retrieval proteins, coat proteins, Rab GTPases, and soluble *N*-ethylmaleimidesensitive factor accessory protein receptor proteins which ensure precise targeting to distinct organelles. Sometimes proteins and lipids at the plasma membrane play a similar role in signaling and pathogenesis as lipid rafts (dynamic cholesterol–sphingolipid assemblies) (Rajendran et al., 2010).

There are several endocytosis mechanisms which are gaining popularity to provide an authentic way to target subcellular systems and two of these are clathrin-mediated endocytosis and raft-mediated endocytic routes (Rajendran et al., 2010).

One of the best examples of subcellular targeting is transferrin (TF) (iron-transporting serum glycoprotein) which is efficiently taken up into cells by the process of receptor-mediated endocytosis through TF receptors (TfRs) which are present on the surface of most proliferating cells, in elevated numbers on erythroblasts, and on many kinds of tumors (Wagner et al., 1994).

In a similar fashion, the cancers which are most difficult to treat via classical methods may be targeted with folate-linked therapeutics (Lu and Low, 2012).

14.2.3 INVERSE TARGETING

The reticuloendothelial system (RES) uptake of a colloidal drug delivery system (Fig. 14.3) in the majority of cases has been found disadvantageous. At the same time, accumulation of the carrier system in the RES may also lead to RES blockade and impairment in defense of the host and accumulation of drug in major RES organs may cause toxicity. But in some cases, the concentration of the drug in the plasma is more advantageous than the RES. Therefore, several attempts have been made to increase the biodistribution by avoiding the RES uptake and such processes can be referred as inverse targeting (Illum et al., 1986). There are various ways of inverse targeting:

1. By reducing the function of RES
2. By modifying the size
3. By modifying the surface charge
4. By modifying the composition and rigidity of the carrier
5. By incorporating hydrophilicity in the carrier

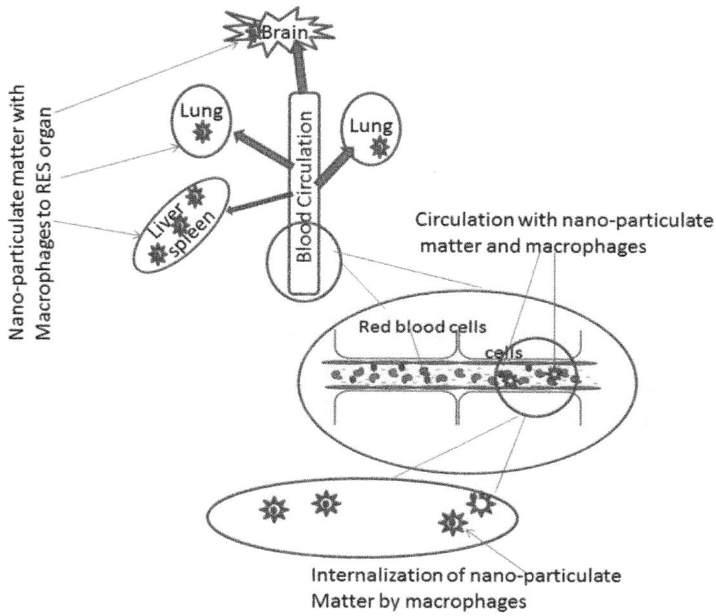

FIGURE 14.3

Active targeting through the RES. *RES*, Reticuloendothelial system.

Preinjection of blank carrier or macromolecules like dextran sulfate before the injection with the final formulation may lead to the super-saturation of the RES, but may lead to the blockade of RES or impair the defense of the host (Patel et al., 1983).

Particle size has the utmost importance in its clearance through RES. Research shows that the blood clearance and organ deposition of polystyrene and cellulose in rabbit models of small (1.27 pm diameter) polystyrene microspheres are easily taken up by the RES, while large polystyrene particles (15.8 pm diameter) are mechanically filtered by capillary beds of the lung. This clearly indicates that by changing the size and surface of the medication, targeting can be modified (Illum et al., 1982).

Koyo et al. found increased liver uptake of the cation-charged molecules rather than the anionic carriers and this phenomenon may be promising in targeting the liver or inverse targeting (Koyo et al., 1990).

Use of surfactant material may also change the pattern of targeting, as it has been observed that coating of microspheres with nonionic surface-active agent poloxamer 338 reduces the uptake in the liver and increases the concentration of carriers in the lungs while uncoated microspheres of 1.27 μm diameter are cleared rapidly from the blood primarily by the RES (Illum and Davis, 1983).

Surface medication of carrier molecules, such as rigidization, may lead to decreased uptake by RES. As in the case of liposomes, it is easily uptaken by

RES due to a high degree of opsonization while the rigidity in the liposomal carrier using sphingomyelin, DSPC, cholesterol, or other rigidifying phospholipids reduces its uptake by RES helping inverse targeting (Allen et al., 1989).

Coating of the colloidal carrier for hydrophilic material like secretory immunoglobulin A (SIgA), egg lecithin, and nonionic surfactants (poloxamer 338), etc., is another approach to avoid the uptake by the RES and movement toward the liver (Ilium et al., 1986).

14.2.4 LIGAND-MEDIATED TARGET

The ligand-mediated target system (LMTS) may be a successful way which can be utilized to increase the safety and efficacy of drug molecules especially cytotoxic drugs and decreasing its toxicity by targeting it at the specific cells. Presently, there are almost 100 such products available in clinical trials and a few of them (such as an immune-toxin and an immune-conjugate) have already received clinical approval. But the success of such targeting is entirely based on the choice of ligand, degree of receptor expression, choice of antibody, antibody fragments or nonantibody ligands, binding affinity of the ligand, and, most importantly, whether the ligand is internalized or not (Garnett, 2001).

The addition of a specific antibody, such as monoclonal antibodies (mAb), at the ligand-mediated targets is one approach used to target the therapeutics at a specific position. Higher binding affinity due to presence of two binding sites on the molecule is a major advantage of using whole mAbs, but also imposes limitations due to binding to normal tissues through Fc receptors particularly on macrophages leading to high liver and spleen uptake of the LMTS. Methotrexate, vinca alkaloid, and anthracylines are some examples of drugs which have been linked to antibodies using various chemical linkers (Allen, 2002).

14.2.5 DUAL TARGETING

Dual targeting is a phenomenon where the inherent property of a carrier molecule is modified in such way that it could synergistically increase the active moiety for delivery at the target site. However, chemotherapy of brain gliome is limited by two major factors, that is, the inability to transport of drug through the BBB and penetration of the drug into the tumor cells. These two hurdles can be minimized by developing the liposomal delivery which is conjugated with p-aminophenyl-α-D-mannopyranoside (MAN) and TF (Ying et al., 2010).

TF is a serum glycoprotein (80 kDa) with isoelectric point 5.5 and exhibits the negative charge in the solution with pH 7.4 leading to a lower zeta potential value. At the same time, MAN helps to delocalize the negative charge centers which results in an increase in the zeta potential value of the liposomes. After dual modification with MAN and TF, the negativity of TF gets weakened by MAN resulting in a higher zeta potential value compared with that of daunorubicin liposomes.

Dual-targeting daunorubicin liposomes are able to improve the therapeutic efficacy of brain glioma in vitro and in animals (Ying et al., 2010).

14.2.6 DOUBLE TARGETING

The combination of drug targeting and is another strategy that may be considered as the basis of double targeting. The combination in double targeting is made by spatial and temporal control (control of drug release prior to targeting) of drug delivery. Spatial control combined with temporal control leads to a better and improved therapeutic index via increased spatial selectivity and activation of drug release in the selected locality with a preferable pattern such as zero order for longer duration (Dunn et al., 2014).

A new polymer–nanoparticle composite-based drug carriers that respond simultaneously to both a change in pH and heat generated by alternating magnetic field forms a unique combination which allows temporal and spatial control to get a burst release of fluorescein amine (a model therapeutic molecule) via bond breakage in a specific region (Dunn et al., 2014).

Photoimmunotherapy of cancer which includes the combination of phototoxic material, hematoporphyrin, and target-oriented M1antibodies is another good example of double targeting. A study revealed the suppression of M-1 tumor (DBA/PJ myosarcoma) growth when the tumor-bearing animals were exposed to incandescent light (Mew et al., 1983).

Several other examples of such conjugates (photosensitizer and monoclonal antibodies) have been reported (Oseroff et al., 1986).

14.2.7 COMBINATION TARGETING

Combination therapy refers to the modification of the molecule which is used for targeting in such a way that it facilitates direct targeting of a specific site. It is a relatively new approach that leads to comparatively better-controlled drug delivery. In 1994, Bonucci et al. tried to utilize combination targeting for the prevention of ovariectomy osteopenia in rats after vaginal administration of benzyl ester of hyaluronic acid (Hyaff 11) microspheres containing salmon calcitonin. They demonstrated that a combination of physical properties of the delivery device and site of delivery may be used to regulate the delivery of the drug. Microspheres are adhered to the mucosal tissue which leads to the protection of calcitonin from proteolytic digestion due to its slow release after the degradation of the microspheres (Bonucci et al., 1995).

The term combination therapy was suggested by Pettit and Gombotz when defining their research related to the site-specific delivery of proteins and peptides, where some of the characteristics of the proteins, peptides, or delivery systems such as molecular weight or particle size, ionic charge, hydrophobicity, membrane or device permeability, susceptibility to biodegradation, and molecular specificities may influence targeting (Pettit and Gombotz, 1998).

Modification of the proteins and peptides with natural or synthetic polymers may change the characteristics to enable targeting. Further use of carrier molecules for these modified protein and peptides may enhance site-specific targeting (Vyas and Khar, 2008).

Studies show that combination targeting has wide utility comprising the modulation in the bioavailability and biodistribution of the drug or carrier molecule, increased transport of the drug and carrier to the tumor cells, significantly higher sensitivity for detection of endothelial injury, and assessment of tumor angiogenesis (Papademetriou et al., 2013; Sun et al., 2010; Bitterman and Du, 2016).

Use of cell adhesion molecules (CAMs) expressed on the vascular endothelium is one of the most important examples of combination targeting. Papademetriou et al. examined nanocarriers functionalized for dual- or triple-targeting to intercellular, platelet-endothelial, and vascular CAMs (ICAM-1, PECAM-1, VCAM-1) differing in expression level, location, pathological stimulation, and endocytic pathway and concluded that multitargeting to CAMs can be controlled by the combination of targets utilized and multiplicity of affinity moieties. This approach may allow manipulation of the targeting performance of nanocarriers, tailored by combining properties of multiple targets rather than optimizing features of the nanocarrier or targeting ligand (Papademetriou et al., 2013).

14.2.8 PHYSICAL TARGETING

The physiology and environment of infected tissues or cells are very different to the normal tissues and cells in several properties like pH and temperature. Cancerous and inflamed tissues are present in the acidified and higher temperature conditions and these two abnormal conditions may be used to develop a stimulus-responsive drug delivery system. These two means are endogenous in origin and lead to developing the carrier for physical targeting. Such developed carriers even move to the entire circulation, but are degraded to the point where a higher temperature or acidic environment leads to the release and accumulation of the drug (Stefanadis et al., 2001; Gerweck and Seetharaman, 1996).

A study shows that doxorubicin (DOX) loaded pH-sensitive poly(ethylene glycol)-poly(L-histidine)-poly(L-lactide) (PEG-PH-PLLA) nanoparticles show better and much faster release of DOX in pH 5.0 than in pH 7.4 (Liu et al., 2011).

A similar study was mentioned by Torchilin in his review regarding pH-sensitive liposomes used to deliver experimental drug and genetic material (Torchilin et al., 1993).

Xu et al. reported that the apoptin (protein) loaded acid-sensitive nanoparticle of crosslinked glycidyl methacrylate-derived dextran shows better release in an acidic environment of the endosomes after endocytosis (Xu et al., 2013).

In several studies, it has been observed that pH or temperature stimuli alone have a limited effect in targeting, but if both are prepared, they will have better effects. Zhang et al. have reported the dual responsive (both to thermo and pH)

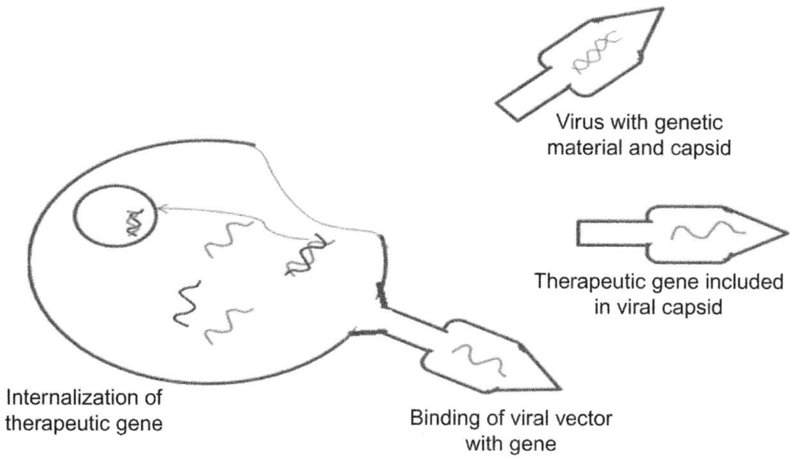

FIGURE 14.4

Diagrammatic representation of gene therapy using viral vectors.

nanoparticles of P(NIPAAm*co*-AA)-*b*-PCL a diblock copolymer comprised of a thermoresponsive hydrophilic P(*N*-isopropylacrylamide-*co*-acrylic acid) block and a biocompatible hydrophobic polycaprolactone (PCL) block, and is used to deliver paclitaxcel to the tumorous cells (Zhang et al., 2007).

Similarly, physical targeting may also be achieved by means which are applied exogenously as in the application of magnetic force for the targeting. Carriers for such types of targeting are materials that have ferromagnetic properties (Fig. 14.4). When a external magnetic field is applied to a specific position it governs the movement of such carriers leading to the accumulation and release of the drug in a specific region. Chomoucka et al. (2010) have researched the use of magnetic nanoparticles for targeted drug delivery and targeted drug therapy to deliver the drug directly to the target cell without having an effect on the body.

A study of iron oxide nanoparticles revealed that magnetic targeting induced a 5-fold increase in the total glioma exposure to magnetic nanoparticles over non-targeted tumors and a 3.6-fold enhancement in the target selectivity index of nanoparticle accumulation in glioma over the normal brain, indicating significant targeting (Chertok et al., 2008).

14.3 VARIOUS CARRIER SYSTEMS FOR TARGETING

Specific use of a carrier system during the targeted delivery is of utmost importance. The choice of carrier system depends on the type of target tissue or organ, drug molecule loaded in the carrier molecule, diseased condition, etc.

(Molema, 2001) Carrier particulates used for targeting can be broadly divided into three basic categories:

1. Particulate-type carrier systems including liposome, nanostructured lipoid carriers, microspheres, nanoparticle, and polymeric micelles, etc.
2. Soluble-type carrier systems including monoclonal antibodies, modified plasma proteins, peptides, and polysaccharides, etc.
3. Biodegradable carriers including various polymeric compositions.

14.3.1 LIPOSOME

Liposomes are microparticulate colloidal carriers composed of different types of lipids existing in the drug in aqueous form entrapped by one or more layers. Depending on the lipophilicity the drug is either entrapped in the lipid vesicles or in the phospholipid bilayer. Since the advent of liposomes almost 40 years ago, it has been converted into one of the most fascinating systems for biological scientists in the development of newer formulations containing liposomes like Doxil and Myocet which have opened a new dimension. The interest in this field is very high considering that more than 2000 published research articles, patents, and review articles detail the different aspects and application of liposome. With the progression of time, one major drawback, that is, elimination of liposome and faster uptake by the RES, has drawn the attention of formulation scientists to develop safer and more efficacious systems like immunoliposomes and stealth liposomes. Immunoliposomes were prepared by attaching some specific ligands at the surface to target specific cells and tissues. These ligands are attached to the surface in such a way that it doesn't affect the integrity of the liposomal system. One of the most common ligands attached to prepare immunoliposome contains immunoglobulin G (Torchilin, 1984).

However, the maximum proportion of the administered immunoliposomes go to the liver due to insufficient time to get attached to the target. Such problems can be eliminated by preparing long circulation liposomes and an attempt was made by attaching the antibody and poly ethylene glycol (PEG) on the surface of the same liposome. It is also be noticed that the antibody should try to attached with the PEG arm to reduce the steric hindrance occurred due to the coating of liposome by PEG (Torchilin et al., 1992; Torchilin, 2005).

PEG transported to the liposomal surface should always be attached in such a fashion that it could easily be removed after the targeting through EPR for better results. Pegylation of the liposome is one of the most popular techniques of stealthing, but researchers are continuously trying to prepare it by using other hydrophilic polymers. Some common examples of polymers being used in the series of such preparations are poly [*N*-(2-hydroxypropyl) methacrylamide], poly-*N*-vinylpyrrolidones, polyvinyl alcohol (PVA), and amino acid-based polymers, etc. (Torchilin, 2005)

Apart from the use of conventional monoclonal antibodies, some other antibodies like antibody CC52 directed against rat colon adenocarcinoma CC531 lines has been used as a ligand to achieve targeting (Kamps et al., 2000).

Similarly doxorubicin-loaded long-circulating liposomes (Doxil, Alza Pharmaceuticals) were modified with the monoclonal nucleosome (NS)-specific 2C5 antibody (mAb 2C5) modified with poly(ethylene glycol)-phosphatidyl ethanolamine conjugate (PEG-PE) with the free PEG terminus activated with the *p*-nitrophenylcarbonyl group (pNP-PEG-PE) that recognizes a broad variety of tumors via the tumor cell surface-bound NSs (Lukyanov et al., 2004).

It has also been observed that tumor cells have overexpressed folate receptors drawing the attention of researchers to develop folate-based delivery systems such as folate-modified liposomes. Li et al. have found that folate-poly (PEG-cyanoacrylate-*co*-cholesteryl cyanoacrylate) (FA-PEG-PCHL)-modified freeze-dried liposomes (FA-PDCT-L) show a higher concentration of docetaxel in the tumors rather than the liver and spleen so that the presence of FA-PEG-PCHL led to better accumulation of liposome in the tumor (Li et al., 2011).

TF receptors are another majorly overexpressed system present in the majority of tumors and so TF and antibodies against TfRs can be used as ligands for liposomal targeting. A study shows that TF-PEG-liposomes were readily internalized by receptor-mediated endocytosis after binding to mouse Colon 26 cells. It has also shown a longer residence time in the circulation and lower uptake by the RES in tumor-bearing mice (Ishida et al., 2001).

In another study, TfR-targeted liposomal formulation of hosphorothioate antisense oligodeoxyribonucleotide (ODN or genasense) was introduced to bcl-2 K562 leukemia cells and more efficient uptake was observed than in the nontargeted controlled system. Targeting of liposome with TfR also led to the significant decrement in the bcl-2 protein (Chiu et al., 2006).

The search for another ligand for the delivery of liposomal systems has opened a new dimension with the use of vasoactive intestinal peptide (VIP) for radionuclide-loaded PEG liposome to VIP receptors present on the tumor leading to a marked decrease in breast cancer in rats (Dagar et al., 2003).

14.3.2 LIPOPROTEINS

Lipoproteins are nanoparticulate carrier systems of endogenous origin and are mainly responsible for the movement of hydrophobic lipids in the blood toward a different cell. Being endogenous in origin, these nanostructure systems are taken up by the RES and are circulated for a longer duration in the systemic flow, so providing a better alternative to other bioincompatible nanocarrier systems (Firestone, 1994).

Studies have also shown that tumors have overexpressed low-density lipoprotein (LDL) receptors especially in cancers, and cholesterol and lipid systems are required for the proliferation and these two characteristics, which may be the key point for targeting therapeutics encapsulated with LDLs (Zheng et al., 2005).

Surprisingly, malignant tumors have a high affinity for the LDLs and this behavior makes LDL useful targeting molecules. Chen et al. have shown enhanced optical imaging in live animals by ligand modified LDL nanoparticles loaded with near infrared dye (Chen et al., 2007).

Corbin et al. (2007) have developed engineered high-density lipoprotein (eHDL) nanoparticles and demonstrated its capability in targeting and use in diagnostics.

Several other attempts have also been made to deliver drugs (e.g., antiviral and antimalarial) to the liver by using lactosylated LDL and HDL nanoparticles. By using the acylated LDL nanoparticles, Kupffer cells, and sinusoidal endothelial cells can be targeted in a more efficient way (Molema, 2001).

14.3.3 NANO-PARTICULATE SYSTEM

In nano-particulate systems, polymeric nanoparticles and dendrimers are the two major categories which are used for targeting. Polymeric nanoparticles are used in the nanosized colloidal delivery system which either contains the drug at the core or the drug is adsorbed on the surface in such a way that produces the controlled release at a specific or targeted site through ligand-driven systems. Polymeric particles behave like a homing device for the therapeutics and the release of the drug occurs by single or miscellaneous mechanism including diffusion, hydrolysis, enzymatic degradation, etc. Some common examples of biocompatible hydrophilic include PVA, polyethylene glycols (PEG), monomethoxy poly-(ethylene glycol) (mPEG), polysorbate, and vitamin E d-alpha tocopheryl polyethylene glycol 1000 succinate (TPGS), among others. These polymers provide a hydrophilic chain-like system at the surface of the nanoparticle preventing its phagocytosis (Masood, 2016).

These polymeric nanoparticles are accumulated in the interstitial space of the tumor cell due to the leaky structure of tumor vessels and the impaired lymphatic system causing an EPR effect.

Studies shows that poly lactic acid (PLA), poly glycolic acid (PGA), and the copolymer of both, that is, PLGA, have proven their efficacy regarding the development of target-oriented controlled-release nanoparticles. PLA/PLGA-sustained release nanoparticles of docetaxel and controlled-release nanoparticles of paclitaxel prepared by the combination of PLGA and vitamin E are two common examples of this category (Mu and Feng, 2003). PLGA copolymers degrade in the body by hydrolytic cleavage of ester linkage to lactic and glycolic acid. These monomers are easily metabolized in the body via the Krebs cycle and eliminated as carbon dioxide and water (Bala et al., 2004; Musumeci et al., 2006).

Dendrimers are one more nanoparticulate delivery system which is used to target the tissues. These are nanosized, hyperbranched, radially symmetric, and nearly mono-dispersed systems. Sometimes these are also referred to as cascade molecules, but this is not a well-established term in comparison to dendrimers. It has been proven as one of the promising tools for the therapy and diagnosis of cancer including the targeted delivery of the antineoplastic agent (Abbasi et al., 2014). Studies have also demonstrated that a well-designed dendrimer structure can be potentially designed for desired biocompatibility, bioavailability, pharmacokinetics, and localized delivery of therapeutics to malignant tissues (Wolinsky and Grinstaff, 2008).

Huang et al. demonstrated the potential use of nucleic acid ligand (aptamers) conjugated gold nanoparticles (AuNPs) for cancer cell detection and in their study they observed intense accumulation of the aptamer-modified gold nanoparticles (Apt-AuNPs) in cancer cells in comparison to the normal cells. They also mentioned that the presence of Apt-AuNPs suppresses the proliferation of MDA-MB-231 cancer cells (Huang et al., 2009).

14.3.4 POLYMERIC MICELLES

Polymeric micelles have emerged as one of the most advantageous tools for the delivery of water-insoluble drugs, although the stability of this carrier system in plasma has raised questions. A polymeric carrier system has a core shell with an A–B diblock structure consisting of a hydrophilic block A and hydrophobic block B. A hydrophobic drug is encapsulated into the core of the micelles while the outer part (hydrophilic block) of the micelles protects the drug from the aqueous environment and avoids the recognition of the micellar system in vivo by the RES (Jones and Leroux, 1999).

Biodegradable, small size, longer circulation time, and accumulation at the required site are some important features needed for better targeting, and polymeric micelles have all these necessary characters. The modification of micelles by using a specific ligand or introducing a pH-sensitive moiety according to the disease condition may be used for active targeting (Kedar et al., 2010). Polymeric micelles have the ability to be internalized into the target cells by endocytosis even without a surface ligand. Paclitaxel-loaded amphiphilic di- and triblock copolymers of poly(ε-caprolactone) (PCL) and monomethoxyl poly(ethylene glycol) (MPEG) show enhanced thermodynamic stability and efficient cellular uptake (Shuai et al., 2004).

14.3.5 CELLULAR AND BIOMIMETIC DRUG DELIVERY CARRIER

Polymeric drug delivery systems have demonstrated their ability for targeting, but they have shown limitations in meeting clinical expectation. This limitation can be avoided by the use of natural cells including pathogenic cells, normal human cells, and immune cells, etc. More intense research related to such carriers is required. Use of immune cells, phagocyte cells (monocyte, macrophages, etc.), lymphocyte, and stem cells, etc., have been proven to be promising drug delivery systems due to their prolonged circulation, decreased immunogenicity and cytotoxicity, better penetrability, and transport to diseased site, etc. (Batrakova et al., 2011) Targeting of methotrexate to hepatocarcinoma cells and cisplatin to ovarian cancer cells through tumor cell-derived microparticles is a good example of such a delivery system (Tang et al., 2012). Resealed erythrocytes are another good example for the targeted and controlled or sustained delivery of bioactive agents, and the advancement regarding such delivery system has been tremendous. There are two basic mechanisms for erythrocyte and therapeutic combination. Either the drug could be loaded to the erythrocyte system or the drug molecule could be attached on the erythrocyte surface, depending up the need (Hamidi et al., 2007).

However, these cellular systems have certain limitations, such as loading the drug being less, disintegration and clearing off the entrapped drug by immune cells and stem cells and unloading of the drug before the target site. Such problems can be minimized by using the polymeric systems coated with cellular system. Steinfeld et al. (2006) demonstrated the use of immune cells (t-lymphocyte) to deliver the therapeutic (doxorubicin) nanoparticles to the cancer cells.

In a similar study, PEG and diacyllipids, such as phosphatidylethanolamine (PE)-based polymeric micelles were initially converted to the immune micelles by attaching various specific antibodies to their surface for enhanced delivery of Taxol to the specific site (Torchilin et al., 2003).

Studies show that the cellular delivery system in the near future will open the new dimension for the targeted delivery of therapeutics at a specific site and will be advantageous for the treatment of different types of cancer.

14.3.6 MODIFIED PROTEIN DELIVERY

A colloidal carrier system designed for targeting is characterized specifically by the modified surface behavior, modified distribution, and increased cellular uptake. Over the past two decades, protein nanoparticles have emerged as a potential tool for the delivery of therapeutics to targeted tissues. Being natural molecules, protein has a particular functional activity and distinct biological application. Protein nanoparticles are nonantigenic, biodegradable, and susceptible to surface modification by attaching the drugs and ligands (Jahanshahi, 2004).

Protein nanoparticles (albumin, gelatin, gliadin, and legumin) by themselves or incorporated into biodegradable polymer microspheres/nanospheres can be utilized for the pulmonary delivery of therapeutics or for controlled release depot systems or oral delivery (Jahanshahi and Babaei, 2008).

Gelatin is a denatured protein, obtained either by partial acid or alkaline hydrolysis or by thermal or enzymatic degradation of structural animal collagen protein. Being biodegradable, antigenic, biocompatible, and easily available makes gelatin a preferred material for preparing drug delivery carriers for controlled and sustained release. Modification of gelatin nanoparticle surfaces could be made available to systemic circulation for a longer time and can be targeted easily to a specific site (Sahoo et al., 2015).

Human serum albumin (HSA) nanoparticles prepared by the desolvation technique has also proven its ability of being a versatile carrier for drug delivery and may form a dimeric and higher aggregate structure due to the presence of a free thiol group (Langer et al., 2008).

In a similar fashion, gliadin (gluten protein found in wheat having bioadhesive properties), elastin (component in connective tissues that is elastic in nature and allows many tissues to maintain their shape), Zein (prolamine-rich protein containing a high proportion of hydrophobic amino acids), soy protein, whey protein (mixture of globular proteins), and milk protein (proteins with unique and diversified functional properties) have also been explored to prepare protein nanoparticles on the basis of cancer cell biology, and improved the efficacy and safety of cancer therapy (Lohcharoenkal et al., 2014).

14.4 GENE THERAPY

Over the past few decades the pharmaceutical and medicine industry has witnessed great changes regarding the treatment of specific diseases with less toxic effects and higher safety and efficacy. The development of targeted drug delivery has changed the scenario drastically. However, several limitations associated with these delivery systems, such as toxic effect, release of the drug before reaching to the target site, and inability to cure diseases of genetic origin. Over the past few years, the number of patients with genetic diseases has increased significantly and the situation is becoming more alarming day by day.

Gene therapy emerged as a potential technique to treat genetic diseases since 1980. The goal of gene therapy is to help to correct genetic disease by providing a functional copy of the gene to make up the genetic defect. The functional copy of the gene is loaded into the vector which acts as the transport vehicle to carry the gene to the required site. Scientists are working continuously to develop a vector which can deliver the gene correctly with no or less toxicity to treat genetic disorders and diseases like cancer, blood diseases, disuse of the central nervous system (CNS), and immune system disease. Studies show that more than 1800 clinical trials have been conducted to date using different type of genes to treat various genetic diseases.

The first clinical study related to gene therapy has been started to treat the defect in a gene which synthesizes adenosine de aminase (ADA) responsible for the generation of immune cells B and T (responsible for cell mediated and humoral immunity) from bone marrow. If under any condition the gene responsible for producing ADA gets faulty, ADA production will be hindered leading to the absence of B and T cell production causing the start of a new progeny without immunity and this condition is referred to as severe combined immunodeficiency (SCID). Scientists have isolated the gene responsible for the production of ADA and introduced in the lymphocytes of patient using retrovirus and these transformed lymphocytes were introduced to the bone marrow leading to temporary synthesis of ADA, during the gene therapy of SCID (Blaese et al., 1995).

As already mentioned, medicine and health science have undergone dramatic changes and gene therapy has emerged as the most advanced tool to treat genetic disease and cancer. Gene therapy includes the transfer of genetic material into cell or tissues to prevent or cure the disease. On the basis of therapy, diseases can be divided into two major categories. The first entails hereditary diseases caused by a single gene defect (muscular dystrophy, hemophilia, etc.) and the second are polygenic or noninherited diseases like cancer, cardiovascular disease, and hepatitis C, etc. These two different conditions may be treated with in vivo gene therapy in which the gene is incorporated into a vector and then administered in the circulation so that the vector transfers the corrected gene to the targeted system. Ex vivo gene therapy, which includes the transfer of genes through vectors, is done in the cultured cells ex vivo and subsequently these genetically modified cells are injected into the circulation.

During the design of gene therapy, some key points should be taken into consideration:

- Identification and isolation of the gene to be targeted.
- Selection of an appropriate vector for the delivery of the gene.
- Identification of the target cell.
- Route of administration to be used to introduce the viral vector.
- A suitable animal model to test the therapy.

It is comparatively easy to find the specific gene and target cells in the case of therapies for mono gene-based diseases and the correct gene could be transferred more easily than for multifactorial diseases like cancer because of the interaction between the genes and the environmental effect. However, the selection of species promoter is always a matter of concern for gene therapy.

Tissue-specific promoters and ubiquitous promoters are concerned in the expression of genes in a specific organ like the liver or heart, etc., while ubiquitous promoters are used for the expression of genes in multiple organs at the same time.

In gene therapy, the therapeutic gene is carried to the target tissue with the help of vectors. An ideal vector should able to transduce the target cells and activate any type of immune response. Since the advent of gene therapy, a large number of vectors have been developed to treat different diseases, but there is still a need for the development of a universal gene vector. The choice pf vector mainly depends on the tissue to be targeted and on the nature of the disease. According to the origin, vectors may be mainly categorized into:

1. Viral vectors
2. Nonviral vectors.

14.4.1 VIRAL VECTORS

Viral vectors are derived from viruses which are highly infectious and transfer their genetic material to the host cells, and this feature is highly useful for

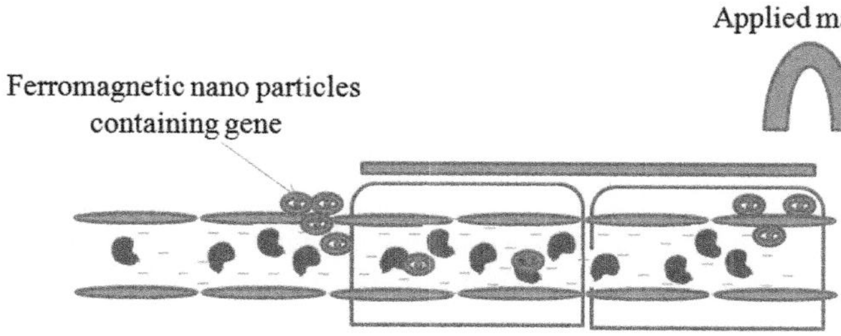

FIGURE 14.5

Representing gene delivery through magnetic particles.

therapeutic gene targeting. For using a virus for gene therapy, the genetic material of the virus is replaced with the therapeutic gene (Fig. 14.5). In the absence of a viral gene, the capsid is nonvirulent and noninfectious, but can efficiently carry the therapeutic gene to the target cell or tissue. The viral structure contains pathogenic genetic material surrounded by capsid (structural protein part with some lipid characteristics) and the common viral vectors used to carry gene therapy include:

1. Retroviral vector
2. Lentiviral vector
3. Adenoviral vector
4. Adeno-associated vectors

14.4.1.1 Retro- and lentiviral vectors

Retroviral vectors and lentiviral vectors are almost similar in structure and features based on their nRNA genomes. When retroviral or lentiviral vectors infect to the target cells, the transcription of RNA genome containing therapeutic gene occurs, leading to the formation new system and incorporation in the cytoplasm DNA which enters to the nucleus and replaces to the faulty gene. The major difference between them is the ability of retroviruses in transfecting the dividing cell while the lentiviral vectors are capable of infecting nondividing cells as well and this characteristic of lentiviruses make them useful in ex vivo and in vivo studies while retroviruses can be used only ex vivo (Anson, 2004).

Retroviral vectors have an important advantage in their ability to transform their single-stranded RNA genomes into double-stranded DNA molecules that stably integrate into the target cell genome and can permanently modify the host cell genome (Verma and Somia, 1997).

Lentiviral vectors can infect both dividing and nondividing cells and have the advantage of long-term stable expression of transgenes, low immunogenicity, and the ability to accommodate larger transgenes (Connolly, 2002; Cockrell and Kafri, 2007).

14.4.1.2 Adenoviral vectors

Adenoviruses contains double-stranded DNA as genetic material which can infect both dividing and nondividing cells. During the use of this viral capsule as a vector, the double-stranded DNA is changed with the nucleic acid containing therapeutic gene. These viral vectors enter the cell by receptor-mediated endocytosis and get attached to the αV integrins of the cell surface through viral pentose base and adenovirus receptor and lead to the infection of host cells (Manjila et al., 2013).

Adenoviruses are able to deliver large molecules of DNA, but in contrast to the retroviruses they do not interact with the host genome which makes its expression for too short a time. Also, the occurrence of immunological responses made them limited to a few tissues during clinical studies. Gene therapy with adenoviruses led to serious side effects and even death in a few cases (Marshall, 1999).

Controlling the replication of the genetic material of these viral vectors by deleting some genes may be beneficial for their use in gene therapy.

14.4.1.3 Adeno-associated vectors

Adeno-associated viruses are single-stranded, nonpathogenic viruses having two genes (cap and rep) sandwiched between inverted terminal repeats defining the start and the end of the virus, with packaging sequence in between. Cap represents the proteins of capsule and rep represents the part involved in the replication and integration, but still need some additional genes for the replication which can be provided by the adeno or herpes simplex viruses (Muzyczka, 1992).

Adenoviral vectors are prepared by replacing the cap and rep genes by a transgene which avoids its integration in the human genome 19. AAVs have been used in the treatment of some diseases, such as CF, hemophilia B, Leber congenital amaurosis, and AAT (Alpha-1 antitrypsin) deficiency. The cumbersome and time-consuming process of its production are the major disadvantages of this vector (Verma and Somia, 1997).

Helper-dependent adenoviral vector, Herpes simplex virus, Poxvirus vectors, and Epstein–Barr virus are some more vectors which may be used as gene therapy vectors for the treatment of neurologic disorders, such as motor neuron diseases, Parkinson's, Alzheimer's, and Huntington's disease as well as lysosomal storage diseases, spinal injury, and different types of cancers (Nayerossadat et al., 2012).

14.4.2 NONVIRAL VECTORS

Gene therapy using viral vectors has set a milestone in the treatment of several diseases, but immunogenic responses and adverse pathological conditions generated by the adenoviruses has brought forward the need of nonviral vectors. Nonviral vectors include all the physical (gene gun, electroporation, particle bombardment, ultrasound utilization, and magnetofection) and chemical means (cationic liposomes and polymers) by which the therapeutic gene can be delivered to the targeted site (Glover et al., 2005). Nonviral vectors have limitations in terms of their effectiveness regarding the gene transduction; however, low immunogenicity, easy availability, and cost effectiveness made them a lucrative approach for gene delivery. Mortality during the clinical trial of viral vectors have also withdrawn the attention of science community to develop nonviral vectors (Lee and Kim, 2005).

14.4.2.1 Physical means of gene delivery

Physical methods of gene delivery include methods which allow transient penetration of genes through the cell membrane. A direct injection of a naked gene to a specific targeted region may lead to easy gene transfer. A single injection of a naked gene is not effective in terms of transgenic expression, but several injections improve this significantly (Nayerossadat et al., 2012).

Bombardment of DNA by using a gene gun is another approach to deliver naked DNA. In this approach, high speed gold or tungsten particles coated with DNA are used to target skin, mucosa, or surgically exposed tissue and is especially effective for DNA-based immunization (Sudha et al., 2001).

Another approach by which DNA could reach the cytoplasm and neoplasm is referred to as electroporation (Fig. 14.6), presenting the condition of temporary destabilization of cell membrane of the targeted tissue by inserting a pair of electrodes (Lurquin, 1997).

Aihara and Miyazaki found that the serum level of interleukin 5(IL-5) increased from 0.2 to 20 ng/mL after electroporation, demonstrating the superiority of this method over a simple intramuscular injection of DNA (Aihara and Miyazaki, 1998).

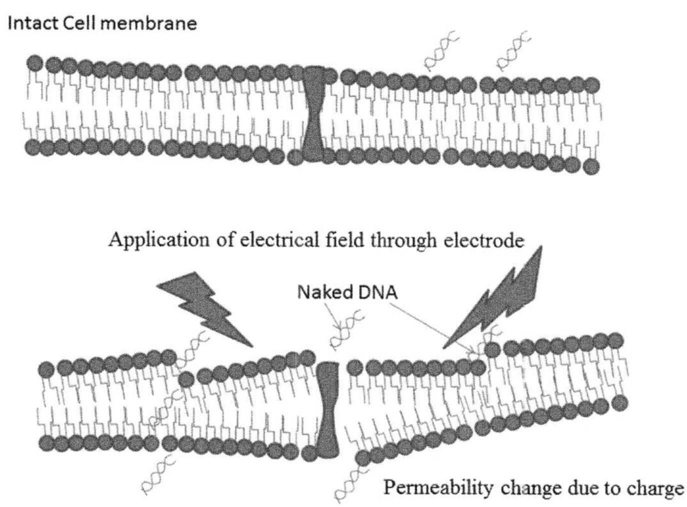

FIGURE 14.6

Diagrammatic representation of electroporation.

In a similar fashion, nanopores in the membrane can be created by ultrasound exposure in the presence of microbubbles to facilitate the movement of DNA into the cells of internal organs or tumors (Newman and Bettinger, 2007).

Magnetofection is another approach for delivering the gene entails a combination of physical and chemical approaches for gene delivery. This newer approach may include magnetic nanoparticles, to which the therapeutic or reporter genes are attached and focused to the target site/cells via high-field/high-gradient magnets. It not only increases the transfection, but also helps to target the specific site with the therapeutic gene. If the DNA is embedded within a matrix-like system such as hydrogels, release of the genes from the magnetic carrier may be achieved by applying alternating fields which will impart heat to the particles containing the gene (Dobson, 2006).

Cai et al. reported the use of magnetic field-driven penetration of nickel-embedded nanotubes (nanotube spearing) into cell membranes. They found unprecedented high transduction efficiency in Bal 17 B lymphoma and concluded that it is a powerful tool for highly efficient gene delivery (Cai et al., 2005).

14.4.2.2 Chemical means of gene delivery vectors

Chemical means of gene delivery vectors are nanosized complexes including compaction of negatively charged nucleic acid by polycationic nanometric particles. It may be cationic liposomes, micelles, or cationic polymers. When a complex is formed between negatively charged nucleic acid and cationic lipid, the combination is called lipoplex (Wasungu and Hoekstra, 2006). This system has several advantages over viral vectors such as the low cost of production, low immunogenic properties, and low toxicity, etc., and cationic lipid-based vectors generally include monovalent cationic lipids, polyvalent cationic lipids, guanidine containing, cholesterol-derivative compounds, and cationic polymers such as poly (ethylenimine) (PEI) poly-L-lysine (PLL) protamine, and a lipid-polymer hybrid (Lv et al., 2006).

Some general mechanisms involved in the gene delivery through cationic lipids include the nonspecific interaction between cationic particles and cell surface, endocytosis into endocytosis vesicles (endosomes), compaction and release of the DNA particle from endosome, translocation of the DNA particle to the nucleus by membrane receptors, and the transgenic expression of it (Nayerossadat et al., 2012).

Cationic liposomes

Cationic liposomes are made up of cationic lipid or positively charged lipid. A cationic lipid exhibits an amphiphilic character having a slight cationic charge which attracts the negatively charged DNA to form a lipoplex. It has been observed that the transfection activity of cationic lipids decreases with the increase in alkyl chain length and saturation. Shorter alkyl chain lengths favor higher rates of inter-membrane transfer of lipid monomers and lipid membrane mixing (Felgner et al., 1994).

The capability to incorporate hydrophilic and hydrophobic drugs, lower immunogenic character, low toxicity, and easy design for targeting are some characteristics of cationic liposomes. The engulfment of these liposomes by RES allows its stability in circulation, which can be enhanced by stealthing the liposome by forming a complex of the lipid material with polyethylene glycol (Son et al., 2000).

N-[1-(2,3-Dioleoyloxy) propyl]-*N,N,N*-trimethylammonium methyl-sulfate (DOTAP), 1,2-di-*O*-octadecenyl-3-trimethylammonium propane (DOTMA), 2,3-dioleyloxy-*N*-[2-(sperminecarboxamido)ethyl]-*N,N*-dimethyl-1-propanaminium (DOSPA), dioctadecylamido-glycylspermine (DOGS) are single synthetic cationic amphiphile while dioleoylphosphatidylethanolamine (DOPE) is combination of a cationic amphiphile and a neutral lipid these newer systems unstabilize the endosomal membrane to facilitate lipid exchange and membrane fusion between lipoplexes and endosomal membrane leading to more gene expression (Mochizuki et al., 2013).

Cationic liposomes can be used for effective gene delivery to the lungs, skeletal muscles, heart, skin cells, spleen, kidney, and liver.

Polymeric gene delivery system

Low immunogenicity, low toxic values, and the possibility of repeated administration are some major factors which have led to the advent of polymer-based, nonviral gene carriers. Currently, there are number of polymeric vectors—PLL-based, PLL-poly ethylene glycol, folate-conjugated PLL, polyethylenimine (PEI)-based, cyclodextrin (CD) etc.—which are being used to deliver therapeutic genes to the target cells (Choi et al., 1998).

Poly(l-lysine) (PLL) is one of the most favored materials used to deliver genes. In physiological environment, protonated primary q-amine groups of lysine in PLL interact negatively charged phosphate groups of DNA to form nanoparticulate polyplex (nanometric complex formed between a cationic polymer and nucleic acids). Poor bioavailability, rapid degradation, and high cytotoxicity are the major limitations of this delivery system. Studies show that the cytotoxicity of this polyplex may be attributed to the presence of a cationic charge present on the surface. However, the presence of this charge may also lead to the specific adsorption of the serum protein leading to rapid clearance (Choi et al., 1998; Ogris et al., 1999).

The stability of such complexes may be enhanced by the conjugation of polyethylene glycol (PEG), which prevents the interparticular aggregation of the system. PEGylation of cationic polymers also reduce the problems of cytotoxicity and nonspecific protein adsorption in vivo. Conjugation of sugar, antibodies, peptides, and folate may also be used to target the specific tissues with these complexes (Park et al., 2006).

Polyethylenimine (PEI) is another very important cationic polymer carrier used to deliver genes due its higher and consistent transfection. Two thirds of the aminic group present in PEI is protonated under physiological conditions

(Garnett, 1999), and the remaining unprotonated amine group works as a buffering system leading to the natural escape from the endosome helping to avoid the degradation of DNA in endosomal compartment during its maturation to lysosome, facilitating to the intracellular trafficking of DNA (Boussif et al., 1995). Although this property is considered to be responsible for its cytotoxicity, grafting of PEI with PEG in different ratios may minimize cytotoxicity and aggregation problems of PEI (Petersen et al., 2002).

Shi et al. have investigated the feasibility of achieving prolonged gene expression in the spinal cord without experiencing any attenuation of gene expression after repeated intrathecal injections of PEI-*g*-PEG/DNA complexes (Shi et al., 2003).

Conjugation of PEI-PEG complex with $\alpha\nu\beta3/\alpha\nu\beta5$ integrin-binding RGD peptide may be used for the targeted delivery of genes to angiogenic endothelial cells. In a similar fashion, monoclonal antibody-conjugated PEIs and folate-conjugated PEIs may be used to deliver the required therapeutic gene to different types of cancer cells (Park et al., 2006).

Low-molecular-weight PEI in conjunction with cholesteryl chloroformate may be used as a water-soluble lipopolymer for gene delivery in the form of micelles with low toxicity (Kunath et al., 2003).

Biodegradable polycations like poly[a-(4-aminobutyl)-L-glycolic acid], poly (b-amino ester)s, poly[2-(dimethylamino)ethyl methacrylate], and poly(2-aminoethyl propylene phosphate) may also be used to deliver genes to specific cells.

Some polysaccharides such as cyclodextarn and chitosan are other major polymers which can be used to deliver genes. CDs are cyclic oligomers with a unique amphiphilic structure with relatively nonimmunogenic and nontoxic properties. Studies show that β CD have similar transfection properties as PEI and lipofectamine.

It has also been observed that low-molecular-weight chitosan (LMWC) may form a complex with the DNA structure leading to its protection from enzymatic degradation. LMWC have also shown superiority on the PLL due to its low cytotoxicity and higher transfection efficiency.

Hydrophobically modified chitosan with deoxycholic acid may acquire a self-aggregating structure with interacted DNA molecules and shows efficient transfection for COS-7 cells (Lee et al., 1998).

Lactosylated chitosan also work as an efficient carrier system for asialoglycoprotein receptor-targeted gene delivery (Erbacher et al., 1998). However, this system is limited in enhancing the transfection in HepG2 cells which may be attributed to the absence of a surface charge after lactosylation. It also shows high interparticulate aggregation behavior.

Cationic polymers have shown their superiority in the development of carrier systems for gene delivery, but neutral polymers like PVA and polyvinyl pyrolidone can also be used to develop carrier systems for gene delivery. It has been observed that these polymers interact with the DNA via the hydrogen bond or Van der Waal forces leading to the protection of DNA from enzymatic degradation and, therefore, facilitate cellular uptake of DNA (Mumper and Rolland, 1998).

14.5 CONCLUSION

The development of new drugs and the existence of new diseases create a new dimension for scientists to develop novel formulations. However, limitations associated with some drug molecules like poor bioavailability, lower half-life, and toxicity have inhibited the attention of researchers. These issues can be overcome by targeted delivery systems. Brain, lung, liver, cancer cell targeting, etc., are the major areas of research. Delivery to a specific site with minimum toxicity is urgently required and present the greatest challenge to the present therapeutic system.

Genetic diseases are a great challenge and gene therapy presents a potential solution. Gene therapy includes the delivery of therapeutic genes to the target cells with the help of specific vectors including viral and nonviral vectors. Because of problems associated with viral vectors like antigenicity, immunogenicity, modified pathological conditions, etc., researchers were motivated to work with nonviral vectors including physical (gene gun, electroporation, particle bombardment, ultrasound utilization, and magnetofection) and chemical (cationic liposomes and polymers) means to deliver the therapeutic genes. In the past few years, the use of polymeric gene delivery approaches have withdrawn the attention of science fraternity due to its nonantigenic, nonimmunogenic nature and cost effectiveness.

14.6 FUTURE PROSPECTS

Targeted delivery systems have proven its position regarding the marked decrement in the adverse and toxic effects of drugs, especially those related to cancer and viral diseases. At the same time, polymeric gene delivery systems for the treatment of cancers and genetically associated diseases has proven its utility, but further research toward the delivery of therapeutic genes and drugs to specific and targeted tissues or cells is urgently needed.

REFERENCES

Abbasi, E., Aval, S.F., Akbarzadeh, A., Milani, M., Nasrabadi, H.T., Joo, S.W., et al., 2014. Dendrimers: synthesis, applications, and properties. Nanoscale Res. Lett. 9 (1), 1–10.

Aihara, H., Miyazaki, J.I., 1998. Gene transfer into muscle by electroporation in vivo. Nat. Biotechnol. 16 (9), 867–870.

Allen, T.M., 2002. Ligand-targeted therapeutics in anticancer therapy. Nat. Rev. Cancer 2 (10), 750–763.

Allen, T.M., Hansen, C., Rutledge, J., 1989. Liposomes with prolonged circulation times: factors affecting uptake by reticuloendothelial and other tissues. Biochim. Biophys. Acta (BBA)—Biomembranes 981 (1), 27–35.

Anson, D.S., 2004. The use of retroviral vectors for gene therapy-what are the risks? A review of retroviral pathogenesis and its relevance to retroviral vector-mediated gene delivery. Genet. Vaccines Ther. 2 (1), 1.

Azzopardi, E.A., Ferguson, E.L., Thomas, D.W., 2013. The enhanced permeability retention effect: a new paradigm for drug targeting in infection. J. Antimicrob. Chemother. 68 (2), 257–274.

Bala, I., Hariharan, S., Kumar, M.R., 2004. PLGA nanoparticles in drug delivery: the state of the art. Crit. Rev. Ther. Drug Carrier Syst. 21 (5), 387–422.

Batrakova, E.V., Gendelman, H.E., Kabanov, A.V., 2011. Cell-mediated drug delivery. Expert Opin. Drug Deliv. 8 (4), 415–433.

Bitterman, D.S., Du, K.L., 2016. Safety and efficacy of combination targeted therapy and radiotherapy. Am. J. Hematol. Oncol. 12 (1).

Blaese, R.M., Culver, K.W., Miller, A.D., Carter, C.S., 1995. T lymphocyte-directed gene therapy for ADA negative SCID: initial trial results after 4 years. Science 270 (5235), 475.

Bonucci, E., Ballanti, P., Ramires, P.A., Richardson, J.L., Benedetti, L.M., 1995. Prevention of ovariectomy osteopenia in rats after vaginal administration of Hyaff 11 microspheres containing salmon calcitonin. Calcif. Tissue Int. 56 (4), 274–279.

Boussif, O., Lezoualc'h, F., Zanta, M.A., Mergny, M.D., Scherman, D., Demeneix, B., et al., 1995. A versatile vector for gene and oligonucleotide transfer into cells in culture and in vivo: polyethylenimine. Proc. Natl. Acad. Sci. U.S.A. 92 (16), 7297–7301.

Cai, D., Mataraza, J.M., Qin, Z.H., Huang, Z., Huang, J., Chiles, T.C., et al., 2005. Highly efficient molecular delivery into mammalian cells using carbon nanotube spearing. Nat. Methods 2 (6), 449–454.

Chen, J., Corbin, I.R., Li, H., Cao, W., Glickson, J.D., Zheng, G., 2007. Ligand conjugated low-density lipoprotein nanoparticles for enhanced optical cancer imaging in vivo. J. Am. Chem. Soc. 129 (18), 5798–5799.

Chertok, B., Moffat, B.A., David, A.E., Yu, F., Bergemann, C., Ross, B.D., et al., 2008. Iron oxide nanoparticles as a drug delivery vehicle for MRI monitored magnetic targeting of brain tumors. Biomaterials 29 (4), 487–496.

Chiu, S.J., Liu, S., Perrotti, D., Marcucci, G., Lee, R.J., 2006. Efficient delivery of a Bcl-2-specific antisense oligodeoxyribonucleotide (G3139) via transferrin receptor-targeted liposomes. J. Control. Release 112 (2), 199–207.

Choi, Y.H., Liu, F., Kim, J.S., Choi, Y.K., Park, J.S., Kim, S.W., 1998. Polyethylene glycol-grafted poly-L-lysine as polymeric gene carrier. J. Control. Release 54 (1), 39–48.

Chomoucka, J., Drbohlavova, J., Huska, D., Adam, V., Kizek, R., Hubalek, J., 2010. Magnetic nanoparticles and targeted drug delivering. Pharmacol. Res. 62 (2), 144–149.

Cockrell, A.S., Kafri, T., 2007. Gene delivery by lentivirus vectors. Mol. Biotechnol. 36 (3), 184–204.

Connolly, J.B., 2002. Lentiviruses in gene therapy clinical research. Gene Ther. 9 (24), 1730–1734.

Corbin, I.R., Chen, J., Cao, W., Li, H., Lund-Katz, S., Zheng, G., 2007. Enhanced cancer-targeted delivery using engineered high-density lipoprotein-based nanocarriers. J. Biomed. Nanotechnol. 3 (4), 367–376.

Dagar, S., Krishnadas, A., Rubinstein, I., Blend, M.J., Önyüksel, H., 2003. VIP grafted sterically stabilized liposomes for targeted imaging of breast cancer: in vivo studies. J. Control. Release 91 (1), 123–133.

Dobson, J., 2006. Gene therapy progress and prospects: magnetic nanoparticle-based gene delivery. Gene Ther. 13 (4), 283–287.

Dunn, A.E., Dunn, D.J., Macmillan, A., Whan, R., Stait-Gardner, T., Price, W.S., et al., 2014. Spatial and temporal control of drug release through pH and alternating magnetic field induced breakage of Schiff base bonds. Polym. Chem. 5 (10), 3311–3315.

Erbacher, P., Zou, S., Bettinger, T., Steffan, A.M., Remy, J.S., 1998. Chitosan-based vector/DNA complexes for gene delivery: biophysical characteristics and transfection ability. Pharm. Res. 15 (9), 1332–1339.

Evans Jr, D.J., Evans, D.G., Gorbach, S.L., 1973. Production of vascular permeability factor by enterotoxigenic *Escherichia coli* isolated from man. Infect. Immun. 8, 725–730.

Fang, J., Nakamura, H., Maeda, H., 2011. The EPR effect: unique features of tumor blood vessels for drug delivery, factors involved, and limitations and augmentation of the effect. Adv. Drug Deliv. Rev. 63, 136–151.

Felgner, J.H., Kumar, R., Sridhar, C.N., Wheeler, C.J., Tsai, Y.J., Border, R., et al., 1994. Enhanced gene delivery and mechanism studies with a novel series of cationic lipid formulations. J. Biol. Chem. 269 (4), 2550–2561.

Firestone, R.A., 1994. Low-density lipoprotein as a vehicle for targeting antitumor compounds to cancer cells. Bioconjug. Chem. 5 (2), 105–113.

Garnett, M.C., 1999. Gene-delivery systems using cationic polymers. Crit. Rev. Ther. Drug Carrier Syst. 16 (2), 147–207.

Garnett, M.C., 2001. Targeted drug conjugates: principles and progress. Adv. Drug Deliv. Rev. 53 (2), 171–216.

Gerweck, L.E., Seetharaman, K., 1996. Cellular pH gradient in tumor versus normal tissue: potential exploitation for the treatment of cancer. Cancer Res. 56 (6), 1194–1198.

Glover, D.J., Lipps, H.J., Jans, D.A., 2005. Towards safe, non-viral therapeutic gene expression in humans. Nat. Rev. Genet. 6 (4), 299–310.

Greish, K., 2010. Enhanced permeability and retention (EPR) effect for anticancer nanomedicine drug targeting. Cancer Nanotechnol.: Methods Protoc. 25–37.

Hamidi, M., Zarrin, A., Foroozesh, M., Mohammadi-Samani, S., 2007. Applications of carrier erythrocytes in delivery of biopharmaceuticals. J. Control. Release 118 (2), 145–160.

Huang, Y.F., Lin, Y.W., Lin, Z.H., Chang, H.T., 2009. Aptamer-modified gold nanoparticles for targeting breast cancer cells through light scattering. J. Nanopart. Res. 11 (4), 775–783.

Ilium, L., Hunneyball, I.M., Davis, S.S., 1986. The effect of hydrophilic coatings on the uptake of colloidal particles by the liver and by peritoneal macrophages. Int. J. Pharm. 29 (1), 53–65.

Illum, L., Davis, S.S., 1983. Effect of the nonionic surfactant poloxamer 338 on the fate and deposition of polystyrene microspheres following intravenous administration. J. Pharm. Sci. 72 (9), 1086–1089.

Illum, L., Davis, S.S., Wilson, C.G., Thomas, N.W., Frier, M., Hardy, J.G., 1982. Blood clearance and organ deposition of intravenously administered colloidal particles. The effects of particle size, nature and shape. Int. J. Pharm. 12 (2–3), 135–146.

Illum, L., Thomas, N.W., Davis, S.S., 1986. Effect of a selected suppression of the reticuloendothelial system on the distribution of model carrier particles. J. Pharm. Sci. 75 (1), 16–22.

Ishida, O., Maruyama, K., Tanahashi, H., Iwatsuru, M., Sasaki, K., Eriguchi, M., et al., 2001. Liposomes bearing polyethyleneglycol-coupled transferrin with intracellular targeting property to the solid tumors in vivo. Pharm. Res. 18 (7), 1042–1048.

Jahanshahi, M., 2004. Re-design of downstream processing techniques for nanoparticulate bioproducts. Iran. J. Biotechnol. 2, 1–12.

Jahanshahi, M., Babaei, Z., 2008. Protein nanoparticle: a unique system as drug delivery vehicles. Afr. J. Biotechnol. 7 (25).

Jones, M.C., Leroux, J.C., 1999. Polymeric micelles–a new generation of colloidal drug carriers. Eur. J. Pharm. Biopharm. 48 (2), 101–111.

Kamps, J., Koning, G., Velinova, M., Morselt, H., Wilkens, M., Gorter, A., et al., 2000. Uptake of long-circulating immunoliposomes, directed against colon adenocarcinoma cells, by liver metastases of colon cancer. J. Drug Target. 8 (4), 235–245.

Kedar, U., Phutane, P., Shidhaye, S., Kadam, V., 2010. Advances in polymeric micelles for drug delivery and tumor targeting. Nanomed.: Nanotechnol. Biol. Med. 6 (6), 714–729.

Kocbek, P., Obermajer, N., Cegnar, M., Kos, J., Kristl, J., 2007. Targeting cancer cells using PLGA nanoparticles surface modified with monoclonal antibody. J. Control. Release 120 (1), 18–26.

Koyo, N., Chiaki, T., Sachi, N., Yoshinobu, T., Mitsuru, H., Hitoshi, S., 1990. Effect of electric charge on the hepatic uptake of macromolecules in the rat liver. Int. J. Pharm. 65 (1–2), 7–17.

Kunath, K., von Harpe, A., Fischer, D., Petersen, H., Bickel, U., Voigt, K., et al., 2003. Low-molecular-weight polyethylenimine as a non-viral vector for DNA delivery: comparison of physicochemical properties, transfection efficiency and in vivo distribution with high-molecular-weight polyethylenimine. J. Control. Release 89 (1), 113–125.

Labhasetwar, V., Song, C., Humphrey, W., Shebuski, R., Levy, R.J., 1998. Arterial uptake of biodegradable nanoparticles: effect of surface modifications. J. Pharm. Sci. 87 (10), 1229–1234.

Langer, K., Anhorn, M.G., Steinhauser, I., Dreis, S., Celebi, D., Schrickel, N., et al., 2008. Human serum albumin (HSA) nanoparticles: reproducibility of preparation process and kinetics of enzymatic degradation. Int. J. Pharm. 347 (1), 109–117.

Laverman, P., Dams, E.T., Storm, G., et al., 2001. Microscopic localization of PEG-liposomes in a rat model of focal infection. J. Control. Release 75, 347–355.

Lee, M., Kim, S.W., 2005. Polyethylene glycol-conjugated copolymers for plasmid DNA delivery. Pharm. Res. 22 (1), 1–10.

Lee, K.Y., Kwon, I.C., Kim, Y.H., Jo, W.H., Jeong, S.Y., 1998. Preparation of chitosan self-aggregates as a gene delivery system. J. Control. Release 51 (2), 213–220.

Li, X., Tian, X., Zhang, J., Zhao, X., Chen, X., Jiang, Y., et al., 2011. In vitro and in vivo evaluation of folate receptor-targeting amphiphilic copolymer-modified liposomes loaded with docetaxel. Int. J. Nanomed. 6 (1167), 289–298.

Liu, R., Li, D., He, B., Xu, X., Sheng, M., Lai, Y., et al., 2011. Anti-tumor drug delivery of pH-sensitive poly (ethylene glycol)-poly (L-histidine-)-poly (L-lactide) nanoparticles. J. Control. Release 152 (1), 49–56.

Lohcharoenkal, W., Wang, L., Chen, Y.C., Rojanasakul, Y., 2014. Protein nanoparticles as drug delivery carriers for cancer therapy. BioMed Res. Int. 2014, 12.

Lu, Y., Low, P.S., 2012. Folate-mediated delivery of macromolecular anticancer therapeutic agents. Adv. Drug Deliv. Rev. 64, 342–352.

Lukyanov, A.N., Elbayoumi, T.A., Chakilam, A.R., Torchilin, V.P., 2004. Tumor-targeted liposomes: doxorubicin-loaded long-circulating liposomes modified with anti-cancer antibody. J. Control. Release 100 (1), 135–144.

Lurquin, P.F., 1997. Gene transfer by electroporation. Mol. Biotechnol. 7 (1), 5–35.

Lv, H., Zhang, S., Wang, B., Cui, S., Yan, J., 2006. Toxicity of cationic lipids and cationic polymers in gene delivery. J. Control. Release 114 (1), 100–109.

Maeda, H., Bharate, G.Y., Daruwalla, J., 2009. Polymeric drugs for efficient tumor-targeted drug delivery based on EPR-effect. Eur. J. Pharm. Biopharm. 71, 409–419.

Mainardes, R.M., Urban, M.C., Cinto, P.O., Khalil, N.M., Chaud, M.V., Evangelista, R.C., et al., 2005. Colloidal carriers for ophthalmic drug delivery. Curr. Drug Targets 6 (3), 363–371.1.

Manjila, S.B., Baby, J.N., Bijin, E.N., Constantine, I., Pramod, K., Valsalakumari, J., 2013. Novel gene delivery systems. Int. J. Pharm. Investig. 3 (1), 1.

Marshall, E., 1999. Gene therapy death prompts review of adenovirus vector. Science 286 (5448), 2244–2245.

Masood, F., 2016. Polymeric nanoparticles for targeted drug delivery system for cancer therapy. Mater. Sci. Eng.: C 60, 569–578.

Melgert, B.N., Olinga, P., Van Der Laan, J., Weert, B., Cho, J., Schuppan, D., et al., 2001. Targeting dexamethasone to Kupffer cells: effects on liver inflammation and fibrosis in rats. Hepatology 34 (4), 719–728.

Mew, D., Wat, C.K., Towers, G.H., Levy, J.G., 1983. Photoimmunotherapy: treatment of animal tumors with tumor-specific monoclonal antibody-hematoporphyrin conjugates. J. Immunol. 130 (3), 1473–1477.

Mochizuki, S., Kanegae, N., Nishina, K., Kamikawa, Y., Koiwai, K., Masunaga, H., et al., 2013. The role of the helper lipid dioleoylphosphatidylethanolamine (DOPE) for DNA transfection cooperating with a cationic lipid bearing ethylenediamine. Biochim. Biophys. Acta (BBA)—Biomembranes 1828 (2), 412–418.

Molema, G., 2001. Drug targeting: basic concepts and novel advances. Methods Princ. Med. Chem. 12, 1–22.

Molema, G., Meijer, D.K., 2001. *Drug Targeting: Organ-Specific Strategies*, Vol. 12. Wiley-VCH.

Mu, L., Feng, S.S., 2003. A novel controlled release formulation for the anticancer drug paclitaxel (Taxol®): PLGA nanoparticles containing vitamin E TPGS. J. Control. Release 86 (1), 33–48.

Mumper, R.J., Rolland, A.P., 1998. Plasmid delivery to muscle: recent advances in polymer delivery systems. Adv. Drug Deliv. Rev. 30 (1), 151–172.

Musumeci, T., Ventura, C.A., Giannone, I., Ruozi, B., Montenegro, L., Pignatello, R., et al., 2006. PLA/PLGA nanoparticles for sustained release of docetaxel. Int. J. Pharm. 325 (1), 172–179.

Muzyczka, N., 1992. Use of adeno-associated virus as a general transduction vector for mammalian cells. Viral Expression Vectors. Springer Berlin Heidelberg, pp. 97–129.

Nayerossadat, N., Maedeh, T., Ali, P.A., 2012. Viral and nonviral delivery systems for gene delivery. Adv. Biomed. Res. 1 (1), 27.

Newman, C.M.H., Bettinger, T., 2007. Gene therapy progress and prospects: ultrasound for gene transfer. Gene Ther. 14 (6), 465–475.

Ogris, M., Brunner, S., Schüller, S., Kircheis, R., Wagner, E., 1999. PEGylated DNA/transferrin-PEI complexes: reduced interaction with blood components, extended circulation in blood and potential for systemic gene delivery. Gene Ther. 6 (4), 595–605.

Oseroff, A.R., Ohuoha, D., Hasan, T., Bommer, J.C., Yarmush, M.L., 1986. Antibody-targeted photolysis: selective photodestruction of human T-cell leukemia cells using monoclonal antibody-chlorin e6 conjugates. Proc. Natl. Acad. Sci. U.S.A. 83 (22), 8744–8748.

Papademetriou, I.T., Garnacho, C., Schuchman, E.H., Muro, S., 2013. In vivo performance of polymer nanocarriers dually-targeted to epitopes of the same or different receptors. Biomaterials 34 (13), 3459–3466.

Park, T.G., Jeong, J.H., Kim, S.W., 2006. Current status of polymeric gene delivery systems. Adv. Drug Deliv. Rev. 58 (4), 467–486.

Patel, K.R., Li, M.P., Baldeschwieler, J.D., 1983. Suppression of liver uptake of liposomes by dextran sulfate 500. Proc. Natl. Acad. Sci. U.S.A. 80 (21), 6518–6522.

Petersen, H., Fechner, P.M., Martin, A.L., Kunath, K., Stolnik, S., Roberts, C.J., et al., 2002. Polyethylenimine-graft-poly(ethylene glycol) copolymers: influence of copolymer block structure on DNA complexation and biological activities as gene delivery system. Bioconjug. Chem. 13 (4), 845–854.

Pettit, D.K., Gombotz, W.R., 1998. The development of site-specific drug-delivery systems for protein and peptide biopharmaceuticals. Trends Biotechnol. 16 (8), 343–349.

Rajendran, L., Knölker, H.J., Simons, K., 2010. Subcellular targeting strategies for drug design and delivery. Nat. Rev. Drug Discov. 9 (1), 29–42.

Sahoo, N., Sahoo, R.K., Biswas, N., Guha, A., Kuotsu, K., 2015. Recent advancement of gelatin nanoparticles in drug and vaccine delivery. Int. J. Biol. Macromol. 81, 317–331.

Shi, L., Tang, G.P., Gao, S.J., Ma, Y.X., Liu, B.H., Li, Y., et al., 2003. Repeated intrathecal administration of plasmid DNA complexed with polyethylene glycol-grafted polyethylenimine led to prolonged transgene expression in the spinal cord. Gene Ther. 10 (14), 1179–1188.

Shuai, X., Merdan, T., Schaper, A.K., Xi, F., Kissel, T., 2004. Core-cross-linked polymeric micelles as paclitaxel carriers. Bioconjug. Chem. 15 (3), 441–448.

Son, K.K., Tkach, D., Hall, K.J., 2000. Efficient in vivo gene delivery by the negatively charged complexes of cationic liposomes and plasmid DNA. Biochim. Biophys. Acta (BBA)—Biomembranes 1468 (1), 6–10.

Song, C., Labhasetwar, V., Cui, X., Underwood, T., Levy, R.J., 1998. Arterial uptake of biodegradable nanoparticles for intravascular local drug delivery: results with an acute dog model. J. Control. Release 54 (2), 201–211.

Stefanadis, C., Chrysochoou, C., Markou, D., Petraki, K., Panagiotakos, D.B., Fasoulakis, C., et al., 2001. Increased temperature of malignant urinary bladder tumors in vivo: the application of a new method based on a catheter technique. J. Clin. Oncol. 19 (3), 676–681.

Steinfeld, U., Pauli, C., Kaltz, N., Bergemann, C., Lee, H.H., 2006. T lymphocytes as potential therapeutic drug carrier for cancer treatment. Int. J. Pharm. 311 (1), 229–236.

Sudha, P.M., Low, S., Kwang, J., Gong, Z., 2001. Multiple tissue transformation in adult zebrafish by gene gun bombardment and muscular injection of naked DNA. Mar. Biotechnol. 3 (2), 119–125.

Sun, D., Nakao, S., Xie, F., Zandi, S., Schering, A., Hafezi-Moghadam, A., 2010. Superior sensitivity of novel molecular imaging probe: simultaneously targeting two types of endothelial injury markers. FASEB J. 24 (5), 1532–1540.

Suzuki, M., Hori, K., Abe, I., Saito, S., Sato, H., 1981. A new approach to cancer chemotherapy: selective enhancement of tumor blood flow with angiotensin II. J. Natl. Cancer Inst. 67, 663–669.

Tang, K., Zhang, Y., Zhang, H., Xu, P., Liu, J., Ma, J., et al., 2012. Delivery of chemotherapeutic drugs in tumour cell-derived microparticles. Nat. Commun. 3, 1282.

Torchilin, V.P., 1984. Liposomes as targetable drug carriers. Crit. Rev. Ther. Drug Carrier Syst. 2 (1), 65–115.

Torchilin, V.P., 2005. Recent advances with liposomes as pharmaceutical carriers. Nat. Rev. Drug Discov. 4 (2), 145–160.

Torchilin, V.P., Klibanov, A.L., Huang, L., O'Donnell, S., Nossiff, N.D., Khaw, B.A., 1992. Targeted accumulation of polyethylene glycol-coated immunoliposomes in infarcted rabbit myocardium. FASEB J. 6 (9), 2716–2719.

Torchilin, V.P., Zhou, F., Huang, L., 1993. pH-sensitive liposomes. J. Liposome Res. 3 (2), 201–255.

Torchilin, V.P., Lukyanov, A.N., Gao, Z., Papahadjopoulos-Sternberg, B., 2003. Immunomicelles: targeted pharmaceutical carriers for poorly soluble drugs. Proc. Natl. Acad. Sci. U.S.A. 100 (10), 6039–6044.

Verma, I.M., Somia, N., 1997. Gene therapy-promises, problems and prospects. Nature 389 (6648), 239–242.

Vyas, S.P., Khar, R.K., 2004. Targeted & Controlled Drug Delivery: Novel Carrier Systems. CBS Publishers & Distributors.

Vyas, S.P., Khar, R.K., 2008. Basis of targeted drug delivery. Targeted And Controlled Drug Delivery. CBS Publishers and Distributors (Reprint), pp. 42–65.

Wagner, E., Curiel, D., Cotten, M., 1994. Delivery of drugs, proteins and genes into cells using transferrin as a ligand for receptor-mediated endocytosis. Adv. Drug Deliv. Rev. 14 (1), 113–135.

Wasungu, L., Hoekstra, D., 2006. Cationic lipids, lipoplexes and intracellular delivery of genes. J. Control. Release 116 (2), 255–264.

Wolinsky, J.B., Grinstaff, M.W., 2008. Therapeutic and diagnostic applications of dendrimers for cancer treatment. Adv. Drug Deliv. Rev. 60 (9), 1037–1055.

Xu, D., Wu, F., Chen, Y., Wei, L., Yuan, W., 2013. pH-sensitive degradable nanoparticles for highly efficient intracellular delivery of exogenous protein. Int. J. Nanomed. 8 (1), 3405–3414.

Ying, X., Wen, H.E., Lu, W.L., Du, J., Guo, J., Tian, W., et al., 2010. Dual-targeting daunorubicin liposomes improve the therapeutic efficacy of brain glioma in animals. J. Control. Release 141 (2), 183–192.

Zhang, L., Guo, R., Yang, M., Jiang, X., Liu, B., 2007. Thermo and pH dual-responsive nanoparticles for anti-cancer drug delivery. Adv. Mater. 19 (19), 2988–2992.

Zheng, G., Chen, J., Li, H., Glickson, J.D., 2005. Rerouting lipoprotein nanoparticles to selected alternate receptors for the targeted delivery of cancer diagnostic and therapeutic agents. Proc. Natl. Acad. Sci. U.S.A. 102 (49), 17757–17762.

FURTHER READING

Matsumura, Y., Maeda, H., 1986. A new concept for macromolecular therapeutics in cancer chemotherapy: mechanism of tumoritropic accumulation of proteins and the antitumor agent smancs. Cancer Res. 46, 6387–6392.

CHAPTER

15 Effects of block copolymer micelles on the pharmacokinetics of encapsulated drugs

Ziyad Binkhathlan[1,2,3] and Afsaneh Lavasanifar[3,4]

[1]Nanomedicine Research Unit, Department of Pharmaceutics, King Saud University, Riyadh, Saudi Arabia [2]King Abdullah Institute for Nanotechnology, King Saud University, Riyadh, Saudi Arabia [3]Faculty of Pharmacy and Pharmaceutical Sciences, University of Alberta, Edmonton, AB, Canada [4]Department of Chemical and Material Engineering, University of Alberta, Edmonton, AB, Canada

15.1 INTRODUCTION

Block copolymers are amphiphilic macromolecules that can self-assemble to form core/shell structures, namely polymeric micelles. In an aqueous media the core of polymeric micelles is hydrophobic and acts as a hydrophobic environment for hosting poorly water-soluble drugs. The shell of polymeric micelles, on the other hand, interfaces the aqueous media restricting the interaction of polymeric micelles with each other. The hydrophilic shell also introduces stealth properties to the polymeric micellar structure in blood, that is, reduces their interaction with opsonins making them unrecognized by the mononuclear phagocytic cells. Owing to their capacity for the encapsulation and slow release of hydrophobic molecules, as well as their small size (<100 nm) and stealth properties, polymeric micelles have been hoped to act as nanocarriers for not only solubilization, but also targeted delivery of poorly soluble drugs. Several studies have focused on the design of polymeric micelles that can hold on to their cargo under sink conditions of systemic circulation and eventually change the normal pharmacokinetics of the encapsulated drug toward that of the nanocarrier. More than two decades of research in this area has resulted in a number of polymeric micellar formulations in clinical trials worldwide. The clinical polymeric micellar formulations that are mostly used by intravenous (IV) route of administration and intended for the delivery of anticancer agents, have all been able to act as efficient solubilizing agents, and in some instances, lower the drug's side effects. However, it is still uncertain whether polymeric micelles can lower drug toxicity by causing a drastic change in the pharmacokinetics of the encapsulated drug in the same way stealth liposomes do. Moreover, the potential benefit of application of polymeric micelles

Nanoarchitectonics in Biomedicine. DOI: https://doi.org/10.1016/B978-0-12-816200-2.00017-7

as vehicles for drug delivery through other routes of administration is still unresolved. It is also unclear whether the role of polymeric micelles as delivery systems for non-IV routes of administration is limited to their high capacity as solubilizing agents for hydrophobic drugs, or whether they can safely penetrate through different sites of drug absorption and be absorbed as intact micelles, because of their small size. The aim of this chapter is to shed light on the answers to these questions by providing an overview of the current information on: (1) the ability of polymeric micellar formulations in clinical trials on changing the pharmacokinetics of the encapsulated drug following IV administration; and (2) the success and role of polymeric micelles as delivery systems for poorly soluble drugs via oral, topical, and mucosal routes of drug administration (Fig. 15.1).

15.2 EFFECTS OF POLYMERIC MICELLES ON THE PHARMACOKINETICS OF ENCAPSULATED DRUGS

15.2.1 INTRAVENOUS ADMINISTRATION

One of the main advantages of the IV route of administration is that the injected drug and its carrier are entirely available at the systemic circulation since there is

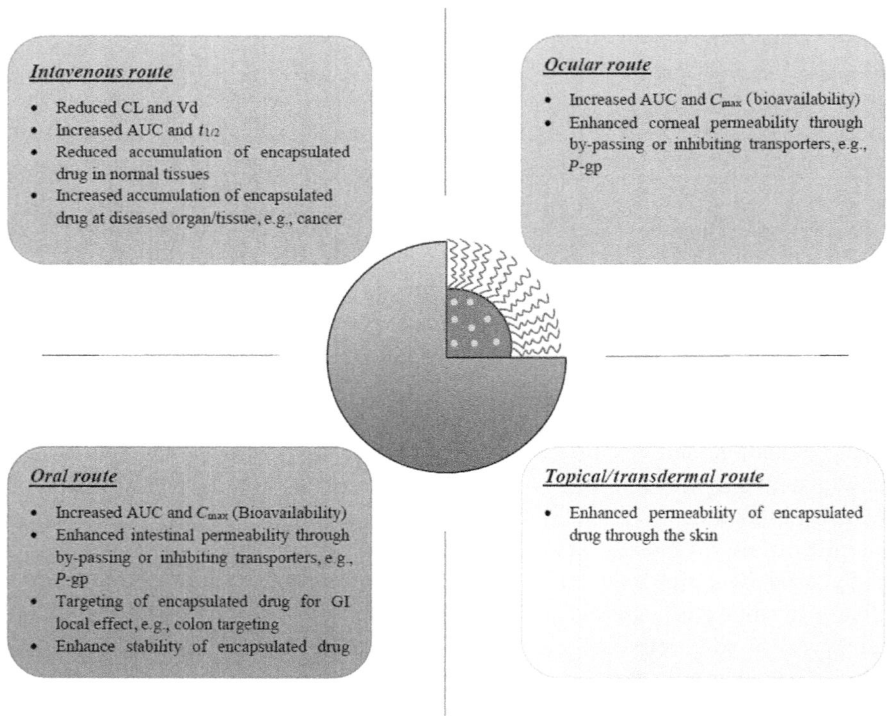

FIGURE 15.1

Effects of block copolymer micelles on the pharmacokinetics of encapsulated drugs after administration through different routes.

no absorption phase. Generally, block copolymer micelles are intended to serve as nanocarriers for hydrophobic drugs. In addition to providing a solution for the poor water solubility of such drugs, the use of a carrier system can modify the toxicity profile of drugs or their instability issues (i.e., rapid clearance from the systemic circulation). Therefore, the main goal of any nano-drug delivery system is to carry the encapsulated drug intact in the blood circulation. If the drug follows the pharmacokinetic and biodistribution of nanocarrier, there is a high likelihood for the delivery of the encapsulated drug to the site of action (the diseased organ/tissue) and for keeping it away from the healthy organs/tissues. A successfully designed drug-loaded nano-delivery system is expected to cause typical changes in the pharmacokinetic profile (and in the tissue distribution profile) of the encapsulated drug. The favorable changes include an increase in the area under the blood or plasma concentration-time curve (AUC), an increase in the elimination half-life ($t_{1/2}$), and decreases in the volume of distribution (Vd) and total body clearance (CL) for the encapsulated drug by its nanocarrier. For optimally designed nanocarriers, the magnitude of change in the drug-related pharmacokinetic parameters by the nanocarrier depend on: (1) the ability of the system to hold on to the encapsulated drug by controlling its release; and (2) the ability of the system to stay intact within the systemic circulation without being quickly recognized by the immune system or eliminated through metabolism and/or excretion.

The majority of the polymeric micelles that have been developed to date or are currently under development and all of the polymeric micellar formulations in clinical trials are administered via the IV route (Table 15.1). For formulations under clinical trials, the reader is referred to Section 15.3 where the influence of polymeric micelles on the pharmacokinetics and tissue distribution of the encapsulated drug after IV administration is discussed in detail. For formulations under preclinical development, this topic has been reviewed extensively (Aliabadi et al., 2008; Aliabadi and Lavasanifar, 2006), although an updated review seems warranted.

15.2.2 ORAL ADMINISTRATION

The oral route of drug administration remains the preferred route for optimum patient compliance, but can be the most challenging one to achieve acceptable bioavailability. Poor water solubility is the major factor that limits the absorption of drugs via the oral route of drug administration. It is estimated that nearly 70% of new drug candidates entering the discovery pipeline fail to proceed because of poor aqueous solubility (Ku and Dulin, 2012). Moreover, around 40% of the currently marketed immediate-release oral drugs are considered practically insoluble in water (<0.1 mg/mL) (Takagi et al., 2006). The second major factor that influences the fate of lead molecules in drug development is intestinal permeability (Amidon et al., 1995). Unfortunately, many of these lead molecules are lipophilic and exhibit low water solubility and/or low permeability. These two

Table 15.1 Polymeric Micellar Formulations in Clinical Trials and Their Changes on the Pharmacokinetics of the Encapsulated Drugs

Product Name	Micelle Size (nm)	Block Copolymer	Encapsulated Drug	Changes in the Pharmacokinetic Profile	References
NK911	40	PEO-*b*-P(Asp)-doxorubicin	Doxorubicin	↑ C_{max}, AUC, and $t_{1/2}$ ↓ Vd_{ss} and CL ↑ Accumulation in tumor tissues	Yokoyama et al. (1991), Nakanishi et al. (2001), Masayuki et al. (1990), Matsumura et al. (2004)
SP1049C	22–27	PEO-*b*-PPO-*b*-PEO	Doxorubicin	In animals: ↑ C_{max}, AUC, and $t_{1/2}$ ↓ Vd_{ss} and CL In humans: No significant changes in these parameters	Alakhov et al. (1999), Danson et al. (2004), Valle et al. (2011)
Paxceed	ND	PEO-*b*-PDLLA	Paclitaxel	In animals: significant ↓ in plasma AUC due to ↑ CL (compared to Taxol) In humans: no pharmacokinetic data reported	Zhang et al. (1997a, 1997b), Burt et al. (1999)
Genexol-PM (Cynviloq; IG001)	ND	PEO-*b*-PDLLA	Paclitaxel	↓ AUC in plasma (compared to Taxol) in animals and humans	Kim et al. (2001, 2004, 2007, 2015), Lee et al. (2008)
NK105	85	PEO-*b*-PPBA	Paclitaxel	↑ C_{max}, AUC, and $t_{1/2}$ ↓ Vd_{ss} and CL ↑ Accumulation in tumor tissues	Hamaguchi et al. (2005, 2007, 2010), Kato et al. (2012), Negishi et al. (2006), Mukai et al. (2016)
NK012	20	PEO-*b*-P(Glu)	SN-38	↑ $t_{1/2}$ of polymer-bound and released drug in plasma	Matsumura (2011), Koizumi et al. (2006), Burris et al. (2016), Hamaguchi et al. (2010)
NC-6004 (Nanoplatin)	30	PEO-*b*-P(Glu)	Cisplatin	↑ C_{max}, AUC, and $t_{1/2}$ ↓ Vd_{ss} and CL ↑ Accumulation in tumor tissues, liver, and spleen	Nishiyama et al. (2003), Plummer et al. (2011), Su et al. (2012), Uchino et al. (2005)
NC-6300 (K-912)	40–80	PEO-*b*-P(Asp)	Epirubicin	↑ C_{max}, AUC, and $t_{1/2}$ ↓ Vd_{ss} and CL ↑ Accumulation in tumor tissues, liver, spleen ↓ Distribution to heart, lungs, and kidneys	Harada et al. (2011), Mukai et al. (2017), Takahashi et al. (2013)
NC-4016	40	PEO-*b*-P(Glu)	Oxaliplatin	↑ AUC and $t_{1/2}$ in plasma ↑ Accumulation in tumor tissues, liver, and spleen	Ueno et al. (2014), Cabral et al. (2005), Cabral et al. (2007), Yamamoto et al. (2014)

↑ *indicates increase;* ↓ *indicates decrease.* PEO, *Methoxy poly(ethylene oxide);* PEO-b-P(Asp)-doxorubicin, *PEO-block-poly(L-aspartic acid)-doxorubicin;* PEO-b-P(Glu), *PEO-block-poly(L-glutamic acid);* PEO-b-PDLLA, *PEO-block-poly(D,L-lactide);* PEO-b-PPBA, *PEO-block-poly(4-phenyl-1-butanoate)-L-aspartamide;* PEO-b-PPO-b-PEO, *PEO-b-poly(propylene oxide)-b-PEO;* AUC, *area under the plasma drug concentration versus time curve;* $t_{1/2}$, *elimination half-life;* Vd_{ss}, *volume of distribution at steady-state;* CL, *total body clearance;* C_{max}, *maximum drug concentration in plasma;* T_{max}, *time to reach the* C_{max}; ND, *not determined.*

parameters (i.e., water solubility and membrane permeability) are used in the Biopharmaceutical Classification System, a tool that allows the prediction of the rate-limiting step in the intestinal absorption process following oral administration (Amidon et al., 1995).

The rate and extent of absorption from the gastrointestinal (GI) tract are important determinants of oral bioavailability of a drug and are affected by several factors. Physicochemical factors are related to the drug itself such as p*K*a, solubility, and stability. There are also physiological factors such as the wide range of pH in the GI, presence of various digestive enzymes, and absorption mechanisms as well as the first-pass effect. Various approaches to overcome these obstacles have been investigated in drug research and development (Kawabata et al., 2011; Stella and Nti-Addae, 2007). The use of new delivery systems such as micro- and nanoemulsions (Li et al., 2016; Fofaria et al., 2016), liposomes (Han et al., 2012; Jung and Han, 2014), polymeric nanoparticles (Joshi et al., 2014; Feng et al., 2009), microcapsules (Gula et al., 2013; Lu et al., 2007), and polymeric micelles has been the focus of much attention in recent years to solve issues associated with poor oral drug delivery.

Polymeric micelles can enhance the oral absorption of encapsulated drugs in multiple ways. This would be evident by apparently increasing the solubility of hydrophobic drugs; protecting the encapsulated drug from the harmful conditions inside the GI, prolonging the residence time in the intestine through inclusion of mucoadhesive shells; facilitating the transport of the encapsulated drug through the epithelial cells by endocytosis; bypassing or inhibiting efflux pumps such as *P*-glycoprotein (*P*-gp); and targeting specific carriers for receptor-mediated transport.

Numerous polymeric micellar formulations have been evaluated as drug delivery systems for oral administration (Table 15.2) (reviewed in Simoes et al., 2015). For instance, Winnik et al. investigated the capability of hydrophobically modified polysaccharide-based micelles to enhance the aqueous solubility of cyclosporine A (CyA) (Francis et al., 2003a, 2003b, 2005b). They showed that dextran-*grafted*-polyethylene glycol *n*-alkyl ether (DEX-*g*-PEG_{10}-C_n) and hydroxypropyl cellulose-*grafted*-polyethylene glycol *n*-alkyl ether (HPC-*g*-PEG_{10}-C_n) were able to load significant amounts of CyA ranging from 2.3% to 8.5% w/w. They also showed that the drug loading efficiency was highly correlated with the alkyl chain grafting density, where higher grafting density was associated with higher drug loading (Francis et al., 2003a,b, 2005b). Moreover, these polymeric micelles showed high stability in simulated gastric and intestinal fluids. In both media, only around 12% of CyA was released within 4 hours, which indicates that the micelles were able to retain the drug (Francis et al., 2005b). Furthermore, both DEX-*g*-PEG-C_{16} and HPC-*g*-PEG-C_{16} micelles were able to enhance the transport of loaded CyA through Caco-2 cell monolayers. Following 4 hours incubation, the apical-to-basolateral permeability of CyA increased by factors of 1.5 and 3 (compared to free CyA) when loaded in DEX-*g*-PEG_{10}-C_{16} and HPC-*g*-PEG_{10}-C_{16} micelles, respectively (Francis et al., 2005b). The superior

Table 15.2 Examples of Block Copolymer Micelles Used for the Oral Delivery of Different Drugs

Drug	Block Copolymer	Diameter Size (nm)	In Vivo Model	Changes in the Pharmacokinetic Profile	Reference
Curcumin	PEO-*b*-PDLLA: TPGS (mixed micelles)	46	Rat	Increased C_{max} and bioavailability compared to curcumin suspension	Duan et al. (2016)
Cyclosporine A	PEO-*b*-PDLLA	55	Rat	C_{max}, T_{max}, and AUC_{0-72h} were similar to those of the optimum oral formulation (Sandimmune Neoral)	Zhang et al. (2010a)
Efavirens	Pluronic F127: Tetronic T904 (mixed micelles)	20–27	Rat	Significantly enhanced the bioavailability of the drug; reduced inter-individual variability	Chiappetta et al. (2010)
Genistein	Pluronic F127	70–90	Rat	Increased C_{max} and T_{max}; enhanced bioavailability	Kwon et al. (2007)
Itraconazole	Mixture of PEO-*b*-PDLLA and PDLLA	23.5	Rat and dog	The pharmacokinetic parameters of the drug and its major metabolite were similar to those of the marketed cyclodextrin formulation (Sporanox)	Yi et al. (2007)
Megestrol	Pluronic L61-PAA	ND	Rat	Significantly increased the bioavailability	Alakhov et al. (2004)
Nifedipine	PEO-*b*-PCL	35	Rat	Relative bioavailability of nifedipine was 246% compared to the commercially available tablets	Yang and Ding (2015)
Paclitaxel	PEG-*b*-PVBODENA	85–90	Rat	Enhanced the oral bioavailability of paclitaxel to reach 12.4%	Lee et al. (2007)
Quercetin	PVCL-PVA-PEG (Soluplus)	79	Dog	The micellar formulation showed a slower T_{max} (7 h vs 5.3 h) and significantly higher AUC compared to quercetin solution (relative bioavailability = 286%)	Dian et al. (2014)
Resperidone	PEG_{750}-poly(CL-*co*-TMC)	24.2	Rat	Similar drug bioavailability, but lower C_{max} compared to aqueous tartaric solution (0.625% v/v)	Ould-Ouali et al. (2005)
Resperidone	^{14}C-labeled PEG_{750}-poly(CL-*co*-TMC)	24.2	Rat	Oral bioavailability of drug was 14%; based on total radioactivity, the oral bioavailability of the copolymer was 40% ($t_{1/2}$ = 29 h); mainly eliminated in urine with low uptake in RES	Mathot et al. (2006)

AUC, *Area under the plasma drug concentration-time curve;* C_{max}, *maximum drug concentration in plasma;* T_{max}, *time to reach the* C_{max}*;* PEG-P(CL-co-TMC), *methoxy poly (ethylene glycol)750-poly(caprolactone-co-trimethylene carbonate);* PEG-b-P(VBODENA), *poly(2-(4-vinylbenzyloxy)-*N,N*-diethylnicotinamide);* PEO-b-PDLLA, *poly(ethylene oxide)-b-poly(D,L-lactide);* PEO-b-PCL, *methoxy poly(ethylene oxide)-block-poly(ε-caprolactone);* PVCL-PVA-PEG, *polyvinyl caprolactam-polyvinyl acetate-polyethylene glycol;* TPGS, *D-a-tocopheryl polyethylene glycol succinate;* PAA, *poly(acrylic acid);* RES, *reticuloendothelial system;* ND, *not determined.*

performance of HPC-*g*-PEG_{10}-C_{16} over DEX-*g*-PEG_{10}-C_{16} was attributed to the better bioadhesive properties of HPC. Interestingly, the results of bidirectional permeability studies using copolymers covalently bound to fluorescein showed that either unimers or intact micelles (or both) were able to transport the Caco-2 cell monolayers as indicated by the detection of fluorescence in the receiving compartment (Francis et al., 2005b). In another study by Winnik et al., the DEX-*g*-PEG_{10}-C_{16} micelles were decorated with vitamin B_{12} on their surface to enhance their uptake by receptor-mediated endocytosis (through the intrinsic factor) (Francis et al., 2005a). Indeed, the results showed that the apical-to-basolateral transport of CyA through Caco-2 cell monolayers was twice as high when incorporated into vitamin B_{12}-decorated micelles compared to unmodified DEX-*g*-PEG_{10}-C_{16} micelles.

Park et al. reported that the use of hydrotropic agents such as *N,N*-diethylnicotinamide (DENA) could enhance the aqueous solubility of paclitaxel (~ 0.3 μg/mL) (Lee et al., 2003). They showed that micelles composed of PEG-*b*-poly(vinylbenzyloxy)-DENA (PEG-*b*-PVBODENA) could load up to 37.4% (w/w) of paclitaxel, thereby increasing the aqueous solubility to reach 39 mg/mL (Lee et al., 2007). In comparison, the highest paclitaxel loading achieved with PEG-*b*-poly(D,L-lactide) (PEG-*b*-PDLLA) was around 25% w/w (Liggins and Burt, 2002; Lee et al., 2007). To date, paclitaxel-loaded PEG-*b*-PVBODENA micelles is perhaps the most impressive example for solubility improvement by polymeric micelles. With such high drug loading levels in the micelles, the oral administration of a therapeutic drug dose became readily achievable. The in vivo pharmacokinetics following oral administration of paclitaxel-loaded PEG-*b*-PVBODENA micelles was studied in rats with catheters placed in their aorta, inferior vena cava, portal vein, or duodenum. The absolute oral bioavailability of paclitaxel-loaded micelles was estimated to be 12.4% (Lee et al., 2007). This is significantly higher than the reported oral bioavailability for the marketed formulation (Taxol) in the same species, which was 6.5% (Peltier et al., 2006). Although PEG-*b*-PVBODENA micelles were able to enhance the solubility and oral absorption of paclitaxel, the drug was still subjected to an extensive first-pass effect (both gastric and hepatic). This was evident by comparing the serum AUC values following a single dose of paclitaxel-loaded micelles administered intravenously via the portal vein and duodenum. The calculated AUC following portal vein administration was less than half of AUC obtained after IV administration (17,139 vs 36,029 ng min/mL) of the same dose of the drug in the micellar formulation (2.5 mg/kg) (Lee et al., 2007). The AUC following duodenal administration of 3.8 mg/kg was the lowest (6774 ng min/mL).

Inhibition of *P*-gp is an approach that has been proven to enhance the oral bioavailability of *P*-gp substrates. For instance, coadministration of CyA (10 mg/kg) with paclitaxel (10 mg/kg) to mice resulted in a fivefold increase in the oral bioavailability of paclitaxel (van Asperen et al., 1998). Instead of administering another therapeutic agent, which increases the incidence of adverse effects and drug–drug interactions, there are several amphiphilic block copolymers that have been shown to inhibit *P*-gp function. D-α-Tocopheryl polyethylene $glycol_{1000}$ succinate (TPGS), poloxamers (Pluronics), methoxy PEO-*b*-PDLLA, and low molecular weight methoxy PEO-*b*-PCL have been identified as potent inhibitors of efflux pumps (Zastre et al., 2008; Kabanov et al., 2002; Wang et al., 2004; Zhang et al., 2010a; Dintaman and Silverman, 1999). Paclitaxel has been loaded into micelles made of methoxy PEO_{2000}-*b*-phosphatidyl ethanolamine (PEG_{2000}-PE) and TPGS at a molar ratio of 1:1. The micelles increased the aqueous solubility of paclitaxel up to 5000 times to reach nearly 5 mg/mL solubilized drug concentration in water. The micelles also exhibited high stability at acidic pH and in the presence of bile acids. Furthermore, the drug-loaded micelles were found to be internalized as intact micelles into Caco-2 cells and bypassed *P*-gp pump. In a separate study, dequalinium was added to the paclitaxel-loaded mixed micelles (PEG_{2000}-PE and TPGS) and similar findings were obtained including the transport of the intact paclitaxel-loaded micelles across the Caco-2 monolayers and the everted gut sac (Fig. 15.2) (Yao et al., 2011). More importantly, following oral administration of these drug-loaded micelles (10 mg paclitaxel/kg), a significant antitumor efficacy was observed in the xenograft resistant tumor (MCF-7/Adr) in mice, which was comparable to the one obtained after administration of the same dose intravenously (Yao et al., 2011).

Although there is always a concern regarding the stability of polymeric micelles in the GI system, there have been several reports that showed polymeric micelles can withstand the challenging conditions in the GI and remain intact. For example, the stability of griseofulvin-loaded methoxy PEO-*b*-PDLLA micelles was evaluated in simulated gastric fluid (SGF) (pH 1.2) and simulated intestinal fluid (SIF) (pH 7.5) at 37°C for 11 days (Pierri and Avgoustakis, 2005). The specific viscosity of the copolymer decreased very slowly with time from day 1 until day 11. Moreover, the drug release profile in SGF and SIF was biphasic with no burst release, with only 10% released within the first 2 days and reaching around 50% at day 6. In another study, it has been shown that the micellar diameter and amount of encapsulated drug did not changed throughout the incubation of CyA-loaded methoxy PEO-*b*-PDLLA micelles in SGF or SIF for 12 hours (at 37°C) (Zhang et al., 2010b). The optimum micellar formulation showed a sustained release of CyA in SIF with only 40% of the drug released within 6 hours and up to around 80% after 24 hours. Furthermore, following a single oral dose of CyA-loaded micelles to rats, the pharmacokinetic parameters of C_{max}, T_{max}, and AUC_{0-t} were comparable to those obtained from the marketed oral CyA formulation (Sandimmun Neoral) (Zhang et al., 2010a).

One of the first mechanistic studies describing the transport of polymeric micelles across the intestinal barrier was reported by Préat et al. (Mathot et al., 2007). They investigated the potential of methoxy poly(ethylene $glycol)_{750}$-poly (ε-caprolactone-*co*-trimethylene carbonate) [PEG_{750}-poly(CL-*co*-trimethylene carbonate (TMC))] micelles as an oral delivery system for risperidone (Mathot et al., 2006; Ould-Ouali et al., 2005). The drug-loaded PEG_{750}-poly(CL-*co*-TMC) micelles were absorbed by pinocytosis (fluid-phase endocytosis), whereas the

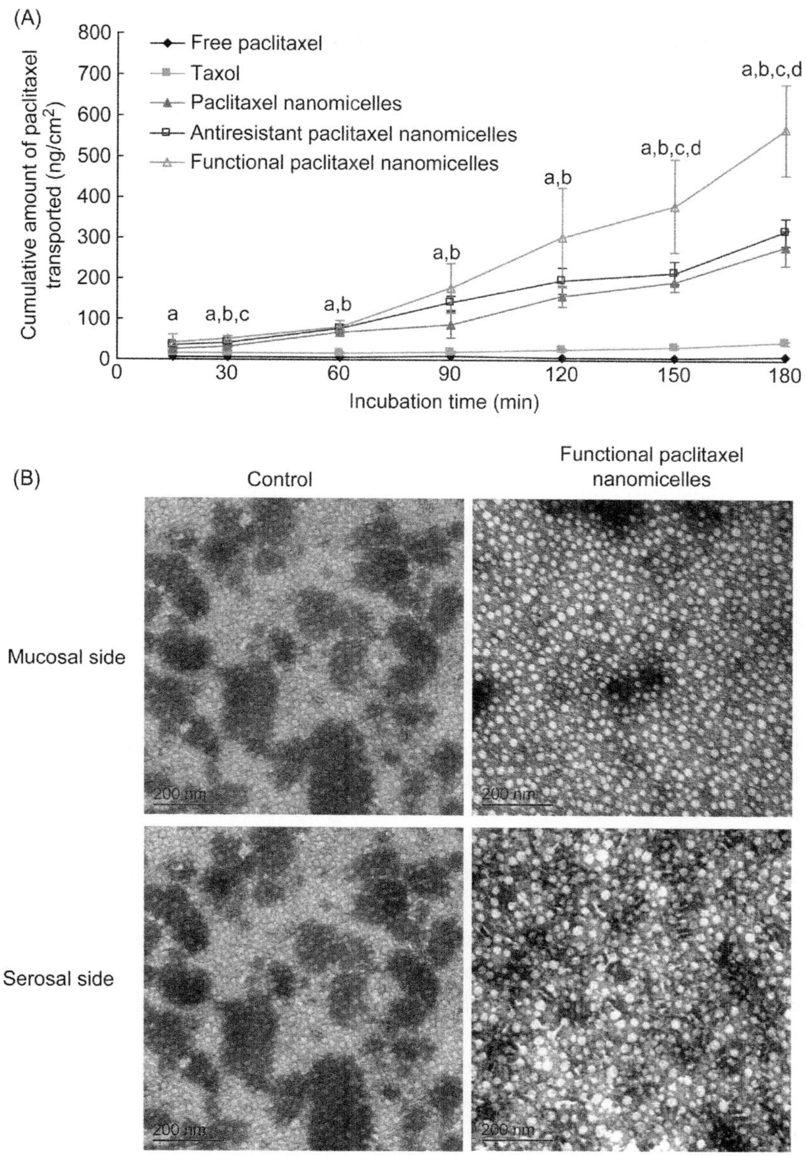

FIGURE 15.2

(A) Cumulative amounts of paclitaxel transported across the everted gut sacs after applying five different paclitaxel-containing formulations. Data are presented as the mean $\pm$ SD ($n = 3$). a, $P < .05$, versus free paclitaxel; b, $P < .05$, versus taxol; c, $P < .05$, versus paclitaxel nanomicelles (PEG_{2000}-DSPE); and d, $P < .05$, versus antiresistant paclitaxel nanomicelles [PEG_{2000}-DSPE:TPGS (1:1)]. (B) Transmission electronic microscopy images of media at the mucosal and serosal side of everted gut sacs incubated with TC199 medium or functional paclitaxel nanomicelles [PEG_{2000}-DSPE:TPGS:dequalinium (1:1:0.4)] for 2 h at 37°C. Samples were stained with uranyl acetate.

Reproduced with permission from Yao, H.J., Ju, R.J., Wang, X.X., Zhang, Y., Li, R.J., Yu, Y., et al., 2011. The antitumor efficacy of functional paclitaxel nanomicelles in treating resistant breast cancers by oral delivery. Biomaterials 32, 3285–3302.

polymeric unimers diffused passively across the membrane concomitantly with micellar endocytosis. This was preceded by a study that investigated the biodistribution of micelles made from [^{14}C]-labeled PEG_{750}-poly(CL-*co*-TMC) following IV and oral administration to rats (Mathot et al., 2006). Based on the total radioactivity obtained in plasma, the dose normalized AUC_{0-72h} after oral administration was around 40% of that obtained following IV administration (Mathot et al., 2006). Further, more than 85% of the total administered PEG_{750}-poly(CL-*co*-TMC)-related radioactivity was detected in the urine within 24 hours after IV administration. Whereas, after oral administration, the percentage urinary excretion of PEG_{750}-poly(CL-*co*-TMC) was around 40% within 24 hours, which matches the estimated oral bioavailability of the polymer. For more information on different mechanisms involved in the transport of different polymeric micelles across the intestinal barrier as well as other epithelial barriers, the reader is referred to the review by Pepić et al. (2013).

15.2.3 OCULAR ADMINISTRATION

Local drug administration to the anterior chamber of the eye by topical application is hampered by clearance mechanisms including lacrimation, tear dilution, and tear turnover (Gaudana et al., 2009; Maurice and Mishima, 1984). In addition, topically administered drugs generally have two pathways for absorption: either through the corneal route (ultimately reaching intraocular tissues through the aqueous humor) or noncorneal route (Gaudana et al., 2009). This explains why only less than 5% of the total administered drug reaches the aqueous humor when conventional eye drops are used (Hughes et al., 2005; Maurice and Mishima, 1984). For instance, after administration of a common eye-drop solution containing 0.05% cyclosporine to treat dry eye syndrome, more than 95% of the drug reaches systemic circulation through transnasal or conjunctival absorption (Peng et al., 2011; Maurice and Mishima, 1984).

Apart from topical administration, the drugs that are currently used to treat anterior and posterior ocular diseases are administered by intravitreal injections, systemic administration, and intraocular implants (Gaudana et al., 2009; Nagarwal et al., 2009). However, these routes have disadvantages and are sometimes associated with complications such as increased intraocular pressure (Maurice and Mishima, 1984) and related retinal detachment (Jager et al., 2004). Numerous delivery systems have shown significant improvement in the bioavailability and activity of ophthalmic drugs upon topical administration. These include nanoemulsions, thermoreversible polymer gels, niosomes, liposomes, nanoparticles, and polymeric micelles (Patel et al., 2013; Cholkar et al., 2013). Certain features such as biocompatibility, small size, potential mucoadhesion, and sustained drug release as well as the relatively transparent appearance would provide block copolymer micelles great potential to be successful carriers for ocular drug delivery.

One of the earliest reports on the use of polymeric micelles for ocular drug delivery was published in 1982 by Miller and Donovan (Miller and Donovan, 1982). They showed that poloxamer 407 (Pluronic F127) gel enhanced miotic activity of pilocarpine nitrate in rabbits compared to the aqueous solution (Miller and Donovan, 1982). Later, Desai and Blanchard (1998a,b) confirmed the previous findings by evaluating the miotic effect of different pluronic F127 formulations of pilocarpine using the albino rabbit eye model. Pluronic F127 formulations enhanced both the duration and the intensity (area under the temporal miotic response curve) of the miotic effect compared to the equal volume and dose of the drug in an isotonic solution (Desai and Blanchard, 1998a).

Numerous studies have shown that block copolymer micelles are safe, biocompatible, and able to enhance the corneal penetration of the loaded drugs (Table 15.3). For instance, Di Tommaso et al. (2011) investigated the in vitro ocular biocompatibility of methoxy poly(ethylene) glycol-hexylsubstituted poly (lactide) (MPEG-hexPLA) micelles using human corneal epithelial cells through different tests. Later, they assessed the in vivo corneal penetration, ocular distribution, and efficacy of CyA-loaded MPEG-hexPLA micelles (Di Tommaso et al., 2012). The objectives of their study were to: (1) investigate the capability of CyA-loaded micelles to overcome the corneal barriers after topical application; (2) test the efficacy of a 0.5% CyA in the micellar formulation for the prevention of corneal allograft rejection in a rat model; and (3) investigate the ocular CyA distribution in rat eyes compared to an oil solution. Following 2 weeks of topical administration, the CyA-loaded micelles were well-tolerated and were associated with a 73% success rate of cornea transplantation compared to a 25% success rate for the control group (physiological saline). The high efficacy observed for local treatment was consistent with the high levels of CyA found in the cornea (Di Tommaso et al., 2012). Furthermore, CyA was also found in the iris-ciliary body, aqueous humor, and vitreous proving the ability of the micelles to deliver CyA into different ocular tissues. The distribution studies on healthy rat eyes showed that the micellar formulation resulted in drug concentrations over 10 times higher in the cornea compared to the control formulation (CyA oil solution) (Di Tommaso et al., 2012).

Guo et al. (2015) studied the in vitro mechanism as well as the in vivo corneal permeation of CyA-loaded micelles following topical application in rabbit eyes. They used a graft copolymer composed of polyvinyl caprolactam-polyvinyl acetate-polyethylene glycol (PVCL-PVA-PEG), which is known commercially as Soluplus, and the micelles were prepared through solvent evaporation/film hydration method (Guo et al., 2015). The in vivo corneal permeation of CyA was evaluated following both single instillation (50 μL) and four instillations (200 μL) of CyA-loaded micelles and the results were compared to an oil-based CyA solution. The pharmacokinetics in the cornea after a single instillation revealed that the drug concentrations of CyA-loaded micelles were significantly higher (6.32%–168.26%) than those obtained with the control oil-based formulation at several time points (Guo et al., 2015). The four-instillation regimen indicated that

Table 15.3 Examples of Block Copolymer Micelles Used for Ocular Delivery of Different Drugs

Drug	Block Copolymer	Diameter Size (nm)	Animal Model	Changes in the Pharmacokinetic Profile	Reference
Curcumin	Pluronic P123: TPGS (mixed micelles)	10.8	Rabbit	Ex vivo corneal penetration study showed 33% higher cumulative permeated amount of curcumin compared with curcumin solution	Duan et al. (2015)
Cyclosporine A	MPEG-hexPLA	54	Rat	Drug concentrations over 10 times higher in the cornea compared to the control formulation (CyA oil solution). CyA was also found in the iris-ciliary body, in the aqueous humor and in the vitreous	Di Tommaso et al. (2012), Di Tommaso et al. (2011)
	PVCL-PVA-PEG (Soluplus)	73	Rabbit	Significantly higher concentration in the cornea compared CyA oil-based ophthalmic solution. The drug was not detected in the aqueous humor	Guo et al. (2015)
Dexamethasone	PEO-*b*-PCL	27.3	Rabbit	The micelles showed a 2.5-fold increase in permeability of the drug across excised rabbit sclera compared to control formulation (dexamethasone suspension)	Vaishya et al. (2014)
	PHEA-PEG-C_{16}	ND	Rabbit	40% higher AUC of drug in aqueous humor compared to dexamethasone suspension	Civiale et al. (2009)
fluorescein diacetate (dye)	PEO-*b*-PCL-*g*-PEI	28	Mouse	The dye in the micelles penetrated into the deep layers of the corneal tissues, while no significant fluorescence was detected in the deep layers of the cornea after administration of control formulations (fluorescein diacetate loaded into either microparticles, nanoparticles, or PEO-*b*-PCL micelles)	Li et al. (2015)
Itraconazole	Pluronic F127	80	Goat	Compared to Itral, the drug-loaded micelles exhibited a 6.67-fold increase in the permeation across goat corneal membrane	Jaiswal et al. (2015)
Pirenzepine	PEO-*b*-PDLLA	152.5	Rabbit	The pharmacokinetic profile of the drug in aqueous humor showed a significantly higher C_{max} and AUC for the PEO-*b*-PDLLA micellar formulation compared to the copolymer-free pirenzepine gel (control); T_{max} was the same for both formulations (~1 h)	Tu et al. (2007)

AUC, *Area under the concentration-time curve;* C_{max}, *maximum drug concentration;* T_{max}, *time to reach the* C_{max}; MPEG-hexPLA, *methoxy poly(ethylene) glycol-hexylsubstituted poly(lactide);* PEO-b-PDLLA, *poly(ethylene oxide)-b-poly(D,L-lactide);* PEO-b-PCL, *methoxy poly(ethylene oxide)-block-poly(ε-caprolactone);* PEO-b-PCL-g-PEI, *methoxy poly(ethylene oxide)-block-poly(ε-caprolactone)-graft-polyethylenmine;* PVCL-PVA-PEG, *polyvinyl caprolactam-polyvinyl acetate-polyethylene glycol;* TPGS, *D-a-tocopheryl polyethylene glycol succinate;* ND, *not determined.*

nanomicelle administration would have a significant advantage over administration of the oil-based ophthalmic solution as the CyA levels were 11.94%–343.35% higher than those obtained using the oil-based ophthalmic solution and the difference was statistically significant ($P < 0.05$) at several time points. Interestingly, the drug was not detected in the aqueous humor of all samples. There is no consensus in the literature whether topical CyA can reach aqueous humor. The various designs of formulations, protocols for animal testing, and different analytical techniques used for CyA detection in the reports are some of the reasons for the contradictory results. Nonetheless, topical formulations of CyA have always been shown to improve the efficacy in treatment of extraocular diseases, but have been inefficient at reaching intraocular targets.

The muscarinic receptor antagonist, pirenzepine was loaded into PEO-*b*-PDLLA micelles and the corneal permeation was evaluated both in vitro and in vivo (Tu et al., 2007). In vitro transcorneal experiments demonstrated that the PEO-*b*-PDLLA micelles significantly enhanced the transcorneal permeation of pirenzepine. The pharmacokinetic profile of pirenzepine in the aqueous humor in the albino rabbits showed a significantly higher peak concentration (C_{max}) and AUC for the PEO-*b*-PDLLA micellar formulation compared to the copolymer-free pirenzepine gel (Tu et al., 2007). The time to reach C_{max} (T_{max}) was similar for both formulations. Furthermore, both the drug-free and pirenzepine-loaded micelles showed no ocular long-term toxicity after twice daily instillations for 9 months (Tu et al., 2007).

Zhai et al. reported on pluronic P123 (P123)/D-*a*-tocopheryl polyethylene glycol succinate (TPGS) mixed micelle in situ gels as an ocular delivery system for curcumin (Duan et al., 2015). Curcumin-loaded mixed micelle in situ gels were formed by mixing micelles with gellan gum solution. The mixed micelles alone were able to control the in vitro release of the drug in phosphate buffered saline (PBS) (pH 7.4; 37°C; 1% Tween 80) over a long period of time. While 80% of curcumin in propylene glycol solution (control) was released within 24 hours, less than 15% of the drug was released from the mixed micelles within the same period (reaching 50% at 96 hours). Presence of the hydrogel further reduced the release rate to only around 5% of drug release within 24 hours and less than 20% after 96 hours. Furthermore, the drug-loaded mixed micelle in situ gels did not cause irritation of the ocular tissues. Moreover, the ex vivo corneal penetration study showed that the cumulative permeated amount of curcumin in the micellar formulation was around 33% higher than that obtained with curcumin solution (Duan et al., 2015). It is worth mentioning that both P123 and TPGS are known to inhibit *P*-gp in different tissues, which is present in several tissues in the body including the ocular tissues (Guan et al., 2011; Dintaman and Silverman, 1999; Kabanov et al., 2002; Guo et al., 2013; Dey et al., 2003).

Salama and Shamma (2015) have reported on single and mixed micelles prepared from linear poly(ethylene oxide)-*b*-poly(propylene oxide)-*b*-poly(ethylene oxide) copolymers (known as poloxamers or commercially as Pluronic or Synperonic) and branched poly(ethylene oxide)-*b*-poly(propylene oxide) copolymers (poloxamines; Tetronic) as a potential ocular delivery system for lornoxicam. Loading lornoxicam into the mixed micelles resulted in more than 70-fold increase in the aqueous solubility of the drug (2.34 vs 0.032 mg/mL). The optimum formulation was composed of T701:P85 at a ratio of (1:1 w/w), had a mean diameter size of nearly 90 nm at room temperature (169.5 nm at 37°C), and showed about 60% drug release within 6 hours. The ability of the mixed micelles to enhance permeability of rhodamine B (lipophilic dye) through a rabbit's cornea was assessed using confocal laser scanning microscopy (CLSM). Compared to dye aqueous solution, the penetration of rhodamine B into the viable cornea was increased following the micellar application. Moreover, the results of in vivo ocular irritation test in rabbits showed no sign of redness or inflammation or increased tear production following repeated instillations of the optimum formulation. However, histopathological studies revealed minimal marked changes and irritation following the instillation of the optimum micellar formulation. The presence of both heterophils and heterophils' exocytosis into the conjunctival epithelium in minimal amounts is usually indicative of subacute inflammation (Salama and Shamma, 2015).

Jaiswal et al. investigated the potential of itraconazole loaded into Pluronic F127 for the treatment of fungal keratitis (Jaiswal et al., 2015). Itraconazole was incorporated into Pluronic F127 micelles and then mixed with a gel base solution (carbopol 934P, 0.5% w/v) to form in situ gel. The optimum micellar formulation had a mean diameter size of 80 nm with a polydispersity index (PDI) of 0.124 and a high drug loading capability that enabled the aqueous solubility of the drug to reach 18.25 mg/mL. The in situ gel had excellent mucoadhesive strength and it further controlled the release of the drug compared to the micelles in solution. Moreover, the histopathological study performed on excised goat cornea revealed that the optimum formulation did not cause any destructive effect on the corneal epithelium as well as on stromal layer, which was similar to the effect obtained with negative control (0.9% NaCl). The in vitro antifungal activity of the itraconazole-loaded micelles was significantly higher compared to that obtained with the drug in the marketed formulation (Itral). Furthermore, the ex vivo transcorneal permeation of the drug in the in situ gel was compared to Itral eye drops using Franz diffusion cell consisting of donor and receptor compartments separated by goat cornea. Interestingly, compared to Itral, the drug-loaded micelles exhibited a 6.67-fold increase, while the in situ gel exhibited a 5.6-fold increase in the permeation across goat corneal membrane. The differences were statistically significant ($P < 0.05$) when compared to Itral and to the pure-drug suspension. Upon fitting the permeability kinetic data of both the micellar formulation alone or in the gel, it followed zero-order kinetics; that is, the flux was independent of the amount of the drug permeated at various time points (Jaiswal et al., 2015).

Recently, Li et al. (2015) demonstrated that methoxy PEG-*b*-PCL-*graft*-polyethyleneimine (PEO-*b*-PCL-*g*-PEI) were able to self-assemble into positively charged micelles in aqueous media. The drug-free PEO-*b*-PCL-*g*-PEI micelles showed a relatively small mean diameter of around 28 nm, a narrow size distribution with a PDI of 0.212, and a moderate positive charge of 12.12 mV. Moreover, the in vitro release of CyA-loaded micelles showed a significant burst release of nearly 35% within the first 2 hours and became steady afterwards (Li et al., 2015). Furthermore, the hydrophobic dye fluorescein diacetate was loaded into the PEO-*b*-PCL-*g*-PEI micelles and the in vivo corneal permeability was assessed in C57BL/6 mice by using two-photon scanning laser microscopy. The images of corneal cross-sections at different time intervals showed that, following topical administration, the dye in the PEO-*b*-PCL-*g*-PEI micelles penetrated into the deep layers of the corneal tissues. In contrast, no significant fluorescence was detected in the deep layers of the ocular cornea (e.g., the stroma), even 60 minutes after the administration of the dye in the control formulations (fluorescein diacetate loaded into either microparticles, nanoparticles, or PEO-*b*-PCL micelles) (Li et al., 2015). It seems that positively charged micelles can enhance the penetration of the payload by prolonging the ocular residence time and by increasing the interaction with the negatively charged cell membrane.

More recently, Prosperi-Porta et al. (2016) developed a series of mucoadhesive block copolymer micelles based on phenylboronic acid (PBA) as an ocular delivery system for CyA. Poly(D,L-lactide)-4-cyano-4-[(dodecylsulfanylthiocarbonyl)-sulfanyl]pentoate (pLA-CDP; MW = 4700) was reacted with methacrylic acid and PBA at different molar feed ratios to get copolymers with a different content of mucoadhesive PBA. The micelles were prepared through a cosolvent evaporation method and their diameter sizes were between 35 and 65 nm. The CyA-loaded micelles showed a typical biphasic release profile with an initial burst release of 35%–45% of the drug within the first 24 hours, followed by a lower rate release reaching around 80% after 14 days. The mucoadhesive properties of the micelles were studied using surface plasmon resonance with chitosan serving as a positive control. The results showed that the mucoadhesion of the micelles increases with increasing PBA content until reaching a limit whereby additional PBA does not significantly increase mucoadhesion, possibly due to the saturation of mucin monolayer. Moreover, the results of an in vivo acute ocular irritation test showed that the micelles had excellent biocompatibility on the ocular surface of Sprague–Dawley rats after 10 days of instillation. Nevertheless, the in vitro human corneal epithelial cell viability results showed that PBA-containing micelles inhibit growth and caused clustered cell growth (Prosperi-Porta et al., 2016).

One of the barriers of drug delivery to the posterior segment of the eye is the presence of transporters that play a critical role in drug transport across ocular tissues including the cornea, conjunctiva, and retina (Dey and Mitra, 2005). Efflux transporters such as *P*-gp and multidrug resistance-associated protein (MRP) are highly expressed on the different layers of the rabbit and human eye (Dey et al., 2003; Dhananjay et al., 2013; Vellonen et al., 2010). Moreover, various influx transporters are expressed in the anterior as well as posterior chambers of the eye (reviewed in Mannermaa et al., 2006; Mandal et al., 2016). Inhibition of one or more of the efflux transporters is expected to enhance the ocular bioavailability of drugs that are substrates to these efflux pumps. Indeed, it has been shown that this strategy has led to an enhancement of the drug transport across the cornea in vitro and an increase of drug concentrations in the aqueous humor in vivo. For instance, the C_{max} and AUC of erythromycin in the aqueous humor of rabbit eyes has been shown to significantly increase when CyA (*P*-gp inhibitor) or MK571 (MRPs inhibitor) was coadministered (Hariharan et al., 2009). Nevertheless, more studies are needed to be done in this area to confirm the applicability and clinical relevance of this approach.

15.2.4 TOPICAL ADMINISTRATION

Topical administration of drugs on skin offers a targeted delivery for treatment of dermatological diseases. Moreover, transdermal delivery is an attractive approach for systemically acting drugs that are subject to an extensive first-pass metabolism upon oral administration. However, the skin is considered a tough defense barrier to protect the body from xenobiotics. Only a few drug molecules are able to passively diffuse across the different layers of the skin at therapeutic rates. Scopolamine, nicotine, nitroglycerin, fentanyl, and estradiol are examples of drugs that are available in transdermal patches. The most challenging permeability barrier in the skin is the uppermost layer, the *stratum corneum*. Over the past few decades, several approaches have been explored to enhance the transdermal permeability of drugs. One of these approaches is to use polymeric micelles as carriers for the delivery of drugs through the skin.

Several polymeric micelles made of diblock and triblock copolymers have been investigated for drug delivery via the skin. In a study by Bachhav et al. (2011) MPEG-hexPLA micelles were used to develop a formulation with an enhanced cutaneous bioavailability of azole antifungals. In this study, three different derivatives of the copolymer as well as three different azole antifungals (clotrimazole, econazole nitrate, and fluconazole) were investigated. The optimum micellar formulation consisted of MPEG-dihexPLA encapsulating econazole with an encapsulation efficiency of 98.3%, which translated to an aqueous concentration of 1.3 mg/mL. The developed micellar formulation (1% econazole nitrate) was evaluated against the commercially available liposomal formulation (Pevaryl 1% econazole nitrate liposomal gel). The micellar formulation showed a significantly higher skin deposition in porcine and human skin. Moreover, the micelle penetration pathway was investigated with CSLM using fluorescein. The CSLM images showed that the micelles were able to deliver higher amounts of the dye to the skin compared to the aqueous fluorescein solution. The same diblock copolymer (MPEG-*b*-dihexPLA) was used by Lapteva et al. (2014) to develop tacrolimus micelles for cutaneous delivery. The micellar formulation was shown to increase the penetration of the encapsulated drug through the skin. Furthermore, while the accumulation of tacrolimus in the deeper skin layers was higher than the marketed formulation (Protopic ointment), the micelles were not able to cross the *stratum corneum* and were deposited in hair follicles. The natural anticancer oridonin was loaded into methoxy PEO-*b*-PCL micelles and the formulation showed about 80% higher permeation of the drug through excised mouse skin compared to a saturated aqueous solution (Xue et al., 2012). Methoxy PEO-*b*-PCL micelles have also shown to increase the permeability of the loaded minoxidil through guinea pig skin by 2.5-fold compared to control formulation (30% ethanol).

15.3 PHARMACOKINETIC PROFILES OF DRUG-LOADED POLYMERIC MICELLES IN CLINICAL TRIALS

There are currently nine micellar formulations in clinical trials, and one of these has recently received marketing status (Genexol-PM or Cynviloq) (Table 15.1). All nine products are formulations of chemotherapeutic agents and were originally developed for cancer. However, one product (Paxceed) was further investigated for other diseases such as rheumatoid arthritis, multiple sclerosis, and psoriasis.

15.3.1 SP1049C

SP1049C is a pluronic-based polymeric micellar formulation of doxorubicin which was primarily developed to overcome multi-drug resistance. It was researched and developed by Professor Kabanov's group and Supratek Pharma, Inc. (Montreal, QC, Canada) (Alakhov et al., 1999; Danson et al., 2004). It consists of Pluronic F127 and Pluronic L61 mixed micelles at a ratio of 1:8 (w/w), which are formulated in an isotonic buffered saline to have final concentrations of 2% and 0.25% (w/v), respectively. Pluronic F127 is used to stabilize the micelles, whereas Pluronic L61 is used for chemosensitization since it has been shown to inhibit the efflux transporter *P*-gp. (Kabanov et al., 2002; Evers et al., 2000). Studies in several animal models have shown superior anticancer efficacy of SP1049C over doxorubicin in the conventional formulation. For instance, in a tissue distribution study performed in mice, SP1049C showed a preferential accumulation of doxorubicin (compared to free doxorubicin; Adriamycin) in the tumor and brain, but not in other tissues such as the liver, kidney, heart, and lung (Alakhov et al., 1999). In agreement with the biodistribution profile, tumor inhibition and survival time were increased in SP1049C-treated animals. In addition, SP1049C has been shown to favorably modify the pharmacokinetic profile of doxorubicin in mice (Alakhov et al., 1999). Specifically, following IV administration of SP1049C to C57BL/6 mice, the mean plasma AUC of doxorubicin was 2.1-fold higher than the free doxorubicin (Adriamycin) (Alakhov et al., 1999). Moreover, SP1049C had lower Vd_{ss} and CL in healthy mice (33% and 52%, respectively) as well as in tumor-bearing mice (28% and 15%, respectively).

In a Phase I clinical trial, 26 patients with advanced cancer were treated with SP1049C administered at doses ranging from 5 to 90 mg/m^2 through IV infusion once per 3 weeks for at least six cycles (Danson et al., 2004). The dose-limiting toxicity (DLT) was neutropenia and was reached at 90 mg/m^2 and the maximum tolerated dose was 70 mg/m^2, which was the recommended dose for the Phase II trials. The hand-foot syndrome reported with the use of liposomal doxorubicin (Doxil) was not observed. The pharmacokinetic profile of SP1049C was similar to the one reported for conventional formulation of doxorubicin although the terminal-phase $t_{1/2}$ obtained with SP1049C (48.8 hours) was longer than the one reported for free doxorubicin (c.30 hours) (Greene et al., 1983). Evidence of antitumor activity was observed in several patients, where three patients showed partial or complete response and eight patients showed stable disease (Danson et al., 2004). In 2005 and 2008, SP1049C was granted orphan drug designation (by the US FDA) for the treatment of esophageal carcinoma and gastric cancer, respectively. Later, a Phase II clinical trial was conducted and the results were published in 2010 (Valle et al., 2011). The antitumor activity of SP1049C was evaluated in 21 patients with advanced adenocarcinoma of the esophagus and gastroesophageal junction. The product was administered at a dose of 75 mg/m^2 as a brief IV infusion every 3 weeks. Out of the 19 evaluable patients, nine patients had a partial response and eight patients had either minor response or stable disease as their best response, with a median overall survival of 10 months (Valle et al., 2011). Moreover, SP1049C showed an acceptable safety profile with neutropenia being the principal toxicity of the product. Although the US FDA has agreed on the design of the study protocol for randomized Phase III pivotal clinical trial, no development news has been reported since 2008 and the trial is not listed in the ClinicalTrials.gov database.

15.3.2 PAXCEED

Paxceed consists of methoxy PEO-*b*-PDLLA micelles that are physically loaded with paclitaxel (25% w/w) (Zhang et al., 1997a,b). This product was developed by Professor Burt's group and Angiotech Pharmaceuticals Inc. (Vancouver, BC, Canada). The molecular weights of methoxy PEO and PDLLA used to make micelles were 2000 and 1333 g/mol, respectively. Paxceed showed a similar in vitro cytotoxicity profile to the conventional formulation of paclitaxel (Taxol) in a variety of tumor cell lines (Kim et al., 2001; Zhang et al., 1997a,b). Moreover, a superior safety profile was shown in vivo for Paxceed compared to Taxol with a fivefold increase in maximum tolerable dose (MTD) upon intraperitoneal (IP) administration to mice (Zhang et al., 1997a). The increase in MTD, however, was modest after IV administration (from 20 to 25 mg/kg). Furthermore, at an equivalent dose of 20 mg/kg, the plasma AUC of Paxceed was about 82% lower than the one obtained with Taxol (7.4 vs 40.6 μg h/mL) (Zhang et al., 1997b). This was later attributed to the rapid dissociation of paclitaxel from methoxy PEO-*b*-PDLLA micelles within minutes after IV administration, which was confirmed by studies on dual-label paclitaxel/copolymers biodistribution (Burt et al., 1999). Following IV administration of radiolabeled paclitaxel/methoxy PEO-*b*-PDLLA to rats, paclitaxel was rapidly and widely distributed into tissues and mainly eliminated in feces, while the methoxy PEO and PDLLA components were rapidly eliminated in urine with a half-life of 2.6 hours. This was supported by a higher degree of radioactivity in the urinary bladder and kidneys relative to the plasma 1 hour following IV administration and the fact that low radioactivity was detected in the GI tract (Burt et al., 1999). Paxceed was investigated in several clinical trials for the treatment of different diseases including psoriasis, rheumatoid arthritis, and multiple sclerosis. However, no news has been reported on this product since 2004.

15.3.3 GENEXOL-PM

Similar to Paxceed, Genexol-PM is a block copolymer micelle formulation of paclitaxel that is composed of methoxy PEO-*b*-PDLLA in which paclitaxel is physically loaded (16.7% w/w) (Kim et al., 2001). The molecular weights of methoxy PEO and PDLLA are 2000 and 1750 g/mol, respectively. It was developed by Samyang Corporation (Seoul, Korea) (Kim et al., 2001). Genexol-PM is the most commonly used name for this product in the literature, but it is known under several other names including Cynviloq, Paclitaxel PM, and IG 001. It is the first polymeric micellar product that made it to the market in several countries around the world. Currently, it is being marketed in the Republic of Korea, India, the Philippines, and Vietnam for the treatment of various types of cancer including metastatic breast cancer, non-small cell lung cancer, and ovarian cancer. In 2013, Sorrento Therapeutics Inc. announced the acquisition of exclusive distribution rights from Samyang to Cynviloq in the European Union and in the United States (after merging with IgDraSol Inc.). In the same year, the FDA Division of Oncology Products agreed that the data available for Cynviloq (Genexol-PM) were sufficient to follow the 505(b)(2) bioequivalence regulatory submission pathway for this product using albumin-bound paclitaxel (Abraxane) and Taxol as the reference listed drugs. Indeed, in 2014, a bioequivalence study was conducted in patients with metastatic or locally recurrent breast cancer using Abraxane as the reference drug (ClinicalTrials.gov identifier: NCT02064829). Although the results of the bioequivalence study were not officially published, the sponsoring company (Sorrento Therapeutics) stated in a press release that "data analysis suggests that Cynviloq meets the bioequivalence criteria for both total and unbound paclitaxel."

Preclinical in vitro and in vivo studies demonstrated that unloaded Genexol-PM micelles are nontoxic and biocompatible and do not have anticancer activity (Kim et al., 2001). Paclitaxel-loaded Genexol-PM micelles showed an in vitro cytotoxicity profile similar to Taxol in several cancer cell lines including breast, ovarian, non-small cell lung cancer, and colon (Kim et al., 2001). Moreover, in vivo studies performed in tumor-bearing mice showed that Genexol-PM had better anticancer activity compared to Taxol when used in their MTDs (Kim et al., 2001). The superior activity of Genexol-PM was most notable in the breast cancer model in which some complete responses were achieved. The pharmacokinetic profile of Genexol-PM is similar to that reported for Paxceed. Following IV administration of 50 mg/kg (MTD) of Genexol-PM to tumor-bearing mice, the calculated plasma AUC of paclitaxel was around 70 μg h/mL (Kim et al., 2001). In contrast, the plasma AUC of paclitaxel was nearly 95 μg h/mL after IV administration of Taxol at a dose of 20 mg/kg (MTD). Although Genexol-PM was administered at a 2.5-times higher dose compared to Taxol, the plasma AUC of the drug was about 26% lower than the value obtained with Taxol. This is in line with the previously reported data on the similar formulation, Paxceed. Nevertheless, the higher MTD for Genexol resulted in two- to threefold higher drug concentrations in all the tissues (in all the time points) including the tumor tissues (Kim et al., 2001).

The first Phase I clinical trial was conducted in patients with advanced tumors in South Korea (Kim et al., 2004). Genexol-PM was administered to 21 patients as a 3-hour infusion once every 3 weeks at doses ranging from 135 to 390 mg/m^2. The MTD was determined to be 390 mg/m^2, with neuropathy, neutropenia, and myalgia as the DLTs. In contrast, the reported MTDs for Taxol and Abraxane were 175 and 300 mg/m^2, respectively (Ibrahim et al., 2005). Beside the advantage of higher MTD obtained with Genexol-PM, no hypersensitivity reactions were reported with its use and for that reason no premedication was needed as was the case with the use of the Cremophor EL formulation (Taxol). The pharmacokinetic profile in patients was found to be linear (C_{max} and $AUC_{0-infinity}$ increases proportionally with dose) and the plasma AUC of paclitaxel obtained with Genexol-PM was lower than that reported for Taxol. Despite that, preliminary signs of clinical efficacy were observed as there were three partial responders and six patients with stable disease. Based on these findings, the recommended dose for Phase II trials was 300 mg/m^2.

In another Phase I trial, 24 patients with solid tumors were treated with Genexol-PM administered as a 1-hour infusion at doses ranging from 80 to 200 mg/m^2 on a weekly basis (followed by a resting week) (Lim et al., 2010). In this dosage regimen, the MTD was determined to be 180 mg/m^2. Although the incidence of neutropenia was similar to that observed in the previously reported study (Kim et al., 2004), the nonhematologic toxic effects were less severe in the weekly regimen of Genexol-PM used in this study. Similar to the previous Phase I trial, the pharmacokinetic profile was linear in the dose range used with both C_{max} and $AUC_{0-infinity}$ values increasing approximately three- and fourfold as the dose of Genexol-PM was increased from 80 to 200 mg/m^2, respectively, with no significant change in total body clearance (median CL = 44 L/h).

Several Phase II trials for Genexol-PM have been conducted on patients with metastatic breast cancer, advanced or metastatic pancreatic cancer, and as a combination therapy with cisplatin in non-small cell lung cancer (Lee et al., 2008; Saif et al., 2010; Kim et al., 2007). Several other Phase I and II trials have been conducted and some are still ongoing for other types of cancer (Lee et al., 2016; Kim et al., 2015; Ahn et al., 2014). The most reported toxicities were hematological (neutropenia and thrombocytopenia), neuropathy, and myalgia. Interestingly, in these trials, hypersensitivity reactions were reported in about 20% of patients. Nevertheless, the results of these studies showed encouraging response rates that led to proceeding with Phase III and IV clinical evaluations in the United States and other countries.

15.3.4 NK911

NK911 is a methoxy PEO-*b*-poly(L-aspartic acid) (PEO-*b*-P(Asp)) micelles that contain both chemically bound and physically loaded doxorubicin. The molecular weights of methoxy PEO and P(Asp) are 5000 and 4000 g/mol, respectively. It was developed by Professor Kataoka's group in Japan (Yokoyama et al., 1991; Nakanishi et al., 2001; Masayuki et al., 1990). The chemically bound drug that is present in the core of the micelles was not intended to be released, but rather to enhance the hydrophobicity of the core and to increase the compatibility and affinity between the core and the physically loaded doxorubicin. Results from preliminary development studies on NK911 micelles revealed that only the physically loaded doxorubicin was responsible for the activity both in vitro and in vivo (Yokoyama et al., 1998). In the optimized formulation (NK911), doxorubicin is conjugated to the side chain of the aspartic acid at a substitution level of 45% (Nakanishi et al., 2001). The micelles are formed upon direct dissolution of the conjugated copolymer in the aqueous medium. Subsequently, free doxorubicin is added to be incorporated inside the core of the formed micelles. Following freeze-drying and reconstitution in PBS, the mean diameter size of NK911 micelles is around 40 nm.

Preclinical assessment of pharmacokinetics, toxicity, and efficacy of NK911 was performed in several tumor-bearing animals (Nakanishi et al., 2001). For instance, in colon adenocarcinoma (C26)-bearing mice, NK911 was clearly shown to have preferential accumulation in the tumor tissues with a 3.4-fold higher $AUC_{0-\text{infinity}}$ compared to free doxorubicin. Moreover, compared to free doxorubicin, NK911 showed a 36.4-fold and 28.9-fold higher C_{max} and AUC_{0-24h} in plasma, respectively (Nakanishi et al., 2001).

A Phase I clinical trial was conducted for NK911 on patients with metastatic or recurrent solid tumors in Japan (Matsumura et al., 2004). NK911 was administered to 23 patients through IV infusion once every 3 weeks at doses ranging from 6 to 67 mg/m^2. The MTD was considered to be 67 mg/m^2 with neutropenia as the DLT. Despite the impressive favorable changes observed in animal models, NK911 showed relatively modest changes in the pharmacokinetic parameters of doxorubicin in humans (Matsumura et al., 2004). The results of a Phase I clinical trial for NK911 showed that the plasma $t_{1/2}$ and $AUC_{0-\text{infinity}}$ were increased by 2.5-fold and 2-fold, respectively, compared to free doxorubicin. These changes were caused by the decrease in CL and Vd_{ss} by 2.2-fold and 1.6-fold, respectively. The C_{max} and $AUC_{0-infinity}$ of NK911 increased proportionally with increasing the dose. When the pharmacokinetic parameters obtained with NK911 were compared to those reported for the liposomal doxorubicin (Doxil), it has clearly shown that NK911 had much lower stability in plasma compared to Doxil. Nonetheless, on the basis of the good tolerability, preliminary signs of efficacy, and the lack of infusion-related reaction (that usually occur with Doxil), it was decided to proceed with clinical development (Matsumura et al., 2004). The recommended dose for the Phase II study was 50 mg/m^2. However, no news has been reported on this product since 2004.

15.3.5 NK105

NK105 appeared to be one of the most promising micellar formulations following preclinical pharmacokinetics and pharmacodynamics studies. This formulation consists of PEO-*b*-poly(4-phenyl-1-butanoate-L-aspartamide) (PEO-*b*-PPBA) micelles containing physically loaded paclitaxel. In this system, half of the carboxylic groups of P(Asp) block were modified with 4-phenyl-1-butanol by esterification reaction to increase the hydrophobicity of the core (Hamaguchi et al., 2005). This product was first developed by Professor Kataoka's group in collaboration with Nanocarrier Co., Ltd (Japan) and Nippon Kayaku Co., Ltd (Japan). The molecular weights of methoxy PEO and PPBA are approximately 12,000 and 8000 g/mol, respectively. NK105 micelles are prepared through self-assembly in an aqueous medium, which is then freeze dried with a paclitaxel content of 23% w/w. The product is supplied in glass vials containing a dose equivalent to 100 mg of paclitaxel per vial, which are usually reconstituted in 100 mL of 5% Dextrose solution prior to administration. The mean diameter size of NK105 micelles is around 85 nm.

Preclinical studies have shown impressive changes on the pharmacokinetic and biodistribution profiles of paclitaxel (Hamaguchi et al., 2005). Following IV administration of NK105 to C-26-bearing mice, there has been an 86-fold increase in paclitaxel's AUC in plasma, a 99% decrease in its CL, and a 93% decrease in its Vd_{ss} compared to Taxol. Moreover, there was approximately a 3-fold increase in the peak concentration of paclitaxel in the tumor tissue as well as a 25-fold increase in the tumor AUC compared to Taxol, which reflects the enhancement of the tumor exposure. A single dose of 100 mg/kg of NK105 to HT-29 colon cancer-bearing mice resulted in the complete disappearance of tumors (Hamaguchi et al., 2005). In another study, NK105 has shown superior antitumor activity compared to free paclitaxel when Lewis Lung carcinoma-bearing mice were treated with radiation in combination with a single IV dose (45 mg/kg) of either NK105 or free paclitaxel (Negishi et al., 2006).

In 2007, a Phase I clinical trial was conducted on 19 patients with solid tumors refractory to conventional chemotherapy (Hamaguchi et al., 2007). NK105 was administered once every 3 weeks as a 1-hour infusion at doses ranging from 10 to 180 mg/m^2 without the antiallergic premedication that is usually given prior to Taxol dosing. The MTD was defined at 180 mg/m^2 with neutropenia as the DLT. Interestingly, there was a significant reduction in neurotoxicity with only grade 1–2 reported for NK105. The pharmacokinetic profile of NK105 at a dose of 150 mg/m^2 showed more than 15-fold increase in AUC, 26-fold decrease in CL, and a 13-fold decrease in Vd_{ss} compared to Taxol at even a higher dose (210 mg/m^2). The magnitude of changes in these pharmacokinetic parameters obtained with NK105 was also higher compared to Genexol-PM and Abraxane. Based on the results of this study, the recommended dose for Phase II trials was 150 mg/m^2.

In 2016, the results of another Phase I study for NK105 were published. In this study, NK105 was administered to 26 patients (with solid tumors refractory to conventional chemotherapy) as a 30-minute infusion at doses ranging from 50 to 100 mg/m^2 on a weekly basis (for 3 weeks followed by a resting week) (Mukai et al., 2016). The study was divided into two parts. The first part was the dose-escalation phase where three to seven patients were enrolled in each dose level ($n = 16$) to determine the recommended dose of NK105. The second part was the expansion phase, where 10 patients were enrolled to evaluate the safety and preliminary efficacy of NK105 at its recommended weekly dose determined in the first part of the study (80 mg/m^2). The pharmacokinetic profile of NK105 was found to be linear in the dosing range used, that is, the C_{max} and AUC were proportionally increasing with the increase in dose, while CL and Vd_{ss} remained constant and independent of dose. Furthermore, using equilibrium dialysis, the estimated plasma $AUC_{0-infinity}$ of free paclitaxel (unbound) was around only 5%–7% of the plasma $AUC_{0-infinity}$ of the total drug (Mukai et al., 2016). Similar to the previous preclinical studies and to the Phase I study, the pharmacokinetic parameters of paclitaxel were substantially changed in NK105 compared to Taxol. Specifically, following the administration of a dose equivalent to 100 mg/m^2 of paclitaxel, NK105 showed around 50-fold decrease in CL (which resulted in a 50-fold increase in plasma $AUC_{0-infinity}$) and a 25-fold decrease in Vd_{ss} compared to those obtained with Taxol (Mukai et al., 2016; Tamura et al., 1995). This, indeed, points to the ability of NK105 micelles to encapsulate paclitaxel and prolong its circulation time in blood, which along with the diameter size of 85 nm provide these micelles with the required features to efficiently target solid tumors through the EPR effect. Moreover, the significantly lower Vd_{ss} of paclitaxel in NK105 compared to Taxol suggests that the drug may have a relatively lower distribution in normal tissues. Regarding the second part of the study, of the 10 patients enrolled, six achieved partial response and four achieved stable disease status (Mukai et al., 2016). Moreover, none of the patients had grade 3 (or more) neuropathy.

A Phase II study was conducted and the results were published in 2011 (Kato et al., 2012). The aim of the study was to evaluate the efficacy and safety of NK105 in patients with advanced gastric cancer after failure of first-line chemotherapy. Fifty-seven patients were enrolled and 56 were evaluable for efficacy. Each patient received NK105 by a 30-minute IV infusion every 3 weeks without antiallergic premedication. Drug-related toxicity was mainly mild and manageable with neutropenia having the most reported toxicity. The incidence of grade 3 neuropathy was only 1.8%, which is considerably lower than those reported for Taxol (10%) and Abraxane (11%) in their Phase II trials. This confirms the findings of the Phase I studies (Hamaguchi et al., 2007; Mukai et al., 2016) where there was no grade 3 peripheral neuropathy reported even in patients who received more than five cycles of NK105. The overall response rate in the Phase II study was 25%, with two complete responses and 12 partial responses (Kato et al., 2012). A multinational Phase III clinical study comparing NK105 versus paclitaxel in patients with metastatic or recurrent breast cancer was started in 2012 and completed in January 2017 (ClinicalTrials.gov identifier: NCT01644890). However, in July 2016, the company announced that "the primary endpoint of the study, progression free survival (PFS), did not meet the prespecified statistical criteria" and no results have yet been published (Nippon Kayaku Co Ltd., 2016).

15.3.6 NK012

NK012 is an SN-38 block copolymer micellar formulation where the drug is chemically attached to the core of methoxy PEO-*b*-poly(L-glutamic acid) (PEO-*b*-P(Glu)) (Matsumura, 2011; Koizumi et al., 2006). SN-38 is an active metabolite of irinotecan (CPT-11), which is an analog of camptothecin, that has shown a 1000-fold more potent in vitro cytotoxicity against various cancer cells compared to irinotecan (Kawato et al., 1991). In NK012 micelles, SN-38 is covalently bound to the P(Glu) segment by the condensation reaction between the carboxylic acid on P(Glu) and the phenol present in SN-38. The molecular weights of methoxy PEO and SN-38-conjugated block are 12,000 and 7000 g/mol, respectively. NK012 has been developed by Nippon Kayaku Co., Ltd (Japan). The final product is a freeze-dried formulation containing 20% (w/w) of SN-38 which is reconstituted in an aqueous media to form micelles with a mean diameter size of 20 nm. The percentage of released SN-38 from NK012 micelles in dextrose 5% solution (at 37°C) was found to be only 1% and 3% at 24 and 48 hours, respectively. The amount of drug released, however, was significantly higher in PBS (at 37°C), where 57% and 74% of SN-38 was released at 24 and 48 hours, respectively (Matsumura, 2011).

In preclinical studies, the anticancer activity of NK012 was evaluated in several orthotopic tumor models including pancreatic cancer, lung cancer, renal cancer, glioma, gastric cancer, and colorectal cancer (reviewed in Matsumura, 2011). In these studies, it has been shown that NK012 had significantly more potent anticancer activity with less intestinal toxicity compared to irinotecan. Furthermore, NK012 has shown preferential distribution and accumulation in solid tumors likely through EPR effect. Following a single IV administration of NK012 to HT-29-bearing mice at a dose of 30 mg/kg, the plasma concentration versus time profile showed a long circulation time with a mean $t_{1/2z}$ of more than 30 hours for the polymer-bound drug and around 62 hours for the free SN-38. Moreover, the mean values of CL and Vd_{ss} for polymer-bound SN-38 were 6 mL/h/kg and 40.4 mL/kg, respectively. More importantly, the peak concentration of polymer-bound and free SN-38 in the tumor were around 14 and 0.2 μg/g, respectively, and were reached 6 hours after administration. Both polymer-bound and free drug were shown to accumulate in the tumor 24 hours after administration of NK012. Specifically, the tumor-to-plasma concentration ratios reached 108 and 11 for polymer-bound and free drug, respectively, 7 days after dosing. Tissue distribution studies showed a relatively high distribution of free SN-38 in organs of the reticuloendothelial system (RES), namely liver and spleen. Nevertheless, in the lung, kidney, and small intestine, the highest concentration of free SN-38 was reached 1 hour after administration of NK012 and the concentrations were negligible at 24 hours.

Several Phase I and Phase II clinical studies were conducted for NK012; however, only the results of two Phase I studies have been published so far. The results of the first study was published in 2010 (Hamaguchi et al., 2010) while the results of the second was published in 2016 (Burris et al., 2016). The first-in-human Phase I study was conducted (in Japan) to determine the DLT, evaluate the pharmacokinetic profile, and document any antitumor activity of NK012 in patients with solid tumors (Hamaguchi et al., 2010). Twenty-four patients were enrolled and received NK012 as a 30-minute infusion once every 3 weeks at a dose range of 2–28 mg/m^2. Neutropenia was the most common hematologic toxicity, while diarrhea (grade 1 or 2) was the predominant nonhematologic toxicity. Two out of nine patients experienced DLTs (mainly neutropenia) in the first cycle at the highest dose level (28 mg/m^2). The pharmacokinetic profiles of total SN-38 (polymer-bound and released) and SN-38 (nonpolymer bound) showed linear pharmacokinetics in the dose range used in the study (2–28 mg/m^2). Similar to the preclinical data, high systemic exposure and slow elimination were observed for SN-38 in this formulation. The mean terminal-phase half-lives ($t_{1/2z}$) for polymer-bound SN-38 and released SN-38 were around 140 and 210 hours, respectively. The values of CL and Vd_{ss} of polymer-bound SN-38 ranged from 99 to 150 mL/h/m^2 and from 2 to 4 L/m^2, respectively. Among the 23 evaluable patients, two had partial responses and nine patients achieved stable disease status. The recommended dose for Phase II studies was 28 mg/m^2 with at least a 3-week interval between treatment cycles.

The second Phase I study was conducted in the United States in which 39 patients with advanced solid tumors were enrolled (Burris et al., 2016). The study was conducted to evaluate the safety, pharmacokinetics, MTD, and recommended dose of NK012 for Phase II studies. Of the 39 patients enrolled, 38 were treated with NK012 as a 30-minute infusion every 3 or 4 weeks at doses ranging from 9 to 37 mg/m^2. The MTD was determined at 28 mg/m^2 where the DLT was myelosuppression. Owing to the delayed neutrophil recovery, the recommended dose of NK012 was 28 mg/m^2 every 4 weeks. Nonhematological toxicities were mild with no grade 3 diarrhea reported. Similar to the findings of the previous study, both polymer-bound SN-38 and polymer-unbound SN-38 showed linear pharmacokinetics within the dose range used in the study (Burris et al., 2016; Hamaguchi et al., 2010). The mean $t_{1/2z}$ for polymer-bound SN-38 and released SN-38 were around 150 and 280 hours, respectively. Following IV infusion of NK012 at a dose of 28 mg/m^2 to six patients, the estimated mean values of CL and Vd_{ss} of polymer-bound SN-38 were around 100 mL/h/m^2 and 2.4 L/m^2, respectively. The calculated pharmacokinetic parameters for SN-38 in this study were comparable to those reported in the previous Phase I study. Nine patients had partial response and 18 achieved stable disease status. The recommended dose for Phase II trials was set at 28 mg/m^2 every 4 weeks (28 days) (Burris et al., 2016). There are several Phase II clinical trials for NK012 (as monotherapy or in combination with other agents) that have been completed in the United States and Japan for the treatment of different types of cancer. However, no results have yet been reported.

15.3.7 NC-6004

NC-6004 (Nanoplatin) is a methoxy PEO-*b*-P(Glu) micellar formulation of cisplatin [cis-diaminodichloroplatinum (II)] developed by NanoCarrier Co., Ltd (Japan) (Nishiyama et al., 2003; Plummer et al., 2011). The molecular weights of methoxy PEO and P(Glu) are 12,000 and 6000 g/mol, respectively. In aqueous media, micelles of NC-6004 are formed through self-assembly and the coordinate complex formation between cisplatin and the carboxylic groups present in P(Glu). NC-6004 micelles have a mean diameter of around 30 nm and about 39% (w/w) drug loading. This product was developed to solve the issues of rapid clearance and dose-limiting nephrotoxicity reported with the conventional formulation of cisplatin (Hartmann and Lipp, 2003). The free platinum is released from NC-6004 in the presence of chloride ions. Interestingly, the release rate in saline (0.9% NaCl solution) at 37°C was found to be relatively slow with only 19.6% and 47.8% of platinum released at 24 and 96 hours, respectively (Uchino et al., 2005).

Several preclinical studies have been performed to assess the pharmacokinetics, biodistribution, efficacy, and toxicity of NC-6004 (Nishiyama et al., 2003; Uchino et al., 2005). The in vitro cytotoxicity studies on various cancer cell lines showed that NC-6004 micelles had lower cytotoxicity compared to the free drug which was likely due to the slower release of cisplatin from the micelles. The in vivo profile, however, showed a superior anticancer effect compared to the control formulation in several tumor models including colon and gastric cancer as well as in a cisplatin-resistant mouse model. Following a single IV dose of 5 mg/kg of free cisplatin (control) or NC-6004 to healthy rats, platinum in the control formulation was only detected until 12 hours post dose, while it was quantified until 48 hours post dose (Uchino et al., 2005). The significant reduction in drug CL resulted in a 65-fold increase in the plasma AUC_{0-12h} of the drug compared to the control. Moreover, the results of tissue distribution showed that the highest platinum concentrations were found in the organs of the RES (i.e., the liver and spleen) especially at late time points (24 and 48 hours post dose). Although presence of PEG molecules on the surface of these micelles, as well as other nano-delivery systems, can provide the system with stealth properties, it seems that it cannot completely escape the recognition and the eventual clearance by the RES. Nevertheless, while a marked increase in platinum concentrations in the kidney obtained immediately after administration of cisplatin control formulation, NC-6004 showed 11-fold and 3-fold lower platinum concentrations in the kidney at 10-minute and 1 hour after IV administration compared to the control. Furthermore, a tissue distribution study conducted in tumor-bearing mice showed a 3.6-fold increase in tumor AUC for total cisplatin after NC-6004 treatment compared to the control formulation. In another study, NC-6004 had 20-fold higher platinum concentrations in the tumor compared to the control formulation at 24 hours after drug administration (Nishiyama et al., 2003). The results of pharmacokinetic and tissue distribution studies were in agreement with the in vivo efficacy and toxicity profiles of NC-6004. Histopathological and biochemical studies revealed that NC-6004 significantly inhibited the nephrotoxicity that was obtained with free cisplatin (control). Moreover, significantly lower neurotoxicity was obtained with NC-6004 compared to the control, which was attributed to the significantly lower concentration of platinum nerve tissues when NC-6004 was administered (Uchino et al., 2005).

Numerous clinical trials have been conducted for NC-6004 (Phases I–III) in different countries. Results of a Phase I clinical trial for NC-6004 were published in 2011 (Plummer et al., 2011). The study was conducted in the United Kingdom in 2006 for patients with advanced solid tumors. Seventeen patients were recruited and treated with NC-6004 administered via a 1-hour IV infusion at doses ranging from 10 to 120 mg/m^2 every 3 weeks. NC-6004 was provided as a sterile solution containing the equivalent of 2.5 mg/mL cisplatin, which could be diluted in 5% dextrose before administration. Generally, NC-6004 was well-tolerated with low nephrotoxicity and no significant neurotoxicity, ototoxicity, myelosuppression, or emesis. One out of six patients dosed at 90 mg/m^2 experienced DLT (grade 3 fatigue). In spite of the prophylactic measures that were implemented (posthydration and premedication regimen), renal impairment and hypersensitivity reactions still developed at 120 mg/m^2 which led to the conclusion that the maximum tolerated dose was 120 mg/m^2 and the recommended dose was set at 90 mg/m^2. The pharmacokinetics of total cisplatin, gel-filterable cisplatin (micelle-encapsulated drug), and ultrafilterable cisplatin (free/released drug) were characterized separately. NC-6004 showed favorable changes on the pharmacokinetics of cisplatin. Pharmacokinetics of total plasma platinum was characterized by longer $t_{1/2}$, higher C_{max} and $AUC_{0-infinity}$, and lower Vd and CL compared to the corresponding values reported for cisplatin (nonmicellar aqueous solution). Specifically, at an equivalent dose of 120 mg/m^2, the C_{max} and $AUC_{0-infinity}$ of NC-6004 were about 11-fold higher than those of the cisplatin aqueous solution (Kitajima et al., 1987; Plummer et al., 2011). Moreover, the terminal-phase $t_{1/2}$ of NC-6004 (158 hours) was longer than that of cisplatin (~74 hours) (Plummer et al., 2011). The CL and Vd values of NC-6004 (48 mL/h and 11 L, respectively) were substantially lower than those reported for cisplatin (350 mL/h and 53 L, respectively) (Calvert et al., 1993; Plummer et al., 2011). The C_{max} and $AUC_{0-infinity}$ of NC-6004 increased linearly with dose escalation, with no apparent change in CL. The pharmacokinetic profile of micelle-encapsulated platinum resembled that of total plasma platinum. At 120 mg/m^2 dose level, the C_{max} and $AUC_{0-infinity}$ represented about 99% and 88% of total platinum in plasma, respectively. Yet, the C_{max} at 120 mg/m^2 of the free/released platinum in plasma reported for NC-6004 was 34-fold lower than that of free (nonprotein-bound) platinum reported for cisplatin at the same dose (Calvert et al., 1993; Plummer et al., 2011). This might be the reason for the lower toxicity reported for NC-6004 compared to the cisplatin solution. On the other hand, the $t_{1/2}$ and $AUC_{0-infinity}$ calculated for the free/released platinum following administration of NC-6004 at 120 mg/m^2 were 230-fold and 8.5-fold larger, respectively, than those reported for nonprotein-bound platinum after administration of an equivalent dose of cisplatin solution. Taken together, NC-6004 micelles were shown to prolong the circulation time of the drug and control the release of the active platinum species. Furthermore, the lower C_{max} and higher $AUC_{0-infinity}$ of the released platinum may lead to a lower toxicity and better efficacy compared to the marketed cisplatin. Based on these findings as well as the stable disease status that was observed in seven patients, the request for further clinical evaluation of NC-6004 was justified.

The results of a Phase I/II study for NC-6004 was published in 2012 in a conference abstract (Su et al., 2012). The study was conducted in Asia in which 19 patients with pancreatic cancer were treated with NC-6004 at doses ranging from 30 to 120 mg/m^2 every 3 weeks, and gemcitabine at 1000 mg/m^2 twice every 3 weeks (on days 1 and 8). Moreover, the patients were given oral steroids as a prophylactic measure to prevent hypersensitivity reactions. The MTD and the recommended dose in combination with gemcitabine were 120 mg/m^2 and 90 mg/m^2, respectively. One patient had a partial response and 10 patients showed stable disease status. In January 2014, a Phase III study was initiated to evaluate the impact of the addition of NC-6004 to gemcitabine in the treatment of patients with locally advanced or metastatic pancreatic cancer in Asian countries (ClinicalTrials.gov identifier: NCT02043288). The estimated study completion date is March 2018. Two more clinical studies (Phase I/II) are ongoing for NC-6004 (ClinicalTrials.gov identifiers: NCT02240238 and NCT02817113).

15.3.8 NC-6300

NC-6300 is a block copolymer micelle formulation consisting of methoxy PEO-*b*-P(Asp), where P(Asp) is partially modified with benzyl groups and epirubicin is covalently bound to P(Asp) through an acid-labile hydrazine bond (Harada et al., 2011). This micellar product has been developed by NanoCarrier Co., Ltd

(Japan). The molecular weight of methoxy PEO used in this system is 12,000 g/mol and the number of Asp residues is 40 in which approximately 20 benzyl groups and eight molecules of epirubicin are conjugated. In aqueous media, epirubicin-conjugated PEO-*b*-P(Asp) self-assemble into micelles with a mean diameter of 40–80 nm. NC-6300 micelles have shown a pH-dependent release of epirubicin. At acidic pH (3.0), around 80% of the drug was released within 1 hour, whereas at neutral pH (7.0) and physiological pH (7.4), only about 20% of epirubicin was released within 48 hours. The in vitro cytotoxicity studies on various cancer cell lines showed that NC-6300 micelles had lower cytotoxicity compared to the free drug due to the slower release of epirubicin from the micelles (Harada et al., 2011).

Preclinical pharmacokinetic, efficacy, and toxicity studies were evaluated in healthy as well as in tumor-bearing animal models (Harada et al., 2011; Takahashi et al., 2013). Following a single IV administration of NC-6300 at a dose equivalent to 1 mg/kg of epirubicin to healthy rats, NC-6300 micelles had an impressive impact on the pharmacokinetics of epirubicin (Harada et al., 2011). There was about 2200-fold decrease in drug CL, which resulted in the increase in plasma $AUC_{0-infinity}$ to the same magnitude compared with the aqueous solution of epirubicin. Furthermore, there was about 63-fold decrease in Vd_{ss} and nearly 9-fold increase in plasma elimination $t_{1/2}$ compared to the epirubicin solution. These findings clearly revealed the ability of NC-6300 micelles to slow down the clearance of the drug and to prolong the circulation time in plasma. Tissue distribution studies on tumor-bearing mice showed that NC-6300, similar to NC-6004, had higher accumulation in the organs of the RES (mainly the liver and spleen) (Takahashi et al., 2013; Harada et al., 2011). Nevertheless, higher concentrations of epirubicin were found in tumor tissues compared to the drug in the conventional formulation. Moreover, NC-6300 decreased the distribution of free epirubicin in kidneys, lungs, and in the heart. The lower concentrations of epirubicin in the heart led to significantly lower cardiotoxicity of NC-6300 compared to the conventional epirubicin (Takahashi et al., 2013).

In January 2017, the results of the first-in-human Phase I study of NC-6300 (currently known as K-912) were published (Mukai et al., 2017). The study was conducted in Japan for patients with advanced or recurrent solid tumors. Nineteen patients were treated with NC-6300 administered as a 10-minute infusion at doses ranging from 15 to 225 mg/m^2. The pharmacokinetic profile of NC-6300 was similar to the one obtained in preclinical animal studies. At each dose level, total epirubicin concentration was steadily high for about 8 hours before it had a mono-exponential decline with a $t_{1/2}$ ranging from 23 to 27 hours. Furthermore, values of AUC_{0-t}, $AUC_{0-infinity}$, and C_{max} showed linear increase upon increasing the dose from 15 to 225 mg/m^2. The calculated Vd_{ss} was approximately equal to the blood volume, which indicates minimal distribution. One patient had partial response and 10 achieved stable disease status. No clinically significant cardiac-related adverse events were reported even in patients who were administered with a cumulative dose exceeding 900 mg/m^2. The MTD and the recommended dose for Phase II studies were determined to be 170 mg/m^2.

15.3.9 NC-4016

NC-4016 is a methoxy PEO-*b*-P(Glu) micelle formulation of oxaliplatin (dichloro-1,2-diaminocyclohexane platinum(II); DACHPt) developed by Professor Kataoka et al. and NanoCarrier Co., Ltd (Japan) (Ueno et al., 2014; Cabral et al., 2005, 2007; Yamamoto et al., 2014). The molecular weights of methoxy PEO and P(Glu) in the optimized NC-4016 formulation are approximately 12,000 and 6000 g/mol, respectively. Before loading of oxaliplatin into the micelles, the drug is mixed with silver nitrate ($AgNO_3$) in distilled water to form the aqueous complex which is then separated from the precipitated silver chloride (AgCl) by centrifugation and filtration. Methoxy PEO-*b*-P(Glu) is added to the solution of oxaliplatin aqueous complex and reacted for 5 days to prepare the oxaliplatin-loaded micelles. In aqueous media, micelles of NC-4016 are formed through self-assembly and the coordinate complex formation between platinum present in oxaliplatin and the carboxylic groups present in P(Glu). NC-4016 micelles have a mean diameter of around 40 nm and about 32% (w/w) of platinum loading (Cabral et al., 2007). In media containing chloride ions, oxaliplatin is released from the micelles by exchange reaction between the carboxylic groups in P(Glu) and the chloride ions. The release profile of NC-4016 in PBS (at 37°C) showed a sustained release of platinum with only 60% of the drug release in 96 hours (Cabral et al., 2005). The in vitro cytotoxicity studies showed that NC-4016 formulation was effective against various cancer cell lines although it was weaker than that of oxaliplatin (control) probably because of the slow release of oxaliplatin from NC-4016 micelles (Cabral et al., 2005, 2007; Ueno et al., 2014).

Preclinical studies of NC-4016 in different animal models demonstrated that NC-4016 micelles had a favorable influence on oxaliplatin's pharmacokinetic, efficacy, and toxicity profiles (Cabral et al., 2005, 2007; Ueno et al., 2014). For instance, following a single IV administration of oxaliplatin-loaded micelles to C-26-bearing bearing mice, the circulation time of oxaliplatin was remarkably prolonged with about 80% of the injected dose still found in the plasma 9 hours after injection. This is in contrast to only 1% of the injected dose found in the plasma 1 hour after IV administration of conventional oxaliplatin (Cabral et al., 2005). Moreover, oxaliplatin-loaded micelles had significantly higher distribution in the tumor tissues compared to the conventional formulation of oxaliplatin (Fig. 15.3). In another study on the same animal model, oxaliplatin-loaded micelles had 20-fold higher concentrations of the drug in tumor tissues compared to the control formulation (Cabral et al., 2007). This preferential accumulation of NC-4016 in tumor tissues led to significant decrease in tumor growth rate compared to the oxaliplatin solution. Nevertheless, like the other formulations discussed, NC-4016 exhibited higher concentrations of the drug in the liver and spleen compared to the conventional formulation of oxaliplatin. In more recent studies, the anticancer

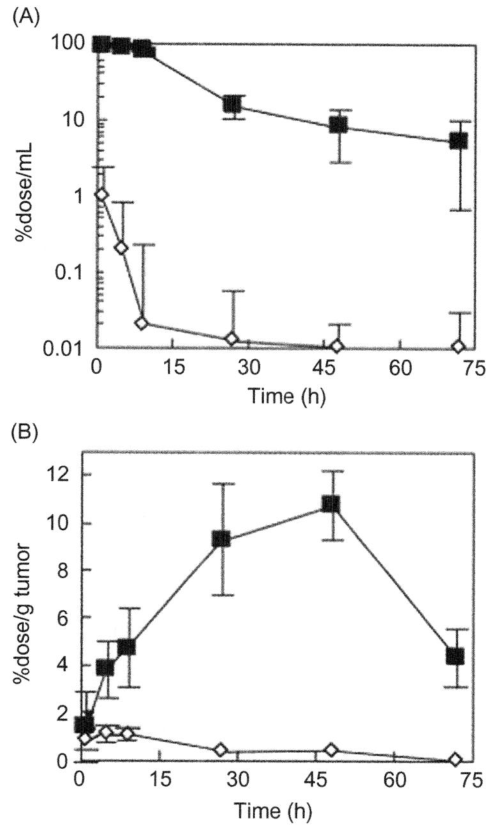

FIGURE 15.3

Time profiles of platinum concentration in the plasma (A) and platinum accumulation in tumor (B) after IV administration of free oxaliplatin (◇) and oxaliplatin-loaded micelles (■). Each drug was administered to C-26-bearing CDF1 mice (female, $n = 6$) at a dose of 100 μg on Pt basis. Data are shown as the mean ± SD.

Reproduced with permission from Cabral, H., Nishiyama, N., Okazaki, S., Koyama, H., Kataoka, K., 2005. Preparation and biological properties of dichloro(1,2-diaminocyclohexane)platinum(II) (DACHPt)-loaded polymeric micelles. J. Control. Release 101, 223–232.

activity of NC-4016 was evaluated in nude mice inoculated subcutaneously with human HeLa-KB cells. NC-4016 micelles were intravenously administered at 14, 21, and 28 days after tumor transplantation. Treatment with NC-4016 resulted in higher anticancer activity which was evident from the significantly lower tumor volume and tumor weight compared to the control groups (oxaliplatin and 5% glucose solution). The percentage reductions in tumor weights after administration of NC-4016 and oxaliplatin solution were 74% and 33%, respectively (Ueno et al., 2014). Moreover, following a single IV administration of NC-4016 at a dose of 4 mg/kg (1.96 mg/kg elemental platinum) to the same xenograft mouse model resulted in significantly higher concentrations of platinum in the plasma as well as in tumor tissues in comparison to the oxaliplatin solution (Ueno et al., 2014). Furthermore, by evaluating peripheral neuropathy (a DLT of oxaliplatin in patients) in a rat model, NC-4016 did not cause cold hyperalgesia/allodynia which were observed with the oxaliplatin solution. The effect of the combined treatment with NC-6300 and NC-4016 on a human gastric model was evaluated in comparison to epirubicin and oxaliplatin solutions (Yamamoto et al., 2014). The micellar formulations showed a significantly higher anticancer efficacy with lower cardiotoxicity and neurotoxicity than the corresponding conventional drugs. In 2013, a Phase I study of NC-4016 was started in patients with advanced solid tumors or lymphoma (ClinicalTrial.gov identifier: NCT01999491). No results have been published yet.

15.4 CONCLUSIONS

Overall, the results of preclinical as well as clinical evaluations of polymeric micellar formulations show great promise for these nanocarriers as safe alternatives for the solubilization of poorly water-soluble drugs. In fact, polymeric micellar formulations have partly been successful in reducing the side effects of intravenously administered drugs simply by replacing relatively toxic solubilizing agents (like Cremophor EL) used in their conventional formulation. A number of polymeric micellar formulations used through IV administration have been able to change the pharmacokinetics of the encapsulated (or complexed or conjugated) drugs at the same direction as stealth liposomes in clinic (e.g., Doxil). These formulations have usually shown a decrease in the Vd and CL and an increase in the AUC and $t_{1/2}$ of incorporated drugs. These changes have resulted in a decreased toxicity for the polymeric micellar formulations compared to the conventional formulations of a number of anticancer drugs allowing the use of higher drug doses in a clinic. The extent of change in the pharmacokinetic parameters by polymeric micellar formulations were the same as that seen for Doxil over free doxorubicin. Nevertheless, this is a very significant accomplishment because the change in pharmacokinetic parameters is shown for hydrophobic drugs, which tend to leak out of liposomal carriers. Despite success in reducing the side effects of medications by polymeric micellar formulations following IV administration, similar to other nanocarrier formulations, there is still no concrete evidence indicating the positive contribution of this formulation approach in increasing the activity of encapsulated drugs in a clinic. Besides, the possibility for polymeric micellar formulation causing a similar change in drug pharmacokinetics through other routes of administration has not been verified yet. Further research is warranted in this context.

REFERENCES

Ahn, H.K., Jung, M., Sym, S.J., Shin, D.B., Kang, S.M., Kyung, S.Y., et al., 2014. A phase II trial of Cremorphor EL-free paclitaxel (Genexol-PM) and gemcitabine in patients with advanced non-small cell lung cancer. Cancer Chemother. Pharmacol. 74, 277–282.

Alakhov, V., Klinski, E., Li, S., Pietrzynski, G., Venne, A., Batrakova, E., et al., 1999. Block copolymer-based formulation of doxorubicin. From cell screen to clinical trials. Colloids Surf. B: Biointerfaces 16, 113–134.

Alakhov, V., Pietrzynski, G., Patel, K., Kabanov, A., Bromberg, L., Hatton, T.A., 2004. Pluronic block copolymers and Pluronic poly(acrylic acid) microgels in oral delivery of megestrol acetate. J. Pharm. Pharmacol. 56, 1233–1241.

Aliabadi, H.M., Lavasanifar, A., 2006. Polymeric micelles for drug delivery. Expert Opin. Drug Deliv. 3, 139–162.

Aliabadi, H.M., Shahin, M., Brocks, D.R., Lavasanifar, A., 2008. Disposition of drugs in block copolymer micelle delivery systems: from discovery to recovery. Clin. Pharmacokinet. 47, 619–634.

Amidon, G.L., Lennernas, H., Shah, V.P., Crison, J.R., 1995. A theoretical basis for a biopharmaceutic drug classification—the correlation of in-vitro drug product dissolution and in-vivo bioavailability. Pharm. Res. 12, 413–420.

Bachhav, Y.G., Mondon, K., Kalia, Y.N., Gurny, R., Möller, M., 2011. Novel micelle formulations to increase cutaneous bioavailability of azole antifungals. J. Control Release. 153 (2), 126–132.

Burris, H.A., Infante, J.R., Anthony greco, F., Thompson, D.S., Barton, J.H., Bendell, J.C., et al., 2016. A phase I dose escalation study of NK012, an SN-38 incorporating macromolecular polymeric micelle. Cancer Chemother. Pharmacol. 77, 1079–1086.

Burt, H.M., Zhang, X., Toleikis, P., Embree, L., Hunter, W.L., 1999. Development of copolymers of poly(D,L-lactide) and methoxypolyethylene glycol as micellar carriers of paclitaxel. Colloids Surf. B: Biointerfaces 16, 161–171.

Cabral, H., Nishiyama, N., Okazaki, S., Koyama, H., Kataoka, K., 2005. Preparation and biological properties of dichloro(1,2-diaminocyclohexane)platinum(II) (DACHPt)-loaded polymeric micelles. J. Control. Release 101, 223–232.

Cabral, H., Nishiyama, N., Kataoka, K., 2007. Optimization of (1,2-diamino-cyclohexane) platinum(II)-loaded polymeric micelles directed to improved tumor targeting and enhanced antitumor activity. J. Control. Release 121, 146–155.

Calvert, H., Judson, I., Van Der vijgh, W.J., 1993. Platinum complexes in cancer medicine: pharmacokinetics and pharmacodynamics in relation to toxicity and therapeutic activity. Cancer Surv. 17, 189–217.

Chiappetta, D.A., Hocht, C., Taira, C., Sosnik, A., 2010. Efavirenz-loaded polymeric micelles for pediatric anti-HIV pharmacotherapy with significantly higher oral bioavailability [corrected]. Nanomedicine (Lond) 5, 11–23.

Cholkar, K., Patel, S.P., Vadlapudi, A.D., Mitra, A.K., 2013. Novel strategies for anterior segment ocular drug delivery. J. Ocul. Pharmacol. Ther. 29, 106–123.

Civiale, C., Licciardi, M., Cavallaro, G., Giammona, G., Mazzone, M.G., 2009. Polyhydroxyethylaspartamide-based micelles for ocular drug delivery. Int. J. Pharm. 378, 177–186.

Danson, S., Ferry, D., Alakhov, V., Margison, J., Kerr, D., Jowle, D., et al., 2004. Phase I dose escalation and pharmacokinetic study of pluronic polymer-bound doxorubicin (SP1049C) in patients with advanced cancer. Br. J. Cancer 90, 2085–2091.

Desai, S.D., Blanchard, J., 1998a. Evaluation of pluronic F127-based sustained-release ocular delivery systems for pilocarpine using the albino rabbit eye model. J. Pharm. Sci. 87, 1190–1195.

Desai, S.D., Blanchard, J., 1998b. In vitro evaluation of pluronic F127-based controlled-release ocular delivery systems for pilocarpine. J. Pharm. Sci. 87, 226–230.

Dey, S., Mitra, A.K., 2005. Transporters and receptors in ocular drug delivery: opportunities and challenges. Expert Opin. Drug Deliv. 2, 201–204.

Dey, S., Patel, J., Anand, B.S., Jain-Vakkalagadda, B., Kaliki, P., Pal, D., et al., 2003. Molecular evidence and functional expression of *P*-glycoprotein (MDR1) in human and rabbit cornea and corneal epithelial cell lines. Invest. Ophthalmol. Vis. Sci. 44, 2909–2918.

Dhananjay, P., Ramya krishna, V., Aswani dutt, V., Mitra, A.K., 2013. Biology of ocular transporters: efflux and influx transporters in the eye. Ocular Transporters and Receptors. Woodhead Publishing.

Di tommaso, C., Torriglia, A., Furrer, P., Behar-Cohen, F., Gurny, R., Moller, M., 2011. Ocular biocompatibility of novel Cyclosporin A formulations based on methoxy poly (ethylene glycol)-hexylsubstituted poly(lactide) micelle carriers. Int. J. Pharm. 416, 515–524.

Di tommaso, C., Bourges, J.L., Valamanesh, F., Trubitsyn, G., Torriglia, A., Jeanny, J.C., et al., 2012. Novel micelle carriers for cyclosporin A topical ocular delivery: in vivo cornea penetration, ocular distribution and efficacy studies. Eur. J. Pharm. Biopharm. 81, 257–264.

Dian, L., Yu, E., Chen, X., Wen, X., Zhang, Z., Qin, L., et al., 2014. Enhancing oral bioavailability of quercetin using novel soluplus polymeric micelles. Nanoscale Res. Lett. 9, 2406.

Dintaman, J.M., Silverman, J.A., 1999. Inhibition of *P*-glycoprotein by D-alpha-tocopheryl polyethylene glycol 1000 succinate (TPGS). Pharm. Res. 16, 1550–1556.

Duan, Y., Cai, X., Du, H., Zhai, G., 2015. Novel in situ gel systems based on P123/TPGS mixed micelles and gellan gum for ophthalmic delivery of curcumin. Colloids Surf. B: Biointerfaces 128, 322–330.

Duan, Y., Zhang, B., Chu, L., Tong, H.H., Liu, W., Zhai, G., 2016. Evaluation in vitro and in vivo of curcumin-loaded mPEG-PLA/TPGS mixed micelles for oral administration. Colloids Surf. B: Biointerfaces 141, 345–354.

Evers, R., Kool, M., Smith, A.J., Deemter, L.V., Haas, M.D., Borst, P., 2000. Inhibitory effect of the reversal agents V-104, GF120918 and Pluronic L61 on MDR1 Pgp-, MRP1- and MRP2-mediated transport. Br. J. Cancer 83, 366–374.

Feng, S.S., Mei, L., Anitha, P., Gan, C.W., Zhou, W.Y., 2009. Poly(lactide)-vitamin E derivative/montmorillonite nanoparticle formulations for the oral delivery of Docetaxel. Biomaterials 30, 3297–3306.

Fofaria, N.M., Qhattal, H.S.S., Liu, X.L., Srivastava, S.K., 2016. Nanoemulsion formulations for anti-cancer agent piplartine—characterization, toxicological, pharmacokinetics and efficacy studies. Int. J. Pharm. 498, 12–22.

Francis, M.F., Lavoie, L., Winnik, F.M., Leroux, J.C., 2003a. Solubilization of cyclosporin A in dextran-*g*-polyethyleneglycolalkyl ether polymeric micelles. Eur. J. Pharm. Biopharm., 56, 337–346.

Francis, M.F., Piredda, M., Winnik, F.M., 2003b. Solubilization of poorly water soluble drugs in micelles of hydrophobically modified hydroxypropylcellulose copolymers. J. Control. Release 93, 59–68.

Francis, M.F., Cristea, M., Winnik, F.M., 2005a. Exploiting the vitamin B12 pathway to enhance oral drug delivery via polymeric micelles. Biomacromolecules 6, 2462–2467.

Francis, M.F., Cristea, M., Yang, Y.L., Winnik, F.M., 2005b. Engineering polysaccharide-based polymeric micelles to enhance permeability of cyclosporin a across Caco-2 cells. Pharm. Res. 22, 209–219.

Gaudana, R., Jwala, J., Boddu, S.H., Mitra, A.K., 2009. Recent perspectives in ocular drug delivery. Pharm. Res. 26, 1197–1216.

Greene, R.F., Collins, J.M., Jenkins, J.F., Speyer, J.L., Myers, C.E., 1983. Plasma pharmacokinetics of adriamycin and adriamycinol: implications for the design of in vitro experiments and treatment protocols. Cancer Res. 43, 3417–3421.

Guan, Y., Huang, J., Zuo, L., Xu, J., Si, L., Qiu, J., et al., 2011. Effect of pluronic P123 and F127 block copolymer on *P*-glycoprotein transport and CYP3A metabolism. Arch. Pharm. Res. 34, 1719–1728.

Gula, A., Ren, L., Zhou, Z., Lu, D.D., Wang, S.Q., 2013. Design and evaluation of biodegradable enteric microcapsules of amifostine for oral delivery. Int. J. Pharm. 453, 441–447.

Guo, Y., Luo, J., Tan, S., Otieno, B.O., Zhang, Z., 2013. The applications of vitamin E TPGS in drug delivery. Eur. J. Pharm. Sci. 49, 175–186.

Guo, C., Zhang, Y., Yang, Z., Li, M., Li, F., Cui, F., et al., 2015. Nanomicelle formulation for topical delivery of cyclosporine A into the cornea: in vitro mechanism and in vivo permeation evaluation. Sci. Rep. 5, 12968.

Hamaguchi, T., Matsumura, Y., Suzuki, M., Shimizu, K., Goda, R., Nakamura, I., et al., 2005. NK105, a paclitaxel-incorporating micellar nanoparticle formulation, can extend in vivo antitumour activity and reduce the neurotoxicity of paclitaxel. Br. J. Cancer 92, 1240–1246.

Hamaguchi, T., Kato, K., Yasui, H., Morizane, C., Ikeda, M., Ueno, H., et al., 2007. A phase I and pharmacokinetic study of NK105, a paclitaxel-incorporating micellar nanoparticle formulation. Br. J. Cancer 97, 170–176.

Hamaguchi, T., Doi, T., Eguchi-Nakajima, T., Kato, K., Yamada, Y., Shimada, Y., et al., 2010. Phase I study of NK012, a novel SN-38-incorporating micellar nanoparticle, in adult patients with solid tumors. Clin. Cancer Res. 16, 5058–5066.

Han, H.K., Shin, H.J., Ha, D.H., 2012. Improved oral bioavailability of alendronate via the mucoadhesive liposomal delivery system. Eur. J. Pharm. Sci. 46, 500–507.

Harada, M., Bobe, I., Saito, H., Shibata, N., Tanaka, R., Hayashi, T., et al., 2011. Improved anti-tumor activity of stabilized anthracycline polymeric micelle formulation, NC-6300. Cancer Sci. 102, 192–199.

Hariharan, S., Gunda, S., Mishra, G.P., Pal, D., Mitra, A.K., 2009. Enhanced corneal absorption of erythromycin by modulating *P*-glycoprotein and MRP mediated efflux with corticosteroids. Pharm. Res. 26, 1270–1282.

Hartmann, J.T., Lipp, H.P., 2003. Toxicity of platinum compounds. Expert Opin. Pharmacother. 4, 889–901.

Hughes, P.M., Olejnik, O., Chang-Lin, J.E., Wilson, C.G., 2005. Topical and systemic drug delivery to the posterior segments. Adv. Drug Deliv. Rev. 57, 2010–2032.

Ibrahim, N.K., Samuels, B., Page, R., Doval, D., Patel, K.M., Rao, S.C., et al., 2005. Multicenter phase II trial of ABI-007, an albumin-bound paclitaxel, in women with metastatic breast cancer. J. Clin. Oncol. 23, 6019–6026.

Jager, R.D., Aiello, L.P., Patel, S.C., Cunningham Jr, E.T., 2004. Risks of intravitreous injection: a comprehensive review. Retina 24, 676–698.

Jaiswal, M., Kumar, M., Pathak, K., 2015. Zero order delivery of itraconazole via polymeric micelles incorporated in situ ocular gel for the management of fungal keratitis. Colloids Surf B: Biointerfaces 130, 23–30.

Joshi, G., Kumar, A., Sawant, K., 2014. Enhanced bioavailability and intestinal uptake of Gemcitabine HCl loaded PLGA nanoparticles after oral delivery. Eur. J. Pharm. Sci. 60, 80–89.

Jung, I.W., Han, H.K., 2014. Effective mucoadhesive liposomal delivery system for risedronate: preparation and in vitro/in vivo characterization. Int. J. Nanomed. 9, 2299–2306.

Kabanov, A.V., Batrakova, E.V., Alakhov, V.Y., 2002. Pluronic block copolymers for overcoming drug resistance in cancer. Adv. Drug Deliv. Rev. 54, 759–779.

Kato, K., Chin, K., Yoshikawa, T., Yamaguchi, K., Tsuji, Y., Esaki, T., et al., 2012. Phase II study of NK105, a paclitaxel-incorporating micellar nanoparticle, for previously treated advanced or recurrent gastric cancer. Invest. New Drugs 30, 1621–1627.

Kawabata, Y., Wada, K., Nakatani, M., Yamada, S., Onoue, S., 2011. Formulation design for poorly water-soluble drugs based on biopharmaceutics classification system: basic approaches and practical applications. Int. J. Pharm. 420, 1–10.

Kawato, Y., Aonuma, M., Hirota, Y., Kuga, H., Sato, K., 1991. Intracellular roles of SN-38, a metabolite of the camptothecin derivative CPT-11, in the antitumor effect of CPT-11. Cancer Res. 51, 4187–4191.

Kim, S.C., Kim, D.W., Shim, Y.H., Bang, J.S., Oh, H.S., Kim, S.W., et al., 2001. In vivo evaluation of polymeric micellar paclitaxel formulation: toxicity and efficacy. J. Control. Release 72, 191–202.

Kim, T.Y., Kim, D.W., Chung, J.Y., Shin, S.G., Kim, S.C., Heo, D.S., et al., 2004. Phase I and pharmacokinetic study of Genexol-PM, a cremophor-free, polymeric micelle-formulated paclitaxel, in patients with advanced malignancies. Clin. Cancer Res. 10, 3708–3716.

Kim, D.W., Kim, S.Y., Kim, H.K., Kim, S.W., Shin, S.W., Kim, J.S., et al., 2007. Multicenter phase II trial of Genexol-PM, a novel Cremophor-free, polymeric micelle formulation of paclitaxel, with cisplatin in patients with advanced non-small-cell lung cancer. Ann. Oncol. 18, 2009–2014.

Kim, H.S., Lee, J.Y., Lim, S.H., Sun, J.M., Lee, S.H., Ahn, J.S., et al., 2015. A prospective phase II study of cisplatin and cremophor EL-free paclitaxel (Genexol-PM) in patients with unresectable thymic epithelial tumors. J. Thorac. Oncol. 10, 1800–1806.

Kitajima, K., Fukuoka, M., Kobayashi, S., Kusunoki, Y., Takada, M., Negoro, S., et al., 1987. [Studies on the appropriate administration of cisplatin based on pharmacokinetics and toxicity]. Gan To Kagaku Ryoho 14, 2517–2523.

Koizumi, F., Kitagawa, M., Negishi, T., Onda, T., Matsumoto, S., Hamaguchi, T., et al., 2006. Novel SN-38-incorporating polymeric micelles, NK012, eradicate vascular endothelial growth factor-secreting bulky tumors. Cancer Res. 66, 10048–10056.

Ku, M.S., Dulin, W., 2012. A biopharmaceutical classification-based Right-First-Time formulation approach to reduce human pharmacokinetic variability and project cycle time from First-In-Human to clinical Proof-of-Concept. Pharm. Dev. Technol. 17, 285–302.

Kwon, S.H., Kim, S.Y., Ha, K.W., Kang, M.J., Huh, J.S., Im, T.J., et al., 2007. Pharmaceutical evaluation of genistein-loaded pluronic micelles for oral delivery. Arch. Pharm. Res. 30, 1138–1143.

Lapteva, M., Mondon, K., Möller, M., Gurny, R., Kalia, Y.N., 2014. Polymeric micelle nanocarriers for the cutaneous delivery of tacrolimus: a targeted approach for the treatment of psoriasis. Mol. Pharm. 11 (9), 2989–3001.

Lee, J., Lee, S.C., Acharya, G., Chang, C.J., Park, K., 2003. Hydrotropic solubilization of paclitaxel: analysis of chemical structures for hydrotropic property. Pharm. Res. 20, 1022–1030.

Lee, S.C., Huh, K.M., Lee, J., Cho, Y.W., Galinsky, R.E., Park, K., 2007. Hydrotropic polymeric micelles for enhanced paclitaxel solubility: in vitro and in vivo characterization. Biomacromolecules 8, 202–208.

Lee, K.S., Chung, H.C., Im, S.A., Park, Y.H., Kim, C.S., Kim, S.B., et al., 2008. Multicenter phase II trial of Genexol-PM, a Cremophor-free, polymeric micelle formulation of paclitaxel, in patients with metastatic breast cancer. Breast Cancer Res. Treat. 108, 241–250.

Lee, S.W., Kim, Y.M., Kim, Y.T., Kang, S.B., 2016. An open-label, multicenter, phase I trial of a cremophor-free, polymeric micelle formulation of paclitaxel combined with carboplatin as a first-line treatment for advanced ovarian cancer: a Korean Gynecologic Oncology Group study (KGOG-3016). J. Gynecol. Oncol. e26.

Li, J., Li, Z., Zhou, T., Zhang, J., Xia, H., Li, H., et al., 2015. Positively charged micelles based on a triblock copolymer demonstrate enhanced corneal penetration. Int. J. Nanomed. 10, 6027–6037.

Li, H.Y., Pan, T.T., Cui, Y., Li, X.X., Gao, J.F., Yang, W.Z., et al., 2016. Improved oral bioavailability of poorly water-soluble glimepiride by utilizing microemulsion technique. Int. J. Nanomed. 11, 3777–3788.

Liggins, R.T., Burt, H.M., 2002. Polyether-polyester diblock copolymers for the preparation of paclitaxel loaded polymeric micelle formulations. Adv. Drug Deliv. Rev. 54, 191–202.

Lim, W.T., Tan, E.H., Toh, C.K., Hee, S.W., Leong, S.S., Ang, P.C., et al., 2010. Phase I pharmacokinetic study of a weekly liposomal paclitaxel formulation (Genexol-PM) in patients with solid tumors. Ann. Oncol. 21, 382–388.

Lu, B., Wen, R., Yang, H., He, Y.J., 2007. Sustained-release tablets of indomethacin-loaded microcapsules: preparation, in vitro and in vivo characterization. Int. J. Pharm. 333, 87–94.

Mandal, A., Agrahari, V., Khurana, V., Pal, D., Mitra, A.K., 2016. Transporter effects on cell permeability in drug delivery. Expert Opin. Drug Deliv. 1–17.

Mannermaa, E., Vellonen, K.S., Urtti, A., 2006. Drug transport in corneal epithelium and blood-retina barrier: emerging role of transporters in ocular pharmacokinetics. Adv. Drug Deliv. Rev. 58, 1136–1163.

Masayuki, Y., Mizue, M., Noriko, Y., Teruo, O., Yasuhisa, S., Kazunori, K., et al., 1990. Polymer micelles as novel drug carrier: adriamycin-conjugated poly(ethylene glycol)-poly(aspartic acid) block copolymer. J. Control. Release 11, 269–278.

Mathot, F., Van beijsterveldt, L., Preat, V., Brewster, M., Arien, A., 2006. Intestinal uptake and biodistribution of novel polymeric micelles after oral administration. J. Control. Release 111, 47–55.

Mathot, F., Des rieux, A., Arien, A., Schneider, Y.J., Brewster, M., Preat, V., 2007. Transport mechanisms of mmePEG750P(CL-*co*-TMC) polymeric micelles across the intestinal barrier. J. Control. Release 124, 134–143.

Matsumura, Y., 2011. Preclinical and clinical studies of NK012, an SN-38-incorporating polymeric micelles, which is designed based on EPR effect. Adv. Drug Deliv. Rev. 63, 184–192.

Matsumura, Y., Hamaguchi, T., Ura, T., Muro, K., Yamada, Y., Shimada, Y., et al., 2004. Phase I clinical trial and pharmacokinetic evaluation of NK911, a micelle-encapsulated doxorubicin. Br. J. Cancer 91, 1775–1781.

Maurice, D.M., Mishima, S., 1984. Ocular pharmacokinetics. In: Sears, M.L. (Ed.), Pharmacology of the Eye. Springer Berlin Heidelberg.

Miller, S.C., Donovan, M.D., 1982. Effect of poloxamer 407 gel on the miotic activity of pilocarpine nitrate in rabbits. Int. J. Pharm. 12, 147–152.

Mukai, H., Kato, K., Esaki, T., Ohsumi, S., Hozomi, Y., Matsubara, N., et al., 2016. Phase I study of NK105, a nanomicellar paclitaxel formulation, administered on a weekly schedule in patients with solid tumors. Invest. New Drugs 34, 750–759.

Mukai, H., Kogawa, T., Matsubara, N., Naito, Y., Sasaki, M., Hosono, A., 2017. A first-in-human Phase 1 study of epirubicin-conjugated polymer micelles (K-912/NC-6300) in patients with advanced or recurrent solid tumors. Invest. New Drugs 35, 307–314.

Nagarwal, R.C., Kant, S., Singh, P.N., Maiti, P., Pandit, J.K., 2009. Polymeric nanoparticulate system: a potential approach for ocular drug delivery. J. Control. Release 136, 2–13.

Nakanishi, T., Fukushima, S., Okamoto, K., Suzuki, M., Matsumura, Y., Yokoyama, M., et al., 2001. Development of the polymer micelle carrier system for doxorubicin. J. Control. Release 74, 295–302.

Negishi, T., Koizumi, F., Uchino, H., Kuroda, J., Kawaguchi, T., Naito, S., et al., 2006. NK105, a paclitaxel-incorporating micellar nanoparticle, is a more potent radiosensitising agent compared to free paclitaxel. Br. J. Cancer 95, 601–606.

Nippon Kayaku Co Ltd., 2016. Results of Phase III study of NK105, a novel macromolecular micelle encapsulating an anticancer drug [Online]. <https://www.nipponkayaku.co.jp/english/news/detail.php?n = 20160705_78TIVWVB> (accessed 10.02.17).

Nishiyama, N., Okazaki, S., Cabral, H., Miyamoto, M., Kato, Y., Sugiyama, Y., et al., 2003. Novel cisplatin-incorporated polymeric micelles can eradicate solid tumors in mice. Cancer Res. 63, 8977–8983.

Ould-Ouali, L., Noppe, M., Langlois, X., Willems, B., Te riele, P., Timmerman, P., et al., 2005. Self-assembling PEG-p(CL-*co*-TMC) copolymers for oral delivery of poorly water-soluble drugs: a case study with risperidone. J. Control. Release 102, 657–668.

Patel, A., Cholkar, K., Agrahari, V., Mitra, A.K., 2013. Ocular drug delivery systems: an overview. World J. Pharmacol. 2, 47–64.

Peltier, S., Oger, J.M., Lagarce, F., Couet, W., Benoit, J.P., 2006. Enhanced oral paclitaxel bioavailability after administration of paclitaxel-loaded lipid nanocapsules. Pharm. Res. 23, 1243–1250.

Peng, C.C., Bengani, L.C., Jung, H.J., Leclerc, J., Gupta, C., Chauhan, A., 2011. Emulsions and microemulsions for ocular drug delivery. J. Drug Deliv. Sci. Technol. 21, 111–121.

Pepić, I., Lovric, J., Filipovic-Grcic, J., 2013. How do polymeric micelles cross epithelial barriers? Eur. J. Pharm. Sci. 50, 42–55.

Pierri, E., Avgoustakis, K., 2005. Poly(lactide)-poly(ethylene glycol) micelles as a carrier for griseofulvin. J. Biomed. Mater. Res. A 75, 639–647.

Plummer, R., Wilson, R.H., Calvert, H., Boddy, A.V., Griffin, M., Sludden, J., et al., 2011. A Phase I clinical study of cisplatin-incorporated polymeric micelles (NC-6004) in patients with solid tumours. Br. J. Cancer 104, 593–598.

Prosperi-Porta, G., Kedzior, S., Muirhead, B., Sheardown, H., 2016. Phenylboronic-acid-based polymeric micelles for mucoadhesive anterior segment ocular drug delivery. Biomacromolecules 17, 1449–1457.

Saif, M.W., Podoltsev, N.A., Rubin, M.S., Figueroa, J.A., Lee, M.Y., Kwon, J., et al., 2010. Phase II clinical trial of paclitaxel loaded polymeric micelle in patients with advanced pancreatic cancer. Cancer Invest. 28, 186–194.

Salama, A.H., Shamma, R.N., 2015. Tri/tetra-block co-polymeric nanocarriers as a potential ocular delivery system of lornoxicam: in-vitro characterization, and in-vivo estimation of corneal permeation. Int. J. Pharm. 492, 28–39.

Simoes, S.M., Figueiras, A.R., Veiga, F., Concheiro, A., Alvarez-Lorenzo, C., 2015. Polymeric micelles for oral drug administration enabling locoregional and systemic treatments. Expert Opin. Drug Deliv. 12, 297–318.

Stella, V.J., Nti-Addae, K.W., 2007. Prodrug strategies to overcome poor water solubility. Adv. Drug Delivery Rev. 59, 677–694.

Su, W., Chen, L., Li, C., Chen, J., Lin, Y., Matsumura, Y., et al., 2012. Phase I/II study of NC-6004, a novel micellar formulation of cisplatin in combination with gemcitabin in patients with pancreatic cancer in Asia. In: ESMO Congress 2012. Vienna, Austria.

Takagi, T., Ramachandran, C., Bermejo, M., Yamashita, S., Yu, L.X., Amidon, G.L., 2006. A provisional biopharmaceutical classification of the top 200 oral drug products in the United States, Great Britain, Spain, and Japan. Mol. Pharmaceutics 3, 631–643.

Takahashi, A., Yamamoto, Y., Yasunaga, M., Koga, Y., Kuroda, J., Takigahira, M., et al., 2013. NC-6300, an epirubicin-incorporating micelle, extends the antitumor effect and reduces the cardiotoxicity of epirubicin. Cancer Sci. 104, 920–925.

Tamura, T., Sasaki, Y., Nishiwaki, Y., Saijo, N., 1995. Phase I study of paclitaxel by three-hour infusion: hypotension just after infusion is one of the major dose-limiting toxicities. Jpn. J. Cancer Res. 86, 1203–1209.

Tu, J., Pang, H., Yan, Z., Li, P., 2007. Ocular permeability of pirenzepine hydrochloride enhanced by methoxy poly(ethylene glycol)-poly(D,L-lactide) block copolymer. Drug Dev. Ind. Pharm. 33, 1142–1150.

Uchino, H., Matsumura, Y., Negishi, T., Koizumi, F., Hayashi, T., Honda, T., et al., 2005. Cisplatin-incorporating polymeric micelles (NC-6004) can reduce nephrotoxicity and neurotoxicity of cisplatin in rats. Br. J. Cancer 93, 678–687.

Ueno, T., Endo, K., Hori, K., Ozaki, N., Tsuji, A., Kondo, S., et al., 2014. Assessment of antitumor activity and acute peripheral neuropathy of 1,2-diaminocyclohexane platinum (II)-incorporating micelles (NC-4016). Int. J. Nanomed. 9, 3005–3012.

Vaishya, R.D., Gokulgandhi, M., Patel, S., Minocha, M., Mitra, A.K., 2014. Novel dexamethasone-loaded nanomicelles for the intermediate and posterior segment uveitis. AAPS PharmSciTech 15, 1238–1251.

Valle, J.W., Armstrong, A., Newman, C., Alakhov, V., Pietrzynski, G., Brewer, J., et al., 2011. A phase 2 study of SP1049C, doxorubicin in *P*-glycoprotein-targeting pluronics, in patients with advanced adenocarcinoma of the esophagus and gastroesophageal junction. Invest. New Drugs 29, 1029–1037.

Van asperen, J., Van tellingen, O., Van Der valk, M.A., Rozenhart, M., Beijnen, J.H., 1998. Enhanced oral absorption and decreased elimination of paclitaxel in mice cotreated with cyclosporin A. Clin. Cancer Res. 4, 2293–2297.

Vellonen, K.S., Mannermaa, E., Turner, H., Hakli, M., Wolosin, J.M., Tervo, T., et al., 2010. Effluxing ABC transporters in human corneal epithelium. J. Pharm. Sci. 99, 1087–1098.

Wang, S.W., Monagle, J., Mcnulty, C., Putnam, D., Chen, H., 2004. Determination of *P*-glycoprotein inhibition by excipients and their combinations using an integrated high-throughput process. J. Pharm. Sci. 93, 2755–2767.

Xue, B., Wang, Y., Tang, X., Xie, P., Wang, Y., Luo, F., et al., 2012. Biodegradable self-assembled MPEG-PCL micelles for hydrophobic oridonin delivery in vitro. J. Biomed. Nanotechnol. 8, 80–89.

Yang, Y.H., Ding, P.T., 2015. Study on the preparation of nifedipine-loaded oral copolymer micelles and its pharmacokinetics in rats. Cell Biochem. Biophys. 71, 155–160.

Yamamoto, Y., Hyodo, I., Takigahira, M., Koga, Y., Yasunaga, M., Harada, M., et al., 2014. Effect of combined treatment with the epirubicin-incorporating micelles (NC-6300) and 1,2-diaminocyclohexane platinum (II)-incorporating micelles (NC-4016) on a human gastric cancer model. Int. J. Cancer 135, 214–223.

Yao, H.J., Ju, R.J., Wang, X.X., Zhang, Y., Li, R.J., Yu, Y., et al., 2011. The antitumor efficacy of functional paclitaxel nanomicelles in treating resistant breast cancers by oral delivery. Biomaterials 32, 3285–3302.

Yi, Y., Yoon, H.J., Kim, B.O., Shim, M., Kim, S.O., Hwang, S.J., et al., 2007. A mixed polymeric micellar formulation of itraconazole: characteristics, toxicity and pharmacokinetics. J. Control. Release 117, 59–67.

Yokoyama, M., Okano, T., Sakurai, Y., Ekimoto, H., Shibazaki, C., Kataoka, K., 1991. Toxicity and antitumor activity against solid tumors of micelle-forming polymeric anticancer drug and its extremely long circulation in blood. Cancer Res. 51, 3229.

Yokoyama, M., Fukushima, S., Uehara, R., Okamoto, K., Kataoka, K., Sakurai, Y., et al., 1998. Characterization of physical entrapment and chemical conjugation of adriamycin in polymeric micelles and their design for in vivo delivery to a solid tumor. J. Control. Release 50, 79–92.

Zastre, J.A., Jackson, J.K., Wong, W., Burt, H.M., 2008. *P*-Glycoprotein efflux inhibition by amphiphilic diblock copolymers: relationship between copolymer concentration and substrate hydrophobicity. Mol. Pharm. 5, 643–653.

Zhang, X., Burt, H.M., Hoff, D.V., Dexter, D., Mangold, G., Degen, D., et al., 1997a. An investigation of the antitumour activity and biodistribution of polymeric micellar paclitaxel. Cancer Chemother. Pharmacol. 40, 81–86.

Zhang, X., Burt, H.M., Mangold, G., Dexter, D., Hoff, D.V., Mayer, L., et al., 1997b. Antitumor efficacy and biodistribution of intravenous polymeric micellar paclitaxel. Anti-Cancer Drugs 8, 696–701.

Zhang, Y., Li, X., Zhou, Y., Fan, Y., Wang, X., Huang, Y., et al., 2010a. Cyclosporin A-loaded poly(ethylene glycol)-*b*-poly(D,L-lactic acid) micelles: preparation, in vitro and in vivo characterization and transport mechanism across the intestinal barrier. Mol. Pharm. 7, 1169–1182.

Zhang, Y., Li, X., Zhou, Y., Wang, X., Fan, Y., Huang, Y., et al., 2010b. Preparation and evaluation of poly(ethylene glycol)-poly(lactide) micelles as nanocarriers for oral delivery of cyclosporine *a*. Nanoscale Res. Lett. 5, 917–925.

CHAPTER

Nano design of extracellular matrix for tissue engineering 16

Jalaledin Ghanavi[1], Poopak Farnia[1,2] and Ali Akbar Velayati[1]

[1]*Mycobacteriology Research Centre (MRC), National Research Institute of Tuberculosis and Lung Disease (NRITLD), Shahid Beheshti University of Medical Sciences, Tehran, Iran*

[2]*Department of Biotechnology, Advanced Technologies in Medicine, Shahid Beheshti University of Medical Sciences, Tehran, Iran*

This chapter provides a comprehensive overview and detailed information concerning tissue engineering for researchers and clinicians. Basically, tissue engineering can be best defined as the interactive effect of cells, scaffold, and growth factors (GFs) for improving or replacing biological functions. Cell in tissue engineering refers to the stem cell or cells that are in different stages of differentiation. These cells seed in tissue culture or are directly injected into the desired organ. Scaffold is a three-dimensional (3D) structure that is made from biodegradable and biocompatible polymers, either natural, synthetic, or a mixture of both. The nanostructural components of the scaffold acts as an extracellular matrix (ECM) and increased cellular attachment to induce shear stress and promotes cell maturation. For enhanced cellular function compared to traditional scaffolds, nanostructural scaffolds deliver signaling molecules such as mitogens, GFs, and morphogens that support cell adhesion, proliferation, differentiation, and control the generation of tissue. Nanostructural scaffold attach to GFs with simple adsorption, electrostatic, or covalent binding. Nanofibrous and nanocomposite scaffolds are used for the regeneration of various tissues. Electrospinning nanofibrous scaffolds are very similar in physical structure to natural nanofibers in ECM. These fibers fabricate a tunable pore structure and the formation of sponge-like scaffolding. Nanocomposite scaffolds are mechanical strength and electrical conductivity and are incorporated into 3D architectures. Drug-delivery systems in nanoscale include nanoparticles (NPs), nanogels, nanocapsules, and dendrimers. Nanocomposite hydrogels combine both nanofillers and hydrogel matrices to improve the mechanical and biological properties for drug delivery and tissue engineering scaffold materials. This system increases solubility and controls release and cellular uptake of signaling molecules. Controlled release of biomolecules in scaffold dramatically enhance the maturity of tissue.

Nanoarchitectonics in Biomedicine. DOI: https://doi.org/10.1016/B978-0-12-816200-2.00010-4

Controlled release can be applied to deliver other biological molecules (plasmid DNA and siRNA) to cells to regulate cell function. Two strategies for the presentation of signaling molecules from nanostructural scaffolds are chemical immobilization and physical encapsulation.

Nanotechnology is the manipulation of atoms and molecules on a nanometer scale (100 nm). In tissue, engineered cells or stem cells are seeded onto prefabricated porous scaffolds or cells can be encapsulated during scaffold formation. The growth and differentiation of stem cells are determined by biochemical and biophysical signals in the extracellular microenvironment. Topography is the main factor in determining and differentiating the stem cells' fate. ECM is the scaffold synthesized by cells. Cells elaborate their ECM by secreting proteins and ,in turn, the ECM regulates cell behavior and influences the remodeling of their matrix (Kaul and Ventikos, 2015; Thorne et al., 2015). Then, the scaffolds are replaced with synthetic biodegradable materials. Polymer scaffolds are used for cell support. This is a basic principle of tissue engineering strategies. Polymeric scaffolds are essentially carriers of cell transplants and act as a temporary structural support until the cells create an adequate matrix. Polymer scaffold interconnectivity should have the capability to transport nutrients, oxygen, GFs and cell produced metabolites (Fig. 16.1). Mechanical shear stress of the scaffold as well as the functionality of targeted organ after implantation of scaffold-cell cause maturation of the cells which are seeded. Nanotechnology in ECM leads to modification of adhesion site, surface and mechanical properties, porosity, biodegradability, and biocompatibility of scaffold.

Specific topographical structures possess the physicochemical variables affecting the cell differentiation, and act as a temporary house in terms of the geometry and cellule size, both of which alter the stem cell response. Stem cell response is induced by the medium topography based on the adhesion in nanometric pressures. During the retraction of cellular structures, the individual forces can reach up to 2,000,000 pN for a whole cell (Dembo and Wang, 1999). The ECM properties include the protein composition of matrices, the availability of binding sites for specific integrin and physical properties like rigidity (Wolfenson et al., 2013). Integrin-mediated adhesion causes cell differentiation and creates ECM adult cells around itself during maturation. The natural structure of tissues or organs are nanometer in dimension and cells directly interact with (and create) nanostructured ECMs and the biomimetic features and the excellent physiochemical properties of nanomaterials play a key role in stimulating cell growth as well as guiding tissue regeneration (Zhang and Webster, 2009); hence, nanostructured biomaterials are used for biomimetics of scaffold design in the molecular level and local delivery of bioactive molecules intimate the cell. The oriented stem cells differentiation is affected by ECM and is in sizes that cells are capable of finding. The sizes should be smaller than the sizes of the cell membrane receptors and physical effects on the cell membrane should have the ability to be interpreted intracellularly. The sensitivity of stem cells to materials is about <50 nm (Mcmurray et al., 2011). Stem cells are defined by essential features as they are undifferentiated cells able to give rise both to differentiated daughter cells and retain their self-renewal (Weissman, 2000). The niche was first described by Schofield as a specialized microenvironment housing stem cells (Schofield, 1977).

The cells, matrix glycoproteins, and the 3D spaces they form provide an ultrastructure for a stem cell niche. The contact between these elements allows molecular interactions that are critical for regulating stem cell function. Secreted proteins offer a paracrine to maintain the stem cell pool or promote expansion of the stem cell pool, but nonprotein components from the local microenvironment also affect stem cell function (Scadden, 2006). Also, the niche dynamically regulates stem cell behavior, maintaining a balance between quiescence, self-renewal, and differentiation (Mitra et al., 2005). The task of niches is to keep the immature and proliferating phenotype of stem cells. The fate of a stem cell is determined by the signals emanating from the niche matrix to the cell membrane. The signals are both biochemical and biophysical. These factors determine the stem cell differentiation pathway. Soluble factors and the interactions between cell and ECM lead to the formation of molecular regulators that determine the stem cell's fate.

The microenvironment is involved in the conversion of pluripotent embryonic stem cells (ESCs) to cell lines, cell-to-cell signaling in the ECM, essential spatial and temporal information for directing cell differentiation and cell migration, and tissue formation. Recapitulating the native ECM is key to providing the 3D structure and biological cues for homing and delivery of stem cells for tissue engineering and organ regeneration purposes (Santos et al., 2012). The biophysical signs regarding the micron sizes of cells are nanotopographical and the pressure on the matrix is in the nano- and pico-Newton (pN) range. For instance, fibroblasts presented with fibronectin-coated beads under restraining forces of 0.02 or 0.18 pN/nm will preferentially couple stiff beads to their cytoskeleton (Jiang et al., 2000) (Fig. 16.1).

Cells and tissue engineering. Tissue engineering ingredients consist of the cells (mesenchymal or stem cell), materials (natural or synthetic matrix), and biochemical factors; while Stem cells exist in every organ and tissue. Naturally, there are two different types of stem cells including ESCs and tissue-specific (or adult) stem cells; however, induced pluripotent stem (iPS) cells can be created artificially in a laboratory by "reprogramming" adult normal cells. These cells differentiate from other cells. Tissue engineering uses living cells seeded on a natural or synthetic extracellular substrate to create implantable pieces of the organism (Nerem, 1992). After an egg is fertilized is becomes totipotent, i.e., it has the potential to create an entire organism. In the initial hours and days following fertilization, this single totipotent cell divides into more totipotent cells that are exact copies of the original. Approximately 4 days after fertilization, the totipotent cells form a cluster of cells known as a blastocyst and then the smaller group of cells, known as the inner cell mass, are formed and these inner pluripotent stem cells will go on to create most of the cells and tissues in the body.

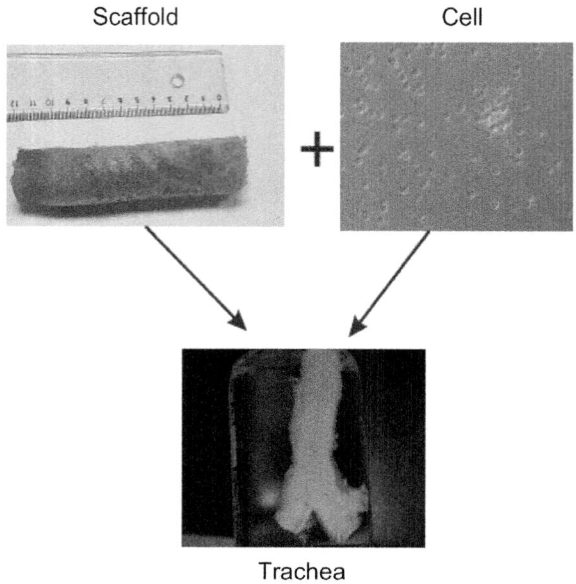

FIGURE 16.1

The cells (differentiated or stem cells) seeded onto a scaffold and cultured in a bioreactor. Physicochemical factors include nutrients, cytotoxins, growth factor, physical factor gradient, and electrical signals combined with the scaffold that provides a structural and mechanical template for cell attachment and tissue formation. The pictures show a tracheal engineered with scaffold made from collagen/chitosan and donor living cells, in the top–down approach of tissue engineering.

Pluripotent stem cells are, therefore, different to totipotent stem cells because they don't develop into a complete organism (Fig. 16.2).

iPS cells are created artificially in a laboratory by "reprogramming" adult normal cells. These cells differentiate from other cells (Deyle, 2015). The original cells (iPSCs) were produced by using viruses to insert extra copies of genes known to be important in embryonic stem cells into the specialized cell. iPSCs are generated from somatic cells that have been reprogrammed by the ectopic expression of defined embryonic transcription factors (Deyle, 2015). Pluripotential cells in the embryo have the capacity to give rise to differentiated progeny representative of all three embryonic germ layers, as well as the extra-embryonic tissues that support development (Reubinoff et al., 2001). Human embryonic stem cells have been shown to differentiate into cells from all three embryonic germ layers in vitro. Embryonic stem cells exhibit two remarkable properties: (1) the ability to proliferate in an undifferentiated but pluripotent state (self-renew); and (2) the ability to differentiate into many specialized cell types (Brivanlou et al., 2003).

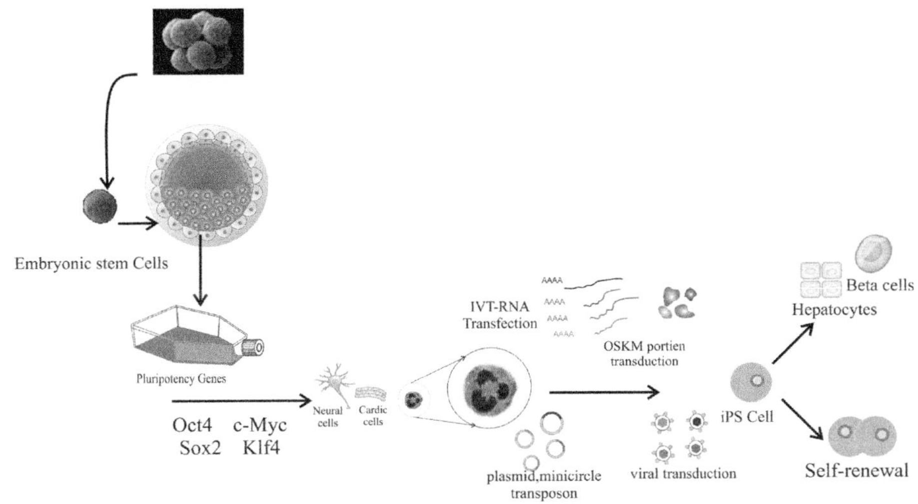

FIGURE 16.2

Pluripotent stem cells include ESCs derived from the inner cell mass of the pre-implantation blastocyst and are able to differentiate to all the three germ layers. The somatic cell are reprogrammed to iPS cells through the reprogramming factors such as OCT4, SOX2, KLF4, RiPS cells, and myelocytomatosis viral oncogene (MYC) (known as OSKM factors) which can reactivate the pluripotency network (Parenti et al., 2016). *ESCs*, Embryonic stem cells; *iPS*, induce pluripotent stem; *RiPS*, RNA-induced pluripotent stem.

16.1 CELL INTERACTION WITH EXTRACELLULAR MATRIX

The self-assembly technique is based on the attachment of oppositely charged polyions onto charged surfaces by electrostatic interactions called Layer-by-Layer (LbL) deposition (Decher, 1997). The encapsulation of living biological cells like platelets and mesenchymal stem cells (MSCs) can serve as a model system in a wide range of biomedical applications including cell targeting and immune protection (Zhao et al., 2011). The modification of the polymer species, type of coating, mechanics, and permeability can be used to control the stem cell microenvironment, allowing directed differentiation toward a desired lineage. Encapsulated stem cells can be expanded in suspension bioreactor systems without being damaged by the presence of hydrodynamic shear forces. Microencapsulation is useful to treat metabolic disorders such as diabetes (Wilson and Mcdevitt, 2013; Hasse et al., 1997; Brun-Graeppi et al., 2011). Cells are suspended in a water-soluble solution with the polymer precursors and appropriate osmolality to prevent cell lyses. Then macromolecular monomers crosslinked covalently for cell encapsulation with radical chain polymerizations and chemical crosslinking. Chemical crosslinking and encapsulation of cells in hydrogels does not require any additional components, such as an initiator (Park et al., 2004) (Fig. 16.3).

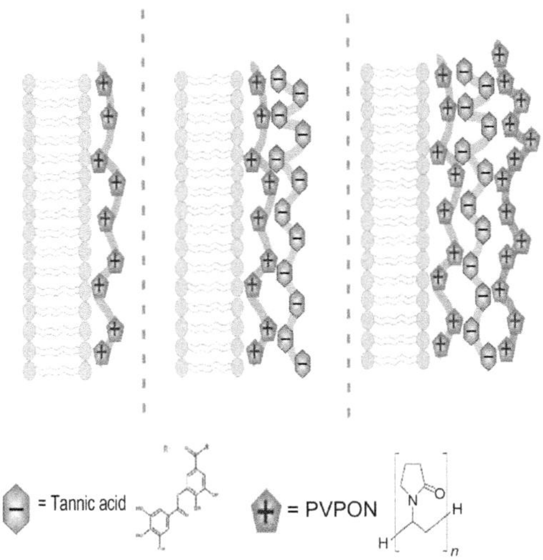

FIGURE 16.3

Chemical structures of TA and poly(*N*-vinyl-pyrrolidone). Islet encapsulation in a hydrogen-bonded $(PVPON/TA)_n$ multilayer using the layer-by-layer assembly. The first PVPON layer was deposited on the cell surfaces through hydrogen-bonded interactions between collagen and PVPON followed by assembly of the TA layer. The layers of PVPON and TA were assembled stepwise until the (poly(N-vinyl-2-pyrrolidone) (PVPON)/TA)$_n$ multilayer was formed (Kozlovskaya et al., 2012). *TA*, Tannic acid.

Human MSCs encapsulated in poly(ethylene glycol) (PEG) hydrogels are enriched with an addition of the oligopeptide sequence Arg-Gly-Asp (RGD) for cell adhesion sites to survive (Salinas et al., 2007). Self-assembling peptide amphiphile (PA) nanofibers have been used in tissue engineering applications (Sargeant et al., 2008). Equal release of GFs in all parts of the matrix is limited in the encapsulation of stem cells. However, matrix design and engineering can prevent cellular density. In addition, release of GFs to these cells and removal of residuals can be controlled based on the properties of the selected materials for the encapsulation of stem cells. Encapsulation is a cellular protection method and provides support to cells in suspension. This also increases the surface to volume ratio to attach an extracellular structure. Cells differentiation is regulated by chemical signals and the biophysical features in the extracellular environment. Biophysical signals that enter the cells from the topography of the extracellular environment, at the micro- and nanoscale, can be changed by engineering the topography of these structures to achieve the intended cell differentiation. The performance of these signaling pathways for the differentiation of stem cells is necessary, but not sufficient, for any cell type. On the one hand, our knowledge of its performance is inadequate since the nanoscale mechanical and chemical effects are unknown and on the other hand, the common methods for generating matrix at large scales are inefficient. Biochemical signals are created within the cell in this process and will eventually lead to adult tissue functions. The micron dimensions of the stem cell, the size and shape of the niche topography at nanoscale are detected so that they react to them in order to differentiate or protect themselves. The main objective of matrix production in tissue engineering is the interaction of cells with the neighboring structures and the stem cell's response to the nanometer scale topography. ECM is where cells interact with each other or with the matrix and it has a nanoscale topography. Cell interaction occurs with its medium, which is a complex structure of nanoscale pores, fine lines, and fibers. A 100–300-nm-thick layer separates the tissue covering the muscles and the nervous system from the connective tissue.

Cellular organization feeds back to regulate signaling and cell function by sensing cell shape and transducing the nanotopography of a surface (Teixeira et al., 2006). All cells interact with the topography of their environment at the nanoscale. Local sensing of force or geometry is transduced into biochemical signals that result in cell responses at the microscale and nanoscale level to regulate cell growth, differentiation, shape changes, and cell death (Vogel and Sheetz, 2006). In the fields of biomimetic materials and tissue implant technology, the importance of nanometric scale surface topography and roughness of biomaterials is, besides chemical surface modifications, increasingly becoming recognized as a crucial factor for tissue acceptance and cell survival (Craighead et al., 2001). Weiss and Garber defined the role of the ECM structure in the orientation and migration of cells (Weiss and Garber, 1952). The topographic pattern regulates signaling for stem cell behaviors, morphology, adhesion, proliferation, clonal expansion, and self-renewal (Chen et al., 2012). Tissue cells have an ability to sense and probe the stiffness of their surroundings as they adhere to, and interact with, the local ECM (Discher et al., 2005). Variables such as stiffness of materials or density of peptide ligands in the matrix and density of the ligands independent of other factors are effective in the interactions between cells and matrix and signal the transduction of specific sequences. For example, alginate with a high density of RGD, fibronectin, fibrin, and vitronectin leads to MSC cartilage-building. The cell movement can be guided by purely physical interactions at the cell–substrate interface (Lo et al., 2000). It has been suggested that during spreading and migration, cells produce initial adhesions with which they "sense" their environment by applying force to these new adhesions. This force application occurs regularly when the leading edge performs protrusion–retraction oscillations (Döbereiner et al., 2005). The cells react to the synthetic surface composition at the nanoscale level, causing the differentiation and normal function of tissue. Natural polymers such as collagen or gelatin resemble most of the natural ECMs and contain specific signals such as RGD motifs for cell adhesion and allow cell infiltration providing a biological advantage (Tian and George, 2011). Using natural biomaterials containing stem cells' niches and enriching them with a combination of GFs, cytokines, or serum proteins lead the stem cells to differentiate into a number of specific types of target cells. Stem cell evolution reacts to the chemical composition, physical appearance, and stiffness of biomaterials. Cellular membrane adhesion to biomaterials can be modified by changing the chemical composition of these materials. In other words, the cellular response to biomaterials can be controlled through modifying the interaction between biomaterials and protoplasmic components. Biomaterials with specific surface properties and extracellular nanostructures are involved in cell proliferation and differentiation. Although the synthetic mimicking of an extracellular environment is quite challenging, synthetic nanostructured polymers are frequently used to fabricate a matrix with different chemical properties usually found in natural ECM for the rapid evolution of stem cells. The cell–ECM connection is a reciprocal interaction in which cells continually remodel the ECM present in their microenvironment and these dynamic modifications of the ECM direct cell behavior (Bershadsky et al., 2003). Mechanical interactions between cells and the environment significantly affect cell differentiation. These mechanical forces are caused by cell–cell and cell–matrix interactions, which direct the differentiation and evolution of a complex tissue pathway. The cell-ECM interface is composed of structural units of nanometer length scales which, in turn, regulate stem cell fate along with other physical factors (Kingham and Oreffo, 2013) (Fig. 16.4).

Cell adhesion to a material with a certain stiffness triggers a signaling cascade, which converts extracellular mechanical signals to intracellular events (Vitillo et al., 2016).

Both genomic and proteomic approaches were examined to determine complex molecular regulation of post-stem–cell-differentiation. Studying cellular receptors at nanoscale and the effect of these receptors within the cell and on the cell nucleus and protein synthesis lead to cell differentiation and create new

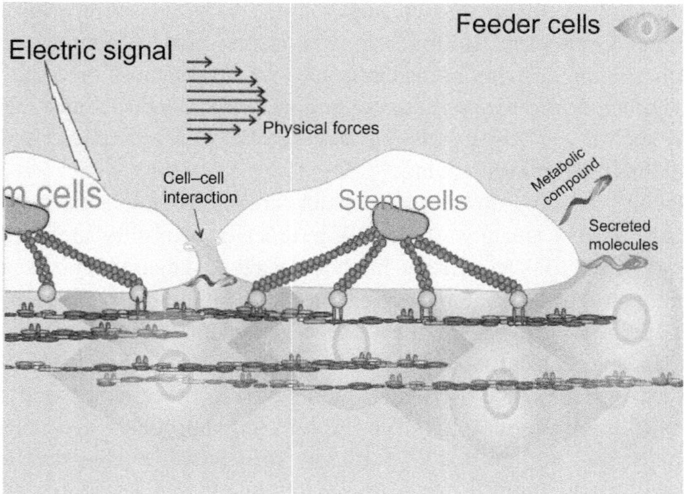

FIGURE 16.4

Schematic of the stem cell microenvironment. Stem cells are regulated by: cell–biomaterial interactions, cell–cell interactions, electrical signals, physical forces, and soluble bioactive molecules such as inhibitors, cytokines, growth factors, and nutrients. The physical characteristics of biomaterials also regulate the stem cell fate (Higuchi et al., 2011).

biomaterials with high functionality for cell adhesion and control over tissue-specific differentiation of stem cells (Wu, 2007).

Accordingly, controlled changes in material properties can lead to the simultaneous differentiation of common progenitor cells to multiple phenotypes on a single scaffold. In addition, construction and evolution of self-repairing materials can control cell adhesion and differentiation based on adhesive parameters, pressure, arrangement, and adhesion of cells to each other to be differentiated into a microvenous network in order to deliver nutrients to the cells within the matrix to complete the tissue structure. Natural or synthetic self-repairing materials support the cells mechanically to repair damaged tissue in order to differentiate stem cells into appropriate phenotypes. These materials should be removed from the surrounding environment simultaneously with cellular differentiation. Porous scaffolds must be interconnected structures, not only for cell seeding and nutrient and metabolite, but also for endothelium and microvascular networks. Diffusion consumption modeling has shown a 200 μm limit in scaffolds for oxygen transport to cells, resulting in a maximum of 400 μm diameter features for cell survival (Dunn et al., 2006). Dynamic interactions between cells and matrix control multiple cell behavior such as cell adhesion, spreading, migration, and formation. Low adhesion could potentially lead to cell quiescence or even apoptosis by anoikis, a type of programmed cell death through "homelessness" (Cavalcanti-Adam et al., 2008).

In nanotechnology there are two approaches for tissue engineering methods, top–down and bottom–up. Bottom–up approaches in developing scaffolds for nano-based tissue engineering are established on self-assembly of small structure blocks into a larger arrangement through noncovalent interactions (Whitesides et al., 1991) such as hydrogen bonding, van der Waals interactions, ionic bonds, and hydrophobic interactions (Zhang, 2003).

However, in bottom–up methods nano or microscale elements such as particles, free ECM molecules, and cells interact or crosslink together to form the "alive 3D scaffold." Numerous methods used to generate nanoscale structures and nanostructured materials are commonly characterized as top–down and bottom–up. The top–down approach uses various methods of lithography to pattern nanoscale structures. This approach technique allows for patterning features in two-dimensions (2D) and include photolithography and scanning beam (or maskless) lithography. The bottom–up approach uses interactions between molecules or colloidal particles to assemble discrete nanoscale structures in two and three dimensions (Gates et al., 2005). Top–down tissue engineering approaches require the use of patterned substrates to fabricate ECM scaffolds and assist the cells in finding their appropriate positions and niching to differentiation states within a tissue (Liu and Gartner, 2012).

The top–down approach mimics ECM by synthesis of porous biocompatible and biodegradable matrix which is combined with biochemical factors into a template, creating an engineered scaffold (Nichol and Khademhosseini, 2009). Another approach is decellularization of an organ to provide all the necessary microarchitecture and microenvironmental factors (Ott et al., 2008). Perfusion decellularization provides a biomimetic natural microenvironment with a biocompatible geometry template for tissue engineering (Ott et al., 2008). This structure is the nanoscale of a 3D scaffold of ECM. For cellular matrices, isolate the ECM and remove cellular components. The source of the matrices can be heterologous, allogeneic, or autologous and in the biomaterial fabric, the scaffold should be biocompatible and biodegradable to confirm cell support.

3D scaffolds for tissue engineering with random micro and nanopores have been fabricated using salt leaching, gas forming, phase separation, and freeze–drying techniques (Loh and Choong, 2013). For producing a porous polymer scaffold, solvent casting and particulate leaching is the widely used method for tissue engineering.

Solvent casting/particle leaching is a simple procedure which begins with the dispersion of a porogen with controlled particle size and porosity into a polymer solution (Quirk et al., 2004).

The nature of the porogen has been demonstrated to play an important role in the interconnectivity of pores and, in turn, the cell–scaffold interactions after seeding (Suh et al., 2002). Particle leaching can be used to fabricate scaffolds for tissue engineering applications (Ma and Langer, 1999). Small particles of salt are transferred into the polymer solution to form the salt-filled mold. The polymer is subsequently hardened and the salt is removed via dissolution in a solvent such as water or alcohol (Lee et al., 2005). After the solvent evaporates, the salt crystals are leached by water to form pores in the scaffold. The pore size and porosity are controlled by the size and number of crystals in the polymer. The polymer is subsequently hardened and the salt is removed via dissolution in a solvent such as water or alcohol. Using salt leaching, based on hydrophilic PEG and hydrophobic poly (epsilon-caprolactone) (PCL) porous hydrogel scaffolds were produced using NaCl as a porogen (pore size distribution: 180–400 μm) and dimethyl sulfoxide as a solvent (Park et al., 2007b). But in the gas-foaming technique, gas is used as a porogen (Nam et al., 2000) and the procedure does not require any organic solvents. Instead, gas bubbles are grown and dispersed throughout a polymer to generate a porous structure (Lips et al., 2005). The gas nucleation and bubbling are formed by chemical reactions (Huh et al., 2005) or by being released from a presaturated gas–polymer mixture at a high pressure (Blacher et al., 2007; Harris et al., 1998).

Highly porous macrocellular foams are obtained from copolymers of lactic and glycolic acid using high pressure CO_2 gas processing. The major problem of the solvent-casting procedure is that the use of organic solvents and the residual solvent in the scaffold after drying caused cell toxicity. Freeze-drying, or lyophilization, can also be used for the fabrication of porous hydrogels for tissue engineering. In this method, thermodynamic instability by rapidly cooling a gel causes phase separation and then the solvent is evaporated under vacuum (Thomson et al., 1995). The freeze–drying procedure is a bio-clean method for formulating biodegradable sponges from natural (Kang et al., 1999) and synthetic polymers.

The structural stability and mechanical properties of the fabricated materials are low and the pore size in the hydrogel architecture are difficult to manage and extremely sensitive to the kinetics of the cooling process (Fig. 16.5).

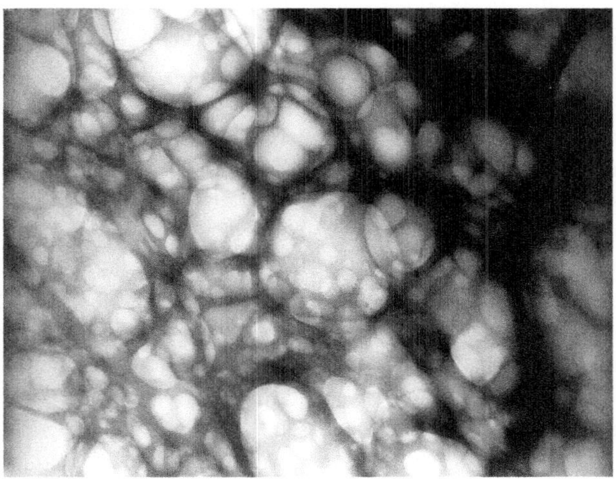

FIGURE 16.5

TEM image of a collagen matrix. *TEM*, Transmission electron microscopy.

These scaffolds are also obtained by manual operation without uniform pore size and low interconnectivity pore structure and nonuniform pore size distribution (Ramay and Zhang, 2003). The pore structure strongly influences cell growth and the drug release profile (Tai et al., 2007). Therefore, scaffold fabrication by automated and computer-controlled processes could offer a great advantage compared to methods. The rapid prototyping (RP) or solid free-form fabrication (SFF) permit automated and full computer-controlled generation of 3D scaffold structure and assembly by controlling the pore size, porosity and architecture. Controlling the essence of ECM pore size provides the facility for cell infiltration and transportation of nutrients and metabolic waste throughout the scaffold. SFF techniques for fabrication of tissue-engineering scaffolds include selective laser sintering (SLS) (Williams et al., 2005; Paul and Baskaran, 1996), fused deposition molding (Zein et al., 2002), stereolithography (Dhariwala et al., 2004), and 3D printing (Hollister, 2005; Wu et al., 2011) with well-defined pore geometry and complex pore architectures at the microscale. SLS constructs scaffolds from 3D digital data by sequentially fusing regions in a powder bed, layer by layer, via a computer-controlled scanning laser beam (Deckard, 1989). The LbL fabrication technique allows for the construction of scaffolds to produce fully interconnected porous scaffolds and does not require organic solvents. The material used is a fine powder (such as hydroxyapatite) whose particles adhere and solidify (or sinter) under laser illumination. SLS are more advantageous to produce scaffolds for tissue engineering than other SFF techniques, such as stereolithography, because the dose not require support structures during the build.

It is a fast, easy to fabrication, cost-effective and efficient method by which scaffolds are constructed to match the complex anatomical geometry (Rutherford et al., 2003; Berry et al., 1997). This LbL fabrication technique allows the construction of scaffolds to produce fully interconnected porous scaffolds with acceptable microstructures and mechanical properties (Williams et al., 2005) without requiring organic solvents. Fused deposition modeling (FDM) is a 3D printing technique based on the deposition of successive layers of thermoplastic materials following their softening/melting. FDM is a RP technology for scaffolds to manufacture honeycomb-like patterns with a fully interconnected channel network and controllable porosity and channel size. A bioresorbable polymer poly (epsilon-caprolactone) (PCL) produces porous scaffolds by computer-controlled extrusion and deposition process (Zein et al., 2002; Melocchi et al., 2016).

Stereolithography can be used for the fabrication of biocompatible complex structures using a LbL approach based on computer-aided design controlled solidification of liquid-based resins by photopolymerization with ultraviolet (UV) light. The first layer of photopolymerized platform supports the structures to fabricated. The process is repeated in each layer individually to form the 3D scaffold (Skoog et al., 2014). Fabrication of photocurable liquid biodegradable copolymers of trimethylene carbonate and poly(e-caprolactone) (PCL) was initially reported to be used to fabricate microscale surface features for tissue engineering (Matsuda et al., 2000).

16.2 THREE-DIMENSIONAL PRINTING

3D printing involves the laying down of successive layers of material to form 3D models, thus, enabling better control of pore sizes, pore morphology, and porosity of the matrix compared with other fabrication methods (Hollister, 2005; Wu et al., 2011).

3D scaffolds are usually fabricated by 3D photoprinting of hydrogels to mimic the properties of ECM in bone and cartilage. The 3D printing of scaffolds was achieved by the methods of material extrusion, material jetting, and sheet lamination (Chia and Wu, 2015; Balogun et al., 2014). In bioprinting, small balls of bioink composed of cells and hydrogel materials (e.g., alginate or decellularized ECM) are printed in a desired shape (Billiet et al., 2012). These methods are rapid and cheap but it is difficult to assure high cell density needed for the fabrication of solid organ structures. Furthermore, due to the high speed of cell deposition, considerable damage is caused to cells, although the latest developments in the field have led to considerable improvement in cell survival (Jakab et al., 2010). Another approach used a direct-write and extruding cells and matrix into spatially organized 3D constructs (Smith et al., 2007) or multicellular spheroids (bio-ink particles) placed into biocompatible environment (bio-paper) that mimics 3D tissue structures by the use of a 3D delivery device (bio-printer) (Jakab et al., 2008).

The disadvantage of the method is associated with the relatively high cost of the printers. Both inkjet and extruder bioprinting are compatible with RP.

A big challenge facing tissue engineering is vascularization of the engineered tissues. A vascular-shaped lumen can be formed by controlling dimensions of lumen size via rapid production of toroid building units (Livoti and Morgan, 2010).

To date, numerous top–down and bottom–up nanofabrication technologies such as electrospinning, phase separation, self-assembly processes, chemical etching, nanoimprinting, photolithography, and lithography are available to synthesize nanomaterials with ordered or random nanotopographies (Bendrea et al., 2011).

16.3 THREE TECHNIQUES FOR ACHIEVING NANOFIBERS FOR TISSUE ENGINEERING

(1) Electrospinning is a method used to generate nanofiber matrices which are manufactured using polymer solutions or molten material. Electrospun fiber diameter ranges is within the 10 nm range. Electrospinning occurs when the electric field reaches a critical value at which the repulsive electric force overcomes the surface tension force of the polymer and a charged jet of the solution is ejected from the tip of the Taylor cone. Since this jet is charged, its trajectory can be controlled by an electric field. As the jet travels in air, the solvent evaporates leaving behind a charged polymer fiber which lays itself randomly on a collecting metal

screen. Thus, continuous fibers are laid to form a nonwoven fabric. This description of the process suggests that the following parameters affect it: solution properties including viscosity, conductivity, and surface tension; controlled variables including hydrostatic pressure in the capillary; electric potential at the tip, and the distance between the tip and the collection screen; and ambient parameters including temperature, humidity, and air velocity in the electrospinning chamber (Doshi and Reneker, 1993).

Flexibility in the spatial orientation of nanofibers is one of the chief advantages of electrospinning. In this fairly complicated process, material properties including solubility, molecular weight, the vapor pressure of the solvent, pH, polymer concentration, and surface tension, on the one hand, and electrodes voltage difference and the geometrical shape of the nozzle, on the other hand, accompanied by the temperature, relative humidity, and plasma between two electrodes, can have an impact on collected fiber diameter and alignment. In order to control the morphology and spatial alignment of nanofibers, all these factors should be designed carefully. The electrospinning process for tissue engineering and cell homing will be improved by the alignment of the nanofiber deposition changed with electric (Li et al., 2003) or magnetic fields (Yang et al., 2007), increasing the distance from tip to collector (Dalton et al., 2008) and reducing the speed of the plate collector (Zhang and Chang, 2007). Electrospinning of collagen, gelatin and laminin, which are natural organic materials and are used as a culture platform for various types of stem cells in tissue engineering causes stem cells to grow due to the existence of natural niches. A significant advantage of electrospinning for studying stem cell nanotopography interactions is its capability to generate 3D nanofibrous architectures that can properly mimic the structure and organization of in vivo ECM (Abrams et al., 2000). This method is cost-effective for preparation of 3D structures from biodegradable synthetic polymers such as PCL and poly(L-lactide-co-glycolide) which are more resistant to degradation, nonimmunogenic, and can be mixed with natural polymers such as collagen and elastin to combine their favorable traits to enhance biocompatibility (Jin et al., 2011). In addition to the physical cues of a native matrix, cells are exposed to an array of biological cues throughout the ECM that direct cellular behavior. Cells are constantly interacting with the surrounding ECM which gives rise to a dynamic transfer of information between the extracellular and intracellular space. Nanofibrous scaffolds mimic native ECM, for example PVA nanofibers as a material to present a biological cue to cartilage repair (Coburn et al., 2012). Biocomposite nanofibrous scaffolds are made from synthetic or natural polymeric blends and electrospinning to provide suitable substrate for tissue engineering. Nanotopography or orientation of the fibers within the scaffolds greatly influence the cell morphology and outgrowth; and the alignment of the fibers ensures better contact guidance of the cells (Kijeńska et al., 2012), the most important being polymer type and solvent composition. The diameter of the nanofiber produced is controlled by the solution concentration, electrical conductivity, addition of surfactant, the amount of solution injection per time unit, as well as the difference between the two electrodes' voltage. The fiber diameter is reduced by decreasing the solution injection velocity and its concentration as well as an increase in solution electrical conductivity. A rotating collector is used to fabricate parallel fibers. In synthetic polymers, the adhesion of fiber surface is improved by using protein molecules, niches, and natural ligands as well as charging polymer surfaces or the surface polymerization by functional groups. The "needle-free electrospinning" concept originates from nanofibers which eject without a needle from the free surface of a polymeric solution (Forward and Rutledge, 2012).

(2) Self-assembled nanostructures are formed according to molecular interactions in 2D sheet structures, 3D networks, and cylindrical shapes. These structures are interconnected to form a nanofiber network after adding opposing ions or titration using pH variables (Carrick et al., 2005).

Self-assembly biopolymers such as peptides and nucleic acids are used for noncovalent interactions to assembling a spontaneous 3D structure. To mimic the human ECM, the PA-based self-assembling system allows for the formation of thermally stable protein-like molecular architectures. The PAs that consist of a dialkyl chain moiety (hydrophobic component/tail group) attach to an *N*-alpha amino group of a peptide chain (hydrophilic component/head group), resulting in a "PA" (Berndt et al., 1995). The nanofibers produced by self-assembly are approximately 5–8 nm in diameter and several microns in length. These nanofibers can be mineralized with hydroxyapatite (HA) crystals. The authors observed the formation of HA crystals oriented along the length of the nanofibers (Hartgerink et al., 2001).

For in situ tissue engineering, these structures are made of amino acid sequences and start formation of architecture of ECM during self-assembling process by consumption of extracellular physiological medium. Moreover, the molecules are interconnected in the form of certain nanofiber networks while producing a hydrogel containing about 99% water. The injectable matrix consists of a synthetic and natural polymer, a stem cell niche, and NPs in the form of cups filled with a GF and physiologic agent used in successful tissue regeneration (Ghanavi, 2014). One of the advantages to self-assembling of nanofiber matrix is their being used as reproduction platform in vivo. Therefore, it provides a strategy for in vivo cell proliferation. Self-assembly is moderated by intermolecular forces, and hydrogels can easily lose their shape (either with or without of cell) and be injected into the body while being reassembled and result in tissue regeneration.

(3) Phase separation is when polymers are dissolved in a solvent at a high temperature which is followed by separation phase by lowering the solution temperature. In this process, the temperature of a polymer solution is controlled to induce the phase separation into two, a polymer-rich phase and a polymer-lean phase. After removal of the solvent, the polymer-rich phase solidifies and forms different pore morphology and structures in polymer foam. Solvent residues in scaffolds fabricated by processes involving organic solvents may damage cells (Lee et al., 2004). Phase separation occurs when the concentration of polymer or polymer blends exceeds its solubility in the solvent. It is convenient to control the

mixture temperature to initiate phase separation in the polymer solution and subsequent generations of micro- and nanoscale polymer-rich and polymer-lean phase domains within the mixture (Smith et al., 2008). Thermally induced phase separation has been used for several years to fabricate synthetic porous scaffolds for tissue engineering. The use of amorphous polymers with a slow cooling rate results in a macroporous open cellular structure, whereas that of semicrystalline polymers with a fast cooling rate generates a microporous closed cellular structure (Nam and Park, 1999). Lithographic methods including colloidal lithography, photolithography, and electron beam lithography are used for formation of extracellular nanomatrix structures. During colloidal lithography, self-assembled NP crystal structures are formed on surfaces to make masks for etching processes (Yang et al., 2006). The photolithographic process uses UV light with wavelengths of 248 and 193 nm to make nanoscale structures with a minimum feature size down to 50 nm (Wong, 2001). Electron beam lithography is a well-defined pattern (sizes $<$10 nm) which is achieved by the reduced wavelength of electron waves (Vieu et al., 2000). In the pattern transfer method, the preexisting nanotopographical structure on a rigid mold transfers to other materials and this technique consists of nanoimprinting and replica molding. Nanoimprinting utilizes hot embossing of thermoplastics to achieve nanoscale surface patterning. The imprint lithography (hard lithography) has high-throughput and low-cost (Chou et al., 1996). But in replica molding (soft lithography), an elastomeric mold and/or stamp such as poly(dimethylsiloxane) with patterned relief structures on its surface is used to generate patterns and structures with feature sizes ranging from 30 nm to 100 μm (Xia and Whitesides, 1998) (Fig. 16.6).

The surface roughening method, referring to physical or chemical etching, generate random nanotopographic on material surfaces. A basic reactive ion etch system is the principal of dry etching method which is currently the most important plasma configuration for creation of nanotopographical structures based on silica glass surfaces (Fracassi and Coburn, 1988) and controlling of nanotopography pattern is valid because of low etching rate and controlling of nanotopography pattern is valid because of low etching rate. Another approach for controlling the cellular microenvironment of stem cells is the microfluidic technique. Microfluidic perfusion culture in particular allows controlled delivery and removal of soluble biochemical molecules in the extracellular microenvironment, and controls the application of mechanical forces exerted via fluid flow. These design considerations include the choice of materials, culture configurations, microfluidic network fabrication, and micro-assays (Kim et al., 2007). Mimicking ECM interactions using peptide sequences leads to the formation of certain tissues. Natural biological processes direct the behavior of stem cells toward the desired tissue and cell under in vivo conditions. For example, RGD peptide is used to enhance cell adhesion, which is a helpful strategy to direct cell differentiation (Qu et al., 2010). The extracellular domains of the integrin subunits recognize ECM and the cytoplasmic domains interact with the cytoskeleton-signaling network through focal adhesion proteins such as vasodilator-stimulated

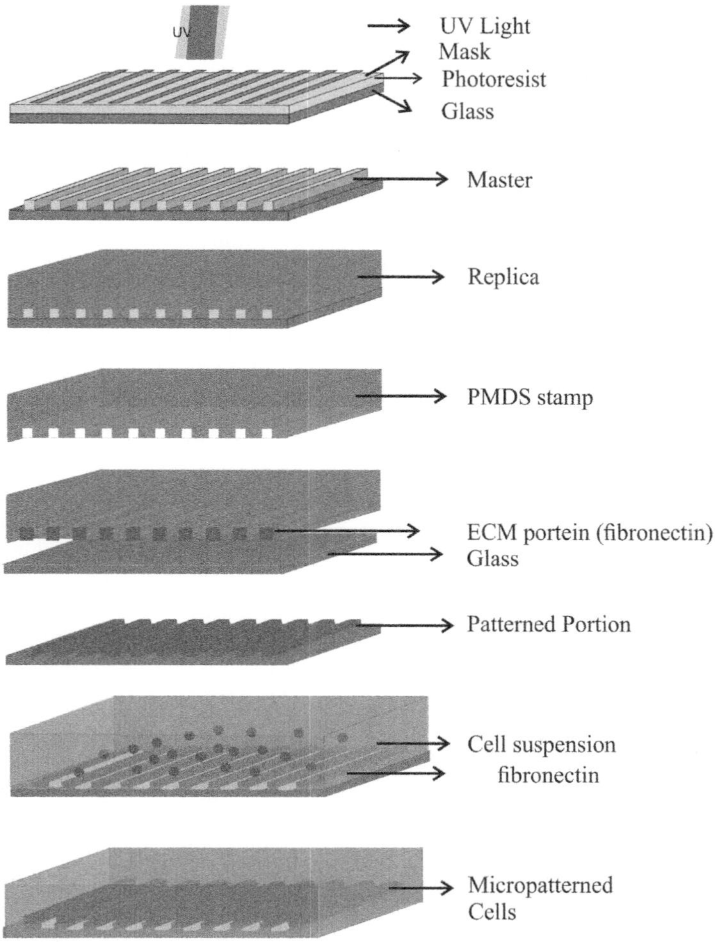

FIGURE 16.6

Microcontact printing is a high-quality 3D nanoscale pattern obtained without deformation of the PDMS stamp. UV light was radiated onto the mask and washes to remove the nonpolymerized gel and expose the underlying glass substrate within the patterned regions. A PDMS stamp is used to transfer ECM onto the supporting materials, creating a specific pattern for cell adhesion. Cells were seeded on fibronectin patterned glass surfaces into the tissue culture. *PDMS*, Poly(dimethylsiloxane); *ECM*, extracellular matrix; *UV*, ultraviolet.

phosphoprotein (VASP), paxillin, tyrosine kinase Src, and focal adhesion kinase (FAK) (Zaidel-Bar and Geiger, 2010). Cells sense the stiffness of their microenvironment by pulling on the ECM and/or adjacent cells. This process is dependent on transcellular cell–ECM or cell–cell adhesions and cell contractility mediated by Rho GTPases to provide a functional linkage through which forces are

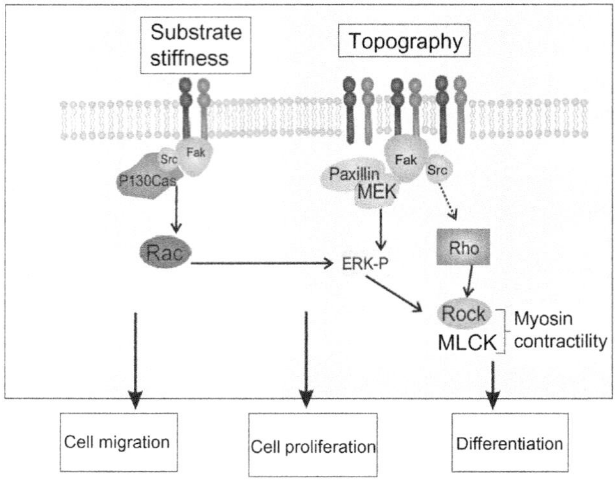

FIGURE 16.7

Focal adhesions regulate some signaling pathways that activate cytoskeletal pathways to modulate cellular responses such as adhesion, migration, proliferation, and differentiation. FAK is a primary regulator of focal adhesion signaling and can be activated by both inside-out and outside-in signaling. Activation of FAK can result in increased cell proliferation through the activation of the ERKs (Provenzano and Keely, 2011). FAK can regulate cell migration by regulation of the RhoA-ROCK pathway. FAK is also able to regulate the differentiation of stem cells (Clemente et al., 2005; Hoon et al., 2016). *FAK*, Focal adhesion kinase; *ERKs*, extracellular signal-regulated kinases.

transmitted through the cytoskeleton by intracellular force-generating proteins to control cell proliferation (Provenzano and Keely, 2011). Also cell adhesion enhanced Grb7 phosphorylation (a physiological substrate of FAK) in cells and regulate the cell migration (Han et al., 2000) and FAK signaling plays a critical role in the control of the cell cycle to terminal differentiation (Clemente et al., 2005) (Fig. 16.7).

According to this strategy, stem cell genes express certain proteins which leads to cell differentiation. However, the multiple matrix affect is a gold guide in 3D structures that directs differentiation of pluripotent stem cells to target cells. The matrix is involved in cell differentiation mechanisms in different pathways. Surfaces modified with methyl, carboxyl, amine, and hydroxyl groups made significant and various structural changes in fibronectin-coated surfaces. These sites can bind to specific integrin. (Lee et al., 2006; Keselowsky et al., 2003; Bai et al., 2013; Capadona et al., 2003).

Binding to a specific integrin can activate a specific signal transduction pathway and, consequently, affect stem cell differentiation. The architectural design of a niche appears to be specifically tailored to suit the particular needs of its resident stem cells and, conversely, stem cells may play an important role in organizing and specifying the niche. Thus, although niches share similarities in activating common signal transduction pathways to achieve the slow-cycling, self-renewing, and undifferentiated state of their residents, each niche is composed of different nonstem cell types as well as stem cell types that establish this microenvironment (Fuchs et al., 2004). The stem cell niche influences stem cell behavior and regulates stem cell fate decisions by providing a variety of signals. These signals may be structural, physical, electrical, or biochemical. ECM is an essential component of the stem cell niche and it is involved in almost all of these signals (Gattazzo et al., 2014).

Each stem cell has a certain performance on its niche; therefore, the number of cellular niches should increase during their proliferation in order to form a functional population from stem cells. The different stem cell types may require different ECM clues to provide the intended benefit. Improved survival of stem cells beyond a critical time may provide these cells with the opportunity to autologously generate an ECM and niche sufficient for viable integration (Kurtz and Oh, 2012). These cells maintain their undifferentiated state even after sequential divisions when grown within unique niches. Ligand and cellular niches connection to cellular culture systems are often required for effective proliferation and differentiation of stem cells. When we employ natural polymers to proliferate and differentiate stem cells due to the elimination of most of the variables, such as attachment type and presence of ligand and niche, the nanofiber diameter is the major variable.

The large diameter of nanofiber changes the mechanical properties of the nanofiber collagen matrix by impacting topography and then changes the intracellular actin structure and makes the proliferation and differentiation process distinct (Raoufi et al., 2016). Retraction forces appear to be the relevant forces since ligand binding causes integrin association to the retrogradely flowing actin (Schmidt et al., 1993) (Fig. 16.8).

It is essential to prepare an environment similar to natural ECM in tissue engineering. Simple polymerized ECM environments similar to natural ECM environments are generated through biomimetic ECM in order to add cellular adhesive peptides such as RGD to the scaffold. Designing a bottom-to-top structure for cell-to-scaffold adhesion is considerably important in nanotechnology. Changes in the molecular structure of ECM and molecular binding sites of cellular membrane to ECM are included as current methods of nanotechnology and tissue engineering. When RGD ligands packed closer than 70 nm to neighboring RGDs, the cells formed focal adhesions and polymerization of contractile actin cytoskeletal stress fibers was observed (Cavalcanti-Adam et al., 2007). Cell adhesion has an important, but incomplete, role in stem cell differentiation. RGD sequences are widely used for binding to integrin (including a2b1, a3b1, a4b1, a5b1, a8b1, avb1, avb3, avb5, and avb6). The main disadvantage of RGD sequence is nonspecificity of sequence-containing ECM proteins. However, RGD peptide contains several peptide motifs with more control over integrin features, cell adhesion, fate, behavior, and evolution (Takagi et al., 2003; Hynes, 2002). Two Gly-Phe-Hyp-Gly-Glu-Arg (GFOGER) (Emsley et al., 2004) and Tyr–Ile–Gly–Ser–Arg (YIGSR)

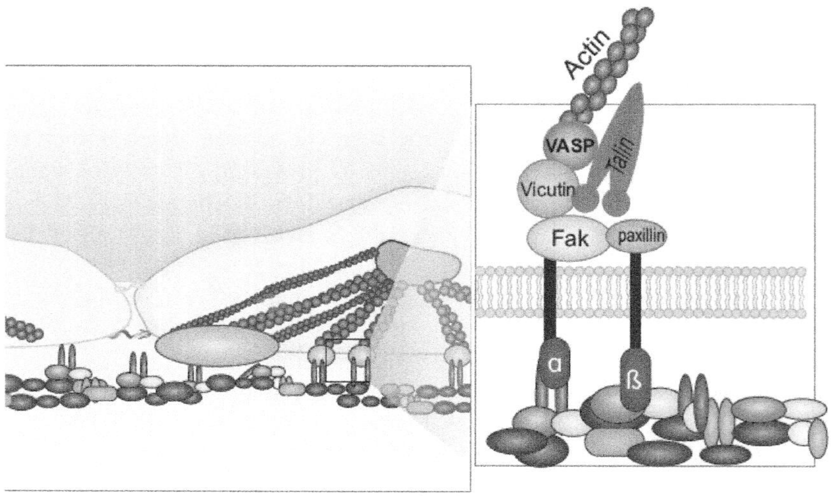

FIGURE 16.8

Integrin-mediated cell–ECM adhesion. Cells residing on their ECM. Integrin-mediated cell adhesion to the ECM requires clustering of multiple integrin receptors and FA complexes for efficient actin fiber assembly. FA consists of proteins including vinculin, talin, Vasodilator-stimulated phosphoprotein (VSAP), paxillin, and FAK. *FAK*, Focal adhesion kinase; *ERKs*, extracellular signal-regulated kinases; *FA*, focal adhesion.

(Graf et al., 1987) peptides are widely and currently used in tissue engineering and microenvironments. These peptides are derived from collagen. These ligands directly bind to the integrin (Fig. 16.9).

Engineering of the RGD-to-RGD nanopatterned adhesive surfaces and use of these surfaces as adhesion substrates indicated that a lateral spacing of 58 nm or less between nanogold-anchored RGD peptides is essential for focal adhesion (FA) formation. There is a critical ligand (RGD) density (interligand spacing of <70 nm) that is essential for the formation of stable integrin-mediated adhesions (Arnold et al., 2004). These engineered protein fragments have several advantages for integrin assessment. ECM molecules are large and complex which not only encompass binding sites to integrin, but also affect many interactions between molecular matrices that can affect cell fate. Integrin interactions can be detected through generation and design of protein fragments. Moreover, specific protein fragments with certain peptide sequences in ECM trigger intercellular molecular mechanisms that significantly contribute to cell differentiation. The peptide nanofibers and synthetic thickening nanofiber were simultaneously functionalized through electrospinning with an arginine-glycine-aspartic acid (RGD) peptide sequence to treat their defects. Moreover, by placing messenger ligands on synthetic nanofibers, or rather by restricted releasing of induction GFs placed in nanocapsule or nanocup, it induced differentiation (Ghanavi, 2014) (Fig. 16.10).

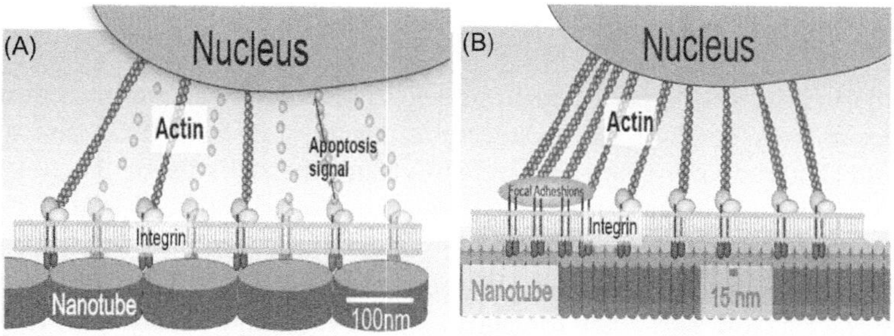

FIGURE 16.9

(A) Schematic illustration of a hypothetical model of nanoscale spacing in directing stem cell fate. (B) A spacing of 15 nm is suitable for integrin clustering and subsequent FA assembly and actin cytoskeletal organization, while a nanotube spacing of >70 nm suppresses FA assembly (Park et al., 2007a). *FA*, Focal adhesion.

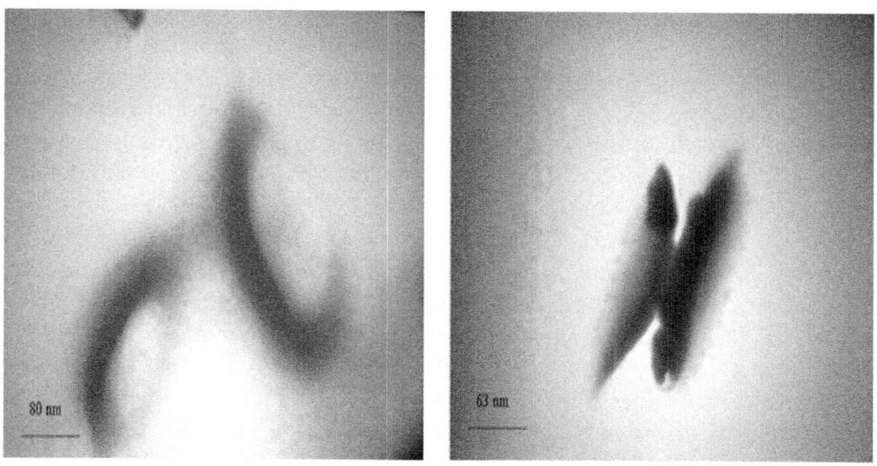

FIGURE 16.10

TEM image of a nanocup with or without a growth factor. *TEM*, Transmission electron microscopy.

Soluble agents are available to be added to a cellular culture to cultivate monolayer cells. These factors can be controlled at different times by substituting the culture solution containing them at certain times. Physical and chemical effects are caused by cellular contacts, soluble molecular signals, and physical interaction of stem cells with ECM. These create future phenotype of the cell. Medium manipulation is used to direct cellular behavior in conventional two-dimensional culture systems. However, the matrix structure restricts homogeneous and free release of materials in 3D structures.

16.4 GROWTH FACTORS

GFs in the ECM must be protected from degradation and deactivation, but degradation occurs very quickly. The architectural design of encrypted GF and storage in scaffold materials must be optimized to ensure the delivery of the functional GF at the right time and place. The tissue-inducing substances (GF) or genes encapsulated in the polymer to obtain NPs, nanospheres, or nanocups as carriers of single or multiple biomolecules. Furthermore, ECM-bound GFs are released into the system by age-related changes in the binding properties of ECM components (Hayashida, 2010). GF incorporation into scaffolds include either noncovalent (surface adsorption, physical entrapment, affinity binding, and ionic complexation) or covalent immobilization on or into the delivery system (chemical conjugation) (Schliephake, 2010). GF entrapment physically causes an initial burst followed by slower release that is controlled by the diffusion and degradation rates of the matrix which are, in turn, dependent on such properties as matrix, porosity, swelling behavior, polymer crosslinking density, and polymer chemistry (Nyberg et al., 2016). Entrapment of GF or gene in NP leads to improved delivery across the cell membrane, reduced toxicity, and lysosomal escape. Nanosized liposomes and dendrimers are also used for the delivery of biomolecules into the scaffold. Particles with sizes of 100 nm are taken up by cells 2.5-times more readily than microparticles 1 μm diameter (Desai et al., 1997). Several mechanisms have been elucidated to describe molecule release from polymer hydrogel systems including diffusion, swelling, and chemically controlled release (Lin and Metters, 2006).

Various approaches used to control GF delivery from scaffolds include physical entrapment, chemical binding, surface modifications, biomineralization, micro- and NP encapsulation, and genetically engineered cells (Nyberg et al., 2016).

However, the continuous and similar effect of GFs is more essential in the 3D culture media due to enhanced cell–cell and cell–matrix interactions. Restrictions on the release of materials complicate the 3D culture of cells. Concentration gradients of soluble factors added to the media (resulting from release of materials in 3D cell structures) depend on the spatial location of cells. Since stem cells are very sensitive to small perturbations in the biochemical composition of their surrounding environment, soluble factors are very effective in cell differentiation. Control and engineering of 3D culture microsystems is essential for stem cell differentiation. The main objective of the strategies for controlling behavior of stem cells in a 3D matrix lies in effective control over the delivery of physical and chemical factors to stem cells at molecular levels. Parameters such as porosity, degradability, and surface coating can be manipulated to deliver materials with the desired properties. Stem cells with biodegradable biomaterials as sponge or fiber scaffolds act as a matrix substrate for the cells, which allow mechanical support and adhesion of the tissue. The synthetic matrix is decomposed as the ECM is synthesized by the cell.

Biomolecular gradients are an important, evolutionarily conserved signaling mechanism for guiding the growth, migration, and differentiation of cells within the dynamic, 3D environment of living tissue (Keenan and Folch, 2008). Chemotaxis, or cell migration directed by an external gradient of chemoattractant, is encountered in various physiological and natural settings and is a means by which cellular processes are coordinated in space and time. In eukaryotes, cell orientation is biased by an external chemoattractant gradient through a spatial contrast in chemoattractant receptor-mediated signal transduction processes that differentially affect cytoskeletal dynamics at the cell front and rear (Schneider and Haugh, 2006). Morphogens are secreted signaling molecules that organize a field of surrounding cells into patterns. They form a gradient of concentration emanating from a localized source and determine the arrangement and fate of responding cells according to the different concentrations of the morphogen perceived by the cells. It is believed that a cell recognizes its circumstance in a concentration gradient of an extracellular signal factor and determines its developmental fate (Gurdon and Bourillot, 2001; Lawrence and Struhl, 1996). Morphogens show a potential economy of signaling, as a single graded signal to which cells respond with serial thresholds can specify several different cell fates (Kay and Thompson, 2009).

16.5 NANOPARTICLES

NPs consist of macromolecular materials and can be used as an adjuvant in vaccines or as drug carriers in which the active principle (drug or biologically active material) is dissolved, entrapped, and encapsulated and to which the active principle is adsorbed or attached (Soppimath et al., 2001). Natural polymers such as gelatin, albumin, chitosan, lectin, alginate, dextran, agarose, and synthetic polymers such as poly(e-caprolactone) PECL, poly(lactic acid), and polystyrene are used for scaffold synthesis, scaffold reinforcement (Yao et al., 2009; Agrawal et al., 2008), and stem cell signaling.

NPs are submicron (10–100 nm) sized materials produced via a number of methods depending on whether approached top–down or bottom–up. Attrition is a top–down method by which macro- or microscale particles are ground in a mill or another size reducing mechanism. Bottom–up approaches include gas phase fabrication (pyrolysis, inert gas condensation) and liquid phase fabrication (solvothermal reaction, sol–gel, and micellar structured media). NPs consist of lipid vesicles (liposomes) and polymeric vesicles (nanocapsules, nanocup, nanospheres), polymeric micelles (based on amphiphilic block copolymers), and dendrimers (highly branched and globular polymeric). Various methods, such as solvent evaporation, spontaneous emulsification, solvent diffusion, salting out/emulsification diffusion, use of supercritical CO_2, and polymerization have been used to prepare NPs (Soppimath et al., 2001). NPs have been one of the most promising devices for improving the delivery of bioactive agents in tissue engineering. NPs exist in the nanosize range, usually <100 nm and, due to their size and surface area, they can be exploited as vectors for the delivery of drugs, GFs, and genetic material (Wahajuddin, 2012). NPs are advantageous to use for gene delivery because of their small size and the specific electrical properties of the surface. The vehicles for GFs delivery take the physical form of porous scaffolds, microspheres, and micro- or nanocapsules and the release profile of a GF or bioactive factor can be adjusted through handling the physical and chemical properties, such as porosity, pore size, degree of crosslinking, and degradation rate (Langer and Tirrell, 2004). The most important NPs used in tissue engineering are liposomes and dendrimers

16.6 LIPOSOMES

Lipids are hydrophobic or amphiphilic small molecules and the most important lipid function in an organism is their role in the plasma membrane. Glycerophospholipids, or phospholipids, are key components of the lipid bilayer of cells (Fig. 16.11).

Phospholipids spontaneously self-assemble into ordered lyotropic liquid-crystalline phases in the presence of water. The formation of these lipid-based structures depends on phospholipid intrinsic factors, such as the nature and size of

FIGURE 16.11

Liposomes are spherical lipid bilayers with diameters ranging from the nanometer to the micrometer scale. Liposomes have been widely used as carriers to encapsulate and protect bioactive agents from the surrounding environments (Banerjee, 2001; Burgess et al., 2010).

the lipid headgroup and the length and degree of unsaturation of the acyl chains, as well as on the extrinsic factors, such as temperature, pH, concentration, and the presence of solutes and other lipids. Micelles are tiny droplets created when the hydrophilic heads of phospholipids face the water while the hydrophobic tails cluster together to "hide" from the water molecules. Phospholipids with compact and short tails tend to form micelles (Monteiro et al., 2014).

Nanoliposome is a technique for the encapsulation and delivery of bioactive agents.

Nanoliposomes are mainly composed of charged molecules, increasing the forces of repulsion or attraction between individual particles and other components within the surrounding environment as well as preventing their unwanted interactions with other molecules (Egorov, 2005). Coating the liposomes with hydrophilic polymers such as PEG changes their surface properties (Woodle and Lasic, 1992). The liposome formulations composed of phospholipids, Chol, and PEG are defined as "sterically stabilized liposomes," also called "stealth" liposomes or PEGylated liposomes. Liposomal formulations can be degraded by lysozymes within the cellular matrix prior to the release of the liposome's therapeutic contents to the cytoplasm (Dijkstra et al., 1984). PEGylated liposomal drug targets $\alpha5\beta1$ integrins and is intended to interact with receptors overexpressed at the surface of the target cells (Shroff and Kokkoli, 2012). This influences the correlation between the integrin and ECM which plays an important role in cell function. Liposomes are bilayer structural vesicles that self-assemble from amphiphilic phospholipids with a size from nanometers to several micrometers and can entrap hydrophilic and hydrophobic biomolecules.

16.7 DENDRIMERS

Dendrimers were first developed under the name cascade polymers by Buhleier et al. (Newkome et al., 1996) and they are synthetic, hyper-branched, tree-like 3D architectures, globular structure macromolecules. The chemical composition and molecular weight of dendrimers can be precisely controlled; hence, it is relatively easy to predict their biocompatibility and pharmacokinetics (Lee et al., 2005).

The permeability of polymeric drug carriers are different depending on their structural features such as hydrodynamic volume, molecular weight, molecular geometry, and charge (Tajarobi et al., 2001). As drug or gene delivery carriers, dendrimers have highly controllable structures, sizes, and terminal functional groups (Haensler and Szoka, 1993). They are higher chemical reactivity compared with other polymers (Frechet, 1994).

Polymers in scaffolds are often crosslinked to achieve the mechanical properties.

Dendrimers are good crosslinkers because they have highly controllable end groups and increase the crosslinking efficiency for fabricated scaffolds (Chan et al., 2008).

Dendrimers have also been used for fabricated dendrimer-peptide scaffolds that self-assemble into fibrillar collagen-like networks (Kinberger et al., 2002; Khew et al., 2008) and conjugated with YIGSR peptide to promote cell adhesion and stimulate nerve growth in corneal tissue (Duan et al., 2007). At the surface of the scaffold there are some electrical charges caused by synthetic or natural polymers in the tissue culture medium and dendrimers are a natural choice to exhibit a uniform charge density or to regularly display certain epitopes to foster the proper interactions between a scaffold and a specific cell-type and faster growth of neurons on polycationic surfaces (Joshi and Grinstaff, 2008). Biomolecules can also be directly incorporated into electrospun nanofibers. GF may be incorporated into the produced fibers during the electrospinning process by mixing the biomolecules into a polymer or by using coaxial electrospinning within the core of the nanofiber surrounded by a shell polymer (Liao et al., 2006). Unification of nanospheres, nanocapsules, or nanocups results in controlled release of the GFs and biomolecules which can be delivered at different times into the scaffold. Physicochemical characteristics of nanospheres, nanocapsules, or nanocups including the kinds of polymer and active molecular agents, molecular weight, and hydrophilicity or hydrophobicity as well as degradation mechanisms determine the controlled release of the GFs and biomolecules. However, matrix design and engineering can prevent cellular density. In addition, release of GFs to these cells and removal of residuals can be controlled based on properties of selected materials for the encapsulation of stem cells. Encapsulation is a cellular protection method and provides support cells in suspension. This also increases the surface-to-volume ratio to attach extracellular structures. Release of GF to these cells depend on the molecular structure of ECM.

Peptide-containing ECM may be involved in cell adhesion and differentiation. Cell binding to integrin generates different versions of ECM and recombinant protein fragments with specific binding to cells, which leads to more specific cell differentiation and generates the desired cell structure. Physical signals traveling from the microenvironment to stem cells change the cellular structure alignment by modifying microfilaments. This causes a change in focal adhesion as well as cellular differentiation.

In the human body, tissue damage often occurs, therefore, stem cells present in blood circulation receive special signals from the damaged tissue microenvironment (ECM, healthy, and damaged somatic cells) and migrate to that area while being differentiated to certain cells and special lineage of that tissue as to target tissue matrix base (Fig. 16.12).

In situ tissue engineering for organ reconstruction is performed by injecting a matrix into the organ or tissue including the polymer, stem cell niche and NPs (containing GFs or physiological agents) which permits the

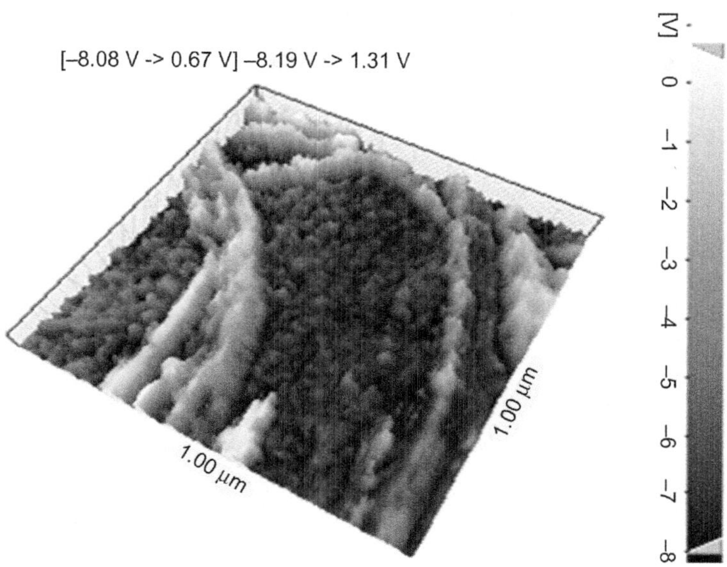

FIGURE 16.12

AFM image of injectable matrix comprising collagen. *AFM*, Atomic force microscopy.

migration of stem cells or tissue-specific progenitor cells into the matrix. GFs and physiological agents are released from NPs within the injected area of the organ or tissue, allowing proliferation and differentiation of the migrated stem cells under mechanical or electrical in vivo stress; whereby the organ or tissue is reconstructed (Ghanavi, 2014). Intravenous or direct injection of stem cells to target tissue significantly improves tissue performance (Gojo et al., 2003).

Therefore, bone marrow transplant is the only therapy using stem cell transplants. The chief reason for its success is the presence of many niches in the previous stem cell media. Therefore, injecting stem cells in the absence of real niches in lost organs and tissues resulted in the death of the transplanted cells and lack of controlling their fate. These scaffolds provide cellular adhesions for the survival of implanted cells as well as instigating the differentiation of stem cells to specialized lineages and providing a proper structure for the maturation of new tissues (Fig. 16.13).

By injecting hydrogels or self-assembling nanofiber hydrogels (either with or without stem cells) and NPs containing GFs and physiological agents, in situ engineered tissue is formed. Nanofiber matrices produced from electrospinning as well as separation of self-assembling and the solution phase provide fertile ground that can be used along with stem cell niches for expansion, differentiation, and transmission of stem and progenitor cells (Fig. 16.14).

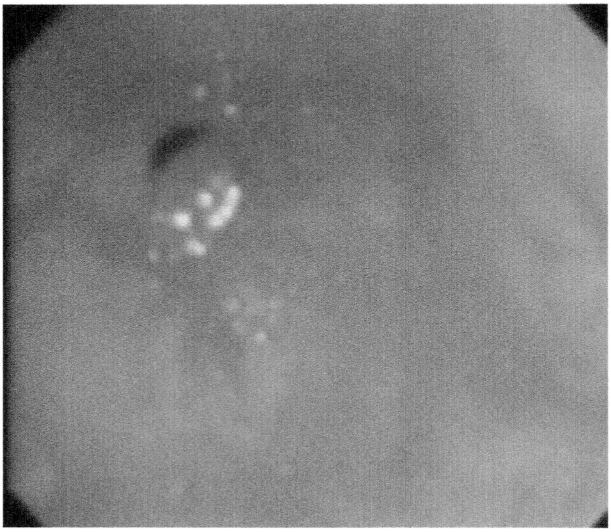

FIGURE 16.13

Bronchoscopic interior view of tracheal-stenosis.

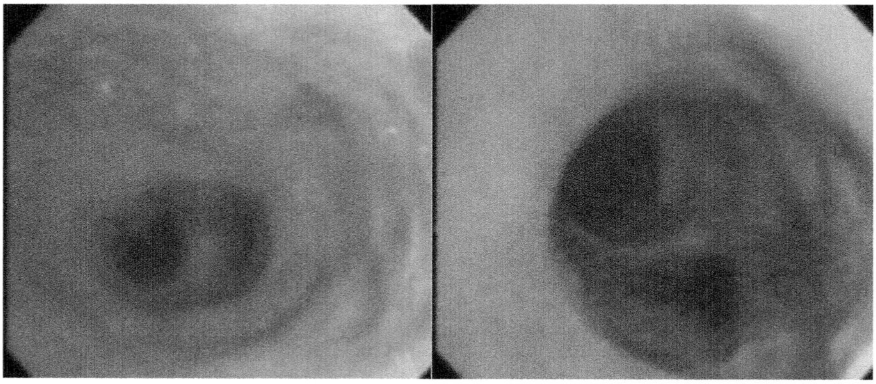

FIGURE 16.14

Bronchoscopic image of trachea 4 months after injection of the injectable matrix.

Nanotopographic features of the nanofiber matrices causes adhesion, migration, proliferation, and differentiation of stem cells while regenerating tissues at nanoscale, molecular message transmission, and cell–cell/cell–substrate interface. The homogenous biochemical and topographical structure at the formed nanofiber scale generated by electrospinning, solution phase separation, and self-assembling are effective methods in tissue engineering.

16.8 CONCLUSION

Cells live and interact in a nanoscale environment. Nano technology in tissue engineering are used for ECM modification, development of the adhesion site, and biochemical transport for temporary structural support until the cells create adequate matrix scaffolds. The nanotopographical architecture of the scaffold interacts with stem cells in the nano- and pico-Newton (pN) range to cause the regulated signaling of behaviors, morphology, adhesion, proliferation, clonal expansion, self-renewal, and differentiation. The stem cell niches regulating cell function and natural ECM signals such as RGD motifs cause cell "sense" that regulates protrusion–retraction oscillations. Top–down and bottom–up are two approaches for tissue engineering methods in nanotechnology. Top–down tissue engineering approaches require the fabrication of ECM scaffolds and assist the cells in finding their appropriate positions or the decellularization of an organ to provide the microarchitecture and microenvironmental for cell attachment. In bottom–up methods, cells and nano- or microscale elements such as particles and free ECM molecules interact or crosslink together to form the "alive 3D scaffold." Another bottom–up approach is the injectable nano-designed matrix to create a 3D scaffold in the area of the desired tissue or organ. Then blood circulatory stem cells or tissue-specific progenitor cells migrate to the area of 3D matrix for organ reconstruction.

ACKNOWLEDGMENT

The authors wish to thank Morteza Karimi for helping with the typography and drawing the illustrations.

REFERENCES

Abrams, G., Goodman, S., Nealey, P., Franco, M., Murphy, C.J., 2000. Nanoscale topography of the basement membrane underlying the corneal epithelium of the rhesus macaque. Cell Tissue Res. 299, 39–46.

Agrawal, S.K., Sanabria-Delong, N., Tew, G.N., Bhatia, S.R., 2008. Nanoparticle-reinforced associative network hydrogels. Langmuir 24, 13148–13154.

Arnold, M., Cavalcanti-Adam, E.A., Glass, R., Blümmel, J., Eck, W., Kantlehner, M., et al., 2004. Activation of integrin function by nanopatterned adhesive interfaces. ChemPhysChem 5, 383–388.

Bai, B., He, J., Li, Y.-S., Wang, X.-M., Ai, H.-J., Cui, F.-Z., 2013. Activation of the ERK1/2 signaling pathway during the osteogenic differentiation of mesenchymal stem cells cultured on substrates modified with various chemical groups. BioMed Res. Int. 1–15.

Balogun, V.A., Kirkwood, N.D., Mativenga, P.T., 2014. Direct electrical energy demand in fused deposition modelling. Procedia CIRP 15, 38–43.

Banerjee, R., 2001. Liposomes: applications in medicine. J. Biomater. Appl. 16, 3–21.

Bendrea, A.-D., Cianga, L., Cianga, I., 2011. Review paper: progress in the field of conducting polymers for tissue engineering applications. J. Biomater. Appl. 26, 3–84.

Berndt, P., Fields, G.B., Tirrell, M., 1995. Synthetic lipidation of peptides and amino acids: monolayer structure and properties. J. Am. Chem. Soc. 117, 9515–9522.

Berry, E., Brown, J., Connell, M., Craven, C., Efford, N., Radjenovic, A., et al., 1997. Preliminary experience with medical applications of rapid prototyping by selective laser sintering. Med. Eng. Phys. 19, 90–96.

Bershadsky, A.D., Balaban, N.Q., Geiger, B., 2003. Adhesion-dependent cell mechanosensitivity. Annu. Rev. Cell Dev. Biol. 19, 677–695.

Billiet, T., Vandenhaute, M., Schelfhout, J., Van vlierberghe, S., Dubruel, P., 2012. A review of trends and limitations in hydrogel-rapid prototyping for tissue engineering. Biomaterials 33, 6020–6041.

Blacher, S., Calberg, C., Kerckhofs, G., Léonard, A., Wevers, M., Jérôme, R., et al., 2007. The porous structure of biodegradable scaffolds obtained with supercritical CO_2 as foaming agent. Stud. Surf. Sci. Catal. 160, 681–688.

Brivanlou, A.H., Gage, F.H., Jaenisch, R., Jessell, T., Melton, D., Rossant, J., 2003. Setting standards for human embryonic stem cells. Science 300, 913–916.

Brun-Graeppi, A.K.A.S., Richard, C., Bessodes, M., Scherman, D., Merten, O.-W., 2011. Cell microcarriers and microcapsules of stimuli-responsive polymers. J. Controlled Release 149, 209–224.

Burgess, P., Hutt, P.B., Farokhzad, O.C., Langer, R., Minick, S., Zale, S., 2010. On firm ground: IP protection of therapeutic nanoparticles. Nat. Biotechnol. 28, 1267.

Capadona, J.R., Collard, D.M., García, A.J., 2003. Fibronectin adsorption and cell adhesion to mixed monolayers of tri(ethylene glycol)-and methyl-terminated alkanethiols. Langmuir 19, 1847–1852.

Carrick, L., Tassieri, M., Waigh, T., Aggeli, A., Boden, N., Bell, C., et al., 2005. The internal dynamic modes of charged self-assembled peptide fibrils. Langmuir 21, 3733–3737.

Cavalcanti-Adam, E.A., Volberg, T., Micoulet, A., Kessler, H., Geiger, B., Spatz, J.P., 2007. Cell spreading and focal adhesion dynamics are regulated by spacing of integrin ligands. Biophys. J. 92, 2964–2974.

Cavalcanti-Adam, E.A., Aydin, D., Hirschfeld-Warneken, V.C., Spatz, J.P., 2008. Cell adhesion and response to synthetic nanopatterned environments by steering receptor clustering and spatial location. HFSP J. 2, 276–285.

Chan, J.C., Burugapalli, K., Naik, H., Kelly, J.L., Pandit, A., 2008. Amine functionalization of cholecyst-derived extracellular matrix with generation 1 PAMAM dendrimer. Biomacromolecules 9, 528–536.

Chen, W., Villa-Diaz, L.G., Sun, Y., Weng, S., Kim, J.K., Lam, R.H., et al., 2012. Nanotopography influences adhesion, spreading, and self-renewal of human embryonic stem cells. ACS Nano 6, 4094–4103.

Chia, H.N., Wu, B.M., 2015. Recent advances in 3D printing of biomaterials. J. Biol. Eng. 9, 1.

Chou, S.Y., Krauss, P.R., Renstrom, P.J., 1996. Imprint lithography with 25-nanometer resolution. Science 272, 85.

Clemente, C.F., Corat, M.A., Saad, S.T., Franchini, K.G., 2005. Differentiation of C2C12 myoblasts is critically regulated by FAK signaling. Am. J. Physiol. 289, R862–R870.

Coburn, J.M., Gibson, M., Monagle, S., Patterson, Z., Elisseeff, J.H., 2012. Bioinspired nanofibers support chondrogenesis for articular cartilage repair. Proc. Natl. Acad. Sci. U.S.A. 109, 10012–10017.

Craighead, H., James, C., Turner, A., 2001. Chemical and topographical patterning for directed cell attachment. Curr. Opin. Solid State Mater. Sci. 5, 177–184.

Dalton, P.D., Joergensen, N.T., Groll, J., Moeller, M., 2008. Patterned melt electrospun substrates for tissue engineering. Biomed. Mater. 3, 034109.

Decher, G., 1997. Fuzzy nanoassemblies: toward layered polymeric multicomposites. Science 277, 1232–1237.

Deckard, C.R., Mcclure, P., 1989. Method and apparatus for producing parts by selective sintering. Google Patents.

Dembo, M., Wang, Y.-L., 1999. Stresses at the cell-to-substrate interface during locomotion of fibroblasts. Biophys. J. 76, 2307–2316.

Desai, M.P., Labhasetwar, V., Walter, E., Levy, R.J., Amidon, G.L., 1997. The mechanism of uptake of biodegradable microparticles in Caco-2 cells is size dependent. Pharm. Res. 14, 1568–1573.

Deyle, D.R., 2015. Generation of induced pluripotent stem cells. Osteoporos Osteoarthritis 43–58.

Dhariwala, B., Hunt, E., Boland, T., 2004. Rapid prototyping of tissue-engineering constructs, using photopolymerizable hydrogels and stereolithography. Tissue Eng. 10, 1316–1322.

Dijkstra, J., Van galen, W., Hulstaert, C., Kalicharan, D., Roerdink, F., Scherphof, G., 1984. Interaction of liposomes with Kupffer cells in vitro. Exp. Cell Res. 150, 161–176.

Discher, D.E., Janmey, P., Wang, Y.-L., 2005. Tissue cells feel and respond to the stiffness of their substrate. Science 310, 1139–1143.

Döbereiner, H.-G., Dubin-Thaler, B.J., Giannone, G., Sheetz, M.P., 2005. Force sensing and generation in cell phases: analyses of complex functions. J. Appl. Physiol. 98, 1542–1546.

Doshi, J., Reneker, D.H., 1993. Electrospinning process and applications of electrospun fibers. In: Industry Applications Society Annual Meeting, Conference Record of the 1993 IEEE, 1993. IEEE, pp. 1698–1703.

Duan, X., Mclaughlin, C., Griffith, M., Sheardown, H., 2007. Biofunctionalization of collagen for improved biological response: scaffolds for corneal tissue engineering. Biomaterials 28, 78–88.

Dunn, J.C., Chan, W.-Y., Cristini, V., Kim, J., Lowengrub, J., Singh, S., et al., 2006. Analysis of cell growth in three-dimensional scaffolds. Tissue Eng. 12, 705–716.

Egorov, S., 2005. Interactions between nanoparticles in supercritical fluids: from repulsion to attraction. Phys. Rev. E 72, 010401.

Emsley, J., Knight, C.G., Farndale, R.W., Barnes, M.J., 2004. Structure of the integrin $\alpha 2\beta 1$-binding collagen peptide. J. Mol. Biol. 335, 1019–1028.

Forward, K.M., Rutledge, G.C., 2012. Free surface electrospinning from a wire electrode. Chem. Eng. J. 183, 492–503.

Fracassi, F., Coburn, J., 1988. Plasma-assisted etching of tungsten films: a quartz-crystal microbalance study. J. Appl. Phys. 63, 1758–1761.

Frechet, J.M., 1994. Functional polymers and dendrimers: reactivity, molecular architecture, and interfacial energy. Science 263, 1710–1714.

Fuchs, E., Tumbar, T., Guasch, G., 2004. Socializing with the neighbors: stem cells and their niche. Cell 116, 769–778.

Gates, B.D., Xu, Q., Stewart, M., Ryan, D., Willson, C.G., Whitesides, G.M., 2005. New approaches to nanofabrication: molding, printing, and other techniques. Chem. Rev. 105, 1171–1196.

Gattazzo, F., Urciuolo, A., Bonaldo, P., 2014. Extracellular matrix: a dynamic microenvironment for stem cell niche. Biochim. Biophys. Acta, Gen. Subj. 1840, 2506–2519.

Ghanavi, J., 2014. Injectable Matrix Having a Polymer and a Stem Cell Niche Composed of Cup-Shaped Nanoparticles Containing Growth Factors or Physiological Agents for Organ Reconstruction. Google Patents.

Gojo, S., Gojo, N., Takeda, Y., Mori, T., Abe, H., Kyo, S., et al., 2003. In vivo cardiovasculogenesis by direct injection of isolated adult mesenchymal stem cells. Exp. Cell Res. 288, 51–59.

Graf, J., Ogle, R.C., Robey, F.A., Sasaki, M., Martin, G.R., Yamada, Y., et al., 1987. A pentapeptide from the laminin B1 chain mediates cell adhesion and binds to 67000 laminin receptor. Biochemistry 26, 6896–6900.

Gurdon, J., Bourillot, P.-Y., 2001. Morphogen gradient interpretation. Nature 413, 797–803.

Haensler, J., Szoka Jr, F.C., 1993. Polyamidoamine cascade polymers mediate efficient transfection of cells in culture. Bioconjugate Chem. 4, 372–379.

Han, D.C., Shen, T.-L., Guan, J.-L., 2000. Role of Grb7 targeting to focal contacts and its phosphorylation by focal adhesion kinase in regulation of cell migration. J. Biol. Chem. 275, 28911–28917.

Harris, L.D., Kim, B.-S., Mooney, D.J., 1998. Open pore biodegradable matrices formed with gas foaming. J. Biomed. Mater. Res. 42, 396–402.

Hartgerink, J.D., Beniash, E., Stupp, S.I., 2001. Self-assembly and mineralization of peptide-amphiphile nanofibers. Science 294, 1684–1688.

Hasse, C., Klöck, G., Schlosser, A., Zimmermann, U., Rothmund, M., 1997. Parathyroid allotransplantation without immunosuppression. Lancet 350, 1296–1297.

Hayashida, T., 2010. Integrins modulate cellular fibrogenesis at multiple levels: regulation of TGF-β signaling. Endocr Metab Immune Disord Drug Targets 10, 302–319.

Higuchi, A., Ling, Q.-D., Ko, Y.-A., Chang, Y., Umezawa, A., 2011. Biomaterials for the feeder-free culture of human embryonic stem cells and induced pluripotent stem cells. Chem. Rev. 111, 3021–3035.

Hollister, S.J., 2005. Porous scaffold design for tissue engineering. Nat. Mater. 4, 518–524.

Hoon, J.L., Tan, M.H., Koh, C.-G., 2016. The regulation of cellular responses to mechanical cues by rho GTPases. Cells 5, 17.

Huh, K.M., Baek, N., Park, K., 2005. Enhanced swelling rate of poly (ethylene glycol)-grafted superporous hydrogels. J. Bioact. Compatible Polym. 20, 231–243.

Hynes, R.O., 2002. Integrins: bidirectional, allosteric signaling machines. Cell 110, 673–687.

Jakab, K., Norotte, C., Damon, B., Marga, F., Neagu, A., Besch-Williford, C.L., et al., 2008. Tissue engineering by self-assembly of cells printed into topologically defined structures. Tissue Eng., A 14, 413–421.

Jakab, K., Norotte, C., Marga, F., Murphy, K., Vunjak-Novakovic, G., Forgacs, G., 2010. Tissue engineering by self-assembly and bio-printing of living cells. Biofabrication 2, 022001.

Jiang, G., Den hertog, J., Hunter, T., 2000. Receptor-like protein tyrosine phosphatase α homodimerizes on the cell surface. Mol. Cell. Biol. 20, 5917–5929.

Jin, G., Prabhakaran, M.P., Ramakrishna, S., 2011. Stem cell differentiation to epidermal lineages on electrospun nanofibrous substrates for skin tissue engineering. Acta Biomater. 7, 3113–3122.

Joshi, N., Grinstaff, M., 2008. Applications of dendrimers in tissue engineering. Curr. Top. Med. Chem. 8, 1225–1236.

Kang, H.-W., Tabata, Y., Ikada, Y., 1999. Fabrication of porous gelatin scaffolds for tissue engineering. Biomaterials 20, 1339–1344.

Kaul, H., Ventikos, Y., 2015. Dynamic reciprocity revisited. J. Theor. Biol. 370, 205–208.

Kay, R.R., Thompson, C.R., 2009. Forming patterns in development without morphogen gradients: scattered differentiation and sorting out. Cold Spring Harbor Perspect. Biol. 1, a001503.

Keenan, T.M., Folch, A., 2008. Biomolecular gradients in cell culture systems. Lab Chip 8, 34–57.

Keselowsky, B.G., Collard, D.M., García, A.J., 2003. Surface chemistry modulates fibronectin conformation and directs integrin binding and specificity to control cell adhesion. J. Biomed. Mater. Res., A 66, 247–259.

Khew, S.T., Yang, Q.J., Tong, Y.W., 2008. Enzymatically crosslinked collagen-mimetic dendrimers that promote integrin-targeted cell adhesion. Biomaterials 29, 3034–3045.

Kijeńska, E., Prabhakaran, M.P., Swieszkowski, W., Kurzydlowski, K.J., Ramakrishna, S., 2012. Electrospun bio-composite P (LLA-CL)/collagen I/collagen III scaffolds for nerve tissue engineering. J. Biomed. Mater. Res., B: Appl. Biomater. 100, 1093–1102.

Kim, L., Toh, Y.-C., Voldman, J., Yu, H., 2007. A practical guide to microfluidic perfusion culture of adherent mammalian cells. Lab Chip 7, 681–694.

Kinberger, G.A., Cai, W., Goodman, M., 2002. Collagen mimetic dendrimers. J. Am. Chem. Soc. 124, 15162–15163.

Kingham, E., Oreffo, R.O., 2013. Embryonic and induced pluripotent stem cells: understanding, creating, and exploiting the nano-niche for regenerative medicine. ACS Nano 7, 1867–1881.

Kozlovskaya, V., Zavgorodnya, O., Chen, Y., Ellis, K., Tse, H.M., Cui, W., et al., 2012. Ultrathin polymeric coatings based on hydrogen-bonded polyphenol for protection of pancreatic islet cells. Adv. Funct. Mater. 22, 3389–3398.

Kurtz, A., Oh, S.-J., 2012. Age related changes of the extracellular matrix and stem cell maintenance. Preventive Med. 54, S50–S56.

Langer, R., Tirrell, D.A., 2004. Designing materials for biology and medicine. Nature 428, 487–492.

Lawrence, P.A., Struhl, G., 1996. Morphogens, compartments, and pattern: lessons from drosophila? Cell 85, 951–961.

Lee, M.H., Ducheyne, P., Lynch, L., Boettiger, D., Composto, R.J., 2006. Effect of biomaterial surface properties on fibronectin–α5β1 integrin interaction and cellular attachment. Biomaterials 27, 1907–1916.

Lee, S.B., Kim, Y.H., Chong, M.S., Hong, S.H., Lee, Y.M., 2005. Study of gelatin-containing artificial skin V: fabrication of gelatin scaffolds using a salt-leaching method. Biomaterials 26, 1961–1968.

Lee, S.H., Kim, B.S., Kim, S.H., Kang, S.W., Kim, Y.H., 2004. Thermally produced biodegradable scaffolds for cartilage tissue engineering. Macromol. Biosci. 4, 802–810.

Li, D., Wang, Y., Xia, Y., 2003. Electrospinning of polymeric and ceramic nanofibers as uniaxially aligned arrays. Nano Lett. 3, 1167–1171.

Liao, I., Chew, S., Leong, K., 2006. Aligned core–shell nanofibers delivering bioactive proteins. Nanomedicine 1, 465–471.

Lin, C.-C., Metters, A.T., 2006. Hydrogels in controlled release formulations: network design and mathematical modeling. Adv. Drug Delivery Rev. 58, 1379–1408.

Lips, P., Velthoen, I., Dijkstra, P., Wessling, M., Feijen, J., 2005. Gas foaming of segmented poly (ester amide) films. Polymer 46, 9396–9403.

Liu, J.S., Gartner, Z.J., 2012. Directing the assembly of spatially organized multicomponent tissues from the bottom up. Trends Cell Biol. 22, 683–691.

Livoti, C.M., Morgan, J.R., 2010. Self-assembly and tissue fusion of toroid-shaped minimal building units. Tissue Eng., A 16, 2051–2061.

Lo, C.-M., Wang, H.-B., Dembo, M., Wang, Y.-L., 2000. Cell movement is guided by the rigidity of the substrate. Biophys. J. 79, 144–152.

Loh, Q.L., Choong, C., 2013. Three-dimensional scaffolds for tissue engineering applications: role of porosity and pore size. Tissue Eng., B: Rev. 19, 485–502.

Ma, P.X., Langer, R., 1999. Fabrication of biodegradable polymer foams for cell transplantation and tissue engineering. Tissue Eng. Methods Protoc. 47–56.

Matsuda, T., Mizutani, M., Arnold, S.C., 2000. Molecular design of photocurable liquid biodegradable copolymers. 1. Synthesis and photocuring characteristics. Macromolecules 33, 795–800.

McMurray, R.J., Gadegaard, N., Tsimbouri, P.M., Burgess, K.V., McNamara, L.E., Tare, R., et al., 2011. Nanoscale surfaces for the long-term maintenance of mesenchymal stem cell phenotype and multipotency. Nat. Mater. 10, 637–644.

Melocchi, A., Parietti, F., Maroni, A., Foppoli, A., Gazzaniga, A., Zema, L., 2016. Hot-melt extruded filaments based on pharmaceutical grade polymers for 3D printing by Fused Deposition Modeling. Int. J. Pharm.

Mitra, S.K., Hanson, D.A., Schlaepfer, D.D., 2005. Focal adhesion kinase: in command and control of cell motility. Nat. Rev. Mol. Cell Biol. 6, 56–68.

Monteiro, N., Martins, A., Reis, R.L., Neves, N.M., 2014. Liposomes in tissue engineering and regenerative medicine. J. R. Soc. Interface 11, 20140459.

Nam, Y.S., Park, T.G., 1999. Porous biodegradable polymeric scaffolds prepared by thermally induced phase separation. J. Biomed. Mater. Res. 47, 8–17.

Nam, Y.S., Yoon, J.J., Park, T.G., 2000. A novel fabrication method of macroporous biodegradable polymer scaffolds using gas foaming salt as a porogen additive. J. Biomed. Mater. Res. 53, 1–7.

Newkome, G.R., Moorefield, C.N., Vögtle, F., 1996. Nomenclature. Dendritic Mol.: Concepts Synth. Perspect. 37–47.

Nichol, J.W., Khademhosseini, A., 2009. Modular tissue engineering: engineering biological tissues from the bottom up. Soft Matter 5, 1312–1319.

Nyberg, E., Holmes, C., Witham, T., Grayson, W.L., 2016. Growth factor-eluting technologies for bone tissue engineering. Drug Deliv. Transl. Res. 6, 184–194.

Ott, H.C., Matthiesen, T.S., Goh, S.-K., Black, L.D., Kren, S.M., Netoff, T.I., et al., 2008. Perfusion-decellularized matrix: using nature's platform to engineer a bioartificial heart. Nat. Med. 14, 213–221.

Parenti, A., Halbisen, M.A., Wang, K., Latham, K., Ralston, A., 2016. OSKM Induce Extraembryonic Endoderm Stem Cells in Parallel to Induced Pluripotent Stem Cells. Stem Cell Rep. 6, 447–455.

Park, J., Bauer, S., Von Dermark, K., Schmuki, P., 2007a. Nanosize and vitality: TiO_2 nanotube diameter directs cell fate. Nano Lett. 7, 1686–1691.

Park, J.S., Woo, D.G., Sun, B.K., Chung, H.-M., Im, S.J., Choi, Y.M., et al., 2007b. In vitro and in vivo test of PEG/PCL-based hydrogel scaffold for cell delivery application. J. Controlled Release 124, 51–59.

Park, Y., Lutolf, M.P., Hubbell, J.A., Hunziker, E.B., Wong, M., 2004. Bovine primary chondrocyte culture in synthetic matrix metalloproteinase-sensitive poly(ethylene glycol)-based hydrogels as a scaffold for cartilage repair. Tissue Eng. 10, 515–522.

Paul, B., Baskaran, S., 1996. Issues in fabricating manufacturing tooling using powder-based additive freeform fabrication. J. Mater. Process. Technol. 61, 168–172.

Provenzano, P.P., Keely, P.J., 2011. Mechanical signaling through the cytoskeleton regulates cell proliferation by coordinated focal adhesion and Rho GTPase signaling. J. Cell Sci. 124, 1195–1205.

Qu, Z., Yan, J., Li, B., Zhuang, J., Huang, Y., 2010. Improving bone marrow stromal cell attachment on chitosan/hydroxyapatite scaffolds by an immobilized RGD peptide. Biomed. Mater. 5, 065001.

Quirk, R.A., France, R.M., Shakesheff, K.M., Howdle, S.M., 2004. Supercritical fluid technologies and tissue engineering scaffolds. Curr. Opin. Solid State Mater. Sci. 8, 313–321.

Ramay, H.R., Zhang, M., 2003. Preparation of porous hydroxyapatite scaffolds by combination of the gel-casting and polymer sponge methods. Biomaterials 24, 3293–3302.

Raoufi, M., Aslankoohi, N., Mollenhauer, C., Boehm, H., Spatz, J.P., Brüggemann, D., 2016. Template-assisted extrusion of biopolymer nanofibers under physiological conditions. Integr. Biol. 8, 1059–1066.

Reubinoff, B.E., Itsykson, P., Turetsky, T., Pera, M.F., Reinhartz, E., Itzik, A., et al., 2001. Neural progenitors from human embryonic stem cells. Nat. Biotechnol. 19, 1134–1140.

Rutherford, R.B., Gu, K., Racenis, P., Krebsbach, P.H., 2003. Early events: the in vitro conversion of BMP transduced fibroblasts to chondroblasts. Connect. Tissue Res. 44, 117–123.

Salinas, C.N., Cole, B.B., Kasko, A.M., Anseth, K.S., 2007. Chondrogenic differentiation potential of human mesenchymal stem cells photoencapsulated within poly (ethylene glycol)-arginine-glycine-aspartic acid-serine thiol-methacrylate mixed-mode networks. Tissue Eng. 13, 1025–1034.

Santos, E., Hernández, R.M., Pedraz, J.L., Orive, G., 2012. Novel advances in the design of three-dimensional bio-scaffolds to control cell fate: translation from 2D to 3D. Trends Biotechnol. 30, 331–341.

Sargeant, T.D., Guler, M.O., Oppenheimer, S.M., Mata, A., Satcher, R.L., Dunand, D.C., et al., 2008. Hybrid bone implants: self-assembly of peptide amphiphile nanofibers within porous titanium. Biomaterials 29, 161–171.

Scadden, D.T., 2006. The stem-cell niche as an entity of action. Nature 441, 1075–1079.

Schliephake, H., 2010. Application of bone growth factors—the potential of different carrier systems. Oral Maxillof. Surg. 14, 17–22.

Schmidt, C.E., Horwitz, A.F., Lauffenburger, D.A., Sheetz, M.P., 1993. Integrin-cytoskeletal interactions in migrating fibroblasts are dynamic, asymmetric, and regulated. J. Cell Biol. 123, 977–991.

Schneider, I.C., Haugh, J.M., 2006. Mechanisms of gradient sensing and chemotaxis: conserved pathways, diverse regulation. Cell Cycle 5, 1130–1134.

Schofield, R., 1977. The relationship between the spleen colony-forming cell and the haemopoietic stem cell. Blood Cells 4, 7–25.

Shroff, K., Kokkoli, E., 2012. PEGylated liposomal doxorubicin targeted to $\alpha5\beta1$-expressing MDA-MB-231 breast cancer cells. Langmuir 28, 4729–4736.

Skoog, S.A., Goering, P.L., Narayan, R.J., 2014. Stereolithography in tissue engineering. J. Mater. Sci.: Mater. Med. 25, 845–856.

Smith, C.M., Christian, J.J., Warren, W.L., Williams, S.K., 2007. Characterizing environmental factors that impact the viability of tissue-engineered constructs fabricated by a direct-write bioassembly tool. Tissue Eng. 13, 373–383.

Smith, L.A., Liu, X., Ma, P.X., 2008. Tissue engineering with nano-fibrous scaffolds. Soft Matter 4, 2144–2149.

Soppimath, K.S., Aminabhavi, T.M., Kulkarni, A.R., Rudzinski, W.E., 2001. Biodegradable polymeric nanoparticles as drug delivery devices. J. Controlled Release 70, 1–20.

Suh, S.W., Shin, J.Y., Kim, J., Kim, J., Beak, C.H., Kim, D.-I., et al., 2002. Effect of different particles on cell proliferation in polymer scaffolds using a solvent-casting and particulate leaching technique. ASAIO J. 48, 460–464.

Tai, H., Mather, M.L., Howard, D., Wang, W., White, L.J., Crowe, J.A., et al., 2007. Control of pore size and structure of tissue engineering scaffolds produced by supercritical fluid processing. Eur. Cell Mater. 14, 64–77.

Tajarobi, F., El-Sayed, M., Rege, B., Polli, J., Ghandehari, H., 2001. Transport of poly amidoamine dendrimers across Madin–Darby canine kidney cells. Int. J. Pharm. 215, 263–267.

Takagi, J., Strokovich, K., Springer, T.A., Walz, T., 2003. Structure of integrin $\alpha5\beta1$ in complex with fibronectin. EMBO J. 22, 4607–4615.

Teixeira, A.I., Mckie, G.A., Foley, J.D., Bertics, P.J., Nealey, P.F., Murphy, C.J., 2006. The effect of environmental factors on the response of human corneal epithelial cells to nanoscale substrate topography. Biomaterials 27, 3945–3954.

Thomson, R., Wake, M., Yaszemski, M.J., Mikos, A., 1995. Biodegradable polymer scaffolds to regenerate organs. Biopolymers II. Springer.

Thorne, J.T., Segal, T.R., Chang, S., Jorge, S., Segars, J.H., Leppert, P.C., 2015. Dynamic reciprocity between cells and their microenvironment in reproduction. Biol. Reprod. 92, 25.

Tian, L., George, S.C., 2011. Biomaterials to prevascularize engineered tissues. J. Cardiovasc. Transl. Res. 4, 685–698.

Vieu, C., Carcenac, F., Pepin, A., Chen, Y., Mejias, M., Lebib, A., et al., 2000. Electron beam lithography: resolution limits and applications. Appl. Surf. Sci. 164, 111–117.

Vitillo, L., Baxter, M., Iskender, B., Whiting, P., Kimber, S.J., 2016. Integrin-associated focal adhesion kinase protects human embryonic stem cells from apoptosis, detachment, and differentiation. Stem Cell Rep. 7, 167–176.

Vogel, V., Sheetz, M., 2006. Local force and geometry sensing regulate cell functions. Nat. Rev. Mol. Cell Biol. 7, 265–275.

Wahajuddin, S.A., 2012. Superparamagnetic iron oxide nanoparticles: magnetic nanoplatforms as drug carriers. Int. J. Nanomed. 7, 3445.

Weiss, P., Garber, B., 1952. Shape and movement of mesenchyme cells as functions of the physical structure of the medium contributions to a quantitative morphology. Proc. Natl. Acad. Sci. U.S.A. 38, 264–280.

Weissman, I.L., 2000. Stem cells: units of development, units of regeneration, and units in evolution. cell 100, 157–168.

Whitesides, G.M., Mathias, J.P., Seto, C.T. 1991. Molecular Self-Assembly and Nanochemistry: A Chemical Strategy for the Synthesis of Nanostructures. DTIC Document.

Williams, J.M., Adewunmi, A., Schek, R.M., Flanagan, C.L., Krebsbach, P.H., Feinberg, S. E., et al., 2005. Bone tissue engineering using polycaprolactone scaffolds fabricated via selective laser sintering. Biomaterials 26, 4817–4827.

Wilson, J.L., Mcdevitt, T.C., 2013. Stem cell microencapsulation for phenotypic control, bioprocessing, and transplantation. Biotechnol. Bioeng. 110, 667–682.

Wolfenson, H., Lavelin, I., Geiger, B., 2013. Dynamic regulation of the structure and functions of integrin adhesions. Dev. Cell 24, 447–458.

Wong, A.K.-K., 2001. Resolution Enhancement Techniques in Optical Lithography. SPIE Press.

Woodle, M.C., Lasic, D.D., 1992. Sterically stabilized liposomes. Biochim. Biophys. Acta, Rev. Biomembr. 1113, 171–199.

Wu, C., 2007. Focal adhesion: a focal point in current cell biology and molecular medicine. Cell Adhes. Migration 1, 13–18.

Wu, C., Luo, Y., Cuniberti, G., Xiao, Y., Gelinsky, M., 2011. Three-dimensional printing of hierarchical and tough mesoporous bioactive glass scaffolds with a controllable pore architecture, excellent mechanical strength and mineralization ability. Acta Biomater. 7, 2644–2650.

Xia, Y., Whitesides, G.M., 1998. Soft lithography. Annu. Rev. Mater. Sci. 28, 153–184.

Yang, D., Lu, B., Zhao, Y., Jiang, X., 2007. Fabrication of aligned fibrous arrays by magnetic electrospinning. Adv. Mater. 19, 3702–3706.

Yang, S.M., Jang, S.G., Choi, D.G., Kim, S., Yu, H.K., 2006. Nanomachining by colloidal lithography. Small 2, 458–475.

Yao, X., Yao, H., Li, Y., Chen, G., 2009. Preparation of honeycomb scaffold with hierarchical porous structures by core-crosslinked core–corona nanoparticles. J. Colloid Interface Sci. 332, 165–172.

Zaidel-Bar, R., Geiger, B., 2010. The switchable integrin adhesome. J. Cell Sci. 123, 1385–1388.

Zein, I., Hutmacher, D.W., Tan, K.C., Teoh, S.H., 2002. Fused deposition modeling of novel scaffold architectures for tissue engineering applications. Biomaterials 23, 1169–1185.

Zhang, D., Chang, J., 2007. Patterning of electrospun fibers using electroconductive templates. Adv. Mater. 19, 3664–3667.

Zhang, L., Webster, T.J., 2009. Nanotechnology and nanomaterials: promises for improved tissue regeneration. Nano Today 4, 66–80.

Zhang, S., 2003. Building from the bottom up. Mater. Today 6, 20–27.

Zhao, Q., Li, H., Li, B., 2011. Nanoencapsulating living biological cells using electrostatic layer-by-layer self-assembly: Platelets as a model. J. Mater. Res. 26, 347–351.

CHAPTER

Polymer-based calcium phosphate scaffolds for tissue engineering applications

17

K. Thanigai Arul[1,*], E. Manikandan[2,3,*] and Rasiah Ladchumananandasivam[4]

[1]Department of Physics, AMET (Deemed to be University), Kanathur, Chennai, India [2]Department of Physics, TUCAS Campus, Thiruvalluvar University, Vellore, India [3]UNESCO UNISA Africa Chair in Nanosciences & Nanotechnology, College of Graduate Studies, University of South Africa, Pretoria, South Africa [4]Post Graduate Programme in Mechanical Engineering, Centre of Technology, Campus Universitário, Natal, Brazil

17.1 INTRODUCTION

For the past 10 years, all the tissues and organs of the human body have been regenerated with the application of tissue engineering. The primary motive is to repair and to restore affected tissue functions using cells, scaffolds and growth factors for treatments. The approach of tissue engineering provides to flourishing new treatments for many diseases. It aims to improve tissue functions which are defective due to pathological conditions. Generally, tissue engineering can be divided into three groups: (1) cell substitutes into the organism; (2) tissue accelerate; and (3) within different matrices. Bone tissue engineering seeks to regenerate bone defects through combining cells, biocompatible scaffolds, and growth factors. Scaffolds with desired biocompatibility, biodegradability, and mechanical properties have been of a great interest to many scientists (Langer and Vacanti, 1993; Arafat et al., 2011; Hutmacher, 2000; Ma, 2004; Hutmacher and Cool, 2007). Recent experimental and clinical results are showed that the high success rate of tissue engineering approaches which enhanced trust perhaps due to their high integration. Next generation scaffolds provide good mechanical and structural properties to control cell attachment, migration, proliferation, and differentiation. The functions of scaffolds are to deliver biological signals and to guide cell functions. A perfect scaffold acquires a three-dimensional (3D) interconnected pore network with microstructure and mechanical strength analogous to those of natural tissues in terms of biocompatibility and resorbability and release of exact bioactive factors for increasing the regeneration process. The mechanical

*Corresponding authors email ids: thanigaiarul.k@gmail;maniphysics@gmail.com

Nanoarchitectonics in Biomedicine. DOI: https://doi.org/10.1016/B978-0-12-816200-2.00011-6

properties were mismatched between bone and scaffolds which inversely affect the implant function in vivo. Scaffolds with weak mechanical strength may fail under load-bearing applications; while scaffolds with higher mechanical strength to that of the bone may cause stress shielding, bone resorption, and weak osseointegration. Another crucial parameter that should be considered in synthetic bone scaffolds is porosity. The range of pore sizes between 100 and 500 mm was required for neovascularization, cell relocation, and the release of nutrients. Hydroxyapatite is one of the most excellent materials for developing bone scaffolds which related to tissue engineering because of its distinctive properties, such as outstanding biocompatibility, osteoconductivity, and chemical content comparable to that of natural bone (Xie et al., 2012, 2013; Kim et al., 2006; Yoshimoto et al., 2003; Loh et al., 2015; Su et al., 2015; Loh et al., 2016).

Biphasic calcium phosphate (BCP) bioceramics have composed of HAp and beta-tricalcium phosphate (TCP) components which fascinated broadly due to their applications in replacement of bone (Macchetta et al., 2009). Beta-TCP has enormous biocompatibility and biodegradability which resorbs easily than hydroxyapatite. Improved bioactivity, osteoconductivity, and controlled degradation rate in BCP ceramics have been further more encouraging properties than hydroxyapatite or beta-TCP in order to repair periodontal defects (Wang et al., 2006; Yamada et al., 1997). However, similar to other calcium phosphate ceramics, their potency is not satisfactory for load bearing. Biodegradable polymer composites with special calcium phosphates are well-known for their appropriateness as bone engineering scaffolds largely due to the hydroxyapatite and other calcium phosphates compounds (Rezwan et al., 2006; Abdal-hay et al., 2014; Wu et al., 2014). In spite of many advantages, the polymeric-ceramics have weak load bearing application in bone replacement (Waizy et al., 2013; Witte, 2015). Magnesium and its alloys are biocompatible and have more suitable mechanical strength together with being closer to that of bone strength (Zhang et al., 2009; Gu and Zheng, 2010). Moreover, magnesium is an important mineral element present in human body which having ability to stimulate the formation of new bone (Hoppe et al., 2011; Diba et al., 2014). Most of the bone implants require to have a good bone bonding ability. The bioactivity of a material can be improved by the addition of calcium phosphate, but it was also proven that magnesium alloys achieve excellent osseointegration (Castellani et al., 2011). Modification of ceramics with magnesium improve bioactivity and stimulate vascularization. Magnesium alloy/metal matrix composites were improved with incorporation of hydroxyapatite particles which were responsible for enhanced corrosion resistance of composites and could be employed for orthopedic applications (Cifuentes et al., 2012) and, recently, it was found that the addition of magnesium improves osseosintegration and further decreases inflammation issues (Cifuentes et al., 2016). A novel composite system for biodegradable intramedullary nails was prepared with polylactide and toughened with magnesium alloy wires and carbon fibers (Morawska-Chochol et al., 2014; Morawska-Chochół et al., 2014). Poly-L-lactide and its composite matrix are biocompatible and broadly used in load-bearing applications (Morawska-Chochół et al., 2015; Rokkanen et al., 1999). It is a semi-crystalline polymer of moderately high mechanical strength; however, there was no chance of using it for load-bearing (e.g., intramedullary nails). Magnesium alloy wires were employed and reinforced in a polylactide matrix, while the rapid degradation of magnesium protected by polymer matrix. In order to achieve better biological properties of the composites, other components such as TCP and an antibiotic gentamicin were also employed (Morawska-Chochół et al., 2014).

Different methods were improved to achieve higher mechanical strength of biphase calcium phosphate scaffolds which included secondary phases such as hydroxyapatite whisker (Ramay and Zhang, 2004) or coating the scaffolds with nanocomposite-based polymers such as polycaprolactone (PCL)/HAp13 or PCL/bioglass (Roohani-Esfahani et al., 2011). In recent years, silicon (Si)−magnesium (Mg) based bioceramics have been drawn interest in the development of bone implant materials (Kharaziha and Fathi, 2010; Shuai et al., 2016). While Mg is an essential element of bone and Si is involved in human bone metabolism (Wu et al., 2010). Diopside with the chemical formula of $CaMgSi_2O_6$ is a silicate-based bioceramic utilized for artificial bone and dental root as a result great ability to form apatite with higher mechanical strength compared to HAp (Wu and Chang, 2007; Nonami and Tsutsumi, 1999). The bending strength and fracture toughness of diopside are 300 and 3.5 MPa $m^{1/2}$, respectively, which are higher than that of HAp (Nonami, 1991; Ashizuka and Ishida, 1997). Furthermore, it has been reported that glass ceramics consisting of a eutectic phase, with the composition of 38 wt.% TCP and 62 wt.% diopside and showed flexural strength as high as 200 MPa (Ghomi et al., 2012). Several techniques have been developed to synthesize porous ceramic scaffolds including gel-casting (Sabree et al., 2015), solid-state sintering (Zia et al., 2015), and replication of polymer foams by impregnation (Roohani-Esfahani et al., 2010). On the other hand, BCP scaffolds have been successfully fabricated by camphene-based freeze-casting (Macchetta et al., 2009) and a combination of gel-casting and polymer sponge (Ramay and Zhang, 2004) techniques. A space-holder approach has been widely used to develop metallic scaffolds. In this technique, spacer particles such as carbamide ($CO(NH_2)_2$), ammonium hydrogen carbonate (NH_4HCO_3), and sodium chloride (NaCl) are mixed with the main powder to enhance porosity during the sintering process. To the best of our knowledge, limited ceramic-based scaffolds have been fabricated using this technique. Depending on their applications, the scaffolds are designed for tissue engineering and regenerative medicine (Langer and Tirrell, 2004). During the past two decades, scientists have been developing potential scaffold materials for tissue engineering. Porous solid biomaterials in 3D form are called scaffoldswhich are designed to: (1) promote cell-biomaterial interactions and adhesion of cells; (2) allow sufficient gases and nutrients to permit cell endurance, abundance, and separation, (3) control biodegradability of tissue regeneration; and (4) allow a minimal degree of inflammation in vivo (Yannas, 1992). The scaffolds were designed with the most favorable characteristics, such as strength, degradation rate, porosity, and microstructure. Shapes and sizes are voluntarily restricted in polymeric scaffolds (Yannas et al., 1977).

A few scaffolds have been shown that the biological activities of them accelerate tissue regeneration and organs. Biological scaffolds are the derivative of human and animal tissues and synthetic scaffolds derived from polymers. Degradation behavior and low antigenicity of biological active scaffolds in vivo, besides, its thromboresistant performance in vitro, have been explained. The main characteristic of biological scaffolds of importance was their rate of degradation (Yannas and Burke, 1980). A number of carriers have been engaged to augment speed fracture or osteotomy revamp in animal models and open fracture in human clinical trials. The administered mixture in an injectable calcium phosphate cement or calcium phosphate medium has been employed to promote faster diaphyseal osteotomy curing in a broad range of animal models including rabbits, dogs, and nonhuman primates. Regeneration of tissues are accelerated in an adult using a scaffold of animals and human beings and it also employed for conjunctiva regeneration (Yannas et al., 1981, 1982, 1985; Burke et al., 1981).

Synthetic frameworks in biomaterials are denoted as supporting systems, scaffolds, and matrices. Over the past few decades, the design of biomaterials has been continuously evolved. The anticipation of biomaterials is to providing benefits in biomedical field by developing synthetic materials that can be used in reinstate purpose and thus by enhance our life span. The experimental results of the biomaterials are in accessible range by confirming their phases with good mechanical strength and thereof employed as a biocompatible implants.

Biomaterials are employed as implants for bone plates, joint replacements, sutures, vascular grafts, heart valves, dental implants, and other medical devices like pacemakers (Ramakrishna et al., 2001; Vert, 2005). Noteworthy advances have been achieved in the development of scaffolds in the biomedical field. This chapter reveals illustration of the different scaffolds on tissue engineering applications. It covers the uses of calcium phosphate-based scaffolds (CPSs). Calcium phosphates are classified into hydroxyapatite, beta-TCP, or the combination of hydroxyapatite and beta-TCP. These materials have been considered in order to attain porous scaffolds in the field of tissue engineering (Karageorgiou and Kaplan, 2005). A polymer duplication route can be used for the synthesis of BCP scaffolds with 70% of interconnected porosity, which effectively allows new bone development in male mice (Teixeira et al., 2010).

Bone is a normal nanoscale composite made of organic components (collagen fibrils) and hydroxyapatite. It is a competent in absolute wound healing and regeneration without scarring as a smart material (Gentili and Cancedda, 2009). When a bone breaks to a exceeding certain size which rejoin its structure by natural healing. Over the past 10 years, damaged bone tissues have been repaired to improve biological function (Amini et al., 2012). Different attempts have been made with the intention to find appropriate scaffolds with comparable structure to bone extracellular matrix. Composite systems combine to compensate polymers and ceramics which are expected alternate for bone tissue engineering provided by many researchers (Maquet et al., 2004; Li and Chang, 2004; Lu et al., 2005; Jiang et al., 2005; Khan et al., 2004; Xu et al., 2004). In a complex biological system, responsive schemes such as individual body with scaffold materials are tremendously involved in tissue engineering. Biocompatibility of the substrate materials is necessary; i.e., the material must not educe an uncertain inciting retort nor reveal immunogenicity. In accumulation, the mechanical strength of the scaffold ought to be satisfactory and not disintegrate during the patient's normal activities. For materials connecting with the human body, scaffolds have to be effortlessly sterilizable for the prevention of infection. This applies particularly for degradable scaffolds where the surface and bulk materials must be sterile. Scaffolds should be predominantly used in bone tissue engineering with controlled interconnected pores providing undeviated growing of cells in the preferred substantial form and to sustain vascularization. A typical porosity (90%) and a pore diameter (100 mm) are known to aid cell penetration and appropriate vascularization (Mikos and Temenoff, 2000; Levenberg and Langer, 2004; Griffith, 2002). Desirable features of scaffolds are to process them at near-net-shape and scalability in a cost-effective way. Various nanostructures in the human body are as shown in Fig.17.1.

The use of bottom–up approaches based on molecular self-assembly are very promising. A class of peptide amphiphile was self-arranged to form a 3D nanofiber. Self-arrangement of peptide amphiphile was permitted for cells in the matrix which lead to a metabolic path. It guides self-arranged nanofibers at the edges. It is possible to use oriented self-arranging supramolecular systems from small molecules. Flourishing of fractures in the form of non-unions and segmental bone defects is frequently raising issues in bone grafting, which intended to limit their contribution in autologous bone . Consequently, orthopedic surgeons have been examined osteoinductive growth and tissue engineering as a potential solutions for bone remedial problems. In the middle of the plentiful growth factors that potentially accelerate bone therapy and bone protein-2 is a potently osteoinductive and has been established with the combination of the lumbar spine and acute tibial fractures. However, administration of exogenous bone morphogenetic protein (BMP-2) provide a great amount of proteins to encourage biologic effects in humans. Therefore, coupling of gene therapy with cell therapy for increased control in the manufacture of osteoinductive within the defect sites to be a smart alternative to increase bone repair. Among these cell sources, mesenchymal stem cells (MSCs) are ahead in the rising curiosity for a promising cell therapy platform, due to their widespread aptitude of self-renewal and multilineage differentiation into adipocytes etc. (Xie et al., 2012). Bone marrow-derived MSCs have been found to heal critical-size femoral defects in a rat model when fixed with hydroxyapatite carrier. Besides, MSCs have been hereditarily customized to coherent BMP-2 to provoke in vivo osteogenic differentiation and bone structure using adenovirals (Hench, 1991; Gamradt and Lieberman, 2004; Bosch et al., 2000; Lieberman et al., 2002; Bruder et al., 1998).

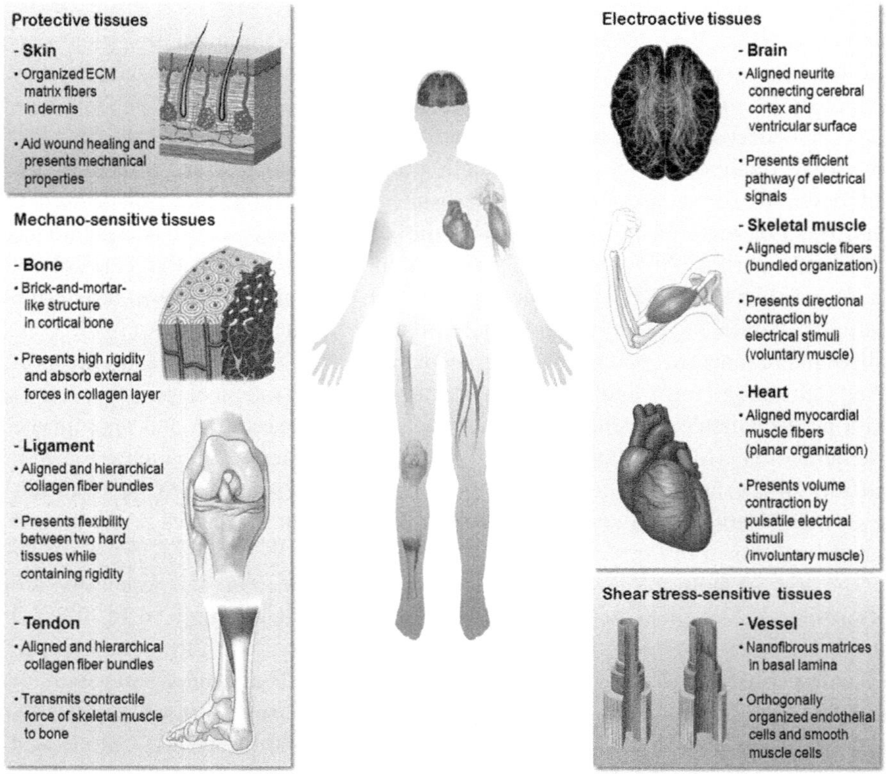

FIGURE 17.1

Various nanostructures in the human body. The tissues are classified into four categories (i.e., protective, mechano-sensitive, electro-active, and shear stress-sensitive tissues) according to the tissue's specific environment and function (Kim et al., 2013).

17.2 TYPES OF CALCIUM PHOSPHATE AND POLYMERS

Calcium phosphate can be classified into monocalcium phosphate, dicalcium phosphate (DCP), TCP, calcium orthophosphate, tetracalcium phosphate, hydroxyapatite, octacalcium phosphate, and BCP. It is an alternative to bone and is employed around dental implants. It is replaced in the place of bone as an implant. A common features of bioactive glasses and ceramics are surface modification that occur after being implanted. The surface forms a biologically dynamic to the hydroxy carbonate apatite (HCA) layer which makes a bonding boundary with tissues. The HCA layer forms on bioactive implants which are chemically and structurally equal to bone which create a bond interfacially (Hench, 1991). The in vivo arrangement of an apatite layer on the surface of a bioactive ceramics can be generated in a protein-free and acellular simulated body fluid. In order to overcome the restrictions , bone-grafts are replaced in the bone defect sites. Ceramics are made from calcium phosphate which have been employed in dentistry and orthopedics. A bone implants must require osteoconductive, osteoinductive, biodegradable, and structurally analogous to bone. Hydroxyapatite and TCP are two ceramics which have been elevated for tissue compatibility and osteoconductivity. Bone tissue engineering is a new approach to regenerate bone tissues. This technique provides a combination of cells for osteogenic and osteoinductive within materials. In order to attain success in bone tissue engineering, scaffolds have osteoconductive property which able to improve the growth of cells. Besides, the scaffolds are able to control the liberation of growth factors. Stimulated metaphyseal bone fix has also been established. Carrier necessities for metaphyseal crack are revamped by creating automatic support and thereby controlling the compression of the trabecular bone which decreasing the fracture by generating a gap in it. Because of mechanical impede with the carrier provides to stop post decrease subsidence. Currently, biodegradable composite scaffolds have been utilized with combination of ceramics and natural polymer.

17.2.1 CALCIUM PHOSPHATE-BASED SCAFFOLDS

Calcium phosphate (CaP)-based scaffolds (CaPs) have been widely examined as a candidate for bone-based tissue engineering. A mixture of macro-porosity (250–350 μm) and micro-porosity (2–8 μm) results in lamellar and woven shaped bone development in HAp scaffolds which is only present in the scaffolds with micro-porosity (Woodard et al., 2007). Microwave-sintered 3D printed β-TCP scaffolds with >60% porosity not only facilitated osteoblast association, but also aided in fresh bone configuration in the pores (Tarafder et al., 2013). High compressive strength (10.95 ± 1.28 MPa) was reported with a porosity of 50% using bioceramics based scaffolds. Present consequences on CaPs allocate, addition of dopant in scaffolds able to systematize closure rates, densification performance (Ramay and Zhang, 2004), mechanical strength (Kwon et al., 1994), and biocompatibility (Roohani-Esfahani et al., 2010). The cells are able to attached to the micropores within the macropores. Zinc and silicon ions were doped in type 1 collagen gene which synchronize kinases discharge that regulates angiogenesis, morphogenesis and differentiation (Shie et al., 2011).

Moreover, hydroxyapatite (HA) and TCP have been clinically examined for bone replacement. During examination, implanted hydroxyapatite and TCP were showed osteoconductive properties. High porosity and low density along with large surface area are used for vascularization and ingrowth of bone. Uniform regular surfaces of hydroxyapatite and TCP were affected cell proliferation and differentiation. In the absence of soft tissues, osteoids are generated on the surface of the ceramics after implantation.

17.2.2 BIOGLASS-BASED BIORESORBABLE SCAFFOLDS

Over the past few years, bioglass of different compositions have been investigated (Jones et al., 2006). Porous 3D bioglass scaffold (70%) with pore size (300–400 μm) revealed that HCA formation significantly enhancing osteoblast activity on in vitro tests. The results of in vivo test showed that the layers of HCA provide protein and growth factors that could assist for the development of new bone (San Miguel et al., 2010). In a further research, cobalt ions (Co) were initiated in the form of mesoporous-based bioglass scaffold to provoke hypoxia that augmented the abundance of bone marrow-based stem cells, separation, and bone-related gene (Wu et al., 2012).

17.2.3 COMPOSITE SCAFFOLDS

Composites consist of two or more various materials (ceramics and polymers). Improved calcium phosphate/polymer scaffolds provide excellent mechanical and physiological nature to host tissues. Addition of calcium phosphate in polymer scaffolds increases the toughness and compressive strength similar to bone. Besides, the properties like mechanical and bioactivity can be enhanced in polymers by the addition of CaP. The weight ratio of PGA to β-TCP in 3D porous composite scaffold was 1:3 having pore size (483.3 ± 113.6 μm) which degrading ($96.2 \pm 3.3\%$) after 90 days of implantation in male rats (Cao and Kuboyama, 2010). Enhanced HA/poly(ester-urethane) scaffolds was established to adsorb large amounts of bovine serum, and fetal calf serum in comparison with poly (ester-urethane) (Laschke et al., 2010). Using conventional salt dischargeor phase inversion was adopted to synthesize porous HA/polyurethane scaffolds (200 μm) with volume fraction porosity ($90 \pm 2\%$) attained. Surface modification of BCP porous/HA/PCL composites is found to enhance the compressive strength by two factors (Roohani-Esfahani et al., 2010; Xue et al., 2009). Polymer/bioglass scaffolds are also used in bone tissue engineering.

17.2.4 POLYMERIC SCAFFOLDS

Natural polymers are more advantages and analogous to macromolecules in biological system which lead to provide good metabolism in it. There are two types of biodegradable polymers, synthetic and natural. Synthetic polymers are also employed in biomaterial. For example, in tissue engineering polymers like polyglycolic acid and their copolymers likes polylactic-*co*-glycolic acid, and polycaprolactone are employed. However, certain clinical complications are observed due to releasing of chemicals from the polymers. Synthetic polymers are able to prevent inflammation or immunological and toxicity which resemble to the extracellular matrix. Natural polymers such as collagen, gelatin, silk, chitosan (CS) etc., have been employed in tissue engineering which are biocompatible and degradable. Collagen is one of the natural polymers which is abundant in the extracellular matrix protein. Osteoblast which is secreted by collagen type 1 which is the main organic component and mineralized soon after the development of bone. Polymers have been extensively employed as a biomaterials for the production of tissue-engineering scaffolds (Vert, 2005; Karageorgiou and Kaplan, 2005; Teixeira et al., 2010; Gentili and Cancedda, 2009; Amini et al., 2012; Maquet et al., 2004; Li and Chang, 2004; Lu et al., 2005; Jiang et al., 2005; Khan et al., 2004; Xu et al., 2004; Mikos and Temenoff, 2000; Levenberg and Langer, 2004; Griffith, 2002; Hench, 1991; Gamradt and Lieberman, 2004; Bosch et al., 2000; Lieberman et al., 2002; Bruder et al., 1998; Kim et al., 2013; Woodard et al., 2007; Tarafder et al., 2013; Kwon et al., 1994; Shie et al., 2011; Jones et al., 2006; San Miguel et al., 2010; Wu et al., 2012; Cao and Kuboyama, 2010; Laschke et al., 2010; Xue et al., 2009; Piskin, 1995).

Biomaterials are selected on the basis of their material chemistry, shape and structure, hydrophilicity and hydrophobicity, lubricity, surface energy, and deterioration. Scaffolds are well known due to their large surface-to-volume ratio, elevated porosity, and mechanical property. Researchers have been attempted to develop skin and cartilage (Ji et al., 2006), bone and cartilage (Eaglstein and Falanga, 1997), liver (Boyan et al., 1999), heart valves and arteries (Mayer et al., 2000), bladder (Mayer et al., 1997), and different other soft tissues (Oberpenning et al., 1999). Scaffold candidates were prepared in synthetic or natural, degradable or nondegradable (Tziampazis and Sambanis, 1995), and can be classified into different types in terms of their structural, chemical, and biological characteristics. Synthetic polymers are increasingly employed in biomedicine and hence, the properties can be engineered for specific applications. Synthetic polymers are generally cheaper and forming in large quantities with longer shelf life. Commercially existing synthetic polymers are demonstrated similar physical, chemical, and mechanical strength as biological tissues. It formed a bigger group in green polymers and are prepared in controlled conditions. The polymers such as polylactic acid, polyglycolic acid, polylactic-*co*-glycolic acid, and related copolymers are the most frequently employed synthetic polymers in tissue engineering (Mohammad et al., 2000). Hydroxyapatite, TCP, and certain compositions of silicate and phosphate glasses are reacted in physiological fluids and then forming cells which tenaciously bonded to them (Germain et al., 1999). Biocompatibility and biodegradability are frequently inadequate properties of the biomaterials which restricted them in clinical applications. The issues in the field of tissue engineering are surmounted by blending synthetic and natural polymers in order to enhance biocompatibility. (DiEdwardo et al., 1999) (Nair and Laurencin, 2007). The grouping of degradable polymers and inorganic bioactive particles demonstrate good mechanical and biological recital in hard tissues (Ma, 2004; Alexander et al., 1996; Gunatillake et al., 2006; Chen and Wang, 2002; Hench, 1998; Cascone et al., 2001; Ciardelli et al., 2005; Roether et al., 2002).

17.2.5 INORGANIC/ORGANIC SCAFFOLDS

Composite materials are highly used in the field of tissue engineering because of their attractive properties and ensemble the mechanical and physiological anxiety in congregation tissues which addressed by scheming the volume fraction and morphology of them (Rezwan et al., 2006). Ceramics were employed for fabricating implants which possess non-absorbable, bioactive or semi-inert (Hench, 1991), and biodegradable or resorbable (non-inert) in nature (Hentrich et al., 1971). Glass based ceramics and dense hydroxyapatite are semi-inert (bioreactive), and some examples of ceramics are aluminum calcium phosphate, hydroxyapatite, TCP etc., (Park and Lakes, 1992). Ceramics are recognized for their superior compatibility, corrosion resistance, and large compression resistance. Some of the drawbacks of ceramics are fragility, low-fracture vigor, complexity to engineer, low-mechanical reliability, lack of elasticity, and high density. Polymers are normally flexible and revealing weak mechanical strength, while inorganic candidates (ceramics and glasses) are recognized to be rigid and fragile. The combined array of polymers and inorganic phases provide enhanced mechanical properties due to their innate superior rigidity and potency. Then, the addition of bioactive phases to polymers lead to decrease deprivation of the scaffolds (Blaker et al., 2003; Kim et al., 2005). Complications in the development of polymer-based bioceramics composite scaffolds are: (1) maintenance of strength and the inflexibility at the interface on the deprivation period and substitution by the usual host tissues; and (2) alike resorption rates with respect to the repair rates of body tissues used in hard tissue implants and tissue engineering due to their excellent biocompatibility, bioactivity, and bioresorption in calcified tissues. Polymer/hydroxyapatite composite scaffolds have outstanding biocompatibility with hard tissues and greater osteoconductivity and bioactivity (Zhang and Ma, 1999). The chief inorganic constituents of natural bone are CP, HAP, and TCP complexes with poly(L-lactic acid) (PLLA) (Woo et al., 2001), collagen (Du et al., 1999), gelatin (Bigi et al., 2002), and CS (Zhang and Zhang, 2001) have been extensively employed as a supporting material for repairing bone.

17.2.6 POROUS AND HYDROGEL SCAFFOLDS

Porous polymeric supporting systems in 3D possess better porosities with consistent interconnected network pores which are extremely valuable in tissue engineering. Sponge like absorbent scaffold have been used in tissue engineering (Zhang and Ma, 1999) predominantly for enlarging host tissues and bone regrowth. This network has a porous nature which stimulates the extracellular matrix building and allowing cells to proficiently interconnect with their environment. Even though foams and sponges are supplementary which having high automatic stability compared to other network structures. Depending on the choice of solvent and phase disconnection, the appearance of foams can be varied. (Ma and Zhang, 2001). Enlargement interconnected pores in porous scaffold is essential to improve synthetic blood vessels. Propose and appeal of biodegradable hydrogels has noticeably augmented to hydrogel materials in the field of biomedical (Freiberg and Zhu, 2004). Gels are fashioned when the network is covalently cross-linked (Mooney et al., 1994). Hydrogels with growth factors improve the growth of the newly created tissues (Wei and Ma, 2004). Hydrogels are frequently positive for promoting cell migration, angiogenesis, and rapid nutrient diffusion (Ouriemchi and Vergnaud, 2000; Hou et al., 2002, 2003; Cabodi et al., 2005; Jhon and Andrade, 1973; Hoffman, 2002; Hubbell, 1999; Lee and Mooney, 2001; Peppas and Khare, 1993; Tabata, 2003; Bryant and Anseth, 2001).

17.3 PHYSICAL, CHEMICAL, AND BIOLOGICAL ANALYSIS OF BIOCERAMIC-BASED SCAFFOLDS

17.3.1 FT-IR ANALYSIS

The characteristic absorption peaks at 2920 and 2875 cm^{-1} are associated to the C–H asymmetric and symmetric stretching vibrations, respectively. The C–H bending vibrations in methylene and methyl groups are 1371 and 1433 cm^{-1}, respectively observed. Also, the existence of bands in the range of 1080^{-1}–1150 cm^{-1} which correspond to the C_O stretching vibrations in glycosidic bonds. The amidation reaction between COOH in PCL and $-NH_2$ groups of CS and sulfonated-CS is appended using 1-ethyl-3-(dimethylaminopropyl)carbodiimide (EDC) and N-Hydroxysuccinimide (NHS) as cross linkers (Ghaee et al., 2017).

The liberation of the poly(ethylene glutarate) (PEG) is exposed to the bone wax samples without starch. The CH stretching and CH-bonding perhaps traced, as overlapped CO-stretching which attribute to the vibrational bonds of the inorganic bone wax component. By concerning the phosphate (PO_4^{3-}) release, all bone waxes are showed a similar burst release up to 4 days. Afterwards, the PO_4^{3-} ions are converged for gaining certain saturation (Fig. 17.2) (Bruckner et al., 2016).

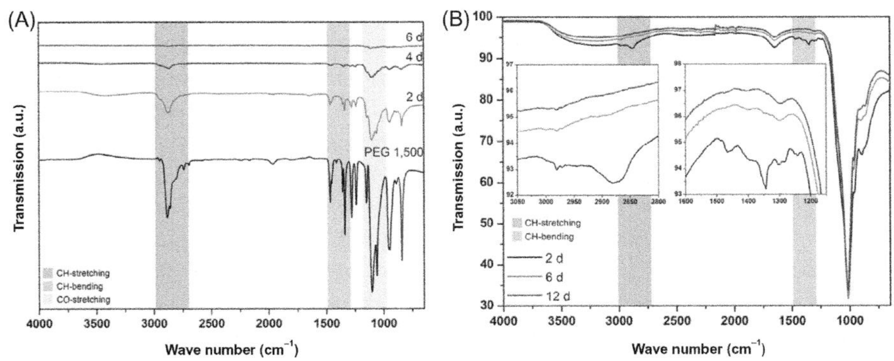

FIGURE 17.2

(A) FT-IR of 10 μL dried PBS in which bone wax without starch was deposited; and (B) FT-IR of dried bone wax samples without starch.

FT-IR spectra of HAp and β-TCP were essentially conquered by the internal modes of PO_4^{3-}. The peak at 3576 cm^{-1} is connected to hydroxyl groups (−OH) in HAp which is deficient in β-TCP. After blending, no shift in the bonded water was observed. Before and after sintering, the absorption peaks of PO_4^{3-} and −OH groups do not shift in their positions. The peaks of -OH and PO_4^{3-} are modified to sharper on sintering, which may be ascribed to the structural conversion of BCP to a crystalline state (Chen et al., 2011).

17.3.2 EXTERNAL GEOMETRY

For reconstruction of tissues , physical characteristics are crucial for scaffolds (Boyan et al., 1996). Scaffolds with an appropriate physical nature are elegant resources that are able to mimic the natural extracellular matrix (ECM). It is played a vital role in architecture of tissues by structure and mechanical strength. There are variety of processes which are relevant to cell differentiation, tissue formation, homeostasis, and others (Tran et al., 2004; Midwood et al., 2004). The production of macro- to nanoscale for structural architectures is established much consideration in medical field. The nano or macroscale structures able to mimic the native state of extracellular matrix in tissues. Scaffolds in 3D form are able to regenerate tissues and organs in their normal condition. Biomaterialshave been employed to mimic the ECM and would be great opportunity to engineer the scaffolds for various types of tissues. As polymeric materials allow flexible assortment of surface characteristics and competent management of processing ECM reconstitution which attained by the interactions of them. The crucial valid points of scaffold geometry are to maintain highly interconnected porous structure with high surface density and high surface-to-volume ratio, which is suitable condition for cell and proliferation (Midwood et al., 2004).

17.3.3 SEM ANALYSIS

CS scaffolds are reinforced by calcium phosphate invert glass and β-TCP and synthesized through a solid−liquid phase separation which result in oriented macropore structure. The density of the scaffolds was increased with the β-TCP content (Zhang and Zhang, 2001). Novel nano-HAp/polymer composite scaffolds with high porosity and controlled pore size were fabricated using thermally induced phase separation technique. By altering the phase separation parameters, the pore size and porosity of the scaffolds can be engineered. The nano-HAp-incorporated PLLA scaffolds showed fibrous morphology which, in turn, enhanced the protein adsorption by three times over the nonfibrous scaffolds. The nano-HAp/polymer composites are showed an open 3D pore structure, whereas the micro-HAp/polymer composite scaffolds showed random, irregular pore structures (Wei and Ma, 2004). Poly(methyl metacrylate) (PMMA) is commonly used as bone cement in the biomedical field. The laminated HAp/PMMA nanocomposites via one-step intercalative bulk polymerization enhanced the thermal stability and tensile strength of the composite (Chu et al., 2004). The PEG–PVA/HAp hydrogel contains interconnected pores with porosity (45–70 vol.%) and pore size (40–100 μm) by varying the concentration of PEG. The molecular weight of PEG plays a vital role in the structure and pore size of the HAp scaffolds (Swain et al., 2015). Yang et al. reported that the PVA/HAp composite membranes with HAp fillers (5–10 wt.%) were homogenous. When the HAp filler contents were increased to 15 wt.%, and depicting large aggregates randomly distributed on the surface of them. (Yang et al., 2008).

Ramli et al. (2011) prepared nano-HAp–CS composite with the average pore size of 26.7 ± 11.7 μm. Garai et al. reported that the HAp–carboxymethyl cellulose (CMC) composites scaffolds depicted the formation of interconnected micro pores which containing aggregates of nano-HAp in the CMC matrix. When the polymer contents were enhanced, the bead-like agglomeration was found to be increased with the reduced size of the beads. The irregular interconnected pore size is formed in 3D scaffolds in the range from 0.7 to 2 μm (Garai and Sinha, 2014). The electronspun poly(L/DL)-lactide (PLDL) and PLDL/nano-HAp membranes showed smooth and uniform fibers with bead formation. The nano-HAp distribution in the polymer matrix was non-homogenous agglomeration. The average diameter of the fiber in the PLDL membrane was 1.7 ± 0.5 μm whereas; the fibers in the PLDL/nano-HAp showed an average diameter ranging from 2.8 ± 1.4 μm. The membranes are showed interconnected pores. The porosity of PLDL/nano-HAp membranes was reduced due to the effective charge dissipation by HAp particles which prevent inter-fiber repulsion (Rajzer et al., 2014). Multiwalled carbon nanotubes–CS–HAp were investigated by Rajesh et al. (2012), and showed uniform dispersion of HAp particles on the CS polymer matrix while the isolated HAp had a flake-like structure with the particle size (50–100 nm).

17.3.4 MECHANICAL PROPERTIES

The pure polymers have been depicted low mechanical strength and modulus. The addition of a small quantity of ceramic particles in the polymer matrix could improve the properties of the polymer. HAp is one of the most common filler in polymers used in the biomedical field. Apart from the HAp, TCP, and DCP were used as fillers (Pazourková et al., 2015). The particle size and the dimension of the HAp nanocomposites play a vital role in the enhancement of properties. Hutmacher et al. reported from a literature review that the compressive strength of the cortical and cancellous bones was approximately 100–230 MPa and 2–12 MPa, respectively. The Young's moduli of the cortical and cancellous bones were in the range of 7–30 GPa and 0.5–0.005 GPa, respectively (Hutmacher et al., 2007).

Weia et al. investigated the influence of nano-HAp on the mechanical properties of PLLA. The compressive modulus of the pure polymer was found to be 4.3 MPa. The compressive modulus of the nano-HAp/PLLA nanocomposites was enhanced with the addition of nano-HAp content. When 50% of nano-HAp was added to the polymer, the mechanical property of the nanocomposites was increased to 8.3 MPa (Wei and Ma, 2004). Kati et al. prepared CS, montmorillonite (MMT), and HAp nanocomposites and CS/MMT/HAp exhibiting good nanomechanical properties compared to the MMT/CS and HAp/CS nanocomposites. The nanoindentation test showed a markable increase in the elastic modulus of the CS/MMT/HAp nanocomposites over the pure polymer composites. The CS/MMT composite exhibited the elastic moduli of 8 GPa. Addition of HAp particles in composites tends to increase the elastic moduli to 9 GPa. A nanomechanical property of the composite was enhanced not only because of the addition of MMT, but also due to the addition of HAp (Katti et al., 2008). The compressive modulus and yield strength of the CS reinforced with calcium phosphate invert glass and β-TCP composites were enhanced when compared to the pure CS scaffolds (Zhang and Zhang, 2001). The addition of HAp particles in the polymer (PLLA) matrix is improved the mechanical properties and adsorption capacity of the scaffolds (Wei and Ma, 2004).

Mao et al. (2003) developed a novel CS gelatin bilayer scaffold by the freeze–drying method and demonstrated the enhanced structural, mechanical and degradable properties. The HAp/silk fibroin coating composites are obtained by matrix-assisted pulsed laser evaporation and showed a decrease in the surface roughness when the concentration of silk was increased (Miroiu et al., 2010). The HAp/PEG biomaterial composites were prepared by ring-opening polymerization of cyclic oligo(ethylene glutarate) in a HAp-based porous scaffold. The compressive strength of the porous HAp scaffolds was observed as 3 MPa and on increasing the PEG concentration, the compressive strength of the hydroxyapatite/poly ethylene glycol composite scaffolds was significantly increased (Siriphannon et al., 2007).

Hossan et al. reported that the tensile strength enhanced linearly with the increasing HAp percentage in the gelatin/HAp composites system which was due to the crosslinking between the gelatin and HAp. It was also observed that the Vickers microhardness increased about six to seven times with the increase in HAp percentage (Hossan et al., 2015). The mechanical properties of HAp/PVA nanocomposite samples are improved due to their interaction between the polymer and the reinforcement material. It has been observed that the HAp/PVA composites prepared from in situ biomimetic HAp crystallization in the presence of PVA showed enhanced tensile strength compared to the direct mixing (Mollazadeh et al., 2007).

17.3.5 SWELLING AND DISSOLUTION STUDIES

Biodegradable polymers such as PGA, PLA, and their copolymers poly(lactic-co-glycolic acid) (PLGA) were extensively used for the fabrication of 3D scaffolds. These synthetic materials are biocompatible and have a controlled in vivo degradation rate (Visscher et al., 1985). Peter et al. reported that the CS—gelatin (CG) and CG/nano-HAp (CG/nHAp) composite scaffolds are exhibited a very high swelling capacity for retaining the water more than its original weight. The addition of nano-HAp, reduced the swelling capability of the CG/nHAp network. The HAp formed between the polymer chains and decrease the hydrophilicity of the gelatin. It eventually binds the calcium and phosphate to the hydrophilic –COOH or NH_2 groups. The swelling and degradation of the CS are involved in the protonation of amino groups and mechanical relaxation of coiled CS chains. Hence, the samples were exhibited a high degree of swelling with large surface-to-volume ratio which allow the samples to have maximum probability of cell infusion (Peter et al., 2010). Chaudhuri et al. compared the swelling behavior of PVA–PVP and PVA/PVP–HAp composites and concluded that the swelling property of PVA/PVP–HAp was increased. In other words, the inclusion of HAp in the polymer matrix is improved the hydrophobicity of the scaffolds which attribute to the strong PVA–PVP–HAp bonding (Chaudhuri et al., 2016). The poly-γ-glutamic acid (γ-PGA)/HAp is showed a good swelling property over the pure HAp powder. The degradation rate of HAp was 10% in 30 days, whereas the γ-PGA/HAp composite showed 30% in 30 days which implies that addition of γ-PGA polymer in the HAp matrix leads to the faster degradation of the material. The accelerated weight loss was observed in the HAp composites incorporated γ-PGA (Shu et al., 2014). A hybrid variety of nanocomposites of PEG, CS, and nano-HAp/zinc oxide (ZnO) with interconnected macroporous structures was prepared. These composites are experienced excellent swelling behavior when the PEG concentration was increased. When the nano-HAp–ZnO nanocomposites were increased and the water uptake abilities of the composite scaffold was drastically reduced (Bhowmick et al., 2015). It is eventually proved that the swelling and degradation properties of the scaffolds could be engineered by altering the concentrations of polymer and calcium phosphate.

17.3.6 CELL VIABILITY

Tissue engineering is a promising tool for creating biological substitutes in the form of implants, where the porosity of the scaffolds play a vital role for cell seeding and assist as a supporting template for the tissue regeneration. The biodegradable polymers from aliphatic polyesters of alpha-hydroxy acid derivatives, such as PLA, indicate a modulus of elasticity similar to natural cortical bone and can retain high strength over time (Daniels et al., 1990). Contradictorily, these materials inhibit the tissue response leading to the delay in bone healing or osteolytic reactions. In order to overcome such complications, calcium phosphate and polyester composites were developed and examined in vitro and in vivo. The developed composites tend to increase the phenotype expression of the cells and improve the bone healing in a dose-dependent fashion compared to the pure polymeric scaffolds. Calcium phosphates such as HAp and β-TCP were used in the composites and evaluated.

Prokop et al. demonstrated that the pure PLLA scaffold showed moderate inflammatory reactions in the sheep femur. When HAp (10% w/w) was added, the bone healing was improved and on higher HAp ratios the inflammatory response of the bone tissue was reduced significantly in the sheep femur bone (Prokop and Chmielnicki, 2013). Similarly, Kikuchi et al. (2004) illustrated PLLA/TCP composites (70/30 w/w) are successively implanted on mandible and tibia bones of dog without any adverse reaction.

HAp-collagen is coated on PCL scaffolds and showed suitable nanostructures and surface roughness for the treatment of osteogensis. The composite enhances the propagation and separation of bone-marrow stromal cells (Wang et al., 2015). Kutikov et al. revealed that the excellent blending of biodegradable, amphiphilic poly(D,L-lactic acid)–poly(ethylene glycol)–poly(D,L-lactic acid) (PLA–PEG–PLA) (PELA) triblock copolymer with hydroxyapatite (HAp) and the fabrication of high-quality rapid prototyped 3D macroporous composite scaffolds using an unmodified consumer-grade 3D printer. The rat bone marrow stromal cells (MSCs) are adhered to HA–PELA which undergoes effective osteogenic differentiation. These cell-adhesion properties can extensively exploited in biphasic constructs with spatially controlled cell adhesion (Kutikov et al., 2015).

The fibrous scaffolds of poly(ε-caprolactone) (PCL) were used as a model polymer for the tissue engineering application, since the rate of degradation rate was relatively slow in vivo. The introduction of calcium phosphate on the PCL scaffolds improves the cytocompatibility (Galperin et al., 2013). The incorporation of HAp in PCL showed better cell viability over the pure PCL scaffolds. The presence of HAp is inhibited the differentiation with preosteoblast cells. Improved cell adhesion and increased protein absorption were obtained on the HAp/PCL nanocomposites. These composites are highly porous having good mechanical properties which was an attractive property in bone tissue regeneration (Kutikov et al., 2015).

Ratner et al. reported that the integrated bi-layer scaffold based on degradable poly(hydroxyethyl methacrylate) hydrogel coated with HAp and layered with hyluronan serve to engineer human cartilage (Galperin et al., 2013). Zhou et al. (2011) examined that the bilayer composite membrane of PCL and gelatin sponge were coated along with HAp by electrospinning technique which in turn enhanced the mechanical properties and porosity thereby, improve the cell affinity of the scaffolds. The PVA:PCL–HAp nanocomposites scaffolds showed densely arranged MG-63 osteoblast cells proliferation (Uma Maheshwari et al., 2014). Zhang and Zhang developed a composite by impregnating HA/β-TCP into polyurethane foams followed by heat treatment of the HA/β-TCP/polyurethane combination to remove the polymer which resulting in macro-porous ceramic material. The macro-porous ceramic material was subsequently soaked in a CS solution and freeze dried to form the CS–HA/β-TCP composite. The composite was examined in vitro and found to support the growth of apatite crystals as well as the attachment and proliferation of osteoblast cells (Zhang and Zhang, 2002).

Khor et al. reported that the porous chitin matrixes from a chitin solution and followed by freeze–drying to produce porous chitin matrixes with pore sizes in the range of 100–500 nm. The human and mouse fibroblast cells are cultured on these matrixes which resulted in excellent cell growth and proliferation (Chow et al., 2001).

17.3.7 DRUG DELIVERY

The development of multifunctionalpolymer-based composites with controlled drug delivery has been fascinated many research groups which induced them to work intensively during the last few decades. The water-soluble drug release was occurred after the penetration of water in polymeric hydrogel network which result in the swelling of the polymer. The water-soluble drug dissolves and diffuses through the polymer matrix. The drug release relies on the swelling properties of the scaffolds and the hydrogel. The addition of nanoparticles in the 3D polymeric structures is a pioneering method for obtaining multicomponent systems with vast applications. The nanoparticle–polymer hydrogel combination results in nanogels as potential carriers for cancer therapy and drug delivery systems (Merino et al., 2015).

The sodium alginate/CS/HAp hydrogel nanocomposites were prepared and subjected to gamma radiation for cross-linking in oral drug delivery. The efficacy of the hydrogel was controlled with the anticancer drug doxorubicin as a model drug. It was showed that the high drug release attained at pH 5 (Abou Taleb et al., 2015). The HAp nanoparticles were synthesized by hydrothermal method and used in drug delivery system with usage of water-soluble anticancer drug paclitaxel (Tax). The Tax-HAp composite was prepared and then suspended in the collagen gel matrix in order to avoid the adverse effect of the drug on normal cells. The HAp could be used as a successful carrier for loading the anticancer drug and the drug-loaded HAp embedded on the collagen gel would be for the chemotherapy (Watanabe et al., 2013). The drug release kinetics of the HAp–PVA nanocomposites loaded with the ciprofloxacin (antibiotic to treat bacterial infections) showed sustained drug release whereas; the pure HAp showed burst release noticed by Sundar Raj et al. (2013). Wang et al. reported that the HAp microspheres and the PLGA coated HAp microspheres were examined for the sustained drug release. Doxycycline hydrochloride was loaded to the models and their adsorption kinetics were studied. Both the systems are agreed well to the Freundlich model at low drug concentrations. The release profile of PLGA–HAp microspheres showed steady and sustained slow release, while the HAp microspheres showed a burst release (Wang et al., 2010).

PLGA/Ibuprofen-loaded HAp were prepared using the conventional electrospinning method for drug delivery application. Ibuprofen is a highly used antiinflammatory drug. Initially, the PLGA/drug-loaded HAp and drug-loaded HAp were showed burst release which was followed by a sustained release pattern. The PLGA/ibuprofen-HAp loaded composite is showed sustained release compared to the drug-loaded HAp and may a potential therapy in bone regeneration and bone tissue engineering (Song et al., 2016).

17.3.8 GENE DELIVERY

Since, the results of baculovirus was competent in transducing mammalian cells. Baculovirus-mediated gene delivery has been oppressed for the progress of cell-based assays, release of small intrusive ribonucleic acid (RNA), detection of major histocompatibility complex (MHC) class I mimotopes, eucaryotic proteins, analysis of gene functions and for liberation of vaccine immunogens (Yang et al., 2007). A different baculoviruses are in vivo gene delivery, except in vivo analysis of genes in animal models (Kaikkonen et al., 2006; Airenne et al., 2000; Wang and Wang, 2005). Although Huser et al. (2001) produced a baculovirus for in vivo appearance of human factor IX (Pieroni et al., 2001). The baculovirus-mediated erythropoietin in the quadriceps and mice which did not showed any healing effects. Wang et al. proved that the baculovirus expresses diphtheria venom A inhibits the glioma xenograft enlargement in the rat brain which revealing the potential usage of baculovirus in cancer therapy. Though, the usages of baculovirus in MSC production and bone refurbish yet to be reported. In this analysis, baculovirus (Bac)-mediated BMP-2 look intended for the in vitro assurance of human mesenchymal stem cells (hMSCs) into osteoblasts at a rate reliant on virus quantity. The BMP-2 term level is mediated by Bac-cannabinoid (CB) which is inferior in the hMSCs than in the rabbit articular chondrocytes. In spite of sending low-level BMP-2, Bac-CB transduction at multiplicity of infection (MOI) 40 actually induced the hMSCs to differentiate into late osteoblast stage, as evidenced by E3–4-fold (Wang et al., 2006).

17.4 PRESENT CHALLENGES AND FUTURE PERSPECTIVES

Tissue engineering is the most stimulating interdisciplinary research field. Scaffolds and production technologies play a vital role in tissue engineering. Various types of polymeric scaffolds were employed in the field of tissue engineering. Scaffolds are supposed altogether for designing their area, despite of their usual property. All these techniques for scaffold manufacture are responsive to the different parameters. Innovations in designing of materials and their production are increasing the option to manufacture implants with the best quality

Scaffolds must be surface compatible and architecturally suitable for host environment. New fabrication principles and theories provide to develop novel design for implants in the field of biomedical. Nanotechnology aids us for creation of scaffolds in dimensional range which may be enough for cells and biomolecules. It is clear to developing novel scaffold structure related to calcium phosphate-based materials for tissue engineering. The research in medical field will be exploring in the scientific frontiers for diagnosing, curing and prevention of diseases at the level of molecules. Significant advances in the clinical usage of medical devices. Currently, the designing of polymeric scaffolds for specific and desired properties which responses surrounding cells. The requirement of different solutions assembles to stipulate for the substitution of organs and parts of tissues will drive to improve tissue engineering. Polymer scaffolds contain all the forthcoming to offer new ways to manage physical and chemical environment in biological system. There are numerous advantages to employ polymers over extensive usage of the synthetic polymerscaffold in the field of tissue engineering.

In future, researchers are able to conquer a lot of magnificent limits in the manufacturing process. We suppose no material will assure all propose parameter in all usage; however, a wide range of materials will locate and employ in an assortment of tissue engineering applications. In general challenges in scaffold are in propose and manufacture of novel stimulating function focused to investigate in scaffold designing which contains polymer congregation, surface morphology, biodegradability, mechanical strength, which able to direct cell functioning and induce the natural tissue formation.

The artificial biodegradable polymer/inorganic bioactive materials are addressed as smart tissue engineering scaffolds due to their bioactivity and adaptable biodegradation kinetics. Conventional materials processing methods have been adapted and extended for incorporation of inorganic bioactive phases into porous and interconnected 3D polymer networks. From the materials science perception, the near confront in tissue engineering for designing and fabricating reliable bioactive and 3D scaffolds bioresorbable for tailoring porosity and pore size, which can maintain the structure and integrity, in load-bearing. The mechanical reliability of composite scaffolds is lesser than the one order of magnitude compared to cortical bone. To achieve the mechanical strength of bone could also permit for replacing damaged bone tissues. The integration of biomolecules such as growth factors with the plan to speed up controlled bone therapeutic. Incorporation of biomolecules during the processing of scaffolds however, it is not simple as at higher temperatures under extreme chemical conditions. A promising approach is the proteins immobilization and enhancement in the growth factors under postprocessing phase through scaffold surface functionalization.

In order to improve clinical applications, abundant in vitro and in vivo studies are expected to be performed in biological systems. Biomaterials have been employed in hard tissue repair which started with bioinertness, and to involve the growth and application of bio-inert materials. It is employed in today clinical use as hip–joint replacements. This period was soon followed by the development of biodegradable materials for bone tissue engineering scaffolds that can stimulate specific cellular responses at the molecular level. The composite scaffolds are explained in this book unite the features of the biomaterials. The developed biomaterials possess increased complexity, from ion interactions and to incorporating growth factor. The biomaterials were composed of purely synthetic materials to be used in biological system, which possess bioactivity and biodegradability. While present research is very much aimed on the interaction between stromal cells and biomaterials, and the fundamentals of biomaterials appear to initiate from stem cells. This new approach shows the surface of scaffold to imitate intricate restricted biological functions and may guide to in vitro and in vivo tissues of growth and organs. In this advance, composite scaffolds to be expected to engineer biodegradability of polymer along with inorganic phases, as considerable action taken for regenerative tissues by developing novel composite biomaterials which could mimic the tissues. A novel approach should be addressed for designing to modulate the surface and bulk properties to tune extracellular matrix by therapeutic molecules in delivery system and 3D scaffolds. The internal modulation was tuned for physicochemical and biological properties to native tissues and to engage in tissue to provide adhesion, proliferation etc.

In fact, the scaffolds are considered to bring multiple bio-factors which deliberately used for the development of vivo tissues. Though, the cellular associations with the pharmaceutical microenvironments changed to design the delivery depots necessitate to improve the composite molecular events and to get better the designs, as well as to attain tissue-engineering construct in vivo situations.

The development of biomaterials is sensitive to temperature, pH etc. Magnetism is very important approach to attain high elegant and multifunctional proceedings of delivery. Nanotechnology has the possible to modify medical research considerably with advances in cell-based technologies. Tissue engineering is a capable therapeutic come close to that associated to cells, biomaterials, for inducing differentiation signals into transplantable formats and encourage tissue revamp. Among various techniques, electrospinning can mimic the electracellular matrix of natural tissues by scaffolds (3D). This technique is a versatile because of its simplicity and control process. In addition, this method is suitable to develop 3D structures for bone tissue engineering. Selection of biomaterials has been considered for the growth of the new bone and electra-cellular microstructure should be mimicked. The technique is so called electrospinning aid to provide bio-signaling in the nanofibers (proteins, antibiotics, gene). Nanofibrous scaffolds is to improving the performance in the application of bone tissue engineering for porosity, diameter and direction of fibers growth. Besides, other advances in surface modification and electrospining which able to provide effective clinical applications. This technology will play a crucial role in the advancement of bone tissue engineering.

The biomaterials will be developed for multiple responses which is considered to be major emerging issues. For next generation smart scaffolds for tissue engineering which composed of calcium phosphate-based polymer nanofibers for multifunctional applications and diagnostics. To realize tissue regenerative biomaterials, the designing rule addressed to provide cell and tissue compatibility. This might mainly limit the option at present obtainable compositions and drives the improvement of novel nanobioactive compositions. Biomaterials will be developed and designed to stimulate and differentiation of bone cells. The calcium phosphate-scaffold systems are refined greatly for the long-term tissue viability. This would further necessitate, creating the best micro-mechanical surrounding. It is important to enhance the bone regeneration to decrease trauma. Currently, micro and nanofabrication method provide to engineer for well-defined stereo-regulated architecture to control spatial organization and which to mimic the microarchitectural in native tissues. Combining material chemistry and technology are employed for regenerated tissue by tuning scaffold degradation rate with respect to the rate of tissue growth.

Bone grafts are required for the replacement of bone which possess the highest success rate for regenerating bone. Harvest of bone tissue is related with many clinical drawbacks which contain healthy bone, morbidity and pain issues. Synthetic bone graft materials with polymer able to promote bone regeneration. Over the past two decades, interdisciplinary efforts emphasis to encourage materials from ceramics to polymer which composed of composite scaffolds, protein loaded mixing both with inorganic and organic phases. The scaffold materials aid unique tunable analogy in crystalline structure and chemistry of materials combined between calcium phosphate and apatite. Analogy of bone tissue is similar from hydroxyapatite to its carbonate. Calcium phosphate has been revealed the enhanced bone development on crystallinity, crystalline phase and ratio of Ca/P which releases calcium and phosphate ions needed for mineralization of bone. This analysis depends uniquely to differentiates from other metal oxides used in orthopedics. Scaffold-type architecture stems with inter-connected porosity is a condition for osteoconductivity and increases angiogenesis. Furthermore, there is more literature revealing that calcium phosphate bioceramic scaffolds enhance both osteogenesis and osseointegration, which are related to surface charge, surface chemistry and topography. However, it should be indicated that the prime application of CPSs for bone replacement. Therefore, the degree of mimicking with respect to bone does not expand beyond chemistry, surface topography, etc. When bone matures and it becomes stiffer and stronger.

A clinically successful tissue engineering move towards a chrono automatic scaffold which capable to trigger molecular agents for bio-integrating tissues. This move toward is part of the prosperous thought of chrono-biotechnology which may be able to manage over dosing and limited signaling molecule which contact in complex, anisotropic and 3D environment. Several attempts have been completed to obtain systems integrating delivery devices and tissue engineering matrix capable to mimic ECM and directionally rearrange tissue. Microsphere-integrated scaffolds, have inadequate protein liberation have been showed. Some related consequences have been achieved in developed scaffolds for the chronological release of multiple and consequent assembling into a porous scaffold. Currently, an anisotropic porous bi-layered scaffold is capable to depict in the spatial region and analogous to polylactic acid microparticles plasticized with polyethylene glycol were sintered into scaffolds created by free protein and loaded protein, thus permitting a discharge of dissimilar bioactive molecules limited to exact regions within the scaffold. These scaffolds may locate usefulness in applications or to increase on tissue growth.

In spite of these supporting results, there are still significant technological boundaries exist. The major problem relies on the use of loaded microspheres which are fairly adapted when fashioned into scaffolds, thus changing their building and as a result liberate features. In addition, engineer dynamic gradients of a signaling molecule, the complete indulgent to release kinetics at the single microsphere level. An alternative move toward to make microsphere-integrated scaffolds capable to adjust both temporally and spatially liberate kinetics which may take benefit of micromanipulation-based techniques. The effortless dispersal of microspheres within gel scaffolds is an ingrained come close to get a temporal control over release. It has been newly established that during the fine alteration of microsphere formulation in scaffold and probable to realize platforms competent for managing the microenvironmental circumstances in terms of time development of bioactive. Probable developments of this result might be advantage from micro and nanotechnologies, which embedded microspheres in the scaffold. In fact, devices performing as single point source might be micro located by 3D printing and to attain highly synchronized structures capable to trigger the scope, and perhaps the planning/structure of tissue arrangement. The combination of micropositioning and mathematical modeling performance in complex and multiple mechanisms leading to release of kinetics from single microspheres within the scaffold and capable to assist scaffolds with extremely controlled architecture by computer-aided scaffold design. In detail, previously pre-programmed in vitro and will not be capable to interactively adjust to release kinetics. As underlined a more effectual biomimicry could be attained if a double reciprocity scheme could be prearranged in the system. In this way, cells can activate the growth of extracellular matrix and engineered cell behaviors through the restricted release of bioactive molecules.

It should also be denoted that a multifunctional scaffold should not only afford a restricted administration of appropriate biomacromolecules, but also nearby such molecules in an appropriate conformation state, mimicking binding. Certainly, it has been exposed that molecularly adorned materials progress tissue ordering through the enunciation of the contact between protein signaling and biomaterials observed to be fundamental to offer an enhanced integration of the scaffold. Rising approaches to propose that the next generation future scaffolds will not be realized by basically integrating within the scaffolds. Certainly, benefit of the current drug discharge and biomaterials, multifunctional scaffolds, where the 3D polymer itself acts as a biomimetic, and multidrug delivery, should be intended.

Unusual progress has been found in the last decade for designing scaffolds with an appropriate hierarchical arrangement and vigorous release of proteins. Addition of progressive scaffold was constructed and lead to important in both therapeutic applications of tissue engineering and knowledge on cell and tissue. Still, scientific challenges have to be surmounted to understand a realistic imitate of the orchestration of structural and accessible by the extracellular matrix. For example, micro and nano technologies should be auxiliary broken in order to systematize biomaterial scaffold with the necessary resolution. On the other hand, to propose of future biomaterial platforms should advantage from a quantitative indulgent for effect of morphogens and gradients on cell. To exceed, the necessary decisive cell density to produce the preferred tissue configuration response. This fast response could be due to an advanced level of local responsive cells at the renovate site or a faster augment in the number of responsive cells. In the second scenario, the retention profile of carrier A does not uphold the minimum concentration for a sufficient period of time and to permit the slow cell population. The slower response may be due to a lesser level of occupant responsive cells or a slower augment in the number of responsive cells. Once the exogenous attentiveness decreases below the least level, the responsive cell density also starts to fall or instead may stay constant.

As a result, the tissue arrangement response is either insufficient for the preferred effect. The augmented preservation now maintains the least amount of exogenous concentration for a long sufficient stage time to permit the slower cell population to reach the critical cell density. Given the potential differences in obtainable receptive cells or differences in response rate, these scenarios assist give facts why certain carriers are efficacious at a given concentration in rodents. These scenarios are also demonstrated that the potential need for higher initial concentrations of carrier essential in large animal models. A wide range of withholding profiles can be achieved with various carrier combinations. The four main types of carrier materials contain natural polymers such as collagen, fibrin, CS, agarose, etc. at lower and higher temperature of calcium orthophosphates (sintered ceramics and inorganic materials).

In addition, combined with natural polymers, calcium orthophosphates and polymers have also been investigated. Allograft and autograft material have also been employed to bring composites of the materials have been employed for optimization for the benefits of each of these materials. Calcium orthophosphate can be incorporated to natural polymers to enhance compression resistance. The restricted release of behavior polymers has been combined with the biocompatibility of natural polymers. Natural and synthetic polymers have also been additional to ceramics to get better conduct uniqueness, injectability, and porosity. Metal and polymer or ceramic composites have also been evaluated.

Bone has a very elevated essential repair and remodeling aptitude, competent of regenerating new tissue the same to normal bone. The periosteum and endosteum can produce appositional novel bone on cortical surfaces. Besides, the two bones trabecular and cortical bone create restore and remodeling that are both recurrent and tremendously movable due to the dual purpose of the skeleton for structural maintain and calcium homeostasis. As a result of the augmented capability for regeneration, the necessities for carriers for bone repair are not as rigorous as for additional tissues.

Carrier provisions for the majority bone conservation are also easy by the simultaneous employ of any internal or exterior or eliminate the necessitate for emotionless or struggle to motion. Segmental diaphyseal repairs, spine fusions and bone loss are noticed and represent exceptions to this simplification. Bone is also extremely vascular and carriers employed for bone repair have to sustain vascular ingrowth.

The major indications for the employ of appendicular skeletal bone repair are inducing of diaphyseal and metaphyseal fracture, bridging defects and enhancing bone combination with metal implants. Extra indications comprise speeding up of interruption osteogenesis, local increase of bone density and osseous repair. The bulk of closed and open fractures have cured suddenly. As a result, quickening of rupture repair requires carriers with the suitable maintenance outline and seat time to permit cell infiltration and devoid of intrusive process, provocative an unsuitable inflammatory or compromising the mechanical repair with carrier. Besides, injectable for closed rupture repair be required and intelligent to administered through needles to reduce patient uneasiness and risk of disease.

Calcium phosphate-based polymer scaffolds for tissue engineering, still there are many issues must be addressed such as resorbability, cell viability etc. Polymer should be chosen in accordance with proper examination of its polymer which was related to enhance tissue ingrowth. Recently, scientist attempt to cure skin cancer in which dissolution and degradation studies are crucial for killing cancerous cells. Foremost thing is to understand the drug interaction with polymers and bioactive ceramics for rapid releasing or sustained delivery, etc. At present, graphene-based materials holding or withstanding higher drugs due to higher surface area. Moreover, graphene-based calcium phosphate polymers will be useful in skin cancer due to its photothermal properties where due to illumination of light heat is generated on the surface which aid to kill or destroy or decrease growth of cells.

Three patterning aids to increase the growth rate due to higher interaction of cells with porous surface of the template. Scientists are fascinated to develop palm-sized mock human organs, to design and to examine drugs and aid to comprehend the basic healthy parts or affected organs. Man-made artificial organic/inorganic based scaffolds for restoring specific function or the group of associated functions may provide or return to a normal life as early as possible. Tissue engineering provides regeneration of diseased tissue in the human body. Researchers continue to improve the generation of tissues in complex. The replacement of tissue is generated by 3D support system or scaffolds for developing functionality. The properties of scaffolds will vary depend on different parameters such as porosity, elasticity, anatomical shapes. The biomaterials are commonly used to restore cell signaling and to meet the required physical nature. Scaffolds have been provided to supply templates with precise topological nature. Major issue is to enable an active interplay with the cells. Extracellular matrix on biomaterial is anticipated to degrade, tissue-like structure which can integrate within the host tissue. Tissue engineering aids to reconstruct different tissues which contain skin, bone, cartilages etc.

17.5 CONCLUSION

Tissue engineering is a highly stimulating multidisciplinary research areas and is budding exponentially over time. Scaffold candidates and production technologies play a vital role in the field of tissue engineering. Wide range of calcium phosphate-based polymeric scaffolds were extensively employed in the region of tissue engineering. Scaffolds must assemble with definite plan parameters in the area and independent of their origin (natural or synthetic). To develop novel materials must be possess strong and have timed bioresorption. However, faster degradation was observed in polymeric materials than ceramics showing osteolysis issues. Innovations in the material propose and production processes are raising the opportunity of manufacture of implants with high performance. The CPS should be surface companionable as well as architecturally appropriate with the host atmosphere. The principles and theories of the manufacture process with calcium phosphate-based polymers would be functional to enlarge a novel plan for implants and also to appreciate the performance of the composite scaffold in the field of biomedical. New research must be addressed towards exact mimicking of the natural process of bone tissue regeneration which produces angiogenesis and osteogenesis for progenitor cell and segregation. The behavior of each biomolecule is organized and needed for the development of bone tissues within the scaffold matrix. Scaffolds have optimized pore size, degradation rate, delivery efficiency, release of biomolecules in time dependent manner much expected requirement of the third-generation composite scaffolds. Some comprehension toward precise signaling pathways and rigid factors can shed light to understand mechanisms at the cellular and molecular level for osteoblastic. However, the response mechanism which influences bone healing process is still an ambiguity.

REFERENCES

Abdal-hay, A., Hamdy, A.S., Khalil, K.A., Lim, J.H., 2014. A novel simple in situ biomimetic and simultaneous deposition of bonelike apatite within biopolymer matrix as bone graft substitutes. Mater. Lett. 137, 260–264.

Abou Taleb, M.F., Alkahtani, A., Mohamed, S.K., 2015. Radiation synthesis and characterization of sodium alginate/chitosan/hydroxyapatite nanocomposite hydrogels: a drug delivery system for liver cancer. Polym. Bull. 72 (4), 725–742.

Airenne, K.J., Hiltunen, M.O., Turunen, M.P., et al., 2000. Baculovirus-mediated periadventitial gene transfer to rabbit carotid artery. Gene Ther. 7 (17), 1499–1504.

Alexander, H., Brunski, J.B., Cooper, S.L., et al., 1996. Chapter 2—classes of materials used in medicine. Biomater Sci. Academic Press, San Diego, CA, pp. 37–130.

Amini, A.R., Laurencin, C.T., Nukavarapu, S.P., 2012. Bone tissue engineering: recent advances and challenges. Crit. Rev. Biomed. Eng. 40 (5), 363–408.

Arafat, M.T., Lam, C.X.F., Ekaputra, A.K., Wong, S.Y., Li, X., Gibson, I., 2011. Biomimetic composite coating on rapid prototyped scaffolds for bone tissue engineering. Acta Biomater. 7 (2), 809–820.

Ashizuka, M., Ishida, E., 1997. Mechanical properties of silicate glass–ceramics containing tricalcium phosphate. J. Mater. Sci. 32 (1), 185–188.

Bhowmick, A., Pramanik, N., Manna, P.J., et al., 2015. Development of porous and antimicrobial CTS–PEG–HAP–ZnO nano-composites for bone tissue engineering. RSC Adv. 5 (120), 99385–99393.

Bigi, A., Boanini, E., Panzavolta, S., Roveri, N., Rubini, K., 2002. Bonelike apatite growth on hydroxyapatite–gelatin sponges from simulated body fluid. J. Biomed. Mater. Res. 59 (4), 709–715.

Blaker, J.J., Gough, J.E., Maquet, V., Notingher, I., Boccaccini, A.R., 2003. In vitro evaluation of novel bioactive composites based on Bioglass-filled polylactide foams for bone tissue engineering scaffolds. J. Biomed. Mater. Res. A 67 (4), 1401–1411.

Bosch, P., Musgrave, D., Ghivizzani, S., Latterman, C., Day, C.S., Huard, J., 2000. The efficiency of muscle-derived cell-mediated bone formation. Cell Transplant. 9 (4), 463–470.

Boyan, B.D., Hummert, T.W., Dean, D.D., Schwartz, Z., 1996. Role of material surfaces in regulating bone and cartilage cell response. Biomaterials 17 (2), 137–146.

Boyan, B.D., Lohmann, C.H., Romero, J., Schwartz, Z., 1999. Bone and cartilage tissue engineering. Clin. Plast. Surg. 26 (4), 629–645. ix.

Bruckner, T., Schamel, M., Kubler, A.C., Groll, J., Gbureck, U., 2016. Novel bone wax based on poly(ethylene glycol)-calcium phosphate cement mixtures. Acta Biomater. 33, 252–263.

Bruder, S.P., Kurth, A.A., Shea, M., Hayes, W.C., Jaiswal, N., Kadiyala, S., 1998. Bone regeneration by implantation of purified, culture-expanded human mesenchymal stem cells. J. Orthop. Res. 16 (2), 155–162.

Bryant, S.J., Anseth, K.S., 2001. The effects of scaffold thickness on tissue engineered cartilage in photocrosslinked poly(ethylene oxide) hydrogels. Biomaterials 22 (6), 619–626.

Burke, J.F., Yannas, I.V., Quinby, W.C., Bondoc, C.C., Jung, W.K., 1981. Successful use of a physiologically acceptable artificial skin in the treatment of extensive burn injury. Ann. Surg. 194 (4), 413–428.

Cabodi, M., Choi, N.W., Gleghorn, J.P., Lee, C.S.D., Bonassar, L.J., Stroock, A.D., 2005. A microfluidic biomaterial. J. Am. Chem. Soc. 127 (40), 13788–13789.

Cao, H., Kuboyama, N., 2010. A biodegradable porous composite scaffold of PGA/beta-TCP for bone tissue engineering. Bone 46 (2), 386–395.

Cascone, M.G., Barbani, N., Cristallini, C., Giusti, P., Ciardelli, G., Lazzeri, L., 2001. Bioartificial polymeric materials based on polysaccharides. J. Biomater. Sci. Polym. Ed. 12 (3), 267–281.

Castellani, C., Lindtner, R.A., Hausbrandt, P., et al., 2011. Bone–implant interface strength and osseointegration: biodegradable magnesium alloy versus standard titanium control. Acta Biomater. 7 (1), 432–440.

Chaudhuri, B., Mondal, B., Ray, S.K., Sarkar, S.C., 2016. A novel biocompatible conducting polyvinyl alcohol (PVA)-polyvinylpyrrolidone (PVP)-hydroxyapatite (HAP) composite scaffolds for probable biological application. Colloids Surf., B: Biointerfaces 143, 71–80.

Chen, L.J., Wang, M., 2002. Production and evaluation of biodegradable composites based on PHB–PHV copolymer. Biomaterials 23 (13), 2631–2639.

Chen, J., Wang, Y., Chen, X., et al., 2011. A simple sol-gel technique for synthesis of nanostructured hydroxyapatite, tricalcium phosphate and biphasic powders. Mater. Lett. 65 (12), 1923–1926.

Chow, K.S., Khor, E., Wan, A.C.A., 2001. Porous chitin matrices for tissue engineering: fabrication and in vitro cytotoxic assessment. J. Polym. Res. 8 (1), 27–35.

Chu, K.T., Oshida, Y., Hancock, E.B., Kowolik, M.J., Barco, T., Zunt, S.L., 2004. Hydroxyapatite/PMMA composites as bone cements. Biomed. Mater. Eng. 14 (1), 87–105.

Ciardelli, G., Chiono, V., Vozzi, G., et al., 2005. Blends of poly-(ε-caprolactone) and polysaccharides in tissue engineering applications. Biomacromolecules 6 (4), 1961–1976.

Cifuentes, S.C., Frutos, E., González-Carrasco, J.L., et al., 2012. Novel PLLA/magnesium composite for orthopedic applications: a proof of concept. Mater. Lett. 74, 239–242.

Cifuentes, S.C., Bensiamar, F., Gallardo-Moreno, A.M., et al., 2016. Incorporation of Mg particles into PDLLA regulates mesenchymal stem cell and macrophage responses. J. Biomed. Mater. Res. A 104 (4), 866–878.

Daniels, A.U., Chang, M.K., Andriano, K.P., 1990. Mechanical properties of biodegradable polymers and composites proposed for internal fixation of bone. J. Appl. Biomater. 1 (1), 57–78.

DiEdwardo, C.A., Petrosko, P., Acarturk, T.O., DiMilla, P.A., LaFramboise, W.A., Johnson, P.C., 1999. Muscle tissue engineering. Clin. Plast. Surg. 26 (4), 647–656. ix–x.

Diba, M., Goudouri, O.-M., Tapia, F., Boccaccini, A.R., 2014. Magnesium-containing bioactive polycrystalline silicate-based ceramics and glass-ceramics for biomedical applications. Curr. Opin. Solid State Mater. Sci. 18 (3), 147–167.

Du, C., Cui, F.Z., Zhu, X.D., de Groot, K., 1999. Three-dimensional nano-HAp/collagen matrix loading with osteogenic cells in organ culture. J. Biomed. Mater. Res. 44 (4), 407–415.

Eaglstein, W.H., Falanga, V., 1997. Tissue engineering and the development of Apligraf®, a human skin equivalent. Clin. Ther. 19 (5), 894–905.

Freiberg, S., Zhu, X.X., 2004. Polymer microspheres for controlled drug release. Int. J. Pharm. 282 (1–2), 1–18.

Galperin, A., Oldinski, R.A., Florczyk, S.J., Bryers, J.D., Zhang, M., Ratner, B.D., 2013. Integrated bi-layered scaffold for osteochondral tissue engineering. Adv. Healthcare Mater. 2 (6), 872–883.

Gamradt, S.C., Lieberman, J.R., 2004. Genetic modification of stem cells to enhance bone repair. Ann. Biomed. Eng. 32 (1), 136–147.

Garai, S., Sinha, A., 2014. Biomimetic nanocomposites of carboxymethyl cellulose–hydroxyapatite: novel three dimensional load bearing bone grafts. Colloids Surf., B: Biointerfaces 115, 182–190.

Gentili, C., Cancedda, R., 2009. Cartilage and bone extracellular matrix. Curr. Pharm. Des. 15 (12), 1334–1348.

Germain, L., Auger, F.A., Grandbois, E., et al., 1999. Reconstructed human cornea produced in vitro by tissue engineering. Pathobiology 67 (3), 140–147.

Ghaee, A., Nourmohammadi, J., Danesh, P., 2017. Novel chitosan-sulfonated chitosan-polycaprolactone-calcium phosphate nanocomposite scaffold. Carbohydr. Polym. 157, 695–703.

Ghomi, H., Fathi, M.H., Edris, H., 2012. Effect of the composition of hydroxyapatite/bioactive glass nanocomposite foams on their bioactivity and mechanical properties. Mater. Res. Bull. 47 (11), 3523–3532.

Griffith, L.G., 2002. Emerging design principles in biomaterials and scaffolds for tissue engineering. Ann. N.Y. Acad. Sci. 961, 83–95.

Gu, X.-N., Zheng, Y.-F., 2010. A review on magnesium alloys as biodegradable materials. Front. Mater. Sci. Chin. 4 (2), 111–115.

Gunatillake, P., Mayadunne, R., Adhikari, R., 2006. Recent developments in biodegradable synthetic polymers. In: El-Gewely, M.R. (Ed.), Biotechnology Annual Review. Elsevier, pp. 301–347.

Hench, L.L., 1991. Bioceramics: from concept to clinic. J. Am. Ceram. Soc. 74 (7), 1487–1510.

Hench, L.L., 1998. Bioceramics. J. Am. Ceram. Soc. 81 (7), 1705–1728.

Hentrich, R.L., Graves, G.A., Stein, H.G., Bajpai, P.K., 1971. An evaluation of inert and resorbable ceramics for future clinical orthopedic applications. J. Biomed. Mater. Res. 5 (1), 25–51.

Hoffman, A.S., 2002. Hydrogels for biomedical applications. Adv. Drug Delivery Rev. 54 (1), 3–12.

Hoppe, A., Guldal, N.S., Boccaccini, A.R., 2011. A review of the biological response to ionic dissolution products from bioactive glasses and glass-ceramics. Biomaterials 32 (11), 2757–2774.

Hossan, M.J., Gafur, M.A., Karim, M.M., Rana, A.A., 2015. Mechanical properties of gelatin–hydroxyapatite composite for bone tissue engineering. Bangladesh J. Sci. Ind. Res. 50 (1), 15–20.

Hou, Q., Grijpma, D.W., Feijen, J., 2002. Preparation of porous poly(ε-caprolactone) structures. Macromol. Rapid Commun. 23 (4), 247–252.

Hou, Q., Grijpma, D.W., Feijen, J., 2003. Porous polymeric structures for tissue engineering prepared by a coagulation, compression moulding and salt leaching technique. Biomaterials 24 (11), 1937–1947.

Hubbell, J.A., 1999. Bioactive biomaterials. Curr. Opin. Biotechnol. 10 (2), 123–129.

Huser, A., Rudolph, M., Hofmann, C., 2001. Incorporation of decay-accelerating factor into the baculovirus envelope generates complement-resistant gene transfer vectors. Nat. Biotechnol. 19 (5), 451–455.

Hutmacher, D.W., 2000. Scaffolds in tissue engineering bone and cartilage. Biomaterials 21 (24), 2529–2543.

Hutmacher, D.W., Cool, S., 2007. Concepts of scaffold-based tissue engineering—the rationale to use solid free-form fabrication techniques. J. Cell. Mol. Med. 11 (4), 654–669.

Hutmacher, D.W., Schantz, J.T., Lam, C.X., Tan, K.C., Lim, T.C., 2007. State of the art and future directions of scaffold-based bone engineering from a biomaterials perspective. J. Tissue Eng. Regen. Med. 1 (4), 245–260.

Jhon, M.S., Andrade, J.D., 1973. Water and hydrogels. J. Biomed. Mater. Res. 7 (6), 509–522.

Ji, Y., Ghosh, K., Shu, X.Z., et al., 2006. Electrospun three-dimensional hyaluronic acid nanofibrous scaffolds. Biomaterials 27 (20), 3782–3792.

Jiang, G., Evans, M.E., Jones, I.A., Rudd, C.D., Scotchford, C.A., Walker, G.S., 2005. Preparation of poly(ε-caprolactone)/continuous bioglass fibre composite using monomer transfer moulding for bone implant. Biomaterials 26 (15), 2281–2288.

Jones, J.R., Ehrenfried, L.M., Hench, L.L., 2006. Optimising bioactive glass scaffolds for bone tissue engineering. Biomaterials 27 (7), 964–973.

Kaikkonen, M.U., Raty, J.K., Airenne, K.J., Wirth, T., Heikura, T., Yla-Herttuala, S., 2006. Truncated vesicular stomatitis virus G protein improves baculovirus transduction efficiency in vitro and in vivo. Gene Ther. 13 (4), 304–312.

Karageorgiou, V., Kaplan, D., 2005. Porosity of 3D biomaterial scaffolds and osteogenesis. Biomaterials 26 (27), 5474–5491.

Katti, K.S., Katti, D.R., Dash, R., 2008. Synthesis and characterization of a novel chitosan/montmorillonite/hydroxyapatite nanocomposite for bone tissue engineering. Biomed. Mater. 3 (3), 034122.

Khan, Y.M., Katti, D.S., Laurencin, C.T., 2004. Novel polymer-synthesized ceramic composite-based system for bone repair: an in vitro evaluation. J. Biomed. Mater. Res. A 69 (4), 728–737.

Kharaziha, M., Fathi, M.H., 2010. Improvement of mechanical properties and biocompatibility of forsterite bioceramic addressed to bone tissue engineering materials. J. Mech. Behav. Biomed. Mater. 3 (7), 530–537.

Kikuchi, M., Koyama, Y., Yamada, T., et al., 2004. Development of guided bone regeneration membrane composed of beta-tricalcium phosphate and poly(L-lactide-*co*-glycolide-*co*-epsilon-caprolactone) composites. Biomaterials 25 (28), 5979–5986.

Kim, H.N., Jiao, A., Hwang, N.S., et al., 2013. Nanotopography-guided tissue engineering and regenerative medicine. Adv. Drug Delivery Rev. 65 (4), 536–558.

Kim, H.W., Lee, E.J., Jun, I.K., Kim, H.E., Knowles, J.C., 2005. Degradation and drug release of phosphate glass/polycaprolactone biological composites for hard-tissue regeneration. J. Biomed. Mater. Res. B: Appl. Biomater. 75 (1), 34–41.

Kim, H.W., Kim, H.E., Knowles, J.C., 2006. Production and potential of bioactive glass nanofibers as a next-generation biomaterial. Adv. Funct. Mater. 16 (12), 1529–1535.

Kutikov, A.B., Gurijala, A., Song, J., 2015. Rapid prototyping amphiphilic polymer/hydroxyapatite composite scaffolds with hydration-induced self-fixation behavior. Tissue Eng. Part C, Methods 21 (3), 229–241.

Kwon, Y.S., Son, G., Suh, J., Kim, K.T., 1994. Densification and grain growth of porous alumina compacts. J. Am. Ceram. Soc. 77 (12), 3137–3141.

Langer, R., Vacanti, J., 1993. Tissue engineering. Science 260 (5110), 920–926.

Langer, R., Tirrell, D.A., 2004. Designing materials for biology and medicine. Nature 428 (6982), 487–492.

Laschke, M.W., Strohe, A., Menger, M.D., Alini, M., Eglin, D., 2010. In vitro and in vivo evaluation of a novel nanosize hydroxyapatite particles/poly(ester-urethane) composite scaffold for bone tissue engineering. Acta Biomater. 6 (6), 2020–2027.

Lee, K.Y., Mooney, D.J., 2001. Hydrogels for tissue engineering. Chem. Rev. 101 (7), 1869–1879.

Levenberg, S., Langer, R., 2004. Advances in tissue engineering. Current Topics in Developmental Biology. Academic Press, pp. 113–134.

Li, H., Chang, J., 2004. Preparation and characterization of bioactive and biodegradable wollastonite/poly(D,L-lactic acid) composite scaffolds. J. Mater. Sci. Mater. Med. 15 (10), 1089–1095.

Lieberman, J.R., Ghivizzani, S.C., Evans, C.H., 2002. Gene transfer approaches to the healing of bone and cartilage. Mol. Ther. 6 (2), 141–147.

Loh, X.J., Abdul Karim, A., Owh, C., 2015. Poly(glycerol sebacate) biomaterial: synthesis and biomedical applications. J. Mater. Chem. B 3 (39), 7641–7652.

Loh, X.J., Lee, T.C., Dou, Q., Deen, G.R., 2016. Utilising inorganic nanocarriers for gene delivery. Biomater. Sci. 4 (1), 70–86.

Lu, H.H., Tang, A., Oh, S.C., Spalazzi, J.P., Dionisio, K., 2005. Compositional effects on the formation of a calcium phosphate layer and the response of osteoblast-like cells on polymer-bioactive glass composites. Biomaterials 26 (32), 6323–6334.

Ma, P.X., 2004. Scaffolds for tissue fabrication. Mater. Today 7 (5), 30–40.

Ma, P.X., Zhang, R., 2001. Microtubular architecture of biodegradable polymer scaffolds. J. Biomed. Mater. Res. 56 (4), 469–477.

Macchetta, A., Turner, I.G., Bowen, C.R., 2009. Fabrication of HA/TCP scaffolds with a graded and porous structure using a camphene-based freeze-casting method. Acta Biomater. 5 (4), 1319–1327.

Mao, J.S., Zhao, L.G., Yin, Y.J., Yao, K.D., 2003. Structure and properties of bilayer chitosan–gelatin scaffolds. Biomaterials 24 (6), 1067–1074.

Maquet, V., Boccaccini, A.R., Pravata, L., Notingher, I., Jerome, R., 2004. Porous poly (alpha-hydroxyacid)/Bioglass composite scaffolds for bone tissue engineering. I: Preparation and in vitro characterisation. Biomaterials 25 (18), 4185–4194.

Mayer, J., Karamuk, E., Akaike, T., Wintermantel, E., 2000. Matrices for tissue engineering-scaffold structure for a bioartificial liver support system. J. Control. Release 64 (1–3), 81–90.

Mayer Jr, J.E., Shin'oka, T., Shum-Tim, D., 1997. Tissue engineering of cardiovascular structures. Curr. Opin. Cardiol. 12 (6), 528–532.

Merino, S., Martín, C., Kostarelos, K., Prato, M., Vázquez, E., 2015. Nanocomposite hydrogels: 3D polymer–nanoparticle synergies for on-demand drug delivery. ACS Nano 9 (5), 4686–4697.

Midwood, K.S., Williams, L.V., Schwarzbauer, J.E., 2004. Tissue repair and the dynamics of the extracellular matrix. Int. J. Biochem. Cell Biol. 36 (6), 1031–1037.

Mikos, A.G., Temenoff, J.S., 2000. Formation of highly porous biodegradable scaffolds for tissue engineering. Electron J. Biotechnol. 3, 23.

Miroiu, F.M., Socol, G., Visan, A., et al., 2010. Composite biocompatible hydroxyapatite–silk fibroin coatings for medical implants obtained by Matrix Assisted Pulsed Laser Evaporation. Mater. Sci. Eng., B 169 (1–3), 151–158.

Mohammad, J., Shenaq, J., Rabinovsky, E., Shenaq, S., 2000. Modulation of peripheral nerve regeneration: a tissue-engineering approach. The role of amnion tube nerve conduit across a 1-centimeter nerve gap. Plast. Reconstr. Surg. 105 (2), 660–666.

Mollazadeh, S., Javadpour, J., Khavandi, A., 2007. Biomimetic synthesis and mechanical properties of hydroxyapatite/poly(vinyl alcohol) nanocomposites. Adv. Appl. Ceram. 106 (4), 165–170.

Mooney, D.J., Organ, G., Vacanti, J.P., Langer, R., 1994. Design and fabrication of biodegradable polymer devices to engineer tubular tissues. Cell Transplant. 3 (2), 203–210.

Morawska-Chochol, A., Chlopek, J., Domalik-Pyzik, P., Szaraniec, B., Grzyska, E., 2014. Magnesium alloy wires as reinforcement in composite intramedullary nails. Biomed. Mater. Eng. 24 (2), 1507–1515.

Morawska-Chochół, A., Domalik-Pyzik, P., Chłopek, J., et al., 2014. Gentamicin release from biodegradable poly-L-lactide based composites for novel intramedullary nails. Mater. Sci. Eng., C 45, 15–20.

Morawska-Chochół, A., Chłopek, J., Szaraniec, B., et al., 2015. Influence of the intramedullary nail preparation method on nail's mechanical properties and degradation rate. Mater. Sci. Eng., C 51, 99–106.

Nair, L.S., Laurencin, C.T., 2007. Biodegradable polymers as biomaterials. Prog. Polym. Sci. 32 (8–9), 762–798.
Nonami, T., 1991. Developmental study of diopside for use as implant material. In: MRS Proceedings, vol. 87. Cambridge Univ Press.
Nonami, T., Tsutsumi, S., 1999. Study of diopside ceramics for biomaterials. J. Mater. Sci. Mater. Med. 10 (8), 475–479.
Oberpenning, F., Meng, J., Yoo, J.J., Atala, A., 1999. De novo reconstitution of a functional mammalian urinary bladder by tissue engineering. Nat. Biotechnol. 17 (2), 149–155.
Ouriemchi, E.M., Vergnaud, J.M., 2000. Processes of drug transfer with three different polymeric systems with transdermal drug delivery. Comput. Theor. Polym. Sci. 10 (5), 391–401.
Park, J.B., Lakes, R.S., 1992. Biomaterials—An Introduction. Plenum Press, New York.
Pazourková, L., Martynková, G.S., Placha, D., 2015. Preparation and mechanical properties of polymeric nanocomposites with hydroxyapatite and hydroxyapatite/clay mineral fillers—review. J. Tissue Eng. Regen. Med. 2 (007), 1–8.
Peppas, N.A., Khare, A.R., 1993. Preparation, structure and diffusional behavior of hydrogels in controlled release. Adv. Drug Delivery Rev. 11 (1–2), 1–35.
Peter, M., Ganesh, N., Selvamurugan, N., et al., 2010. Preparation and characterization of chitosan–gelatin/nanohydroxyapatite composite scaffolds for tissue engineering applications. Carbohydr. Polym. 80 (3), 687–694.
Pieroni, L., Maione, D., La Monica, N., 2001. In vivo gene transfer in mouse skeletal muscle mediated by baculovirus vectors. Hum. Gene Ther. 12 (8), 871–881.
Piskin, E., 1995. Biodegradable polymers as biomaterials. J. Biomater. Sci., Polym. Ed. 6 (9), 775–795.
Prokop, A., Chmielnicki, M., 2013. [Management of dislocated radial head fractures with biodegradable implants]. Z. Orthop. Unfall. 151 (6), 565–568.
Rajesh, R., Hariharasubramanian, A., Senthilkumar, N., Ravichandran, Y.D., 2012. A biocompatible and load bearing composite of multiwalled carbon nanotubes chitosan and natural hydroxyapatite derived from the chicken bones wasted in the slaughter houses. Int. J. Pharm. Pharm. Sci. 4 (4), 716–720.
Rajzer, I., Menaszek, E., Kwiatkowski, R., Chrzanowski, W., 2014. Bioactive nanocomposite PLDL/nano-hydroxyapatite electrospun membranes for bone tissue engineering. J. Mater. Sci. Mater. Med. 25 (5), 1239–1247.
Ramakrishna, S., Mayer, J., Wintermantel, E., Leong, K.W., 2001. Biomedical applications of polymer-composite materials: a review. Compos. Sci. Technol. 61 (9), 1189–1224.
Ramay, H.R.R., Zhang, M., 2004. Biphasic calcium phosphate nanocomposite porous scaffolds for load-bearing bone tissue engineering. Biomaterials 25 (21), 5171–5180.
Ramli, R.A., Adnan, R., Bakar, M.A., Masudi, S.M., 2011. Synthesis and characterisation of pure nanoporous hydroxyapatite. J. Phys. Sci. 22 (1), 20–37.
Rezwan, K., Chen, Q.Z., Blaker, J.J., Boccaccini, A.R., 2006. Biodegradable and bioactive porous polymer/inorganic composite scaffolds for bone tissue engineering. Biomaterials 27 (18), 3413–3431.
Roether, J.A., Boccaccini, A.R., Hench, L.L., Maquet, V., Gautier, S., Jerjme, R., 2002. Development and in vitro characterisation of novel bioresorbable and bioactive composite materials based on polylactide foams and Bioglass for tissue engineering applications. Biomaterials 23 (18), 3871–3878.
Rokkanen, P., Böstman, O., Hirvensalo, E., et al., 1999. Bioabsorbable implants in orthopaedics. Curr. Orthop. 13 (3), 223–228.

Roohani-Esfahani, S.-I., Nouri-Khorasani, S., Lu, Z., Appleyard, R., Zreiqat, H., 2010. The influence hydroxyapatite nanoparticle shape and size on the properties of biphasic calcium phosphate scaffolds coated with hydroxyapatite–PCL composites. Biomaterials 31 (21), 5498–5509.

Roohani-Esfahani, S.I., Nouri-Khorasani, S., Lu, Z.F., Appleyard, R.C., Zreiqat, H., 2011. Effects of bioactive glass nanoparticles on the mechanical and biological behavior of composite coated scaffolds. Acta Biomater. 7 (3), 1307–1318.

Sabree, I., Gough, J.E., Derby, B., 2015. Mechanical properties of porous ceramic scaffolds: influence of internal dimensions. Ceram. Int. 41 (7), 8425–8432.

San Miguel, B., Kriauciunas, R., Tosatti, S., et al., 2010. Enhanced osteoblastic activity and bone regeneration using surface-modified porous bioactive glass scaffolds. J. Biomed. Mater. Res. A 94 (4), 1023–1033.

Shie, M.-Y., Ding, S.-J., Chang, H.-C., 2011. The role of silicon in osteoblast-like cell proliferation and apoptosis. Acta Biomater. 7 (6), 2604–2614.

Shu, X., Shi, Q., Feng, J., Xie, X., Chen, Y., 2014. Design and in vitro evaluation of novel γ-PGA/hydroxyapatite nanocomposites for bone tissue engineering. J. Mater. Sci. 49 (22), 7742–7749.

Shuai, C., Liu, T., Gao, C., et al., 2016. Mechanical and structural characterization of diopside scaffolds reinforced with graphene. J. Alloys Compd. 655, 86–92.

Siriphannon, P., Monvisade, P., Jinawath, S., Hemachandra, K., 2007. Preparation and characterization of hydroxyapatite/poly(ethylene glutarate) biomaterials. J. Biomed. Mater. Res. Part A 81A (2), 381–391.

Song, B., Xu, Q., Wang, C., Xu, S., Zhang, H., 2016. Fabrication of polymer/drug-loaded hydroxyapatite particle composite fibers for drug sustained release. J. Appl. Polym. Sci. 133 (3), 42871.

Su, X., Tan, M.J., Li, Z., et al., 2015. Recent progress in using biomaterials as vitreous substitutes. Biomacromolecules 16 (10), 3093–3102.

Sundar Raj, M., Arkin, V.H., Adalarasu, Jagannath, M., 2013. Nanocomposites based on polymer and hydroxyapatite for drug delivery application. Indian J. Sci. Technol. 6 (55), 4653–4658.

Swain, S.K., Bhattacharyya, S., Sarkar, D., 2015. Fabrication of porous hydroxyapatite scaffold via polyethylene glycol–polyvinyl alcohol hydrogel state. Mater. Res. Bull. 64, 257–261.

Tabata, Y., 2003. Tissue regeneration based on growth factor release. Tissue Eng. 9 (Suppl 1), S5–S15.

Tarafder, S., Balla, V.K., Davies, N.M., Bandyopadhyay, A., Bose, S., 2013. Microwave-sintered 3D printed tricalcium phosphate scaffolds for bone tissue engineering. J. Tissue Eng. Regen. Med. 7 (8), 631–641.

Teixeira, S., Fernandes, H., Leusink, A., et al., 2010. In vivo evaluation of highly macroporous ceramic scaffolds for bone tissue engineering. J. Biomed. Mater. Res. A 93 (2), 567–575.

Tran, K.T., Griffith, L., Wells, A., 2004. Extracellular matrix signaling through growth factor receptors during wound healing. Wound Repair Regen. 12 (3), 262–268.

Tziampazis, E., Sambanis, A., 1995. Tissue engineering of a bioartificial pancreas: modeling the cell environment and device function. Biotechnol. Prog. 11 (2), 115–126.

Uma Maheshwari, S., Samuel, V.K., Nagiah, N., 2014. Fabrication and evaluation of (PVA/HAp/PCL) bilayer composites as potential scaffolds for bone tissue regeneration application. Ceram. Int. 40 (6), 8469–8477.

Vert, M., 2005. Aliphatic polyesters: great degradable polymers that cannot do everything. Biomacromolecules 6 (2), 538–546.
Visscher, G.E., Robison, R.L., Maulding, H.V., Fong, J.W., Pearson, J.E., Argentieri, G.J., 1985. Biodegradation of and tissue reaction to 50:50 poly(DL-lactide-*co*-glycolide) microcapsules. J. Biomed. Mater. Res. 19 (3), 349–365.
Waizy, H., Seitz, J.-M., Reifenrath, J., et al., 2013. Biodegradable magnesium implants for orthopedic applications. J. Mater. Sci. 48 (1), 39–50.
Wang, C.Y., Wang, S., 2005. Adeno-associated virus inverted terminal repeats improve neuronal transgene expression mediated by baculoviral vectors in rat brain. Hum. Gene Ther. 16 (10), 1219–1226.
Wang, C.Y., Li, F., Yang, Y., Guo, H.Y., Wu, C.X., Wang, S., 2006. Recombinant baculovirus containing the diphtheria toxin A gene for malignant glioma therapy. Cancer Res. 66 (11), 5798–5806.
Wang, S., Wang, X., Xu, H., et al., 2010. Towards sustained delivery of small molecular drugs using hydroxyapatite microspheres as the vehicle. Adv. Powder Technol. 21 (3), 268–272.
Wang, T., Yang, X., Qi, X., Jiang, C., 2015. Osteoinduction and proliferation of bone-marrow stromal cells in three-dimensional poly (ε-caprolactone)/hydroxyapatite/collagen scaffolds. J. Transl. Med. 13, 152.
Wang, X., Fan, H., Xiao, Y., Zhang, X., 2006. Fabrication and characterization of porous hydroxyapatite/β-tricalcium phosphate ceramics by microwave sintering. Mater. Lett. 60 (4), 455–458.
Watanabe, K., Nishio, Y., Makiura, R., Nakahira, A., Kojima, C., 2013. Paclitaxel-loaded hydroxyapatite/collagen hybrid gels as drug delivery systems for metastatic cancer cells. Int. J. Pharm. 446 (1–2), 81–86.
Wei, G., Ma, P.X., 2004. Structure and properties of nano-hydroxyapatite/polymer composite scaffolds for bone tissue engineering. Biomaterials 25 (19), 4749–4757.
Witte, F., 2015. Reprint of: the history of biodegradable magnesium implants: a review. Acta Biomater. 23 (Suppl), S28–S40.
Woo, B.H., Kostanski, J.W., Gebrekidan, S., Dani, B.A., Thanoo, B.C., DeLuca, P.P., 2001. Preparation, characterization and in vivo evaluation of 120-day poly(D,L-lactide) leuprolide microspheres. J. Control. Release 75 (3), 307–315.
Woodard, J.R., Hilldore, A.J., Lan, S.K., et al., 2007. The mechanical properties and osteoconductivity of hydroxyapatite bone scaffolds with multi-scale porosity. Biomaterials 28 (1), 45–54.
Wu, C., Chang, J., 2007. Degradation, bioactivity, and cytocompatibility of diopside, akermanite, and bredigite ceramics. J. Biomed. Mater. Res. B: Appl. Biomater. 83 (1), 153–160.
Wu, C., Ramaswamy, Y., Zreiqat, H., 2010. Porous diopside ($CaMgSi_2O_6$) scaffold: a promising bioactive material for bone tissue engineering. Acta Biomater. 6 (6), 2237–2245.
Wu, C., Zhou, Y., Fan, W., et al., 2012. Hypoxia-mimicking mesoporous bioactive glass scaffolds with controllable cobalt ion release for bone tissue engineering. Biomaterials 33 (7), 2076–2085.
Wu, S., Liu, X., Yeung, K.W.K., Liu, C., Yang, X., 2014. Biomimetic porous scaffolds for bone tissue engineering. Mater. Sci. Eng., R: Rep. 80, 1–36.
Xie, J., Blough, E.R., Wang, C.H., 2012. Submicron bioactive glass tubes for bone tissue engineering. Acta Biomater. 8 (2), 811–819.

Xie, J., Zhong, S., Ma, B., Shuler, F.D., Lim, C.T., 2013. Controlled biomineralization of electrospun poly(epsilon-caprolactone) fibers to enhance their mechanical properties. Acta Biomater. 9 (3), 5698–5707.

Xu, H.H.K., Quinn, J.B., Takagi, S., Chow, L.C., 2004. Synergistic reinforcement of in situ hardening calcium phosphate composite scaffold for bone tissue engineering. Biomaterials 25 (6), 1029–1037.

Xue, W., Bandyopadhyay, A., Bose, S., 2009. Polycaprolactone coated porous tricalcium phosphate scaffolds for controlled release of protein for tissue engineering. J. Biomed. Mater. Res. B: Appl. Biomater. 91 (2), 831–838.

Yamada, S., Heymann, D., Bouler, J.M., Daculsi, G., 1997. Osteoclastic resorption of calcium phosphate ceramics with different hydroxyapatite/beta-tricalcium phosphate ratios. Biomaterials 18 (15), 1037–1041.

Yang, C., Lee, Y., Chiu, S., et al., 2008. Preparation of a PVA/HAP composite polymer membrane for a direct ethanol fuel cell (DEFC). J. Appl. Electrochem. 38 (10), 1329.

Yang, D.-G., Chung, Y.-C., Lai, Y.-K., Lai, C.-W., Liu, H.-J., Hu, Y.-C., 2007. Avian influenza virus hemagglutinin display on baculovirus envelope: cytoplasmic domain affects virus properties and vaccine potential. Mol. Ther. 15 (5), 989–996.

Yannas, I., Burke, J., Orgill, D., Skrabut, E., 1982. Regeneration of skin following closure of deep wounds with a biodegradable template. Trans. Soc. Biomater. 5, 24–27.

Yannas, I.V., 1992. Tissue regeneration by use of collagen-glycosaminoglycan copolymers. Clin. Mater. 9 (3–4), 179–187.

Yannas, I.V., Burke, J.F., 1980. Design of an artificial skin. I. Basic design principles. J. Biomed. Mater. Res. 14 (1), 65–81.

Yannas, I.V., Burke, J.F., Gordon, P.L., Huang, C., 1977. Multilayer membrane useful as synthetic skin. In: Google Patents.

Yannas, I.V., Burke, J.F., Warpehoski, M., et al., 1981. Prompt, long-term functional replacement of skin. Trans. Am. Soc. Artif. Intern. Organs 27, 19–23.

Yannas, I.V., Orgill, D.P., Silver, J., Norregaard, T.V., Zervas, N.T., Schoene, W.C., 1985. Polymeric template facilitates regeneration of sciatic nerves across 15-mm gap. Trans. Soc. Biomater. 8, 145.

Yoshimoto, H., Shin, Y.M., Terai, H., Vacanti, J.P., 2003. A biodegradable nanofiber scaffold by electrospinning and its potential for bone tissue engineering. Biomaterials 24 (12), 2077–2082.

Zhang, E., Xu, L., Yu, G., Pan, F., Yang, K., 2009. In vivo evaluation of biodegradable magnesium alloy bone implant in the first 6 months implantation. J. Biomed. Mater. Res. A 90 (3), 882–893.

Zhang, R., Ma, P.X., 1999. Porous poly(L-lactic acid)/apatite composites created by biomimetic process. J. Biomed. Mater. Res. 45 (4), 285–293.

Zhang, Y., Zhang, M., 2001. Synthesis and characterization of macroporous chitosan/calcium phosphate composite scaffolds for tissue engineering. J. Biomed. Mater. Res. 55 (3), 304–312.

Zhang, Y., Zhang, M., 2002. Three-dimensional macroporous calcium phosphate bioceramics with nested chitosan sponges for load-bearing bone implants. J. Biomed. Mater. Res. Part A 61 (1), 1–8.

Zhou, Z., Zhou, Y., Chen, Y., et al., 2011. Bilayer porous scaffold based on poly-(ε-caprolactone) nanofibrous membrane and gelatin sponge for favoring cell proliferation. Appl. Surf. Sci. 258 (5), 1670–1676.

Zia, R., Riaz, M., Maqsood, S., Anjum, S., Kayani, Z., Hussain, T., 2015. Titania doped bioactive ceramics prepared by solid state sintering method. Ceram. Int. 41 (7), 8964–8972.

CHAPTER

Metal nanoparticles-based colorimetric methods for drug analyses 18

Suresh Kumar Kailasa[1,2], Tae-Jung Park[2] and Janardhan Reddy Koduru[3]

[1]*Department of Applied Chemistry, S. V. National Institute of Technology, Surat, India*

[2]*Department of Chemistry, Institute of Interdisciplinary Convergence Research, Research Institute of Halal Industrialization Technology, Chung-Ang University, Seoul, Republic of Korea*

[3]*Department of Environmental Engineering and, Kwangwoon University, Seoul, Republic of Korea*

18.1 INTRODUCTION

Metal nanoparticles (NPs)-based colorimetric approaches have been proven to be successful analytical strategies for the recognition and quantification of a wide variety molecules in complex samples due to their simplicity, rapidity, and cost-effectiveness (Li et al., 2013; Bhamore et al., 2015; Jans and Huo 2012). In recent years, great research has been performed on the use of metal NPs as colorimetric sensors for the selective and sensitive recognition of various species including drugs, biomolecules, pesticides, and inorganic species at low volume of samples (Vilela et al., 2012; Upadhyayula, 2012; Kailasa and Wu, 2012). Recent reports have revealed that the promising roles of metal NPs in analytical sciences for the detection of various analytes can be described via localized surface plasmon resonance (SPR) (LSPR), fluorescence resonance energy transfer, surface enhanced Raman scattering (SERS), and nonlinear optical properties (Ghosh and Pal, 2007 and Saha et al., 2012). It is well-known that metal NPs exhibit plasmonic (SPR and LSPR) that have significantly received much attention in assaying of various chemical species due to their remarkable enhancements in selective and sensitivity (Li et al., 2015; Wang and Astruc, 2014). The SPR phenomenon of metal NPs and their applications have already been described in the review articles (Mayer and Hafner, 2011; Klinkova et al., 2014). The metal NPs possess unique optical properties that is, LSPR and surface plasmon polaritons, which make them as good candidates to integrate with UV–visible spectrometric methods for the quantification of various molecules (Kelly et al., 2003; Gwo et al., 2016). Notably, the use of metal NPs in UV–visible spectrometric methods has lift-up the performance of UV–visible spectrometry for assaying of various molecules (inorganic species, pesticides, biomolecules and toxic molecules), just like as

Nanoarchitectonics in Biomedicine. DOI: https://doi.org/10.1016/B978-0-12-816200-2.00003-7

sophisticated instruments (Ringe et al., 2013; Vilela et al., 2012; Upadhyayula, 2012; Borisov and Klimant, 2008).

Importantly, drug assays in their pharmaceutical and biological fluids play an essential role in evaluating the physiological action of a drug. Generally, drug analysis is required for several sample preparation steps prior to their identification by an analytical instrument (Rani et al., 2016). The presence of other components seriously interferes with target drugs, which results in generating analytical data with significant errors. In view of this, various analytical methods have been utilized for the analysis of various drugs including gas chromatography/mass spectrometry (GC/MS), and liquid chromatography (LC) coupled with various detectors (UV, fluorescence, and MS) (Kostiainen et al., 2003). Although these methods effectively quantified the drugs with more accuracy and precision, unfortunately they are expensive and time consuming as well as requiring tedious sample preparations. Furthermore, the complexity in pharmaceutical and biological fluids requires efficient chromatographic techniques for the separation of drugs from their samples.

The integration of Au and Ag NPs with UV−visible spectrometry has proven to miniaturize the colorimetric strategy for cost-effective, in situ identification of various trace level target molecules without the aid of any chromatographic techniques (Mehta et al., 2013; Modi et al., 2014). Interestingly, Au and Ag NPs acted as promising colorimetric probes for instantaneous identification of trace level analytes, which showed instantaneous color change that can be observed with the naked eye. The Au and Ag NPs-based colorimetric strategies rely on two points, that is, (1) the specific interaction of the target analytes with surface ligands of metal NPs; and (2) the significant change in the SPR band and visual color change (visual region 390−750 nm) (Mehta et al., 2013; Modi et al., 2014). These sensing strategies are classified into two categories, that is, (1) analyte-induced aggregation of Au and Ag NPs (aggregation mechanism); and (2) either in the morphology transition or in the surface integrations of a single particle.

Due to their significant achievements in recognizing the trace level analytes, several reports have explored the optical properties and analytical applications of Au and Ag NPs (Liu et al., 2014; Li et al., 2017a,b; Xie et al., 2017). These reviews explain the fabrication, characterization, optical properties, and applications of metal NPs (Peng et al., 2015; Li et al., 2012). Further, colorimetric sensing applications of metal NPs have been well-documented and summarized in the literature (Sun and Xia, 2003; Upadhyayula, 2012; Li et al., 2015). In this chapter, we will explore the analytical applications of Au and Ag NPs for simple and sensitive colorimetric sensing of various drugs in their pharmaceutical and biological samples. Apart from Au and Ag NPs, we will also summarize the use of Cu and Pt NPs for the detection of drugs. Due to space limitations, we will only explore the use of metal NPs (Ag, Au, Cu, and Pt) in spectrometric methods for the analysis of various drugs with minimized sample preparations at low volume of samples. Finally, we will discuss the challenges and future prospects of metal NPs for label-free and sensitive detection of drugs in real samples.

18.2 AU NANOPARTICLES-ASSISTED UV–VISIBLE SPECTROMETRIC METHODS FOR DRUG ANALYSIS

Among several analytical applications of Au NPs, molecular sensing is an exciting area as it deals with the detection of various chemical species through specific interactions between the surfaces of Au NPs and target analytes. As a result, the dispersed Au NPs began to aggregate which can cause significant changes in their color (red to blue) and spectral changes (red-shift). These specific characteristics make them promising colorimetric probes for the selective recognition of various molecular species including inorganic ions (Mehta and Kailasa, 2015; Mehta et al., 2014), pesticides (Rohit et al., 2014; Rohit et al., 2016), and biomolecules (Rawat et al., 2016; D'souza et al., 2015; Kasibabu et al., 2015). In recent years, Au NPs have been used as legendary colorimetric probes for the selective and sensitive identification of various drugs in their pharmaceutical formulations and biofluids. For example, adrenaline was successfully detected by using molecularly imprinted polymer (MIP) immobilized Au NPs as a novel color indicator (Matsui et al., 2004). The MIP-Au NPs were selectively interacted with adrenaline, causing a blue shift in their SPR band, which can use for the detection of adrenaline. Four neurotransmitters [L-3,4-dihydroxyphenylalanine-L-DOPA, dopamine (DA), adrenaline, and noradrenaline] were assayed by the degree of generation of Au NPs through the activity of tyrosinase (Baron et al., 2005). It was noticed that the absorption maxima of Au NPs is gradually increased with increasing tyrosinase concentration, and a blue shift was also observed due to the formation of both larger Au NPs and small Au nanoclusters. This approach was effectively applied to estimate the activity of tyrosinase in melanocytes and melanoma cells, which enables that this can be used as a simple analytical protocol for clinical diagnostics. Similarly, Kong et al. (2011) developed a facile and new spectrophotometric method for the detection of L-DOPA and tyrosinase activity using Au NPs/SiO_2 as a probe. In this approach, the growth of Au NPs was proportional to the concentration of L-tyrosine increases, confirming the formation of L-DOPA via catalytic reactions between tyrosinase and $AuCl_4^-$. As a result, L-DOPA was successfully detected by measuring the growth of Au NPs via UV–visible spectrometry. Furthermore, tyrosinase acted as a biocatalyst to hydroxylation of L-tyrosine and then forming L-DOPA (Fig. 18.1). Finally, dopaquinone was formed due the oxidation of L-DOPA either catalyzed by tyrosinase or oxidized by $AuCl_4^-$. Based on these specific reactions, this strategy was applied to evaluate tyrosinase activity in clinical and food samples.

5-Fluorouracil (5-FU) acts as an anticancer drug in cancer chemotherapy. In view of this, Au NPs were used as probes for the quantification of 5-FU and the interactions of Au NPs with 5-FU were characterized by UV–visible, fluorescence, and Fourier transform infrared (FTIR) spectroscopy (Selvaraj and Alagar, 2007). Importantly, 5-FA (folic acid)-Au NPs have shown higher antibacterial activity than that of pure 5-FU. A simple optical sensing method was developed

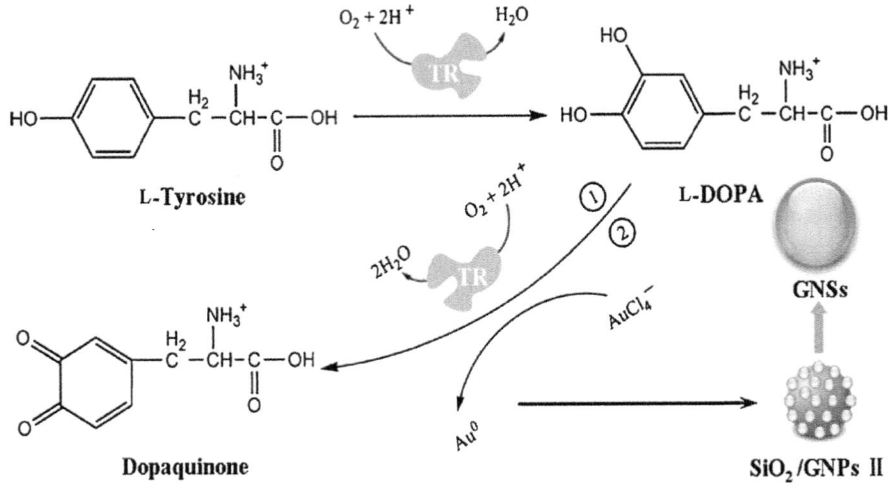

FIGURE 18.1

Sensing mechanism for detection of L-DOPA by the formation of Au NPs. *NP*, Nanoparticle.

Reprinted from Kong, F., Liu, H., Dong, J., Qian, W., 2011. Growth-sensitive gold nanoshells precursor nanocomposites for the detection of L-DOPA and tyrosinase activity. Biosens Bioelectron. 26, 1902–1907. With permission.

for the analysis of polyionic drugs (protamine and heparin) (Jena and Raj, 2008). This approach was based on the aggregation of Au NPs induced by protamine and de-aggregation of Au NPs induced by heparin (Fig. 18.2A). The citrate-Au NPs exhibited the SPR band at 514 nm, however it was red-shifted with the addition of protamine (red to blue color). Interestingly, the aggregated Au NPs were de-aggregated with the addition of heparin, yielding a blue shift in the SPR band (Fig. 18.2B). The strong electrostatic interactions were observed between negative charges of citrate-Au NPs and positive charges of protamine (polycation), resulting the aggregation of citrate-Au NPs. The citrate-Au NPs-protamine aggregates were de-aggregated with the addition of heparin, yielding a blue shift in the SPR band and a color change from blue to red. This reveals the dissociation of antithrombin III–heparin complex by protamine, which results in strong electrostatic interactions between protamine and heparin.

Glutathione is also used as a drug in antioxidant cellular defense (Mates et al., 2000). Copolymers [poly(*N*-*n*-isopropylacrylamide-*co*-acryloyldiethyletriamine)] were fabricated onto gold cores and used as a probe for visual sensing of glutathione in eye drops and health supplements (Uehara et al., 2010). This strategy was based on the de-aggregation of Au NPs with the addition of glutathione, favoring the increase of the interparticle distance of the Au NPs, which cause a red shift and a color change. Cetyl trimethyl ammonium bromide (CTAB) was decorated on Au NPs surfaces and utilized as a probe for colorimetric determination of

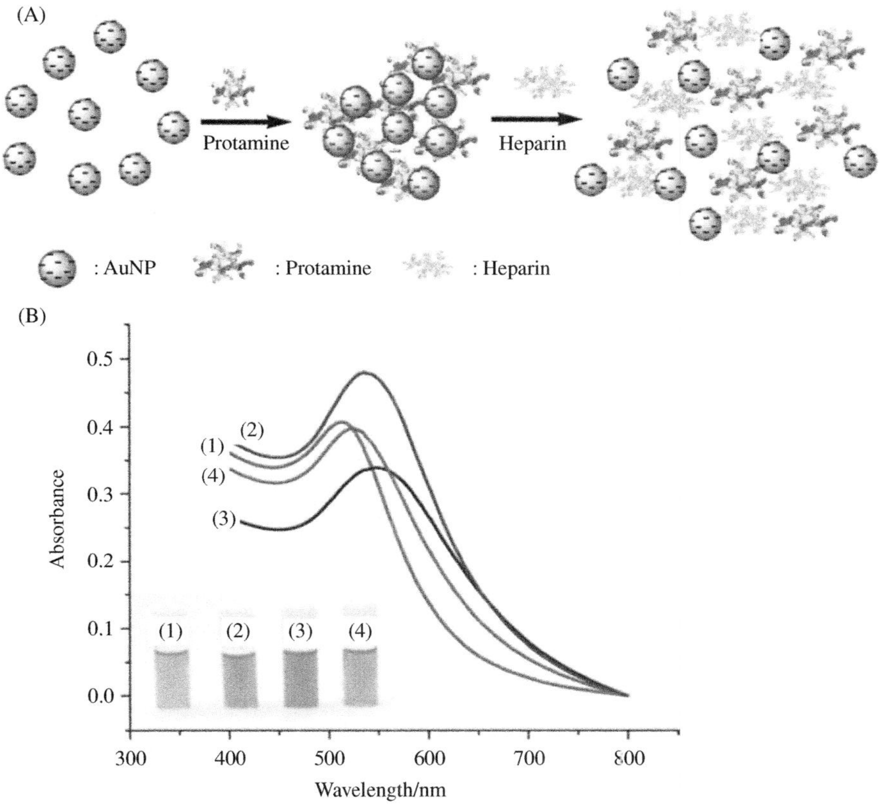

FIGURE 18.2

(A) Colorimetric principle for the detection of protamine and heparin via Au NPs aggregation and de-aggregation. (B) UV–visible absorption spectra of (1) Au NPs; (2) Au NPs with (2) 0.7 μg/mL; (3) 1.6 μg/mL of protamine; and (4) Au NPs with heparin (10.2 μg/mL). Inset: Photographic image of corresponding Au NPs solutions. *NP*, Nanoparticle.

Reprinted from Jena, B.K., Raj, C.R., 2008. Optical sensing of biomedically important polyionic drugs using nano-sized gold particles. Biosens. Bioelectron. 23, 1285–1290. With permission.

Penicillin G acylase (PGA) enzyme (Tiwari et al., 2010). The detection principle was based on an absorption ratio between 545 and 527 nm and the color intensity was progressively changed from red to blue with the increasing concentration of PGA. This strategy was successfully applied to analyze Pen G in pharmaceutical samples. Zhang et al. (2010) quantified DA by integrating citrate-Au NPs with UV–visible spectrometry. This method was based on the aggregation of citrate-Au NPs induced by DA via hydrogen bonding. To enhance the selectivity of the probe and aggregation, the authors used Cu^{2+} ion as a linker for binding of DA molecule and citrate-Au NPs, favoring to form Cu^{2+} ion-DA-citrate-Au NPs tetragonal geometry of complexes. As a result, this method exhibited good selectivity toward DA and provided a lower limit of detection (LOD) for DA. Similarly, gemcitabine HCl (GMCT) was successfully quantified in its bulk drug and pharmaceutical formulations by the integration of citrate-Au NPs with extractive UV spectrophotometry (Menon et al., 2012). The authors noticed that GMCT has strong tendency to aggregate citrate-Au NPs via electrostatic interactions, causing a SPR band shift from 522 to 688 nm which offers the detection of GMCT at micromolar concentrations.

Atenolol is widely used as a blocking agent for cardioselective β-adrenergic receptor. Shamsipur et al. (2014) estimated that the concentration of atenolol in pharmaceutical and biological (urine and plasma) samples by various analytical instruments (differential pulse voltammetry, cyclic voltammetry, and chronoamperometry) through its electrochemical oxidation on Au NPs/multiwall carbon nanotubes. The probe exhibited a wider linear range and detected atenolol concentrations as low as 59 nM. Santos et al. (2014) illustrated the use of chitosan as a reducing and capping agent for the synthesis of Au NPs. The chitosan-Au NPs were coated on the glass slide and used as a promising substrate for the sensing of paracetamol by SERS. This method has proven to be a facile, rapid, and cost-effective strategy for the quantification of paracetamol in pharmaceutical samples.

Recently, plasmonic metal nanocomposites have been prepared by using various organic moieties and fabricated on the surfaces of electrodes. For example, Alarfaj and El-Tohamy (2016) described the use of gelatin capped (Ag and Au) bimetallic NPs as a potential probe for the analysis of raloxifene hydrochloride in pharmaceutical samples. Under optimized conditions, the developed strategy exhibited good linearity in the range of $1.0 \times 10^{-9}-1.0 \times 10^{-1}$ M, which allows the detection of raloxifene hydrochloride even at nanomolar concentrations (1.0×10^{-9} M). This method exhibited higher selectivity toward raloxifene hydrochloride over the other chemical species.

Neurotransmitters exhibit numerous arrays of functions in the central as well as the peripheral nervous systems (Yusoff et al., 2015). In view of this, a facile and accurate analytical strategy is essential for the quantification of DA in various samples. For example, dithiobis(succinimidylpropionate) (DSP) was decorated on the surfaces of Au NPs for colorimetric assay of DA using Fe^{3+} ions as a binding agent (Liu et al., 2012). The authors observed the formation of supramolecular networks by forming bonds between $-NH_2$ group of DA and $-COOH$ group of DSP-Au NPs via coupling reaction. At the same time, the Fe^{3+} ions greatly reduced the DA-DSP-Au NPs interparticle distance via the formation of $DA-Fe^{3+}$ ion complexes. As a result, the calibration graph of the absorption ratio versus DA concentration shows a linear range from 5.0 to 500 nM with a detection limit of 2.0 nM. Similarly, DA was successfully detected in the presence of other competitive chemical species [α-amino acids, glutathione, glucose, uric acid, and ascorbic acid (AA)] using Au NPs as a probe and Cu^{2+} ions as a discriminator and linker (Su et al., 2012). Good linearity was observed between the absorption ratio at 650 and 525 nm and a DA concentration in the range of 5×10^{-7}–1×10^{-5} M, which allows to detect DA at 2×10^{-7} M. This probe has the ability to detect DA in human serum within 30 min, which confirms that sample preparation steps are greatly minimized. Independent studies have been performed on the functionalization of Au NPs with melamine (MA) for the simple and rapid colorimetric assay of DA in biofluids (human serum and urine) (Chen et al., 2015a,b). These sensing strategies were based on the formation of larger DA-MA-Au NPs supramolecular networks via hydrogen bonding, resulting in a change of color from blue to purple-blue, which can be noticed with the naked eye. As a result, DA was effectively estimated in human biofluids (urine and plasma) with an impressive detection limit at nanomolar concentrations. These methods have simplified the analytical protocols for the detection of DA in biological samples at a small volume of samples and the results well-agreed with sophisticated analytical instruments. Furthermore, a novel non-aggregation method was developed for the detection of DA by the formation of Au-Ag shell nanorods (Au@AgNRs) (Liu et al., 2013). This method was based on the reduction of Ag^{+} ion to Ag^{0} by DA as a reducing agent facilitating the formation of Au@AgNRs nanostructured networks, which cause a significant blue shift in the SPR band of Au NPs. The intensity of the blue shift showed good linear enhancement in the range of 0.2–12.0×10^{-6} M, offering a detection limit of 47×10^{-8} M. The SPR band changes (blue shift) were relatively negligible with the addition of other drugs and chemical species, signifying the selectivity of the method for DA analysis.

To improve the selectivity of Au NPs toward DA, Leng et al. (2015) developed a simple and sensing UV–visible spectrometric method for the detection of DA in urine and serum samples by using Ag NPs as a probe. In this approach, DA has shown a strong tendency to aggregate Au NPs at basic pH. The authors used thioglycolic acid (TGA) to modify DA-Au NPs through the hydrolysate of TGA, which results in the formation of Au–S bonding. Furthermore, strong hydrogen bonding was observed between TGA and DA, promoting a high degree of aggregation with Au NPs. The color of Au NPs was gradually changed, that is, red→purple→gray→yellow, which is attributed to the increasing concentration of DA (Fig. 18.3A and B). These visual changes led to drastic changes in the SPR peak of Au NPs, resulting in a decrease in the intensity with increasing DA concentration, which provides good linearity in the range of 0 and 10^{-9}–10^{-5} M (Fig. 18.3C). Common interfering species (aminoacids, drugs, metal ions, and anions) do not induce the visual and spectral changes in an Au NPs solution, enabling the successful identification of DA in biofluids. Furthermore, Altintas et al. (2015) synthesized NPs using molecular imprinting (MIP) with diclofenac. The synthesized MIP was decorated on Au nanochips via carbodiimide 1-ethyl-3-(3-dimethylaminopropyl)-carbodiimide (EDC)/*N*-hydroxysuccinimide coupling reaction. It was observed that protein kinases have been used as drug targets in pharmaceutical industry. Since protein kinases act as cellular logic gates for

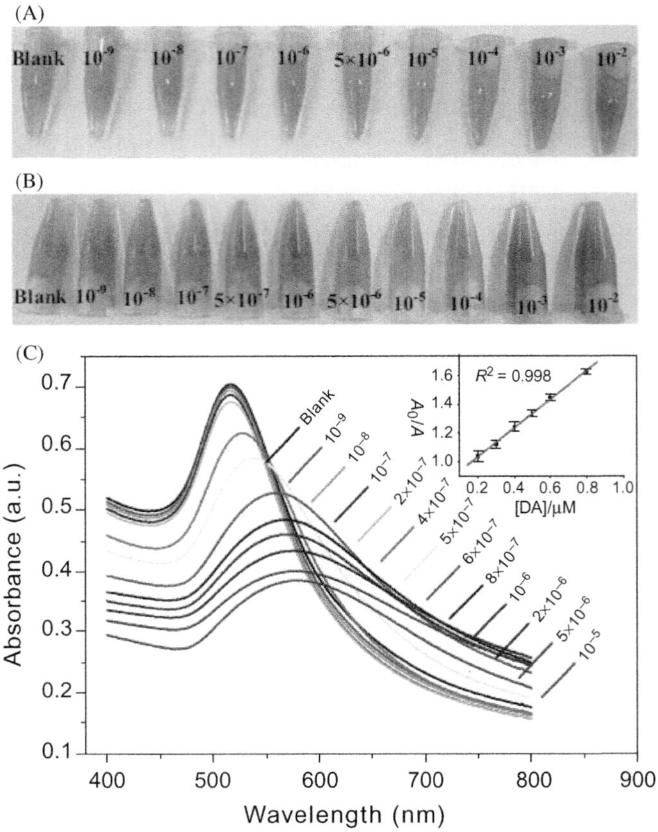

FIGURE 18.3

Colorimetric detection of concentrations of DA based on Au NPs (A) without and (B) with TGA^{2-}, (C) the corresponding UV–vis absorption spectra of (B), and the inset of (C): absorbance ratio (A_0/A) of (C) vs. DA concentrations. (The concentrations of DA: 0, 10^{-9}, 10^{-8}, 10^{-7}, 2×10^{-7}, 4×10^{-7}, 5×10^{-7}, 6×10^{-7}, 8×10^{-7}, 10^{-6}, 2×10^{-6}, 5×10^{-6}, 10^{-5}, 10^{-4}, 10^{-3}, and 10^{-2} M, respectively). *NP*, Nanoparticle; *DA*, dopamine; *TGA*, thioglycolic acid.

Reprinted from Leng, Y., Xie, K., Ye, L., Li, G., Lu, Z., He, J., 2015. Gold-nanoparticle-based colorimetric array for detection of dopamine in urine and serum. Talanta 139, 89–95. With permission.

phosphorylation of proteins and the abnormal phosphorylation of proteins causes lethal diseases. For example, Bhalla et al. (2015) developed a versatile analytical strategy for the evaluation of protein phosphorylation using Au metal–insulator–semiconductor capacitor nanostructures. In this approach, the authors fabricated Au NPs with thio-phosphorylated proteins, which was confirmed by the SPR band change while the amount of thio-phosphorylated proteins was estimated with quartz crystal microbalance.

Testosterone is an important chemical substance that was used in various clinical applications. Therefore, clinical analytical chemistry requires highly selective and sensitive analytical platforms for the accurate quantification of testosterone in various biological samples. To meet these merits, Au NPs have been combined with UV–visible spectrometry for the trace level identification of testosterone in various samples. Mitchell and Lowe (2009) synthesized an Au NPs-based sensor by conjugating oligoethylene glycol with testosterone. The authors noticed that the fabricated sensor exhibited an ultrasensitive nature toward testosterone in the presence of a secondary antibody. Interestingly, the use of a secondary antibody plays key role in improving the sensitivity to 12.5-fold compared to a pure primary antibody. This method exhibited a LOD at 3.7 pg/mL and complete testosterone analysis was completed within 13 min. Similarly, Yockell-Lelièvre et al. (2015) synthesized 70 nm size branched Au NPs and again they reduced size of Au NPs with 25 nm via Ostwald ripening process. The Au NPs were decorated with testosterone-biotin or mercapto carboxylic acids—11-mercaptoundecanetriethylglycol acid and 11-mercaptoundecanoic acid—causing a bathochromic shift from 525 to 538 and 556 nm, which resulted in visual changes (red to purple). The authors noticed that testosterone-biotin/folate-Au NPs acted as a superior sensing platform for the accurate detection of testosterone, which provides a detection limit of 0.17 nM.

In recent years, the use of Au NPs in UV–visible spectrometric techniques has boosted its performance with regard to selectivity and sensitivity, similarly to sophisticated instruments. To enhance their role in UV–visible spectrometric-based platforms, Au NPs have been synthesized by using various novel organic molecules as reagents and successfully integrated with an UV–visible spectrometric technique for the quantification of various drugs. Generally, triptan-family drugs (i.e., sumatriptan, rizatriptan, naratriptan, and zolmitriptan) are widely used as antimigraines to control various paroxysmal disorders (headache, nausea, photophobia, vomiting, and malaise) (Belvis et al., 2009). In view of this, Au NPs were synthesized using 4-aminoantipyrine (AAP) as a new reducing and capping reagent for parallel colorimetric detection of sumatriptan, naratriptan, zolmitriptan, and rizatriptan in their pharmaceutical samples (Rawat et al., 2014). The SPR band and size of Au NPs were tuned by optimizing the AAP concentration, reaction time, and temperature, respectively. Interestingly, each drug has shown a different degree of Au NPs aggregation, resulting in the appearance of new SPR bands at different wavelengths, which reveals selective visual color changes. The above four drugs were selectively induced by the aggregation of AAP-Au NPs through $\pi-\pi$ and electrostatic interactions. The other competitive chemical species and drugs did not interfere with the detection of the four drugs, providing the limits of detections at nanomolar concentrations. Due to the specific degree of aggregations, this method was successfully applied to detect the four drugs in their pharmaceutical formulations. Four other drugs (venlafaxine, amlodipine, imipramine, and alfuzosin) have been successfully detected by using citrate-Au NPs as a probe (Rawat et al., 2015). The four drugs greatly induced the aggregation of citrate-Au NPs and red-shifted with different values at 653, 695, 688 and 636 nm, accompanied by a color change from red to blue. Upon the exposure of different concentrations of the four drugs, the calibration graph showed good linearity in the range of 0.001–100 μM, which enables the detection of the four drugs even at nanomolar range. Because of their selectivity and sensitivity towards target drugs, the authors applied them as promising colorimetric probes to detect four drugs in their pharmaceutical samples and biofluids.

A novel UV–visible spectrometric method was developed for the quantification of ascorbic acid (AA), acetaminophen (ACET), and FA using Au NPs-based hydrogel as a sensor (Bahram et al., 2014). In this approach, the Au NPs-based hydrogel was fabricated by using poly(styrene-alt-maleic acid) as a template. This approach exhibited wider linear ranges with LODs of 1.8×10^{-6}, 2×10^{-6} and 2.7×10^{-7} M for ACET, FA, and AA, respectively. The practical application of this method was assessed by quantifying the above drugs in serum samples. Trovafloxacin mesylate (TVX-M) acts as an antibacterial agent for oral administration and is widely used for various treatments, e.g., gynecologic, pneumonia, and pelvic infections. Alarfaj and El-Tohamy (2015a) quantified trovafloxacin mesylate in its formulations by the development of two analytical strategies, that is, sequential injection chemiluminescence analysis and fluorescence spectrometric methods. Utilizing these systems, trovafloxacin mesylate was detected even at 1.7×10^{-10} by chemiluminescence and at 3.3×10^{-9} M by fluorescence methods. These methods were effectively applied to detect trovafloxacin mesylate in its bulk and pharmaceutical formulations. Table 18.1 illustrates the developed works for drug analysis using Au NPs-based UV–visible, fluorescence, SERS, and voltammetric techniques.

Bromelain is an enzyme that exhibits numerous applications in pharmaceutical as well as other industries (Rowan et al., 1990). It acts as an anti-tumor agent and shows antiedematous, anti-inflammatory, antithrombotic, and fibrinolytic activities (Maurer, 2001). In view of this, bromelain was detected by using carbon dots (CDs) functionalized Au NPs as a probe in two spectrometric (UV–visible and fluorescence) methods (Rawat et al., 2016). In the colorimetric method, bromelain was quantified based on the degree of aggregation of CDs-Au NPs, exhibiting good linearity between at the absorption ratio (A_{576}/A_{523}) and concentration (1.0–260 μM), which allows a detection limit of 18.9 nM. Similarly, bromelain was detected based on the progressive enhancement of the fluorescence signal at 425 nm, resulting good linear relationship at I/I_0 verses concentration 1.0–100 nM with a detection limit of 0.52 nM. The concentration of bromelain was successfully estimated in pineapple and biofluids. Similarly, Zhang et al. (2016) developed Au NPs-based dual-mode analytical strategies, that is, colorimetric and fluorometric methods for the analysis of DA. In this approach, the authors treated fluorophore with aptamers and then conjugated it with Au NPs. It was observed that the inter-particle distance of Au NPs was greatly reduced with

Table 18.1 Overview of Au Nanoparticles (NPs)-Based Analytical Techniques for the Analysis of Various Drugs

NPs	SPR Peak (nm)	Size (nm)	Name of Drug	Linear Range (M)	LOD (M)	Technique	Reference
Au-MIP	533	5.3	Adrenaline	5×10^{-6}–2×10^{-3}	–	UV–visible	Matsui et al. (2004)
Au NPs	–	10–20	DA, L-DOPA, adrenaline, and noradrenaline	–	2.5×10^{-6}	UV–visible	Baron et al. (2005)
SiO_2/Au NPs	520		L-DOPA and tyrosinase		2.5×10^{-6}	UV–visible	Kong et al. (2011)
Citrate-Au NPs	518	18	5-FU	–	–	UV–visible	Selvaraj, and Alagar (2007)
Citrate-Au NPs	514	5	Protamine and heparin	–	0.1 and 0.6 μg/mL	UV–visible	Jena and Raj (2008)
Citrate-Au NPs	520	13	Glutathione	6×10^{-6}	2.9×10^{-8}	UV–visible	Uehara et al. (2010)
CTAB-Au NPs	527	15	Penicillin G acylase	–	–	UV–visible	Tiwari et al. (2010)
Citrate-Au NPs	520	13	DA	3.3×10^{-8}–1.0×10^{-7} and 3.0×10^{-7}–4.5×10^{-6}	30×10^{-9}	UV–visible	Zhang et al. (2010)
Citrate-Au NPs	522	41	GMCT	2.0–40 μg/mL	0.44 μg/mL	UV–visible	Menon et al. (2012)
Au NPs/MWCNTs		80	Atenolol	1–60×10^{-6}	0.5×10^{-6}	Voltammetry	Shamsipur et al. (2014)
Au NPs–chitosan	537	1.5–12.5	Paracetamol		1×10^{-3}	SERS	Santos et al. (2014)
Gelatin/Au–Ag NPs	480	30	Raloxifene HCl	1.0×10^{-9}–1.0×10^{-1}	5.0×10^{-10}	Chemiluminescence	Alarfaj and El-Tohamy (2016)
DSP-Au NPs	520		DA	5–600 nM	2 nM	UV–visible	Liu et al. (2012)
Au NPs	525	–	DA	5×10^{-7}–1×10^{-5}	2×10^{-7}	UV–visible	Su et al. (2012)
MA-Au NPs	530	15	DA	33×10^{-9}–3.33×10^{-3}	33×10^{-9}	UV–visible	Chen et al. (2015a)

(*Continued*)

Table 18.1 Overview of Au Nanoparticles (NPs)-Based Analytical Techniques for the Analysis of Various Drugs *Continued*

NPs	SPR Peak (nm)	Size (nm)	Name of Drug	Linear Range (M)	LOD (M)	Technique	Reference
Citrate-MA-Au NPs	530	15	DA	0.3×10^{-6}–10×10^{-9}	30×10^{-9}	UV–visible	Chen et al. (2015b)
Au@AgNRs	–	–	DA	0.2–12.0×10^{-6}	47×10^{-8}	UV–visible	Liu et al. (2013)
TGA-Au NPs	525	–	DA	0 to $\times 10^{-6}$	1.0×10^{-7} and 9.4×10^{-8}	UV–visible	Leng et al. (2015)
Functionalized Au NPs	–	25	Testosterone	–	0.17×10^{-9}	UV–visible	Yockell-Lelièvre et al. (2015)
4-AAP-Au NPs	544	<13.0	Naratriptan, sumatriptan, rizatriptan and zolmitriptan	2.5–1250×10^{-6}	0.031, 0.033, 0.043 and 0.019×10^{-6}	UV–visible	Rawat et al. (2014)
Citrate-Au NPs	521	5.3	Venlafaxine, imipramine, amlodipine and alfuzosin	0.01–100×10^{-6}	6.3, 0.9, 8.4 and 9.3×10^{-6}	UV–visible	Rawat et al. (2015)
Au NPs hydrogel	–	13	Acetaminophen, ascorbic acid, and folic acid	6–38×10^{-6}, 1–7×10^{-6}, 5–38×10^{-6}	1.8×10^{-6}, 2.7×10^{-7}, 2×10^{-6}	UV–visible	Bahram et al. (2015)
Au NPs–luminol–potassium ferricyanide	–	5	Trovafloxacin	1.0×10^{-8}–1.0×10^{-3} and 1.0×10^{-9}–1.0×10^{-2}	3.3×10^{-9} and 1.7×10^{-10}	Fluorescence and SI-CL	Alarfaj and El-Tohamy (2015a)
CDs-Au NPs	523	8.2	Bromelain	1.0–260×10^{-6} and 1.0–100×10^{-9}	18.9×10^{-9} and 0.52×10^{-9}	UV–visible and fluorescence	Rawat et al. (2016)
Aptamer-Au NPs	524	13	DA	0.17–4.0×10^{-6}	78.7×10^{-9}	UV–visible	Zhang et al. (2016)
Aptamer-Au NPs	520	14	Methamphetamine	–	5.0×10^{-3}	UV–visible	Yarbakht and Nikkhah (2016)
BSA-Au NPs	–	–	6-MP	1.0×10^{-7}–1.2×10^{-4}	1.98×10^{-8}	Fluorescence	Chen et al. (2013)
CTAB-Au NPs	–	–	6-MP	0.0681–1.702 μg/mL	3.32 ng/mL	RLS	Li et al. (2017)

SPR, *Surface plasmon resonance;* LOD, *limit of detection;* MIP, *molecularly imprinted polymer;* CTAB, *cetyl trimethyl ammonium bromide;* DA, *dopamine;* GMCT, *gemcitabine HCl;* MWCNT, *multiwall carbon nanotube;* DSP, *dithiobis(succinimidylpropionate);* MA, *melamine;* TGA, *thioglycolic acid;* AAP, *aminoantipyrine;* CD, *carbon dot;* BSA, *bovine serum albumin;* MP, *mercaptopurine;* FU, *fluorouracil.*

the addition of DA, which leads to a remarkable visual change (red to blue) and a red shift in their SPR band. Notably, emission intensity of fluorophore was quenched by its modification with an aptamer; however, it was gradually restored with the increasing concentration of DA. As a result, both the strategies have the ability to detect DA even at 78.7 nM. These sensing platforms have proved to be rapid tools for the quantification of DA in plasma samples. Yarbakht and Nikkhah (2016) designed a sensitive bioassay strategy for the detection of methamphetamine (MA) and 3,4-methylenedioxy-*N*-methylamphetamine using Au NPs as a probe. In this approach, aptamer (APtMeth) was assembled on Au NPs where aptamers did not induce the aggregation. Interestingly, MA greatly implies the degree of aggregation, yielding a decrease in the SPR band due to the formation of larger Au NPs agglomerates, for which a visual change was observed even at 5.0 mM. This method provides a LOD as low as 10^{-4} M. The analytical results of the method are well-agreed with the GC-mass spectrometric method, which signifies the potential application of this method.

6-Mercaptopurine (6-MP) is used as a drug for various treatments, that is, Crohn's disease and acute lymphoblastic leukemia (Nielsen et al., 2001). A fluorescence sensor was established for the detection of 6-MP using bovine serum albumin (BSA)-Au NPs (Chen et al., 2013). In this approach, Cu^{2+} ions were used as a binder to quench the fluorescence of BSA-Au NPs; however, the fluorescence intensity of BSA-Au NPs was gradually restored with the addition of 6-MP (1.0×10^{-7}–1.2×10^{-4} M). As a result, the authors reported an impressive LOD of 1.98×10^{-8} M for 6-MP. The authors applied this sensor to detect 6-MPA in serum samples. Recently, Au NPs were decorated with CTAB and then integrated with Rayleigh light scattering (RLS) technique for assays of 6-MP in synthetic samples (Li et al., 2017a,b). This method is based on the RLS intensity enhancement on the addition of 6-MP. Since the mercapto group of 6-MP has a strong tendency to replace CTAB, which results to form Au–S bond. As a result, RLS intensity was progressively increased with the addition of 6-MP (0.0681–1.702 μg/mL), with a detection limit of 3.32 ng/mL.

18.3 AG NANOPARTICLES-ASSISTED SPECTROMETRIC METHODS FOR DRUG ANALYSES

In recent years, great efforts have been made with the use of Ag NPs as sensors for the analysis of various molecules due to their size-dependent color and optical properties (Shim et al., 2008). Importantly, Ag NPs exhibit remarkable brilliant colors based on their size variations that make them potential candidates in assaying various molecules (Vilela et al., 2012). In this section, we explore Ag NPs-assisted spectrophotometric methods for the analysis of various drugs in pharmaceutical samples. For example, Ling et al. (2008) integrated Ag NPs with UV–visible spectrophotometric method for the analysis of berberine hydrochloride in pharmaceutical samples with high selectivity. The detection principle was based on the aggregation of Ag NPs induced by berberine hydrochloride via strong electrostatic interactions at pH 5.0. Since, the surfaces of Ag NPs have negative charges due to citrate ions (pK_a 3.13), strong electrostatic interactions were generated with berberine hydrochloride. Upon the increasing concentration of berberine hydrochloride, the color of Ag NPs solutions turned from yellow to dark yellow to green-blue, resulting a good linear relationship between SPR intensities and berberine hydrochloride ($0.5–5.0 \times 10^{-7}$ M), which provides a detection limit of 1.3×10^{-8} M. This method has proven to be the simplest protocol for the assay of berberine hydrochloride in tablets. Das et al. (2009) developed a prototype Ag NPs biosensor to detect the drug binding ability with cytochrome P450 3A4 (CYP3A4) and their interactions were evaluated by UV–visible spectrometry. In this work, Ag NPs were fabricated with CYP3A4 via nanosphere lithography. According to Alarfaj et al. (2015b), the percentage of cefditoren pivoxil was estimated in its pharmaceutical formulations (Meiact 200 mg/tablet) by integrating Ag NPs with a sequential injection analysis chemiluminescence technique. In the presence of other drugs, cefditoren pivoxil strongly influenced the chemiluminescence signal intensity, confirming the selectivity of the method, which provides a detection limit of 0.5 pg/mL.

Similarly, introducing citrate on the surfaces of Ag NPs enables a simple "visual read out" analytical strategy for the analysis of various drugs using UV–visible spectrometry. For example, Laliwala et al. (2014) developed a simple and rapid UV–visible spectrometric method for the parallel detection of the four triptan-family drugs (rizatriptan, sumatriptan, zolmitriptan, and naratriptan) in their pharmaceutical samples. The sensing mechanism is based on the aggregation of Ag NPs induced by the four drugs via electrostatic interactions, resulting in a remarkable SPR band-shift toward longer wavelengths, which comes with visual changes (Fig. 18.4). Importantly, the SPR band shifts of Ag NPs with other drugs are negligible, suggesting the selectivity of the method. The Ag NPs-assisted UV–visible spectrometric method showed better performance than traditional analytical techniques and was successfully applied to detect their concentrations in their pharmaceutical samples—tablets–Rizatriptan benzoate (rizatriptan–5.0 mg); sumatriptan and naproxen (sumatriptan—85 mg) and nasal spray (zolmitriptan—5.0 mg)—respectively. Further, Ag NPs have been integrated with optical spectroscopic techniques (i.e., SERS, chemiluminescence, and UV–visible spectroscopic techniques) for the detection of various drugs, including acetylsalicylic acid (Sallum et al., 2014), entecavir (Gao et al., 2014), 6-MP (Biparva et al., 2015), gamma-aminobutyric acid (Jinnarak and Teerasong, 2016), timolol (Amirjani et al., 2016), and azithromycin (Chavada et al., 2017) in their pharmaceutical formulations and biological samples. The developed sensing strategies were based on the aggregation of Ag NPs induced by drug, favoring drastic

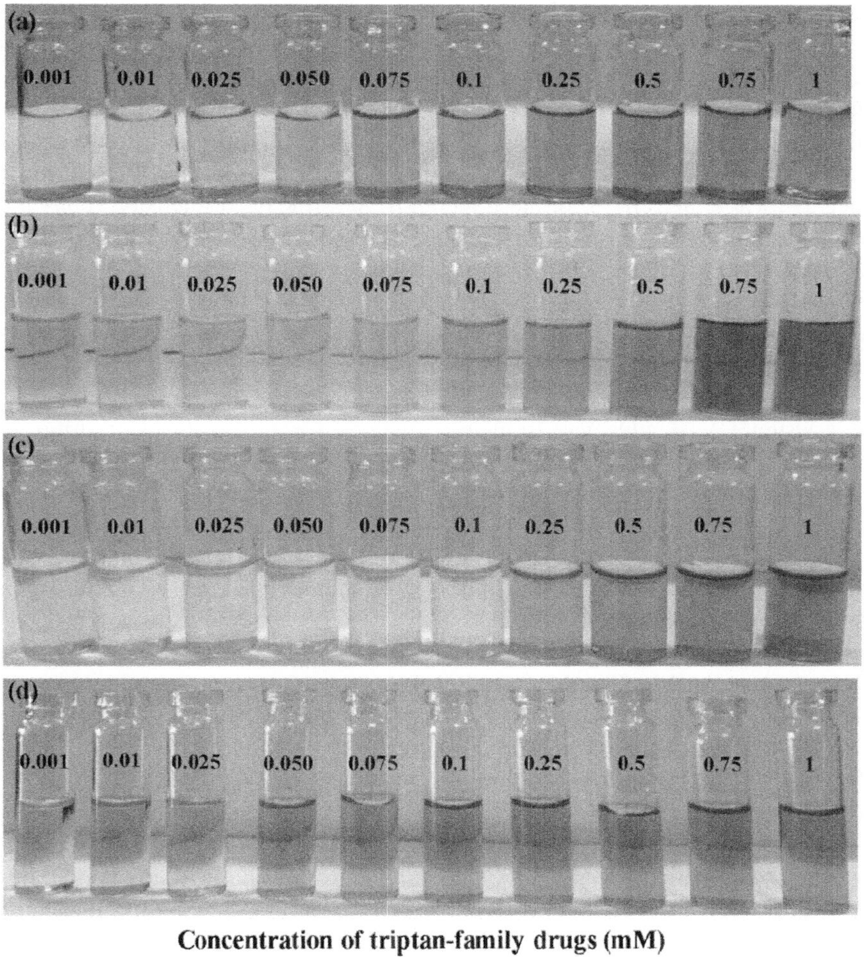

FIGURE 18.4

Photographic images of citrate-capped Ag NPs aggregation with (A) rizatriptan, (B) naratriptan, (C) sumatriptan, and (D) zolmitriptan in concentration range of 0.001 to 1.0 mM, respectively. *NP*, Nanoparticle.

Reprinted from Laliwala, S.K., Mehta, V.N., Rohit, J.V., Kailasa, S.K., 2014. Citrate-modified silver nanoparticles as a colorimetric probe for simultaneous detection of four triptan-family drugs. Sens. Actuators, B: Chem. 197, 254–263. With permission.

spectral changes in the UV–visible absorption spectra of Ag NPs, which cause remarkable visual changes. These changes are mainly due to strong interactions (electrostatic, $\pi-\pi$, and hydrogen bonding) and the spectral intensities were progressively enhanced with the increasing concentration of the drug, which enables the construction of calibration graphs with good correlation coefficients. The authors noticed that other common chemical species showed negligible effects during their quantifications, and impressive LODs at nanomolar concentrations were achieved with these methods. Finally, these methods were successfully applied to quantify the drugs in their pharmaceutical formulations and biofluids with good analytical merits. Recently, Ag NPs were fabricated by using FA as a reducing agent for dual detection (UV−visible and fluorescence) of 6-MP in its tablets (Rawat et al., 2017). In these methods, the SPR band intensity at 399 nm was gradually decreased with the addition of 6-MP concentration (20−1000 nM), whereas the emission peak intensity at 448 nm was progressively enhanced with addition of 6-MP (5−300 nM). As a result, the SPR band was blue-shifted with the restoration of the FA emission peak at 448 nm, allowing to quantify 6-MP in tablets and biofluids, which offers detection limits of 13.2 and 2.3 nM by both methods. Table 18.2 summarizes the developed Ag NPs-assisted UV−visible spectrometric methods for the analysis of drugs in their pharmaceutical samples.

Apart from these NPs in drug analysis, few reports have also illustrated the potential use of platinum NPs (Pt NPs) and copper NPs (Cu NPs) as probes for drugs analysis. Briefly, Derikvand et al. (2016) fabricated a novel nanocomposite by mixing Pt NPs with carbon nanotubes-polyethyleneimine and used as an aptamer sensor for the detection of diclofenac (DIF). Under optimal conditions, the concentration of DIF was effectively estimated by impedance spectroscopy, showing good linearity (10−200 nM), which is more suitable for the quantitative analysis of DIF. This method showed LOD as low as 2.7 nM. Similarly, Wang et al. (2015) quantified *N*-acetylcysteine (NAC) drug in NAC tablets by integrating poly(thymine) (poly T)-capped Cu NPs with fluorescence spectrometry. The emission intensity of poly-T-Cu NPs was remarkably quenched by the addition of NAC, thereby a calibration graph was constructed between the fluorescence intensity and concentration of NAC (0.2−100 μM), which allows a lower LOD at 50 nM to be achieved. The authors noticed that the emission intensity quenching by NAC was not influenced over the other competitive drugs, which reveals the selectivity of the method for assaying NAC.

A simple and rapid analytical strategy was developed for the analysis of penicillin antibiotics (penicillin G, penicillin V, oxacillin, dicloxacillin, nafcillin, and cloxacillin) using Pt NPs combined with magnetic nanocomposites (Kwon et al., 2015). The detection principle was based on the color change of 3,3′5,5′-tetramethylbenzidine (TMB) by the catalytic function of Pt NPs in the presence of hydrogen peroxide, which results in quantifying penicillin drugs in meat samples by UV−visible spectrometry. A similar approach was applied to detect penicillin G by decorating penicillin G-antibody-dendritic Pt NPs on a glass vial (Kwon et al., 2018). This method was based on the oxidation of TMB in the presence of antibody-conjugated Pt NPs-H_2O_2. A linear relationship was observed between color intensity and penicillin G concentration, displaying high selectivity toward penicillin G over other drugs, which represents its application for penicillin G analysis in milk samples.

Table 18.2 Summary of Ag Nanoparticles (NPs)-Based Analytical Methods for the Analysis of Various Drugs

NPs	SPR Peak (nm)	Size (nm)	Name of the Drug	Linear Range (M)	LOD (M)	Technique	Reference
Citrate-Ag NPs	–	–	Berberine-HCl	$0.5–5.0 \times 10^{-7}$	1.3×10^{-8}	UV–visible	Ling et al. (2008)
Ag NPs-CYP3A4 nanodisk	415	–	Testosterone, ketoconazole	–	–	LSPR spectroscopy	Das et al. (2009)
Citrate-Ag NPs	–	–	Cefditoren pivoxil	0.001–5000 ng/mL	0.5 pg/mL	Chemiluminescence	Alarfaj et al. (2015b)
Citrate-Ag NPs	396	5.0	Rrizatriptan, naratriptan, sumatriptan, and zolmitriptan	$0.001–1.0 \times 10^{-6}$	$7.3–84.0 \times 10^{-9}$	UV–visible	Laliwala et al. (2014)
Ag NPs-filter paper		180	Acetylsalicylic acid	$0.1–1.0 \times 10^{-3}$	–	SERS	Sallum et al. (2014)
Ag NPs-citrate	398	22	Entecavir	5.04–25.2 μg/mL	1.51 μg/mL	UV–visible	Gao et al. (2014)
Ag NPs–TCPO-H_2O_2	425	18.23	6-Mercaptopurine	$5.5 \times 10^{-7}–5.5 \times 10^{-5}$	1.6×10^{-7}	Chemiluminescence	Biparva et al. (2015)
Citrate-Ag NPs	390	17	Gamma-aminobutyric acid	100–500 mg/L	57.7 mg/L	UV–visible	Jinnarak and Teerasong (2016)
Ag NPs-polyvinylpyrrolidone	400	10–15	Timolol	0.043–432 μg/mL	0.519 μg/mL	UV–visible	Amirjani et al. (2016)
Citrate-Ag NPs	394	12	Azithromycin	$0.2–20.0 \times 10^{-6}$	0.053×10^{-6}	UV–visible	Chavada et al. (2017)
FA-Ag NPs	399	28 ± 6.6	6-MP	$20–1000 \times 10^{-9}$ and $5–300 \times 10^{-9}$	13.2×10^{-9} and 2.3×10^{-9}	UV–visible and fluorescence	Rawat et al. (2017)

SPR, *Surface plasmon resonance;* LOD, *limit of detection;* LSPR, *localized surface plasmon resonance;* SERS, *surface enhanced Raman scattering;* TCPO, *bis-(2,4,6-trichlorophenyl) oxalate;* CILE, *carbon ionic liquid electrodes;* FA, *folic acid;* MP, *mercaptopurine.*

18.4 SUMMARY

In this chapter, we have explored the progress of metal (Au, Ag, Cu, and Pt) NPs-based spectrometric approaches for the analyses of various drugs in real samples (pharmaceutical formulations, urine and plasma). The analytical merits (selectivity, sensitivity, accuracy, and precision) of UV–visible spectrometry have been remarkably boosted by using metal (Au, Ag, Cu, and Pt NPs) as probes. These approaches possess a remarkable selectivity and sensitivity over other competitive drugs. Noticeably, the analytical data of these methods were well-agreed with sophisticated analytical instruments (GC–MS and liquid chromatography–MS), exhibiting LODs at nanomolar concentrations. In addition, the selectivity of metal NPs was tuned by functionalizing them with different organic ligands. Despite the remarkable achievements of metal NPs in drug analysis, significant efforts must be paid on the development of metal NPs-assisted colorimetric methods for routine analysis of drugs in their formulations and biofluids. Therefore, modifications of metal NPs will continue to draw attention in the metal NPs-based UV–visible spectrometric approaches for accurate and real-time quantification of drugs in real samples.

ACKNOWLEDGMENTS

SKK and TJP greatly acknowledge the financial support from Brain Pool Program through the Korean Federation of Science and Technology Societies (KOFST) funded by the Ministry of Science, ICT and Future Planning (172S-4-3-1923). SKK acknowledges the financial support (37(2)/14/07/2015/BRNS/10401) from the Board of Research in Nuclear Science (BRNS), Government of India. SKK acknowledges the Director, SVNIT for granting sabbatical leave for the Brain Pool Korea Program.

REFERENCES

Alarfaj, N.A., El-Tohamy, M.F., 2015a. Utility of gold nanoparticles in luminescence determination of trovafloxacin: comparison of chemiluminescence and fluorescence detection. Luminescence 30, 1403–1408.

Alarfaj, N.A., El-Tohamy, M.F., 2016. Eco-friendly synthesis of gelatin-capped bimetallic Au–Ag nanoparticles for chemiluminescence detection of anticancer raloxifene hydrochloride. Luminescence 31, 1194–1200.

Alarfaj, N.A., Aly, F.A., El-Tohamy, M.F., 2015b. Application of silver nanoparticles to the chemiluminescence determination of cefditoren pivoxil using the luminol–ferricyanide system. Luminescence 30, 91–97.

Altintas, Z., Guerreiro, A., Piletsky, S.A., Tothill, I.E., 2015. Nano MIP based optical sensor for pharmaceuticals monitoring. Sens. Actuators, B: Chem. 213, 305–313.

Amirjani, A., Bagheri, M., Heydari, M., Hesaraki, S., 2016. Colorimetric determination of timolol concentration based on localized surface plasmon resonance of silver nanoparticles. Nanotechnology. 27, 375503–375511.

Bahram, M., Hoseinzadeha, F., Farhadia, K., Saadata, M., Najafi-Moghaddama, P., Afkhami, A., 2014. Synthesis of gold nanoparticles using pH-sensitive hydrogel and its application for colorimetric determination of acetaminophen, ascorbic acid and folic acid. Colloids Surf., A: Physicochem. Eng. Aspects 441, 517–524.

Baron, R., Zayats, M., Willner, I., 2005. Dopamine-, l-DOPA-, adrenaline-, and noradrenaline-induced growth of Au nanoparticles: assays for the detection of neurotransmitters and of tyrosinase activity. Anal. Chem. 77, 1566–1571.

Belvis, R., Pagonabarraga, J., Kulisevsky, J., 2009. Individual triptan selection in migraine attack therapy. Recent Pat. CNS Drug Discovery 4, 70–81.

Bhalla, N., Formisano, N., Miodek, A., Jain, A., Lorenzo, M.D., Pula, G., et al., 2015. Plasmonic ruler on field-effect devices for kinase drug discovery applications. Biosens. Bioelectron. 71, 121–128.

Bhamore, J., Rawat, K.A., Basu, H., Singhal, R.K., Kailasa, S.K., 2015. Influence of molecular assembly and NaCl concentration on gold nanoparticles for colorimetric detection of cysteine and glutathione. Sens. Actuators, B: Chem. 212, 526–535.

Biparva, P., Abedirad, S.M., Kazemi, S.Y., 2015. Silver nanoparticles enhanced a novel TCPO–H_2O_2–safranin O chemiluminescence system for determination of 6-mercaptopurine. Spectrochim. Acta, A: Mol. Biomol. Spectrosc. 145, 454–460.

Borisov, S.M., Klimant, I., 2008. Optical nanosensors smart tools in bioanalytics. Analyst 133, 1302–1307.

Chavada, V.D., Bhatt, N.M., Sanyal, M., Shrivastav, P.S., 2017. Surface plasmon resonance based selective and sensitive colorimetric determination of azithromycin using unmodified silver nanoparticles in pharmaceuticals and human plasma. Spectrochim. Acta, A: Mol. Biomol. Spectrosc. 170, 97–103.

Chen, Z., Zhang, G., Chen, X., Chen, J., Liu, J., Yuan, H., 2013. A fluorescence switch sensor for 6-mercaptopurine detection based on gold nanoparticles stabilized by biomacromolecule. Biosens. Bioelectron. 41, 844–847.

Chen, Z., Zhang, C., Wang, C., 2015a. A colorimetric assay of dopamine utilizing melamine modified gold nanoparticle probes. Anal. Methods 3, 838–841.

Chen, Z., Zhang, C., Zhou, T., Ma, H., 2015b. Gold nanoparticle based colorimetric probe for dopamine detection based on the interaction between dopamine and melamine. Microchim. Acta 182, 1003–1008.

Das, A., Zhao, J., Schatz, G.C., Sligar, S.G., Duyne, R.P.V., 2009. Screening of type I and II drug binding to human cytochrome P450-3A4 in nanodiscs by localized surface plasmon resonance spectroscopy. Anal. Chem. 81, 3754–3759.

Derikvand, H., Roushani, M., Abbasi, A.R., Derikvand, Z., Azadbakht, A., 2016. Design of folding-based impedimetric aptasensor for determination of the nonsteroidal anti-inflammatory drug. Anal Biochem. 513, 77–86.

D'souza, S.L., Pati, R.K., Kailasa, S.K., 2015. Ascorbic acid-functionalized Ag NPs as a probe for colorimetric sensing of glutathione. Appl. Nanosci. 5, 747–753.

Gao, M., Lin, R., Li, L., Jiang, L., Ye, B., He, H., et al., 2014. Label-free silver nanoparticles for the naked eye detection of entecavir. Spectrochim. Acta, A: Mol. Biomol. Spectrosc. 126, 178–183.

Ghosh, S.K., Pal, T., 2007. Interparticle coupling effect on the surface plasmon resonance of gold nanoparticles: from theory to applications. Chem. Rev. 107, 4797–4862.

Gwo, S., Chen, H.-Y., Lin, M.-H., Sun, L., Li, X., 2016. Nanomanipulation and controlled self-assembly of metal nanoparticles and nanocrystals for plasmonics. Chem. Soc. Rev. 45, 5672–5716.

Jans, H., Huo, Q., 2012. Gold nanoparticle-enabled biological and chemical detection and analysis. Chem. Soc. Rev. 41, 2849–2866.

Jena, B.K., Raj, C.R., 2008. Optical sensing of biomedically important polyionic drugs using nano-sized gold particles. Biosens. Bioelectron. 23, 1285–1290.

Jinnarak, A., Teerasong, S., 2016. A novel colorimetric method for detection of gamma-aminobutyric acid based on silver nanoparticles. Sens. Actuators, B: Chem. 229, 315–320.

Kailasa, S.K., Wu, H.F., 2012. One-pot synthesis of dopamine dithiocarbamate functionalized gold nanoparticles for quantitative analysis of small molecules and phosphopeptides in SALDI and MALDI-MS. Analyst 137, 1629–1638.

Kasibabu, B.S.B., Bhamore, J.R., D'souza, S.L., Kailasa, S.K., 2015. Dicoumarol assisted synthesis of water dispersible gold nanoparticles for colorimetric sensing of cysteine and lysozyme in biofluids. RSC Adv. 5, 39182–39191.

Kelly, K.L., Coronado, E., Zhao, L.L., Schatz, G.C., 2003. The optical properties of metal nanoparticles: the influence of size, shape, and dielectric environment. J. Phys. Chem. 107, 668–677.

Klinkova, A., Choueiri, R.M., Kumacheva, E., 2014. Self-assembled plasmonic nanostructures. Chem. Soc. Rev. 43, 3976–3991.

Kong, F., Liu, H., Dong, J., Qian, W., 2011. Growth-sensitive gold nanoshells precursor nanocomposites for the detection of L-DOPA and tyrosinase activity. Biosens Bioelectron. 26, 1902–1907.

Kostiainen, R., Kotiaho, T., Kuuranne, T., Auriola, S., 2003. Liquid chromatography/atmospheric pressure ionization–mass spectrometry in drug metabolism studies. J. Mass Spectrom. 38, 357–372.

Kwon, D., Lee, W., Kim, W., Yoo, H., Shin, H.-C., Jeon, S., 2015. Colorimetric detection of penicillin antibiotic residues in pork using hybrid magnetic nanoparticles and penicillin class-selective, antibody-functionalized platinum nanoparticles. Anal. Methods 7, 7639–7645.

Kwon, D., Yoo, H., Lee, H., Jeon, S., 2018. Colorimetric detection of penicillin G in milk using antibody-functionalized dendritic platinum nanoparticles. Sens. Actuators, B: Chem. 255, 552–556.

Laliwala, S.K., Mehta, V.N., Rohit, J.V., Kailasa, S.K., 2014. Citrate-modified silver nanoparticles as a colorimetric probe for simultaneous detection of four triptan-family drugs. Sens. Actuators, B: Chem. 197, 254–263.

Leng, Y., Xie, K., Ye, L., Li, G., Lu, Z., He, J., 2015. Gold-nanoparticle-based colorimetric array for detection of dopamine in urine and serum. Talanta 139, 89–95.

Li, M., Gou, H., Al-Ogaidi, I., Wu, N., 2013. Nanostructured sensors for detection of heavy metals: a review. ACS Sustain. Chem. Eng. 1, 713–723.

Li, M., Cushing, S.K., Wu, N., 2015. Plasmon-enhanced optical sensors: a review. Analyst 140, 386–406.

Li, Y., Jing, C., Zhang, L., Long, Y.T., 2012. Resonance scattering particles as biological nanosensors in vitro and in vivo. Chem. Soc. Rev. 41, 632–642.

Li, T., Wu, X., Liu, F., Li, N., 2017a. Analytical methods based on the light-scattering of plasmonic nanoparticles at the single particle level with dark-field microscopy imaging. Analyst 142, 248–256.

Li, Z., Bi, S., Wang, T., Wang, Y., Zhou, H., Wu, J., 2017b. Investigation of the binding of AuNPs-6-mercaptopurine and the sensitive detection of 6-mercaptopurine using resonance Rayleigh light scattering. Luminescence 32, 502–508.

Ling, J., Sang, Y., Huang, C.Z., 2008. Visual colorimetric detection of berberine hydrochloride with silver nanoparticles. J. Pharm. Biomed. Anal. 47, 860–864.

Liu, J.-M., Wang, X.-X., Cui, M.-L., Lin, L.-P., Jiang, S.-L., Jiao, L., et al., 2013. A promising non-aggregation colorimetric sensor of AuNRs–Ag^{+} for determination of dopamine. Sens. Actuators, B: Chem 176, 97–102.

Liu, L., Li, S., Liu, L., Deng, D., Xia, N., 2012. Simple, sensitive and selective detection of dopamine using dithiobis(succinimidylpropionate)-modified gold nanoparticles as colorimetric probes. Analyst 137, 3794–3799.

Liu, M.M., Chao, J., Deng, S., Wang, K., Li, K., Fan, C.H., 2014. Dark-field microscopy in imaging of plasmon resonant nanoparticles. Colloids Surf., B 124, 111–117.

Mates, J.M., Perez-Gomez, C., Blanca, M., 2000. Chemical and biological activity of free radical 'scavengers' in allergic diseases. Clin. Chim. Acta 296, 1–15.

Matsui, J., Akamatsu, K., Nishiguchi, S., Miyoshi, D., Nawafune, H., Tamaki, K., et al., 2004. Composite of Au nanoparticles and molecularly imprinted polymer as a sensing material. Anal. Chem. 76, 1310–1315.

Maurer, H.R., 2001. Bromelain: biochemistry, pharmacology and medical use. Cell. Mol. Life Sci. 58, 1234–1245.

Mayer, K.M., Hafner, J.H., 2011. Localized surface plasmon resonance sensors. Chem. Rev. 111, 3828–3857.

Mehta, V.N., Kailasa, S.K., 2015. Malonamide dithiocarbamate functionalized gold nanoparticles for colorimetric sensing of Cu^{2+} and Hg^{2+} ions. RSC Adv. 5, 4245–4255.

Mehta, V.N., Kumar, M.A., Kailasa, S.K., 2013. Colorimetric detection of copper in water samples using dopamine dithiocarbamate-functionalized Au nanoparticles. Ind. Eng. Chem. Res. 52, 4414–4420.

Mehta, V.N., Solanki, J.N., Kialsa, S.K., 2014. Selective visual detection of Pb(II) ion via gold nanoparticles coated with a dithiocarbamate-modified 4′-aminobenzo-18-crown-6. Microchim. Acta 181 (Issue 15–16), 1905–1915.

Menon, S.K., Mistry, B.R., Joshi, K.V., Sutariya, P.G., Patel, R.V., 2012. Analytical detection and method development of anticancer drug Gemcitabine HCl using gold nanoparticles. Spectrochim. Acta, A: Mol. Biomol. Spectrosc. 94, 235–242.

Mitchell, J.S., Lowe, T.E., 2009. Ultrasensitive detection of testosterone using conjugate linker technology in a nanoparticle-enhanced surface plasmon resonance biosensor. Biosens. Bioelectron. 24, 2177–2183.

Modi, R.P., Mehta, V.N., Kailasa, S.K., 2014. Bifunctionalization of silver nanoparticles with 6-mercaptonicotinic acid and melamine for simultaneous colorimetric sensing of Cr^{3+} and Ba^{2+} ions. Sens. Actuators, B: Chem. 195, 562–571.

Nielsen, O.H., Vainer, B., Rask-Madsen, J., 2001. The treatment of inflammatory bowel disease with 6-mercaptopurine or azathioprine. Alimentary Pharmacol. Ther. 15, 1699–1708.

Peng, Y.H., Xiong, B., Peng, L., Li, H., He, Y., Yeung, E.S., 2015. Recent advances in optical imaging with anisotropic plasmonic nanoparticles. Anal. Chem. 87, 200–215.

Rani, S., Malik, A.K., Heena, Kaur, R., Kaur, R., 2016. A review for the analysis of antidepressant, antiepileptic and quinolone type drugs in pharmaceuticals and environmental samples. Crit. Rev. Anal. Chem 46, 424–442.

Rawat, K.A., Surati, K.R., Kailasa, S.K., 2014. One-pot synthesis of gold nanoparticles by using 4-aminoantipyrine as a novel reducing and capping agent for simultaneous colorimetric sensing of four triptan-family drugs. Anal. Methods 6, 5972–5980.

Rawat, K.A., Basu, H., Singhal, R.K., Kailasa, S.K., 2015. Simultaneous colorimetric detection of four drugs in their pharmaceutical formulations using unmodified gold nanoparticles as a probe. RSC Adv. 5, 19924–19932.

Rawat, K.A., Singhal, R.K., Kailasa, S.K., 2016. Colorimetric and fluorescence "turn-on" methods for the sensitive detection of bromelain using carbon dots functionalized gold nanoparticles as a dual probe. RSC Adv. 6, 32025–32036.

Rawat, K.A., Singhal, R.K., Kailasa, S.K., 2017. One-pot synthesis of silver nanoparticles using folic acid as a reagent for colorimetric and fluorimetric detections of 6-mercaptopurine at nanomolar concentration. Sens. Actuators, B: Chem. 249, 30–38.

Ringe, E., Sharma, B., Henry, A.-I., Marks, L.D., Duyne, R.P.V., 2013. Single nanoparticle plasmonics. Phys. Chem. Chem. Phys. 15, 4110–4129.

Rohit, J.V., Solanki, J.N., Kailasa, S.K., 2014. Surface modification of silver nanoparticles with dopamine dithiocarbamate for selective colorimetric sensing of mancozeb in environmental samples. Sens. Actuators, B: Chem. 200, 219–226.

Rohit, J.V., Singhal, R.K., Kailasa, S.K., 2016. Dithiocarbamate-calix[4]arene functionalized gold nanoparticles as a selective and sensitive colorimetric probe for assay of metsulfuron-methyl herbicide via non-covalent interactions. Sens. Actuators, B: Chem. 237, 1044–1055.

Rowan, A.D., Buttlet, D.J., Barrett, A., 1990. The cysteine proteinases of the pineapple plant. J. Biochem. 266, 869–875.

Saha, K., Agasti, S.S., Kim, C., Li, X., Rotello, V.M., 2012. Gold nanoparticles in chemical and biological sensing. Chem. Rev. 112, 2739–2779.

Sallum, L.F., Soares, F.L., Ardila, J.A., Carneiro, R.L., 2014. Determination of acetylsalicylic acid in commercial tablets by SERS using silver nanoparticle-coated filter paper. Spectrochim. Acta, A: Mol. Biomol. Spectrosc. 133, 107–111.

Santos, E.D.B., Lima, E.C.N.L., Oliveira, C.S.D., Sigoli, F.A., Mazali, I.O., 2014. Fast detection of paracetamol on a gold nanoparticle–chitosan substrate by SERS. Anal. Methods 6, 3564–3568.

Selvaraj, V., Alagar, M., 2007. Analytical detection and biological assay of antileukemic drug 5 fluorouracil using gold nanoparticles as probe. Int. J. Pharm. 337, 275–281.

Shamsipur, M., Saber, R., Emami, M., 2014. A highly sensitive electrochemical sensor based on gold nanoparticles/multiwall carbon nanotubes modified glassy carbon electrode for selective determination of traces of atenolol. Anal. Methods 6, 7038–7045.

Shim, S.-Y., Lim, D.-K., Nam, J.-M., 2008. Ultrasensitive optical biodiagnostic methods using metallic nanoparticles. Nanomedicine 3, 215–232.

Su, H., Sun, B., Chen, L., Xu, Z., Ai, S., 2012. Colorimetric sensing of dopamine based on the aggregation of gold nanoparticles induced by copper ions. Anal. Methods 4, 3981–3986.

Sun, Y., Xia, Y., 2003. Gold and silver nanoparticles: a class of chromophores with colors tunable in the range from 400 to 750 nm. Analyst 128, 686–691.

Tiwari, N.R., Rathore, A., Prabhune, A., Kulkarni, S.K., 2010. Gold nanoparticles for colorimetric detection of hydrolysis of antibiotics by penicillin G acylase. Adv. Biosci. Biotechnol. 1, 322–329.

Uehara, N., Ookubo, K., Shimizu, T., 2010. Colorimetric assay of glutathione based on the spontaneous disassembly of aggregated gold nanocomposites conjugated with water-soluble polymer. Langmuir 26, 6818–6825.

Upadhyayula, V.K.K., 2012. Functionalized gold nanoparticle supported sensory mechanisms applied in detection of chemical and biological threat agents: a review. Anal. Chim. Acta 715, 1–18.

Vilela, D., González, M., Escarpa, C.A., 2012. Sensing colorimetric approaches based on gold and silver nanoparticles aggregation: Chemical creativity behind the assay. A review. Anal. Chim. Acta 751, 24–43.

Wang, C., Astruc, D., 2014. Nanogold plasmonic photocatalysis for organic synthesis and clean energy conversion. Chem. Soc. Rev. 43, 7188–7216.

Wang, H.-B., Zhang, H.-D., Chen, Y., Ou, L.-J., Liu, Y.-M., 2015. Poly(thymine)-templated fluorescent copper nanoparticles for label-free detection of *N*-acetylcysteine in pharmaceutical samples. Anal. Methods 7, 6372–6377.

Xie, T., Jinga, C., Long, Y.-T., 2017. Single plasmonic nanoparticles as ultrasensitive sensors. Analyst 142, 409–420.

Yarbakht, M., Nikkhah, M., 2016. Unmodified gold nanoparticles as a colorimetric probe for visual methamphetamine detection. J. Exp. Nanosci. 11, 593–601.

Yockell-Lelièvre, H., Bukar, N., Keating, K.S.M., Arnaud, M., Cosin, P., Guo, Y., et al., 2015. Plasmonic sensors for the competitive detection of testosterone. Analyst 140, 5105–5111.

Yusoff, N., Pandikumar, A., Ramaraj, R., Lim, H.N., Huang, N.M., 2015. Gold nanoparticle based optical and electrochemical sensing of dopamine. Microchim. Acta 182, 2091–2114.

Zhang, Y., Li, B., Chen, X., 2010. Simple and sensitive detection of dopamine in the presence of high concentration of ascorbic acid using gold nanoparticles as colorimetric probes. Microchim. Acta 168, 107–113.

Zhang, Y., Qi, S., Liu, Z., Shi, Y., Yue, W., Yi, C., 2016. Rapid determination of dopamine in human plasma using a gold nanoparticle-based dual-mode sensing system. Mater. Sci. Eng., C 61, 207–213.

CHAPTER

Source, effect, and risk assessment of nanoparticles with special reference to occupational exposure 19

Venugopal Dhananjayan*, Beerappa Ravichandran, Somnath Sen and Karunamoorthy Panjakumar

Industrial Hygiene & Toxicology Division, ICMR-Regional Occupational Health Centre (Southern), Indian Council of Medical Research, Bangalore, India

19.1 INTRODUCTION

The growing industrial revolution and urbanization releasing large amounts of hazardous substance into the surrounding environment and consequently threaten the human. The industrial workplace is often a major source of nanosized materials, or "nanoparticles" (NPs), that influence different biological systems of the body and may cause adverse effects (Coogan et al., 2012; Kelishadi, 2012). NPs find their way into the human body either through inhalation, ingestion, or absorption (Ibrahim et al., 2016; Yildirimer et al., 2011). Worldwide epidemiological studies have shown that human exposure to respirable particulate matter is correlated with an increase in cardiac and respiratory morbidity and mortality. In addition to environmental exposure, workers are exposed to NPs through their intentional use in engineering applications. The increasing use of NPs to improve consumer products and medical treatments may significantly increase the potential for human occupational and environmental exposure to NPs. Some recent studies have demonstrated the potential health impacts of such exposure (DeLoid et al., 2014; Rastogi et al., 2017; Weidemann et al., 2016).

NPs generated may play a major role in many chronic diseases where infectious pathogens have not been suspected, including diseases that were previously attributed only to genetic factors and lifestyle. These small particles have the ability to enter and travel within the circulatory systems and finally affect the host (Masserini, 2013). The smallest particles contain tens or hundreds of atoms, with dimensions in the scale of nanometers, hence NPs. The toxicity of size-dependent

*Corresponding author (dhananjayan_v@yahoo.com)

Nanoarchitectonics in Biomedicine. DOI: https://doi.org/10.1016/B978-0-12-816200-2.00004-9

materials depends largely on the bonding arrangement of its atoms (Powers et al., 2006). However, many emerging novelties of NPs would expedite the rise of potential risks to industrial workers, consumers, and the environment. Continuing exposure to NPs may cause serious damage to the human respiratory tract, lung diseases (Lu et al., 2014), heart diseases, and premature death (Pui et al., 2014).

The fast propagation of nanotechnologies into different industries and consumer products is causing the exponential growth of nanomaterial production. Hence, increasing amounts of NPs reach occupational settings and the indoor and outdoor environments, thus, representing a potentially serious chronic hazard to human health (Castranova, 2011). While a significant number of biological processes have been identified, only a few case studies have reported on risk assessment methods for the exposure of workers to NPs. Although the route of entry of NPs into the cell is well-documented, their interactions with biological systems are relatively unknown (Wang et al., 2014).

19.1.1 NATURE OF NANOPARTICLES

NPs are defined as minute objects with sizes varying from 1 to 100 nm and behave as a whole unit with respect to transport and properties (Buzea et al., 2007). Ultrafine particles are similar to NPs with particle sizes in the range of 100–2500 nm and coarse particles cover the range of 2500–10,000 nm. Nanoclusters have at least one dimension between 1 and 10 nm and a narrow size distribution. NPs are agglomerates of ultrafine particles, NPs, or nanoclusters (Fahlman, 2007). A representation of the various sizes of nano- and microscale materials is given in Fig. 19.1. The chemical processing and synthesis of high-performance technological components in various industrial sectors requires the use of high-purity raw products and composite materials. In condensed bodies formed from fine powders, the irregular particle sizes and shapes often lead to nonuniform packing morphologies that result in variations in the powder compact and production of fine particles. NPs often possess unexpected optical properties

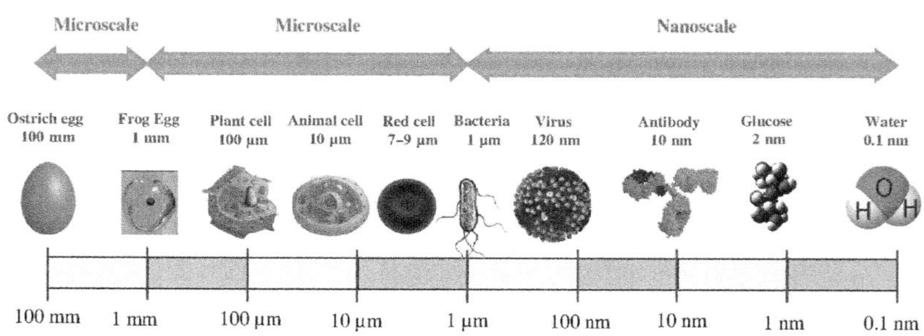

FIGURE 19.1

Representation of the size of nanomaterials compared to biological components.

as they are small enough to confine their electrons and produce quantum effects (Hewakuruppu et al., 2013).

NPs possess various new properties and their industrial use creates new opportunities, but they also present new risks and uncertainties. NPs are broadly classified into different categories depending on their morphology size and chemical nature. Based on physical and chemicals properties, some of the known classes of NPs which synthesized by various process are carbon-banded NPs [carbon nanotubes (CNTs), metal NPs, ceramics NPs, semiconductor NPs, polymeric NPs, lipid-based NPs, etc.] (Khan et al., 2017) During the synthesis of products a variety of NPs are released and workers are, thus, exposed to them. Very a few studies have reported on workplace exposure to nanomaterials (Peixe et al., 2015). NPs present potential harm to humans, both medically and environmentally (Shi et al., 2012).

NPs have very high surface area to volume ratio with more inner body surface and quick reactive or catalytic properties (Biener et al., 2009). They are also able to pass through cell membranes in organisms, but their interactions with biological systems are relatively unknown. It is unlikely the particles would enter the cell nucleus, golgi complex, endoplasmic reticulum, or other internal cellular components due to the particle size and intercellular agglomeration (Greulich et al., 2011). A recent study looking at the effects of different NPs on humans found varying levels of susceptibility to cytotoxicity (Foldbjerg et al., 2011). There are concerns that pharmaceutical companies, seeking regulatory approval for nanoreformulations of existing medicines, are relying on safety data produced during clinical studies of earlier, prereformulation versions of the medicine. The number of nanotechnology-based consumer products listed in the Nanotechnology Consumer Products Inventory (Woodrow Wilson International Center for Scholars, 2011) is currently >1300. The expansion of the nanotechnology-based consumer products market is cause for concern regarding potential human exposure to nanomaterials and possible health risks.

19.1.2 SOURCES OF NANOPARTICLES

19.1.2.1 Natural sources

NPs are abundant in nature, as they are produced in many natural processes including photochemical reactions, volcanic eruptions, forest fires, and simple erosion as well as by plants and animals. Aerosols generated by human activities are estimated to be about 10% of the total NPs, the remaining 90% having a natural origin (Taylor, 2002). These large-scale phenomena are visible from satellites and produce particulate matter and airborne particles of dust and soot ranging from micro- to nanoscales. Small particles suspended in the atmosphere, often known as aerosols, affect the entire planet's energy balance because they both absorb radiation from the sun and scatter it back to space (Houghton, 2005). Many organisms also produce nanoparticulate inorganic materials through intracellular and extracellular processes.

19.1.2.2 Anthropogenic sources

Humans have created nanomaterials for millennia as by-products of simple combustion and cooking food, and more recently, chemical manufacturing, welding, ore refining and smelting, combustion in vehicles and airplane engines, combustion of treated pulverized sewage sludge, and combustion of coal and fuel oil for power generation (Linak et al., 2000). While engineered NPs have been on the market for some time and are commonly used in cosmetics, sporting goods, tires, stain-resistant clothing, sunscreens, toothpaste, and food additives, etc., these nanomaterials—and new more deliberately fabricated NPs, such as CNTs—constitute a small minority of environmental nanomaterials. Diesel and automobile exhaust fumes are the primary source of atmospheric nano- and microparticles in urban areas (USEPA, 2002). NPs constitute 20% of the particles' mass, but more than 90% of the number of diesel-generated particles. Humans and their activities generate considerable amounts of particulate matter in indoor environments. NPs are generated through common indoor activities, such as cooking, smoking, cleaning, and combustion. NPs have also been shown to enter buildings from outdoors through ventilation systems (Sioutas et al., 2005). As humans generally spend much of their time indoors ($>80\%$), indoor pollution directly affects our health and immune systems. As a combustion product, tobacco smoke is composed of NPs with sizes ranging from around 10 nm up to 700 nm, with a maximum of these being around 150 nm. Environmental tobacco smoke has a very complex composition, with more than 100,000 chemical components and compounds (Ning et al., 2006). Several studies have demonstrated increased toxic effects per mass dose of ultrafine metal exposure compared to their larger counterparts (Mittal and Pandey, 2014; Zhang et al., 2012) and a better correlation of the effects to surface area rather than mass. Metals in ultrafine particles associated with known sources are listed in Table 19.1.

19.1.2.3 Engineered nanomaterials

Researchers evaluating toxicity and human exposure potential of engineered nanomaterials are challenged by the rapid development of novel materials for new applications as the nanotechnology industry drives forward. Nanomaterials can be synthesized by many methods including gas phase processes, vapor deposition synthesis, colloidal, or liquid phase methods. A critical fact to consider with engineered nanomaterials is that they can be synthesized in almost any shape and size. Nanostructured materials do not pose a health risk as long as they do not detach from the substrate. However, published toxicity studies clearly show that particle size alone is not a good criterion for differentiating between more or less hazardous materials. For instance, inhalation studies using rodents have demonstrated that 20 nm diameter TiO_2 particles had a greater impact on the animals' lungs than pigment-grade particles with the same composition, even though both particle sizes were administered as micrometer–diameter agglomerates (Bermudez et al., 2004). Engineered nanomaterials are notoriously difficult to

Table 19.1 Metals in Ultrafine Particles Associated With Known Sources

Source	Elements Associated	References
Diesel	Al, Ca, Cu, Fe, Mg, Mn, V, Zn	Pulles et al. (2012), Corbin et al. (2018)
Gasoline (emissions), lubricating oil, fuel borne catalysts	Sr, Cu, Mn, Fe, Ca, P, Zn, Mg Fe, Mn, Ce	Zmozinski et al. (2010), Chrastný et al. (2018), Heeb et al. (2013)
Diesel fuel (marine engines)	Al, Ca, Fe, Ni, V, Zn	Corbin et al. (2018)
Nonexhaust Traffic Sources		
Brakes	Fe, Cu, Sn, Zn	Malachova et al. (2016)
Tires	Cd, Co, Cr, Cu, Fe, Mn, Pb	Shakya et al. (2006), Galvagno et al. (2009)
Road dust	Sb, As, Cu, Zn, Cr, Cd, Ni, Co, V, Mn, Pb	Massadeh and Snook (2002), Lanzerstorfer (2018)
Industrial Sources		
Metal working	Fe, K, Na, Pb, Zn etc.,	Ružičková et al. (2018)
Power generation	Fe, Mn, Cr, Zn, Ni, Cu, Cd, and Pb	Kisku et al. (2018)
Incinerators	As, Be, Cd, Cr, Pb, Mn, Hg, Ni, and V	Miyake et al. (2017), Mauriello et al. (2017)
Mines	Cr, As, Cd, Pb, Hg, Zn	Singh et al. (2018)

characterize; even two materials that are notionally the same may have subtle but significant differences that determine their behavior. For instance, introducing a small percentage of impurities to the surface of nano-TiO_2 particles may fundamentally alter their propensity to generate free radicals under UV radiation (Wakefield et al., 2004). Without rigorous nanomaterials characterization it will be near impossible to interpret toxicity studies, compare similar studies, and develop predictive models of nanomaterials hazard. Instead, it is more practical as a first step to consider the three key physical exposure metrics: number concentration, surface area concentration, and mass concentration.

19.1.3 EXPOSURE ROUTES

Many investigations have been carried out to determine the mechanisms in which NPs can cause damage to humans in the industrial workplace. The primary routes of entry to the human body from the outside environment are inhalation, skin exposure, and ingestion, where the toxicity targets are the respiratory, integumentary, and gastrointestinal systems, respectively. Human skin, lungs, and the gastrointestinal tract are in constant contact with the outer environment and exposed varieties of particles. While the skin is generally an effective barrier to foreign substances, the lungs and gastrointestinal tract are more vulnerable. These three ways are the most likely points of entry for natural or anthropogenic NPs (Chan-Remillard et al., 2009). Injections and implants are other possible routes of exposure and are primarily limited to engineered materials.

19.1.3.1 Inhalation exposure

NPs settle in different parts of the respiratory tract depending on particle size and inertial, gravitational, and diffusional behavior. Exposure via inhalation is the most significant route in terms of potential toxicity, followed by dermal contact with chemicals, and then ingestion. NP deposition in the respiratory tract increases with decreasing particle size. After inhalation, particles can be phagocytosed by macrophages, cleared via the mucociliary escalator, and swallowed (Sajid et al., 2015) or retained in the lungs, aggregated or metabolized, and further transported to different tissues through the bloodstream (Sharifi et al., 2012). Studies have explored the potential human health effects resulting from exposure to NPs and nanomaterials. About one third of the particles is cleared through the defensive system in the upper airways (mucociliary clearance) and lower airways of the lung, the alveoli (macrophage clearance), which transports soluble and degradable particles to the larynx, where they are swallowed, digested, and excreted (Sajid et al., 2015).

The lack of information on risks emanating from exposure to nanomaterials has led to concerns that the nanotechnology sector is not fully prepared to deal with these (Gewin, 2006). A recent study has explored the risk of prostate cancer among a population exposed to ultrafine particles (Weichenthal et al., 2017). Studies have also reported the clinical responses of metal fume fever after inhalation exposure to welding fumes (Greenberg and Vearrier, 2015). This is accompanied by changes in the composition of bronchoalveolar lavage fluid, including early increase in tumor necrosis factor α followed by numbers of polymorphonuclear leukocytes involved in ZnO NP-induced human bronchial epithelial cells (Wu et al., 2010).

19.1.3.2 Dermal exposure

Despite the fact that a number of commercial products containing NPs, such as cosmetics and suntan lotions, are already available, there is considerable uncertainty about whether this is a significant exposure route. Studies have suggested that it is an important route of entry for NPs in occupational settings and it should be considered for risk evaluation (Pinaki et al., 2010; Senzui et al., 2010). Most existing studies appear to indicate that few NPs penetrate through the stratum corneum and the dermis. The major pathways of penetration across the skin have been identified depending on physicochemical properties of the compound: intercellular, trans-cellular, and transappendageal like through hair follicles and sweat glands (Desai et al., 2010). Various factors that influence the dermal absorption of NPs include: (1) the location and skin conditions at the application site; (2) the physicochemical properties of the penetrating molecule; and (3) the physicochemical properties of the vehicle dispersing the penetrating molecule. Persistent organic pollutants bound to TiO_2 in cosmetics, sun creams, and toothpaste could be absorbed via the dermis of skin, hair, and follicles. However, this is generally unlikely in healthy skin as only fatty compounds with a molecular weight <500 should penetrate the epidermis (Senzui et al., 2010). This penetration is of interest due to the direct contact of the skin with potentially risky consumer products. Researchers also caution that leaching of selected components of the particles through the skin and into the bloodstream is possible.

19.1.3.3 Ingestion

Several studies have shown that NPs are efficiently absorbed through the gastrointestinal tract (Bouwmeester et al., 2018; Kaiser et al., 2013). The route of entry is both directly through intentional ingestion and indirectly via dissolution from food contact materials or by secondary ingestion of inhaled particles. Additionally, growing use of nanomaterials may lead to increased environmental contamination and unintentional ingestion (Bergin and Witzmann, 2013). A number of studies on humans have highlighted the association between occupations and substances where inadvertent ingestion may be significant, typically through case reports or from theoretical analysis. These scenarios involve exposure to various types of metals and metal compounds, pesticides, infectious agents, pharmaceuticals, unsealed radioactive sources, and high molecular weight allergens. A theoretical investigation of ingestion exposure highlights the primary routes of exposure are through hand-to-mouth and object-to-mouth events. While more emphasis is placed by employers on controlling inhalation and dermal exposures, the ingestion route of exposure has tended to be considered unimportant and has rarely been considered (Cherrie et al., 2006). The reasons for less consideration of ingestion exposure are: (1) the common belief that ingestion of hazardous substances can only occur by intentional means or acts of gross negligence, and hence can be avoided; (2) the recognition that many materials are only very poorly absorbed from the gut (i.e., they have low bioavailability) and as such are unlikely to produce toxic effects when swallowed in small quantities and because ingested materials may be metabolized in the liver and excreted before having the opportunity to exert any toxic effect; and (3) the presumption that where a worker is exposed by inhalation, dermal contact, and ingestion, the mass of material taken into the body by ingestion may be small in comparison with other routes.

19.1.4 HEALTH EFFECTS OF NANOPARTICLES

NPs translocate from the entry portals into the circulatory and lymphatic systems and, ultimately, to body tissues and organs and can produce irreversible damage to cells by oxidative stress and/or organelle injury. The toxicity of NPs depends on various factors, including size, aggregation, composition, crystallinity, and surface functionalization, etc. (Sajid et al., 2015; Shin et al., 2015).

Diseases associated with inhaled NPs are asthma, bronchitis, emphysema, lung cancer, and neurodegenerative diseases, such as Parkinson's and Alzheimer's diseases (Kumar, 2006; NIOSH, 2009; WHO, 1999). NPs in the gastrointestinal tract have been linked to Crohn's disease and colon cancer. It's entry into the circulatory system are related to the occurrence of arteriosclerosis, blood clots, arrhythmia, heart diseases, and ultimately cardiac death. Translocation to other organs, such as the liver or spleen, etc., may lead to diseases of these organs as well. Exposure to some NPs is associated with the occurrence of autoimmune diseases, such as systemic lupus erythematosus, scleroderma, and rheumatoid arthritis. The types of NPs, sources, and their health effects on human are listed in Table 19.2.

NPs are detrimental to human health and several studies have reported the various health effects associated with a variety of nanomaterials. NP exposure and health effects on inflammatory response depends on the surface area of deposition (Schmid and Stoeger, 2016), superoxide radicals and stimulation of tyrosine (Sisler et al., 2016), reactive oxygen species generation (USEPA, 2002), alteration in immune cells and process (DeLoid et al., 2016; Roy et al., 2014), reduction in cholinesterase, and exceedance of residues in blood (Dhananjayan et al., 2012a,b; Dhananjayan and Ravichandran, 2013; Šinko et al., 2014). Inflammatory response closely matches the number of ultrafine particles reaching the airways and alveoli (Churg, 2000) indicating that oxidative stress plays an important role in fine particulate matter induced augmentation of functional and morphological alterations of Clara cells in allergic lung inflammation (Alessandrini et al., 2010). Furthermore, there is increasing evidence that NPs, if deposited in the upper airways, can be translocated to the central nervous system and ganglia along axons and dendrites of olfactory neurons (Oberdörster et al., 2005a). Recently, it has been demonstrated that toxicity profiles of NPs are more toxic than those of larger particles and show cytotoxicity and cell death (Albanese et al., 2012; Hwang et al., 2015). In vivo research has shown that NPs could penetrate the skin, reach different tissues, and then induce the inflammation and vascular homeostasis disorder (Nemmar et al., 2016).

19.1.5 RISK ASSESSMENT PARADIGM

The development of a risk assessment is a process to systematically characterize the scientific evidence of potential adverse health effects from human exposures to hazardous agents (NRC, 1983). This assessment process is one of the essential strategies to support policy development and decision making. Different approaches to assess the health risk for NPs were recently proposed by many researchers. However, a quantitative risk assessment of NPs is not yet accomplished. The traditional risk assessment process involves the use of toxicology data to attain the derived no effect level (DNEL) of exposure. Risk is assessed by comparing the DNEL with the exposure levels obtained from the different scenarios in the exposure assessment. The major steps involved in the risk assessment

Table 19.2 Health Effects and Sources of Different Nanoparticles

Source/Nanoparticles	Health Effects	References
CoO and La_2O_3 metal oxide	CoO nanoparticles produce superoxide radicals and caused stimulation of total tyrosine and threonine phosphorylation at both 6 and 24 h when compared with La_2O_3 nanoparticles. These data provide evidence that different toxicological modes of action were involved in CoO and La_2O_3 metal oxide nanoparticle-induced cellular toxicity.	Sisler et al. (2016)
Diesel soots and silicate dusts	Effects of particle surface conditioning by a primary component of phospholipids pulmonary surfactant, diacyl phosphatidyl choline.	Wallace et al. (2007)
Silica NPs	Toxic effect mechanisms of silica nanoparticles on endothelial cells was through DDR via Chk1-dependent G2/M checkpoint signaling pathway.	Duan et al. (2013)
Fumed silica	Plasma membrane perturbation, potassium efflux, Reactive Oxygen Species (ROS) generation.	Zhang et al. (2012), Sun et al. (2015)
Diesel exhaust	Potential or probable human carcinogen.	USEPA (2002)
Hard metal, cobalt metal, tungsten carbide	Heightened levels of catalytically generated reactive oxygen species in aqueous media.	Stephens et al. (1998), Keane and Wallace (2002)
Metal oxides (SiO_2, Fe_2O_3, ZnO, CeO_2, TiO_2, and an Ag/SiO_2 composite)	Induced cytotoxicity is mitigated in the presence of serum, potentially through masking of reactive surface groups by serum proteins, whereas ZnO nanoparticles were cytotoxic both in the presence and in the absence of serum. Alter innate and adaptive immune cells and process, possibly causing impaired protection or disruption of immune homeostasis, leading to or exacerbating existing allergic and inflammatory condition, lysosomal damage, cathepsin B release.	Shi et al. (2012), Smith et al. (2014), Li et al. (2014), Roy et al. (2014), DeLoid et al. (2016)

(*Continued*)

Table 19.2 Health Effects and Sources of Different Nanoparticles *Continued*

Source/Nanoparticles	Health Effects	References
Super fine particles	Exposure of macrophages include changes in phagocytic function, gene expression, cytokine secretion, reactive oxygen and nitrogen species production, and surface markers. Cause diverse diseases, including lung cancer, pulmonary and cardiovascular disease, diabetes, and neurodegenerative disorders. Produce abiotic ROS which potentially can result in cell death.	Liu et al. (2010), Kodali et al. (2013), Sahu et al. (2014), Lin et al. (2014), Zheng et al. (2016)
Carbon nanotubes	Lysosomal damage, cathepsin B release.	Nel et al. (2013), Lin et al. (2014)
Amino-polystyrene	Toxic to macrophage, hepatoma, epithelial cells, cardiovascular endothelial, mitochondrial uptake, cell death, Adenosine triphosphate (**ATP**) depletion	Kim et al. (2013), Ware et al. (2014)
Graphene oxides	TLR4 activation, lysosomal damage, cathepsin B release, ROS generation.	Ma et al. (2015), Trommer and Bergmann (2015)
Lipid and polymeric nanoparticles in the presence of cationic surfactants	Cationic nanoparticles interact with the cell membrane, triggering membrane disruption and the following Ca^{2+} influx. The elevation of intracellular Ca^{2+} induces degranulation and oxidative stress. The consequence of these effects is cytotoxicity and cell death.	Hwang et al. (2015)
Fibers and metals	Chronic lung fibrosis or fibrotic reaction in the airways or the lung interstitium.	Dostert et al.(2008)
Silver nanoparticles	Most potent inhibitor of both AChE and BChE.	Šinko et al. (2014)
Polychlorinated biphenyls (PCBs), dibenzo-p-dioxins (PCDDs), dibenzofurans (PCDFs)	Sick building syndrome.	Schön et al. (2016)
Exposure to agriculture pesticides	Reduction in AChE and BChE activities among farmers, exceeding organochlorine pesticide residues in human blood and other environmental matrix.	Dhananjayan et al. (2012a, 2012b), Dhananjayan and Ravichandran (2013)

(Continued)

Table 19.2 Health Effects and Sources of Different Nanoparticles *Continued*

Source/Nanoparticles	Health Effects	References
Si NPs and BaP	Co-exposure enhanced the MDA contents and reduced SOD and GSH-Px activity.	Wu et al. (2016)
CdO	Alterations in lung and liver morphology and damage to their tissue. Modulation in glutathione levels and increased lipid peroxidation after the greater chronic exposure.	Lebedová et al. (2016)
Cu, Cd, Pb	Toxicity to microorganism and soil microbe activity, cytotoxicity, membrane damage, inflammatory response, and oxidative stress.	Guo et al. (2011)
Au, Ni	Affects cellular micromotility, mitochondrial damage, autophagy, oxidative stress, acute inflammation and apoptosis in the liver, sperm motility, penetration of gold nanoparticles into sperm head and tail.	Li et al. (2010), Lasagna-Reeves et al. (2010)
Ag	Cytotoxicity and chromosome instability, oxidative stress, apoptosis, intracellular calcium transients, cell cycle arrest, interference with DNA replication fidelity, mammalian cells, free radical-induced oxidative stress and alteration of gene expression, blood–brain barrier destruction and astrocyte swelling, neuronal degeneration, and induce brain edema formation.	Asharani et al. (2009), Foldbjerg et al. (2011)
Quantum dots (QD)	oxidative stress in plants, endothelial toxicity, through activation of mitochondrial cells, cell viability decreased with increasing concentration of QD.	Valizadeh et al. (2012)

NP, *Nanoparticle;* DDR, *DNA damage response;* MDA, *malondialdehyde;* SOD, *superoxide dismutase.*

process are identification of hazards associated with NPs. If hazards identified are defined as having the potential to cause harmful effects, risk is the likelihood of this harm. This process of identification is highly difficult due to the very complex mixture of particles and dependent on the route of exposure, and the specific physicochemical properties of the individual particles, etc. (Hoet et al., 2004). Thus, NPs assessment methods necessarily focus on all the body systems that are considered potential targets for NPs. Currently, there is a data gap in this field and further data showing the interaction of NPs with cells will help in assessing the risk of NPs exposure to the developmental systems. In the occupational framework, hazard identification needs to be restated. According to NIOSH no conclusive data on engineered NPs exists for answering the question of what levels these particles create risk to the work force. Workers within certain industries have the potential to be exposed to exclusively engineered materials with fine particles and chemical properties at levels far exceeding ambient concentrations (Pietroiusti and Magrini, 2014). Occupational hygienists can contribute to the knowledge and understanding of Engineering Nano-Material (ENM) safety and health effects by the thorough documentation of exposures and effects. The major components of a risk assessment program include: exposure assessment, hazard identification; risk characterization, and risk management. A summary of the state-of-the-art for these components is discussed in Section 19.3.3.

19.2 EXPOSURE MONITORING

19.2.1 OBJECTIVES

Exposure assessment through various advance monitoring techniques is required to protect workers and the environment from toxic contaminants that can be released during industrial processes. The United States Environmental Protection Agency (USEPA) has imposed strict regulations on the concentrations of many environmental contaminants. However, current monitoring methods are costly and time-intensive while limitations in sampling and analytical techniques exist. In addition, the integrity of offsite laboratory analyses can be compromised during sample collection, transport, storage, and analysis, which can span over several days or more. Clearly, a need exists for accurate, inexpensive, long-term monitoring of environmental contaminants and their exposure status using special devices that can be operated onsite continuously. A wide ranging review described the field of nanotoxicology including some mechanisms of toxicity, portals of entry, their translocation, and the state of their risk assessment at the time (Oberdörster et al., 2005a). More recent reviews have focused on the major challenges, key questions, and research needs to assess ENM toxicity and risk (Stone et al., 2010). However, the ability to deploy and use emerging techniques for these applications is uncertain due to many technological barriers. The objective of this review is to assess the source, health effects, and risk assessment techniques in the identification of health hazards associated with certain occupations. This critical review was elucidated for the potential benefits of using various techniques used for environmental monitoring for NPs exposure.

19.2.2 INSTRUMENTATION

Traditionally, several techniques have been used to assess workplace pollutants and nanosized particles. The measurement techniques differ for assessing the chemical content of a NP sample, the reactions on the surface of the NPs, or for the interactions with other chemical species present. Almost all the techniques depend on quantity, size, and the nature of samples acquired for analysis. However, NPs are difficult to assess by traditional gravimetric sampling methods because they are very small to affect mass concentration and the difficulty of finding their emission source (Ham et al., 2016). Differential mobility analyzers (DMAs) or electrical mobility analyzers were combined with other instruments to generate different measurement techniques. Some techniques were used for getting the measurements and then required mathematical modeling to calculate the NPs. Electrostatic Low Pressure Impactor was used to calculate the mass concentration of an aerosol sample if the particle charge and density are known. The Nanomechanical Resonator was used to measure the mass of biological molecules of NPs and single bacterial cells.

19.2.3 REAL-TIME MONITORING OF NANOPARTICLES

Measurement techniques are continuously evolving as they are challenged and improved by research. There are different measurements of NPs and it is not clear which measurement relates closest to the risk posed by NPs. Therefore, various real-time monitoring devices are proposed by researchers to provide different information and accuracy on NPs in the risk assessment process. A comparison study of real-time NP monitoring instruments in the workplace demonstrated the need for source identification of emission and caution in comparing the data generated from different NP measurement devices (Ham et al., 2016). The accuracy of the methods cannot always be determined, but the comparison of results on the same sample through different techniques can provide an indication of accuracy. The details of available techniques given here are a general overview. Instruments will vary in their accuracy, sensitivity, and application ability depending on the manufacturer.

Microscopy Methods, Photon Correlation Spectroscopy, Surface Area Monitor, Dust-track Monitor (DTM), Condensation Particle Counter, DMA, Scanning Mobility Particle Sizer, NP Tracking Analysis, X-Ray Diffraction, Aerosol Time of Flight Mass Spectroscopy, and Aerosol Particle Mass Analyzer are the various techniques used for the monitoring of NPs. Experimentally, these approaches use online, real-time, and in situ methods for measuring NP retention dynamics, eliminating the laborious and less accurate sampling and offline analyses. The data analysis part is a process simulator which provides both kinetic properties of the retention process as well as the overall capacity and loading. This integrated method is useful for both comparison of the filtration effectiveness of various porous materials as well as for process optimization and scale-up for industrial applications (Rottman et al., 2013). The developed concepts of real-time monitoring can be applied to other types of NPs such as Au, Ag, and Si—which are relevant in medical and industrial applications. Real-time monitoring provides a powerful means for fast and effective optimization of synthesis conditions. It was also shown that simultaneous measurements can be performed on NPs even in low concentration.

19.2.4 AREA MONITORING OF NANOPARTICLES

19.2.4.1 Surface aerosol monitor

A NP aerosol monitor is used to measure the surface area of NPs (below 1.0 μm size) likely to be deposited in the lung in various industrial shop-floors. Workers are exposed to submicrometer aerosols in many occupations. The deposition of efficiency can be measuring the surface concentration of an aerosol that would be deposited either in the tracheobronchial or alveolar section of the airway of the human respiratory tract (Fissan et al., 2007). Studies have reported the indoor aerosol concentration during different activities (Mokhtar et al., 2013) and aerosol emission from a NP manufacturing process (Mäkelä et al., 2009). Therefore, aerosol monitoring at workplaces has become a major, and necessary, task.

19.2.4.2 Dust-track monitor

DTM is a compact instrument that includes an impaction inlet, laser diode, focusing optics, photo detector, vacuum pump, and control electronics. Aerosol dust monitor is currently used to monitor the different size fractions of dust (total dust, PM_{10}, PM_4, $PM_{2.5}$, and PM_1) simultaneously in different work environment for the duration of 8 hours with a calculation of Time Weighted Average (TWA). This size-segregated mass fraction measurement technique is superior to either a basic photometer or optical particle counter. It delivers the mass concentration of a photometer and the size resolution of an optical particle counter. The DTM cannot detect particles which are smaller than 0.1 μm in diameter. This measurement differs from gravimetric measurements of airborne particulate matter when the size distribution of the airborne particles differs significantly from the size distribution of the test aerosol (Chung et al., 2001).

19.2.4.3 Five-stage cascade impactor

Five-stage Sioutas Cascade Impactor with a high-efficiency personal air pump can collect NPs through a preseparator allowing particles into five different size categories with upper cut points ranging from 10 μm down to <0.25 μm. The mass concentrations of aerodynamic particle diameter ranges of <0.25, 0.25–0.5, and 0.5–1.0 will be measured. The equipment is capable of measuring the NPs size fraction gravimetrically, after which it is possible to find out the characteristics of the NPs.

19.3 DISCUSSION

19.3.1 OCCUPATIONAL EXPOSURE LIMITS

To prevent occupational diseases or other adverse effects of workers exposed to hazardous chemicals in the workplace, occupational exposure limit (OEL) values have been developed. OELs presume that exposed workers are adults, although in some cases it also aims to protect vulnerable groups (pregnant women or more susceptible people). OELs are several tools available to protect the health of workers who may be exposed to physical or chemicals hazards in the working environment. Generally, OELs are set for single substances, although in some cases they are also produced for a mixture of chemicals in the workplace. As the key criterion for making distinctions between exposures that are excessive and those that are not expected to result in adverse health effects, selection of the OEL is a critical and required step in the exposure assessment process.

There are well-developed methodologies by which formal OELs can be established. The conditions that need to be met are: (1) the criteria for exposure assessment should be established (e.g., exposure metrics); (2) the exposure assessment strategy should specify whether one needs to measure short-term or long-term exposures; (3) the instrumentation and analytical methods for measuring these metrics should be available; and (4) a dose–response relationship should be established by means of toxicity data and quantitative risk assessment. Only when these conditions are met can an exposure limit be identified. Because of the broad spectrum of nanomaterials, there is difficulty in setting an OEL for specific components. Expections include amorphous silicon dioxide with an OEL in Germany (AGS, 2006) and titanium dioxide with a proposed drat OEL from NIOSH of 1.5 mg/m^3 for fine particles and 0.1 mg/m^3 for ultrafine particles as TWA concentrations for up to 10 h/day during a 40 hours work week (NIOSH, 2005). Bayer Materials Science derived an in-house OEL of 0.05 mg/m^3 for its multiwalled CNTs (MWCNTs) based on subchronic inhalation studies (Bayer Material Science, 2010).

Additionally "benchmark levels" for nanomaterials have been developed by the British Standard Institution (BSI, 2007) to represent a pragmatic guidance level. The general benchmark level for insoluble nanomaterials is 0.066 × OEL of the corresponding microsized bulk material (mass concentration) is proposed. This factor of 0.066 is in line with the potency difference of microscale and nanoscale titanium dioxide as described by the American National Institute for Occupational Safety and Health (NIOSH, 2005). For fibrous nanomaterials, a benchmark level of 0.01 fibers/mL is proposed. This level is derived from the current limit value in asbestos removal activities in the United Kingdom. The benchmark of 0.5 × OEL is proposed for highly soluble nanomaterials. A benchmark of 0.1 × OEL of the corresponding microsized bulk material (mass concentration) for CMAR (carcinogenetic, mutagenic, asthmogenic, or reproductive) nanomaterials has been suggested. The benchmarks proposed were 20,000 particles/cm^3 for biopersistent metal oxides and other granular materials with a density between 6000 and 40,000 kg/m^3 for CNTs.

In case of respirable dust there is good evidence from epidemiology and toxicology studies that dust exposures may present a risk to workers and that for some of those who are affected, there are devastating health consequences. It is proposed that until regulators agree on the safe OELs for low-toxicity dusts, health and safety professionals should consider 1 mg/m^3 of respirable dusts as a more appropriate guideline than the value of 4 mg/m^3 currently used in Britain (Cherrie et al., 2013). Besides these exceptions, most nanomaterials have no OELs. The Institute for Occupational Safety and Health of the German Social Accident Insurance recommended benchmark limits for an 8 hours work shift that can be used for monitoring the effectiveness of protective measures in the workplace. Based on the similar physical attributes for CNTs, refractive ceramic fibers (RCFs), and asbestos fibers several OELs can be used to develop an ad-hoc OEL for CNTs. For asbestos fibers >5 μm in length, NIOSH recommends a Recommended Exposure Limit (REL) of 0.1 fibers/cm^3. For RCFs, REL and the ACGIH TLV can be used. The NIOSH REL of 0.5 fibers/cm^3 is intended to prevent the development of lung cancer, mesothelioma, or impaired pulmonary function. Current proposed nanomaterial OELs are generally low-mass concentrations compared with existing OELs for larger respirable particles of the same chemical composition. Among the proposed nanomaterial OELs, the differences among the same or similar materials may be largely due to different methods and assumptions used to derive the OELs. Despite these differences, there is relative consistency in the OELs within particle types and these generally fall within the same exposure control bins (e.g., order-of-magnitude categories) (Hewett et al., 2006). Each of these bins is associated with specific engineering control options, based on proven control technologies from other industries (e.g., pharmaceuticals, cosmetics, and dry powder processes) (Naumann et al., 1996). The CNT OEL estimates fall within either 1–10 or 10–100 μg/m^3 (8 hours TWA), which may be achievable with containment systems or ventilated enclosures, respectively (Naumann et al., 1996; Paik et al., 2008).

The applicability of existing OELs for larger particles to those with nanoscale forms has generally not been evaluated, except for example, NIOSH (2011). In general, the health basis for existing OEL values will vary largely based on the type of data and methods adapted to derive the OEL. Thus, standardized risk assessment methods are needed to provide a more consistent health-based approach for setting OELs for nanomaterials. The steps involved in using rodents for dose–response studies of inhaled particles in an occupational risk assessment are shown in Fig. 19.2. This method of process was applied for multiwalled carbon nanotubes and the exposure–response data obtained from subchronic inhalation studies provide data of relevance of workers exposure and quantitative measures of dose and response.

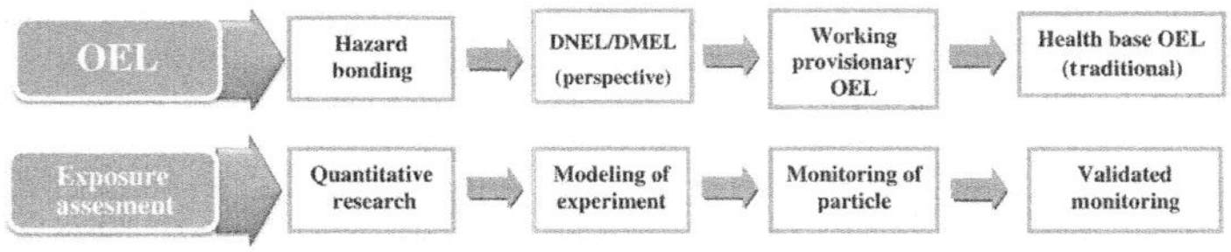

FIGURE 19.2

General risk assessment steps using airborne particles to develop occupational exposure limits.

19.3.2 EFFECT EVALUATION

Public concerns about NPs on health may arise with reports of some effects in a laboratory study or their presence in tissues of animals and humans. Results obtained from the research report must be interpreted carefully before concluding NP risk for health. The risk is always defined as a combined function of a chemical's ability to produce an adverse effect and the likelihood of exposure to those chemicals.

The advances in analytical chemistry have led to highly sensitive techniques that can detect chemicals at remarkably low levels. The detectable level may be far lower than any dose shown to produce an adverse effect. Further, a single finding in the literature may gather public attention and it may be statistically significant, but its scientific importance remains uncertain until it is replicated, preferably in another laboratory. In this regard, a follow-up study may be warranted to characterize the relevant parameters of dose, duration, and route of exposure as outlined in the risk assessment. The toxicity from ultrafine materials and NPs is not always the same (Madl and Pinkerton, 2009). For example, toxicity was greater from cadmium-containing quantum dots than the free cadmium ion (Priester et al., 2009). Therefore, one cannot always predict NPs toxicity from the known effects of the mixture of particles. Common findings of many studies are induction of inflammatory processes and oxidative stress using various model organisms.

19.3.3 RISK ASSESSMENT

In general, health risk assessment is a tool to evaluate the hazard and exposure to characterize the risk of workers during their job-related activities. A proposed life risk assessment framework was used for precommercial screening of selected applications of NPs as a method for systematically identifying and assessing potential risks from occupational, consumer, and environmental exposures. The Organization for Economic Cooperation and Development (OECD) has selected fullerenes (C60), single-walled CNTs, MWCNTs, silver NPs, iron NPs, carbon black, titanium dioxide, aluminum oxide, cerium oxide, silicon dioxide, polystyrene, dendrimers, and nanoclays as a list of representative manufactured nanomaterials (OECD, 2008) for further review on exposure assessment. Exposure to engineered nanomaterials can be found mainly in two ranges: one at production, formulation, and application in industrial process and the other, at research and university laboratories. The methods of risk assessment have been harmonized for industrial chemicals in general or on a national level. Currently there is no consensus what testing strategies and methods of risk assessment can be applied for nanomaterials. A general identified methods for characterization of physicochemical properties are in vitro- and in vivo-toxicity as relevant for testing of nanomaterials (Oberdörster et al., 2005b). Additionally, a newly identified method evaluated by SCENIHR (2006) for risk assessment has been described in the Technical Guidance Document. Rocks et al. (2008) suggested that the available risk assessment approaches for manufactured nanomaterials are appropriate. The analysis identifies potential exposure scenarios, evaluates toxicity, and assesses the adequacy of available data to characterize risk and highlighting data needs and gaps that must be filled to reduce the current uncertainty about safety. There are four basic steps involved in a comprehensive risk assessment process, namely exposure assessment, hazard identifications, risk characterization, and risk management.

19.3.3.1 Exposure assessment

The major key element of the risk assessment framework is exposure assessment, which includes the most likely routes of NP exposure. Exposure assessment is an estimation of the concentration to which workers are exposed directly during their work or indirectly via the surrounding environment and its compartments. Workers may be exposed to NPs while manufacturing these materials, formulating them into products, transporting them, or handling them in the storage facilities. Generally higher concentrations of nonmaterial and higher frequency of exposure to them are more likely to happen in workplace settings. Not much is known about the extent of occupational exposure to NPs. There are very few studies available on these aspects (Brouwer, 2010), however due to the absence of solid exposure data, no solid risk evaluation can be conducted (Schmid and Riedike, 2008). There is obvious value in conducting exposure assessments in the workplace to identify the routes, extent, and frequency of NP exposure. In assessing worker exposure, the traditional industrial hygiene sampling method of collecting samples in the breathing zone of workers (personal sampling) is preferred over area sampling. Only a few studies (Brouwer, 2010; Nazarenko et al., 2011) have conducted breathing zone measurements. The primary route of exposure for workers to NPs is considered to be through inhalation and/or dermal contact, since most of the NPs in manufacturing processes are performed in closed reaction chambers (Shi et al., 2013; Yokel and MacPhail, 2011). Contamination and exposure of workers are more likely to happen during handling and are associated with the reuse of personal protective equipments (PPEs) during the operations (Tsai, 2015). Exposure assessment in the risk assessment framework focuses on the initial portion of environmental health paradigm: from sources, to environmental concentrations, to exposure, to dose. It includes environmental sources, concentrations, exposure, dose, etc. The major purpose of an exposure assessment is to develop a qualitative and quantitative description of the environmental agents and their contact with humans and the exposed dose. The major emphasis is placed on estimating the magnitude, duration, and frequency of exposures along with the number of the population exposed to various concentrations of the pollutants (Callahan and Bryan, 1994).

Occupational exposures in the product formulation phase are most likely to occur while handling the materials from various processes. According to NIOSH, high exposures are likely to occur during product machining, repair, destruction, and recycling. Additionally, area samples and real-time exposure measurements are useful for evaluating engineering controls and their efficacy and work practices. When monitoring potential workplace exposure to NPs it is vital that background particle measurements are conducted before their production and/or processing in order to obtain baseline data. The study by Tsai et al. (2008) reported on the production of a nanocomposite containing alumina in a polymer by a twin-screw extrusion process caused release of 5–20 and 50–200 nm alumina in the worker's breathing zone.

19.3.3.2 Hazard identification

Hazard identification of NPs among occupationally exposed workers in industries is urgently needed; however, there is no conclusive data available on the effect of NPs in occupational contexts for many of the chemicals. Workers within industries have the potential to be exposed to uniquely engineered materials with novel sizes, shapes, and chemical properties at levels far exceeding ambient concentrations (Pietroiusti and Magrini, 2014). Even at low ENP exposure, increases in cytokines were observed among those who handled NPs compared with those who did not (Glass et al., 2017).

The most extensive exposures to NPs likely occur in workplaces (Brouwer, 2010; Yokel and MacPhail, 2011). Occupational hygienists can contribute to the knowledge and understanding of NPs safety and health effects by thorough documentation of exposures and effects. The US Occupational Safety and Health Administration (OSHA) is responsible for making and enforcing the regulations. NPs of less than 100 nm size demonstrated greater uptake and are more biologically active than larger sized particles of the same chemistry, due to their greater surface area per mass (Jiang et al., 2008). Aggregated particles are a cohesive mass consisting of particulate subunits tightly bound by covalent or metallic bonds due to a surface reconstruction, often through melting or annealing on surface impact, and often having an external surface area significantly smaller than the sum of calculated surface areas of the individual components (BSI, 2007). The optimal particle radius to accelerate adhesion to a cell-surface lipid bilayer is 15 and 30 nm for cylindrical and spherical particles, respectively (Nel et al., 2009). While a growing number of lipid vesicles, protein, and nucleic acid systems have been explored with nanoplasmonic sensors, there has been only very limited investigation of the interactions between nanomaterials and supported lipid membranes (Ferhan et al., 2017). Therefore, NPs sized between 10 and 30 nm have a spherical appearance to have the potential for more profound biological effects than other particles. The application of the continued understanding of the influence of physicochemical properties on biological responses can similarly enhance the benefit/risk ratio of future NPs. Additionally, observations of workers exposed to NPs can greatly add to this understanding, to increase confidence in the predicted effects of future NPs.

19.3.3.3 Risk characterization

Risk characterization for NPs would not, in principle, differ from that followed for soluble chemicals or the micro/macroscale materials. However, use of uncertainty factors must be considered. Japan's Ministry of Health, Labour and Welfare funded studies starting in 2005 to establish health risk assessment methodologies of manufactured nanomaterials and concluded that these studies were not yet sufficient to evaluate engineered nanomaterials risk (Hiroyuki, 2010). However, a new study incorporated a physiology-based lung model estimated that NP (anatase titania) adversely affects the lung cells. The authors concluded that workers exposed to relatively high airborne NPs ranging in 10–30 nm size are unlikely to have substantial risk for lung inflammatory responses, but are also a risk for cytotoxicity (Liao et al., 2009). The toxicological and exposure data on NPs are generally less well-developed than for other particles (Singh, 2016). Therefore, characterizing the uncertainty associated with NPs may require special attention to risk characterization. Risk characterization is the final phase of the risk assessment method. The different types of results obtained from the actual exposure assessment and the effects assessments will be combined to estimate the human health risks from the occupational exposure.

19.3.3.4 Risk management

Three types of standards are relevant for NPs under the Occupational Safety and Health Act (Balbus et al., 2007). (1) Substance-specific standards, for which there are none for nanomaterials. (2) General respiratory protection standards, under which inhalable nanomaterials would be considered particulates not otherwise regulated, for example, "nuisance dust", with a 5 mg/m^3 TWA air exposure limit, determined by breathing-zone air samples. The respiratory protection standard requires employers provide workers with NIOSH-certified respirators or other PPE to protect health. (3) The hazard communication standard states producers and importers of chemicals must develop Material Safety Data Sheets (Balbus et al., 2007). The EPA, using the legislative authority of the Toxic Substances Control Act has taken steps to limit the use and exposure to NPs, including CNTs. EPA has required the use of PPE and limitation on nanomaterials use and environmental exposures (Kagan et al., 2010). NIOSH recommends an 8-hour TWA exposure limit of 7 μg CNTs and nanofibers/m^3 air, and that employers minimize exposure to these materials. Suggested implementation includes many of the primary prevention measures discussed in this review and an occupational health surveillance program of exposure and medical monitoring. The goal in managing the potential risks from NPs is to minimize exposure. In the absence of specific information on NPs, the extensive scientific literature on airborne, respirable aerosols, and fibers has been used to develop interim guidance for working safely with nanomaterials (NIOSH, 2009) as the general approach to minimizing exposure.

19.3.4 CONTROLLING NANOPARTICLES EXPOSURE

Field investigations by the NIOSH team have reported relatively large reductions in the airborne particle exposures after installation of standard engineering controls (NIOSH, 2010). A study of laboratory fume hoods showed that the hood design affects the NP release and an air-curtain hood design significantly reduced workers' exposure (Tsai et al., 2010). There are different types of NPs exposure controlling methods that have been proposed by several studies (Hewett et al., 2006; Paik et al., 2008; Sajid et al., 2015). Since the use of nanomaterials continues to expand and information about potential health effects and exposure limits for nanomaterials is still being developed, industrial workers should be protected using a combination of measures (OSHA, 2013). The general steps involved in risk assessment and control methods are illustrated in Fig. 19.3.

19.3.4.1 Engineering controls

NP exposure can be reduced through the use of engineering controls, such as process changes, material containment, and enclosures operating at negative pressure compared to the workers' breathing zone, worker isolation, separated rooms, and local exhaust ventilation. Given the lack of occupational exposure standards to provide guidance, the most prudent approach is needed to minimize exposure. An international survey of NP generating industry managers conducted during 2009–10 by the University of California Center for Environmental Implications of Nanotechnology focused on industry controls of NPs exposure reported on the use of PPE, environmental risks, and perceptions revealed 46% of the companies had a nano-specific environmental health and safety program, compared to 58% in a 2006 survey (Conti et al., 2008). Some companies were using inappropriate occupational environmental cleanup methods, such as sweeping and compressed air (Harthorn, 2010). These results suggest more widespread adoption of nano-specific environmental health and safety programs and the use of PPE in the absence of superior controls measures.

19.3.4.2 Administrative controls

Administrative control should be implemented when engineering controls are not feasible for reducing exposure. Implementation of new policies and procedures

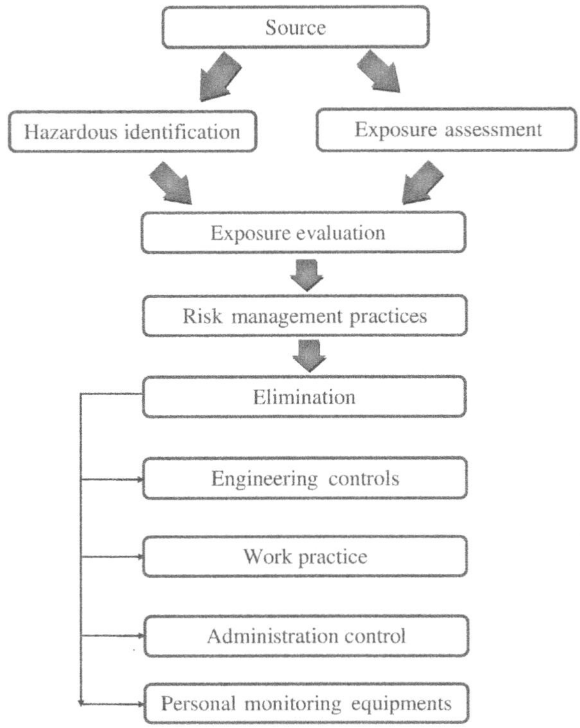

FIGURE 19.3

Flowchart of the hierarchy of source and control methods of nanoparticles.

are considered as part of the control methods in limiting worker exposure to hazards (Schulte et al., 2008). The other possible methods include hygiene plan for training and monitoring of standard operating procedures, modification of work practices and housekeeping practices as well as establishing procedures to address cleanup of nanomaterials spills and decontaminating surfaces to minimize worker exposure. Another vital part of administrative control is biological monitoring and medical examination.

19.3.5 PERSONAL PROTECTIVE EQUIPMENTS

PPEs should be provided appropriately to workers as per guidelines. NIOSH initiated a study of US workers exposed to engineered carbon NPs which included identification of the eligible companies and the most feasible population for study, followed by characterization of workers' exposures to NPs. These exposure data include respirable particle mass, number, and active surface area; personal full-shift daily exposure; and targeted sampling of the tasks associated with the highest exposures. Assessing the health effect of workers exposed to NPs is a critical task for the risk assessment of technology development (Schulte et al., 2011, 2012). The last line of defense in the hierarchy of exposure control is the use of PPE, such as respirators, protective clothing, and gloves.

19.3.6 BIOLOGICAL MONITORING AND MEDICAL EXAMINATION

Secondary prevention in the continuum of the prevention of exposure control includes biological monitoring and medical examination, the early detection of asymptomatic disease, and prompt intervention when the disease is preventable or more easily treatable (Halperin, 1996). An appropriate medical screening and surveillance program should be conducted for workers exposed to NPs. An occupational health surveillance is the process by which information obtained from any activity in the continuum of prevention and hierarchy of exposure control is collected and used to support or alter what is done at a step higher in the hierarchy as discussed (Trout and Schulte, 2010). Occupational health surveillance is the ongoing systematic collection, analysis, and dissemination of exposure and health data on groups of workers for the purpose of early detection and injury. It includes hazard surveillance, the periodic identification of potentially hazardous practices or exposures in the workplace, assessing the extent to which they can be linked to workers, the effectiveness of controls, and the reliability of exposure measures. A goal is to help define effective elements of the risk management program for exposed workers. Occupational health surveillance also includes medical surveillance, which examines health status to determine whether an employee is able to perform their essential job function (NIOSH, 2009). NIOSH and other studies recommend assessment of the need for medical monitoring of nanotechnology workers, as well as consideration of exposure registries, and development of epidemiological studies (NIOSH, 2009; Schulte et al., 2011).

19.4 CONCLUSION

19.4.1 RESEARCH NEEDS AND FUTURE DIRECTIONS

The objective of this review was to contribute to the various risk assessment methods for NPs exposure and, in particular, exposure risk assessment methods. The current data literature gives reason for concern due to the potentially and considerable toxicity of NPs. It is currently difficult to determine the most relevant risk indicator for nanomaterials. Data scarcity is profound and obstructs risk assessments of nanomaterials. In order to obtain more risk-relevant data and develop exposure models for nanomaterials, more collaboration between risk assessors namely industrial hygienists, environmental scientists, and chemists would be helpful, since these specialists know best what can be measured and how, whereas risk assessors know best what would be interesting to measure from a risk perspective. These contributions should be seen as a number of small steps toward a harmonized method for the risk assessment of nanomaterials. The profound data scarcity in the field of risk assessment of nanomaterials must be remedied. Deriving emission coefficients and collecting data on the production of nanomaterials constitute two important goals for future research. Regardless of which modeling approach is considered most relevant, data on emissions to the environment will always constitute important input to such models. Further studies on proxy risk indicators for nanomaterials would help to reveal the relevance of such an approach as an alternative to risk assessment. Further studies by risk assessing specialists on the relevance of different risk indicators for nanomaterials should be conducted.

19.5 SUMMARY AND RECOMMENDATIONS

This chapter demonstrates that the exposure in workplaces may cause health impairments if prevention is not taken. The observations from epidemiological studies show that exposure to fine dust is associated not only with diseases of the respiratory tract, but also with an elevated incidence and increased mortality of cardiovascular diseases. Considering the available literature guidelines, it is advisable for manufactures/industries to substitute hazardous nanomaterials with non-hazardous materials wherever feasible and also special attention should be given on ensuring low-dust and dust-free working methods. Protective measures such as encapsulation or exhaust systems may be effective for nanomaterials emissions. One of the key inputs in risk assessment is the potential exposure to a substance, yet evaluation of nanomaterial exposure is a particularly challenging endeavor. In general, this chapter demonstrated that engineering controls should be considered for effective control in NP emissions. The use of PPEs such as a mask, gloves, or eye protection should be strictly followed by workers. There is a small amount of exposure assessment information and a paucity of information required for a risk characterization. Therefore, more research and workplace monitoring information are needed to refine the knowledge of NPs exposure and associated risks in order to curtail these hazards in workplaces.

ACKNOWLEDGMENTS

The authors gratefully acknowledge the support of the Director, ICMR-National Institute of Occupational Health, Ahmedabad, India for NP research. The authors thank the editor of this book Dr. Alexandru Mihai Grumezescu and all the staff of ROHC(S), Bangalore for their help on a prior version of this chapter.

REFERENCES

AGS, 2006. TRGS 900 – Technische Regeln für Gefahrstoffe. Ausschuss für Gefahrstoffe – AGS-Geschäftsführung – BAuA. Available from <www.baua.de>.

Albanese, A., Tang, P.S., Chan, W.C., 2012. The effect of nanoparticle size, shape, and surface chemistry on biological systems. Annu. Rev. Biomed. Eng. 14, 1–16.

Alessandrini, F., Weichenmeier, I., van Miert, E., Takenaka, S., Karg, E., Blume, C., et al., 2010. Effects of ultrafine particles-induced oxidative stress on Clara cells in allergic lung inflammation. Part. Fibre Toxicol. 7, 11. Available from: https://doi.org/10.1186/1743-8977-7-11.

Asharani, P.V., Hande, M.P., Valiyaveettil, S., 2009. Antiproliferative activity of silver nanoparticles. BMC Cell Biol. 10, 65.

Balbus, J.M., Florini, K., Denison, R.A., Walsh, S.A., 2007. Protecting workers and the environment: an environmental NGO's perspective on nanotechnology. J. Nanopart. Res. 9, 11–22.

Bayer Material Science, 2010. Occupational Exposure Limits (OEL) for Baytubes defined by Bayer Material Science. http://www.baytubes.com/news_and_services/news_09116_oel.html.

Bergin, I.L., Witzmann, F.A., 2013. Nanoparticle toxicity by the gastrointestinal route: evidence and knowledge gaps. Int. J. Biomed. Nanosci. Nanotechnol. 3, 1–2. Available from: https://doi.org/10.1504/IJBNN.2013.054515.

Bermudez, E., Mangum, J.B., Wong, B.A., Asgharian, B., Hext, P.M., Warheit, D.B., et al., 2004. Pulmonary responses of mice, rats, and hamsters to subchronic inhalation of ultrafine titanium dioxide particles. Toxicol. Sci. 77, 347–357.

Biener, J., Wittstock, A., Baumann, T.F., Weissmüller, J., Bäumer, M., Hamza, A.V., 2009. Surface chemistry in nanoscale materials. Materials 2, 2404–2428.

Bouwmeester, H., van-der-Zande, M., Jepson, M.A., 2018. Effects of food-borne nanomaterials on gastrointestinal tissues and microbiota. Wiley Interdiscip. Rev. Nanomed. Nanobiotechnol. 10 (1). Available from: https://doi.org/10.1002/wnan.1481.

Brouwer, D., 2010. Exposure to manufactured nanoparticles in different workplaces. Toxicology 269, 120–127.

BSI, 2007, British Standards, Terminology for Nanomaterials. Publicly Available Specification, PAS. p. 16 ICS 01.040.71; 71.100.99; PAS 136:2007. http://shop.bsigroup.com/upload/Shop/Download/Nano/PAS136.pdf.

Buzea, C., Pacheco, I.I., Robbie, K., 2007. Nanomaterials and Nanoparticles: Sources and Toxicity. Biointerphases Vol. 2, 17–71.

Callahan, M.A., Bryan, E.F., 1994. Exposure assessment. In: Milman, H.A., Weisburger, E.K. (Eds.), Handbook of Carcinogen Testing. Noyes Publications, Park Ridge, NJ, pp. 651–671.

Castranova, V., 2011. Overview of current toxicological knowledge of engineered nanoparticles. J. Occup. Environ. Med. 53 (6 suppl), S14–S17.

Chan-Remillard, J.S., Kapustka, L., Goudey, S., 2009. Nanotechnology. In: Linkov, I., Steevens, J. (Eds.), Nanomaterials: Risks and Benefits, NATO Science for Peace and Security Series-C, Environmental Security. Springer, Dordrecht, pp. 53–66.

Cherrie, J.W., Semple, S., Christopher, Y., Saleem, A., Hughson, G.W., Philips, A., 2006. How important is inadvertent ingestion of hazardous substances at work? Ann. Occup. Hyg. 50, 693–704.

Cherrie, J.W., Brosseau, L.M., Hay, A., Donaldson, K., 2013. Low-toxicity dusts: current exposure guidelines are not sufficiently protective. Ann. Occup. Hyg. 57, 685–691.

Chrastný, V., Šillerová, H., Vítková, M., Francová, A., Jehlička, J., Kocourková, J., et al., 2018. Unleaded gasoline as a significant source of Pb emissions in the subarctic. Chemosphere 193, 230–236.

Chung, A., Chang, D.P., Kleeman, M.J., Perry, K.D., Cahill, T.A., Dutcher, D., et al., 2001. Comparison of real-time instruments used to monitor airborne particulate matter. J. Air Waste Manage. Assoc. 51, 109–120.

Churg, A., 2000. Particle uptake by epithelial cells. In: Gehr, P., Heyder, J. (Eds.), Particle–Lung Interactions. Marcel Dekker, New York, pp. 401–435. , ISBN 0-8247-9891-0.

Conti, J.A., Killpack, K., Gerritzen, G., Huang, L., Mircheva, M., Delmas, M., et al., 2008. Health and safety practices in the nanomaterials workplace: results from an international survey. Environ. Sci. Technol. 42, 3155–3162.

Coogan, P.F., White, L.F., Jerrett, M., Brook, R.D., Su, J.G., Seto, E., et al., 2012. Air pollution and incidence of hypertension and diabetes mellitus in black women living in Los Angeles. Circulation 25, 767–772.

Corbin, J., Mensah, A.A., Pieber, S.M., Orasche, J., Michalke, B., Zanatta, M., et al., 2018. Trace metals in soot and PM2.5 from heavy-fuel-oil combustion in a marine engine. Environ. Sci. Technol. Available from: https://doi.org/10.1021/acs.est.8b01764.PubMedPMID: 29688717.

DeLoid, G., Casella, B., Pirela, S., Filoramo, R., Pyrgiotakis, G., Demokritou, P., et al., 2016. Effects of engineered nanomaterial exposure on macrophage innate immune function. NanoImpact 2, 70–81.

DeLoid, G., Cohen, J.M., Darrah, T., Derk, R., Rojanasakul, L., Pyrgiotakis, G., et al., 2014. Estimating the effective density of engineered nanomaterials for in vitro dosimetry. Nat. Commun. 5, 3514. 10.1038/ ncomms4514.

Desai, P., Patlolla, R.R., Singh, M., 2010. Interaction of nanoparticles and cell-penetrating peptides with skin for transdermal drug delivery. Mol. Membr. Biol. 27, 247–259.

Dhananjayan, V., Ravichandran, B., 2013. Organochlorine pesticide residues in foodstuffs, fish, wildlife and human tissues from India: historical trend and contamination status. In: Malik, A., Grohmann, E. (Eds.), Environmental Deterioration and Human Health: Natural and Anthropogenic Determinants. Chapter 10, 2013. Springer, The Netherlands, pp. 229–262. , ISBN: 978-94-007-7889-4.

Dhananjayan, V., Ravichandran, B., Anitha, N., Rajmohan, H.R., 2012a. Assessment of acetylcholinesterase and butyrylcholinesterase activity in blood plasma of agriculture workers. Indian J. Occup. Environ. Med. 16, 127–130.

Dhananjayan, V., Ravichandran, B., Rajmohan, H.R., 2012b. Organochlorine pesticide residues in blood samples of agriculture and sheep wool workers in Bangalore (rural), India. Bull. Environ. Contam. Toxicol. 88 (4), 497–500.

Dostert, C., Petrilli, V., Van Bruggen, R., Steele, C., Mossman, B.T., Tschopp, J., 2008. Innate immune activation through Nalp3 inflammasome sensing of asbestos and silica. Science 320, 674–677.

Duan, J., Yu, Y., Li, Y., Yu, Y., Li, Y., Zhou, X., et al., 2013. Toxic effect of silica nanoparticles on endothelial cells through DNA damage response via Chk1-dependent G2/M checkpoint. PLoS One 8 (4), e62087. Available from: https://doi.org/10.1371/journal.pone.0062087.

Fahlman, B.D., 2007. Materials Chemistry. Springer, pp. 282–283, ISBN 1-4020-6119-6.

Ferhan, A.R., Gamaliel, J.M., Jackman, J.A., Sut, T.N., Park, J.H., Cho, N.J., 2017. Probing the interaction of dielectric nanoparticles with supported lipid membrane coatings on nanoplasmonic arrays. Sensors (Basel) 17, 1484.

Fissan, H., Neumann, S., Trampe, A., Pui, D., Shin, W.G., 2007. Rationale and principle of an instrument measuring lung deposited nanoparticle surface area. J. Nanopart. Res. 9, 53–59.

Foldbjerg, R., Dang, D.A., Autrup, H., 2011. Cytotoxicity and genotoxicity of silver nanoparticles in the human lung cancer cell line, A549. Arch. Toxicol. 85, 743–750.

Galvagno, S., Casciaro, G., Casu, S., Martino, M., Mingazzini, C., Russo, A., et al., 2009. Steam gasification of tyre waste, poplar, and refuse-derived fuel: a comparative analysis. Waste Manage. 29, 678–689.

Gewin, V., 2006. Nanotech's big issue. Nature 443, 137.

Glass, D.C., Mahjabeen, M., Sue, X., Pamela, D., Pamela, S., Brian, P., et al., 2017. Immunological effects among workers who handle engineered nanoparticles Deborah Catherine. Occup. Environ. Med. 74, 868–876.

Greenberg, M., Vearrier, D., 2015. Metal fume fever and polymer fume fever. Clin. Toxicol. (PA) 53, 195–203.

Greulich, C., Diendorf, J., Simon, T., Eggeler, G., Epple, M., Köller, M., 2011. Uptake and intracellular distribution of silver nanoparticles in human mesenchymal stem cells. Acta Biomater. 7, 347–354.

Guo, G.Y., Yuan, T., Wang, W.H., Li, D., Cheng, J.P., Gao, Y., et al., 2011. Bioavailability, mobility, and toxicity of Cu in soils around the dexing Cu mine in China. Environ. Geochem. Health 33, 217–224.

Halperin, W.E., 1996. The role of surveillance in the hierarchy of prevention. Am. J. Ind. Med. 29, 321–323.

Ham, S., Lee, N., Eom, I., Lee, B., Tsais, P.-J., Lee, K., et al., 2016. Comparison of real time nanoparticle monitoring instruments in the workplaces. Saf. Health Work 7, 381–388.

Harthorn, B., 2010. Characterization of the Nanotechnology Workforce. Presentation: Nanomaterials and Worker Health. Medical Surveillance, Exposure Registries, and Epidemiologic Research, Keystone, CO.

Heeb, N.V., Zennegg, M., Haag, R., Wichser, A., Schmid, P., Seiler, C., et al., 2013. PCDD/F formation in an iron/potassium-catalyzed diesel particle filter. Environ. Sci. Technol. 47, 6510–6517.

Hewakuruppu, Y.L., Dombrovsky, L.A., Chen, C., Timchenko, V., Jiang, X., Baek, S., et al., 2013. Plasmonic "pump–probe" method to study semi-transparent nanofluids. Appl. Opt. 52, 6041–6050.

Hewett, P., Logan, P., Mulhausen, J., Ramachandran, G., Banerjee, S., 2006. Rating exposure control using Bayesian decision analysis. J. Occup. Environ. Hyg. 10, 568–581.

Hiroyuki, T., 2010. Risk assessment studies of nanomaterials in Japan and other countries. Asia-Pac. J. Cancer Prevent. 11, 13–14.

Hoet, P.H.M., Brüske-Hohlfeld, I., Salata, O.V., 2004. Nanoparticles – known and unknown health risks. J. Nanobiotechnol. 2, 12.

Houghton, J., 2005. Global warming. Rep. Prog. Phys. 68, 1343–1403.

Hwang, T.L., Aljuffali, I.A., Lin, C.F., Chang, Y.T., Fang, J.Y., 2015. Cationic additives in nanosystems activate cytotoxicity and inflammatory response of human neutrophils: lipid nanoparticles versus polymeric nanoparticles. Int. J. Nanomed. 10, 371–385.

Ibrahim, R.K., Hayyan, M., AlSaadi, M.A., Hayyan, A., Ibrahim, S., 2016. Environmental application of nanotechnology: air, soil, and water. Environ. Sci. Pollut. Res. 23, 13754–13788.

Jiang, J., Oberdörster, G., Elder, A., Gelein, R., Mercer, P., Biswas, P., 2008. Does nanoparticle activity depend upon size and crystal phase? Nanotoxicology 2, 33–42.

Kagan, V.E., Shi, J., Feng, W., Shvedova, A.A., Fadeel, B., 2010. Fantastic voyage and opportunities of engineered nanomaterials: what are the potential risks of occupational exposures? J. Occup. Environ. Med. 52, 943–946.

Kaiser, J.P., Roesslein, M., Diener, L., Wick, P., 2013. Human health risk of ingested nanoparticles that are added as multifunctional agents to paints: an *in vitro* study. PLoS One 8, e83215. Available from: https://doi.org/10.1371/journal.pone.0083215.

Keane, M.J., Wallace, W.E., 2002. A quantitative in vitro fluorescence imaging method for phospholipid loss from respirable mineral particles. Inhal. Toxicol. 17, 287–292.

Kelishadi, R., 2012. Environmental pollution: health effects and operational implications for pollutants removal. J. Environ. Public Health. Available from: https://doi.org/10.1155/2012/341637. Article ID 341637.

Khan, I., Saeed, K., Khan, I., 2017. Nanoparticles: properties, applications and toxicities. Arab. J. Chem. Available from: https://doi.org/10.1016/j.arabjc.2017.05.011.

Kim, J.A., Åberg, C., de Cárcer, G., 2013. Low dose of amino-modified nanoparticles induces cell cycle arrest. ACS Nano 7, 7483–7494.

Kisku, G.C., Kumar, V., Sahu, P., Kumar, P., Kumar, N., 2018. Characterization of coal fly ash and use of plants growing in ash pond for phytoremediation of metals from contaminated agricultural land. Int. J. Phytorem. 20, 330–337.

Kodali, V., Littke, M.H., Tilton, S.C., Teeguarden, J.G., Shi, L., Frevert, C.W., et al., 2013. Dysregulation of macrophage activation profiles by engineered nanoparticles. ACS Nano 7, 6997–7010.

Kumar, C., 2006. Nanomaterials—Toxicity, Health and Environmental Issues. Wiley-VCH.

Lanzerstorfer, C., 2018. Heavy metals in the finest size fractions of road-deposited sediments. Environ. Pollut. 239, 522–531.

Lasagna-Reeves, C., Gonzalez-Romero, D., Barria, M.A., Olmedo, I., Clos, A., Sadagopa Ramanujam, V.M., et al., 2010. Bioaccumulation and toxicity of gold nanoparticles after repeated administration in mice. Biochem. Biophys. Res. Commun. 393, 649–655.

Lebedová, J., Bláhová, L., Večeřa, Z., Mikuška, P., Dočekal, B., Buchtová, M., et al., 2016. Impact of acute and chronic inhalation exposure to CdO nanoparticles on mice. Environ. Sci. Pollut. Res. 23, 24047–24060.

Li, J.J., Hartono, D., Ong, C.N., Bay, B.H., Yung, L.Y., 2010. Autophagy and oxidative stress associated with gold nanoparticles. Biomarker 31, 5996–6003.

Li, R., Ji, Z., Chang, C.H., Dunphy, D.R., Cai, X., Meng, H., et al., 2014. Surface interactions with compartmentalized cellular phosphates explain rare earth oxide nanoparticles hazard and provide opportunities for safer design. ACS Nano 8, 1771–1783.

Liao, C.M., Chiang, Y.H., Chio, C.P., 2009. Assessing the airborne titanium dioxide nanoparticle-related exposure hazard at workplace. J. Hazard. Mater. 162, 57–65.

Lin, C.D., Kou, Y.Y., Liao, C.Y., Li, C.H., Huang, S.P., Cheng, Y.W., et al., 2014. Zinc oxide nanoparticles impair bacterial clearance by macrophages. Nanomedicine (Lond.) 9, 1327–1339.

Linak, W.P., Miller, C.A., Wendt, J.O., 2000. Comparison of particle size distributions and elemental partitioning from the combustion of pulverized coal and residual fuel oil. J. Air Waste Manage. Assoc. 50, 1532–1544.

Liu, R., Yin, L.H., Pu, Y.P., Li, Y.H., Zhang, X.Q., Liang, G.Y., et al., 2010. The immune toxicity of titanium dioxide on primary pulmonary alveolar macrophages relies on their surface area and crystal structure. J. Nanosci. Nanotechnol. 10, 8491–8499.

Lu, X., Zhu, T., Chen, C., Liu, Y., 2014. Right or left: the role of nanoparticles in pulmonary diseases. Int. J. Mol. Sci. 15, 17577–17600.

Ma, J., Liu, R., Wang, X., Liu, Q., Chen, Y., Valle, R.P., et al., 2015. Crucial role of lateral size for graphene oxide in activating macrophages and stimulating pro-inflammatory responses in cells and animals. ACS Nano 9, 10498–10515.

Madl, A.K., Pinkerton, K.E., 2009. Health effects of inhaled engineered and incidental nanoparticles. Crit. Rev. Toxicol. 39, 629–658.

Mäkelä, J.M., Rostedt, A.A., Krinke, T.J., Janka, K., Marjamäki, M., Keskinen, J., 2009. Liquid flame spray for generating metal and metal oxide nanoparticle test aerosol. Hum. Exp. Toxicol. 28, 6–7.

Malachova, K., Kukutschova, J., Rybkova, Z., Sezimova, H., Placha, D., Cabanova, K., et al., 2016. Toxicity and mutagenicity of low-metallic automotive brake pad materials. Ecotoxicol. Environ. Saf. 131, 37–44.

Massadeh, A.M., Snook, R.D., 2002. Determination of Pb and Cd in road dusts over the period in which Pb was removed from petrol in the UK. J. Environ. Monit. 4, 567–572.

Masserini, M., 2013. Nanoparticles for brain drug delivery. ISRN Biochem. Available from: https://doi.org/10.1155/2013/238428. Article ID 238428.

Mauriello, M.C., Sbordone, C., Montuori, P., Alfano, R., Triassi, M., Iavicoli, I., et al., 2017. Biomonitoring of toxic metals in incinerator workers: a systematic review. Toxicol. Lett. 272, 8–28.

Mittal, S., Pandey, A.K., 2014. Cerium oxide nanoparticles induced toxicity in human lung cells: role of ROS mediated DNA damage and apoptosis. BioMed. Res. Int. Available from: https://doi.org/10.1155/2014/891934. Article ID 891934.

Miyake, Y., Tokumura, M., Iwazaki, Y., Wang, Q., Amagai, T., Horii, Y., et al., 2017. Determination of hexavalent chromium concentration in industrial waste incinerator stack gas by using a modified ion chromatography with post-column derivatization method. J. Chromatogr. A 1502, 24–29.

Mokhtar, M.-A., Jayaratne, R.R., Morawska, L., Mazaheri, M., Surawski, N., Buonanno, G., 2013. NSAM-derived total surface area versus SMPS-derived "mobility equivalent" surface area for different environmentally relevant aerosols. J. Aerosol Sci. 66, 1–11.

Naumann, B.D., Sargent, E.V., Starkman, B.S., Fraser, W.J., Becker, G.T., Kirk, D.L., 1996. Performance-based exposure control limits for pharmaceutical active ingredients. Am. Ind. Hyg. Assoc. J. 57, 33–42.

Nazarenko, Y., Han, T.W., Lioy, P.J., Mainelis, G., 2011. Potential for exposure to engineered nanoparticles from nanotechnology-based consumer spray products. J. Exp. Sci. Environ. Epidemiol. 21, 515–528.

Nel, A.E., Madler, L., Velegol, D., Xia, T., Hoek, E.M., Somasundaran, P., et al., 2009. Understanding biophysicochemical interactions at the nano-bio interface. Nat. Mater. 8, 543–557.

Nel, A., Xia, T., Meng, H., Wang, X., Lin, S., Ji, Z., et al., 2013. Nanomaterial toxicity testing in the 21st century: use of a predictive toxicological approach and high-throughput screening. Acc. Chem. Res. 46, 607–621.

Nemmar, A., Yuvaraju, P., Beegam, S., Yasin, J., Kazzam, E.E., Ali, B.H., 2016. Oxidative stress, inflammation, and DNA damage in multiple organs of mice acutely exposed to amorphous silica nanoparticles. Int. J. Nanomed. 11, 919.

Ning, Z., Cheung, C.S., Fu, J., Liu, M.A., Schnell, M.A., 2006. Experimental study of environmental tobacco smoke particles under actual indoor environment. Sci. Total. Environ. 367, 822–830.

NIOSH, 2005. NIOSH Testimony on the Occupational Safety and Health Administration's Proposed Rule on Occupational Exposure to Hexavalent Chromium, January 5, 2005, OSHA Docket No. H-054A. NIOSH Policy Statements. Cincinnati, OH: U.S. Department of Health and Human Services, Public Health Service, Centers for Disease Control and Prevention, National Institute for Occupational Safety and Health.

NIOSH, 2009. Approaches to Safe Nanotechnology. Managing the Health and Safety Concerns Associated With Engineered Nanomaterials. NIOHS, CDCP, DHHS; DHHS (NIOSH) Publication 2009-125. Available from: <http://www.cdc.gov/niosh/topics/nanotech/safenano/>.

NIOSH, 2010. Progress Toward Safe Nanotechnology in the Workplace: A Report From the NIOSH Nanotechnology Research Center. U.S. Department of Health and Human Service, Centers for Disease Control and Prevention, National Institute for Occupational Safety and Health, DHHS (NIOSH). Publication No. 2010-104, Cincinnati, OH.

NIOSH, 2011. Current Intelligence Bulletin 63. Occupational Exposure to Titanium Dioxide. U.S. Department of Health and Human Services, Public Health Service, Centers for Disease Control and Prevention, National Institute for Occupational Safety and Health, NIOSH, DHHS. Publication No. 2011-160, Cincinnati, OH.

NRC, 1983. Committee on the Institutional Means for Assessment of Risks to Public Health. National Research Council: Risk Assessment in the Federal Government: Managing the Process, p. 191.

Oberdörster, G., Maynard, A.D., Donaldson, K., Castranova, V., Fitzpatrick, J., Ausman, K., et al., 2005a. A report from the ILSI Research Foundation/Risk Science Institute Nanomaterial Toxicity Screening Working Group, principles for characterizing the potential human health effects from exposure to nanomaterials: elements of a screening strategy. Part Fibre Toxicol. 2.

Oberdörster, G., Oberdörster, E., Oberdörster, J., 2005b. Nanotoxicology: an emerging discipline evolving from studies of ultrafine particles. Environ. Health Perspect. 113, 823–839.

OECD, 2008. Organization for Economic Co-operation and Development, Series on the safety of manufactured nanomaterials, number 6, List of manufactured nanomaterials and list of endpoints for phase one of the OECD testing programme. <www.olis.oecd.org/olis/2008doc.nsf/LinkTo/NT00003282/$FILE/JT03246895.PDF> (accessed 06.02.18.).

OSHA, 2013. OSHA factsheet working safely with nanomaterials. Available from: <https://www.osha.gov/Publications/OSHA_FS-3634.pdf>.

Paik, S.Y., Zalk, D.M., Swuste, P., 2008. Application of a pilot control banding tool for risk level assessment and control of nanoparticle exposures. Ann. Occup. Hyg. 52, 419–428.

Peixe, T.S., de-Souza-Nascimento, E., Schofield, K.L., Arcurid, A.S.A., Bulcão, R.P., 2015. Nanotoxicology and exposure in the occupational setting. Occup. Dis. Environ. Med. 3, 35–48.

Pietroiusti, A., Magrini, A., 2014. Engineered nanoparticles at the workplace: current knowledge about workers' risk. Occup. Med. (Lond.) 64 (5), 319–330.

Pinaki, D., Patlolla, R.R., Singh, M., 2010. Interaction of nanoparticles and cell-penetrating peptides with skin for transdermal drug delivery. Mol. Membr. Biol. 27, 247–259.

Powers, K.W., Brown, S.C., Krishna, V.B., Wasdo, S.C., Moudgil, B.M., Roberts, S.M., 2006. Part VI. Characterization of nanoscale particles for toxicological evaluation. Toxicol. Sci. 90, 296–303.

Priester, J.H., Stoimenov, P.K., Mielke, R.E., Webb, S.M., Ehrhardt, C., Zhang, J.P., et al., 2009. Effects of soluble cadmium salts versus CdSe quantum dots on the growth of planktonic *Pseudomonas aeruginosa*. Environ. Sci. Technol. 43, 2589–2594.

Pui, D.Y.H., Chen, S.C., Zuo, Z.L., 2014. PM2.5 in China: measurements, sources, visibility and health effects, and mitigation. Particuology 13, 1–26.

Pulles, T., van der Gon, H.D., Appelman, W., Verheul, M., 2012. Emission factors for heavy metals from diesel and petrol used in European vehicles. Atmos. Environ. 61, 641–651.

Rastogi, A., Zivcak, M., Sytar, O., Kalaji, H.M., He, X., Mbarki, S., et al., 2017. Impact of metal and metal oxide nanoparticles on plant: a critical review. Front. Chem. 5, 78. Available from: https://doi.org/10.3389/fchem.2017.00078.

Rocks, S., Polland S., Dorey, R., Levy, L., Harrison, P., Handy, R., 2008. Comparison of risk assessment approaches for manufactured nanomaterials. <http://randd.defra.gov.uk/Document.aspx?Document = CB0403_7306_ABS.doc> (accessed 05.02.18.).

Rottman, J., Sierra-Alvarez, R., Shadman, F., 2013. Real-time monitoring of nanoparticle retention in porous media. Environ. Chem. Lett. 11, 71–76.

Roy, R., Kumar, S., Tripathi, A., Das, M., Dwivedi, P.D., 2014. Interactive threats of nanoparticles to the biological system. Immunol. Lett. 158, 79–87.

Ružičková, S., Remeteiová, D., Mičková, V., Dirner, V., 2018. Sediment matrix characterization as a tool for evaluating the environmental impact of heavy metals in metal mining, smelting, and ore processing areas. Environ. Monit. Assess. 190, 158. Available from: https://doi.org/10.1007/s10661-018-6551-4.

Sahu, D., Kannan, G.M., Vijayaraghavan, R., 2014. Size-dependent effect of zinc oxide on toxicity and inflammatory potential of human monocytes. J. Toxicol. Environ. Health A 77, 177–191.

Sajid, M., Ilyas, M., Basheer, C., Tariq, M., Daud, M., Baig, N., et al., 2015. Impact of nanoparticles on human and environment: review of toxicity factors, exposures, control strategies, and future prospects. Environ. Sci. Pollut. Res. Int. 22, 4122–4143.

SCENIHR, 2006. Scientific Committee on Emerging and Newly Identified Health Risks, the synthesis report on the public consultation of the SCENIHR opinion on the appropriateness of existing methodologies to assess the potential risks associated with engineered and adventitious products of nanotechnologies. <http://ec.europa.eu/health/ph_risk/documents/synth_report.pdf> (accessed 13.11.08.).

Schmid, K., Riedike, R.M., 2008. Use of nanoparticles in Swiss industry: a targeted survey. Environ. Sci. Technol. 42, 2253–2260.

Schmid, O., Stoeger, T., 2016. Surface area is the biologically most effective dose metric for acute nanoparticle toxicity in the lung. J. Aerosol Sci. 99, 133–143.

Schön, P., Ctistis, G., Bakker, W., Luthe, G., 2016. Nanoparticular surface-bound PCBs, PCDDs, and PCDFs—a novel class of potentially higher toxic POPs. Environ. Sci. Pollut. Res. Int. 24, 12758–12766.

Schulte, P., Geraci, C., Zumwalde, R., Hoover, M., Kuempel, E., 2008. Occupational risk management of engineered nanoparticles. J. Occup. Environ. Hyg. 5, 239–249.

Schulte, P.A., Kuempel, E., Zumwalde, R., Geraci, C.L., Schubauer-Berigan, M., Castranova, V., et al., 2012. Focused actions to protect carbon nanotube workers. Am. J. Ind. Med. 55, 395–411.

Schulte, P.A., Trout, D., Hodson, L., 2011. Introduction to nanomaterials and worker health: medical surveillance, exposure registries, and epidemiologic research. J. Occup. Environ. Med. 53 (6S), 1.

Senzui, M., Tamura, T., Miura, K., Ikarashi, Y., Watanabe, Y., Fujii, M., 2010. Study on penetration of titanium dioxide (TiO_2) nanoparticles into intact and damaged skin in vitro. J. Toxicol. Sci. 35, 107–113.

Shakya, P.R., Shrestha, P., Tamrakar, C.S., Bhattarai, P.K., 2006. Studies and determination of heavy metals in waste tyres and their impacts on the environment. Pak. J. Anal. Environ. Chem. 7, 70–76.

Sharifi, S., Behzadi, S., Laurent, S., Forrest, M.L., Stroeve, P., Mahmoudi, M., 2012. Toxicity of nanomaterials. Chem. Soc. Rev. 41, 2323–2343.

Shi, J., Karlsson, H.L., Johansson, K., Gogvadze, V., Xiao, L., Li, J., et al., 2012. Microsomal glutathione transferase 1 protects against toxicity induced by silica nanoparticles but not by zinc oxide nanoparticles. ACS Nano 6, 1925–1938.

Shi, H., Magaye, R., Castranova, V., Zhao, J., 2013. Titanium dioxide nanoparticles: a review of current toxicological data. Part. Fibre Toxicol. 10, 15.

Shin, S.W., Song, I.H., Um, S.M., 2015. Role of physicochemical properties in nanoparticle toxicity. Nanomaterials 5, 1351–1365.

Singh, A.K., 2016. Human and environmental risk characterization of nanoparticles. In: Engineered Nanoparticles. doi: 10.1016/B978-0-12-801406-6.00009-1.

Singh, U.K., Ramanathan, A.L., Subramanian, V., 2018. Groundwater chemistry and human health risk assessment in the mining region of East Singhbhum, Jharkhand, India. Chemosphere 204, 501–513.

Šinko, G., Vinković Vrček, I., Goessler, W., Leitinger, G., Dijanošić, A., Miljanić, S., 2014. Alteration of cholinesterase activity as possible mechanism of silver nanoparticle toxicity. Environ. Sci. Pollut. Res. Int. 21, 1391–1400.

Sioutas, C., Delfino, R.J., Singh, M., 2005. Exposure assessment for atmospheric ultrafine particles (UFPs) and implications in epidemiologic research. Environ. Health Perspect. 113, 947–955.

Sisler, J.D., Pirela, S.V., Shaffer, J., Mihalchik, A.L., Chisholm, W.P., Andrew, M., et al., 2016. Toxicological assessment of CoO and La_2O_3 metal oxide nanoparticles in human small airway epithelial cells. Toxicol. Sci. 150, 418–428.

Smith, M.J., Brown, J.M., Zamboni, W.C., Walker, N.J., 2014. From immunotoxicity to nanotherapy: the effects of nanomaterials on the immune system. Toxicol. Sci. 138, 249–255.

Stephens, J.W., Harrison, J.C., Wallace, W.E., 1998. Correlating auger electron spectroscopy with scanning electron microscopy-energy dispersive spectroscopy for the analysis of respirable particles. Scanning 20, 302–310.

Stone, V., Nowack, B., Baun, A., van-den-Brink, N., Kammer, F., Dusinska, M., et al., 2010. Nanomaterials for environmental studies: classification, reference material issues, and strategies for physico-chemical characterisation. Sci. Total Environ. 408, 1745–1754.

Sun, B., Pokhrel, S., Dunphy, D.R., Zhang, H., Ji, Z., Wang, X., et al., 2015. Reduction of acute inflammatory effects of fumed silica nanoparticles in the lung by adjusting silanol display through calcination and metal doping. ACS Nano 9, 9357–9372.

Taylor, D.A., 2002. Dust in the wind. Environ. Health Perspect. 110, A80–A87.

Trommer, R.M., Bergmann, C.P. (Eds.), 2015. Flame Spray Technology: Method for Production of Nanopowders, VIII. Springer-Verlag, Berlin Heidelberg. Available from: http://dx.org/10.1007/978-3-662-47162-3.

Trout, D.B., Schulte, P.A., 2010. Medical surveillance, exposure registries, and epidemiologic research for workers exposed to nanomaterials. Toxicology 269, 128–135.

Tsai, C.S., 2015. Contamination and release of nanomaterials associated with the use of personal protective clothing. Ann. Occup. Hyg. 59, 491–503.

Tsai, S.J., Huang, R.F., Ellenbecker, M.J., 2010. Airborne nanoparticle exposures while using constant-flow, constant-velocity, and air-curtain-isolated fume hoods. Ann. Occup. Hyg. 54, 78–87.

Tsai, S.-J., Ashter, A., Ada, E., Mead, J.L., Barry, C.F., Ellenbecker, M.J., 2008. Airborne nanoparticle release associated with the compounding of nanocomposites using nanoalumina as fillers. Aerosol Air Qual. Res. 8, 160–177.

USEPA, U.S. Environmental Protection Agency, 2002. Health Assessment Document for Diesel Engine Exhaust. USEPA EPA/600/8-90/057F. 01 May (2002). U.S. EPA, Office of Research and Development, National Centre for Environmental Assessment, Washington, DC.

Valizadeh, A., Mikaeili, H., Samiei, M., Farkhani, S.M., Zarghami, N., Kouhi, M., et al., 2012. Quantum dots: synthesis, bioapplications, and toxicity. Nanosc. Res. Lett. 7 (1), 480. Available from: https://doi.org/10.1186/1556-276X-7-480.

Wakefield, G., Green, M., Lipscomb, S., Flutter, B., 2004. Modified titania nanomaterials for sunscreen applications—reducing free radical generation and DNA damage. Mater. Sci. Technol. 20, 985–988.

Wallace, W.E., Keane, M.J., Murray, D.K., Chisholm, W.P., Maynard, A.D., Ong, T.-M., 2007. Phospholipid lung surfactant and nanoparticles surface toxicity: lessons from diesel soots and silicate dusts. J. Nanopart. Res. 9, 23–38.

Wang, B., Zhang, Y., Mao, Z., Yu, D., Gao, C., 2014. Toxicity of ZnO nanoparticles to macrophages due to cell uptake and intracellular release of zinc ions. J. Nanosci. Nanotechnol. 14, 5688–5696.

Ware, M.J., Godin, B., Singh, N., Majithia, R., Shamsudeen, S., Serda, R.E., et al., 2014. Analysis of the influence of cell heterogeneity on nanoparticle dose response. ACS Nano 8, 6693–6700.

Weichenthal, S., Lavigne, E., Valois, M.F., Hatzopoulou, M., Ryswyk, K.V., Shekarrizfard, M., et al., 2017. Spatial variations in ambient ultrafine particle concentrations and the risk of incident prostate cancer: a case-control study. Environ. Res. 156, 374–380.

Weidemann, E., Patrik, L., Andersson, Bidleman, T., Boman, C., Carlin, D.J., et al., 2016. 14th congress of combustion by-products and their health effects-origin, fate, and health effects of combustion-related air pollutants in the coming era of bio-based energy sources. Environ. Sci. Pollut. Res. 23, 8141–8159.

WHO, 1999. Hazard prevention and control in the work environment: airborne dust. Available from: <http://www.who.int/occupationalhealth/publications/en/oehairbornedust3.pdf>.

Woodrow Wilson International Center for Scholars, 2011. The project on emerging nanotechnologies: consumer products: an inventory of nanotechnology-based consumer products currently on the market. Available from: <http://www.nanotechproject.org/inventories/consumer/>.

Wu, W., Samet, J.M., Peden, D.B., Philip, A., Bromberg, P.A., 2010. Phosphorylation of p65 is required for zinc oxide nanoparticle–induced interleukin 8 expression in human bronchial epithelial cells. Environ. Health Perspect. 118, 982–987.

Wu, J., Shi, Y., Asweto, C.O., Feng, L., Yang, X., Zhang, Y., et al., 2016. Co-exposure to amorphous silica nanoparticles and benzo[*a*]pyrene at low level in human bronchial epithelial BEAS-2B cells. Environ. Sci. Pollut. Res. Int. 22, 23134–23144.

Yildirimer, L., Thanh, N.T.K., Loizidou, M., Seifalian, A.M., 2011. Toxicology and clinical potential of nanoparticles. Nano Today 6, 585–607.

Yokel, R.A., MacPhail, R.C., 2011. Engineered nanomaterials: exposures, hazards, and risk prevention. J. Occup. Med. Toxicol. 6, 7. Available from: https://doi.org/10.1186/1745-6673-6-7.

Zhang, H.Y., Ji, Z.X., Xia, T., Meng, H., Low-Kam, C., Liu, R., et al., 2012. Use of metal oxide nanoparticle band gap to develop a predictive paradigm for oxidative stress and acute pulmonary inflammation. ACS Nano 6, 4349–4368.

Zheng, R., Tao, L., Duan, T., Liua, N., Cheng, Y., Chang, Y., et al., 2016. Potential hazards of superfine particles to human bronchial epithelial cells through inducing oxidative stress. Nano Impact 2, 93–98.

Zmozinski, A.V., de Jesus, A., Vale, M.G., Silva, M.M., 2010. Determination of calcium, magnesium and zinc in lubricating oils by flame atomic absorption spectrometry using a three-component solution. Talanta 83, 637–643.

Index

Note: Page numbers followed by "*f*" and "*t*" refer to figures and tables, respectively.

C

D

E

M

N

Q

R

S

T

U

V

W

X

Z

CPI Antony Rowe
Eastbourne, UK
April 02, 2019